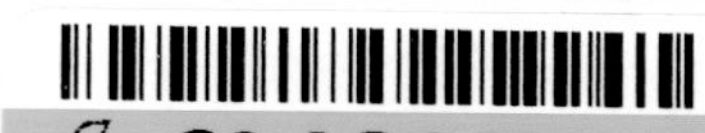

Quick Reference Guide to SuperCalc3 Release 2

SORCIM/IUS MICRO SOFTWARE

<E> signifies <Enter> or <Return>

/ Arrange
- Row
  - row number --- → <E> for entire row; ascending sort; no adjust / , col. range --- , → Ascend / Descend → Yes adj. / No adj. → <E> primary / , secondary → row number , / col. letter , → Ascend / Descend
  - <E> current row
- Col
  - col. number --- → , row range --- , / <E> for entire column; ascending sort; no adjust
  - <E> current col.

/ Blank
- range --- <E>
- <E> for current cell
- * graph range

/ Copy
- from range --- , to upper/left cell of destination range → <E> adjust / , options → No adjust / Ask for adjust / Values only / + - * /
- from * graph number (1-9) , to graph number (1-9) <E>

/ Delete
- Row — row range (to delete one or more rows of data) ---
- Column — column range (to delete one or more columns of data) ---
- File
  - *filename* (to delete a file from disk in data drive or drive specified)
  - ESC for current filename
  - <E> for Directory options

/ Edit
- any cell --- <E>
- <E> for current cell

/ Format
- Global level
- Column level - column range --- ,
- Row level ---- row range to 254 or Remaining
- Entry level - any range --- ,
- Define table (User-defined formats: $n,nnn; (neg. #); 0=blank; %; dec. places; scaling
- → Integer for no decimals
- General (num. with best fit)
- Exponential numbers only
- $ for two decimal places
- Right numeric justification
- Left numeric justification
- TR text right justification
- TL text left justification
- TC text centered
- * for asterisk linear display
- User-defined format - (1-8)
- Hide values
- Default settings (G,R,TL,9)
- (0-127) column width <E>

/ Global
- Optimum Spreadsheet Conditions menu → memory usage (max. speed or spreadsheet data-space) / spreadsheet boundary (up to 127 col. by 9999 rows) / screen options
- Keep (settings at Global menus, Output Setup, & Directory
- Graphics menus → Colors menu (graph component colors) / Fonts menu (graph type styles) / Layout menu (graph or sheet size & graph position) / Options menu (graph appearance & device settings) / Device Selection menu (select graphics printer or plotter)
- Formula display (on/off)
- " for text entry (on/off)
- Next to auto-advance cursor (on/off)
- Border display (on/off)
- Tab to skip blank & protect cells (on/off)
- Row, Column, or Dependency order of calc.
- Iteration control → (1-99) fixed max. number of iterations <E> / Solve (SuperCalc3 controls iter's) → Delta → - for Delta = 0.01 / cell containing Delta--- <E>; Range to converge to Delta --- <E>
- Manual or Automatic recalculation

/ Insert
- Row — row range (to insert one or more empty rows) --- <E>
- Column — column range (to insert one or more empty columns) --- <E>

/ Load
- *filename* ,
- ESC for current name ,
- <E> for Directory options
- → All
- Part - from range --- , - to upper/left cell --- → <E> adjust / , options → No adjust / Ask for adjust / Values only / + - * /
- Consolidate
- * - from graph range --- , - to fist graph number (1-9)

/ Move
- Row from row range ---, to row number --- (will be top row if move is up; bot. row if move is down)
- Column - from col. range ---, to col. letter --- (will be left col. if move is left; right col. if move is right)

/ Output
- Display / Contents → range --- , → Printer / Console / Disk → *filename* , / ESC for current name , / <E> for Directory options → Change name / Backup / Overwrite
- Setup menu
  - Length (lines)
  - Width (characters)
  - New border character
  - Border printing (on/off/match)
  - Auto-form-feed (on/off)
  - Double space (on/off)
  - End-line-feed (on/off)
  - Set printer control codes
  - Retain printer control codes
  - Print report

/ Protect
- range --- <E>
- <E> for current cell

/ Quit
- Yes to exit from SuperCalc3 (does not save current work)
- No to cancel this command
- To quit & re-load SC# or load another program → program *filename* <E> / <E> for Directory options

/ Replicate
- from cell ---, to cell/partial row/partial column ---
- from partial row ---, to left partial column ---
- from partial column ---, to top partial row ---
- → <E> adjust / , options → No adjust / Ask for adjust / Values only / + - * /

/ Save
- *filename* , / Esc for current name , / <E> for directory options → Change name / Backup / Overwrite → All / Values only / Part → All / Values only → range --- <E>

/ Title
- Horizontal lock
- Vertical lock
- Both
- Clear

/ Unprotect
- range --- <E>
- <E> for current cell

/ View
- <E> for view current graph
- [?] for current graph description summary
- (1-9) graph #
- Data — Variable range ---, or <E>
- Graph-Type — Pie / X-Y / Bar / Stacked-Bar / Line / Area / Hi-Lo
- Time-Labels — range --- <E>
- Variable-Labels — range --- <E>
- Point-Labels — range --- <E>
- Headings — Main / Sub / X-axis / Y-axis → Cell <E>
- Options
  - Format — Axis scale labels / Time-Labels / Variable-Labels / Point-Labels / % pie segment labels - (0-9) or Default
  - Explosion — All pie segments / None / (1-8) pie segment #
  - Pie-Mode — One Variable / All Variables
  - Scaling — X-axis / Y-axis

/ Window
- Horizontal split
- Vertical split
- Clear to right or below split
- Synchronize split-wise scroll
- Unsynchronize split-wise scroll

/ Xecute)
- *filename* for execute file <E>
- ESC for current name
- <E> for Directory options

/ Zap
- Yes to delete current spreadsheet; retains settings of Global menus, Output Setup, & Directory
- No to cancel this command
- Contents, same as Yes, but also retains User-defined format table settings

// Data mgt.
- Input / Criterion / Output → range --- <E>
- Find → ↓ next record / ↑ previous record / → next field / ← previous field / <E> to cancel
- Extract records specified
- Select records or veto → Yes to accept / No to reject / → next field / ← previous field / <E> to cancel
- Remain at current location

INTRINSIC FUNCTIONS AND OPERATORS

Logical Functions
- AND($cond_a$,$cond_b$)
- FALSE
- IF(cond,c/v_1,c/v_2)
- NOT(cond)
- OR($cond_a$,$cond_b$)
- TRUE

Financial Functions
- FV(pmt,int,per)
- IRR([guess,]range)
- NPV(disc,range)
- PMT(prin,int,per)
- PV(pmt,int,per)

Mathematical Functions
- ABS(c/v)
- ACOS(c/v)
- ASIN(c/v)
- ATAN(c/v)
- COS(c/v)
- EXP(c/v)
- INT(c/v)
- LN(c/v)
- LOG(c/v)
- MOD(c/v_1,c/v_2)
- PI
- RAN
- ROUND(c/v,places)
- SIN(c/v)
- SQRT(c/v)
- TAN(c/v)

Special Functions
- ERROR
- LOOKUP(c/v,range)
- NA

Statistical Functions
- AV(range)
- COUNT(range)
- MAX(range)
- MIN(range)
- SUM(range)

+	Plus
-	Minus
*	Multiply
/	Divide
^ (or **)	Raise to power
%	Percent
=	Equal to
<>	Not equal to
<	Less than
<=	Less than or equal
>	Greater than
>=	Greater than or equal

c/v - a Cell reference or Value
cond - some algebraic Condition (e.g., A5>5.)
disc - Discount rate
guess - a Guess
int - Interest rate
per - Period or term
places - number of Places after the decimal
pmt - Payment
prin - Principal
range - a Range of cells (e.g., B3.C6)

FUNCTION KEY	ACTION
F1 or ?	Help
F2 or Ctrl Z	Clears/Aborts
F9 or Ctrl Y	Graph on Plotter
F10 or Ctrl T	Graph on Screen

PREFIXES FOR CELL ENTRIES

Letters and most symbols begin text mode
" begins forced text mode
' begins repeating entry
Numbers, functions, cell references, +, -,., cell addresses begin formulas

RANGES CELL, COLUMN, ROW

SINGLE CELL
D5 is column D row 5

BLOCK OF CELLS
Two Opposite Corner Cells
D5.F48
H1.A17

FULL COLUMN
F

MULTI-COLUMNS
B.G

PARTIAL COLUMN
C3.C10

FULL ROW
39

MULTI-ROWS
45.83

PARTIAL ROW
G25.M25

5TH EDITION

ENGINEERING

An Introduction to a Creative Profession

GEORGE C. BEAKLEY

DONOVAN L. EVANS

JOHN BERTRAND KEATS

Arizona State University

Macmillan Publishing Company NEW YORK Collier Macmillan Publishers LONDON

Macmillan Publishing Company
866 Third Avenue, New York, New York 10022

Collier Macmillan Canada, Inc.

Library of Congress Cataloging-in-Publication Data

Beakley, George C.
Engineering: an introduction to a creative profession.
Includes index.
1. Engineering—Vocational guidance. 2. Engineering.
3. Engineering design. I. Evans,
Donovan L. II. Keats, John Bertrand. III. Title.
TA157.B394 1986 620′.0023 85-15808

Printing: 1 2 3 4 5 6 7 8 Year: 6 7 8 9 0 1 2 3 4 5

ISBN 0-02-307090-0

Preface

Today we live in an era of rapid technological change. The abundance of updated content in this new edition is a reflection of this dynamic condition. Although this book is correctly designated as the 5th edition under the present title, in actuality it is the 8th edition in an unbroken sequence of printings. Three earlier editions of this same text appeared in 1949, 1951, and 1960 under different titles. The 1949 edition was a collection of the original authors' lecture notes. It was one of the first elementary texts to flavor classroom problem solving techniques with the realism of contemporary industrial practices. Almost instantly the book was a best seller. The 1967 edition was boldly innovative. It was one of the first introductory texts to include a strong introduction to engineering design—content that had traditionally been reserved for the junior or senior years in most curricula. Again this new content proved to be highly popular with adopters, and other authors began to emphasize engineering design in their own texts.

This 1986 edition also charts some previously unexplored territory. It contains approximately two-thirds new content and one-third updated and revised material. As with earlier editions, this text contains many solved examples and hundreds of problems for class and homework assignment. The appendixes contain an abundance of information useful in problem solving and not found in most other elementary texts. For example, the interest tables in Appendix III, would be especially useful later in the student's career for solving engineering economics problems of the type found in the national Engineering in Training (EIT) examination. As an aid in engineering analysis, the 1949 edition featured an in-depth study of uses of the slide rule; the 1977 edition featured instruction in hand-held calculators. This edition is the first elementary text in the field to feature the use of microcomputers—particularly with regard to the use of electronic spreadsheets—in problem solving and design. The authors' intent is to smooth the transition into the rapidly approaching era of mainframe power in desktop computers and text book size computers with built in spreadsheets, equation solvers, and graphics. The sequence of chapters in the text is such that only the last two chapters of the text are microcomputer oriented. A handheld calculator is sufficient for the other chapters. For those schools that do not have the resources to teach the last two chapters, they can be used by the students on a self-study basis.

Today most high schools teach one or more programming languages, BASIC being the most widely used. The authors have assumed that students studying the material in this text will have completed high school level algebra and trigonometry and have some experience in computer programming. Chapter 17 builds upon this experience and offers the student further instruction in two computer languages: BASIC and FORTRAN. The presentation serves as a bridge between these two languages and provides, to our knowledge, one of the few multilingual programming language textual presentations. FORTRAN continues to be the favored language of most engineers; BASIC is the language most students use on their home computer.

Chapter 18 contains instructional material for two of the most widely used spreadsheets, Supercalc3[1] and Lotus 1-2-3[2]. Both chapters 17 and 18 contain side-by-side *TOPICS* (computer examples) to assist the student in mastering and reviewing the subject material. Both chapters are written to accommodate a self-study format.

Chapter 19 provides a climax for modern problem solving using the microcomputer. It includes computer codes and solution procedures for design and optimization problems such as projectile motion, solar cell models, engineering economic analysis, the shortest route algorithm, PERT, and Kirchhoff's laws.

The authors have prepared handy quick-reference guides for the languages BASIC and FORTRAN, and for the spreadsheets, Supercalc3 and Lotus 1-2-3. For convenience these are printed on the front and back endpapers of the text. Some students may find it more convenient to remove these reference sheets from the text and to either encase them in plastic or triple-fold them to fit into a shirt or blouse pocket.

The authors believe that an engineering education must present an early awareness of the full responsibility to society of the engineer-designer. Previous editions have addressed this point, but in the brief span of time

[1]© SORCIM/IUS Micro Software, San Jose, CA
[2]© Lotus Development Company, Cambridge, MA

that has elapsed since their introduction, society's problems seem only to have become more critical. This newest edition both updates and places increased emphasis on these responsibilities. Chapters 2, 3, and 6 are both new and unique in the urgency of their contemporary messages. Also, throughout the text a special effort has been made to impart to the student a professional attitude and personal concern to produce designs compatible with the occupational health and safety of everyone who might be affected. In this regard the authors are especially appreciative of the advice and time devoted to manuscript review afforded them by William N. McKinnery, Jr. and his staff of the National Institute for Occupational Safety and Health.

Chapter 3, *Welcome to the 21st Century,* reflects the research of Jack Stadmiller, who has attempted to expand the horizons of our imagination and at the same time help us to understand that frequently, through innovative engineering design, the seemingly impossible can be made possible. In Chapter 6, *Professional Responsibilities of the Engineer,* Keith Roe has used a number of contemporary examples to enhance the student's sensitivity to the importance of accepting a moral and professional responsibility in achieving designs that are entirely compatible with safe and healthful operation.

Those who create, artists-designers-architects-engineers, have always found it expedient to communicate to others the essence of their ideas through drawings, sketches, diagrams, or other graphic means. Until the latter part of the 18th century most design drawings were made freehand. Leonardo da Vinci's middle 16th century drawings depicting his creative ideas are still easy to comprehend today. Beginning about 1820, engineers in this country began to be taught projection drawing, based on the French system that was first developed by Gaspard Monge (1746–1818). Projection drawings are most useful in the manufacturing process, particularly with regard to the production of mechanical components. Courses in engineering drawing and descriptive geometry have been standard components of most engineering curricula until the past few years. However, in the last decade the situation has changed drastically.

Traditional engineering drawing methods stressing orthographic projection no longer serve as a common medium of graphic communication in many engineering disciplines such as microelectronics, advanced chemical processes, nuclear power, and computing systems. However, the methods of freehand drawing using by da Vinci are still effective today in enabling engineers from all disciplines to clearly communicate the essence of their creative thought. Chapter 7, *Freehand Drawing and Visualization,* has been developed to enable students (by class instruction or by self-study) to enhance spatial visualization and to acquire those freehand drawing skills that will be universally useful to engineers of all disciplines for the visual communication of ideas and design concepts. Martha Heier, who has extensive experience in teaching freehand drawing skills to university students, prepared most of the textual material for the chapter, including the student assignments. Del Bowers developed the microcomputer portion of the chapter.

A recent nationwide study by the Educational Testing Service has shown that during the last two decades, the competence of high school students in spatial visualization has diminished at an alarming rate. This is an essential skill for engineers, architects, and medical doctors. We believe that since drawing is "seeing," the practice of drawing stimulates the ability to visualize, something that the more creative persons in our society appear to do naturally. Exercising visualization skills of engineering students can only improve their performance in the creative aspects of design.

The popular in-depth design content of previous editions has been further strengthened with the addition of a new chapter, Chapter 13, *Engineering Design Philosophy.* The importance of topics such as aesthetics, life-cycle cost, and logistics to the design process is emphasized. System effectiveness measures such as reliability, availability, and serviceability are related to the design phase, which is discussed in depth in Chapter 12.

James Pruett and Kenneth Case provided helpful critiques of Chapter 8, and Phillip Wolfe assisted in preparing the PERT program of Chapter 19. Improvements in content organization and the accuracy of textual materials have been achieved because of the dedicated efforts of the text reviewers, and the authors are most appreciative of their work. They were William N. Anderson, Rickey Brouillette, W. George Devens, Donald K. Jamison, and Larry Simmons.

A number of friends and colleagues have also reviewed or supplied information for manuscript preparation, and the authors are most grateful for their assistance. These are:

Donald D. Autore	Louise Majeres
Carl Bailey	John H. Matson
John Gambrell	Carl Mueller
Kenneth Geiser	Michael J. Nielsen
Andrew L. Hopper	Robert Raudebaugh
John E. Kelly	John M. Sims
I. Herbert Lundgren	Gregory P. Wilson

The authors are also most grateful to many of the adopters of the 4th edition and to other professional colleagues who have made suggestions regarding content to be included in this book.

G.C.B.
D.L.E.
J.B.K.

Acknowledgments

Front Cover	Computerized color-coded topographic map of the surface of the Space Telescope Primary Mirror at the start of the polishing process. The pattern is based on precise interferometric measurements. White represents the optimum surface shape while the blues and reds identify the highs and lows. NASA photo.
Frontispiece	TIAA-CREF
Figure 1-1	Duane Michals; Eli Lilly & Company
Figure 1-4	Maddox and Hopkins
Figure 1-6	W.S. Dickey Clay Manufacturing Company
Figure 1-7	Ewing Galloway
Figure 1-8	Maddox and Hopkins
Figure 1-9	Maddox and Hopkins
Figure 1-13	Maddox and Hopkins
Figure 1-16	Hercules Incorporated
Figure 1-17	Fischer Scientific Company
Figure 1-18	Texas Instruments Incorporated
Figure 1-19	General Electric FORUM
Figure 1-21	National Society of Professional Engineers
Figure 1-22	Matheson Gas Products
Figure 1-23	National Society of Professional Engineers
Figure 1-24	National Society of Professional Engineers
Figure 1-25	National Society of Professional Engineers
Figure 1-26	Medtronic, Inc.
Figure 1-27	National Society of Professional Engineers
Figure 1-28	National Society of Professional Engineers
Figure 1-29	National Aeronautics and Space Administration
Figure 1-30	National Society of Professional Engineers
Figure 1-31	Westinghouse Electric Corporation
Figure 1-32	American Society for Engineering Education
Figure 2-1	Western Electric
Figure 2-2	United States Navy
Figure 2-3	Exxon Chemical Company, USA
Figure 2-4	Uniroyal, Inc.
Figure 2-5	Bethlehem Steel Corporation
Figure 2-6	H. Armstrong Roberts Photo
Figure 2-7	Stennett Heaton Photo, Courtesy Neil A. Maclean Co., Inc.
Figure 2-9	Floyd A. Craig, Christian Life Commission, Southern Baptist Convention
Figure 2-10	Planned Parenthood—World Population
Figure 2-12	Planned Parenthood—World Population
Figure 2-13	Ministers Life and Casualty Union
Figure 2-14	Monsanto Company
Figure 2-15	Ambassador College
Figure 2-16	Adolph Coors Company
Figure 2-18	Ford Motor Company
Figure 2-19	Citizens for Clean Air
Figure 2-21	Southcoast Air Quality Management District
Figure 2-22a.	Torit Corporation
Figure 2-22b.	*Arizona Republic*
Figure 2-23	George Kranse; Presbyterian Ministers Fund Life Insurance
Figure 2-24	Planned Parenthood—World Population
Figure 2-25	United States Department of Agriculture
Figure 2-26	Floyd A. Craig, Christian Life Commission, Southern Baptist Convention
Figure 2-27	The Carborundum Company
Figure 2-28	Shell Oil Company
Figure 2-29	Phil Stitt, *Arizona Architect*
Figure 2-30	Floyd A. Craig, Christian Life Commission, Southern Baptist Convention
Figure 2-32	Planned Parenthood—World Population
Figure 2-34	Xerox Corporation
Figure 2-35	Hewlett-Packard
Figure 2-36	Hewlett-Packard
Figure 2-37	*Kaiser News*
Figure 2-38	Southern California Edison Company
Figure 2-39	Pacific Gas and Electric Company
Figure 2-40	Southern California Edison Company
Figure 2-41	Humble Oil and Refining Company
Figure 2-43	Los Angeles County Air Pollution Control District
Figure 2-44	Shell Oil Company
Figure 2-45	American Telephone and Telegraph Company
Figure 2-46	*Arizona Republic*
Figure 2-47	Peter Kiewit Son's Company
Figure 2-48	Institute of Traffic Engineers
Figure 2-49	American Iron & Steel Institute
Figure 2-50	American Express Company
Figure 2-51	Polaroid Corporation
Figure 2-52	General Telephone & Electronics
Figure 2-55	Floyd A. Craig, Christian Life Commission, Southern Baptist Convention
Figure 2-58	The Carborundum Company
Figure 2-59	General Electric Company
Figure 3-1	Wausau Insurance Company
Figure 3-2	Schott Glaswerke, Design: Luerzer, Conrad & Leo Burnett
Figure 3-3	Champion International Corporation
Figure 3-4	Shell Oil Company
Figure 3-5	Champion International Corporation
Figure 3-9	National Aeronautics and Space Administration
Figure 3-10	National Aeronautics and Space Administration
Figure 3-11	National Aeronautics and Space Administration
Figure 3-13	Blue Cross Association
Figure 3-15	Champion International Corporation
Figure 3-17	Floyd A. Craig, Christian Life Commission, Southern Baptist Convention
Figure 3-18	General Electric Company
Figure 3-19	Northrop Corporation
Figure 3-20	Northrop Corporation
Figure 3-21	General Electric Company
Figure 3-22	National Aeronautics and Space Administration

Figure 3-23 Artist Robert McCall
Figure 3-24 Lockheed Missiles & Space Company
Figure 3-25 Champion International Corporation
Figure 3-27 General Motors Corporation and American Society of Civil Engineers
Figure 3-28 Vehicle Systems Development Corporation
Figure 3-29 Gates Learjet Corporation
Figure 3-30 Symbion, Inc.
Figure 3-31 Champion International Corporation
Figure 3-32 EG&G, International Geodyne Division
Figure 3-33 Champion International Corporation
Figure 3-34 Ameron
Figure 3-35 Photo by Eric Lantz, *Walnut Grove Tribune*
Figure 3-36 Paramount Die Casting
Figure 3-37 Exxon Company, USA
Figure 3-38 National Aeronautics and Space Administration
Figure 3-39 U.S. Air Force Photo
Figure 4-1 McDonnell Douglas
Figure 4-3 International Harvester Company
Figure 4-4 Aluminum Company of America
Figure 4-5 General Motors Research Laboratories
Figure 4-7 Glass Container Manufacturers Institute
Figure 4-8 American Institute of Plant Engineers
Figure 4-9 Bethlehem Steel Corporation
Figure 4-10 *Engineering Times*
Figure 4-11 General Electric Research and Development Center
Figure 4-12 General Electric Research and Development Center
Figure 4-13 RCA Electric Corporation
Figure 4-14 Cessna Aircraft Company
Figure 4-15 Information Handling Services
Figure 4-16 Phillips Petroleum Company
Figure 4-17 Renault USA, Inc.
Figure 4-18 Union Carbide Corporation
Figure 4-19 Cities Service Company
Figure 4-20 Phelps Dodge Corporation
Figure 4-22 Southern California Edison Company
Figure 4-23 Atlantic Richfield Company
Figure 4-24 Exxon Company, USA
Figure 5-1 AT&T Bell Laboratories
Figure 5-2 J.E. Cermack, Director, Fluid Dynamics and Diffusion Laboratory, Colorado State University
Figure 5-3 Fafnir Bearing Division of Textron
Figure 5-6 Information Handling Services
Figure 5-7 Information Handling Services
Figure 5-8 Naval Undersea Center
Figure 5-9 *Industrial Engineering*
Figure 5-12 American Institute of Plant Engineers
Figure 5-13 Bethlehem Steel Corporation
Figure 5-14 NCR Corporation
Figure 5-15 International Paper Company
Figure 5-16 Arkansas Best Corporation
Figure 5-17 Union Carbide Corporation
Figure 5-18 Fisher Scientific Company
Figure 6-1 Ernst and Ernst
Figure 6-2 General Dynamics
Figure 6-4 Federal Highways Administration
Figure 6-8 Modine Manufacturing Company
Figure 6-9 *American Youth Magazine*
Figure 6-12 *Industrial Research*
Figure 6-13 American Lung Association
Figure 6-14 Merrill, Lynch, Pierce, Fenner & Smith, Inc.
Figure 6-16 Goulds Pumps, Inc.
Figure 6-17 Johns-Manville Corporation
Figure 6-19 Rockwell International
Figure 6-20 Fireman's Fund American Insurance Companies
Figure 6-21 The Foxboro Company
Figure 6-22 Ford Motor Company
Figure 6-24 Sperry Corporation
Figure 6-25 NASA, Ames Research Center
Figure 6-26 NASA, Ames Research Center
Figure 6-27 NASA, Ames Research Center
Figure 6-28 Kansas City Star Company
Figure 6-29 Kansas City Star Company
Figure 6-30 Texas Instruments Incorporated
Figure 6-31 NCR Corporation
Figure 6-32 Floyd Clark, California Institute of Technology
Figure 6-33 FAG Bearings Corporation
Figure 7-2 Bruning Division, Addressograph Multigraph Corporation
Figure 7-3 U.S. Air Force Photo
Figure 7-5 Collection of IBM Corporation
Figure 7-6 Campus Crusade of Christ International
Figure 7-7 *Sperry Rand Engineering Review*
Figure 7-12 "Waterfall," M.C. Escher (1898–1972), Escher Foundation, Gementemuseum, The Hague
Figure 7-13 After a drawing by M. Gardner, *Scientific American,* 1970, May, 124-7
Figure 7-14 Photograph by William G. Hyzer, Reprinted from *Research & Development,* August 1971 © 1971 by Technical Publishing Company
Figure 7-18 Andrea Raab Corporation
Figure 7-21 Union Electric Company
Figure 7-25 After a drawing by J. Frazer, *British Journal of Psychology,* 1980, Vol. 2 (1980) p.307 Photo by Gordon Smith from National Audubon Society
Figure 7-27 Carl Zeiss, Inc. New York
Figure 7-28 A. W. Faber—Castell
Figure 7-30 Juan Gris, Portrait of Max Jacob, 1919
Figure 7-32 Thomas F. Brennan
Figure 7-33 A/P Wide World Photos
Figure 7-36 "Drawing Hands", M.C. Escher (1898–1972), Escher Foundation, Gementemuseum, The Hague
Figure 7-47 Photo by Mike Jakub
Figure 7-48 Nicholas Orsini, *The Language of Drawing,* Reprinted by permission of Doubleday & Co., Inc.
Figure 7-49 Nicholas Orsini, *The Language of Drawing,* Reprinted by permission of Doubleday & Co., Inc.
Figure 7-51 Honoré Daumier (1808–1879)
Figure 7-52 Formsprag Company
Figure 7-56 John A. Lytle and Bank of America
Figure 7-57 John A. Lytle and Bank of America
Figure 7-58 Barbara Banthien
Figure 7-61 Pete Cowgill Photo
Figure 7-61 Modified Pete Cowgill Photo
Figure 7-64 Dallas/Ft. Worth Airport
Figure 7-66 Albert Lorenz Studio
Figure 7-67 Gary Meyer
Figure 7-69 Jan Vredeman de Vries (1527–after 1604) From *Perspective* (Leiden) 1604
Figure 7-70 Jan Vredeman de Vries (1527–after 1604) From *Perspective* (Leiden) 1604
Figure 7-72 Albrecht Dürer (1471–1528)
Figure 7-73 Andrea Mantegna (1431–1506), Brera, Milan, Italy
Figure 7-74 William G. Reynolds
Figure 7-75 Barbara Banthien
Figure 7-78 Jan Vredeman de Vries (1527–after 1604) From *Perspective* (Leiden) 1604
Figure 7-79 Ibis-headed Figures, Tomb of Tethmosis III

Figure 7-82	*Arizona Republic*
Figure 7-84	Research Laboratories, General Motors Corporation
Figure 7-87	International Business Machines
Figure 7-90	Robert G. Steele, Illustrator
Figure 7-97	"Day and Night", M.C. Escher (1898–1972), Escher Foundation, Gementemuseum, The Hague
Figure 7-98	Jim Sharpe
Figure 7-99	"Beckie King" (1949), Andrew Wyeth, Dallas Museum of Art
Figure 7-100	Nicholas Orsini, *The Language of Drawing,* Reprinted by permission of Doubleday & Co., Inc.
Figure 7-102	Alfred Leslie, Landfall Press Inc.
Figure 7-106	Ri Kaiser
Figure 7-109	© Murray Tinkelman
Figure 7-110	Bob Peters
Figure 7-111	Bob Peters
Figure 7-112	Mark Adams
Figure 7-113	Bob Peters
Figure 7-114	Edward Gazzi
Figure 7-115	Newsday, Inc.
Figure 7-116	"The Persistence of Memory" (1931), Salvador Dali
Figure 7-117	James McConnell Illustration
Figure 7-118	Bruce Schluter
Figure 7-119	Gerry Gersten
Figure 7-120	Vickers, Incorporated
Figure 7-121	General Dynamics, Convair Division
Figure 7-124	Formsprag Company
Figure 7-126	Calcomp
Figure 8-1	Texas Instruments Incorporated
Figure 8-2	United California Bank and Foote, Cone & Belding
Figure 8-36	Beckman Instruments Company
Figure 9-1	*Chicago Tribune*
Figure 9-2	Midas, Inc.
Figure 9-3	Ford Motor Company, Design Center
Figure 9-13	*Engineering Graphics*
Figure 10-3	Brown Brothers
Figure 10-11	Monarch Machine Tool Company
Figure 10-17	Westinghouse Electric Corporation
Figure 11-1	Ingersoll-Rand Company
Figure 12-2	Robert Shaw Controls Company
Figure 12-3	First National Bank of America
Figure 12-4	United Aircraft
Figure 12-5	Modine Manufacturing Company
Figure 12-6	Can-Tex Industries
Figure 12-7	American Tel and Tel, Madison Ave., New York
Figure 12-8	Encyclopaedia Britannica
Figure 12-11	The Falk Corporation
Figure 12-12	Eaton Company
Figure 12-13	Al Capp
Figure 12-14	National Aeronautics and Space Administration
Figure 12-28	New York Life Insurance Company
Figure 12-19	Frank Roberge and *Arizona Republic*
Figure 12-30	College of Engineering, Colorado State University
Figure 12-31	Research Laboratories, General Motors Corporation
Figure 12-35	Renault USA, Inc.
Figure 12-38	Sandia National Laboratory
Figure 12-42	Pontiac Corporation
Figure 12-43	Pontiac Corporation
Figure 12-44	Pontiac Corporation
Figure 12-45	Citicorp Leasing, Inc.
Figure 12-46	Mercedes-Benz of North America
Figure 12-47	National Aeronautics and Space Administration
Figure 12-49	Walker/Parkersburg
Figure 12-50	Walker/Parkersburg
Figure 12-51	National Aeronautics and Space Administration
Figure 12-52	Somerset Importers, Ltd.
Figure 13-2	Fisher Scientific Company
Figure 13-4	U.S. Navy
Figure 13-5	International Business Machines Corporation
Figure 13-6	The Boeing Company
Figure 13-7	GTE Sylvania Incorporated
Figure 13-9	USDA Photo
Figure 13-11	Bell Telephone Laboratories
Figure 13-14	North American Rockwell
Figure 13-19	General Dynamics Convair Division
Figure 13-20	Eastman Kodak Corporation
Figure 13-21	Swanson Analysis Systems, Inc.
Figure 13-22	Photo by Karl H. Maslowski from National Audubon Society
Figure 13-24	American Welding Society
Figure 13-25	Reynolds Metals Company
Figure 13-26	Reynolds Metals Company
Figure 13-27	Western Division, GTE Government Systems
Figure 13-29	Enjay Chemical
Figure 13-30	General Radio Corporation
Figure 13-31	Renault USA, Inc.
Figure 13-34	Ford Motor Company
Figure 13-35	National Aeronautics and Space Administration
Figure 13-37	A. H. Robins Company
Figure 13-38	Ambassador College
Figure 13-39	Reprinted from the January 19, 1970 issue of *Design News,* a Cahners Publication
Figure 13-40	International Paper Company
Figure 13-41	American Motors Corporation
Figure 13-42	Boeing Commercial Airplane Company
Figure 13-43	Mercedes-Benz of North America
Figure 13-45	The Foxboro Company
Figure 13-46	Texas Instruments Incorporated
Figure 13-51	McDonnell Douglas-Douglas Aircraft Company
Figure 13-52	National Aeronautics and Space Administration
Figure 13-53	National Aeronautics and Space Administration
Figure 14-3	*Arizona Republic*
Figure 14-4	INA Corporation
Figure 14-11	The Boeing Company, Vertol Division
Figure 14-21	Phoenix Newspapers, Inc.
Figure 15-5	The Timken Company
Figure 15-6	Automatic Switch Company
Figure 15-7	Trustees of the Science Museum (London)
Figure 15-8	Trustees of the Science Museum (London)
Figure 15-9	Trustees of the Science Museum (London)
Figure 15-10	International Business Machines
Figure 15-11	International Business Machines
Figure 15-12	Texas Instruments Incorporated
Figure 15-13	International Business Machines
Figure 15-14	National Semiconductor Corporation
Figure 15-17	Adapted from Drawing, *Kaiser News*
Figure 15-21	Plessey Peripheral Systems
Figure 15-22	Pertec Peripherals Corporation
Figure 15-23	Keytronic
Figure 15-24	International Business Machines
Figure 15-25	International Business Machines
Figure 15-26	Calcomp
Figure 15-27	O'Connell & Associates, Inc.
Figure 15-28	Hewlett-Packard
Figure 15-29	Apple Computer, Inc.
Figure 15-30	Digital Equipment Corporation
Figure 15-31	Control Data
Figure 16-8	Naval Surface Weapons Center

Contents

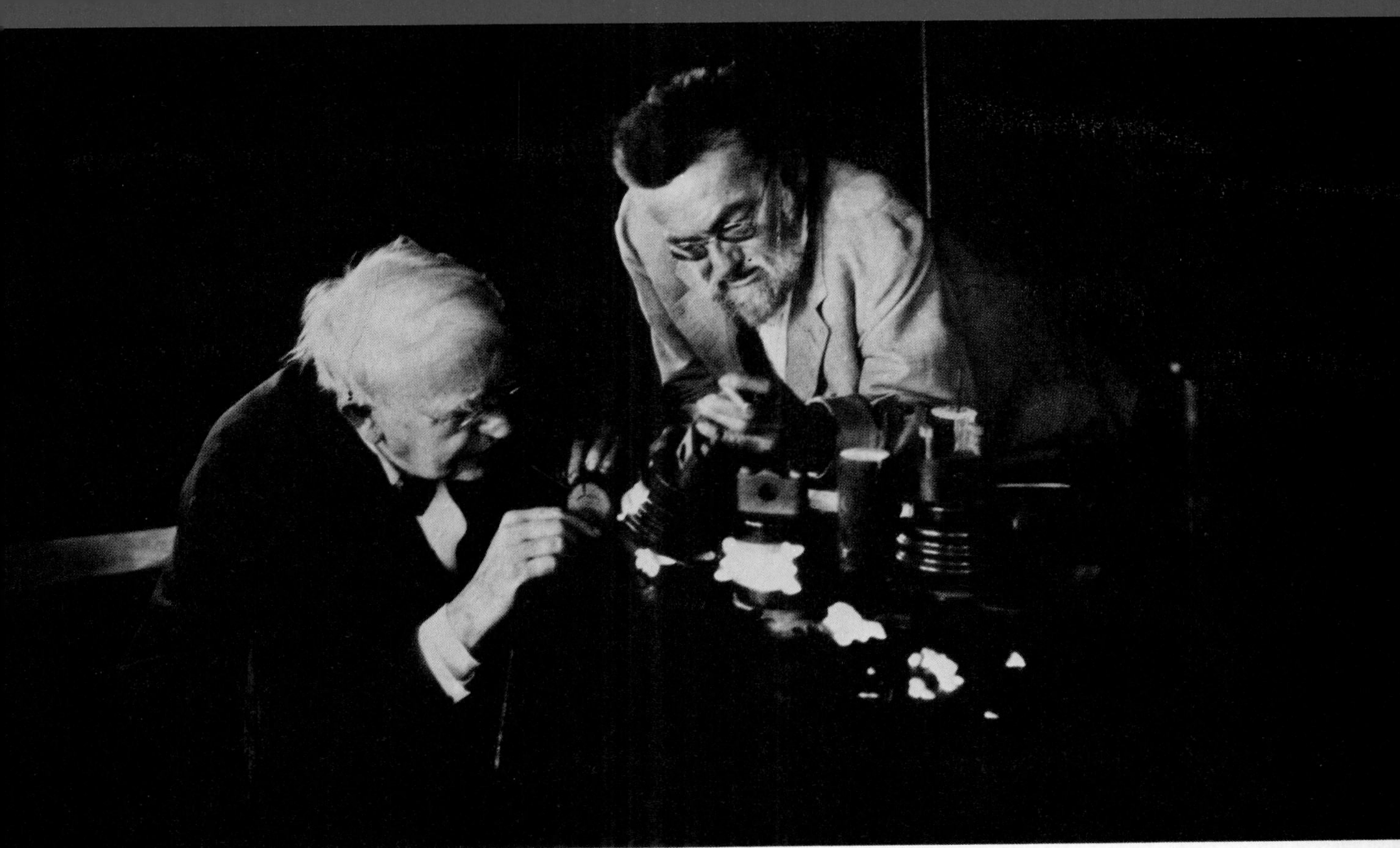

Section One

The Engineering Profession

Chapter 1

Engineering Achievements of the Past

If we open a quarrel between the past and the present, we shall find that we have lost the future.

Sir Winston Churchill, 1874–1965
Speech, House of Commons, 1940

When did engineering begin? Who were the first engineers? What were the first engineering designs? Answers to these questions and others concerning the beginning of engineering appear in fragments of historical information available to us. In fact, the beginnings of civilization and the beginnings of engineering were the same. As prehistoric man[1] emerged from caves to make homes in communities, he adapted rocks and sticks as tools to serve his needs. Simple as these items may seem to us today, their useful employment suggests that the creative ideas which emerged in the mind of early man were developed into useful products. Some served as tools in the struggle for existence, and others were used for protection against wild animals or warlike neighbors.

Through the ages, the engineer has been in the forefront as a maker of history. His creative designs have had as much impact on world history as have political, economic, or social development. Sometimes these designs have developed from the pressures of need from evolving civilizations. At other times abilities to produce and meet needs have led the way for civilizations to advance. It is still true today that engineers create designs to serve the needs of people and their culture.

The role of the engineer has not changed through the centuries. The primary task has always been to make practical use of converting scientific theory into useful application. In so doing, the engineer helps to provide for mankind's material needs and well-being. From era to era, only the objectives pursued, the techniques of solution used, and the tools available for analysis have changed.

It is helpful to review the past to gain insight into the driving forces of science and to learn of the individuals who developed and applied these principles. A review will also reveal important facts concerning the discovery and use of fundamental scientific principles. Primarily, science builds its store of knowledge on facts which, once determined, are available from then on for application and further discovery. This principle is in contrast to the arts, where, for example, the ability of one person to produce a beautiful painting does not increase the skill of others who want to produce paintings.

[1-1] Everyone should learn about the past.

To be ignorant of the lives of the most celebrated men of antiquity is to continue in a state of childhood all our days.

Plutarch, ca. A.D. 106

[1] In discussing activities of mankind—the human species—the authors use the generic word *man* to represent both male and female.

[1-2] Primitive man fashioned tools to sustain his existence. The stone-bladed axe was one of his first inventions.

Characteristics of engineers through the centuries have included willingness to work and an intellectual curiosity about the behavior of things. Their queries about "Why?," "How?," "With what?," and "At what cost?" have all served to stimulate an effort to find desirable answers to many types of problems.

Another characteristic associated with engineers is the ability to "see ahead." The engineer must have a fertile imagination, must be creative, and must be ready to accept new ideas. Whether an engineer lived at the time of construction of the pyramids or has only recently graduated in computer engineering, these characteristics have been an important part of his or her intellectual makeup.

The following sections present a brief picture of the development of engineering since the dawn of history and outline the place that the engineer has held in various civilizations.

THE BEGINNINGS OF ENGINEERING: 6000–3000 B.C.

The beginning of engineering probably occurred in Asia Minor or Africa some 8000 years ago. About this time, man began to cultivate plants, domesticate animals, and build permanent houses in community groups. With the change from a nomadic life came requirements for increased food production. The stone-bladed axe was one of man's first inventions [1-2]. Among the first of his major engineering projects were irrigation systems to promote crop growing. Increased food production permitted time to engage in other activities. Some men became rulers, some priests, and many became artisans, whom we may call the first engineers.

Early achievements in this era included methods of producing fire at will, melting certain rocklike materials to produce copper and bronze tools, invention of the plow, the yoke for animals, wheel and axle, animal-, water-, and wind-driven mills, development of a system of symbols for written communication, origination of a system of mathematics, and construction of irrigation works [1-3].

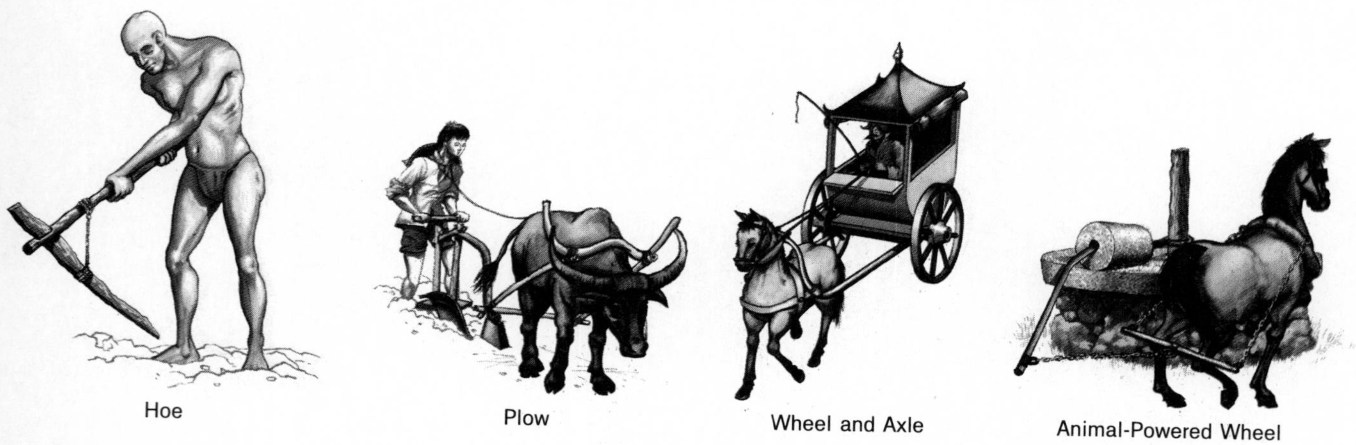

[1-3] History is an account of the ever-increasing complexity of man's inventions.

Early records are so fragmentary that only approximate dates can be given for any specific discovery, but evidence of the impact of early engineering achievements is readily discernible. For example, in setting up stable community life in which land was owned, there had to be provisions both for irrigation and for accurate location and maintenance of boundaries. This necessity stimulated the development of surveying and of mathematics. The moving of earth to make canals and dams required computations. To complete the work the efforts of many men had to be organized and directed. As a result, a system of supervisors, foremen, and workers was established that formed the beginnings of a class society. In this society, craftsmen became a distinct group producing useful items such as pottery, tools, and ornaments that were desired by others. As a result, trade and commerce were stimulated and roads were improved.

In order to record the growing accumulation of knowledge about mathematics and engineering, the early engineer needed a system of writing and some type of writing material. In the Mesopotamian region, soft clay was used on which cuneiform characters were incised. When baked, the clay tile material was used for permanent documents, some of which are legible even today [1-4]. In the Nile

[1-4] Mesopotamia, often called the "cradle of civilization," helped to nurture engineering in its infancy. Clay tablets have been unearthed which show city plans, irrigation, water supply systems, and road maps. Although no engineering tools have been discovered among the remains of ancient Mesopotamia, the evidence unearthed of their remarkable architectural construction indicates that they used measuring tools, which, even though primitive, aided in producing engineering of a high degree for this period. Their cities, with their water supply, irrigation systems, and road networks, were among the wonders of the ancient world.

Many outstanding contributions of mathematics were made by the Mesopotamians. It has been proven that they had knowledge of the sexagesimal system, in which they divided the circle into 360 degrees, the hour into 60 minutes, and the minute into 60 seconds.

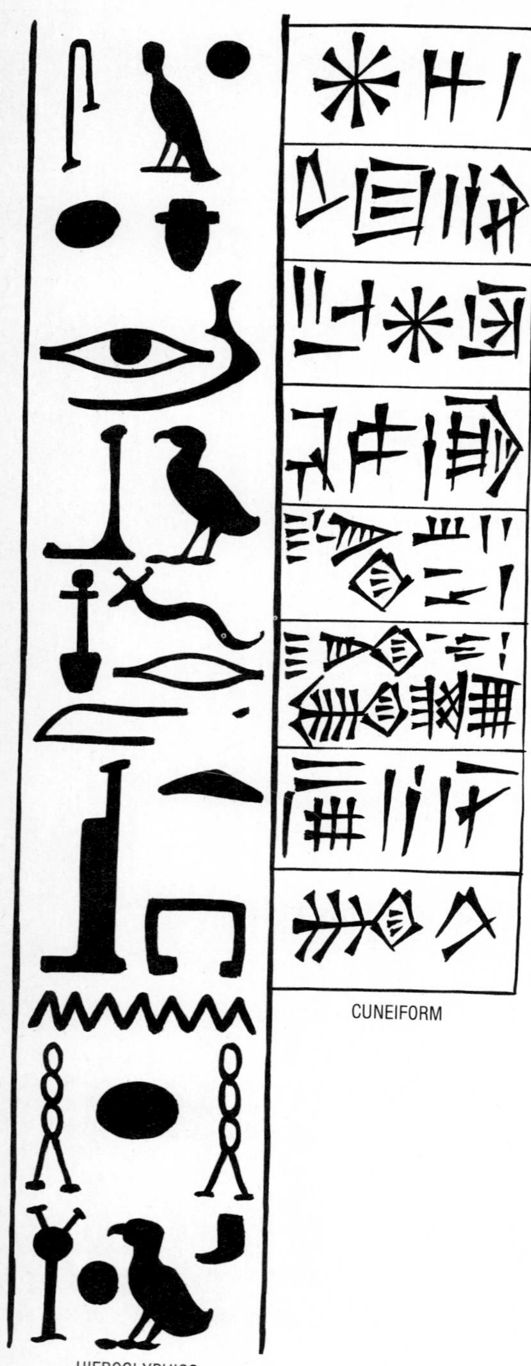

[1-5] Cuneiform and hieroglyphics were two forms of writing used by ancient engineers.

Valley, a paperlike material called *papyrus* was made from the inner fibers of a reed. In other parts of Asia Minor, treated skins of animals were used to form parchment. Occasionally, slabs of stone or wood were used as writing materials. The type of writing that developed was strongly influenced by the writing material available. For example, the incised characters in soft clay differed significantly from the hieroglyphics (picture writing) used to write on papyrus [1-5].

Engineering work requires a source of energy. Unfortunately, this requirement has from time to time led to the enslavement and use of numbers of human beings as primary sources of energy. The construction of all early engineering works, whether Oriental, Mediterranean, or American Indian, was accomplished principally by human labor. It was not until near the end of the period of history known as the Middle Ages that mechanical sources of power were developed.

ENGINEERING IN EARLY CIVILIZATIONS: 3000–600 B.C.

After about 3000 B.C., enough records were made on clay tablets, on papyrus and parchment, on pottery, and as inscriptions on monuments and temples to provide us with information about ancient civilization. These records show that urban civilizations existed in Egypt, Mesopotamia, and the Indus Valley, and that a class society of craftsmen, merchants, soldiers, and government officials was a definite part of that civilization.

In Mesopotamia, clay tablets have been uncovered which show that Babylonian engineers were familiar with basic arithmetic and algebra. From these writings we know that they routinely computed areas and volumes of land excavations. Their number system, based on 60 instead of 10, has through the centuries been handed down to us in our measures of time and angle. Their buildings were constructed using basic engineering principles still in use today [1-6]. Primitive arches were used in some of their early hydraulic works. Bridges were built with stone piers carrying wooden stringers for the roadway. Some roads were surfaced with a naturally occurring asphalt, a construction method that was not used again until the nineteenth century.

It was in Egypt that some of the world's most remarkable engineering was performed. Beginning about 3000 B.C. and lasting for about 1000 years, the Pyramid Age flourished in Egypt. The first pyramids were mounds covered with stone. Building techniques progressed rapidly until the Great Pyramid was begun about 2900 B.C. Stones for the structures were cut by workmen laboriously chipping channels in the native rock. A ball made of a harder rock was sometimes used as a tool. At other times grooves were cut in the stone with mallet and bronze chisel. Using wedges, the stones could then be split apart from the bedrock. By this method, blocks weighing 15 tons or more were cut for use in building. Over 2,300,000 building stones weighing about 5000 pounds each were used to construct the Great Pyramid of Cheops. Its 206 courses stand 481 feet high. Egyptian masons anchored facing stones, weighing up to 15 tons each, in a thin mortar of gypsum and sand. Many of the joints separating the facing stones are imperceptible, less than 0.02 inch in width. The Egyptian engineers apparently used only the lever, the inclined plane, the wedge, and the roller in their construction efforts [1-7].

Although early construction tools were primitive, the actual structures, even by today's standards, are outstanding examples of engineering skill in measurement and layout. For example, the base of the Great Pyramid is square within about 1 inch in a distance of 755 feet, and its angles are in error by only a few minutes despite the fact that the structure was built on a sloping rocky ledge.

The Egyptian engineers and architects held a high place in the Pharaoh's court. Imhotep, a designer of one of the large pyramids, was so revered for his wisdom and ability that he was included as one of the Egyptian gods after his death. Egyptian engineers were skilled in building and land measurement. Annual overflows of the Nile River obliterated many property lines and a resurvey of the valley was frequently necessary. Using geometry and primitive measuring equipment, they restored markers for land boundaries after the floods receded [1-8].

The Egyptians also were skilled in irrigation work. Using a system of dikes and canals, they reclaimed a considerable area of desert. An ancient engineering contract to build a system of dikes about 50 miles long has recently been discovered.

Although the skill and ingenuity of the Egyptian engineers were outstanding, the culture lasted only a relatively short time. Reasons that may account for the failure to maintain leadership are many, but most important was the lack of pressure to continue development. The civilization was affluent. Little influence was brought to bear on the engineers to cause them to continue their creative efforts. Also, after an agricultural system was established, living conditions were favorable and little additional engineering was required. The lack of urgency to do better finally stifled most of the creativity of the engineers and the civilization fell into decay.

[1-6] Ancient builders employed engineering principles in the construction of their structures. Clay plumb bobs were used by Babylonian builders.

[1-7] The pyramids of Egypt exemplify man's desire to create and build enduring monuments.

[1-8] In ancient Egypt warfare and strife delayed the development of engineering; however, with the unification of Upper and Lower Egypt, the science of measurement and construction made rapid progress. Buildings, city planning, and irrigation systems show evidence of this development. The Pyramids are still engineering marvels today both in design and construction.

The Egyptians also advanced mathematics. This is attested to by papyrus scrolls, dating back to 1500 B.C., which show that the Egyptians had knowledge of the triangle and were able to compute areas and volumes. They also had a device to obtain the azimuth from the stars.

The annual floods of the Nile afforded ample practice in measurement surveying. This may well have been the first example of the importance of resurveys. The rope used as a measure was first soaked in water, dried, and then coated heavily with wax to ensure constant length.

SCIENCE OF THE GREEKS AND ROMANS: 600 B.C.–A.D. 400

The history of engineering in Greece had its origins in Egypt and the East. With the decline of the Egyptian civilization, the center of learning shifted to the island of Crete and then about 1400 B.C. to the ancient city of Mycenae in Greece.

To the engineers of Mycenae were passed not only the scientific discoveries of the Egyptians but also a knowledge of structural building materials and a language that formed the basis of the early Greek language. These engineers subsequently developed the corbeled arch and made wide use of irrigation systems.

The Greeks of Athens and Sparta borrowed many of their developments from the Mycenaean engineers. In fact, the engineers of this period were better known for the intensive development of borrowed ideas than for creativity and invention. Their water system, for example, modeled after Egyptian irrigation systems, showed outstanding skill in the use of labor and materials. Their engineering

When looms weave by themselves man's slavery will end.

Aristotle, 287—212 B.C.

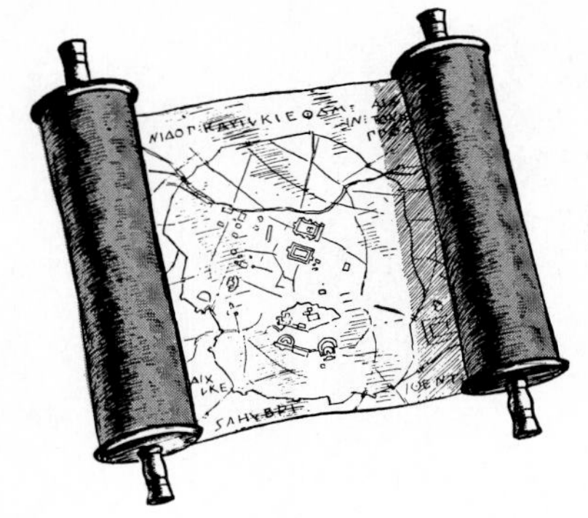

[1-9] The outstanding progress made by the Ancient Grecians in architecture and mathematics and their contribution to the advancement of engineering demand our admiration.

Aristotle contended that the world was a spheroid. He stated that observations of the various stars showed the circumference of the earth to be about 400,000 stadia (4600 miles).

Erathosthenes of Cyrene observed that the sun's rays, when perpendicular to a well at Alexandria, cast a shadow equal to one-fiftieth of a circle at Syene (Aswan) 500 miles away. From this he established that the circumference of the earth was 50 times 500 miles or 25,000 miles.

The Greeks constructed many buildings and structures of large size, which show engineering skill and excellent architectural design. One tunnel, which was built to bring water to Athens, measured 8 feet by 8 feet and was 4200 feet in length. The construction of such a tunnel necessitated extremely accurate alignment both on the surface and underground.

efforts were much like those of today, where the vast amount of technological progress can be classified as *development* rather than research.

Greece was also famous for its outstanding philosophers [1-9]. Significant contributions were made by men such as Socrates, Plato, Aristotle (acclaimed as the greatest physical scientist of the period), and Archimedes. In the realm of abstract thought, they perhaps have never been equaled. However, at that time extensive use of their ideas was retarded because of the belief that verification and experimentation, which required manual labor, were fit only for slaves. Archimedes is best known as one of the greatest mathematicians of antiquity. He developed mathematically based ideas, particularly as related to the statics of solid and liquid bodies. Archimedes (well known for buoyancy and flotation) is also thought to have invented the water screw during a visit to Egypt. This type of machine is still in wide use there today for irrigation [1-10]. Of all the contributions of the Greeks to the realm of science, perhaps the greatest was the

[1-10] The Archimedes water screw is still used today to lift water.

[1-11] The Roman army used a gear-driven odometer on the axle of the unit commander's chariot to measure distance traveled. For each measured distance a round stone was discharged into a cup on the tail of the chariot.

It chanced that Archimedes was by himself, working out some problem with the aid of a diagram. Having fixed his thoughts and his eyes as well upon the matter of his study, he was not aware of the incursion of the Romans or of the capture of the city. Suddenly a soldier came upon him and ordered him to go with him to Marcellus. This Archimedes refused to do until he had worked out his problem and established his demonstration, whereupon the soldier flew into a passion, drew his sword, and despatched him.

Friedrich Klemm, A History of Western Technology, *1979*

Eureka! (I have found it!)

Archimedes, 287–212 B.C.

discovery that nature has general laws of behavior which can be described with words.

The Great Wall of China, over 1500 miles long, was completed about 2200 years ago. It averaged 25 feet in height, 15 feet in width at the top, and 25 feet in width at the base. It was built by human labor to protect the Chinese against Mongolian raids.

The best engineers of antiquity were the Romans. They were called *architectus,* and they were greatly respected. Within a century after the death of Alexander, Rome had conquered many of the eastern Mediterranean countries, including Greece. Within two more centuries Rome had dominion over most of the known civilized areas of Europe, Africa, and the Middle East. Roman engineers liberally borrowed scientific and engineering knowledge from the conquered countries for use in warfare and in their public works. Although in many instances they lacked originality of thought, Roman engineers were superior in the application of ideas and techniques that they learned from their conquests of neighboring countries. Hero invented one of the earliest calculating devices, the odometer

[1-11], and the steam turbine (aeolipile), which is the forerunner of today's reaction turbines [1-12]. He also described a hydraulic clock and a hydraulic organ, a force pump, a fire engine, an air gun, and a most spectacular theater scene with steam- and hot-air-operated puppets revolving and dancing (a skilled engineering effort later perfected at Disneyland).

From experience, Rome had learned the necessity for establishing and maintaining a system of communications to hold together the great empire. Thus Roman roads became models of engineering skills. By first preparing a deep subbase and then a compact base, the Romans advanced the technique of road construction to such a level that some Roman roads are still in use today. At the peak of Roman sovereignty, the network of roads comprised over 180,000 miles stretching from the Euphrates Valley to Great Britain.

In addition, Roman engineers were famous for the construction of aqueducts for water supply, systems of drainage, heated houses, and public baths. Using stone blocks in the constructing of arches, they exhibited unusual construction skills. An outstanding example of this construction is the famous Pont du Gard near Nîmes, France,

[1-12] Hero's turbine was a forerunner of today's reaction turbine.

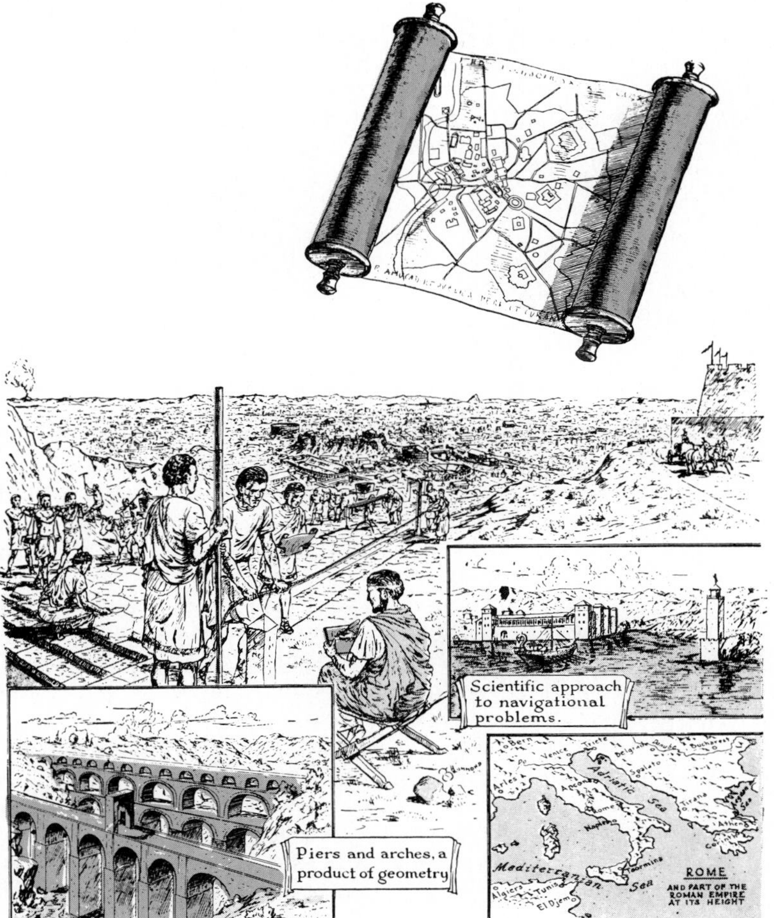

[1-13] The Romans excelled in the building of aqueducts. Many carried water for great distances with perfect grade and alignment. The key design element in this type of construction was the arch, which was also used in bridges, tunnels, buildings, and other construction.

Evidence of the Romans' knowledge and understanding of basic geometric principles is further shown by their river and harbor construction and scientific approach to navigational problems.

Sanitary systems, paved roads, magnificent public buildings, water supply systems, and other public works still in evidence today stand as monuments to the Roman development of engineering as a key to the raising of the standard of living.

The rise of the Roman Empire was attributable to the application of engineering principles applied to military tactics. The invincibility of the Roman legions was the result not only of the valor of the fighting men but also, and perhaps more strongly, to the genius of the Roman military engineers.

which is 160 feet high and over 900 feet long. It carries both an aqueduct and a roadway [1-13].

By the time of the Christian era, iron refining had developed to the extent that iron was being used for small tools and weapons. However, the smelting process was so inefficient that over half of the metallic iron was lost in the slag. Except in the realm of medicine, no interest was being shown in any phase of chemistry.

Despite their outstanding employment of construction and management techniques, the Roman engineers seemed to lack the creative spark and imagination necessary to provide the improved scientific processes required to keep pace with the expanding demands of a farflung empire. The Romans excelled in law and civil administration but were never able to bring distant colonies fully into the empire. Finally, discontent and disorganization within the empire led to the fall of Rome to a far less cultured invader.

ENGINEERING IN THE MIDDLE AGES: FIRST TO SIXTEENTH CENTURIES

Vessels can be made which row without the force of men, so that they can sail onward like the greatest river or sea-going craft, steered by a single man; and their speed is greater than if they were filled with oarsmen. Likewise, carriages can be built which are drawn by no animal but travel with incredible power . . . flying machines can be constructed, so that a man, sitting in the middle of the machine, guides it by a skillful mechanism and traverses the air like a bird.
Roger Bacon, ca. 1220–ca. 1292

After the fall of Rome, scientific knowledge was dispersed among small groups, principally under the control of religious orders. In the east, an awakening of technology began among the Arabs but little organized effort was made to carry out any scientific work. Rather, it was a period in which isolated individuals made new discoveries or rediscovered earlier known scientific facts.

It was during this time that the name *engineer* was first used. Historical writings of about A.D. 200 tell of an *ingenium*, an invention, which was a type of battering ram used in attacks on walled defenses. Some thousand years later, we find that an *ingeniator* was the man who operated such a device of war—the beginning of our modern title, *engineer*.

Several technical advances were made late in this period. One important discovery involved the use of charcoal and a suitable air blast for the efficient smelting of iron. Another advance was made when the Arabs began to trade with China and a process of making paper was secured from the Chinese. Within a few years the Arabs had established a paper mill and were making paper in large quantities. With the advent of paper, communication of ideas began to be reestablished. Also in Arabia, the sciences of chemistry and optics began to develop. Sugar refining, soap making, and perfume distilling became a part of the culture. The development of a method of making gun powder, probably first learned from China about the fourteenth century, also had rapid and far-reaching results [1-14].

[1-14] The hand cannon ended the superiority of armor.

After centuries of inaction, the exploration of faraway places began again, aided greatly by the development of a better compass. With the discovery of other cultures and the uniting of ideas, there gradually emerged a reawakening of scientific thought.

With the growth of Christianity, an aversion arose to the widespread use of slaves as primary sources of power. This led to the development of waterwheels and windmills and to a wider use of animals, particularly horses, as power sources.

About 1454, Johann Gutenberg, using movable type, produced

the first books printed on paper. This meant that the knowledge of the ages, which previously had been recorded laboriously by hand, could now be disseminated widely and in great quantities. Knowledge, which formerly was available only to a few, now was spread to scholars everywhere. Thus the invention of paper and the development of printing served as fitting climaxes to the Middle Ages.

Seldom has the world been blessed with a genius such as that of Leonardo da Vinci (1452–1519). Although still acclaimed today as one of the greatest of all artists, his efforts as an engineer, inventor, and architect are even more impressive. The extent of his observations and creative and innovative ideas covered the whole field of science from psychology to mathematics. Long after his death his designs of a steam engine, a machine gun, a camera, conical shells, a submarine, a parachute, and a helicopter have been proven to be workable [1-15].

[1-15] Leonardo da Vinci . . . artist, architect, engineer . . . many of the ideas that he had 500 years ago have only been proven to be feasible in the past 100 years.

Galileo (1564–1642) was also a man of great versatility. He was an excellent writer, artist, and musician, and he is also considered one of the foremost scientists and innovators of that period. One of his greatest contributions was his formulation of what he considered to be a new method of gaining knowledge, the interplay between calculable forces *(theory)* and measurable bodies *(experimentation)*.

THE REVIVAL OF SCIENCE: SEVENTEENTH AND EIGHTEENTH CENTURIES

When Newton saw an apple fall, he found . . .
A mode of proving that the earth turn'd round
In a most natural whirl, called "gravitation";
And thus is the sole mortal who could grapple,
Since Adam, with a fall or with an apple.

George Byron, 1788–1824
Don Juan

Following the invention of printing, the self-centered medieval world changed rapidly. At first, the efforts to present discoveries of nature's laws met with opposition and in some cases even hostility. Slowly, however, freedom of thought was permitted and a new concept of *testing to evaluate a hypothesis* replaced the early method of establishing a principle solely by argument.

Four men in this period made discoveries and formulated laws that have proved to be of great value to engineering. They were Robert Boyle (1627–1691), who formulated a law relating pressures and volumes of gases; Christian Huygens (1629–1695), who invented the pendulum clock and investigated the effects of gravitational pull; Robert Hooke (1635–1703), who experimented with the elastic properties of materials and invented the universal joint; and Isaac Newton (1642–1727), who is famous for his three basic laws of motion. All the early experimenters were hampered by a lack of a concise vocabulary to express their ideas. Because of this, many of the principles were expressed in a maze of wordy statements.

During this period, significant advancements were made in communication and transportation. Canals and locks were built for inland water travel and docks and harbors were improved for ocean commerce. Advances in ship design and improved methods of navigation permitted a wide spreading of knowledge that formerly had been isolated in certain places.

The search for power sources to replace human labor continued. Water power and wind power were prime sources, but animals began to be used more and more. About this time, the first attempts to produce a steam engine were made by Denis Papin (1647–1712) and Thomas Newcomen (1663–1729). Although these early en-

[1-16] Not all innovative ideas are operational. For example, this concept of man's attempt to harness birdpower, in an attempt to fly, first appeared in *The Man in the Moon,* published in Paris in 1648.

gines were very inefficient, they did mark the beginning of power from heat engines.

An important industry was made possible in this period by the development of spinning and weaving machinery by such men as Jurgen, Hargreaves, Crampton, and Arkwright. This period also marked a general awakening of science after the Dark Ages. Individual discoveries, although usually isolated, found their way into useful products within a short period of time because of the development of printing and the improvements in communication.

The basic discoveries in this era were made by men who were able to reject old, erroneous concepts and search for principles that were more nearly in accord with nature's behavior. Engineers in any age must be equally discerning if their civilization is to advance.

BEGINNINGS OF MODERN SCIENCE: NINETEENTH CENTURY

Early in the nineteenth century, two developments provided an impetus for further technological discoveries. The two developments were the introduction of a method, developed by Henry Cort (1740–1800), of refining iron and the invention of an efficient steam engine by James Watt (1736–1819) [1-17]. These developments provided a source of iron for machinery and power plants to operate the machinery.

As transportation systems began to develop, both by water and by land, a network of railroads and highways was built to tie together the major cities in Europe and in the United States.

In this period, the awakening of science and engineering truly had begun. Now, although people were slow to accept new ideas, knowledge was not rejected as it had been in earlier centuries. Colleges began to teach more and more courses in science and engineering, and it was here that the fuse was lighted for an explosion of discoveries in the twentieth century.

One of the most important reasons for the significant develop-

[1-17] This engraving of James Watt, a Scottish engineer, first appeared in 1860. It portrays an experimental design which was to become the forerunner of the modern condensing steam engine. Watt had been given a Newcomen engine to repair, in 1764, and, noting its extreme wastefulness of fuel, set about the task of building a better machine.

[1-18] Henry Ford's assembly line of 1915 is acclaimed today as a milestone toward modern manufacturing practices.

ment of technology in this period was the increasingly close cooperation between science and engineering. It began to become more and more evident that discoveries by research scientists could be used to develop new articles for commerce. Industry soon began to realize that money spent for research and development eventually returned many times its value.

I know this world is ruled by Infinite Intelligence. It required Infinite Intelligence to create it and requires Infinite Intelligence to keep it on its course. Everything that surrounds us—everything that exists—proves that there are Infinite Laws behind it. There can be no denying this fact. It is mathematical in its precision.

Thomas Alva Edison, 1847–1931

No man really becomes a fool until he stops asking questions.

Charles P. Steinmetz, 1865–1923

TWENTIETH-CENTURY TECHNOLOGY

As the twentieth century came into being, a number of inventions emerged that were destined to have far-reaching effects on our civilization. The automobile began to be more widely used as better roads were made available. Henry Ford designed and built a successful "gasoline buggy" in 1896 and in the early part of the twentieth century he began to build and sell automobiles. He provided American industry with a new dimension in manufacturing. More than any other person he has been universally acclaimed as the individual responsible for bringing modern mass production into being [1-18]. Today much of industry follows his industrial philosophy—to reduce the price of the product while improving its quality, to increase the volume of sales, to improve production efficiency, and then to increase output to sell at still lower prices. Just prior to the turn of the century, Frederick Taylor (1856–1915) originated principles of scientific management that served as the principles on which modern industrial engineering has been built. The inventions of Thomas Alva Edison (1847–1931) and Lee DeForest (1873–1961) of electrical equipment and electron tubes started the widespread use of power systems and communication networks [1-19]. One of the most creative individuals of this era was Nikola Tesla (1856–1943), who introduced the first practical application of alternating current, the polyphase induction motor. Following the demonstrations by Orville and Wilbur Wright that man could build a machine that would fly, aircraft of many types developed rapidly [1-20].

These inventions, typical of many basic discoveries that were made early in the century, exemplify the spirit of progress of this

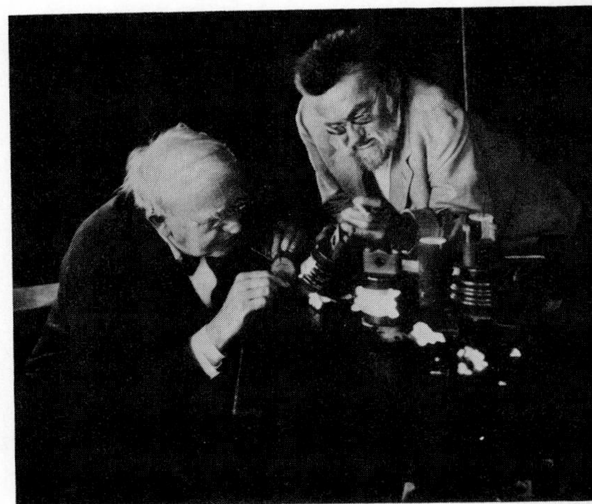

[1-19] Nations the world over acknowledge their indebtedness to Thomas Alva Edison, inventor, and Charles Proteus Steinmetz, electrical engineer, for their significant inventions relating to electricity.

[1-20] Orville and Wilbur Wright, U.S. inventors and aviation pioneers, achieved in 1903 the world's first successful powered, sustained, and controlled flights of an airplane.

period. So fast was the pace of discovery, with one coming on the heels of another, that it is difficult to evaluate properly their relative importance, although we certainly can realize their impact on our way of life. However, in a number of instances the practicality of an engineering invention had been demonstrated many years in advance of its implementation.

The past 50 years has been a period of remarkable technological development. In 1984, the National Society of Professional Engineers selected the 10 most outstanding engineering achievements of the previous 50 years as follows:[2]

How Nylon Got Its Name

First, it was called "No Run." But there was an objection on sound grounds. Then NoRun was turned backward to form NURON. But, wouldn't this make nylon appear to be a backward child? So the Y was substituted for the U and the R became an L, and NYLON came into the American language as a generic term for a polyamide fiber.

E.I. duPont de Nemours & Company

Synthetic Fibers: Nylon

Wallace Carothers led a team of organic chemists and chemical engineer researchers at the duPont Company to develop *nylon,* a new polymer. This synthetic resin, characterized by elasticity, strength, and resistance to abrasion and chemicals, was announced October 28, 1938. Subsequent research brought about thousands of other polymers, each with unique characteristics [1-21]. These were used extensively for the war effort and today they make up more than 70 percent of the fibers produced from all sources, being used in a countless variety of applications—from medical supplies to consumer products.

[1-21] One of the first uses for nylon was for women's hose.

Nuclear Energy: The First Nuclear Pile

An outstanding characteristic of this century is the increased use of power. Albert Einstein, a world-renowned physicist at Princeton University, hypothesized an equivalence of mass and energy, $E = mc^2$ [1-22]. Using Einstein's model, a group of scientists from Europe and the United States worked in secrecy at the University of Chicago in the early 1940s to produce a working nuclear pile composed of graphite, uranium, and cadmium control rods. On December 2, 1942, the first test of the pile was successfully completed. The age of controlled nuclear reaction was born. Today, nuclear power is used

[2] "The Ten Outstanding Engineering Achievements of the Past 50 Years," *Professional Engineer,* Summer 1984, p. 18.

for a large number of applications, from powering ships to metropolitan utility systems. In the last few years public concern for enhanced safety and reliability, and for improved disposal processes for nuclear waste, has slowed the pace of development of nuclear power plants in the United States.

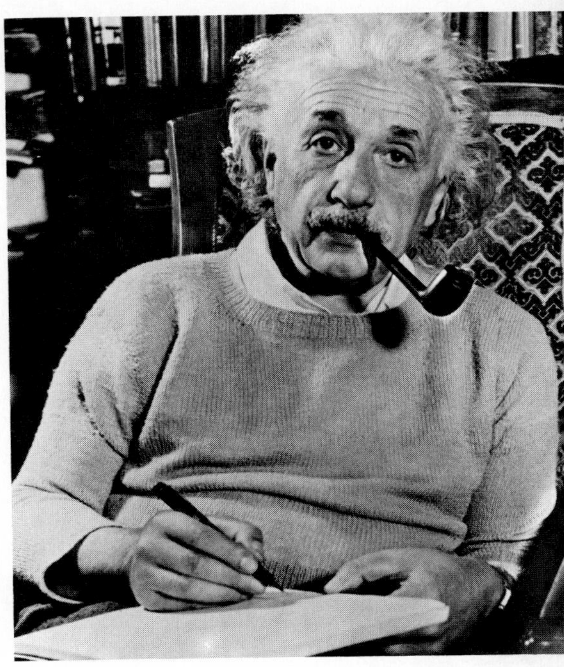

[1-22] Albert Einstein, father of the nuclear age, emphasized the importance of simplicity in all matters. He believed that harmony will result if people act in accordance with principles founded on consciously clear thinking and experience.

Computers: ENIAC

In 1943, a federal development contract was awarded to the University of Pennsylvania's Moore School of Engineering to develop with existing electronics a super-high-speed calculating machine. Under the direction of John Brainerd as project supervisor, and after $2\frac{1}{2}$ years of research, development, and manufacture, a computer (called ENIAC) was designed and constructed [1-23]. It weighed well over 30 tons and occupied a floor space of some 1500 square feet. Although slow by today's standards, the ENIAC was a remarkable breakthrough that established the major concepts on which high-speed electronic computers are based. In the past 40 years the computer has become an essential part of the fabric of our society—in research, development, business, and even the home.

Solid-State Electronics: The Transistor

In December 1947, while working on radio diodes, a team of Bell Laboratory researchers, led by John Bardeen, Walter Brattain, and William Shockley, discovered that current changes in one part of a diode caused current changes in another part [1-24]. Thus the transistor—an efficient and economical way to regulate current—was discovered. The development of commercial uses for transistors soon followed and made possible a modern revolution in many fields, including communications, computers, and medicine. Important refinements were discovered in 1959 at Texas Instruments and Fairchild Semiconductors, when the transistor's silicon crystal was made to serve as its own circuit board. For over two decades the transistor has been the switch that controls the world.

[1-23] Today's desktop microcomputers are more powerful than the ENIAC.

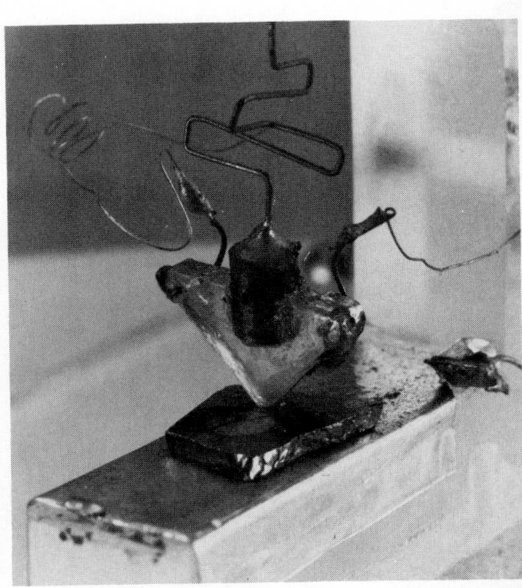

[1-24] Over 1,000,000 transistors today can occupy less space than that required for the world's first transistor.

[1-25] Jet aircraft have brought world travels to the masses.

Jet Aircraft: Boeing 707

Using new jet technology and other improvements in plane design developed during the war, in the mid-1950s Boeing Airplane Company designed a plane capable of carrying over 180 passengers at speeds exceeding 600 miles per hour [1-25]. The result was the highly successful Boeing 707. It was powered by four Pratt & Whitney turbojet engines, pod-mounted on wings that swept back at a 35° angle. Six other jetliners—from the 707-120 to the 747 superjet—have been developed from the basic 707 configuration.

Biomedical Engineering: The Implantable Pacemaker

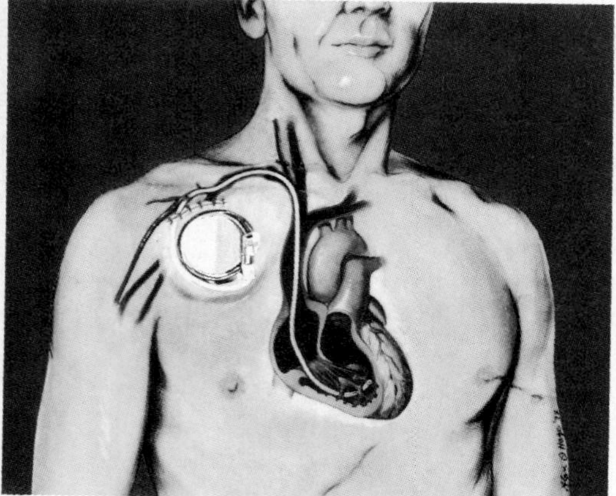

[1-26] Over 2,000,000 lives have been saved by the implanted pacemaker.

In 1960, engineers at Medtronic, working with Wilson Greatbatch and several physicians, developed a new technology to diminish the toll from heart disease. Using high-performance plastics, metals, and electronics, they produced the first implantable pacemaker, a small-signal generator to keep the heart pumping. By 1981, over 122,000 pacemaker implants were being performed each year [1-26]. The development of the self-contained pacemaker served as a medical breakthrough and underlined the growing influence of engineers in medical technology.

Lasers: Helium–Neon Laser

The basic principles of the laser (*l*ight *a*mplification by *s*timulated *e*mission of *r*adiation), a totally new light source, was first conceptually envisioned by Albert Einstein in 1917. In 1958, A. L. Schawlow and C. H. Townes proposed the first theoretical basis for a laser, the first coherent light source. In 1960, Theodore Maiman focused the glare of a flash lamp in a rod of synthetic ruby, producing the first working laser on earth. From this laser he produced a burst of crimson light so brilliant it outshone the sun. This laser beam could be transmitted over great distances, for example, to the moon. Laser light is bright, intense, pure, and directional. Its power is rated in watts and can easily concentrate 10 billion watts into an area no larger than a dime. Such a device has the potential to revolutionize communications and influence a host of diverse fields. In 1962, A. Javan, W. R. Bennett, Jr., and D. R. Herriott, working at Bell Laboratories, were able to design the first laser that emitted a continuous beam of light—a helium–neon laser [1-27].

[1-27] Practical uses of lasers seem to be limitless.

Today, lasers are used in a variety of ways and account annually for over $2 billion worth of industrial goods. By using robots to direct laser tools and sensors, future factories will be much safer for the workers than has been the case in the past. Even now they are used extensively to assist surgeons, to transmit telephone calls, to track storms, to checkout in supermarkets, to weld steel, to cut fabric, and to produce holograms. Holography itself is a virtual miracle of modern technology. A hologram can be created by the intersecting beams of a laser. This allows us to view a recorded scene or object from different directions, all appearing to be three-dimensional. Holography has great potential to produce three-dimensional movies and television, and it can vastly improve our ability to measure small changes in heat, vibration, and deformation.

Communications Satellites: Telstar

In 1945, science fiction writer A. C. Clarke (author of *2001: A Space Odyssey*) proposed satellite communications. During the 1950s, J. L. Pierce and other scientists and engineers at Bell Laboratories worked to make Clarke's dream come true. Many engineering problems had to be solved. These included the development of special materials, electronics, and rocket technology. This dream materialized on July 10, 1962, with the successful launching of Telstar I [1-28]. This satellite would remain in a geosynchronous orbit and relay signals from one location to another. Telstar I was the first of a series of increasingly complex satellites to be used in today's global communications network. Communications satellites now handle more than half of all transoceanic telephone traffic, television, and audio network program distribution, and high-speed data services for government and industry. No other form of communication has had such a profound and immediate impact on people's ability to exchange information.

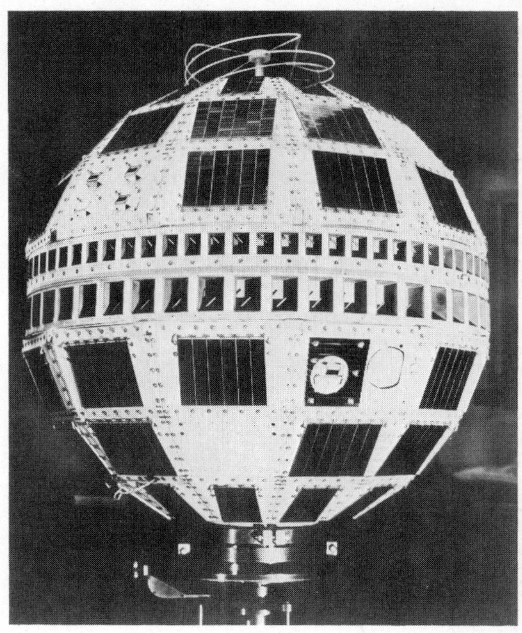

[1-28] Communication satellites have simplified man's ability to communicate with man.

The U.S. Space Program: Apollo Eleven

Perhaps the greatest united engineering effort of all times was set into motion by President John F. Kennedy in May 1961 when he challenged the United States to place a man on the moon before the end of the decade. To coordinate these complex efforts he initiated the Apollo Space Program. The next eight years recorded more advances in aerospace engineering than the world had seen since the first primitive rocket was launched in the thirteenth century. Virtually every element in the mission had to be imagined, built, tested, and matched with millions of other components. On July 20, 1969, the world watched in awe as Apollo Eleven astronaut Neil Armstrong set foot on the lunar surface. It was indeed "One giant step for mankind" [1-29]. The modules, the thruster engines, the launch-

Today the spectrum of laser types extends from those as large as football fields to others smaller than a pinhead. Their light ranges from invisible—ultraviolet and infrared—through all colors of the rainbow. A few fire in pulses lasting but quadrillionths of a second; many could beam steadily for decades . . . some lasers can focus light to a fine point bright enough to vaporize iron or any other earthly material, concentrating energy on it a million times faster and more intensely than a nuclear blast. Others do not emit enough energy to coddle an egg. . . . Lasers weld car parts, husk peanuts, and cut teeth in saws. They also drill holes in baby-bottle nipples. Often lasers are the preferred tools of surgeons in the operating room. . . . Growing more powerful every year, high energy lasers are bringing the world ever closer to the energy source of the stars—nuclear fusion.

"The Laser: A Splendid Light for Man's Use,"
National Geographic, *March 1984, p. 335*

One day, about noon, going towards my boat, I was exceedingly surprised with the print of a man's naked foot on the shore.

Daniel Defoe, 1660—1731
Robinson Crusoe

[1-29] Man's safe exploration of the moon will long stand as one of his greatest engineering achievements.

[1-30] Inertial guidance systems have been greatly responsible for man's successful conquest of space.

ing rockets, the control and guidance equipment, the radio and television equipment—all had to be engineered to work together perfectly. And for 195 hours, 18 minutes—from lift-off to splash-down—everything worked. This incredible achievement was made possible by the combined efforts of thousands of engineers from all the disciplines.

Automation and Control Systems: Inertial Navigational Guidance

The space program could not have succeeded without the development of a highly sophisticated navigational system. Working at his MIT Instrumentation Laboratory, Charles Stark Draper developed such a system. He was the first to succeed in combining gyroscopes, small computers, and accelerometers to produce the first true inertial guidance system. The system was installed in the Apollo project's command and lunar excursion modules [1-30]. It allowed astronauts to target their splashdowns from over a quarter of a million miles away. Other inertial guidance systems developed in the MIT laboratory now guide ships and submarines, and are used in airplane autopilots.

[1-31] Bertha Lamme, the nation's first woman graduate engineer (Ohio State University, 1893), had an outstanding professional career with Westinghouse. She was the author of many patents.

ENGINEERING TODAY

Broadly speaking, modern engineering in this country had its beginnings about the time of the close of the Civil War. Within the last century, the pace of discovery has been so rapid that it can be classed as a period within itself. In these modern times, engineering endeavor has changed markedly from procedures used in the time of Imhotep, Galileo, or Edison. Formerly, engineering discovery and development were accomplished principally by individuals. With the increased store of knowledge available and the widening of the field of engineering to include so many diverse branches, it is usual to find groups or teams of engineers and scientists working on a single project. Where formerly an individual could absorb and understand practically all of the scientific knowledge available, now the amount of information available is so vast that an individual can retain and employ at best only a part of it.

Today there are more engineers and scientists alive than have ever lived in all of history. Since 1900 the ratio of engineers and scientists in the United States in comparison to the total population has been steadily increasing. Predictions based on past increases indicate the following trends:

Year	Ratio of U.S. engineers and scientists to population
1900	1 to 1800
1950	1 to 190
1960	1 to 130
1980	1 to 65
2000	1 to 30

If this is the case, there will be even greater increases in technological advances in the next 20 years than there have been in the past 20 years.

In the past decade a larger percentage of women have begun to study engineering. Traditionally, particularly in the United States, women have not been so free to select engineering as a career choice. In this mechanized age physical strength is of little importance to the engineer, while mental ability, creativity, curiosity, and interest in problem solving are more important attributes. Both men and women possess these traits [1-31 and 1-32].

In this age, as in any age, engineers must be creative and able to visualize what may lie ahead. They must possess fertile imaginations and a knowledge of what others have done before them. As Sir Isaac Newton is reputed to have said, "If I have seen farther than other men, it is because I have stood on the shoulders of giants." The giants of science and engineering still exist. All any person must do to increase his or her field of vision is to climb up on their shoulders.

[1-32] In 1984, Eleanor Baum became the first woman to be named dean of an engineering school, Pratt Institute, Brooklyn, N.Y.

We should all be concerned about the future because we will have to spend the rest of our lives there.
Charles Franklin Kettering, 1876–1958
Seed for Thought

It's just as sure a recipe for failure to have the right idea fifty years too soon as five years too late.
J. R. Platt

PROBLEMS

1-1. Prepare a chart as a series of columns, showing happenings and their approximate dates in a vertical time scale for various civilizations, beginning about 3000 B.C. and extending to about A.D. 1200.

Chinese	*Middle East*	*Egyptian*	*Greek*	*Roman*	*Western Europe*

1-2. Prepare a brief essay on the possible circumstances surrounding the discovery that an iron needle when rubbed on a lodestone and then supported on a bit of wood floating on water will point to the north.

1-3. Determine from historical references the approximate number of years that major civilizations existed as important factors in history.

1-4. What were the principal reasons for the lack of advancement of discovery in Greek science?

1-5. Explain why the development of a successful horsecollar was a major technological advancement.

1-6. Draw to some scale a typical cross section of the Great Wall of China, and estimate the volume of rock and dirt required per mile of wall.

1-7. Trace the development of a single letter of our alphabet from its earliest known symbol to the present.

1-8. Describe the details of preparing papyrus from reedlike plants that grew in Egypt.

1-9. Describe the patterns of behavior and accomplishments of ancient engineers that seem to have made successful civilizations and to have prolonged their existence.

1-10. Prepare lists of prominent persons who contributed outstanding discoveries and developments to civilization during the period from A.D. 1200 to 1900 in the fields of (a) mathematics, (b) astronomy, (c) electricity, (d) mechanics, and (e) light.

Person	***Date***	***Major contribution***

1-11. List the 10 most significant engineering achievements of the nineteenth century. Explain your reasons for making these selections.

1-12. Beginning with 3000 B.C., list the 25 most significant engineering achievements. Explain your reasons for making these selections.

1-13. Based on your knowledge of world history, describe the probable changes that might have occurred had the airplane not been invented until 1970.

1-14. Describe the precision with which the pyramids of Egypt were constructed. How does this precision compare with that of modern office buildings of more than 50 stories in height?

1-15. Trace the development of the power-producing capability of man from 3000 B.C. to the present.

1-16. Write a short essay about the engineering accomplishments of each of three women engineers.

1-17. Write an essay on the positive and negative aspects of nuclear power.

1-18. List the 10 most significant problems that had to be overcome in successfully completing the Apollo Eleven space project.

1-19. How many U.S.-launched communication satellites (nonmilitary) are in orbit today? When was the last one successfully launched? How many are working satisfactorily?

1-20. List 20 current uses of lasers.

1-21. List five inventions of Nikola Tesla.

1-22. List five commercial uses of nylon.

1-23. Write a 50-word paragraph concerning the creative engineer Charles F. Kettering.

1-24. Leonardo da Vinci made many sketches of ideas that he envisioned for engineering inventions. Describe one such idea-sketch that has not yet been made to work.

1-25. Make a list of 10 inventions that would improve our quality of life if they were made to work economically.

Chapter 2

Challenges Surrounding Today's Engineer

Our earth is a magnificent planet, unlike others that we know about. The more we learn about the earth, the more wonderful it becomes. As we move about on its surface, we recognize that we are actually living in two worlds at the same time. One is a *natural world,* the other, a *man-made* world. The former, a creation of God, is manifested by certain laws of nature, the latter by designs of man—engineering designs. These designs are the things that have been made by man since his habitation of the earth—his creations, inventions, tools, and products—which have been the means of his survival on this earth.

The natural world existed long before man first made his appearance on the earth, and it will continue to exist, with or without man. We learn about the laws of nature through a study of the various sciences. If we use these laws wisely, we can make our lives safe, comfortable, and productive. However, when we defy these laws, or ignore them, we invariably suffer unpleasant and sometimes deadly consequences.

Biologists have named us *Homo sapiens*—Man the Thinker. All that man has created from and in the world is a product of his thoughts . . . his mind. Thought without action is a mere pastime of philosophers. So man might more appropriately be called *Homo faber*—Man the Skillful, or perhaps Man the Maker. We might even describe him as Man the Tool User, or Man the Fabricator. The authors prefer to use the term *Man*[1] *the Engineer.*

[1] In discussing activities of mankind—the human species—the authors use the generic word *man* to represent both male and female.

[2-1] Where does the firefly get its light?

There is another design that is far better. It is the design that nature has provided. . . . It is pointless to superimpose an abstract, man-made design on a region as though the canvas were blank. It isn't. Somebody has been there already. Thousands of years of rain and wind and tides have laid down a design. Here is our form and order. It is inherent in the land itself—in the pattern of the soil, the slopes, the woods—above all, in the patterns of streams and rivers.

William H. Whyte
The Last Landscape

[2-2] In the beginning God created the heavens and the earth.

Genesis 1:1
The Holy Bible

[2-3] Creative effort is a function of the mind, not the age of the individual.

Imagination is more important than knowledge.
Albert Einstein
On Science, 1931

My toys were all tools—they still are.
Henry Ford, 1863–1947

[2-4] The system.

WE ARE ALL ENGINEERS

The statement above may seem farfetched to you at first. But, consider, if you will, your world as you first discovered it as a child. Like other children, when you first became aware of the world outside yourself, you were curious. You wanted to know about things. You asked many questions of older persons. This attribute of curiosity is also one of the strongest characteristics of engineers. Next, as a child indulging in one of the activities that children like best, you played "make believe"—and you built or made things. You constructed your designs out of such things as big and little boxes, building blocks, Tinker Toys, Lincoln Logs, or perhaps Erector Sets. When you didn't have manufactured toys, you "pretended," using sticks, rocks, dirt, sand on the beach, or even an available puddle of water. You splashed it, diverted it, dammed it, or maybe changed its flow in as many ways as you could imagine.

In other words, you were a builder of play houses, tree houses, forts, castles, mountains (of dirt, snow, or sand), cars, rafts and boats, and other childhood fantasies. As you grew older, your building projects became more sophisticated, as did your ability to repair, modify, and invent more complex devices or ways of doing things. Actually, many of these actions on your part were elementary forms of engineering design.

Most individuals do not choose to focus their interests, ambitions, and education to become occupational practicing engineers. However, for most people their innate urge to create never completely leaves them. Most continue to find other ways to make creative contributions. They may paint a landscape, write a poem, compose a piece of music, devise a new accounting system, create a new hair style, develop a new computer program . . . or engage themselves in a host of other mind-stimulating design activities.

TODAY'S ENGINEERING ENVIRONMENT

Chapter 1 considered man's history in innovation and engineering design. It covered the simplest (the axe) to the most complex (safe travel to the moon and return) of design activities. We found that man has always been able to meet the challenges placed before him. Today's engineers must lead our society in being equally successful. This chapter explores some of the conditions and challenges that confront today's engineers. Some important questions will be asked such as: How can we be sure that our natural environment continues to provide an enhanced quality of life for all people in every part of the earth? And . . . what is my own responsibility?

The earth is a constantly changing system. Until man arrived on the scene the changes that took place over the geological ages were the result of natural processes. However, through time, as man's numbers have grown, he has exerted his own influence on the system. During the past century his population increased at such an alarming rate that his "artificial" changes have taken precedence over the slower natural changes of the system.

Perhaps more than the members of any other profession, engineers throughout history have made possible the incremental changes in our way of life—some desirable, some undesirable. However, too often the engineer has not been involved in the political, religious, economic, demographic, and military decisions that have determined the ultimate use and effect of his creations—his designs.

The system—the earth—has undergone many artificial but significant changes in the last 100 years, many of which are detrimental to maintaining a desirable quality of life and too many of which are irreversible in their influence. Once man questioned, "Am I my brother's keeper?" Now the answer is no longer one of uncertainty; it is, "Yes, I am!" To survive on this earth we *must* be concerned about our neighbor's well-being. For example, when insecticides can be plowed into farmlands in the midwest and a few months later their traces found in animals in such remote locations as the Antarctic, man can no longer ignore the ultimate consequences of his actions—his designs. By analyzing particulate matter in the upper atmosphere, scientists at the Mauna Loa Observatory in Hawaii can tell with great accuracy when spring plowing starts in northern China.[2] The situation is becoming critical and people in all parts of the world *must* begin to show a special concern for the welfare of their fellowmen and of the plants and animals that share this planet with man.

The engineer of today must be more aware of the effects that his designs could have on the total system, on the lives of his fellowmen, both now and in the future, and on his own well-being. In this regard his designs must always reflect a consciousness for the safety and health of all people. He is responsible also to warn those in government and business of the consequences and possible misuse of his designs.

Before we can realistically evaluate the effects of our designs, we must first have a good understanding of the natural order of the system in which we and our fellowmen live. Since the whole realm of the physical and life sciences is devoted to gaining such understanding, no in-depth attempt will be made in this book to address that part of the task. Rather, the discussion here will be devoted to an exploration of the natural constraints of the system and particularly to the major challenges facing the engineer today—especially with regard to restoring the system to a state of change more nearly in harmony with its natural rate of change. However, lasting solutions to complex problems of this magnitude are possible *only* if the general populace demands and supports a national priority sufficient to supply engineers with resources adequate to sustain their designs.

THE BIOSPHERE

The earth and its inhabitants form a complex system of constantly changing interrelationships. From the beginning of time until a few thousand years ago, the laws of nature programmed the actions of all

[2]Josef R. Parrington et al., "Asian Dust: Seasonal Transport to the Hawaiian Islands," *Science,* April 8, 1983.

[2-5] We must have a good understanding of the natural order of the earth.

The act of creation and the act of appreciation of beauty are not in essence, distinguishable.
H. E. Huntley
The Divine Proportion, 1970

Engineering is the art of directing the great sources of power in nature for the use and convenience of man.
Thomas Tredgold, 1783–1829

It [the Universe] is an infinite sphere whose centre is everywhere, its circumference nowhere.
Blaise Pascal, 1623–1662
Pensées

Beauty is in the eyes of the beholder.
Margaret Wolfe Hungerford, 1855–1897
Molly Brown

The chess-board is the world, the pieces are the phenomena of the universe, the rules of the game are what we call the laws of Nature. The player on the other side is hidden from us. We know that his play is always fair, just, and patient. But also we know, to our cost, that he never overlooks a mistake or makes the smallest allowance for ignorance.
Thomas H. Huxley, 1825–1895
A Liberal Education

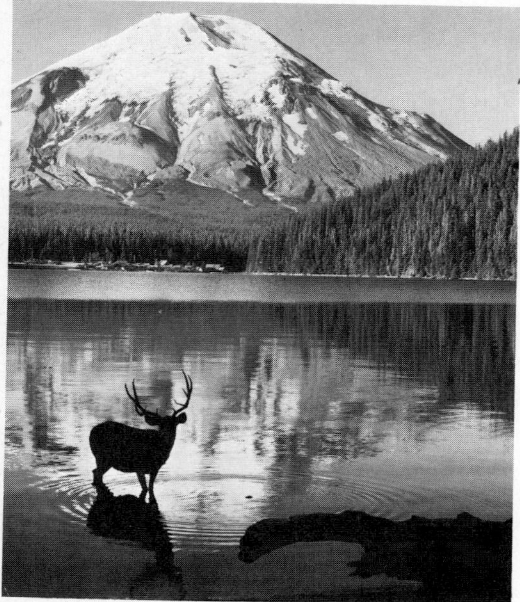

[2-6] There are too few beautiful areas to be found today that have been spared the ravages of man.

living things in relation to each other and to their environments. From the beginning, however, change was ever present. For example, as glaciers retreated, new forests grew to reclaim the land, ocean levels and coastlines changed, the winding courses of rivers altered, lakes appeared, and fish and animals migrated, as appropriate, to inhabit the new environments. Changes in climate and/or topography always brought consequential changes in the distribution and ecological relationships of all living things—plants and animals alike. Almost invariably these changes brought about competitive relationships between the existing and migrating species. In this way certain competing forms were forced to adapt to new roles, while others became extinct. As a result of this continual change in the ecological balance of the earth over eons of time, there currently exist some 1,300,000 different kinds of plants and animals that make their homes in rather specific locations. Only a few, such as the cockroach, housefly, body louse, and house mouse, have been successful in invading a diversity of environments—this because (willingly and unwillingly) they followed man in his travels. Presumably, even these would be confined to specific regions if man did not exist on the earth.

No man is an island, entire of itself.
John Donne, 1573–1631
Devotions

Primitive man was concerned with every facet of his environment and he had to be acutely aware of many of the existing ecological interrelationships. For example, he made it his business to know those places most commonly frequented by animals that he considered to be good to eat or whose skins or pelts were valued for clothing. He distinguished between the trees, plants, and herbs, and he knew which would provide him sustenance. Although by today's standards of education men of earlier civilizations might have been classified as "unlearned," they certainly were not ignorant. The Eskimo, for example, knew long ago that his sled dogs were susceptible to the diseases of the wild arctic foxes, and the Masai of east Africa have been aware for centuries that malaria is caused by mosquito bites.[3]

[2-7] And when the span of man has run its course
sitting upon the ruins of his civilization will be
a cockroach,
calmly preening himself
In the rays of the setting sun.

From century to century man has continued to add to his store of knowledge and understanding of nature. In so doing he has advanced progressively from a crude nomadic civilization in which he used what he could find useful to him in nature, to one sustained by domestication and agriculture in which he induced nature to produce more of the things that he wanted, and currently to one in which he is endeavoring to use the technologies of his own design to control the forces of nature.

For thousands of years after man first inhabited the earth, populations were relatively small and, because man was mobile, the cumulative effect of his existence on his environment was negligible. If, perchance, he did violence to a locality (for example, caused an entire forest to be burned), it was a relatively simple matter for him to move to another area and to allow time and nature to heal the wound. Because of the expanded world population, this alternative is no longer available. Now he affects not only his immediate environment but that of the whole earth.

[3] Peter Farb, *Ecology* (New York: Time, Inc., 1963), p. 164.

Man's existence is a part of, not independent of, nature—and specifically it is most concerned with that part of nature that is closest to the surface of the earth known as the *biosphere*. This is the wafer-thin skin of air, water, and soil comprising only a thousandth of the planet's diameter and measuring less than 8 miles thick.[4]

It might be said to be analogous to the skin of an apple. However, this relatively narrow space encompasses the entire fabric of life as we know it—from virus to field mouse, man, and whale. Most life forms live within a domain extending from ½ mile below the surface of the ocean to 2 miles above the earth's surface, although very few creatures can live in the deep ocean or above a 20,000-foot altitude. A number of processes of nature provide a biosphere with a delicate balance of characteristics that are necessary to sustain life. The life cycles of all living things, both fauna and flora, are interdependent and inextricably interwoven to form a delicately balanced *ecological system* that is as yet not completely understood by man. We do know, however, that not only is every organism affected by the environment of the "world" in which it lives, but it also has some effect on this environment.[5] The energy necessary to operate this system comes almost entirely from the sun and is utilized primarily through the processes of photosynthesis and heat. These processes are cyclic and are often referred to as the *chain of life cycle* [2-8]. Elements in the soil are combined with carbon dioxide (by photosynthesis) to produce plants. In turn, they serve as the energy basis for other life forms. The entire cycle is activated by energy from the sun. In accordance with the second law of thermodynamics, there is a loss of energy at each conversion in the cycle. Any changes that man exerts on any part of the system will affect its tenuous balance and cause internal adjustment of either its individual organisms, its environment, or both. The extent and magnitude of the modifications that man has exerted on this ecological system have increased immeasurably within the past few years, particularly as a consequence of his rapid population growth. Certain of these modifications are of particular concern to today's engineer.

Even during the period of the emergence of agriculture and domestication of animals, man began to alter the ecological balance of his environment. Eventually, some species of both plants and animals became extinct, while the growth of others was stimulated artificially. All too often man has not been aware of the extent of the consequences of his actions and, more particularly, of the irreversibility of the alterations and imbalances that he may have caused in nature's system. His concerns have more often been directed toward *subduing* the earth than to *replenishing* it. Over the period of a few thousand years nature's law of "survival of the fittest" was gradually replaced by man's law of "survival of the most desirable." From man's short-term point of view, this change represented a significant

[4] R. C. Cook, ed., "The Thin Slice of Life," *Population Bulletin*, Vol. 24, No. 5 (1968), p. 101.

[5] Marston Bates, "The Human EcoSystem," in National Academy of Sciences–National Research Council, *Resources and Man* (San Francisco: W. H. Freeman, 1969), p. 25.

I have long believed that our Nation has a God-given responsibility to preserve and protect our natural resource heritage. Our physical health, our social happiness, and our economic well-being will be sustained only to the extent that we act as thoughtful stewards of our abundant natural resources.

Ronald Reagan
Message to Congress, July 11, 1984

The first law of ecology is that everything is related to everything else.

Barry Commoner, 1917–

The air, the water, and the ground are free gifts to man and no one has the power to portion them out in parcels. Man must drink and breathe and walk and therefore each man has a right to his share of each.

James Fenimore Cooper
The Prairie, 1827

Ecology—the study of plants and animals in relation to their natural environment.

Harper's Encyclopedia of Science, 1967

The health of nations is more important than the wealth of nations.

Will Durant, 1885–1981

We must and will be sensitive to the delicate balance of our ecosystems, the preservation of endangered species, and the protection of our wilderness lands. We must and will be aware of the need for conservation, conscious of the irreversible harm we can do to our natural heritage, and determined to avoid the waste of our resources and the destruction of the ecological systems on which these precious resources are based.

Ronald Reagan
The White House, July 11, 1984

Edible.—Good to eat, and wholesome to digest, as a worm to a toad, a toad to a snake, a snake to a pig, a pig to a man, and a man to a worm.

Ambrose Bierce
The Devil's Dictionary

> And God blessed them, and God said unto them, be fruitful, and multiply, and *replenish* the earth, and *subdue* it: and have dominion over the fish of the sea, and over the fowl of the air, and over every living thing that moveth upon the earth.
>
> *Genesis 1:28*
> The Holy Bible

"AND ON THE SEVENTH DAY"

In the end,
There was Earth, and it was with form and beauty.
And Man dwelt upon the lands of the Earth, the meadows and trees, and he said,
"Let us build our dwellings in this place of beauty."
And he built cities and covered the Earth with concrete and steel.
And the meadows were gone.
And Man said, "It is good."
On the second day, Man looked upon the waters of the Earth.
And Man said, "Let us put our wastes in the waters that the dirt will be washed away."
And Man did.
And the waters became polluted and foul in smell.
And Man said, "It is good."
On the third day, Man looked upon the forests of the Earth and saw they were beautiful.
And Man said, "Let us cut the timber for our homes and grind the wood for our use."
And Man did.
And the lands became barren and the trees were gone.
And Man said, "It is good."
On the fourth day, Man saw that animals were in abundance and ran in the fields and played in the sun.
And Man said, "Let us cage these animals for our amusement and kill them for our sport."
And Man did.
And there were no more animals on the face of the Earth.
And Man said, "It is good."
On the fifth day, Man breathed the air of the Earth.
And Man said, "Let us dispose of our wastes into the air for the winds shall blow them away."
And Man did.
And the air became heavy with dust and all living things choked and burned.
And Man said, "It is good."
On the sixth day, Man saw himself and seeing the many languages and tongues, he feared and hated.
And Man said, "Let us build great machines and destroy these lest they destroy us."
And Man built great machines and the Earth was fired with the rage of great wars.
And Man said, "It is good."
On the seventh day, Man rested from his labors and the Earth was still, for Man no longer dwelt upon the Earth.
And it was good.

New Mexico State Land Office

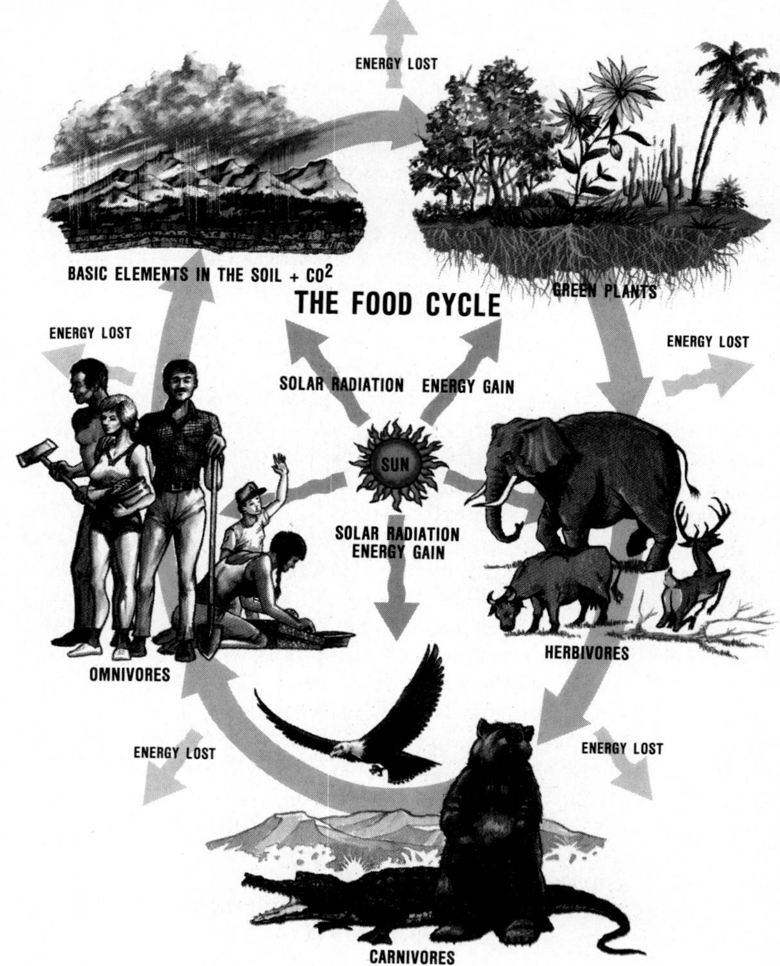

[2-8] The chain of life cycle.

advantage to him. Whereas once he was forced to gather fruit and nuts and to hunt animals to provide food and clothing for himself and his family—a life filled with uncertainty, at best—now he could simplify his food-gathering processes by increasing the yield of such crops as wheat, corn, rice, and potatoes. In addition, certain animals, such as cows, sheep, goats, and horses, were protected from their natural enemies and, in some instances, their predators were completely annihilated. Such eradication seemed to serve man's immediate interests, but it also eliminated nature's way of maintaining an ecological balance. Man, in turn, was also affected by these changes. At one time only the strongest of his species survived, and the availability or absence of natural food kept his population in balance with the surroundings. With domestication these factors have become less of a problem and "survival of the fittest" no longer governs his increase in numbers. In general, today both strong and weak live and procreate. Because of this condition the world's population growth has begun to mount steadily and *alarmingly* . . . because of the manifold problems that accompany large populations and for which solutions are still to be found.

It many respects the young engineer of today lives in a world that is vastly different from the one known to his or her grandfather or great grandfather. Without question we enjoy a standard of living unsurpassed in the history of mankind; yet, in spite of the significant agricultural and technological advances made in this generation, over half the world's population still lives in perpetual hunger. Famine, disease, and war continue to run rampant throughout portions of the earth, and wastes from our own technology continue to mount steadily. Nevertheless, the world's population continues to grow and further compounds these problems.

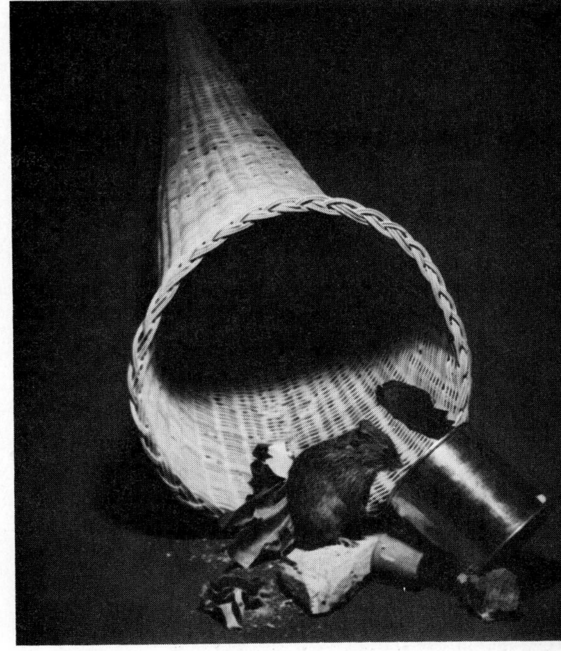

[2-9] Man's continued misuse of the world's resources can only lead to a depletion of the essentials for life.

THE POPULATION EXPLOSION: A RACE TO GLOBAL FAMINE

Until recently the growth of the human race was governed by the laws of nature in a manner similar to the laws of nature controlling the growth of all other living things. However, as man's culture changed from nomadic to agrarian to technological, he began to alter nature's population controls significantly. Through control of disease and pestilence his average life span has been extended by a factor of three. His ability to supply his family consistently with food and clothing has also been improved immeasurably. Combinations of these two factors have caused his population to increase in geometric progression: 2-4-8-16-32-64-128, and so on.[6] Initially it took hundreds of thousands of years for a significant change to occur in the world population. However, as the population numbers became larger, and particularly in more recent times as man's life span began to lengthen as a result of his gaining some control over starvation, disease, and violence, the results of geometric growth began to have a profound effect. Whereas it has taken an estimated 2 million years for the world population to reach slightly over 3 billion persons, it will take only 30 years to add the next 3 billion *if present growth rates remain unchanged*. The significance of this problem is dramatically illustrated by Figure [2-11].

As in the case of man's altering his ecological environment, he has not always been wise enough to anticipate all the various effects of his changes. In the case of his own propagation, he has managed to introduce "death control," but birth rates have continued to climb, particularly in the underdeveloped countries. It is estimated that the average annual increase in world population in 1650 was only 0.3 percent.[7] In 1900 this annual growth rate had increased to 0.9 percent; in 1930, 1.0 percent; in 1960, 2.0 percent; but in 1983

"THAT'S NOT A PLANET — IT'S AN INCUBATOR."

[2-10]

[6] T. R. Malthus, *An Essay on the Principle of Population As It Affects the Future Improvements of Mankind,* 1798. Facsimile reprint in 1926 for J. Johnson, Macmillan, London.

[7] J. M. Jones, *Does Overpopulation Mean Poverty?* (Washington, D.C.: Center for International Economic Growth, 1962), p. 13.

By the close of this century the world may have to feed as many as 2 billion additional people. Most of them will be born in developing countries, especially in marginal lands ill-suited for food production.

Donald L. Plucknett and Nigel J. H. Smith
Science, *July 16, 1982, p. 215*

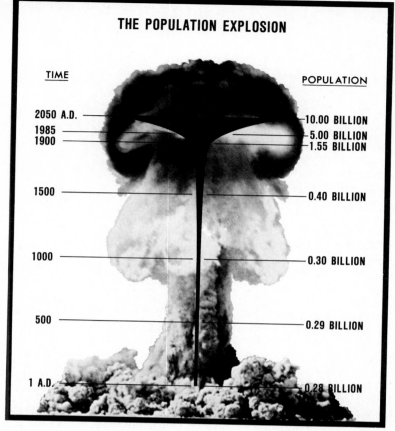

The arithmetic of population growth is awesome, and sobering: The earth gains 150 new persons per minute, 9,100 per hour, 219,100 per day and 79.6 million per year and does so mostly in the nations least able to cope with the burden of added people to feed and clothe, and minds to nourish.

John Collins
U.S. News & World Report, *July 23, 1984*

THE POPULATION EXPLOSION

TIME	POPULATION
2050 A.D.	10.00 BILLION
1985	5.00 BILLION
1900	1.55 BILLION
1500	0.40 BILLION
1000	0.30 BILLION
500	0.29 BILLION
1 A.D.	0.28 BILLION

[2-11]

[2-12] Traditionally, one person signifies loneliness; two persons—companionship; three persons—a crowd. In more recent times a new concept has been added: multitudes signify pollution and loneliness.

had decreased slightly to 1.7 percent. This is approximately double the United States annual growth rate. Unless the world rates decline significantly, which does not now seem likely, a worldwide crisis is fast approaching. Unfortunately, 92 percent of the increase in the world's population between 1986 and 2000 is predicted to occur in the less-developed countries.

The earth's land area, only 10 percent of which appears to be arable, is fixed and unexpandable, and a shortage of food and water is already an accepted fact of life in many countries. Although conditions in some slum areas of the United States are very bad, they bear little resemblance to many areas of the world where people grovel in filth and live little better than animals. Unfortunately, Paul Ehrlich's description of a visit to India could just as well have referred to a similar visit to a multitude of other countries.[8]

> I have understood the population explosion intellectually for a long time. I came to understand it emotionally one stinking hot night in Delhi a couple of years ago. My wife and daughter and I were

[8] P. R. Ehrlich, *The Population Bomb* (New York: Ballantine Books, 1968), p. 15.

> returning to our hotel in an ancient taxi. The seats were hopping with fleas. The only functional gear was third. As we crawled through the city, we entered a crowded slum area. The temperature was well over 100°, and the air was a haze of dust and smoke. The streets seemed alive with people. People eating, people washing, people sleeping. People visiting, arguing, and screaming. People thrusting their hands through the taxi window, begging. People defecating and urinating. People clinging to buses. People herding animals. People, people, people, people. As we moved slowly through the mob, hand horn squawking, the dust, noise, heat, and cooking fires gave the scene a hellish aspect.

Over one-fourth of the human beings now living on the earth are starving, and another one-third are ill fed. The underdeveloped countries of the world are incapable of producing enough food to feed their populations. For example, the growth in Africa's food supply compares favorably with that for the world as a whole. However, its increase in human numbers is far more rapid. The continent has the world's fastest population growth, as well as widespread soil erosion and desertification. The food production in Africa has fallen 11 percent since 1970.[9] This deficiency is 20 million tons of food each year and will grow to a staggering 100 million tons of food per year by 1990. For these people to be fed an adequate diet, the current world food production would have to double by 1990, *which appears to be an impossible task*. The quantity of food available is not the only problem; it must also be of the proper quality. Again, in central Africa every other baby born dies before the age of 5 *even*

[9] U.S. Department of Agriculture, Economic Research Service, *World Indices of Agricultural and Food Production, 1950–82* (Washington, D.C.: USDA, 1983).

Very few Americans, picking and choosing among the piles of white bread in a supermarket, have ever appreciated the social standing of white bread elsewhere in the world. To be able to afford white bread is a dream that awaits fulfillment for billions of the world's population. To afford it signifies that one enjoys all the comforts of life.

Isabel Cary Lundberg
Harper's Magazine

[2-13]

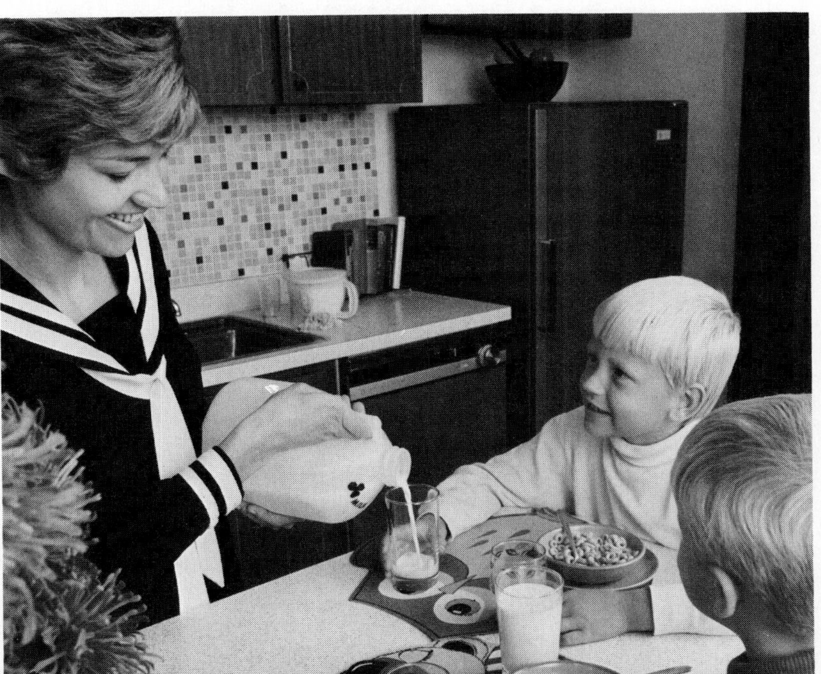

[2-14] One-half of the people in the world had food to eat this morning. . .

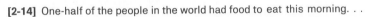

[2-15] . . . the other one-half went hungry.

[2-16] Wheat is one of the world's most important food crops because each grain contains a relatively large amount of useful protein. Even though engineering designs and advances in agricultural science have made possible a substantially increased production, the disparity between food supply and world population continues to increase. The problem is not merely one of increasing the production of food products. The economic, social, political, and engineering aspects of the logistics of distribution must also receive attention.

U.S. POPULATION GROWTH

1900 - 76 MILLION

1940 - 132 MILLION

1980 - 231 MILLION

2000 - 280 MILLION

SOURCE: U.S. BUREAU OF THE CENSUS

[2-17]

though food is generally plentiful. They die from a disease known as *kwashiorkor,* which is caused by a lack of sufficient protein in the diet. The magnitude of the problem continues to grow as the world population increases. More population means more famine. It also means more crowding, more disease, less sanitation, more waste and garbage, more pollution of air, water, and land, and ultimately . . . the untimely death of millions.

In the United States, just as in the underdeveloped countries of the world, the population has tended to migrate to cities and some of the cities have grown to monstrous size, often called *megalopolises.* For example, if current world trends continue, Mexico City will be the world's largest city by the year 2000, bulging with 25 to 30 million people.[10] In this country, these migrations have been brought about by the widespread use of mechanized agriculture and the impoverishment of the soil. With these migrations special problems have arisen. No longer is a city dweller self-sufficient in his ability to provide sustenance, shelter, and security for his family. Rather, his food, water, fuel, and power must all be brought to him by others, and his wastes of every kind must be taken away. Most frequently his work is located many miles from his home, and his reliance on a transportation system becomes critical. He finds himself vulnerable to every kind of public emergency, and the psychological pressures of city life often lead to mental illness or escape into the use of alcohol and drugs. The incidence of crime increases, and his clustering invites rapid spread of disease and pestilence. In general, cities are enormous consumers of electrical and chemical energy and producers of staggering amounts of wastes and pollution. Today 70 percent of the people in our country live on 1 percent of the land, and the exodus from the countryside has diminished only slightly. For the past decade, there have been regional shifts in population from the north central and northeastern states to the South and the West.

What are the implications for the engineer of these national and international sociological crises? The engineer is particularly affected because he or she is an essential participant among those whose creative efforts should be directed *to improving* man's physical and economic lot. First, we must learn all we can about the extent and causes of the technological problems that have resulted, and then direct the necessary energies and abilities to solve them. In general, we must recognize our responsibility to restore the equilibrium to the ecological system of nature in those cases in which it has become unbalanced. This requires that we be cognizant of the manifold effects of our designs *prior to their implementation.* All engineering designs are subject to failure under both predictable and unpredictable conditions. Some such failures could have catastrophic effects with regard to the health and safety of people, while others are of lesser import. It is our responsibility to design redundant engineering control systems for those failure situations that could be hazardous (such as in the design of processing systems and in the control of hazardous

[10] United Nations, Department of International Economics and Social Affairs, "Estimates and Projections of Urban, Rural, and City Populations (1950–2025: The 1980 Assessment)," *ST-ESA-R-45* (New York: UN, 1982), Table 8, p. 61.

emissions from processes and operations; see p. 122). The task is not an easy one, and the challenge is great, but the consequences are too severe to be disregarded.

Just as you learn that the physical laws of nature *do* govern the universe, so also must you be aware that nature's laws governing the procession and diversity of life on this planet are equally valid and unyielding. In the remainder of this chapter we consider the severity and complexity of several problems that confront today's society and, more particularly, the engineer's social and humanitarian responsibility for their solution.

[2-18] Go west, young man.

John Babsone Lane Soule, 1815–1891
Article in the Terre Haute, Indiana Express, *1981*

OUR POLLUTED PLANET

In the last few years the average U.S. citizen has become aware that our "spaceship earth" is undergoing many severe and detrimental ecological changes, which may take hundreds of years to repair. Some genetic changes may be irreversible. Pathological effects may be delayed in an individual for many years (such as the contraction of lung cancer 20 years after exposure to asbestos); also, insidious, genetic mutations (from ionizing radiation or mutagenic chemicals) that become manifest several generations later. Unfortunately, man is not always able to distinguish effects that are of a temporary nature from those with long-term consequences. Frequently, man's most damaging actions to the environment are either of an incremental or visually indistinguishable nature, and for this reason he participates willingly in them. In some measure man's reactions are dulled by the slowness of deterioration. This is somewhat analogous to the actions of a frog that will die rather than jump out, when placed in a bucket of water that is being *slowly* heated. In constrast, if the frog is pitched into a bucket of boiling water, he will immediately jump out and thereby avoid severe injury. The engineer in particular must learn to understand such cause-and-effect relationships so that his designs will not become detrimental to the orderly and natural development of all life. In the past two decades significant progress has been made in improving the working environment. The landmark Occupational Safety and Health Act (OSHA) established as national policy that American workers should be provided with a safe and healthful place of employment. The Clean Air Act of 1970 and 1972 amendments to the Federal Water Pollution Control Act (since renamed the Clean Water Act) focused primarily on cleaning up "conventional" pollutants—smoke and sulfur oxides in the air, oxygen-depleting discharges to the water, and solid wastes on the land.

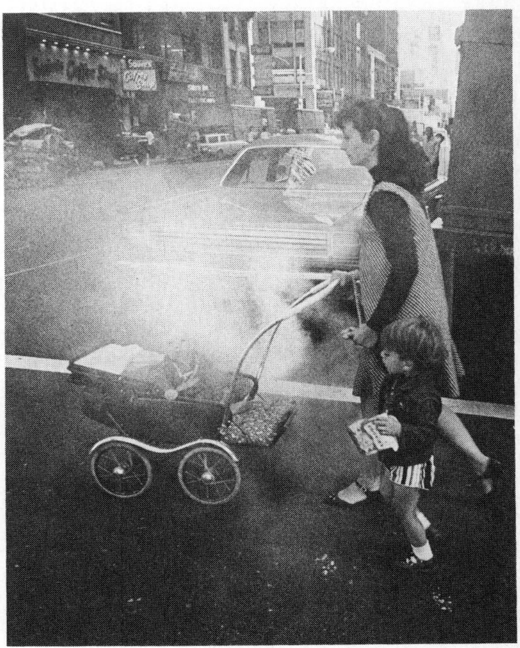

[2-19] Chronic exposure to lead in food or air can cause anemia, convulsions, kidney damage, and brain damage. In the United States, approximately 675,000 children between six months and five years old have high blood-lead levels.

Joseph L. Annest et al.
Blood-Lead Levels for Persons 6 Months–74 Years of Age, United States, 1976–1980, *Department of Health and Human Services*

THE AIR ENVIRONMENT

The atmosphere, which makes up the largest fraction of the biosphere, is a dynamic system that absorbs continuously a wide range of solids, liquids, and gases from natural and man-made sources. These substances often travel through the air, disperse, and react with one another and with other substances. Eventually, most of these constituents find their way into a depository, such as the ocean,

The sky is the daily bread of the eyes.

Ralph Waldo Emerson
Journal, May 25, 1843

Hell is a city much like London—a populous and smoky city.

Percy Bysshe Shelley
Peter Bell the Third, 1819

I durst not laugh, for fear of opening my lips and receiving the bad air.

Shakespeare
Julius Caesar

or to a receptor, such as man. A few, such as helium, escape from the biosphere. Others, such as carbon dioxide, may enter the atmosphere faster than they can be absorbed and thus gradually accumulate in the air.

Clean, dry air contains 78.09 percent nitrogen by volume and 20.94 percent oxygen. The remaining 0.97 percent is composed of carbon dioxide, helium, argon, krypton, and xenon, as well as very small amounts of some other inorganic and organic gases whose amounts vary with time and place in the atmosphere. Varying amounts of contaminants continuously enter the atmosphere from natural and man-made processes that exist on the earth. That portion of these substances which interacts with the environment to cause toxicity, disease, aesthetic distress, physiological effects, or environmental decay has been labeled by man as a pollutant. Environmental contaminants may be in the form of a chemical (or biological) agent or physical agent, such as noise, electromagnetic radiation, and so on. In general, the actions of people are the primary cause of pollution and, as population increases, the attendant pollution problems

[2-20] We can't agree on what to call Noah's new invention. He wants to call it *fire* . . . I want to call it *pollution*.

increase at the same geometric rate. This is not a newly recognized relationship, however. The first significant change in man's effect on nature came with his deliberate making of a fire. *No other creature on earth starts fires.* (The authors concede that the infamous Chicago fire of 1871 was blamed on Mrs. O'Leary's cow!) Prehistoric man built a fire in his cave home for cooking, heating, and to provide light for his family. Although the smoke was sometimes annoying, no real problem was perceived to exist with regard to pollution of the air environment. However, when his friends or neighbors visited him and also built fires in the same cave, even prehistoric man recognized that he then had an *air-pollution problem.* People in some nineteenth-century cities with their hundreds of thousands of smoldering soft-coal grates coughed amid a thicker and deadlier smog than any modern city can concoct. Today the natural terrain that surrounds large cities is recognized as having a significant bearing on the air-pollution problem. However, this is not an altogether new concept either. Historians tell us that the present Los Angeles area, which in recent years has become a national symbol of comparison for excessive smog levels,[11] was known as the "Valley of Smokes" when the Spaniards first arrived.[12] In recent years air pollution has become a problem of world concern.

In the United States the most common air pollutants are carbon monoxide, sulfur oxides, hydrocarbons, ozones, nitrogen oxides, and suspended particulates. Their primary sources are motor vehicles, industry, electrical power plants, space heating, and refuse disposal, with approximately 60 percent of the bulk being contributed by motor vehicles and 17 percent by industry. It seems probable that by the year 2000 America's streets will contain 20 percent more automobiles than the current 125 million.

In 1969, every man, woman, and child in the United States was producing an average of 1400 pounds per year of air pollutants. The problem was one of serious proportions. The National Air Quality Standards Act of 1970, which specified that motor vehicle exhaust emissions should be reduced by 90 percent by January 1, 1975, has provided the impetus for a sincere and necessary national commitment. In the 15 years since that time, over $300 billion has been spent for air pollution control. Industry supplied approximately 60 percent of these funds, the federal government 30 percent, and the consumer public 10 percent. The results have been a 50 percent decrease in measurable outdoor toxic air contaminants. But by focusing on outside air, we have missed many serious problems created by pollutants indoors, often referred to as occupational environmental contamination. Since on an average, American workers spend about one-fourth of their total life (about one-half of their waking hours) in the workplace, it is of utmost importance that this environment be free of contamination.

[2-21] Air pollution can make your eyes water and your throat burn. It can cause dizziness, blurred vision, coughing, chest discomfort and impaired breathing. During episodes of heavy air pollution, scores of people come to hospital emergency rooms with serious breathing problems, and premature deaths from heart and lung diseases jump dramatically. Furthermore, many scientists are convinced that air pollution contributes to three major types of chronic disease that kill millions of people annually—heart disease, lung disease and cancer.

Besides these health effects, dirty air has other impacts. It injures crops, flowers, shrubs and forests; it corrodes and dirties buildings, statues, fabrics and metals. When air pollution emissions return to the earth in the form of acid rain, populations of fish and other organisms in sensitive lakes can be decimated. Air pollution can impair visibility on the highway and in national parks. And, it may even be altering the earth's climate.

Deborah A. Sheiman, 1981
LWVEF Environmental Quality Department

---

[11] The term *smog* was coined originally to describe a combination of smoke and fog, such as was common in London when coal was widely used for generating power and heating homes. More recently it has come to mean the accumulation of photochemical reaction products that result largely from the action of the radiant energy of the sun on the emissions of internal combustion engines (automobile exhaust).

[12] H. C. Wohlers, *Air Pollution—The Problem, the Source, and the Effects* (Philadelphia: Drexel Institute of Technology, 1969), p. 1.

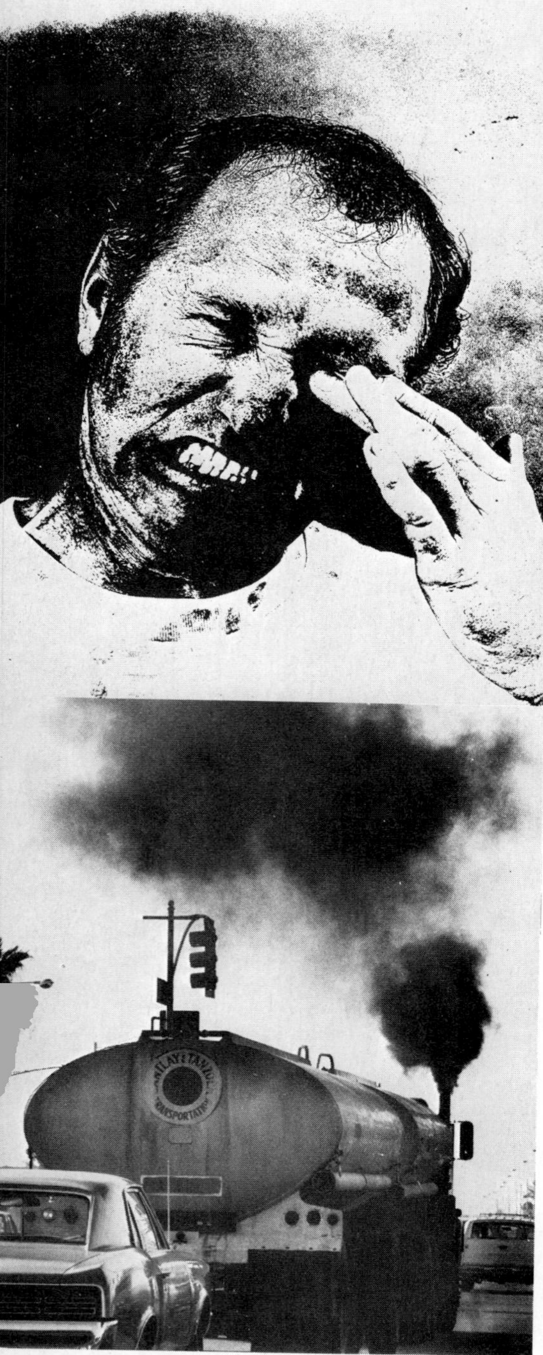
[2-22] Air pollution is a health hazard.

It has been found that the significantly increasing volume of particulate matter entering the atmosphere scatters the incoming sunlight. This reduces the amount of energy that reaches the earth and lowers its temperature. The decreasing mean global temperature of recent years has been attributed to the rising concentrations of airborne particles in the atmosphere.[13] A counteracting phenomenon, commonly referred to as the *greenhouse effect,* is caused by the presence of water vapor in the atmosphere, and to a lesser extent carbon dioxide and ozone, which combine to act in a manner similar to the glass in a greenhouse. Light from the sun arrives as short-wavelength radiation (visible and ultraviolet) and passes through it to heat the earth, but the relatively long-wavelength infrared radiation (heat radiation) that is emitted by the earth is absorbed by the carbon dioxide and water vapor—thereby providing an abnormal and additional heating effect to the earth. Although carbon dioxide occurs naturally as a constituent of the atmosphere and is not normally classified as an air pollutant, man generates an abnormally large amount of it in those combustion processes that utilize coal, oil, and natural gas. Some have estimated that if the carbon dioxide content in the atmosphere continues to increase at the present rate, the mean global temperature could rise by almost 4°C in the next 40 to 50 years. This might become a matter of great importance, because small temperature increases could cause a partial melting of the ice caps of the earth (causing flooding of coastal land, towns, and cities) with consequential and devastating effects to man.[14] A recent study by the National Academy of Sciences concludes that the impact of increasing CO_2 may not be as serious as others fear, although there are many factors that still need further investigation.[15]

Air pollution can cause death, impair health, reduce visibility, bring about vast economic losses, and contribute to the general deterioration of both our cities and countryside. Even though significant improvements have been made in the last two decades to reduce pollution levels, the overall situation has continued to deteriorate because in many cases the composition of the pollutants has changed also. Today, many pollutants that are released are geneotoxic and will cause genetic damage that will affect future generations. The problem must be viewed as one of protecting the health (genetic or otherwise) of the entire biosystem. It is therefore a matter of grave importance that engineers of all disciplines consciously incorporate in their designs sufficient constraints, controls, and safeguards to ensure that they do not contribute to the pollution of the atmosphere. In addition, they must apply their ingenuity and problem-solving abilities to eliminating air pollution where it exists and restoring the natural environment. The preferable course of action is for the engineer to produce designs that do not contribute to pollution in any form.

[13] R. E. Newell, "The Global Circulation of Atmospheric Pollutants," *Scientific American,* January 1971, p. 40.

[14] The Conservation Foundation, *State of the Environment,* (Washington, D.C.: CF, 1984), p. 103.

[15] U.S. Environmental Protection Agency, Office of Policy and Resource Management, Strategic Studies Staff, "Can We Delay a Greenhouse Warming?" (Washington, D.C.: EPA, September 1983).

THE QUEST FOR WATER QUALITY

Water is the most abundant compound to be found on the face of the earth and, next to air, it is the most essential resource for man's survival. The per capita daily water withdrawal in the United States is over 1000 gallons per day, and this demand continues to grow. Early man was most concerned with the quality (purity) of his drinking water, and even he was aware that certain waters were contaminated and could cause illness or death. In addition, modern man has found that he must be concerned also with the quantity of the water available for his use. An abundant supply of relatively pure water is no longer available in most areas. Today, water pollution, *the presence of toxic or noxious substances or heat in natural water sources,* is considered to be one of the most pressing social and economic issues of our time. Unlike the nation's relatively consistent and successful improvements in air quality in the last decade, success in cleaning up surface waters has not been as impressive. Some streams and rivers have improved and the "dying" Great Lakes, Erie and Ontario, are reviving, with fewer algae blooms and growing fish populations.[16] Unfortunately, many streams and lakes have been degraded or have shown no significant change since the early 1970s.

The processes of nature have long made use of the miraculous ability of rivers and lakes to "purify themselves." After pollutants find their way into a water body they are subject to dilution or settling, the action of the sun, and to being consumed by beneficial bacteria. The difficulty arises when man disturbs the equilibrium of the ecosystem by dumping large amounts of his wastes into a particular water body, thereby intensifying the demand for purification. In time the body of water cannot meet the demand, organic debris accumulates, anaerobic areas develop, fish die, and putrification is

[16]The Conservation Foundation, *State of the Environment* (Washington, D.C.: CF, 1982), p. 97.

[2-23] Oil and grease discharges are responsible for polluting many of the rivers and lakes of America.

[2-24] Pollution does not necessarily need to accompany poverty, but it most frequently does.

Modern man . . . has asbestos in his lungs, DDT in his fat, and strontium 90 in his bones.

Today's Health, April 1970

the result. This process also occurs in nature, but it may take many thousands of years to complete the natural processes of deterioration. Man can alter nature's time scale appreciably.

Many city water-treatment plants merely remove the particulate matter and disinfect the available water with chlorine to kill bacteria, since they were not originally designed to remove pesticides, herbicides, and other organic and inorganic chemicals that may be present.[17] Industrial liquid wastes frequently carry large quantities of dissolved material which should be recycled. Silver salts from photographic processes are but one example. This problem has become acute in a number of areas in recent months as hundreds of new contaminants have been discovered in deep wells, streams, and lakes: bacteria and viruses, chemicals from solid-state electronics processes, detergents, municipal sewage, acid from mine drainage, pesticides and weed killers, radioactive substances, phosphorus from fertilizers, trace amounts of metals and drugs, and other organic and inorganic chemicals. As the population continues to increase, the burden assumed by the engineer to design more comprehensive wastewater treatment plants also mounts. Indeed, the well-being of entire communities may depend on the engineer's design abilities, because it is now a recognized condition of population increase that "everyone cannot live upstream."

Every citizen has a personal stake in the quality of the nation's waters. We need water that is safe to drink, safe to swim in, habitable for aquatic life, free of nuisance conditions, and usable for agriculture and industry. Health, jobs, and the quality of our lives are thus affected in many ways by water quality.

Environmental Quality—1977
The Eighth Annual Report of the Council on Environmental Quality

The presence of radioactive wastes and excess heat are relatively new types of pollution to water bodies, but they are of no less importance for the engineer to take into account in designing. All radioactive materials are biologically injurious. Therefore, radioactive substances that are normally emitted by nuclear power plants are suspected of finding their way into the ecological food chain, where they could cause serious problems. For this reason all radioactive wastes should be isolated from the biological environment during

[17] Gene Bylinsky, "The Limited War on Water Pollution," *The Environment* (New York: Harper & Row, 1970), p. 20.

Soil conservation must be given a higher priority in the future. This will mean not only obvious measures such as terracing and tree planting, but also less conventional approaches including a much greater emphasis on organic fertilizers (including biological nitrogen fixation), minimum tillage techniques, and, in the tropics, farming methods that preserve negative cover and protect the soil from baking by sun and battering by rain, such as use of dead or living mulches and agro-forestry.

Paul Harrison
Ambio, 1984, p. 167

[2-25] Soil erosion is a major source of water pollution in agricultural areas.

the "life" of the isotopes (as much as 600 years or longer in some cases), and this must be incorporated in the design.

The heat problem arises because electrical power generating plants require great quantities of water for cooling. Although the heated discharge water from a nuclear power plant is approximately the same temperature as that from a fossil-fuel power plant, the quantity of water must be increased by approximately 40 percent. The warmer water absorbs less oxygen from the atmosphere, and this decelerates the normal rate of decomposition of organic matter. This abnormal heat also unbalances the life cycles of fish, which, being cold blooded, cannot regulate their body temperatures correspondingly. If the number of large power plants increases, this problem could loom larger than ever.

Engineering designs of the future must take into account all these factors to ensure for all the nation's inhabitants a water supply that is both sufficient in quantity and unpolluted in quality. This is not only the engineer's challenge, but his responsibility as well.

[2-26] The character of a nation is revealed by examining its garbage.

SOLID-WASTE DISPOSAL AND RECYCLING MATERIALS

We are living in a most unusual time—a time when possibly the most valuable tangible asset that a person could own would be a "bottomless hole." Never before in history have so many people had so much garbage, refuse, trash, and other wastes to dispose of, and at the same time never before has there been such a shortage of "dumping space." The proliferation of refuse, however, is only partly attributable to the population explosion. A substantial portion of the blame must be assumed by an affluent society that is careless with its increasing purchasing power and that has demonstrated a decided distaste for secondhand articles.

The most popular method of solid-waste disposal has long been "removal from the immediate premises." For centuries man has been aware of the health hazards that accompany the accumulations of garbage. Historians have recorded that a sign at the city limits of ancient Rome warned all persons to transport their refuse outside the city or risk being fined. Also, it has been recognized that in the Middle Ages the custom of dumping garbage in the streets was largely responsible for the proliferation of disease-carrying rats, flies, and insects that made their homes in piles of refuse.

Sooner or later closed-cycle manufacturing, materials recycling, energy recovery from wastes, and the return to the land of fiber and nutrients will certainly occur.

Ariel Parkinson
Environment, *December 1983*

On a per capita basis the United States has established an unenviable record with regard to the generation of solid waste, refuse, and garbage:

1920: 2.75 pounds per person per day.

1970: 5.5 pounds per person per day.

1985: 10.0 pounds per person per day.

Industries in this country generate about 2500 pounds of hazardous waste per capita annually. From New York to Los Angeles, cities throughout the nation are rapidly depleting their disposal space, and there is considerable concern that too little attention has

[2-27] The world will little note nor long remember what we say here; but it can never forget what they did here. . . .

Abraham Lincoln
Address at Gettysburg, 1863

been given to what is fast becoming one of man's most distressing problems—solid-waste disposal. Why is trash becoming such a problem? The answer seems to lie partially in man's changing value system. Just a generation or two ago, thrift and economy were considered to be important tenets of American life and few items with any inherent value were discarded. Today, we live in a "throwaway society." More and more containers of all types are being made of inert plastic or glass, or nondegradable aluminum, and everything from furniture to clothing is being made from disposable paper products, which are sold by advertising that challenges purchasers to "discard when disenchanted." The truth is that most "consumers" in America are becoming "users and discarders," but this fact is not always recognized.

Unfortunately, most current waste-disposal practices make no attempt to recover any of the potential values that are in solid wastes. The problem is not just one of wasting materials. We also throw away energy when we discard products. Only 4 percent as much energy is required to recycle aluminum as to produce it originally from bauxite. To recycle copper would save 90 percent and to produce steel from scrap would save almost 50 percent. The methods of disposing of refuse in most common use today are dumping and/or burning in the open, sanitary land fills, burial in abandoned mines, dumping at sea, and grinding in disposal systems followed by flushing in sewers. Some edible waste, such as garbage, is fed to hogs.

Waste is a human concept. In nature, nothing is wasted, for everything is part of a continuous cycle [2-8]. Even the death of an animal provides nutrients that will eventually be reincorporated in the chain of life. For example, a newspaper has been "consumed"

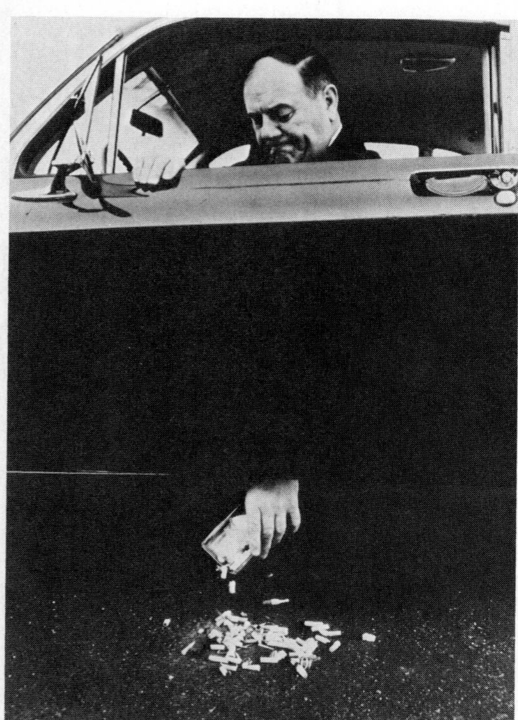

[2-28] Pollution is a personal matter.

[2-29] I think that I shall never see
A billboard lovely as a tree
Indeed, unless the billboards fall
I'll never see a tree at all.

Ogden Nash
Song of the Open Road, *1945*

when its purchaser has finished reading it. But one person's waste newspapers are another person's recyclable fibers or cellulose insulation.[18] Also, a few cities have begun to convert their municipal garbage into energy. Landfill gases are created through bacterial decomposition of organic waste, and consist of approximately 55 percent methane (CH_4), 45 percent CO_2, and trace amounts of other gases [2-31]. At present less than 1 percent of the solid waste generated in the United States each year is used to generate energy. Several Scandinavian countries convert up to 40 percent of their garbage into useful energy.[19]

Since almost all engineering designs will eventually be discarded due to wear or obsolescence, it is imperative that consideration for disposal be given to each design *at the time that it is first produced.* In addition, the well-being of society as we know it appears to depend in some measure on the creative design abilities of the engineer to devise new processes of recycling wastes by either changing the physical form of wastes or the manner of their disposal, or both. Such designs must be accomplished within the constraints of economic considerations and without augmenting man's other pollution problems: air, water, and sound. Basically, the solution of waste disposal is a matter of attitude, ingenuity, and economics—all areas in which the engineer can make significant contributions.

In addition to reducing the cost of increasing solid-waste disposal, recycling of discardables will save energy and expensive raw

[2-30] As the earth becomes more crowded, there is no longer an "away." One person's trash basket is another's living space. (National Academy of Sciences)

[18] Denis Hayes, *Repairs, Reuse, Recycling—First Steps Toward a Sustainable Society* (Washington, D.C.: Worldwatch Institute, 1978), p. 6.

[19] *Two Energy Futures* (Washington D.C.: American Petroleum Institute, 1980), p. 108.

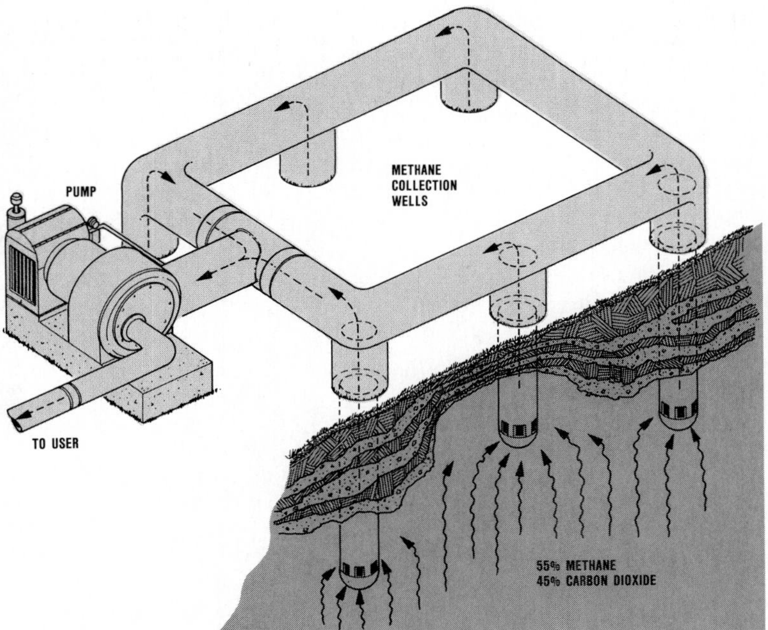

At least two-thirds of the material resources that we now waste could be reused without important changes in our life-styles. With products designed for durability and for ease of recycling, the waste streams of the industrial world could be reduced to small trickles. And with an intelligent materials policy, the portion of our resources that is irretrievably dissipated could eventually be reduced to almost zero.

Dennis Hayes, Repairs, Reuse, Recycling—First Steps Toward a Sustainable Society *(Washington, D.C.: Worldwatch Institute, September 1978)*

Feedlots now produce more organic waste than the total sewage from all U.S. municipalities.

Barry Commoner
The Closing Circle, 1971

The paper packaging for McDonald's first eight billion hamburgers used up 890 square miles of forest.

Bruce Hannon
Letters to the Editor,
Not Man Apart,
September 1972

[2-31] The recovery of methane from garbage has become a profitable practice.

[2-32] Adequate housing is a concern for people everywhere.

Woe unto them that join house to house that lay field to field, till there be no place, that they may be placed alone in the midst of the earth!

Isaiah 5:7–8
The Holy Bible

materials, as well as protecting the environment. It has been estimated that throwing away an aluminum beverage container wastes an amount of energy equivalent to that which would be wasted by pouring out such a container half-filled with gasoline. Failing to recycle a weekday edition of the *Washington Post* or the *Los Angeles Times* wastes just about as much.[20]

Some progress in recycling aluminum has been made in the past two decades, growing from 16 percent to 32 percent of the total production.[21] This represents 54 percent of all aluminum beverage cans used each year. Paper consumption has more than doubled during this period, but less than one-fourth is recycled paper. The recovery rate of this country's annual steel consumption is a mere 35 percent.

The placement of debris in space (nose cones, pieces of disintegrated satellites, nuts, and bolts) is becoming a matter of concern to NASA. It is estimated that over 15,000 man-made objects are circling the earth in space, with at least 5000 of these being tracked continuously. Perhaps it is time to consider the design of a space debris scavenger satellite.[22]

THE RISING CRESCENDO OF UNWANTED SOUND

A silent world is not only undesirable but impossible to achieve. Man's very nature is psychologically sensitive to the many sounds that come to his ears. For example, he is pleased to hear the gurgle and murmur of a brook or the soothing whispering wind as it filters through overhead pine trees, but his blood is likely to chill if he recognizes the whirring buzz of a rattlesnake or hears the sudden screech of an automobile tire as it slides on pavement. He may thrill to the sharp bugle of a far-off hunting horn, but his thoughts often tend to lapse into dreams of inaccessible places as a distant train whistle penetrates the night.[23] Yes, sounds have an important bearing on man's sense of well-being. Although the average city dweller's ears continue to alert him to impending dangers, their sensitivity is far less acute than that of people who live in less densely populated areas. It is said that, even today, aborigines living in the stillness of isolated African villages can easily hear each other talking in low conversational tones at distances as great as 100 yards, and that their hearing acuity diminishes little with age.[24] Even as man's technology has brought hundreds of thousands of desirable and satisfying inno-

Noise has only recently been recognized as a serious environmental pollutant. In comparison to efforts to control air and water pollution, the control of noise pollution is still in its infancy.

Ronald M. Buege, Director of Environmental Health
Journal of Environmental Health, *Vol. 45, No. 5*

[20] William V. Chandler, *State of the World 1984,* Worldwatch Institute (New York: W. W. Norton, 1984), p. 95.

[21] Aluminum Association, *Aluminum Statistical Review for 1981,* Washington, D.C.

[22] Peter T. White, "The Fascinating World of Trash," *National Geographic Magazine,* April 1983, p. 456.

[23] *Noise: Sound Without Value* (Washington, D.C.: Committee on Environmental Quality of the Federal Council for Science and Technology, 1968), p. 1.

[24] The Editors of *Fortune, The Environment* (New York: Harper & Row, 1970), p. 136.

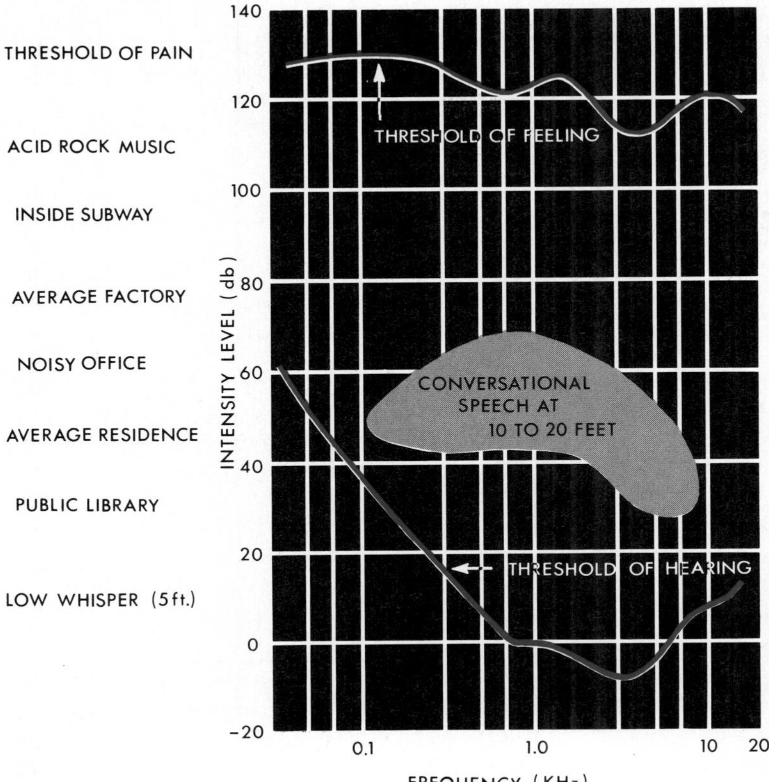

[2-33] Sound levels.[26]

Noise is the only pollutant that some people actually want. Truckdrivers want the meanest sounding rig. A company tried to sell a quiet vacuum cleaner years ago, but few wanted them. They would never sell because they were too quiet.

Edward DiPolmere, Association of Noise Control Officials
Quoted in U.S. News & World Report, July 16, 1984

Not many sounds in life, and I include all urban and all rural sounds, exceed in interest a knock at the door.

Charles Lamb
Valentine's Day, 1823

It will be generally admitted that Beethoven's Fifth Symphony is the most sublime noise that has ever penetrated into the ear of man.

Edward Morgan Forster, 1879–1970
Howards End

vations, it has also provided the means for the retrogression of his sense of hearing—for deafness caused by a deterioration of the microscopic hair cells that assist in transmitting sound from the ear to the brain. It has been found that prolonged exposure to intense sound levels will produce permanent hearing loss, and it matters not that such levels may be considered to be pleasing. Some people purport to enjoy "rock" music concerts at sound levels exceeding 110 decibels[25] [2-33]. Today noise-induced hearing loss looms as one of America's major health hazards.

Noise is generally considered to be any annoying or unwanted sound. Noise (like sound) has two discernible effects on man. One causes a deterioration of his sensitivity of hearing, the other affects his psychological state of mind. The adverse effects of noise have long been recognized as a form of environmental pollution. Julius Caesar was so annoyed by noise that he banned chariot driving at night, and, prior to 1865, studies in England reported substantial hearing losses among blacksmiths, boilermakers, and railroad men.[27]

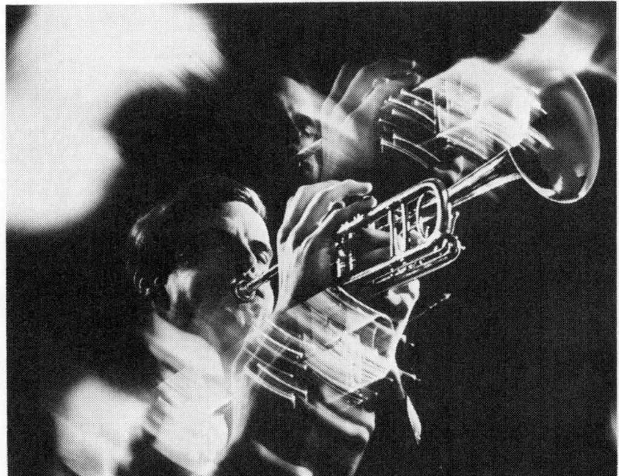

[2-34] Man is affected psychologically by the sounds that he hears.

[2-35]

---

[25] A *decibel* (abbreviated dB) is a measure of sound intensity or pressure change on the ear.

[26] Adapted from W. E. Woodson and D. W. Conover, *Human Engineering Guide for Equipment Designers* (Berkeley: University of California Press, 1964), pp. 4–10.

[27] Aram Glorig, in *Noise as a Public Health Hazard,* W. D. Ward and J. E. Fricke, eds. (Washington, D.C.: The American Speech and Hearing Association, 1969), p. 105.

He who sleeps in continual noise is wakened by silence.

William Dean Howells, 1837–1920
Pordenone

Men trust their ears less than their eyes.

Herodotus
ca. 485–ca. 425 B.C.

The crescendo of noise—whether it comes from truck or jackhammer, siren or airplane—is more than an irritating nuisance. It intrudes on privacy, shatters serenity and can inflict pain. We dare not be complacent about this ever mounting volume of noise. In the years ahead, it can bring even more discomfort—and worse—to the lives of people.

Lyndon Baines Johnson

Unnecessary noise is the most cruel absence of care which can be inflicted either on sick or well.

Florence Nightingale
1820–1910

Nature has given man one tongue and two ears, that we may hear twice as much as we speak.

Epictetus
Fragments, A.D. 90

Psychologists have found that music does things to you whether you like it or not. Fast tempos invariably raise your pulse, respiration, and blood pressure; slow music lowers them.

Doron K. Antrim

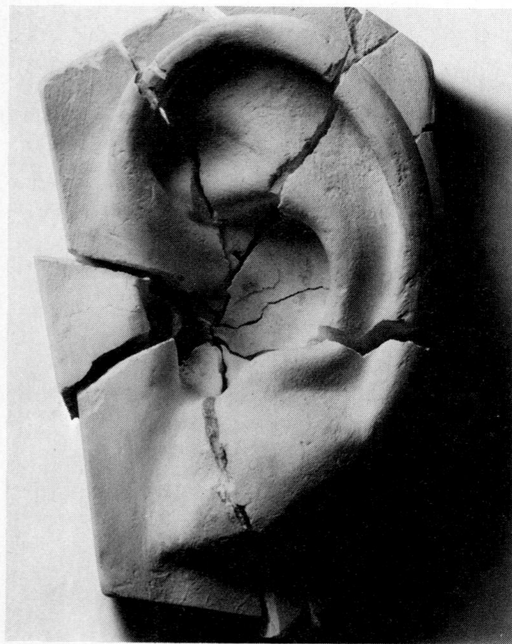

[2-36] If the ear were to shatter or bleed profusely when subjected to abuse from intense or prolonged noise, we might be more careful of its treatment.

TABLE 2-1 PERMISSIBLE NOISE EXPOSURES[a]

Duration per day (hr)	*Sound level (dBA slow response)*
8	90
6	92
4	95
3	97
2	100
$1\frac{1}{2}$	102
1	105
$\frac{1}{2}$	110
$\frac{1}{4}$ or less	115

[a] Noise measured on the A scale of a sound level meter is given in dB(A) units. The A scale measures noise processed by the human ear. It gives greatest weight to noise in the octave band centered on 2 kHz. Noise of low frequency receives less weight.
Source: OSHA-29CFR 1910.956.

It has only been in the last 50 years or so, however, that noise has been recognized naturally as an occupational health hazard.[28]

It is estimated that the average background noise level throughout the United States has been doubling each 10 years. At this rate of increase, living conditions will be intolerable within a few years. Such a crescendo of sound results from the steady increase of population and the concomitant growth of the use of power on every hand—from the disposal in the kitchen and the motorcycle in the street to power tools in the factory. Buses, jet airliners, television sets, stereos, dishwashers, tractors, mixers, waste disposers, air conditioners, automobiles, jackhammers, power lawn mowers, vacuum cleaners, typewriters, and printers are but a few examples of noise producers that are deemed desirable to today's high standard of living, but which may very well also prevent man from fully enjoying the fruits of his labors, unless the sound levels at which they operate are altered significantly.

OSHA has established permissible levels of noise exposure, as shown in Table 2-1. When employees are subjected to sound exceeding the exposures listed, feasible administrative or engineering controls must be utilized.

Except in the case of minimizing aircraft noise, the United States lags far behind many countries in noise prevention and control. Virtually all manmade noise can be suppressed, and the same engineer who formulates the idea for a new type of kitchen aid or designs an improved family vehicle must also be capable of solving the acoustical problems that are associated with its manufacture and use. In this regard, as an engineer, you are responsible to generations yet unborn for the consequences of your actions.

[28] "Effect of Noise on Hearing of Industrial Workers," State of New York Department of Labor, *Special Bulletin, 166* (New York: Bureau of Women in Industry, 1930).

MAN'S INSATIABLE THIRST FOR ENERGY

In man's earliest habitation of the earth he competed for energy with other members of the earth's ecological environment. Initially his energy requirements were primarily satisfied by food—probably in the range of 2000 calories per person per day.[29] However, as he has been able to make and control the use of fire, domesticate the plant and animal kingdoms, and initiate technologies of his own choosing, his per capita consumption of energy has increased appreciably.

In 1940, it was estimated that the total energy generated in the United States would be equivalent in "muscle-power energy" of 153 slaves working for every man, woman, and child in this country. Today a similar calculation would show that over 500 "slaves" are available to serve each person. In this respect every person is a monarch. This demand for increasing forms of energy has followed an exponential pattern of growth similar to the growth of the world population, *except that the annual rate of increase for nonnutrient energy utilization is growing at a rate* (approximately 4 percent per year) *considerably in excess of the world's growth in popultion* (slightly less than 2 percent per year). This is brought about by man's appetite for more gadgets, faster cars and airplanes, heavier machinery, and so on.

The principal sources of the world's energy prior to about A.D. 1200 were solar energy, wood, wind, and water. At about this time in England it was discovered that certain "black rocks" found along the seashore would burn. From this there followed in succession the mining of coal (the black rocks) and the exploration of oil and natural-gas reservoirs.[30] More recently, nuclear energy has emerged as a promising source of power. The safe management and disposal of radioactive wastes, however, continue to present problems for the engineer. Renewable energy sources appear to have considerable promise in providing the additional energy needs of the country that will develop over the next two decades. These include wind power, geothermal energy, solar energy, fuel cells, hydropower, wood fuel, ocean thermal energy differences, and energy crops [2-38 to 2-40].

The graph [2-41] provides a record of the history of energy consumption in the United States since 1850. Of course, the future is unknown, and a prediction of our energy sources for the year 2000 and beyond is mere conjecture. It depends to a large extent on the background and experience of the predictor. External factors may also intervene. It may well be, for example, that although fossil fuels seem to be sufficient in quantity, they might be undesirable for expanded use because of their combustive pollutant effects. The solving of such problems represents a number of challenges for the engineer.

[2-37] Man stands at the end of a long cycle of energy exchanges in which there is a calculable and irreversible loss of energy at each exchange. A grown adult irradiates heat equivalent to that of a 75 watt bulb. His total energy output, in 12 hours of hard physical work, is equivalent to only 1-kilowatt hour. He requires daily 2,200 calories of food intake, 4½ pounds of water and 30 pounds of air, and he discards 5 pounds of waste. Considered as an energy converter, man is the least efficient link in his particular "food chain," and for this reason the most vulnerable to catastrophic ecologic change. Such a change can be caused by overloading the energy circuit. There are two new humans added to the globe's population *every second.*

$E = mc^2$

Albert Einstein
Annalen der Physik, *1905*
Statement of the mass–energy equivalence relationship

F. D. Roosevelt
President of the United States
White House
Washington, D.C.

Sir:

Some recent work by E. Fermi and L. Szilard, which has been communited to me in manuscript, leads me to expect that the element of uranium may be turned into a new and important source of energy in the immediate future . . . that it may be possible to set up a nuclear chain reaction in a large mass of uranium by which vast amounts of power and large quantities of new radium-like elements would be generated. . . .

Albert Einstein
Old Grove Road, Nassau Point
Peconic, Long Island
August 2, 1939

[29] The calorie is an energy unit. However, the food industry conventionally uses a "calorie" that is 1000 times the size of the "calorie" used by the general scientific community. The 2000 food calories referred to here would be equivalent to 2000 kilocalories expressed in the more common scientific units.

[30] There are evidences that coal was used in China, Syria, Greece, and Wales as early as 1000 to 2000 B.C.

[2-38] Wind power's potential value is in its ability to substitute for the use of oil, coal, or nuclear energy. The most promising form of wind machine is the large, horizontal-axis wind turbine with propeller-type rotor blades.

[2-39] Geothermal energy includes the harnessing of both natural steam and hot water. Some 2400 quadrillion (2.4×10^{18}) Btu of geothermal resources have been identified in this country. This is over 30 times the nation's present total energy consumption.

In the United States, the energy consumption is distributed approximately as follows:

Residential	21 percent
Commercial	14
Automobiles	16
Transportation	8
Industrial and other	41

Considering the fact that currently the five most common air pollutants (carbon monoxide, sulfur oxides, hydrocarbons, nitrogen oxides, and solid particles) are primarily by-products of the combustion of fossil fuels, it behooves the engineer to design and utilize energy sources that are as free from pollution as possible.

In the next two decades the United States will consume more energy than it has in the past 200 years combined. The artful manipulation of energy is an essential component of man's ability to survive in a competitive world environment.

The most accepted index for measuring a nation's aggregate economic output, the total market value of the goods and services produced by a nation's economy during a specific period of time (usually a year) is the gross national product (GNP). Although the GNP is not necessarily a measure of the quality of life, it does represent in the broadest sense a measure of a nation's standard of living. If one plots the GNP of the nations of the world versus their total energy consumption per year per capita, a linear relationship will result [2-42]. A compelling case can be made, therefore, for the hypothesis that the productive utilization of energy has played a primary role in shaping the science and culture of the world.[31]

It would appear that in the long run the earth can tolerate a significant increase in man's continuous release of energy (perhaps as much as 1000 times the current U.S. daily consumption—or more) without deleterious effect.[32] Such increases would, of course, be necessary to accommodate a constantly increasing population. However, extrapolations and statements of this type concerning the future are meaningless unless the short-range problems—the problems of today—are solved. Our society has invested the engineer with a responsibility for leadership that must not fail.

GO-GO-GO

At the present time American motorists are traveling over 1.5 million million miles per year on the nation's highways—an equivalent distance of almost 3 million round trips to the moon. More than one-half of this travel is in urban areas [2-43], where for the most part the physical layouts—the planning, the street design, and basic service systems—were created over 100 years ago. As the population of the nation continues to shift to the urban and suburban areas,

[31] Chauncey Starr, "Energy and Power," *Scientific American,* September 1971.

[32] A. M. Weinberg and R. P. Hammond, "Limits to the Use of Energy," *American Scientist,* August 1970, p. 413.

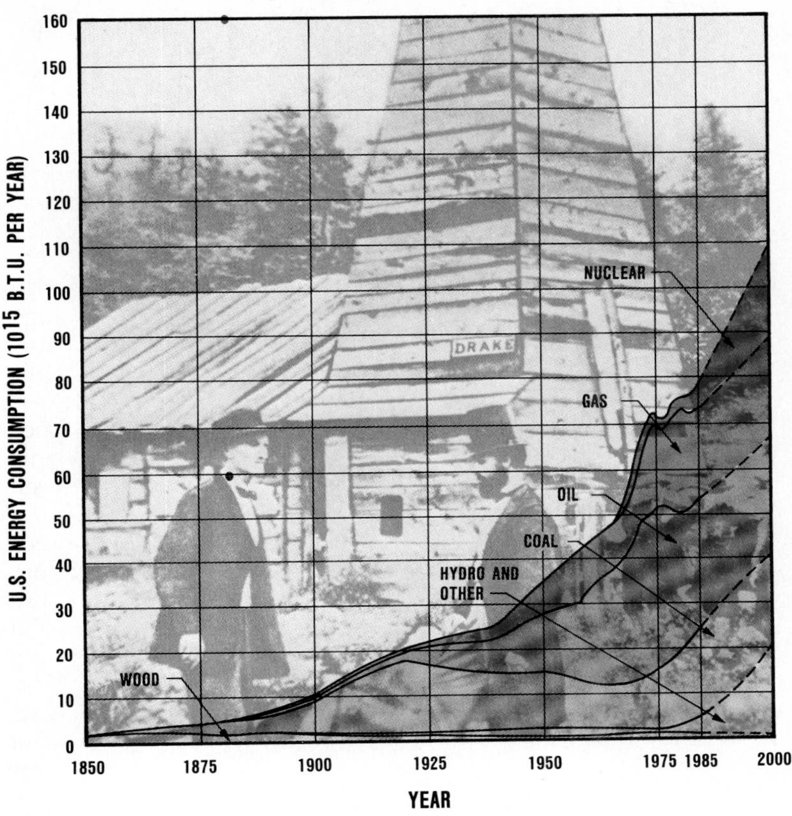

[2-41] Energy consumption in the United States.

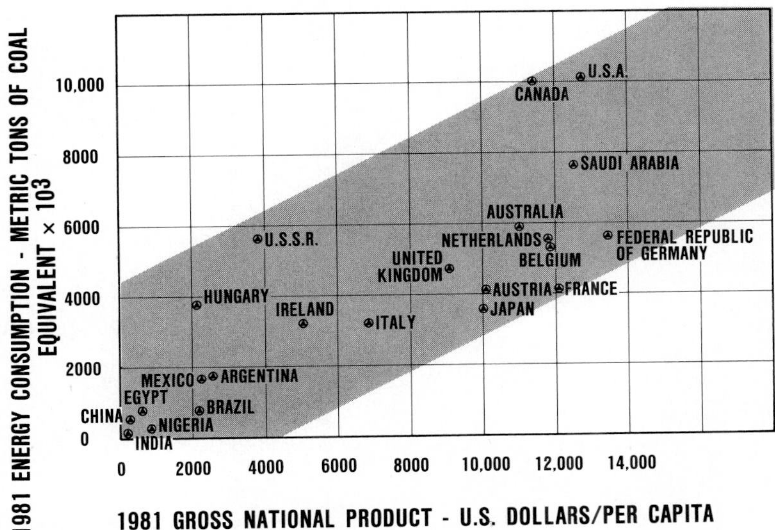

[2-42] The relationship of energy consumption to gross national product.

[2-40] Solar One, a 10-megawatt installation located near Barstow, California, is the nation's first solar-thermal central receiver electric generating station. A field of 1818 heliostats, controlled by computer, tracks the sun throughout the day and reflects its energy onto the receiver, where steam is produced.

"If we can build an atomic bomb, if we can put a man on the moon, why can't we solve our energy problems?" The answer, in a word, boils down to "complexity." Although building the first bomb and putting the first man on the moon were admittedly complicated efforts involving the examination of many alternative approaches, and the coordination of many institutions, there is a certain essential unity about both these efforts. There was, in each case, one mission, one developer, one user. The objective could be clearly, simply, and unambiguously defined. There were no degrees of success short of the goal. Success could be clearly—and dramatically—measured.

By constrast, the energy problem is a multi-headed one. Virtually every person and institution in the country is directly affected. There are multiple possible goals, degrees of success, and pathways to the goal. It might not even be clear just when the goal is reached. There are endless interactions between energy decisions and the environment, the economy, and public welfare. Different regions have different needs, different resources, different limitations, different rules and regulations, and competing interests. There are many questions for public debate. There is, in short, no easy way to identify a goal and reach it.

Nuclear power development seems to have stagnated. Synthetic fuel plants are not yet proven commercially. Renewable resources remain too highly priced to be competitive in most applications. Declining domestic oil and gas discoveries portend decreasing supplies in coming years. . . .

Gail H. Marcus
The Changing Role for Federal Energy R&D, *1983*
The Library of Congress

It is better to more effectively insulate our homes than to compete for oil. It is better to design around a need for a scarce metal than to go to war over it.

Myron Tribus
Roy V. Wright Lecture, ASME, *1971*

It is our task in our time and in our generation to hand down undiminished to those who come after us, as was handed down to us by those who went before, the natural wealth and beauty which is ours.

John Fitzgerald Kennedy
March 3, 1961

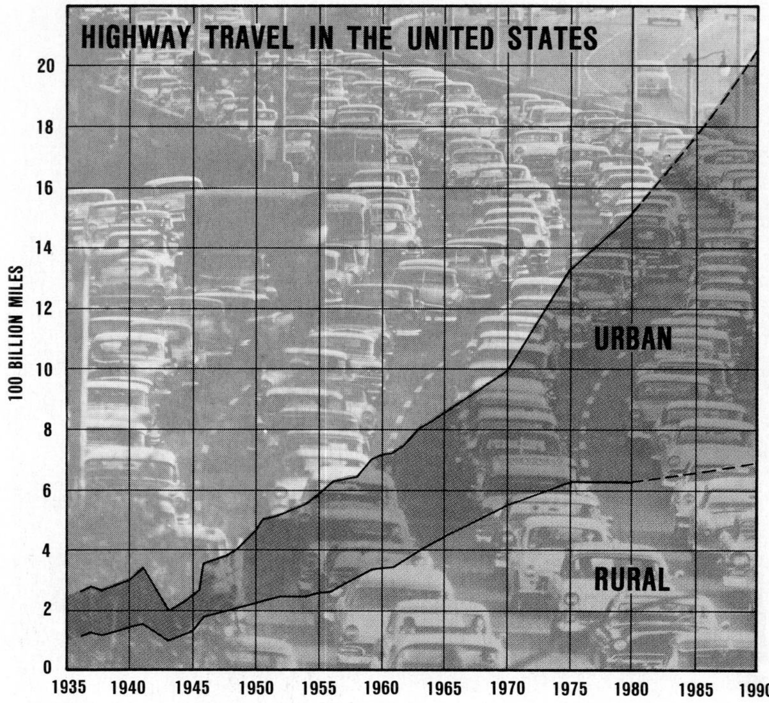

[2-43] Highway travel in the United States.

The family which takes its mauve and cerise, air-conditioned, power-steered, and power-braked automobile out for a tour passes through cities that are badly paved, made hideous by litter, blighted buildings, and posts for wires that should long since have been put underground. They pass on into a countryside that has been rendered largely invisible by commercial art . . . . They picnic on exquisitely packaged food from a portable ice box by a polluted stream and go on to spend the night at a park which is a menace to public health and morals. Just before dozing off on an air mattress beneath a nylon tent, amid the stench of decaying refuse, they may reflect vaguely on the curious unevenness of their blessings. Is this, indeed, the American genius?

John K. Galbraith, 1908–
The Affluent Society

many of the *frustrating* problems of today will become *unbearable* in the future. Since 1896, when Henry Ford built his first car, the mores of the nation have changed gradually from an attitude of "pioneer independence" to a state of "apprehensive dependence"—to the point where one's possession of a means of private transportation is now considered to be a *necessity*.

[2-44] Automobile travel and parking problems of yesterday!

[2-45] Automobile travel today.

Over 80 percent of the families in the United States own at least one automobile, and over one-fourth can boast of owning two or more. However, because of inadequate planning, this affluence has brought its share of problems for all concerned. Beginning about 3500 B.C. and until recent times, roads and highways were used primarily as trade routes for the transport of commerce between villages, towns, and cities. The Old Silk Trade Route that connected ancient Rome and Europe with the Orient, a distance of over 6000 miles, was used extensively for the transport of silk, jade, and other valuable commodities. The first really expert road builders, however, were the Romans, who built networks of roads throughout their empire to enable their soldiers to move more quickly from place to place. In this country early settlers first used the rivers, lakes, and oceans for transportation, and the first communities were located at points easily accessible by water. Then came the railroads. A few crude roads were constructed, but until 1900 the railroad was generally considered to be the most satisfactory means of travel, particularly when long distances were involved. With the advent of the automobile, individual desires could be accomodated more readily, and many road systems were improvised to connect the railroad stations with frontier settlements. At first these roads existed mainly so that farmers could market their produce, but subsequent extensions were the direct result of public demands for an improved highway system. People in the cities wanted to visit the countryside and people in the outlying areas were eager to "get a look at the big city." Within a few years we became a *mobile* society, but the road and

Motor trucks average some six miles an hour in New York traffic today, as against eleven for horse-drawn trucks in 1910—and the cost to the economy of traffic jams, according to a *New York Times* business survey, is five billion dollars yearly!

The Poverty of Abundance

In the next 40 years, we must completely renew our cities. The alternative is disaster. Gaping needs must be met in health, in education, in job opportunities, in housing. And not a single one of these needs can be fully met until we rebuild our mass transportation systems.

Lyndon Baines Johnson
1968

[2-46] Parking problems today.

[2-47] Rapid-transit systems have proved their usefulness in many major metropolitan areas.

[2-48] Highway-guideway systems would relieve the driver of the tedious and tiring task of maneuvering his vehicle from one destination to another.

highway system in use today was designed primarily to accommodate the transfer of goods rather than large volumes of people. Because of this, many of these "traffic arteries" are not in the best locations nor of the most appropriate designs to satisfy *today's* demands. Thus attempts to *drive* to work, *drive* downtown to shop, or take a leisurely *drive* through the countryside on a Sunday afternoon are apt to be "experiences in frustration." Vehicle parking is also becoming a critical problem. It takes an acre of parking area, for example, to accommodate 200 subcompact vehicles.

Most cities have made only half-hearted attempts to care for the transportation needs of their most populous areas. Although those owning automobiles do experience annoying inconveniences, those without automobiles suffer the most—especially the poor, the handicapped, the elderly, and the young. Too often the public transit services that do exist are characterized by excessive walking distances to and from stations, poor connections and transfers, infrequent service, unreliability, slow speed and delays, crowding, noise, lack of comfort, and a lack of information for the rider's use. Moreover, passengers are often exposed to dangers to their personal safety while awaiting service. Not to be minimized are the more than 18 million vehicle accidents and the 54,000 fatalities that result annually from motor vehicle accidents. (For perspective, in the past seven years highway fatalities have exceeded the total loss of American lives in the Vietnam War.)

Traditionally, people have moved into a locality, built homes, businesses, and schools and then demanded that adequate transportation facilities be brought to them. We may now live in an era when this independence is no longer feasible; rather, people eventually may be required to settle around previously designed transportation systems. (This is much the same as it was 100 years ago. However, during this period of time the transportation systems have changed greatly.) Engineers can provide good solutions for all these problems *if they are allowed to do so by the public.* However, there will be a cost for each improvement—whether it be a better vehicle design, computerized control of traffic flow, redesigned urban bus systems, rapid-transit systems, highway-guideway systems for vehicles, or some other entirely new concept. In some instances, city, state, or federal taxes must be levied; in others, the costs must be borne by each person who owns private transportation. The quality of our life-style depends on a unified commitment to this end.

THE CHALLENGE OF CRIME

Crime, one form of social pollution, is increasing rapidly in the United States in particular and throughout the world in general. The rate of increase in this country can be attributed variously to the population explosion, the increasing trend to urbanization, the changing composition of the population (particularly with respect to such factors as age, sex, and race), the increasing affluence of the populace, the diminishing influence of the home, and the deterioration of previously accepted value systems, mores, and standards of morality. A survey made by the National Opinion Research Center

CRIMES REPORTED IN THE U.S.

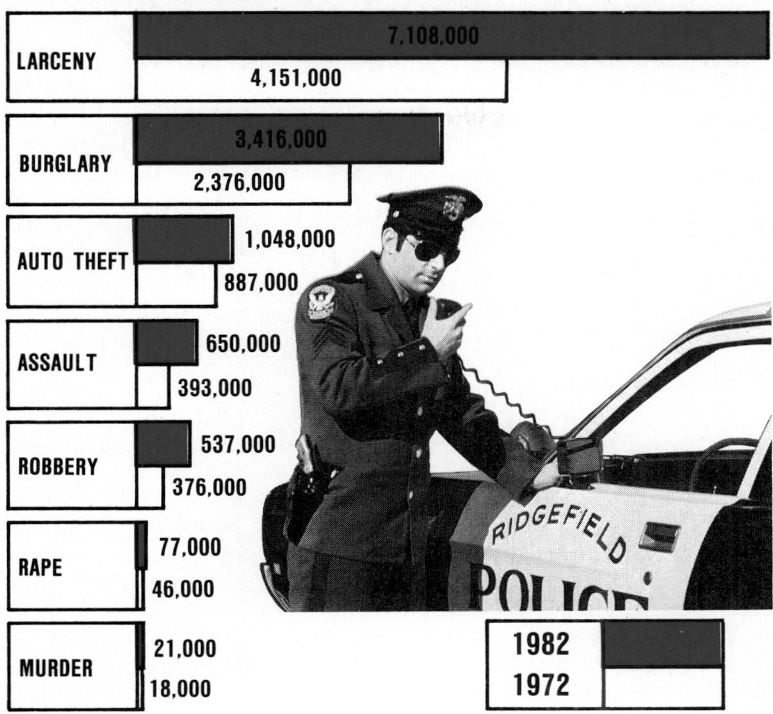

[2-49] The growth of reported crime in the United States.

of the University of Chicago indicates that the actual amount of crime in the United States is known to be several times that reported. A comparison of recent increases in the seven forms of crime considered to be the most serious in this country is shown in Figure [2-49]. A brief examination of these data indicates that the rate of increase of crime is now several times greater than the rate of increase of the population. In fact, crime is becoming such a serious social issue as to challenge the very fabric of our American way of life.

Not all people react in the same way to the threat of crime. Some are inclined to relocate their residences or places of business; some become fearful, withdrawn, and antisocial; some are resentful and revengeful; and a large percentage become suspicious of particular ethnic groups whom they believe to be responsible. A number, of course, seize the opportunity to "join in," and they adopt crime as an "easy way" to get ahead in life. The majority, however, merely display moods of frustration and bewilderment. In all cases, consequential results are detrimental to everyone concerned, because a free society cannot long endure such strains on public and private confidences, nor tolerate the continual presence of fear within the populace.

Traditionally, the detection, conviction, punishment, and even the prevention of crime have been functions of local, state, or federal agencies. Only in rare instances has private enterprise been called on to assist in any significant way, and there has been no concentrated effort to bring to bear on these situations the almost revolutionary

The state of Arizona uses space-age engineering innovation to catch "water rustlers." (Zane Grey and the *Hashknife Outfit* would be jealous with envy!) By state law, water usage in some areas is regulated. Only land that was irrigated from 1975 to 1980 is allowed irrigation rights. No new land can be irrigated for agriculture. Anyone who does so is prosecuted for water rustling. Instead of using a deputy sheriff to seek out those who break the water law, today's law enforcement officers use Landsat, a satellite that orbits the earth 14 times each day. Although it is 560 miles above the ground, its television sensors transmit pictures of the land area. Agricultural lands that have been irrigated show up as a bright, deep-red color. Then it is a simple matter to overlay the satellite map with a computer map that shows only the authorized water areas. The water thieves are revealed.

[2-50] Many of us are unconcerned about the mounting epidemic of crime—until it gets personal.

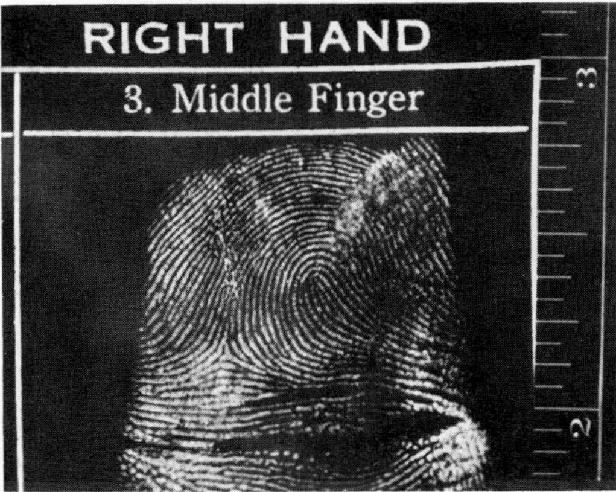

[2-51] Two decades ago fingerprints were symbols associated only with criminals. Today, parents record their children's fingerprints to protect them from criminals.

[2-52] Using a computer, the location of accidents can be quickly identified, license-plate checks made in a matter of minutes, and verification of criminal records greatly simplified.

We have met the enemy . . . and he is us.

Pogo

The world is very different now. For man holds in his mortal hands the power to abolish all forms of human poverty . . . .

John F. Kennedy
January 1961

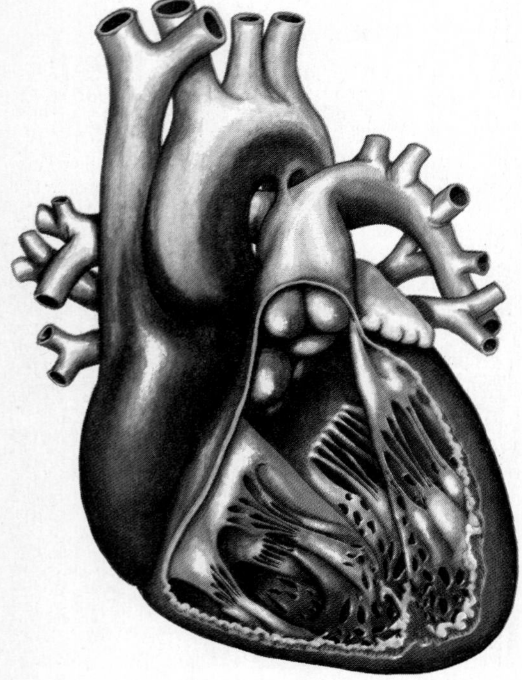

[2-53] If one could substitute for the heart a kind of injection [of blood], one would succeed easily in maintaining alive indefinitely any part of the body. (Julian-Jean César La Gallois, 1812)

advances that have been made in recent years in engineering, science, and technology. Rather, a few of the more spectacular developments have been modified or adapted for police operations or surveillance.

What is needed, and needed now, is a delineation of the vast array of problems that relate to the prevention, detection, and punishment of crime, with particular attention being directed toward achieving *general* rather than *specific* solutions. In this way technological efforts can be concentrated in those areas where they are most likely to be productive. For example, petty thefts may occur more frequently in one area of the city, murders more frequently in some other area, burglaries in another, and so on. What may be needed is a systems analysis of the city to delineate the contributing factors—rather than, for example, equipping all homes with burglary alarms. The engineer can make a significant contribution in such an endeavor.

OTHER OPPORTUNITIES AND CHALLENGES

So far we have discussed primarily the societal environment as an area of challenge for the engineer today. Of necessity, many very important challenges have not been discussed, such as the mounting congestion caused by the products of communication media and the threatening inundation of existing information-processing systems, ocean exploration with all of its varied technical problems and yet almost unlimited potential as a source of material, the expertise that the engineer can contribute to the entire field of health care and biological and medical advance, and the attendant social problems that are closely related to urbanization and population growth—such as mass migration, metropolitan planning, improved housing, and unemployment caused by outmoded work assignment.

It is axiomatic that technological advance always causes sociocultural change. In this sense the engineers and technologists who create new and useful designs are also "social revolutionaries." After all, it was they who brought about the obsolescence of slave labor, the emergence of transportation machines that allowed redistribution of the population, the radio and television sets that provide "instant communication," and every convenience of liberation for the housewife—from mixers, waste disposers, dishwashers, ironers, and dryers to frozen foods. They have most recently brought robotics into the forefront of modern manufacturing methods, and no family has been untouched by the new information society, brought about by dramatic advances in solid-state electronics and the computer revolution. The wonderful thing about our modern sophisticated technology is that it gives us options. Unlike our ancestors, whose limited technical means gave them no choice but to do and make things in the same way as their ancestors, we have many different choices we can make regarding where and how and when we should apply our informational expertise. And we can make the decisions on the basis of the perceived social consequences as well as on the technical elements involved.[33] Frequently, society is not pre-

[33] Melvin Kranzberg, "Technological Revolutions," *National Forum,* Summer 1984, p. 10.

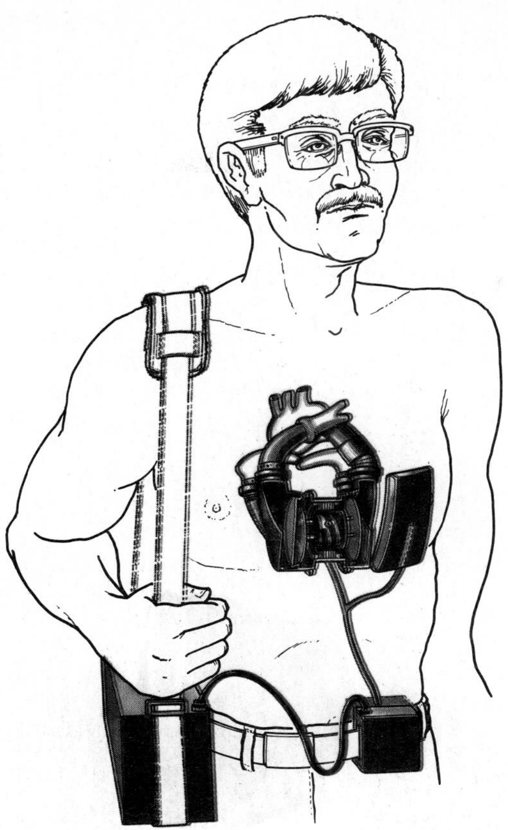

[2-54] An artificial human heart must be many things. It has to be the appropriate size, about a kilogram or less, to fit into a small area of the chest cavity. It must work without failing, beating about 60 times a minute, 40 million times a year. The human heart beats about 3 billion times in an average life span, pumping some 250 million liters of blood.

pared to accept such abrupt changes—even though it is generally agreed that they are for the overall betterment of mankind. Because of this, as a future engineer, you have a dual responsibility to society. You must not only continue to bring about improvements for the benefit of society, but you must exert every possible means to acquaint society with its responsibility for continual change. Without such an active voice in community and governmental affairs, irrational forces and misinformation can prevail.

AN ENVIRONMENT OF CHANGE

Man has always lived in an environment of change. As he has been able to add to his store of technical knowledge, he has also been able to change his economic structure and his sociological patterns. For centuries the changes that took place during a lifetime were hardly discernible. Beginning about 1600, the changes became more noticeable; and today technological change is literally exploding at an exponential rate. It is interesting to contemplate one's future if a growth-curve relationship such as $k = a(i)^t$ is followed [2-57] (both a and i are constants).

In Figure [2-57] engineering and scientific knowledge is assumed to be doubling every 15 to 20 years. Experience with other growth curves of this nature indicates that at some point a limit will

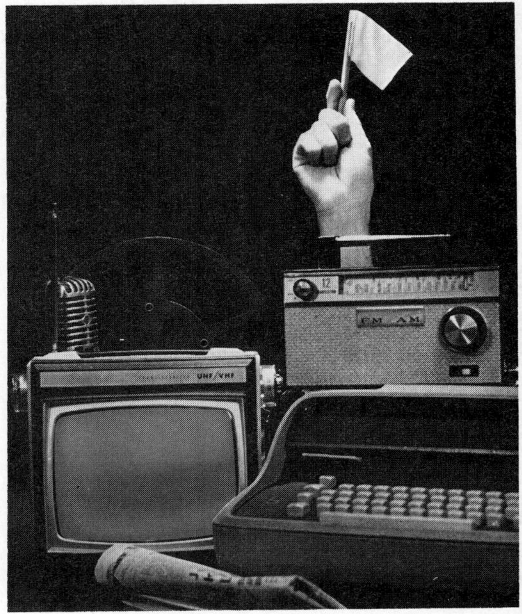

[2-55] Lest we alter our course, we may soon become captives to the media of communication we have created.

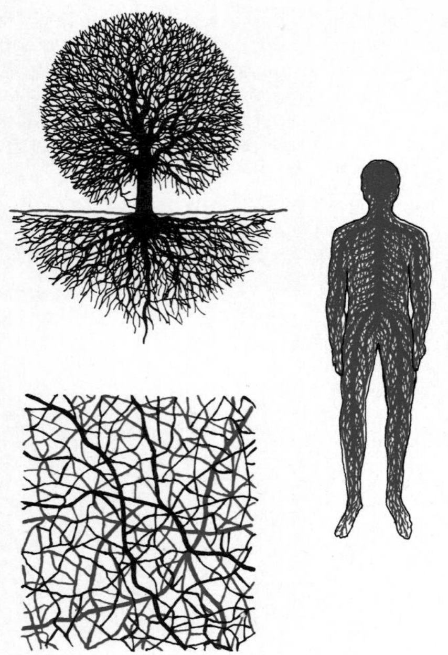

[2-56] In human communities, networks are the interfaces along which the interaction takes place between organic systems—nature, man, society, and, of course, other networks. Rural villages have fairly simple systems of networks, but urban communities interweave many systems of networks at various levels—water supply, sewage and waste disposal, electrical and natural-gas systems, movement of people and goods, telephone, radio, television, and mass printed media. What is important about the vitality of an urban community is not its size or complexity but the degree to which the networks function efficiently. Shown here are analogous network systems in nature, man, and an urban community.

There is no excuse for western man not to know that the scientific revolution is the one practical solution to the three menaces which stand in our way—H-bomb war, overpopulations, and the gap between the rich and the poor nations.

Lord C. P. Snow, 1905

The man who graduates today and stops learning tomorrow is uneducated the day after.

Newton D. Baker

be reached and the rate will begin to level off and then decline. However, in considering the expansion of technology, no one can say with certainty that there is a limit or when this slowing down might occur.

Similar factors are working to provoke changes in educational goals and patterns. In 1900, for example, the engineering student studied for four years to earn a baccalaureate degree, and relatively little change took place in the technological environment during this period. Today, however, due to the accelerated growth pattern of engineering and scientific knowledge, many significant changes will have taken place between the freshman and senior years in college. In fact, complete new industries will be bidding for the services of the young graduate that were not even in existence at the time of his or her enrollment as an engineering student. This is particularly true of the engineering student who continues graduate studies for a master's or a doctorate. It is also interesting to contemplate that at the present rate of growth, engineering and scientific knowledge will have doubled within 20 years after graduation. This places a special importance on continuing lifetime studies for all levels of engineering graduates.

These growth patterns, which are promoting change in all phases of society, are also causing educators and leaders in industry to reappraise educational practices with a view to increasing their scope and effectiveness. From time to time these changes, although not revolutionary, often provoke a sense of progress that shocks those who received their formal education a scant generation before.

THE EDUCATION OF THE ENGINEER

Engineering students who will be best prepared for a career of change should have better-than-average abilities:[34]

[34] Joseph Kestin, Brown University *Engineer,* No. 7 (May 1965), p. 11.

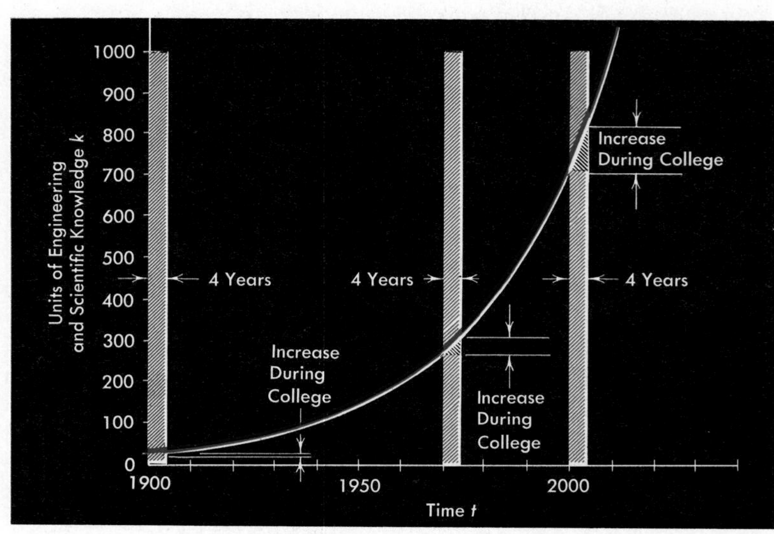

[2-57] Growth of engineering and scientific knowledge.

1. To think with imagination and insight.
2. To understand scientific principles and apply analytical methods to the study of natural phenomena.
3. To conceive, organize, and carry to completion appropriate experimental investigations.
4. To synthesize and to design.

In general, engineering programs in colleges and universities have concentrated on providing a broad-based education that is not closely aligned to a specific state of the art. This has been necessary because for one to acquire even a small part of all the factual knowledge now available, a continuous memorization process would be required. It is, therefore, more appropriate to learn the basic laws of nature and certain essential facts that contribute to an understanding of problem solving. Emphasis must be placed on developing mature minds and in educating engineers who can both *think* and get things done. A means of condensing and concentrating the material to be learned is also of paramount importance. A powerful way of doing this is to employ mathematical techniques that can describe technical situations. For this reason mathematics is a most effective tool of the engineer and its mastery early in one's college career will allow for more rapid progress in such subjects as engineering, mechanics, physics, and electrical circuit analysis. In a similar way, if a student learns the principles of physics, this knowledge will bind together such diverse engineering developments as magnetic materials, gas

[2-58] History is the accumulation of the ideas of man that have burned for varying lengths of time.

The average person puts only 25 percent of his energy and ability into his work. The world takes off its hat to those who put in more than 50 percent of their capacity, and stands on its head for those few and far between souls who devote 100 percent.

Andrew Carnegie, 1835–1919

Education is what you have left over after you have forgotten everything you have learned.

Education is the instruction of the intellect in the laws of nature.

Thomas Huxley, 1825–1895

[2-59] The challenge of engineering lies in providing light for man in the search for truth and happiness—without also bringing about some undesirable side effects.

discharges, semiconductors, and dielectric and optical properties of materials. Similarly, there is no substitute for a mastery of the fundamental principles of other sciences, such as chemistry and biology.

Naturally, the education of an engineer must not end upon graduation from college. The pace of discovery is too great to consider any other course of action than to study and keep abreast of the expanding realm of science and technology. Therefore, in addition to learning fundamental principles of science and engineering in college, the student must develop an intellectual and technical curiosity that will serve as a stimulus for continuing study after graduation.

For the foreseeable future, opportunity for the engineer will continue to expand. Barring a national catastrophe or world war, available knowledge, productivity, and the living standard will probably continue to increase. It is essential that man's appreciation for moral and esthetic values will continue to deepen and keep pace with this technological explosion.

PROBLEMS

2-1. Describe one instance in which the ecological balance of nature has been altered unintentionally by man.

2-2. Plot the rate of population growth for your state since 1900. What is your prediction of its population for the year 2000? Explain the reasons underlying your predictions.

2-3. What can the engineer do that would make possible the improvement of the general "standards of living" in your home town?

2-4. Investigate world conditions and estimate the number of people who need some supplement to their diet. How can the engineer help to bring such improvements about?

2-5. From a technological and economic point of view, what are the fundamental causes of noise in buildings?

2-6. Borrow a sound-level meter and investigate the average sound level in decibels of (a) a busy freeway, (b) a television "soap opera," (c) a college classroom lecture, (d) a library reading room, (e) a personal computer, (f) a riverbank at night, (g) a "rock" combo, (h) a jackhammer, (i) a chain saw, and (j) a kitchen mixer.

2-7. Which of the air pollutants appears to be most damaging to man's longevity? Why?

2-8. Explain the greenhouse effect.

2-9. Investigate how the *smog intensity level* has changed over the past 10 years for the nearest city of over 100,000 population. With current trends, what level would you expect for 1995?

2-10. Describe some effects that might result from a continually increasing percentage of carbon dioxide in the atmosphere.

2-11. Investigate the methods used in purifying the water supply from which you receive your drinking water. Describe improvements that you believe might be made to improve the quality of the water.

2-12. What are the apparent sources of pollution for the water supply serving your home?

2-13 Seek out three current newspaper accounts in which man has caused pollution of the environment. What is your suggestion for remedy of each of these situations?

2-14. Where do some of the highest chemical exposures occur? Give some examples.

2-15. Why should society be concerned with uncontrolled emissions (solids,liquids, or gases) of toxic chemicals into the workplace and environment?

2-16. Estimate the amount of electrical energy consumed by the members of your class in one year.

2-17. Considering the expanding demand for energy throughout the world, list 10 challenges that require better engineering solutions.

2-18. What are the five most pressing problems that exist in your state with regard to transportation? Suggest at least one engineering solution for each.

2-19. List five new engineering designs that are needed to help suppress crime.

2-20. In the United States, what are the most pressing communications problems that need solving?

2-21. List five general problems that need engineering solutions.

2-22. Which three of the renewable energy sources show the greatest promise of development? Why?

2-23. Two decades ago it was widely predicted that solar energy would be used to supply most of the world's energy needs. Why has this not happened?

2-24. Discuss the impact of robotics on the manufacture of automobiles.

2-25. Discuss the obligation of the engineer to check carefully the environmental impact of his designs.

Chapter 3

Welcome to the Twenty-first Century

In times past, foretelling the future was a role delegated to prophets, astrologers, soothsayers, and stock market analysts. Practical men and women did not delve into such matters. Times have changed, and that is the basic reason for this chapter. Today we live in an environment of rapid change. It is not uncommon for an industry to have seemingly "cornered the market" with a new semiconductor device, only to find that their wealth of knowledge (and warehouse filled with products ready for the customer) has been made insignificant by still another device or product that will do even more marvelous things—and at one-tenth the cost. It has been estimated that over one-half of the larger industries in this country now have technological forecasting units to keep abreast of developing research concepts and breakthroughs in pertinent fields. Often technological forecasting is done by "trend analysis" and is done by one or two persons at the corporate level. Today, the average planning horizon for technological forecasting is about 10 years. Although this is an elementary text, it is important to recognize that engineering graduates of the future must be more perceptive than they have been in the past in "looking to the future." To neglect this dimension is to flirt with economic suicide. Now, so much for the rational, let's put our imagination to work.

Anything that is theoretically possible, will be achieved in practice, no matter what the technical difficulties, if it is desired greatly enough.

Arthur C. Clarke

[3-1] I like the dreams of the future better than the history of the past.

Thomas Jefferson

[3-2] The limits to man's environment ranges from 36,000 feet below sea level to 36,000,000,000,000 feet above.

The greatest treasure in the world is knowledge. You are being given an opportunity to gain a greater amount of it than has been possible at any time in the past. Each era of history has been a new, exhilarating, and sometimes seemingly dangerous adventure.

In retrospect, a thumbnail sketch of the phases of man's existence and his goals on this planet (at least in the Western world) might be ordered as follows:

Pre-civilized man: concern for survival.

Greek civilization: pursuit of excellence.

The Roman world: social law and order.

Christianity: salvation of the soul.

The Renaissance: individualism and self-expression.

The Age of Enlightenment: rational truth (science and human quality).

The nineteenth and twentieth centuries: materialism (a product of industrialization).

The twenty-first century: ????? You will help determine its accomplishments.

For many people in this country this is truly a magnificent world, especially for the younger people who are preparing for their life's work. The opportunities and challenges have never been greater at any time. The frontiers of space, computerization, information processing, robotics, health care, transportation, environmental en-

We recognize as cultural all activities and resources which are useful to men for making the earth serviceable to them, and for protecting them against the violence of the forces of nature. As regards this side of civilization, there can be scarcely any doubt. If we go back far enough, we find that the first acts of civilization were the use of tools, the gaining of control over fire, and the construction of buildings.
Sigmund Freud, 1856–1939
Civilization and Its Discontents

[3-3] We must welcome the future, remembering that soon it will be the past; and we must respect the past, remembering that once it was all that was humanly possible.

George Santayana

[3-4] What will my viewpoint be in the year 2020?

[3-5] The road to an improved future is paved with the results of many dedicated and imaginative efforts.

hancement, search for new energy sources, and feeding of the world population are all challenges which are more exciting, and more challenging than ever. This is because today they require a higher level of knowledge, expertise, and risk to accomplish.

Through man's ingenuity, imagination, and nobleness of purpose, he has made possible a world that is more wonderful for more people in all parts of the earth, and throughout the history of mankind, than has ever been the case before. But the task has really only begun. There still remains much to be done to raise both quality and quantity standards of human life everywhere [3-5].

Within the last century, man's science and engineering efforts have been combined into what has generally come to be known as "technology." But the advances of man's technology have been growing more and more complex [3-6]. In the last few decades, man has learned how to produce marvelous machines, devices, methods, and systems to provide the highest level of comfort, safety, health, convenience, and mobility that has ever existed. These wonders of our age—the invention and development of such technologies as radio, TV, air travel, space vehicles, automobiles, radar, lasers, computers (the list goes on)—have enhanced our lives to such a degree that it is almost impossible for those born into such a world to imagine life without them. Quite remarkably, all these things have come about in roughly the last 0.5 percent of the time that man has existed on the earth. The reason for this is that knowledge begets knowledge. In more than 300,000 years of man's history, most of his inventions have been made in the last 250 years.

While we have provided ourselves with the fruits of technology (which is really only another name for man's intellect, creativity, and resourcefulness), we have also brought about other problems that still await solution.

As André Maurois,[1] one of France's best-known writers today, has said:

> In the last fifty years man has discovered more of nature's secrets than in the last twenty thousand. He has found sources of energy so abundant they make him almost too strong. He has embarked upon the exploration of the cosmos and floated in the space between the stars. He flies from city to city at a speed three times greater than that of sound and builds machines that calculate better than the human brain. . . . Your generation will make new discoveries at an even faster pace, but there is plenty left for you to do. You must endow biology with the same precision as physics, unlock the secrets of heredity, make economics into an exact science. The tasks of the future are unending. The more we discover the more we realize that we know nothing.

He goes on to say,

> Man must learn to wield power . . . not to be all powerful. To make the leap to the moon, to Mars and Venus, or to the galaxies infinitely farther away may seem to magnify man's courage and ingenuity. But in the perspective of the universe, it is nothing. If the inhabitant of an electron were to manage to move from his electron to another, all the "Electronians" would clap their hands.

[1] André Maurois, *Open Letter to a Young Man* (London: Heinemann, 1967).

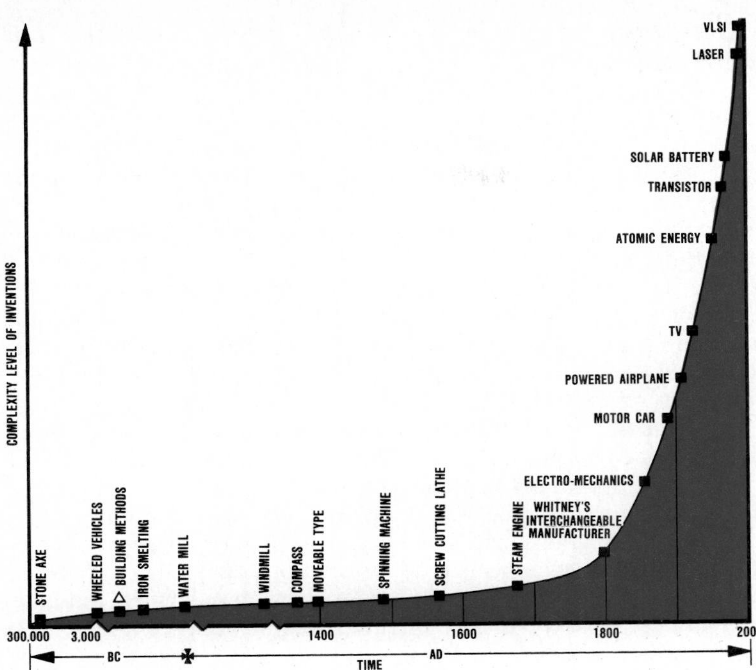

[3-6] The complexity level of inventions has increased with time.

> But a feat on such a small scale would have no cosmic importance. We have taken a few steps in space, but what are these few steps in relation to the infinite? We believe that we know all about the chains of molecules that transmit hereditary traits, but every one of these molecules is a world of its own and we do not know what goes on within it.
>
> Pascal's two infinities are and always will be beyond our comprehension. We are gods. On our small scale and on our drop of mud we have acquired a certain power. We have yet to measure up to it.

The challenge for us, then, is to build on what has been done, correct the errors we have made, and create a world that allows every human being the circumstances and environment to reach his or her potential: to create a "technology" that is both human and humanitarian, and to make war obsolete. As William Faulkner said in his Nobel Prize acceptance speech,

> I believe that man will not merely endure; he will prevail.

As human beings, we have a moral imperative: to control our technology.

FORETELLING THE FUTURE

> You can never plan the future by the past.
> *Edmund Burke, 1729–1797*
> Letter to a member of the National Assembly, 1791

Thus far we have spoken of the past and present, with their problems and solutions. We have considered man's ambitions and achievements. Before we begin a short trip into the future, we must understand that our guides, the authors, are not psychics and they are not even the slightest bit clairvoyant. However, as engineers they can plot the past and present achievements of man on the scale of time; then just as you can, they can dream of man's achievements a few decades into the future. This chapter, then, is a register of events that *might* occur—not a chronicle of achievements that *will* occur. Now let's have some fun and let our imagination investigate the future.

It is not possible to predict the future accurately. When we attempt to do so in any detail, we are bound to look ridiculous in a few year's time. What lies ahead? Who can tell? The future is not fixed nor predetermined by external forces. However, in a sense we can determine our future and, in a sense, the future of the world and mankind. The "we," of course, are those who have lived in the past, those who are alive now, and those who will be here in the future—all playing their parts.

> Where there is no vision, the people perish.
> *Proverbs 29:18*
> The Holy Bible

> The greatest task before civilization at present is to make machines what they ought to be, the slaves, instead of the masters of men.
> *Havelock Ellis, 1859–1939*
> Little Essays of Love and Virtue, *Chapter 3*

The science fiction writers of the last part of the twentieth century have increasingly concerned themselves with the sociological, political, and moral consequences of man's technological and scientific achievements, rather than dreaming about the development of new systems, weapons, and machines. This may be because so many of the intriguing inventions they wrote about earlier—rocket belts, space travel, time warps, robots, antigravitational devices, and talking computers—have in our lifetime become realities and are no longer fiction. In fact, it has become increasingly difficult to keep our dreams ahead of the realities of our daily activities. Figure [3-7] shows how man's inventive ingenuity continues to expand at an ever-increasing rate.

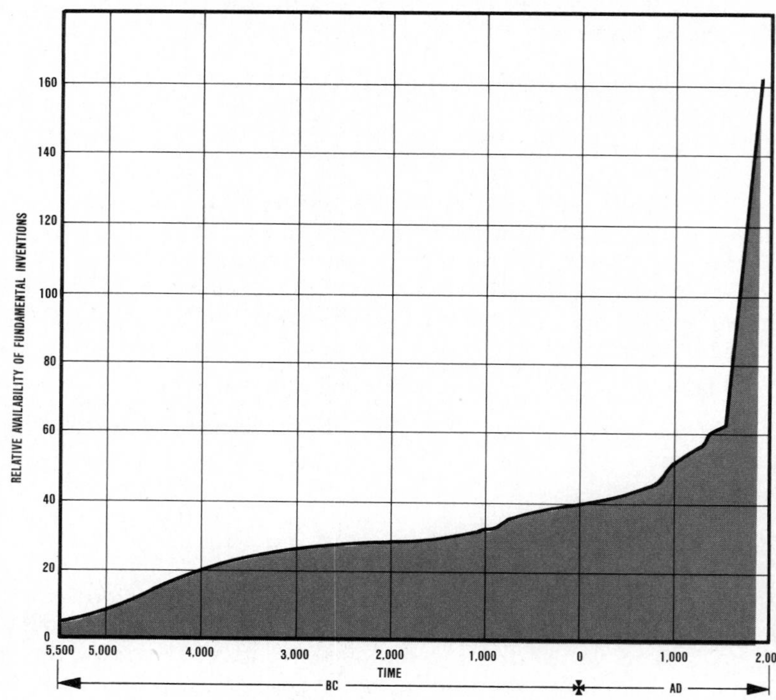

[3-7] The relative availability of fundamental inventions is increasing.

[3-8] Dreams yesterday . . . realities today.

In the late nineteenth century, Jules Verne filled his books with fantasy dreams of submarines, aircraft, and television. You may have read his book or perhaps you saw the movie *Voyage to the Center of the Earth*. To name just a few giants of modern science fiction—Ray Bradbury, Frederick Pohl, Isaac Asimov, Arthur C. Clarke *(Space Odyssey)*, and even Tarzan's author, Edgar Rice Burroughs, who wrote a series of books about people and the society that was imagined to inhabit the planet Mars—all wrote fantasies of science fiction which have had an interesting way of turning into realities.

If you had been a teenager growing up in the 1930s and 1940s, you undoubtedly would have awaited each week's adventures of Buck Rogers (and his lady friend, Wilma) in the Sunday comics or on the radio [3-8]. These fantasies are still being reprinted in some newspapers today, and Buck Rogers comic books have become collector's items. The rocket ships, ray guns, and individual jet packs have all become realities; so have computers and space communications that even Buck Rogers did not have in his fantasy world.

As we consider our "high-technology" world today, we can glance back at the past and see that in the pages of history we will receive little direction. Only the imagination of man can help us to envision the future. There are those who make it their business to dream about the future. They are called *futurists*.

[3-9] Freedom to move about . . .

[3-10] . . . in space . . .

[3-11] . . . is now a reality.

FUTURISTS

Today, a growing area of study and research is being undertaken by persons who are known professionally as futurists. They are scientifically and technologically oriented persons (including social scientists and psychologists) who attempt to identify trends and developments in all fields of human endeavor, particularly in the technologies. Their purpose is to anticipate the direction—and the consequences, good or bad—of man's innovative efforts to create an improved society. Many of their predictions are and will be wrong, particularly in the time frame in which they are predicted to come about.

A review of some of the technological innovations that have emerged within the last hundred years, and which were unforeseen, suggests to us that the unpredictable may be more likely to come about than those things which might be predicted. Some of these ideas that defied prediction are nuclear energy, transistors, TV, sound recordings, x rays, photography, superfluids, night-vision instruments, and radio astronomy.

Man's creative genius, however, is no recent phenomenon. Since mankind first made an appearance on the earth, he has used his mind to ensure his survival. Much of what we now consider as "new" technology had its beginning in antiquity. Man's imagination projected ideas, but for the most part his lack of mastery of the physical laws of the universe and particularly the basic laws of economics prohibited him from bringing his dreams into reality. As our understanding of the basic laws of science developed, so did their application. New technology development is increasing at an ever-increasing rate. In fact, the length of time that it takes for a new technology to find its way into practical application is constantly decreasing [3-12]. Each new "discovery" today makes possible a host of other applications that were not originally anticipated. Consequently, the "seeding effects" of new technology produce results that frequently amaze us all. The possibilities seem infinite. Each is the result of someone else's imaginative thoughts.

COMPUTERS AND INFORMATION

The world we live in today is being altered in spectacular fashion by a new technology, computer technology, whose total history is slightly longer than one generation of man, 25 to 30 years. Not only is its dramatic entrance on the scene of modern life startling, but so is its almost overwhelming impact on virtually every aspect of our daily lives. The consequential effects that it has already had on us individually and on societies around the world are not recognized by most people. In time this impact will be better understood, as the adaptability of the computer, with its seemingly inexhaustible capabilities and applications, becomes an integral part of the lives of more and more people on the earth.

As we view our rapidly changing world of high technology, we must also consider several other aspects of the information processing world which are just variations or extensions of the computer and computer technology in general. In particular, we should con-

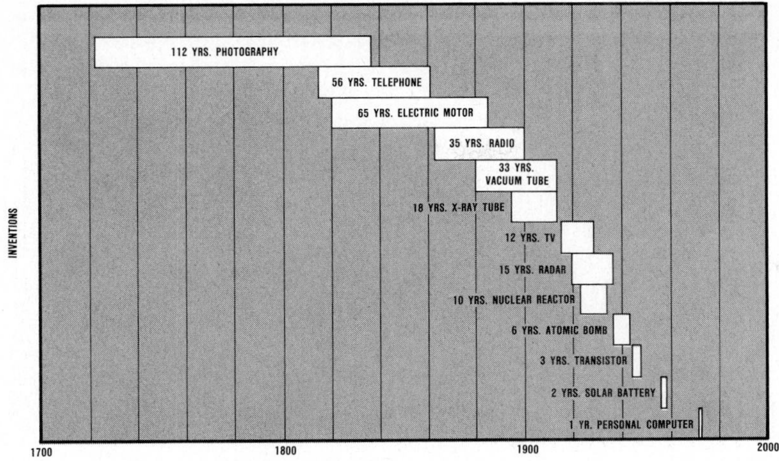

[3-12] The elapsed time between the origin of an invention and its product application is decreasing.

sider the development of industrial automation, robotics, computer-aided design and manufacturing (CAD/CAM), expert systems, and artificial intelligence.

We can trace man's efforts to express and record his inventory of facts, thoughts, and ideas by examining pictograph drawings on the walls of ancient smoky and dank caves. If we follow this path it continues from engraved stone tablets through the hieroglyphics of Egypt—the alphabets and number systems of Greece, Rome, China, and Arabia—through the illuminated, calligraphic manuscripts of medieval Europe—to the printing press—and now to the word processor, a recent capability brought about by the digital computer.

Man's efforts and ability to record and pass on information to future generations is one element of his being that distinguishes him from other mammals and makes him the dominant being on earth. The abacus, the slide rule, the simple adding machine, and now the calculator have been designed as tools to assist him in manipulating the facts that he has developed. These organized data in turn become information that is needed to increase knowledge, some of which will lead to wisdom.

Lest we forget, it is important to keep in mind that the computer is merely another tool of man [3-13]. It is, however, an unusual tool of immense versatility and vast potential. We can speculate endlessly about what we might accomplish with this tool. In advertising a recently published book, *The Micro Millenium* written by Christopher Evans and published by the Viking Press, New York, the following dramatic statements were made:

- If the efficiency and cheapness of the car had improved at the same rate as has the computer's over the last two decades, a Rolls-Royce automobile would cost about $3.00, would get three million miles to the gallon, and would deliver enough power to drive the *Queen Elizabeth II*.
- Today a single silicon chip can store 10,000 words; by the end of the '80s, a whole set of books—and at a tenth of the production cost.

The more information we have, the more our imagination is excited—we can tend to look a little further ahead.

Edward Cornish
President of the World Future Society

Human knowledge and human power meet in one; for where the cause is not known, the effect cannot be produced. Nature to be commanded must be obeyed, and that which in contemplation is as the cause is in operation as the rule.

Francis Bacon, 1561–1626
Novum Organum, *Aphorism III*

Computers, far from robbing man of his individuality, enable technology to adapt to human diversity.

John McCarthy
"Information," Scientific American, September 1966

Animals live without knowing how they live, and they communicate without knowing how they communicate. By and large, so do we. Unlike animals, however, we speculate about how we live and how we communicate. Our better brains and our unique means of communication—language—makes such speculation possible.

John R. Pierce
"Communications," Scientific American, *September 1972*

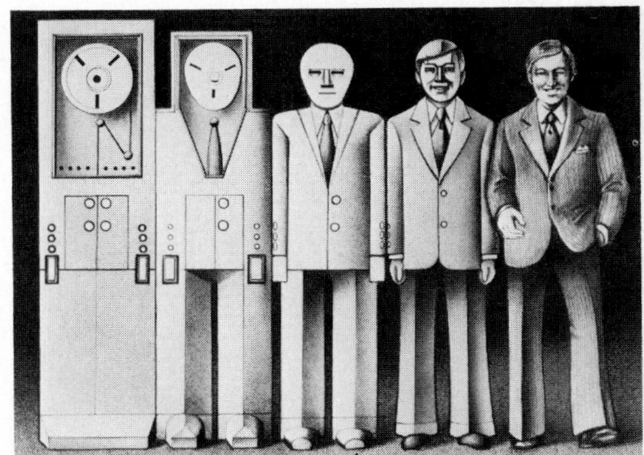

[3-13] The computer has made possible the greatest breakthrough in office technology in history, and has even promised to eliminate paper. Oddly enough, it has become our most ravenous consumer of paper. Offices in the United States generate three-quarters of a million pages of computer printout every day, along with 250 million photocopies and 80 million letters.

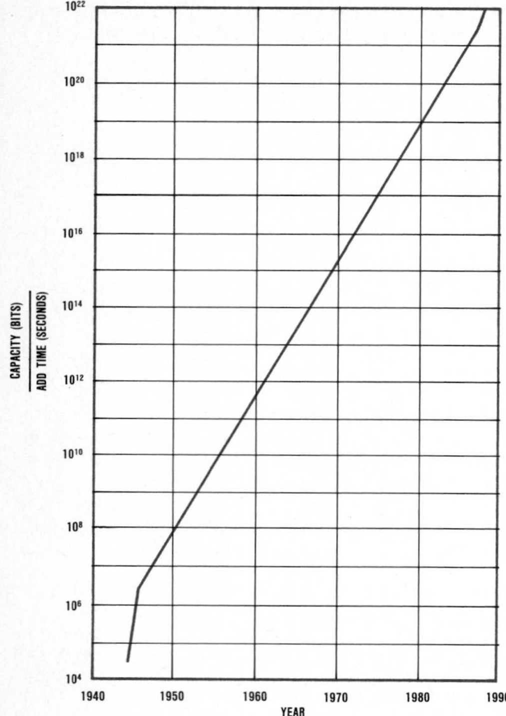

[3-14] Computer performance is continuing to increase.

[3-15] Today, electronic information gathering and analysis techniques could be used to find out more about you, and what you think and feel, than you may want others to know. How will we prevent our minds from turning into open books?

- Fifteen years ago experts said no computer could be programmed to play a decent game of chess; today there are computers which can defeat 99.5 percent of the world's chess players, and very soon computers will reach the International Master level.
- Thirty years ago a computer with the same number of functions as a human brain would have had to be the size of New York City and would have used more power than the subway system; today, that computer would be the size of a TV set; by the end of the '80s it will be as small as a human brain and will run on a transistor radio battery.

It was originally envisioned that the computer would free man's mind for higher-order thought. This has not yet happened. Although marvelous indeed, the computer has primarily performed clerical work. In the past 40 years the computer's performance has increased exponentially, and there is little doubt that this rate will continue for the foreseeable future [3-14].

As we continue to amass and process information (which is also growing at an exponential rate), we are running out of storage space. However, because of recent advances in computer microchip technology, it will soon be possible to store the contents of an entire book on a microchip. If such storage capacity advances continue at their present rate, it will soon be possible to store a number of books on a microchip and perhaps in the future even entire libraries can be stored in a space no larger than the size of a paperback book [3-15]. An added attraction to this scenario will be that the microchip of the future will be much less costly than even one book costs now. What we will have then is an "electronic book." An electronic machine will be used to translate the code on the chip, and another will be used to project it for viewing. However, this is not a dream. These designs are already in the development stage.

The implications of this type of potential are almost limitless. Schools, as we know them, could become obsolete. Through the computer the accumulated knowledge of mankind could be almost instantaneously available to anyone anywhere. In the future, school may very well be wherever your computer terminal is located. Let's continue this line of reasoning a bit. The same might be said for your job. No longer would it be necessary to *go to* your job and suffer the transportation inconveniences attendant to commuting. You might be able to do much of your work at home, at your own pace and on your own schedule (your boss permitting, of course).

Even today, personal, portable computers allow you to carry "school" around with you. Do you need an instant translation of a Spanish phrase? Just say it to your computer and listen for the translation. Would you like to learn a foreign language? A computer-aided instructional program is available for you to use whenever you wish. One of the problems that has been brought about by the tremendous increase in the generation of information is that it is impossible for an individual to stay current with the publications in his or her professional field [3-16]. Computerized data banks make it possible to provide more information.

What are some of the other possibilities? Seemingly, they are endless. In 1958, a paperless banking system was proposed, but although extensive use is being made of credit cards, it has yet to be implemented to any significant extent. For many transactions,

[3-16] Because of the increasing number of scientific journals published, it is becoming more and more difficult to keep abreast of new developments throughout the world.

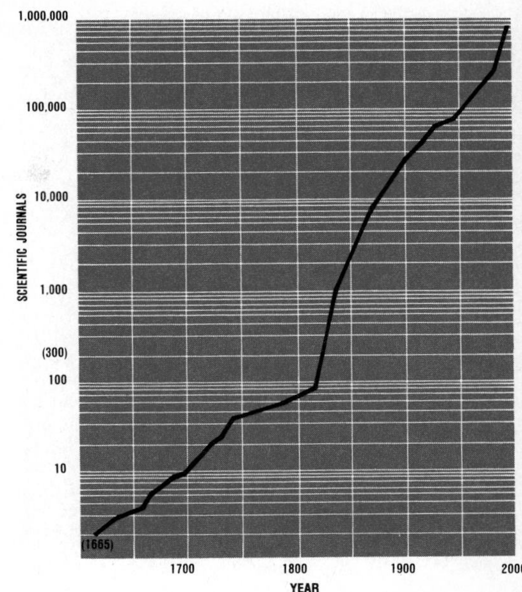

money (cash) could largely become obsolete and all banking would be accomplished electronically. You would deposit, withdraw, or transfer funds by computer, yours or those at establishments of your choosing [3-17]. This banking system will become available in all parts of the world.

Are you having a medical problem? Medical doctors, as with engineers, will not become obsolete as a consequence of the computer revolution. They will, however, be immeasurably assisted and their expertise will be enhanced by the use of expert systems on a computer. This will include using the computer for such tasks as patient medical history interviewing, case history compilation, diagnostic procedures, and, finally, computer-assisted diagnosis.

Microprocessors are already installed in some cars. We have the capability today to develop a transportation system that will allow you to enter a program into your car's control system that would get you to and from your destination safely and expeditiously. In the future, microprocessors will assure you an almost accident-proof vehicle. It would have sensors to prevent collisions, plot alternative routes, detect impending danger, and take evasive action if necessary. It may even have an electronic sensor that will decide when the driver has had too much alcohol in his system to allow him to drive. Simply, in such a case, the ignition would not activate.

The development of telecommunications also has many beneficial possibilities. Again, if the microprocessor is linked to a satellite, potentially it can give everyone in the world his or her own personal communication channel. Some satellites are already earmarked for person-to-person paging—nothing less than the two-way wrist radio (or TV) that Dick Tracy (the comic strip detective) pioneered in the 1930s. Communication by teleconference has the potential to reduce energy demands for fuel, automobiles, transportation networks, and the expense of providing such utilities.

We have been focusing attention on information as it is used to facilitate the business, commercial, and educational segments of our lives. Let us now take a look at the productive side—how we are likely to provide ourselves with the goods and services that we will require. It has been estimated that by the year 2000, 90 percent of all of the production (assembly line) jobs in the industrialized nations will be done by robots. These will not be the humanlike machines described in science fiction stories. Rather, they will be computerized, programmed machines that can endlessly and efficiently do tasks that human beings find boring. They will also do other tasks for which human beings become progressively inefficient as they attempt to sustain a high productivity level during a typical workday [3-18].

In England in 1847, the first labor law was passed; it established the 10-hour per day limit on work. However, it did not limit the

[3-17] In a paperless banking system, expenditures will be charged to your account almost instantaneously . . . much the same as occurred centuries ago when purchases were made by barter.

[3-18] Robots are especially useful in executing manufacturing processes that are hazardous, require close precision, or are fatiguing.

[3-19] Robots have simplified many manufacturing tasks, thereby reducing the cost of the product to the consumer.

number of allowable workdays per week. Now, less than 150 years later, we live in the eight-hour per day range (36 to 40 hours per week). In the near future the normal work time will probably be reduced even more. This is predicted to come about inevitably as a consequence of advances in information processing, automation, and robotics. The workplaces of tomorrow, whether they be in offices or factories, will be vastly different—and probably would not be recognizable to those people who performed similar jobs during the early part of this century.

In general, engineering designs should have as their ultimate purpose not only a superior solution with economic advantages but also the advancement of the human condition and the overall betterment of the human race. However, the design of robots raises some interesting questions. For example, should we create machines (robots) that will reduce or eliminate much of the need for manual and skilled labor? Is there anything that we should not design a robot to do? Isaac Asimov's "Laws of Robotics"[2] addresses this question.

1. A robot must not harm, or through inactivity allow harm to come to a human being.
2. A robot must obey all commands given to it by a human being except when these are in conflict with the first law.
3. A robot must preserve itself at all times unless by doing so it contradicts the first two laws.

Robots could conduct many high-risk jobs, such as defusing bombs, fighting fires, and conducting deep-sea rescues.

Although the future holds great promise, challenges, and prospects for the creative, imaginative, and inventive engineer, new and difficult ethical problems will also make their appearance. Not the least of these may come with the advent of the biological computer.

Engineers are working with life scientists to produce computer circuitry in biological laboratories from living bacteria. It will theoretically then be possible to produce microprocessors with up to 10 million times the memory of today's powerful computers.

If this new technology is successful, future circuits will be created from groups of organic proteins the size of molecules to act in the place of electronic memory and switching devices in chips.

An almost infinite list of applications exist for these tiny supercomputers. A desktop device of this design could hold all the information ever recorded by the human race. It is possible that such a computer could be connected to the human nervous system and serve as artificial eyes, ears, and vocal chords. Tiny computers could even be implanted in the bloodstream to monitor body chemistry and correct imbalances.

> Fooling around with alternating current is just a waste of time. Nobody'll use it, ever. It's too dangerous . . . it could kill a man as quick as a bolt of lightning. Direct current is safe.
>
> *Thomas Edison to Nikola Tesla*

THE FACTORY OF THE FUTURE

The twenty-first century, if not before, will see the paperless, peopleless office and factory. Electrotechnology is creating an era where many of the work activities that people have engaged in previously will be done by and through electronic processes. As data-

[2] Isaac Asimov, *I, Robot* (New York: Doubleday, 1950).

processing software is developed to expand the capabilities of computer-integrated manufacturing (CIM), we will be able to design factories that will be fully automated. Many major manufacturers, both in this country and abroad, are working toward that goal.

In such a factory all the manufacturing operations, from the inception of an idea to the production of a new device or the modification of an old product, can be done through the use of the computer. Market and cost analyses can be extracted from the computer, as can the location and accessibility of raw materials and component parts. The product design can be produced using the data base containing previous designs, or a new design can be generated through the computer-aided design (CAD) process. Production will be carried out through use of computer-aided manufacturing (CAM), and inspection can be done by robots. Consequently, the factory of the future will, to an ever-greater extent, be automated. Process computers and robots will assume many of the tasks of production—from the ordering and assembling of raw materials to fabricating, testing, and moving the finished product into the consumer's hands.

Ultimately, computers will be designing computers, and the engineer's job will be directed more and more to the creative aspects of design. The engineer will continue to direct the computer . . . but the computer will be a much more valuable servant than at present.

Much work remains to be done in the next decade or so to integrate various computer systems into the manufacturing process. Present capabilities allow a video camera to track the movement of automatically guided vehicles (AGVs) to integrate the robot with its environment. Robots also have the capability of tracking the moving load and maintaining a precise orientation with it.

Each of our five senses—sight, sound, taste, smell, and touch—can be duplicated in robots, but not necessarily in the same way. For example, robots can be designed to "see" in normal light, or in darkness, using infrared or ultraviolet light. Even depth perception can be simulated with an electronic sensor.

The vision-coordinated robotic system will be used in manufacturing to determine locations, and for inspection, recognition, counting, and motion applications. The human machine operator as we know him today will become largely superfluous.

Robot control via sensory feedback using a vision system can be used to inspect a variety of products ranging from complete assemblies (such as instrument panels, distributors, and fully assembled appliance) to printed circuit boards.

The future of utilizing robotics for manufacturing appears to be moving from hardware-intensive mechanical systems as stand-alone units to software-intensive machines with a high degree of intelligence and integration with their environment and surrounding equipment.[3] The handling of toxic materials is a particularly appropriate function for robots.

It is easy to see that in the future there will be an even greater need for engineers who will develop, design, and put into place the manufacturing systems of tomorrow. Eventually, engineers will be able to design self-replicating robots—robots designed to build other robots.

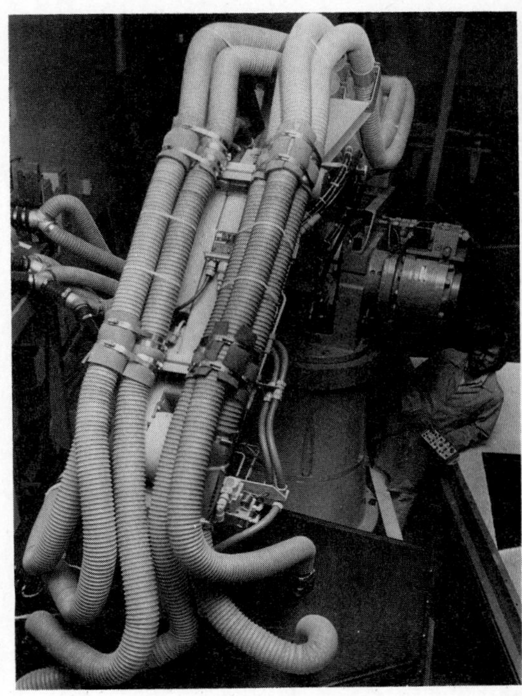

[3-20] The "tentacles" of this squid-like industrial robot are especially designed suction hoses that pick up pieces of fabric-like graphite material and place them in tool forms used to produce finished parts.

[3-21] Robots can be programmed to accomplish almost any manufacturing or industrial task.

---

[3] *Robotics Today,* October 1983.

[3-22] During the Challenger's second orbital mission in 1983, astronaut Sally K. Ride floats freely on the flight deck as she communicates with ground controllers in Houston.

Factories in Space

While the greatest part of manufacturing in the future will concern itself with producing goods here on earth, another distinct possibility is the use of space as a site for manufacturing. The properties of space—high radiation, low and high temperatures, unlimited vacuum, and zero gravity—will allow both production of products and research experiments that would otherwise be dangerous to conduct on earth (remember the gas leak in India in 1984, for example, that killed over 1700 people and affected thousands more). These orbiting factories would also make possible the manufacture of devices that could either not be produced at all, or not be produced economically on earth.

Space manufacturing will probably be rather limited for some time to come, but the future offers tremendous advantages for it. For example, because cooling metals do not form convection currents in zero-gravity environment, strong alloys can be formed from metals that would crumble if they were combined on earth. Because the weight factor of materials and processing equipment is the most critical factor in the economics of space manufacturing, the high "value to weight" ratio of such products as electronic devices, pharmaceuticals, specialty glasses, and certain alloys will be the first items manufactured in space.

General Technology

As we look at the world of engineering, high technology, manufacturing, and industry, we can speculate concerning new products, processes, devices, materials, and systems that we might expect in the coming century. Here is just a sampling of feasible possibilities that will be brought into being when it is demonstrated that they are economically feasible.

- A worldwide fully-automated air traffic system.
- Superconducting magnets for use in electrical power generation.
- Ceramic automobile engines.
- Fiberglass springs for automobiles.
- Epoxy-reinforced plywood for auto or truck bodies.
- Ultrasonics in manufacturing for cleaning, metalworking, assembly, and testing materials.
- Robots to repair live electrical lines.
- Mass drivers to move large objects through space (for example, to bring asteroids closer to the earth for mining).
- Holographic projectors and receivers for home entertainment.
- TV pictures projected on the entire wall of a room.
- GEMs (ground-effect machines). These vehicles may some day allow us to move thousands of tons of material across the sky by using antigravity freighters.
- A gravity control unit that could be strapped on a person's back.

As we advance in our abilities to create a new work environment we will be creating many new jobs for people to maintain and service our electronic environment. In this new age people will be trained or retrained to fill positions that require a greater degree of technical understanding. Many of the jobs that were formerly "production" oriented will evolve into the servicing of a new "high-tech" society. Engineers and scientists will find that these new "electronic tools" will increase their creative abilities and productivity at an ever-increasing rate, and the twenty-first-century creations of man will be marvelous indeed.

SPACE: THE INFINITE FRONTIER

Space is to place as eternity is to time.
Joseph Joubert

Not quite 500 years ago, an Italian explorer set out on a voyage that was to forever alter the history of the world. In some ways he might be likened to the Soviet astronaut who became the first man in space in 1961. The nationalities are not important. In any case, both were stretching man's knowledge and technology of the time. Today we are in the infancy of space travel, exploration, and ultimately, perhaps, even space colonization. From 1957, with the launching of the artificial satellite, Sputnik, we have witnessed a sequence of events as follows:

Interplanatary rockets.
Humans in space.
Man on the moon.
Robot to Venus.
Robot to Mars.
Probe to Jupiter and Saturn.
Reusable spacecraft (space shuttle).
Space walk.
Retrieval of satellites in space.
Disaster (the Challenger) in space.

[3-23] Space exploration and habitation of the moon offer exciting and rewarding opportunities.

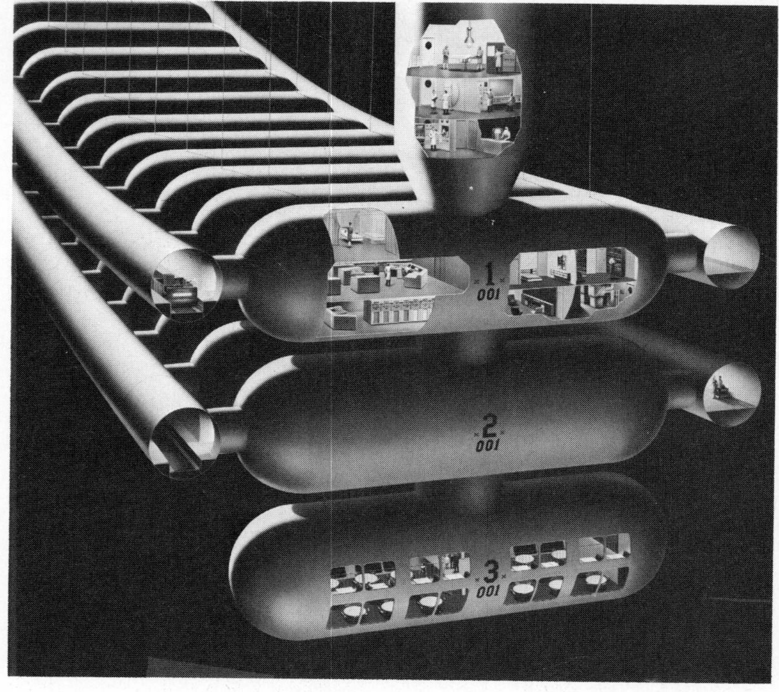

[3-24] This drawing shows how people might live and work in space within the next decade. The modules containing work areas, living spaces, and a restaurant area, are part of a giant wheel silently circling Earth. People might live in this space city for six month periods, and vacationers from Earth could treat themselves to the city's resort hotel.

The geological engineers of the not-so-far future will be concerned with the discovery of precious minerals and metals which will come from other planets and asteroids. Once such materials have been located and assessed, several engineering activities will be involved in their mining and processing. Then it will be necessary to transport and deliver them to the laboratories and manufacturing facilities—either to earth or to space laboratories, which can utilize the environment of zero gravity, high and low temperatures, and an unlimited vacuum in which to manufacture useful goods [3-24].

The January 28, 1986 Challenger disaster, with resulting loss of the crew of seven, had a sobering effect on manned space exploration. Before this catastrophic failure, there had been 24 successful shuttle space flights. This accident has brought about an even greater recognition of the importance of risk analysis (see page 121).

The speculation . . . is interesting, but the impossibility of ever doing it is so certain that it is not practically useful.

The editor of Popular Astronomy *in a rejection letter to Robert H. Goddard concerning his article "On The Possibility of Navigating Interplanetary Space," in which he proposed the idea of nuclear energy (1907)*

The first great space project may very well be a solar power station. It would probably consist of several square miles of photovoltaic cells that would convert sunlight into electrical current, which might be radiated in a light beam to the earth. Although this design has already been envisioned, the challenges to perfect such a system and operate it efficiently will challenge engineers and technicians for years to come. Eventually, we can have solar power stations orbiting the earth which could supply the world's need for electricity.

Communications satellites are another phase of the space program that has intriguing possibilities. As we develop and improve them, perhaps by substituting laser light for microwaves, we can create so many millions of TV channels that people and/or businesses all over the earth can each have their own. Such an eventuality would revolutionize the educational system as we know it today. The person being educated will have access to and control of all the information needed for his or her education.

No matter what your interests may be, the infinite possibilities of space seem to have something for everyone. Observation satellites can further our knowledge of climatology, oceanography, and geology; all of these are linked with such human concerns as earthquake warning, hurricane predictions, weather patterns, pollution assessment, fishery potential, and mineral exploration.

In the area of medicine and health, space laboratories may create pharmaceuticals more effectively and less expensively than can be done in the earth's atmosphere.

Space colonization is another idea that offers intrigue. Someone has even coined an interesting name, "terra forming." This is the concept of space exploration with the idea of creating environments throughout space that can be made compatible with human beings and their physiological and psychological limitations. As has been said, "Life in space for the foreseeable future will be like that in a submarine, an offshore oil platform, or an Antarctic mining camp—dangerous, cramped, isolated, and uneventful." It is likely, then, that large-scale space colonization will not be feasible until the latter part of the twenty-first century, but for those who dream and are inspired by what man has done so far, the challenge is real and compelling [3-25].

[3-25] With change must also come an improvement in the quality of life.

TRANSPORTATION

"One small step for man: one giant step for mankind." That statement is appropriately associated with the moon landing and the space age, but it could also be considered as a description of the beginning of man's journey from prehistoric times to the present in the context of movement, travel, or locomotion. At first we had only the power of our muscles. We then discovered that we could drag what we could not carry. Later someone found that a round rock with a hole in its middle could be made into a wheel—"something that would roll." In fact, put two of them together, and, with a pole in between, we could perch a bundle on top (later a box), and we had a cart in which we could carry things too heavy to drag or carry.

Man's ingenuity led to steam carriages, locomotives, automobiles, airplanes, and finally space travel by rockets. What lies ahead?

Our purpose here is to speculate concerning the future transportation devices and systems that are waiting to be developed. (We "perfect" designs only to find that they can still be improved.) The saga of transportation has been linked with the idea of speed; from the slow pace of walking to the speed of supersonic aircraft and space vehicles that travel thousands of miles an hour, man has always wanted to go faster. Figure [3-26] tells the story of man's quest for speed. We have already surpassed the speed of sound. Is there a possibility that we will ultimately travel through space, exceeding the speed of light in some modes of transportation? Is there a limit? This

Relativity

There was a young lady named Bright,
Who could travel much faster than light.
She started one day
In the relative way
And came back on the previous night.

Anonymous

[3-26] In the past 150 years the speed with which man can be transported has increased dramatically.

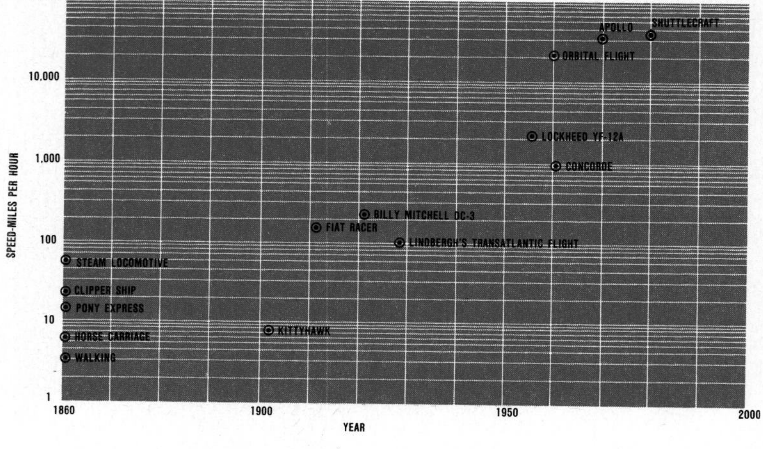

is a marvelous challenge. However, there are other earthly challenges that could bring even greater dividends.

Congestion, traffic snarls, pollution, frustration, and the potentially dangerous overloading of our urban traffic system are reasons in themselves for engineers to design safer, more efficient, and nonpolluting transportation systems and to create an environment that is essentially clean and healthful. This will mean that some aspects of today's cities will need to be redesigned. It is an accepted fact that mass-transit systems can be greatly improved to move people more comfortably and efficiently [3-27]. Automated roadways with programmed route and interval spacing may be able to let the driver pick a destination, enter it into the computer, activate the car, and then sit back and relax while being transported to a destination with no interruptions for traffic lights, snarls, or chance of accident.

The efficiency of home furnaces, steam turbines, automobile engines, and light bulbs helps in fixing the demand for energy. A major need is a kind of energy source that does not add to the earth's heat load.
Claude M. Summers
"The Conversion of Energy," Scientific American, *September 1971*

A great challenge also exists to design automobiles using lighter-weight materials and which have more fuel-efficient engines. This will probably eliminate the metal internal combustion engine. The Japanese are presently working on a ceramic engine and several companies in this country are also experimenting with them as well. Ceramics, made of clay hardened by fire, are extremely hard, resist heat and wear, insulate effectively, and use plentiful raw materials such as silicon and carbon. Mass-produced electric cars are also becoming a potential reality. This would depend on the development of a long-life, low-cost, rechargeable battery. Research in progress includes using a lithium–silicon system. Research is presently under way to produce a zinc chloride (or some other type) battery that could carry a car at least 500 miles on a single charge—about five times the distance achieved by conventional batteries and one-third the cost of fuel for an internal combustion engine.

The actual building of roads devoted to motor cars is not for the near future, in spite of many rumors to that effect.
Harper's Weekly, *1902*

Engineers will be increasingly involved with city planning to create metropolitan areas which have fewer cars and encourage more pedal and footborne traffic. In Gothenburg, Sweden, for example, the city is divided into five cells, each shaped something like a piece of pie. From a surrounding ring road, motor vehicles can enter or leave any cell—but they cannot drive from one cell to an adjacent

[3-27] Traffic problems of the future can be minimized if cities are properly designed.

one. Only streetcars, buses, cyclists, and pedestrians are allowed to make these crossings. The system works. Motor vehicle traffic has been cut 20 percent. Can you devise a better plan?

A new engineering discipline is developing. It is called *macroengineering*. This is the study, preparation, and management of extremely large and complex engineering projects. Examples of the past of this type of engineering have been such things as the Great Wall of China, the Roman aqueducts, Gothic cathedrals, and the Panama Canal, to name just a few. The future may offer such macroengineering projects as a Great Lakes reservoir that would convert part of Hudson Bay into a freshwater lake and divert water south by a series of canal-rivers. Water could be delivered as far south as Mexico.

Another idea being considered by macroengineers is the construction of a transcontinental trench or subway that would allow magnetically levitated trains to travel up to several thousand miles per hour. Impossible? Don't be too sure. American, German, French, and Japanese engineers are already working on a train that has been referred to as "Maglev" (for magnetic levitation). The train will have no moving parts or wheels. Floating above its track on a magnetic cushion and propelled by linear induction motors, such a train has already reached a cruising speed of 310 mph. Like other technological developments, this progress appears to be only the beginning.

[3-28] Devise a transportation system incorporating the use of a "mother" tractor-truck that is used to transport mini electric cars where great distances are involved.

[3-29] Would you like to own this "dream airplane" by the year 2000?

Thanks to science, you can now fly almost anywhere in half the time it will take you to wait for your luggage after you get there.

Bill Vaughan

Rail travel at high speeds is not possible, because passengers, unable to breathe, would die of asphyxia.

Dr. Dionysys Lardner, 1793–1859

Even if the propeller had the power of propelling a vessel, it would be found altogether useless in practice, because of the power being applied in the stern it would be absolutely impossible to make the vessel steer.

Sir William Symonds,
Surveyor of the British Navy, 1837

With air transportation, the greatest problems confronting the industry today involve serious overcrowding, both in the air and on the ground. More air traffic controllers will only temporarily alleviate the problem. However, a better way is to develop and refine aircraft routing approach and landing systems. Automation, computerization, and mechanical and electrical design can produce an air-traffic system of tomorrow that will make today's seem antiquated.

Ticketing, baggage handling, and airport facilities, including surface transportation getting in and out of airports, are all challenges and responsibilities of tomorrow's engineer. What good is it to be able to fly from Paris to New York in three hours only to find that it takes half that much time to get from the airport to the city's center? Helicopters, high-speed monorail trains, and auto expressways must be designed to solve these problems.

Atomic powered, ocean-going ships with sails, dirigibles pulling other freight-carrying dirigibles, the rocket backpack (such as used by both Buck Rogers and the Space Shuttle astronauts), and turbine or jet-powered automobiles are all ideas whose time is coming—possibly within the next two decades. What forms will they take, who will develop them in an optimal way, and what will they mean to the betterment of mankind? These are all questions for which you may be working to find the answers. The "need" is there, but the "demand" lags behind.

MEDICINE AND HEALTH: THE ENGINEER'S ROLE

"God grant you long life and the good health to enjoy it." This toast has been offered, in one form or another, down through the centuries. Priests, shamans, witch doctors, medicine men, and today's physicians have all been involved with curing disease and restoring or maintaining the physical (and in modern times) the mental well-being of mankind. As with so many other areas of technology, advancements in medicine and the healing arts have grown exponentially and are almost unbelievable. Only in the past 150 years has the acceleration of the techniques of immunology, antiseptic methods, vaccination, anesthetics, and infection control made it possible for one to survive serious illness or operations. Pasteur's understanding of the relationship of bacteria to infectious diseases, Lister's antiseptic method of prevention of infection of wounds, and the work of various scientists in the mid-nineteenth century to develop successful anesthetics all made it possible for man to survive illness and injuries that had ultimately caused his death—a death that we at the end of the twentieth century would certainly call "premature."

The elimination of innumerable diseases in the relatively recent past—from chickenpox to polio, from anthrax to yellow fever—has been truly astounding. For example, in 1980, smallpox was officially declared to be eliminated on a global basis. This was a remarkable achievement. However, even these achievements are overshadowed by what is currently being done in the medical fields. Today's accomplishments will pale in contrast to what the future will hold.

What new challenges, then, does tomorrow's young engineer face? What additional work is to be done? The truth is, the future is forever unknowable. However, its challenges can be mastered by those who are unafraid to enter the unknown. Ironically, as you may have heard, *the more you know, the more there is to know.*

In the past two decades we have seen successful open-heart surgery, heart transplants, several models of the artificial heart, and even the implantation of animal hearts into human patients. Other organs are becoming transplantable as we learn more about immune systems, rejection, and regeneration. Hands and feet have been reattached, corneas have been replaced, artificial joints have been implanted, and laser surgery is now becoming commonplace. A new concern focusing on holistic health, preventive medicine, and genetic manipulation can go far to fulfilling predictions that the life span of the average teenager today could be lengthened as much as 20 to 30 years.

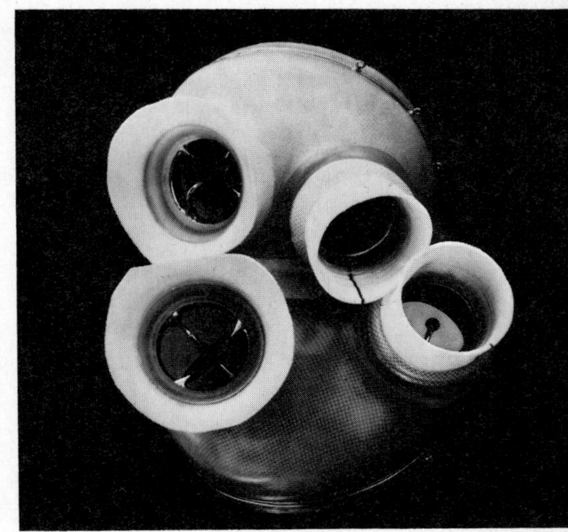

[3-30] The Jarvik 7 artificial heart.

Physicians are increasingly joining engineers in design efforts. This new design team, by cross-fertilizing knowledge, skills, and concerns for the betterment of mankind, will produce a myriad number of new procedures, methods, devices, and materials that will make the idea of a "bionic" man a partial reality in your lifetime.

Bioengineers will face particular challenges in the effort to produce "new" body parts to replace those that through illness or accident have lost their effectiveness [3-30]. We now have available artificial limbs, arteries, ears, intraocular eye lenses, and the oldest replicated human part—false teeth.

The future is destined to bring efficient, functional, portable, and long-lasting replacements for hearts, kidneys, and other vital organs. As with surgical replacements of these parts, the engineer will be vitally concerned with developing improved artificial joints and limbs that can be controlled by computers and affected by miniaturization.

Another procedure that the engineer will be involved with is microsurgery. In microsurgery, doctors use powerful microscopes to work with tissue and vessels thinner than a human hair—vessels of 1 millimeter in diameter or less and structures of 50 micrometers or less—the size of a human hair. Research in this field has brought other breakthroughs: for example, surgery on a fetus in the mother's womb; reverse sterilization of people who have been voluntarily sterilized; computer-guided surgery using holograms, laser beams, and miniature, multipurpose instruments. It is not uncommon to talk of "bench surgery" in which a defective organ is removed from the patient, repaired outside the body, and then replaced.

Chemotherapy, often the province of hemotologists, may alter the treatment of mental diseases as well as provide a means of boosting human learning, attention, and memory. Some drugs already have been demonstrated to cause improvements from a few percent to as much as 20 percent on various tests of learning and memory. Other drugs have increased cognitive power by boosting alertness or attention. Drugs are also being developed to combat obesity, increase sex drive, induce sleep in insomniacs, calm those suffering from mental and emotional illness, and alter moods. Much more is yet to be done.

It Couldn't Be Done

Somebody said that it couldn't be done,
But he with a chuckle replied
That "maybe it couldn't," but he would be one
Who wouldn't say so til he'd tried.
So he buckled right in with the trace of a grin
On his face. If he worried he hid it.
He started to sing as he tackled the thing
That couldn't be done, and he did it.

Somebody scoffed: "Oh, you'll never do that;
At least no one ever has done it";
But he took off his coat and he took off his hat,
And the first thing we knew he'd begun it.
With a lift of his chin and a bit of a grin,
Without any doubting or quiddit.
He started to sing as he tackled the thing
That couldn't be done, and he did it.

There are thousands to tell you it cannot be done,
There are thousands to prophesy failure;
There are thousands to point out to you, one by one,
The dangers that wait to assail you.
But just buckle in with a bit of a grin,
Just take off your coat and go to it;
Just start to sing as you tackle the thing
That "cannot be done," and you'll do it.

Edgar A. Guest,
The Path To Home

[3-31] If life extension were to become a national priority, we could produce dramatic results within the foreseeable future. Life spans of 100 to 200 years and more are feasible.

Researchers are working on several approaches to longevity, which include:

- *Transplantation,* which might allow us to continue replacing organs until almost our entire bodies are new.
- *Regeneration,* a process by which deactivated genes are switched back to renew cell tissue.
- *Prevention of lipofuscin buildup.* Lipofuscins are a form of destructive cellular garbage produced by the body, and are thought by many scientists to contribute to aging.
- *Restrictive diet,* which in the young delays maturity and increases longevity; and in the middle-aged seems to rejuvenate the immune system.
- *Prosthetics and cyborgs,* machine–human combinations of which the $6,000,000 Man is an almost credible preview.
- *Lowering body temperature,* which alone might add many years to human life.

Champion International Corporation

The future? An improved artificial ear! It would enable the stone-deaf to hear. Artificial blood vessels a few millimeters in diameter to be implanted in any part of the body? Geriatric engineering—engineering aimed at the problems of helping old and aging people live lives which are satisfying and relatively pain free [3-31].

As they struggle with perfecting these capabilities, engineers will be faced with the ethical problems of how much effort should be expended to develop solutions to keep the comatose alive. Although it will not be the engineer's role to make this decision for society, still the question must be answered, "Is my design adding to the quality of human existence?"

Another challenge facing tomorrow's engineers is perfection of the mind–machine linkup (cybernetics). Researchers are already developing an advanced aircraft which uses an aiming system that follows the pilot's eyes. They may soon be joined by sensors that read a pilot's brain waves to control flight. The next step could be a computer–brain linkup that would operate by placing electrodes that are tied to a minicomputer on the neck and temples. Perhaps this could be expanded to provide a biochip which would allow the ultimate in mind–machine hookups, a direct link between the brain and microprocessors.

These are just a few possibilities and challenges in the field of medicine and health that will concern the engineer of tomorrow.

THE CHALLENGE OF THE OCEAN

The world's oceans form an interconnecting body of salt water that lies between the continents. They cover over 71 percent of the earth's surface. An *ocean* is one of the major subdivisions of this expanse of water, with *seas* being smaller partially enclosed subdivisions. Since man first appeared on the earth, the ocean has played a dominant role in the development of his history. It was first used as a source of food and later as a means of transportation. These uses are still important today [3-32]. In recent years we have recognized that because of its vastness, its benefits are often contradictory. On the one hand it stores heat and water that greatly influence weather and climate. Yet castaways at sea may die from cold or thirst. It is unforgiving and many who have attempted to use it as a highway have perished. On the other hand, to those who master its technology, the ocean offers a safe and economical way to travel.

The ocean continues to be an important source of food, particularly animal proteins and fats. However, billions of tons of refuse, garbage, and sewage of seaside communities are dumped into it. In recent years this has even included radioactive wastes. It serves as a recreational playground, and as a source for coal, oil, and gas, mineral deposits, precious metals and jewels, important raw materials, and oxygen. For example, the annual release of oxygen by marine plants provides 70 percent of the oxygen in the atmosphere. In turn, the organic substances of plants are a virtual banquet table for species living near the surface of the water. These then feed deeper-living organisms.

In general terms the ocean's reserves of minerals have been estimated at 6×10^{15} tons.[4] Today over one-third of the world's production of salt is obtained from seawater. The ocean is also a repository for over 70 different and highly useful minerals and materials. The United States is especially interested in harvesting minerals from the sea, because we presently are dependent completely or in part on foreign countries for 69 of these [3-33]. The Soviet Union, in contrast, is presently self-sufficient in all but two of these resources.[5] However, let's not let paranoia take over our thinking. The world is not being depleted of minerals, at least, not for the most of them. The amount of minerals in the earth's crust is awesome. In fact, our entire planet is composed of minerals. The main exception is oil and gas, which are exhaustible (at least in terms of practical economics).

Anyone who has seen the sea in its variety of personalities has to be impressed by its energy potential. Energy from the sea could come from several major sources, such as tides, ocean currents, waves, thermal or salinity differences, and marine biomass. Each of these has its advantages and disadvantages as well as a host of possible applications. For example, the energy in ocean waves might be extracted by one of the following procedures:

- Use of the vertical rise and fall of the crests and troughs of successive waves to drive an air- or water-powered turbine.
- Use the rolling motion of waves to move vanes or cams to turn turbines.
- Converge wave forms into channels and concentrate their energy.

In the future, in addition to food and energy problems, certain countries, and cities in particular, will face critical shortages of fresh water. Los Angeles is but one example in this country that needs an additional supply of pure water. However, the amount of water used for human consumption is but a fraction of that needed for agriculture and industry. There are three major problems that must be overcome in converting seawater for aquaculture use:

1. A substantial energy source must be available.
2. The salt must be removed from the seawater.
3. A transportation mechanism must be available to move the water economically to the user.

The vastness of the ocean and its unique physical characteristics suggest that it may offer even more profitable solutions to our pressing problems than does space [3-34].

WEATHER FORECASTING AND CONTROL

"Everyone talks about the weather, but no one does anything about it." This statement has been made for many, many years, but it need

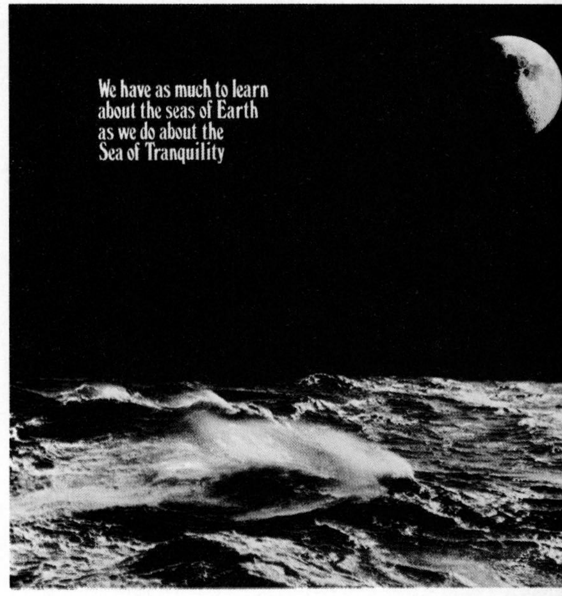

We have as much to learn
about the seas of Earth
as we do about the
Sea of Tranquility

[3-32]

[3-33] The seas—the last, largely untapped resources left on earth—have the potential for supplying far more energy, raw materials, and food than man has ever imagined.

[4] *The Ocean and Its Resources* (Moscow: Progress Publishers, 1977), p. 12.

[5] David A. Ross, *Opportunities and Uses of the Ocean* (New York: Springer-Verlag, 1978), p. 103.

[3-34] The vastness of the ocean and its unique physical characteristics suggest that it may offer even more profitable solutions to our pressing problems than does space.

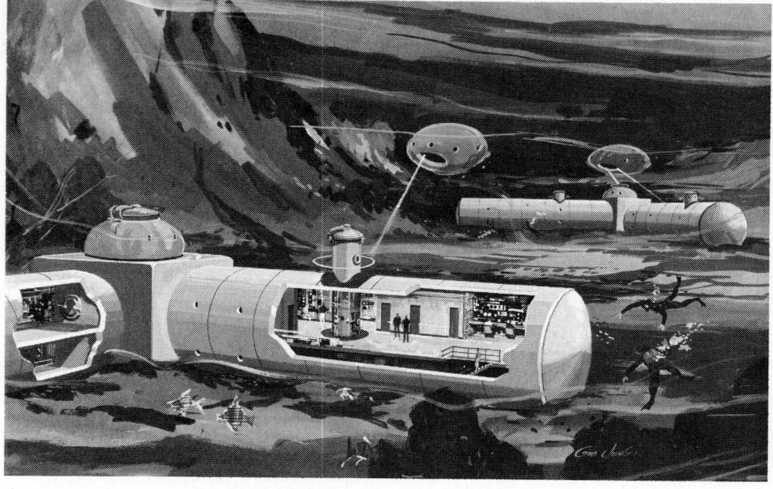

not be true forever. More and more of the "guesswork" is going out of predicting the weather, and we can look forward to the time when we will be able to use technology to control some facets of the weather. Weather prediction is still an inexact science, but great improvements have been made both in the short-term prediction accuracy and the long-term challenge of climate modification.

The lightning is his slave; heaven's utmost deep gives up her stars, and like a flock of sheep they pass before his eye, are numbered and roll on! The tempest is his steed, he strides the air; and the abyss from her depth laid bare Heaven, hast thou secrets? Man unveils me; I have none.

Percy Bysshe Shelley, 1792–1822
Prometheus Unbound

As our computers, and in turn our satellites, become more sophisticated, we can look forward to an end of much of man's suffering by being able to put an end to famine, floods, deserts, and drought. Climate has a great effect on our lives and on world economies. Consequently we must try to understand much more about it, the factors that comprise it, and to be able to solve the problems it creates. We still do not know enough about weather forecasting and control. Before future climates can be accurately predicted and controlled, we must understand enough about them to build realistic, quantitative models that relate cause and effect. At present such models are primitive and incomplete. The challenges of the future are to improve data collection through more sophisticated equipment that will be designed by the engineers. An "Automatic Weather Integrating Prediction System" and an "Automatic Surface Observation System" are currently being developed. Other improvements will follow.

Several scenarios have been advanced for the next decade in an effort to assess and understand possible climatic trends. These solutions include such possibilities as a study of global cooling and global warming. Particularly, we must be able to anticipate the effects that these possibilities would have on the world and its various geographic regions [3-35 and 3-36].

Here are just a few challenges with regard to the weather that will occupy the creative efforts of engineers in the coming years:

- Crops forecasting from space satellites.
- Assessing the quantitative and qualitative state of forests around the world.

[3-35] Some weather brings destruction and death . . .

- Remote sensing for measuring crop production, supplies of ores, and the effects of pollution.
- Assessing the world climate by examining such climatic subsystems as sea-surface temperature changes and their interaction with the overlying air to produce seasonal forecasts a season in advance, as well as to identify long-term climatic trends.
- Using computer analysis in weather forecasting to provide an early warning system for such impending crises as typhoons, hurricanes, and other difficulties. Besides saving lives, and resources, this will have a great economic impact.
- Predicting or estimating such occurrences as: How much deforestation is taking place per year? How much soil is lost to cultivation through erosion, salinization, desertification, and so on? We want to measure the gains in crop yields as a result of improved agricultural technologies.

We should have the capability to put into orbit huge metal shades and reflectors that could focus sunlight onto the earth's surface to direct tropical storm systems, relieve droughts, and alter deserts. Today's desert regions of the earth could become gardens! The engineer, working as a team member with meteorologists and other scientists, will play an important role in enhancing the well-being of people throughout the world.

[3-36] . . . while other weather brings the moisture necessary for life.

PREDICTIONS FOR THE TWENTY-FIRST CENTURY, AS DESCRIBED IN THE 1985 LITERATURE

- Develop artificial eyesight for blind people.
- Arrest senility through chemicals.
- Neighborhood computer terminals to print and bind any book while you wait.
- Illnesses treated electromagnetically. Cells are deceived into producing antibodies, coagulants, new tissue, and chemicals when they are exposed to certain kinds of electric and magnetic fields.
- Cure for cancer if detected early enough.
- Football games played entirely by robots.
- An open market for used and reconditioned body parts.
- Earthquake prevention by injecting water into wells along fault lines.
- Elimination of nighttime from selected portions of the earth through the use of solar satellites.
- Solar-powered satellites to supply a majority of the world's energy.
- Melting of antarctic icebergs on demand to relieve water shortages.
- Advances in human understanding of the growth of crystal structures, enabling us to grow buildings.

There is therefore much ground for hoping that there are still laid up in the womb of nature many secrets of excellent use, having no affinity or parallelism with anything that is now known, but lying entirely out of the best of the imagination which have not yet been found out. They too no doubt will some time or other, in the course and revolution of many ages, come to light themselves, just as the others did; only by the method of which we are now treating can they be speedily and suddenly and simultaneously presented and anticipated.

Francis Bacon, 1561–1626
Novum Organum, *Aphorism CIX*

[3-37] Synthetic fuels will be developed.

The marvels of modern technology include the development of a soda can which, when discarded, will last forever—and a $7,000 car which, when properly cared for, will rust out in two or three years.

Paul Harwitz
The Wall Street Journal

[3-38] Photovoltaic panels will serve as energy collectors in space.

- Faster-than-light travel through use of unified-field theory (first proposed by Einstein).
- People living in high-orbital mini-earth homes.
- DNA research to create larger plants, increased crop yields, grow larger cows to produce more milk, and to produce square tomatoes.
- Using protein to replace silicon in computer chips which would allow for more information to be contained in the same space.
- Domestic robots in many homes.
- Increase the average life span to 100 plus years.
- Regeneration of human limbs and organs.
- Drugs to facilitate learning.
- Lunar colonies.
- Undersea cities.
- Synthetic fuel [3-37].
- Reliable weather forecasting.
- Nonnarcotic drugs for altering personality.
- Mining of the ocean floor.
- Regional weather control.
- Synthetic food.
- Total immunization against disease.
- Space telescopes launched.
- Operating particle beam accelerators.
- Flushless toilets that compost wastes.
- Home TV sets with 300 or more channels.
- Digital synthesizers to enable a person to play many instruments if he can play at least one.
- The replacement of gasoline for vehicle fuel.
- Paying passengers for space shuttle voyages.
- Three-dimensional pictures through the use of laser holography.
- 1000-seat jet liner.
- Most assembly-line tasks accomplished by robots.
- A computer makes an "original" scientific discovery and is awarded a Nobel Prize.
- Autos equipped with microcomputer, sensor, and control actuator for self-operation by voice commands.
- Autos equipped with collision-avoidance electronic devices.
- Daily body checkups by computer to provide warning of impending illness.
- Energy collectors in space [3-38].

PROBLEMS

3-1. Imagine a machine, a device, or a system that has never existed before—that you have just developed or invented. Describe its features.

3-2. Assuming that your new invention (Problem 3-1) can become a reality, how will it benefit humankind?

3-3. What would be the possible dangers to society or negative effect that your invention might have (Problem 3-1)?

3-4. The further development of which particular technology (e.g., communication, transportation, computers, automation, medical, energy, or food, would, in your opinion, contribute the most to the enhanced welfare of mankind? Why did you make it your choice?

3-5. Other than the examples in this chapter, what other inventions have come into being even though some "expert" said they were "impossible"?

3-6. Make a list of the advantages of space colonization.

3-7. Make a list of the disadvantages of space colonization.

3-8. When robots can accomplish so many manufacturing and production tasks, including building other robots, what role will people play in the factory of the future?

3-9. In which "frontier" would you rather work, outer space or the undersea world? Why?

3-10. We are entering a world where even elementary school students are familiar with and can operate a computer. What do you predict will be the computer's effect in 10 years on: (a) the school; (b) society; (c) science and engineering?

3-11. If you have chosen to become an engineer, why did you make this choice? What kind of an engineer do you want to be? Why? What challenges interest you most?

3-12. Advances in medicine have often been called "a two-edged sword." Modern medicine helps us cure illnesses and keeps us alive longer. On the other hand, medical advances have contributed to the population explosion problem by helping the unproductive elderly to live and keep the "weak" alive to procreate. What are your views about this?

3-13. Can you think of some "new" energy sources that have never been developed?

3-14. What "people" problems do you foresee arising in space travel and/or space colonization? What solutions might be found?

3-15. As the world population increases, more and more people are moving to the cities. In the year 2000 we will have a number of "megacities" throughout the world. What will be the best types of transportation systems to use in these cities? Why?

3-16. The ocean contains recoverable resources. It can also be easily polluted. If you were an oceanographer-engineer, which activity would you rather be involved in: production of ocean products, or devising pollution controls to preserve the oceans? Why did you make this choice?

3-17. It has been said, "What mankind can dream, technology can achieve." What is your dream? What "new" technology would you like to help develop? How will it benefit mankind?

3-18. We can envision a day when all financial transactions will be processed electronically—through the use of "plastic" money—credit or transaction cards. Is a "cashless" society a good or a bad idea? Why?

3-19. When much of our production process can be accomplished by machines (robotics), what will the people do whose work is replaced by robots?

3-20. Speculate and imagine what will replace the automobile and/or its power supply by the year 2050.

3-21. Assuming that commercial space travel is available at an affordable price (Space Shuttle, for example), would you make a trip? Why? Why not?

3-22. It is believed that plants, and eventually animals, can be cloned. Should we attempt to clone human beings? Why?

3-23. What are the advantages of controlling the weather? What are some potential problems in doing so?

3-24. List the present energy sources presently being used in the industrialized world.

3-25. List potential energy sources that have either not been perfected or exist only in your imagination.

[3-P21] Have you bought your ticket yet for the 6:30 A.M. flight to Mars?

Chapter 4

Engineering Career Fields

When a person alibis that he could have amounted to something if it had not been for his race, creed, or religion, one should call attention to Epictetus, the slave who lived in the first century in Greece, and became one of the world's most profound scholars and philosophers. He should be reminded that Disraeli, the despised Jew, became Prime Minister of Great Britain; that Booker T. Washington, who was born in slavery in this country, became one of the nation's greatest educators; and that another slave, George Washington Carver, became one of the greatest scientists of his generation. Lincoln, born of illiterate parents in a log cabin in Kentucky, lived to be acclaimed one of the greatest statesmen of all times.
Phil Conley

Much of the change in our civilization in the past 100 years has been due to the work of the engineer. We hardly appreciate the changes that have occurred in our environment unless we attempt to picture the world of a few generations ago, without electric lights, automobiles, telephones, radios, computers, transportation systems, supersonic aircraft, robotic-operated machine tools, television, and all the modern appliances in our homes. In the growth of all these things the role of the engineer is obvious.

Development in the field of science and engineering is progressing so rapidly at present that within the last 10 years we have acquired materials and devices that are now considered commonplace but which were unknown to our parents. Through research, development, and mass production, directed by engineers, ideas are made into realities in an amazingly short time.

The engineer is concerned with more than research, development, design, construction, and the operation of technical industries, however, since many are engaged in businesses that are not concerned primarily with production. Formerly, executive positions were held almost exclusively by persons whose primary training was in the field of law or business, but the tendency now is to utilize engineers more and more as administrators and executives.

No matter what kind of work the engineer may wish to do, there will be opportunities for employment not only in purely technical fields but also in other functions, such as general business, budgeting, rate analysis, purchasing, marketing, personnel, labor relations, and industrial management. Other opportunities also exist in such specialized fields of work as teaching, writing, patent practice, and work with the national defense.

Although college engineering curricula contain many basic courses, there will be some specialized courses available that are either peculiar to a certain curriculum or are electives. These specializations permit each student to acquire a particular proficiency in certain technical subjects so that, for example, professional identification can be acquired in such fields as chemical, civil, electrical, mechanical, or industrial engineering.

Education in the application of certain subject matter to solve technological problems in a particular engineering field constitutes engineering specialization. Such training is not for manual skills as in trade schools, but rather is planned to provide preparation for research, design, operation, management, testing, maintenance of projects, and other engineering functions in any given specialty.

The principal engineering fields of specialization that are listed in college curricula and that are recognized in the engineering profession are described in the following sections.

AEROSPACE AND ASTRONAUTICAL ENGINEERING

Powered flight began in 1903 at Kitty Hawk, North Carolina. Perhaps no other single technological achievement has had such an influence on our way of life. Through faster transportation, space exploration, and improved communications, almost every aspect of our daily life has been affected. However, not all challenges are associated with spaceflight. Problems associated with conventional aircraft, and the development of special vehicles such as hydrofoil ships, ground-effect machines, and deep-diving vessels for oceanographic research are all concerns of the industry.

Within the past few years many changes have taken place which have altered the work of the aeronautical engineer—not the least of which is man's successful conquest of space. Principal types of work vary from the design of guided missiles and spacecraft to analyses of aerodynamic studies dealing with the performance, stability, control, and design of various types of planes and other devices that fly.

Although aerospace engineering is one of the newer fields, it offers a great variety of opportunities for employment. Continued exploration and research in previously uncharted areas is needed in the fields of propulsion, materials, thermodynamics, cryogenics, navigation, cosmic radiation, and magnetohydrodynamics. It is predicted that within the near future the chemically fueled rocket engine, which has enabled man to explore lunar landscapes, will become obsolete as the need increases to cover greater and greater distances over extended periods of time.

[4-1] Aerospace engineering includes a broad range of engineering activities.

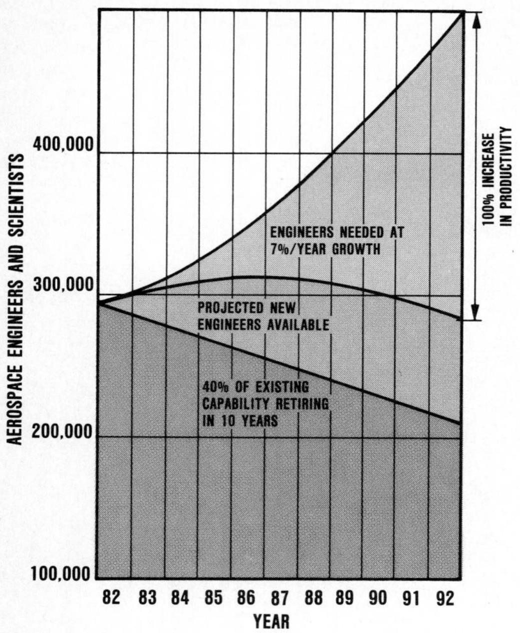

[4-2] In the future, one aerospace engineer or scientist must do the work that two do now.[1]

There will be fewer aerospace engineers and scientists available in ten years than we have today. . . . Meanwhile, other major nations press the training of engineers. The Soviet Union, Japan, and even the Peoples Republic of China are graduating more engineers than the United States . . . In view of this trend, we will have to be twice as productive in aerospace ten years from now as we are today.

Robert W. Hager, Vice President, Engineering,
Boeing Aerospace
Astronautics & Aeronautics, *May 1983*

[4-3] Agricultural engineers at a research center test the safety features of a farm tractor. Agricultural engineers apply fundamental engineering principles of analysis and design to improve our methods of food production and land utilization.

The rapidly expanding network of airlines, both national and international, provides many openings for the engineering graduate. Since the demand for increasing numbers of aircraft of various types exists, there are opportunities for work in manufacturing plants and assembly plants and in the design, testing, and maintenance of aircraft and their component parts. The development of new types of aircraft, both civilian and military, requires the efforts of well-trained aeronautical engineers, and it is in this field that the majority of positions exists. Employment opportunities exist for specialists in the design and development of fuel systems using liquid and solid propellants. Control of the newer fuels involves precision valving and flow sensing at very low and very high temperatures. The design of ground and airborne systems that will permit operation of aircraft under all kinds of weather conditions is also a part of the work of aeronautical engineers.

The aerospace engineer works on designs that are not only challenging and adventuresome but also play a major role in determining the course of present and future world events.

AGRICULTURAL ENGINEERING

Agricultural engineering is that discipline of engineering that spans the area between two fields of applied science—agriculture and engineering. It is directly concerned with supplying the means whereby food and fiber are supplied in sufficient quantity to fill the basic needs of all mankind. By the year 2000 almost 2 billion additional people will populate the earth—a number roughly equal to the world's total population in 1940. This factor, plus the increasing demands of people throughout the world for increased standards of living, provides unparalleled challenges to the agricultural engineer. Not only must the quantity of food and fiber be increased, but the efficiency of production must also be steadily improved in order that personnel may be released for other creative pursuits. Through applications of engineering principles, materials, energy, and machines may be used to multiply the effectiveness of man's effort. This is the agricultural engineer's domain.

In order that the agricultural engineer may understand the problems of agriculture and the application of engineering methods and principles to their solution, instruction is given in agricultural subjects and the biological sciences as well as in basic engineering. Agricultural research laboratories are maintained at schools for research and instruction using various types of farm equipment for study and testing. The young person who has an analytical mind and a willingness to work, together with an interest in the engineering aspects of agriculture, will find the course in agricultural engineering an interesting preparation for his or her life's work.

Many agricultural engineers are employed by companies that serve agriculture and some are employed by firms that serve other industries. Opportunities are particularly apparent in such areas as (1) research, design, development, and sale of mechanized farm

[1]*Astronautics & Aeronautics,* May 1983, p. 66.

equipment and machinery; (2) application of irrigation, drainage, erosion control, and land and water management practices; (3) application and use of electrical energy for agricultural production, and feed and crop processing, handling, and grading; (4) research, design, sale, and construction of specialized structures for farm use; and (5) the processing and handling of food products.

He gave it for his opinion, that whoever could make two ears of corn or two blades of grass to grow upon a spot of ground where only one grew before, would deserve better of mankind, and do more essential service to his country than the whole race of politicians put together.

Jonathan Swift, 1667–1745
Gulliver's Travels *II, vi*

ARCHITECTURAL ENGINEERING

The architectural engineer is interested primarily in the selection, analysis, design, and assembly of modern building materials into structures that are safe, efficient, economical, and attractive. The education received in college is designed to teach one how best to use modern structural materials in the construction of tall buildings, manufacturing plants, and public buildings.

The architectural engineer is trained in the sound principles of engineering and at the same time is given a background supportive of the point of view of the architect. The architect is concerned with the space arrangements, proportions, and appearance of a building, whereas the architectural engineer is more nearly a structural engineer and is concerned with safety, economy, and sound construction methods.

Opportunities for employment will be found in established architectural firms, in consulting engineering offices, in aircraft companies, and in organizations specializing in building design and construction. Excellent opportunities await the graduate who may be able to associate with a contracting firm or who may form a partnership with an architectural designer. In the field of sales an interesting and profitable career is open to the individual who is able to present ideas clearly and convincingly.

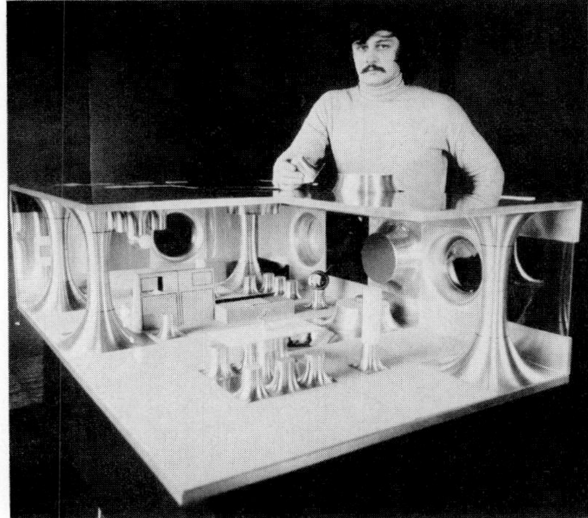

[4-4] Architectural engineers must be equally cognizant of aesthetic and structural design considerations.

BIOMEDICAL ENGINEERING

Biomedical engineering encompasses all aspects of the application of engineering methods to the use and control of biological systems. It bridges the engineering, physical, and life sciences in identifying and solving medical and health-related problems. Biomedical engineers are team players in much the same way as many athletes. For example, engineers, physicists, chemists, and mathematicians routinely join with the biologist and physician in developing techniques, equipment, and materials.

One of the earliest applications of bioengineering in the use of prosthetic devices was made by maimed ancient warriors who made their own wooden limbs. In the Middle Ages "experts" got into the field when armorers fashioned artificial limbs for knights injured in battle. It was a natural by-product of their work since they had to create suits of armor that fit the human form and joints that moved with ease. Goethe wrote of one German knight, Goetz von Berlichengin, who had an artificial hand.[2]

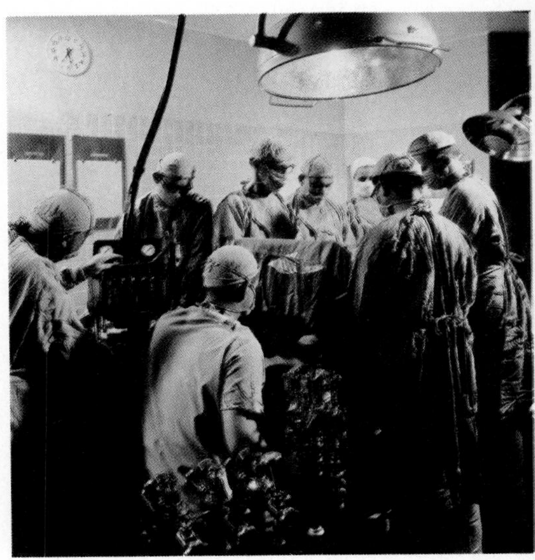

[4-5] The work of bioengineers has made possible the development of many life-lengthening and life-enhancement systems.

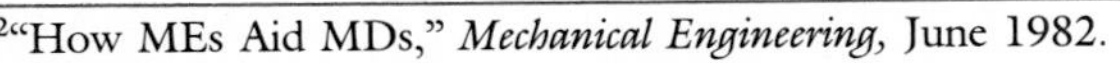

[2]"How MEs Aid MDs," *Mechanical Engineering,* June 1982.

ARTIFICIAL KIDNEY

TO VEIN
FROM ARTERY
HEMODIALYZER CONTAINING
SEMIPERMEABLE MEMBRANE
DIALYZING FLUID
BLOOD LEAK DETECTOR
CONDUCTIVITY MONITOR
WATER
PROPORTIONING PUMP
35 : 1
ELECTROLYTE CONCENTRATE
UREA & METABOLIC WASTE

[4-6] A schematic drawing of the artificial kidney.

[4-7] Ceramic engineers hold in their hands a very important answer to the impending crisis in the shortage of metallic materials.

Today, the range of the bioengineers' interests is very broad. It would involve, for example, the development of highly specialized medical instruments and devices—including artificial hearts and kidneys and the development of lasers for surgery and cardiac pacemakers that regulate the heartbeat. Other biomedical engineers may specialize more particularly in the adaptation of computers to medical science, such as monitoring patients or processing electrocardiograph data. Some will design and build systems to modernize laboratory, hospital, and clinical procedures.

Those selecting a career in biomedical engineering should anticipate earning a graduate degree since advanced study beyond the bachelor's degree is acutely needed to attain a depth of knowledge from at least two diverse disciplines.

At present biomedical engineering is a small field because few engineers have attained the necessary depth of academic training and experience in the life sciences. Therefore, job opportunities for graduates are excellent. Here indeed is a promising new field for those so inclined.

CERAMIC ENGINEERING

Today, our technological world is amazingly dependent on ceramics of all types. Unlike many other products they appear in every part of the spectrum of life, from beautiful but commonplace table settings, to the protective coatings of electrical transducers or the refractories of space exploratory rocket nozzles, to the spark plugs of a farmer's tractor. Exactly what are ceramics? When did man first find a use for them?

Ceramics are nonmetallic, inorganic materials that require the use of high temperatures in their processing. In the earliest form, clay pottery of 10,000 B.C. has been found to be excellently preserved. The most common of ceramics, glass—an ancient discovery of the Phoenicians (about 4000 B.C.), is a miracle material in every sense. It may be made transparent, translucent, or opaque, weak and brittle or flexible and stronger than steel, hard or soft, water soluble or chemically inert. Truly, it is one of the most versatile of engineering materials. Ceramic engineers are employed by a variety of industries, from the specialized raw material and ceramic product manufacturers to the chemical, electrical and electronic, automotive, nuclear, and aerospace industries.

CHEMICAL ENGINEERING

Chemical engineering is responsible for new and improved products and processes that affect every person. This includes materials that will resist extremities of heat and cold, processes for life-support systems in other environments, new fuels for reactors, rockets, and booster propulsion, medicines, vaccines, serum, and plasma for mass distribution, and plastics and textiles to serve a multiplicity of human needs. Consequently, chemical engineers must be able to apply scientifically the principles of chemistry, physics, and engineering to

[4-8] Chemical engineers frequently make use of scale models in the design of petrochemical installations.

the design and operation of plants for the production of materials that undergo chemical changes during their processing.

The courses in chemical engineering cover inorganic, analytical, physical, and organic chemistry in addition to the basic engineering subjects; and the work in the various courses is designed to be of a distinctly professional nature and to develop capacity for original thought. The industrial development of our country makes large demands on the chemical engineer. The increasing uses for plastics, synthetics, and building materials require that a chemical engineer be employed in the development and manufacture of these products. Although well trained in chemistry, the chemical engineer is more than a chemist in that he or she applies the results of chemical research and discovery to the use of mankind by adapting laboratory processes to full-scale manufacturing plants.

The chemical engineer is instrumental in the development of the newer fuels for turbine and rocket engines. Test and evaluation of such fuels and means of achieving production of suitable fuels are part of the work of a chemical engineer. This testing must be carefully controlled to evaluate the performance of engines before the fuel is considered suitable to place on the market.

Opportunities for chemical engineers exist in a wide variety of fields. Not only are they in demand in strictly chemical fields but also in nearly all types of manufacturing. The production of synthetic rubber, the uses of petroleum products, the recovery of useful materials from what was formerly considered waste products, and the better utilization of farm products are only a few of the tasks that will provide work for the chemical engineer.

CIVIL AND CONSTRUCTION ENGINEERING

Civil engineers plan, design, and supervise the construction of facilities essential to modern life in both the public and private sectors—

[4-9] This civil engineer is marking a structural member that will be used in the erection of a building that she designed.

facilities that vary widely in nature, size and scope, space satellites and launching facilities, offshore structures, bridges, buildings, tunnels, highways, transit systems, dams, airports, irrigation projects, treatment and distribution facilities for water, and collection and treatment for wastewater.

Construction engineering is concerned primarily with the design and supervision of construction of buildings, bridges, tunnels, and dams. The construction industry is America's largest industry today. Geotechnic investigations, such as in soil mechanics and foundation investigations, are essential not only in civilized areas but also for successful conquest of new lands such as Antarctica and extraterrestrial surfaces. Transportation systems include the planning, design, and construction of necessary roads, streets, thoroughfares, and superhighways. Engineering studies in water resources are concerned with the improvement of water availability, harbor and river development, flood control, irrigation, and drainage. Pollution is an ever-increasing problem, particularly in urban areas. The environmental engineer is concerned with the design and construction of water supply systems, sewerage systems, and systems for the reclamation and disposal of wastes. City planning and municipal engineers are concerned primarily with the planning of urban centers for the orderly, comfortable, and healthy growth and development of business and residential areas. Surveying and mapping are concerned with the measurements of distances over a surface (such as the earth or the moon) and the location of structures, rights-of-way, and property boundaries.

Civil engineers engage in technical, administrative, or commercial work with manufacturing companies, construction companies, transportation companies, and power companies. Other opportunities for employment exist in consulting engineering offices, in city and state engineering departments, and in the various bureaus of the federal government.

Civil engineering is the oldest and one of the most exciting branches of engineering. Civil engineers are the "earth changers."

The Book of Knowledge

[4-10] Man has always suspected that some state of perfection exists where function, beauty, truth, and everlastingness converge. Perhaps the closest he has come to reaching this state is with the bridges he has built.

From "An Essay on Bridges," A *CBS-TV Reports* program, February 1965

ELECTRICAL AND COMPUTER ENGINEERING

It is well into the 1990s, a time when computers can understand and reply to human voice commands. The place is the Pentagon war room. Suddenly, an alarm goes off, signaling that the country is under attack. One of the generals shouts to a computer, "Is the attack coming from land, air or sea?"
"Yes," the computer responds.
"Yes what?" the general snaps.
"Yes, sir!" the machine replies.

Boston Globe

Electrical engineering is concerned, in general terms, with the utilization of electric energy. It is divided into broad fields, such as information systems, automatic control, and systems and devices. Electricity used in one form or another reaches nearly all our daily lives and is truly the servant of mankind.

The electrical engineer applies engineering principles, both mechanical and electrical, in the design and construction of computers and auxiliary equipment. The basic requirements of a computer constantly change and new designs must provide for these necessary capabilities. In addition, a computing machine must be built that will furnish solutions of greater and greater problem complexity and at the same time have a means of introducing the problem into the machine in as simple a manner as possible.

There are many companies that build elaborate computing machines, and employment possibilities in the design and construction part of the industry are not limited. Many industrial firms, colleges, and governmental branches have set up computers as part of their capital equipment, and opportunities exist for employment as computer applications engineers, who serve as liaison between computer programmers and engineers who wish their problems evaluated on the machines. Of course, in a field expanding as rapidly as computer design, increasing numbers of employment opportunities become available. More and more dependence will be placed on the use of computers in the future, and an engineer specializing in this work will find ample opportunity for advancement.

The automatic control of machines and devices, such as autopilots for spacecraft and missiles, has become a commonplace requirement in today's technically conscious society. Automatic controlling of machine tools is an important part of modern machine shop operation. Tape systems are used to furnish signals to serve units on automatic lathes, milling machines, boring machines, and other types of machine tools so that they can be programmed to perform repeated operations. Not only can individual machines be controlled but also entire power plants can be operated on a programmed system. The design of these systems is usually performed by an electrical or mechanical engineer.

Energy-conversion systems, where energy is converted from one form to another, are also a necessity in almost every walk of life. Power plants convert heat energy from fuels into electrical energy for transmission to industry and homes. In addition to power systems, communication systems are a responsibility of the electrical engineer. Particularly in communications the application of modern electronics has been most evident. The electrical engineer who specializes in electronics will find that the majority of communication devices employs electronic circuits and components.

Other branches of electrical engineering that may include power or communication activities, or both, are illumination engineering, which deals with lighting using electric power; electronics, ultrasound, which has applications in both power and communications; and such diverse fields as x-ray, acoustics, and seismograph work.

[4-11] Electrical engineers learn the fundamentals of designing and fabricating microelectronic circuit chips.

[4-12] A human hair is laid across a semiconductor chip that includes transistors and other circuit components which are smaller than 1 micrometer.

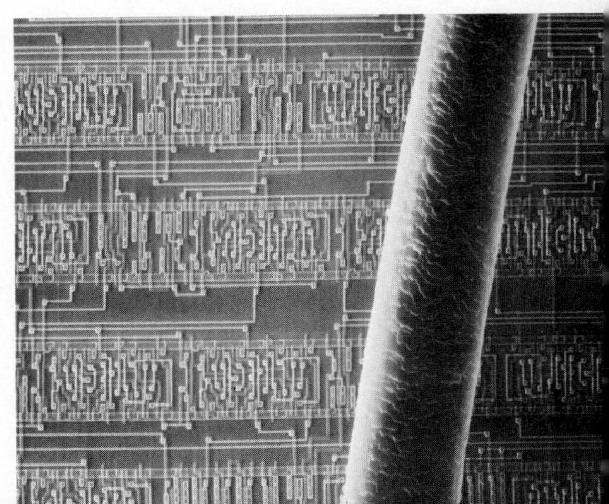

[4-13] Industrial engineers are proficient in discovering ways to economize the manufacturing process.

Employment opportunities in electrical engineering are extremely varied. Electrical manufacturing companies use large numbers of engineers for design, testing, research, and sales. Electrical power companies and public utility companies require a staff of qualified electrical engineers, as do the companies that control the networks of telegraph and telephone lines and the radio systems. Other opportunities for employment exist with oil companies, railroads, food processing plants, lumbering enterprises, biological laboratories, and chemical plants. The aircraft and missile industries use engineers who are familiar with circuit design and employment of flight data computers, servomechanisms, computers, and solid-state devices. There is scarcely any industry of any size that does not employ one or more electrical engineers as members of its engineering staff.

INDUSTRIAL ENGINEERING

Industrial engineers are concerned with the design, improvement, and installation of integrated systems of people, materials, equipment, and energy in a production environment. Whereas other branches of engineering tend to specialize in some particular phase of science, the realm of industrial engineering may include parts of all engineering fields. The industrial engineer then will be more concerned with the larger picture of management of industries and production of goods than with the detailed development of processes.

The work of the industrial engineer is rather wide in scope. Generally, work is with people and machines, and because of this it is important that one be educated in both personnel administration and in the relations of people and machines to production.

The advent of the digital computer and other solid-state support equipment has revolutionized the business world. Many of the resultant changes have been made as a result of industrial engineering designs. Systems analysis, operations research, statistics, queuing theory, information theory, symbolic logic, and linear programming are all mathematics-based disciplines that are used in industrial engineering work.

The industrial engineer must be capable of preparing plans for the arrangement of plants for best operation and then organizing the workers so that their efforts will be coordinated to give a smoothly functioning unit. In such things as production lines, the various processes involved must be timed perfectly to ensure smooth operation and efficient use of the worker's efforts. In addition to coordination and automating of manufacturing activities, the industrial engineer is concerned with the development of data-processing procedures and the use of computers to control production, the development of improved methods of handling materials, the design of plant facilities and statistical procedures to control quality, the use of mathematical models to simulate production lines, and the measurement and improvement of work methods to reduce costs.

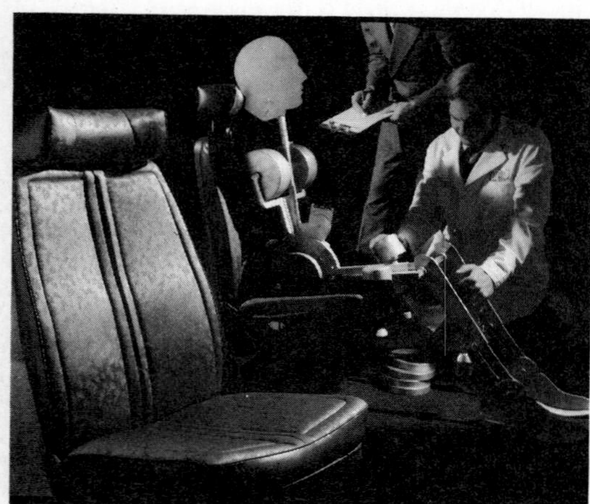

[4-14] The matching of the human being to the machine is a special area of study within industrial engineering.

Opportunities for employment exist in almost every industrial plant and in many businesses not concerned directly with manufacturing or processing goods. In many cases the industrial engineer may be employed by department stores, insurance companies, consulting companies, and as a city engineer. The industrial engineer is trained in fundamental engineering principles, and as a result may also be employed in positions that would fall in the realm of the civil, electrical, or mechanical engineer.

MARINE ENGINEERING, NAVAL ARCHITECTURE, AND OCEAN ENGINEERING

"How thick do you judge the planks of our ship to be?" "Some two good inches and upward," returned the pilot. "It seems, then, we are within two fingers' breadth of damnation."

Rabelais, Works, Book IV, *1548, Chapter 23.*

For many centuries the sea has played a dominant role in the lives of people of all cultures and geographical locations. For this reason in every era the designers of ships have been held in the highest regard for their knowledge and understanding of the sea's physical influences and for their artistry and ability in marine craftsmanship. As our civilization increases in complexity, all peoples of future generations will depend to an even greater extent on vessels of the sea to keep food, materials, and fuel flowing.

Naval architects design marine vehicles as total systems, especially the internal layout. Ship design is a refined art as well as an exacting science since most ships are custom built—one at a time. Many large ships are virtually floating cities containing their own power sources, sanitary facilities, food-preparation center, and recreational and sleeping accommodations. Every service that would be provided to city dwellers must also be provided for the ship's crew. As with aircraft design, the ship's structural members and intricate networks of piping and electrical circuits must fit together harmoniously in the minimum space possible.

Marine engineers must have a broad-based engineering educational background. They will design the mechanical systems that go into the ship, especially the propulsion and auxiliary power machinery.

Ocean engineers are designers of those systems that cannot be called a ship or a boat but still must operate in the marine environment. Some examples are oil rigs that drill offshore, man-made islands, and offshore harbor facilities.

The basic design of seaworthy cargo ships changed very little from 1900 to 1960. However, with the advent of nuclear power and sophisticated electronic computers a new era in ship design has begun to develop. Surface-effect or air-cushion-type vehicles, submarine tankers, and deep-submergence research vehicles have all emerged from the realm of science fiction to enter one of engineering reality. The application of these newer ideas for the shipbuilding industry awaits only a more positive commitment to the task by government and industry. With the shrinking world supply of food and energy, this commitment is certain to come.

[4-15] The marine engineer's role is significant in helping to solve the world's transportation problems.

MECHANICAL ENGINEERING

Mechanical engineering is concerned with the design of machines and processes used to generate power and apply it to useful purposes. These designs may be simple or complex, inexpensive or expensive, luxuries or essentials. Such items as the kitchen food mixer,

[4-16] The mechanical engineer is a specialist in designing and testing a great variety of instruments and machines.

the automobile, air-conditioning systems, nuclear power plants, and interplanetary space vehicles would not be available for human use today were it not for the mechanical engineer. In general, the mechanical engineer works with systems, subsystems, and components that exhibit motion. The range of work that may be classed as mechanical engineering is wider than that in any of the other branches of engineering, but it may be grouped generally under two heads: work that is concerned with power-generating machines, and work that deals with machines that transform or consume this power in accomplishing their particular tasks. Design specialists may work with parts that vary in size from the microscopic part of the most delicate instrument to the massive parts of heavy machinery. Included are mass transit systems that are rapidly becoming a part of our nationwide transportation system. Automotive engineers work constantly to improve the vehicles and engines. Heating, ventilating, air-conditioning, and refrigeration engineers are concerned with the design of suitable systems for making our buildings more comfortable and for providing proper conditions in industry for good working environments and efficient machine operation.

Employment may be secured by mechanical engineering graduates in almost every type of industry. Manufacturing plants, power-generating stations, public utility companies, transportation companies, airlines, and factories, to mention only a few, are examples of organizations that need mechanical engineers. Experienced engineers are needed in the missile and space industries in the design and development of such items as gas turbine compressors and power plants, air-cycle cooling turbines, electrically and hydraulically driven fans, and high-pressure refrigerants. Mechanical engineers are also needed in the development and testing of airborne and missile fuel systems, servovalves, and electro-mechanical control systems.

METALLURGICAL ENGINEERING

[4-17] Industrial robots, such as this "elephant trunk" painting robot in a Renault assembly plant, are designed by mechanical engineers.

In many respects the past 25 years may be said to be an "age of materials"—an age which has seen the maturing of space exploration, nuclear power, digital computer technology, and ocean conquest. None of these engineering triumphs could have been achieved without the contributions of the metallurgical engineer. Metals are found in every part of the earth's crust, but rarely in immediately usable form. It is the metallurgical engineer's job to separate them from their ores and from other materials with which they exist in nature.

Metallurgical engineering may be divided into two branches. One branch deals with the location and evaluation of deposits of ore, the best way of mining and concentrating the ore, and the proper method of refining the ore into the basic metals. The other branch deals with the fabrication of the refined metal or metal alloy into various machines or metal products.

The metallurgist performs pure and applied research on vacuum melting, arc melting, and zone refining to produce metallic materials having unusual properties of strength and endurance. In addition, the metallurgist in the aircraft and missile industries is often called upon to give an expert opinion on the results of fatigue tests of metal parts of machines.

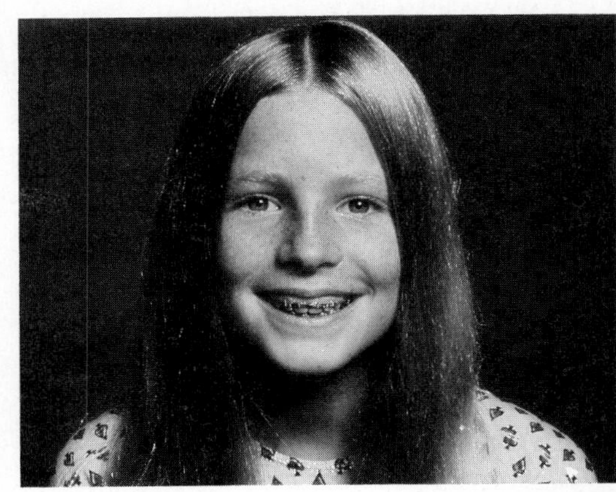

[4-18] Almost every aspect of our life is affected by advances in metallurgy and materials science. For example, teeth can now be straightened because of the development of a special type of steel . . . rust-proof, strong yet ductile, and hard yet smooth . . . unchanged through ice-cold sodas and red-hot pizzas.

MINING AND GEOLOGICAL ENGINEERING

The mining and geological engineer of today who searches the earth for hidden minerals is necessarily a person of quite different stature than the traditional explorer of yesteryear. These engineers must possess a combination of fundamental engineering and scientific education and field experience to enable them to understand the composition of the earth's crust. They must be experts in utilizing very sensitive instruments as they seek to locate new mineral deposits and to anticipate the problems that might arise in getting them out and transporting them to civilization. For this reason it is not unusual to find a mining or geological engineer in a modern office building in New York one week, and the next in Arizona or Zambia— or commuting between an expedition campsite and technical laboratories.

The work of mining and geological engineers lies generally in three areas: finding the ore, extracting it, and preparing the resulting minerals for manufacturing industries to use. These engineers design the mine layout, supervise the construction of mine shafts and tunnels in underground operations, and devise methods for transporting minerals to processing plants. Mining engineers are also responsible for mine safety and the efficient operation of the mine, including ventilation, water supply, power, communications, and

[4-19] Geological engineers investigate new sources of essential mineral bodies.

[4-20] The recovery of certain minerals can best be accomplished by open-pit mining.

[4-21] Disposal of spent nuclear fuel requires careful monitoring by the nuclear engineer.

equipment maintenance. Geological engineers are more directly concerned with locating and appraising mineral deposits.

An important part of the mining and geological engineer's work is to keep in mind inherent air- and water-pollution problems that might develop during the mining operation. This involves establishing efficient controls to prevent harmful side effects of mining and designing ways whereby the land will be restored for people to use after the mining operation terminates.

NUCLEAR ENGINEERING

Nuclear engineering is one of the newest and most challenging branches of engineering. Although much work in the field of nucleonics at present falls within the realm of pure research, a growing demand for people educated to utilize recent discoveries for the benefit of humankind has led many colleges and universities to offer courses in nuclear engineering. The nuclear engineer is familiar with the basic principles involved in both fission and fusion reactions; and by applying fundamental engineering concepts, is able to direct the enormous energies involved in a proper manner. Work involved in nuclear engineering includes the design and operation of plants to utilize heat energy from reactions, and the solution of problems arising in connection with safety to persons from radiation, disposal of radioactive wastes, and decontamination of radioactive areas.

The wartime uses of nuclear reactions are well known, but of even more importance are the less spectacular peacetime uses of controlled reactions. These uses include such diverse applications as electrical power generation and medical applications. Other applications are in the use of isotopes in chemical, physical, and biological research, and in the changing of the physical and chemical properties of materials in unusual ways by subjecting them to radiation.

Recent advances in our knowledge of controlled nuclear reactions have enabled engineers to build power plants that use heat

[4-22] Nuclear power offers a great potential to supplying this country's energy needs. Safety of operation and disposal of nuclear wastes are still problems that must be solved in a more reassuring way. The San Onofre Nuclear Power Plant, located along the coast midway between Los Angeles and San Diego, began operation in 1968 and supplies 2650 megawatts of power.

from reactions to drive machines. Nuclear energy plays an important role in our national energy supply. More than 12 percent of today's electrical power is nuclear generated, and this is projected to be 20 percent of the national total by the year 2000. Submarine nuclear power plants, long a dream, are now a reality, and experiments are being conducted on smaller nuclear power plants that can be used for airborne or railway applications.

At present, there are opportunities for employment of nuclear engineers in both privately owned and government-operated plants, for the generation of electricity, and where radioactive waste disposal or processing of nuclear materials is performed. Nuclear engineers are also needed by companies that may use radioactive materials in research or processing involving agricultural, medical, metallurgical, and petroleum products.

[4-23] Petroleum engineers specialize in discovering and recovering petroleum products.

PETROLEUM ENGINEERING

In early America, wood was the primary source of energy. Today the major source of energy is petroleum. It is the most widely used of all energy sources because of its mobility and flexibility in utilization. Approximately three-fourths of the total energy needs of the United States are currently supplied by petroleum products, and this condition will likely continue for many years. Petroleum engineering is the practical application of the basic sciences (primarily chemistry, geology, and physics) and the engineering sciences to the development, recovery, and field processing of petroleum.

Petroleum engineering is concerned with all phases of the petroleum industry, from the location of petroleum in the ground to the ultimate delivery to the user. Petroleum products play an important part in many phases of our everyday life in providing our clothes, food, work, and entertainment. Because of the complex chemical structure of petroleum, we are able to make an almost endless number of different articles. Owing to the wide demand for petroleum products, the petroleum engineer strives to satisfy an ever-increasing demand for oil and gas from the ground.

The petroleum engineer is concerned first with finding deposits of oil and gas in quantities suitable for commercial use, in the extraction of these materials from the ground, and in the storage and processing of the petroleum above ground. The petroleum engineer is also concerned with the location of wells in accordance with the findings of geologists, the drilling of wells and the myriad problems associated with the drilling, and the installation of valves and piping when the wells are completed. In addition to the initial tapping of a field of oil, the petroleum engineer is concerned with practices that will provide the greatest recovery of the oil, considering all possible factors that may exist many thousand feet below the surface of the earth.

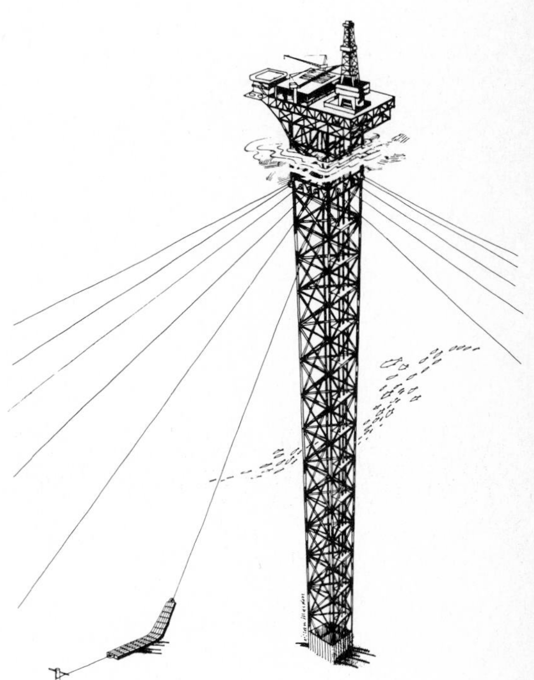

[4-24] Exxon's 1300-foot Lena Guyed Tower is slightly higher than New York City's Empire State Building. It is held in place by 14 piles and 20 guy lines. This permits the tower to move slightly with the wind and waves.

After the oil or gas has reached the surface, the petroleum engineer will provide the means of transporting it to suitable processing plants or to places where it will be used. Pipelines are providing an ever-increasing means of transporting both oil and gas from field to consumer.

Many challenges face the petroleum engineer. Some require pioneering efforts, such as with the rapidly developing Alaska field. Other opportunities lie closer at hand. For example, it is known that because of excessive costs in recovery, more than one-half of the oil already discovered in the United States *has yet to be brought to the surface of the earth*. It is estimated that even a 10 percent increase in oil recovery would produce 3 billion barrels of additional oil, a worth of over $60 billion.

Owing to the expanding uses for petroleum and its products, the opportunities for employment of petroleum engineers are widespread. Companies concerned with the drilling, producing, and transporting of oil and gas will provide employment for the majority of engineers. Because of the widespread search for oil, employment opportunities for the petroleum engineer exist all over the world; and for the young person wishing a job in a foreign land, oil companies have crews in almost every country over the globe. Other opportunities for employment exist in the field of technical sales, research, and as civil service employees of the national government.

PROBLEMS

4-1. Discuss the changing requirements for aerospace and astronautical engineers.

4-2. Investigate the opportunities for employment in agricultural engineering. Discuss your findings.

4-3. Write a short essay on the differences in the utilization and capability of the architectural engineer and the civil engineer who has specialized in structural analysis.

4-4. Interview a chemical engineer. Discuss the differences in her work and that of a chemist.

4-5. Assume that you are employed as an electrical engineer. Describe your work and comment particularly concerning the things that you most like and dislike about your job.

4-6. Explain why the demand for industrial engineers has increased significantly during the past 10 years.

4-7. Write a 200-word essay describing the challenging job opportunities in engineering that might be particularly attractive for an engineering graduate.

4-8. Explain the importance of mechanical engineers in the electronics industry.

4-9. Describe the changes that might be brought about to benefit humankind by the development of new engineering materials.

4-10. Investigate the need for nuclear and petroleum engineers in your state and report your findings.

Chapter 5

Engineering Work Opportunities

During the college years, engineering students will study courses in many subject areas. English courses make it easier to organize and present ideas effectively; mathematics courses improve the analysis or computational skills useful in modeling processes and devices; social science courses assist in finding a place in society as informed citizens; and various technical courses help to gain an understanding of natural laws. In your study of technical courses, you will become familiar with a store of factual information that will form the basis for your engineering decisions. The nature of these technical courses, in general, will influence your choice of a major field of interest. For example, you may decide to concentrate your major interest in a particular field, such as civil, chemical, electrical, industrial, or mechanical engineering.

College courses also provide training in learning facts and in developing powers of reasoning. Since it is impossible to predict what kind of work a practicing engineer will be doing after graduation, the objective of an engineering education is to provide a broad base of facts and skills on which the engineer can rely.

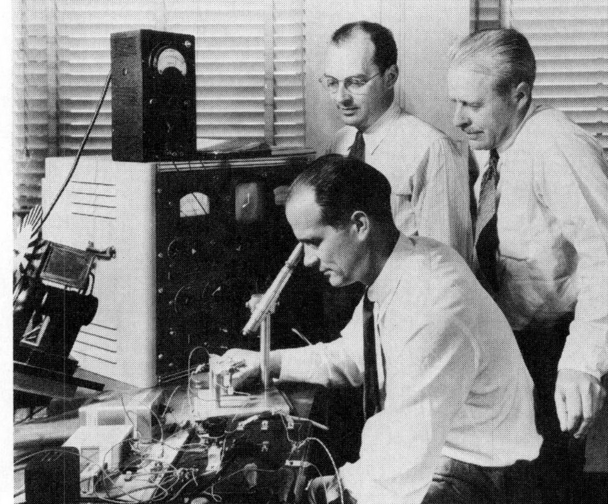

[5-1] Bell Telephone laboratories researchers William Shockley, John Bardeen, and Walter Brattin received the Nobel Prize in 1956 for their invention of the transistor.

It usually is not sufficient to say that an engineer is working as a *civil engineer*. The work experience may vary over a wide spectrum. As a civil engineer, for example, one may be performing research on materials for surfacing highways, or be employed in government service and be responsible for the budget preparation of a missile launch project. In fact, there are many things that a practicing engineer will be called on to do which are not described by a specialized course of study. The *type* of work that the engineer may do, as differentiated from a major field of specialization, can be called "engineering function." The major functions are research, development, design, production, construction, operations, sales, and management.

It has been found that in some engineering functions, such as in the management of a manufacturing plant, specialization is of lesser importance, whereas in other functions, such as research in microelectronics, specialization may be extremely important. To understand more fully the activities of a practicing engineer, let us examine some of these functions.

RESEARCH

Research is an organized method of trying to find out what you are going to do after you find that you cannot do what you are doing now.

Charles F. Kettering, 1876–1958

In some respects, research is perceived to be one of the more glamorous functions of engineering. In this type of work the engineer

[5-2] Research is an important type of work performed by the engineer. Engineers in research employ basic scientific principles in the discovery and application of new knowledge. This engineer is experimenting with a wind tunnel model of the city of Denver, Colorado, during a simulated atmospheric inversion.

delves into the nature of matter, exploring processes to use engineering materials and searching for reasons for the behavior of the things that make up our world. In many instances the work of the scientist and the engineer who are engaged in research will overlap. The work of scientists is usually closely allied with research. The objective of the researcher is to *discover truths*. The objective of the research engineer, on the other hand, usually is directed toward the practical side of the problem: not only to discover but also *to find a use for the discovery*.

The research engineer must be especially perceptive. Patience is also required, since most tasks have never before been accomplished. It is also important to be able to recognize and identify phenomena previously unnoticed. As an aid to educating an engineer to do research work, some colleges offer specific courses in research techniques. However, the life of a research engineer can be quite disheartening. In addition to probing and exploring new areas, much of the work is trial and error, and outstanding results of investigation usually occur only after long hours of laborious and painstaking work. On occasion, the final result is entirely different from the original mission.

Observation, not old age, brings wisdom.
Plublilius Syrus
Sententiae

The greatest invention of the nineteenth century was the invention of the method of invention.
Alfred North Whitehead, 1861–1947
Science and the Modern World

But in science the credit goes to the man who convinces the world, not to the man to whom the idea first occurs.
Sir Francis Darwin, 1848–1925
First Galton Lecture before the Eugenics Society

Until the last few decades, almost all research was solo work by individuals. However, with the rapid expansion of the fields of knowledge of chemistry, physics, and biology, it became apparent that groups or "research teams" of scientists and engineers could accomplish better the aims of research by pooling their efforts and knowledge. Within the teams, the enthusiasm and competition provide added incentive to push the work forward. Since each person is able to contribute from a particular specialty, discovery is accelerated.

As has been indicated, a thorough training in the basic sciences and mathematics is essential for a research engineer. In addition, an inquiring mind and a great curiosity about the behavior of things is desirable. Most successful research engineers have a fertile and uninhibited imagination and a knack of observing and questioning phenomena that the majority of people overlook. For example, one successful research engineer has worked on such diverse projects as an automatic lawn mower, an electronic biological eye to replace natural eyes, and the use of small animals as electrical power sources.

Most research engineers secure advanced degrees because they need additional background in the basic sciences and mathematics. In addition, this study usually gives them an opportunity to acquire useful skills in research procedures.

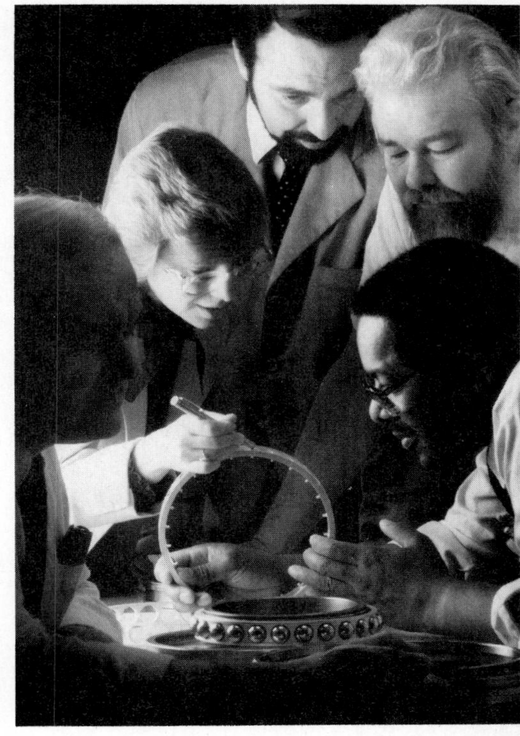

[5-3] Today, significant research is most often accomplished by teams of engineers and technicians.

DEVELOPMENT

After a basic discovery in natural phenomena is made, the next step in its utilization involves the development of processes or machines that employ the principles involved in the discovery. In the research and development fields, as in many other functions, the areas of activity overlap. In many organizations, the functions of research and development are so interrelated that the department performing this work is designated simply as a research and development (R&D) department.

The engineering features of development are concerned principally with the actual construction, fabrication, assembly, layout, and testing of scale models, pilot models, and experimental models for pilot processes or procedures. Where the research engineer is concerned more with making a discovery that will have commercial or economic value, the development engineer will be interested primarily in producing a process, an assembly, or a system *that will work.*

The development engineer does not deal exclusively with new discoveries. Actually, the major part of work assignments will involve using well-known principles and employing existing processes or machines to perform a new or unusual function. It is in this region that many patents are granted. In times past, the utilization of basic machines, such as a wheel and axle, and fundamental principles, including Ohm's law and Lenz' law, have eventually led to patentable machines, such as the electric dynamo. On the other hand, within a very short time after the announcement of the discovery of the laser in 1960, a number of patents were issued on devices employing this new principle. Thus the lag between the discovery of new knowledge and the use of that knowledge has been steadily decreasing through the years.

A patent is a legal document granting its owner the sole right to exclude others from using, manufacturing, or selling the invention described by the patent claims. In the United States the rights granted by the patent endure for a period of 17 years from its date of issue. Upon expiration of this grant, the information disclosed passes into the public domain and then can be used by anyone. As an exam-

Nothing is invented and perfected at the same time.
John Ray

The zipper, invented in 1879, was so shrewdly and strongly opposed by button manufacturers that it didn't get on the market for 30 years.
E. V. Durling

Invention, strictly speaking, is little more than a new combination of those images which have been previously gathered and deposited in memory. Nothing can be made of nothing; he who has laid up no materials can produce no combinations.
Sir Joshua Reynolds, 1723–1792

[5-4] A schematic drawing of a conventional refrigeration system.

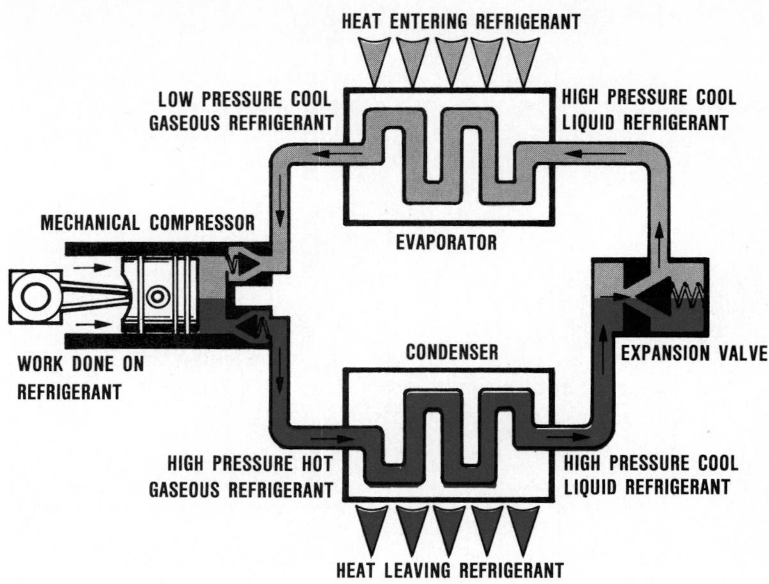

In 1899, the director of the U.S. Patent Office urged President McKinley to abolish the Patent Office along with his own job because "everything that can be invented has been invented."

United States Patent [19] [11] **4,157,015**

Hosterman et al. [45] **Jun. 5, 1979**

[54] **HYDRAULIC REFRIGERATION SYSTEM AND METHOD**

[75] Inventors: **Craig Hosterman,** Scottsdale; **Warren Rice,** Tempe, both of Ariz.

[73] Assignee: **Natural Energy Systems,** Tempe, Ariz.

[21] Appl. No.: **862,119**

[22] Filed: **Dec. 19, 1977**

[51] **Int. Cl.**² **F25B 1/00;** F25D 15/00; F25B 1/06

[52] **U.S. Cl.** **62/115;** 62/119; 62/500

[58] **Field of Search** .... 62/115, 116, 119, 122, 62/260, 498, 500, 514 R, 467 R; 126/400

[56] **References Cited**

U.S. PATENT DOCUMENTS

1,882,256	10/1932	Randel	62/500
2,152,663	4/1939	Randel	62/500
2,191,864	2/1940	Schaefer	62/119
3,789,617	2/1974	Rannow	62/115
3,848,424	11/1974	Rhea	62/115

Primary Examiner—Lloyd L. King
Attorney, Agent, or Firm—Cahill, Sutton & Thomas

[57] **ABSTRACT**

A refrigerant fluid is entrained within a down pipe of a closed loop water flow circuit to compress the refrigerant fluid from a gaseous state to a liquid state. A separation chamber at the lower extremity of the down pipe separates the liquid refrigerant fluid from the water and the water is drawn off. The water flows upwardly through a return pipe and pump, through a pipe for reintroduction to the down pipe at the lower end thereof. The drawn off liquid refrigerant flows upwardly through a return pipe and through an expansion valve. The refrigerant fluid, converted to a mixture of vapor and liquid, called a "quality mixture of the refrigerant" by the expansion valve, flows through an evaporator to cool a medium, such as air, passing therethrough. The refrigerant fluid, flowing from the evaporator and in a gaseous state at a higher temperature, is introduced to the upper end of the down pipe for reentrainment in the water flowing into the down pipe.

18 Claims, 5 Drawing Figures

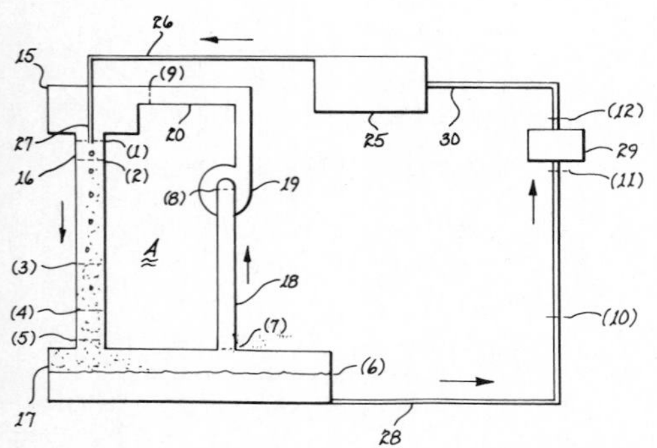

[5-5]

ple of how the process works, let us examine a new entry into the continuing development of mechanical refrigeration systems.

Conventional refrigeration systems require four major components: an *evaporator* (to allow heat to enter the liquid refrigerant and change it to gaseous state), a mechanical *compressor* (to increase the pressure and temperature of the refrigerant), a *condenser* (to cool the refrigerant and change its state from gas to liquid), and an *expansion valve* (to allow the liquid refrigerant to change to a partly gaseous state). Such a system is illustrated in schematic form in Figure [5-4]. Two Arizona inventors have been issued a patent, No. 4,157,015, *Hydraulic Refrigeration System and Method* (HRS), that eliminates the need for a mechanical compressor in the refrigeration cycle.[1] This novel invention uses a pump-driven descending liquid column (such as water) to accept and entrain the vaporized refrigerant (such as Freon) coming from the evaporator. As the water and Freon mixture descends, the gaseous Freon is compressed to its liquid state. This natural compression of the refrigerant eliminates the need for a conventional refrigerant compressor, which is a mechanically complex machine that needs maintenance and is expensive to operate. The patent drawing [5-5] shows schematically how the invention has been designed to function. The numbers on the drawing refer to the specific locations in the system where various changes of state occur. For example, the substantial differences in densities of the condensed Freon and the water at the bottom of the system (17) allows for separation of the two fluids (7 and 28). Heat rejection from the water column at 3 (the condenser function) can be accomplished by heat absorption by the earth, a water body, or by a conventional cooling tower. The expansion valve (29) and the evapora-

[1] This patent is owned by Natural Energy Systems, 2042 East Balboa Drive, Tempe, AZ 85282.

tor (25) are standard units. This is a novel refrigeration system because it requires a single moving part for it to operate—the water pump (19). (It is not unusual for submersible water pumps to operate for 20 years or more without maintenance.) The processes that occur at each numbered point on the patent drawing are described in turn in the patent document.

Considerable research has been conducted on this invention, including the construction of a full-scale experimental model. Further development is currently under way. What might be the benefits to humankind when such a new concept in refrigeration can be fully and economically developed? Let us consider some possibilities.

The natural compression of the HRS provides a 15 to 50 percent energy conservation advantage over conventional industrial and residential units that use mechanical compression. Perhaps of even greater significance, however, the HRS greatly simplifies the construction and maintenance of refrigeration systems. It has long been predicted that the introduction of a simple, relatively inexpensive method of producing refrigeration would greatly accelerate both the quality of life and the industrialization of Third World countries, particularly large population areas such as those found in India, China, Africa, Mexico, and South America. The HRS may be the answer these people are seeking.

In most instances the tasks of the development engineer are dictated by immediate requirements. For example, a new type of device may be needed to determine at all times the position in space of an airplane. Let us suppose that the development engineer does not know of any existing device that can perform the task to the desired specifications. Should he or she immediately attempt to invent such a device? The answer, of course, is "usually not." First, the files of available literature should be searched for information pertaining to existing designs. Such information may come from two principal sources. The first source is library material on processes, principles, and methods of accomplishing the task or related tasks. The second source is manufacturers' literature. It has been said humorously that "There is no need to reinvent the wheel." A literature search may discover a device that can accomplish the task with little or no modification. If no device is available that will do the work, a system of existing subassemblies may be set up and joined to accomplish the desired result. Lacking these items, the development engineer must explore further into the literature, and, using results from experiments throughout the world, formulate plans to construct a model for testing. Previous research may point a way to go, or perhaps a mathematical analysis will provide clues as to possible methods.

The development engineer usually works out ideas on a trial or "breadboard" basis, whether it be a machine or a computer process [5-8]. Having the parts or systems somewhat separated facilitates changes, modifications, and testing. In this process, improved methods may become apparent and can be incorporated. When the system or machine is in a workable state, the development engineer must then refine it and package it for use by others. Here again, ingenuity and a knowledge of human needs are important. A device that works satisfactorily in a laboratory when manipulated by skilled technicians may be hopelessly complex and unsuited for field use.

[5-6] Literature searches can be frustrating and nonproductive . . .

If you want to be happy for an hour, get drunk.
If you want to be happy for three days, get married.
If you want to be happy for eight days, kill your pig and eat it.
If you want to be happy forever, invent a machine useful to your fellowmen.

Old Chinese Proverb (Revised)

Never get to the point where you will be ashamed to ask anybody for information. The ignorant man will always be ignorant if he fears that by asking another for information he will display ignorance. Better once display your ignorance of a certain subject than always know nothing of it.

Booker T. Washington, 1856–1915

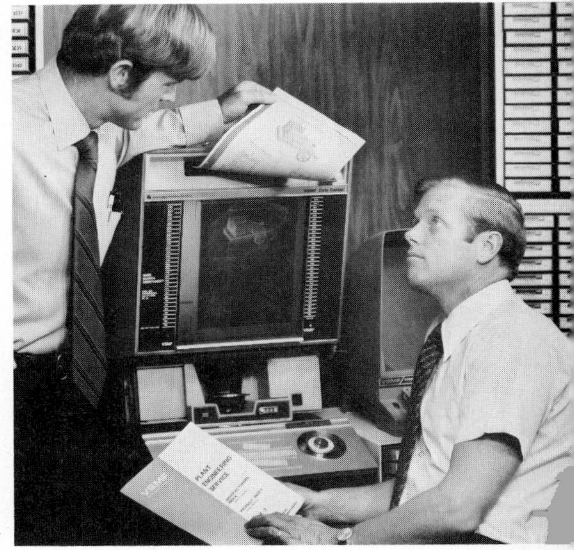

[5-7] . . . or satisfying and successful.

[5-8] The development engineer frequently uses a "breadboard" model to test the design's operational characteristics.

Were it not for imagination, Sir, a man would be as happy in the arms of a chambermaid as of a Duchess.
Samuel Johnson
Boswell's Life, *May 9, 1778*

The development engineer is the important person behind every pushbutton.

The education of an engineer for development work is similar to the education that the research engineer will expect to receive. However, creativity and innovation are perhaps of more importance, since the development engineer is standing between the scientist or the research engineer and the members of management who provide money for the research effort. The economic value of certain processes over others to achieve a desired result must be recognizable. It is also important to be able to convince others of the soundness of any conclusions reached. A comprehensive knowledge of basic principles of science and an inherent cleverness in making things work are also essential skills for the development engineer.

DESIGN

In our modern way of life, mass production has given us less expensive products and has made more articles available than ever before in history. In the process of producing these articles, the design engineer enters the scene just before the actual manufacturing process begins. After the development engineer has assembled and tested a device or a process and it has proved to be one that it is desirable to produce for a mass market, the final details of making it adaptable for production will be handled by a design engineer.

Thousands of engineers can design bridges, calculate strains and stresses, and draw up specifications for machines, but the great engineer is the one who can tell whether the bridge or the machine should be built at all, where it should be built, and when.
Eugene G. Grace

The design engineer must anticipate all manner of problems that the user may create in the application of the machine, or use of the structure. The design must prevent user errors, accidents, and dissatisfaction. This is especially true for products that will be integrated into larger systems, or be used by customers under widely varied circumstances. For example, a car designed in Detroit must survive and operate in arctic Alaska, desert Arizona, humid Florida, and atop Pike's Peak.

In bridging the gap between the laboratory and the production line, the design engineer must be a versatile individual. This requires a mastery of basic engineering principles and mathematics, and an understanding of the capabilities of machines. It is also important to understand the temperament of the people who operate them. The design engineer must also be conscious of the relative costs of producing items, for it will be the design that will determine how long the product will survive in the open market. Not only must the device or process work; in many cases, it must also be made in a style and at a price that will attract customers.

As an example, let us take a clock, a simple device widely used to indicate time [5-10]. It includes a power source, a drive train, hands, and a face. Using these basic parts, engineers have designed spring-driven clocks, weight-driven clocks, and electrically driven clocks with all variations of drive trains. The basic hands and face have been modified in some models to give a digital display. The case has been made in many shapes and, perhaps in keeping with the slogan "time flies," it has even been streamlined! In the design of each modification the design engineer has determined the physical structure of the assembly, its aesthetic features, and the economics of producing it.

Of course, the work of the design engineer is not limited solely to performing engineering on mass-produced items. Design engineers may work on items such as bridges or buildings where only one item is made. However, in such work they are still fulfilling the design process of adapting basic ideas to provide for making a completed product for the use of others. In this type of design, engineers must be able to use their training, in some cases almost intuitively, to arrive at a design solution that will provide for adequate safety without excessive redundancy. The more we learn about the behavior of structural materials, the better we can design without having to add additional materials to cover the "ignorance factor" area. Particularly in the aircraft industry, design engineers have attempted to use structural materials with minimum excess being allowable as a safety factor. Each part must perform without failure, and every ounce of weight must be saved.

Of course, to do this, fabricated parts of the design must be tested and retested for resistance to failure due either to static loads or to vibratory fatiguing loads. Also, since surface roughness has an important bearing on the fatigue life of parts which are subjected to high stress or repeated loads, much attention must be given to specifying that surface finishes meet certain requirements.

Since design work involves a production phase, the design engineer is always considering costs as a factor in our competitive economy. One way in which costs can be minimized in manufacture or construction is to use standard parts, and standard sizes and dimensions for raw material. For example, if a machine were designed using nonstandard bolt threads or a bridge designed using nonstand-

(a)

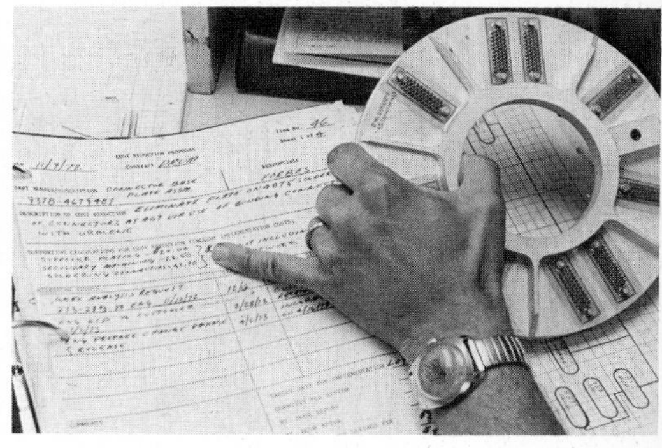

(b)

[5-9] The design engineers pictured in part (a) are evaluating a cost-reduction proposal for the connector base plate assembly shown in part (b).

[5-10] The availability of solid state circuitry has revolutionized the design of clocks and watches.

The fact is, that civilization requires slaves. The Greeks were quite right there. Unless there are slaves to do the ugly, horrible, uninteresting work, culture and contemplation become almost impossible. Human slavery is wrong, insecure, and demoralizing. On mechanical slavery, on the slavery of the machine, the future of the world depends.

Oscar Fingle O'Flahertie Wills Wilde, 1854–1900
The Soul of Man under Socialism

Laws of Thermodynamics

1. You cannot win.
2. You cannot break even.
3. You cannot get out of the game.

Anonymous

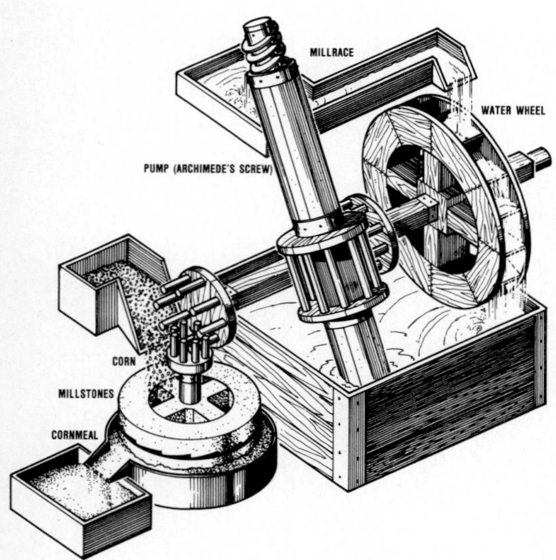

[5-11] In 1618, an English physician named Robert Fludd proposed a perpetual motion machine to grind grain. The design violates the principle of energy conservation, more formally stated in the first law of thermodynamics.

ard steel I-beams, the design probably would be more expensive than needed to fulfill its function. Thus the design engineer must be able to coordinate the parts of a design so that it functions acceptably and is produced at minimum cost.

The design engineer soon comes to realize also that there usually are many acceptable ways to solve a design problem. Unlike an arithmetic problem with fixed numbers which gives one answer, real design problems can have many answers and many ways of obtaining a solution, *and all may be acceptable*. In such a case the engineer's decision becomes a matter of experience and judgment. At other times it may become just a matter of making a decision one way or the other. Regardless of the method used, the final solution to a problem should be a conscious effort to provide the *best* method, considering fabrication, costs, and sales.

What are the qualifications of a design engineer? Here creativity and innovation are key elements. Every design will embody a departure from what has been done before. However, all designs must be produced within the constraints of the reality of the physical properties of available materials and by economic factors. Therefore, design engineers must be thoroughly knowledgeable in fundamental engineering in a rather wide range of subjects. They must be able to apply the natural laws of nature appropriately to ascertain whether proposed ideas are feasible. In addition, they must be familiar with basic principles of economics, both from the standpoint of employing people and using machines. As they progress upward into supervisory and management roles, the employment of principles of psychology and economics becomes of even more importance. For this reason they usually will have more use for management courses than will research or development engineers.

PRODUCTION AND CONSTRUCTION

In the fields of production and construction, the engineer is more directly associated with the technician, mechanic, and laborer. The production or construction engineer must take the design engineer's drawings and supervise the assembly of the object as it was conceived and illustrated by drawings or models.

Usually, production or construction engineers are associated closely with the process of estimating and bidding for competitive jobs. In this work they employ their knowledge of structural materials, fabricating processes, and general physical principles to estimate both time and cost to accomplish tasks. In construction work the method of competitive bidding is usually used to award contracts, and the ability to reduce an appropriate amount from an estimate by skilled engineering practices may mean the difference between a successful bid and one that is either too high or too low.

Once a contract has been awarded, it is usual practice to assign a "project engineer" as the person who assumes overall responsibility and supervision of the work from the standpoint of materials, labor, and money. The engineer will supervise other production or construction engineers, who will be concerned with more specialized features of the work, such as civil, mechanical, electrical, or chemical engineering. Here the project engineer must complete the details of the designers' plans. Provision must be made to provide the specialized construction tools needed for the work. Schedules of production and/or construction must also be set up and questions that technicians or workers may raise concerning features of the design must be answered. Design engineers will need to be advised concerning desirable modifications that will aid in the construction or fabrication processes. In addition, the project engineer must be able to work effectively with construction or production crafts and labor unions.

Preparation of a schedule for production or construction is an important task of the engineer. In the case of an industrial plant, all planning for the procurement of raw materials and parts will be based on this production schedule. An assembly line in a modern electronics plant is one example that illustrates the necessity for scheduling the arrival of parts and subassemblies at a predetermined time. As another example, consider the construction of a multistory office building. The necessity for parts and materials to arrive at the right time is very important. If they arrive too soon, they probably will be in the way, and if they arrive too late, the building is delayed, which will cause an increase in costs to the builder.

Qualifications for production or construction engineers includes a thorough knowledge of engineering principles. In addition, they must have the ability to visualize the parts of an operation, whether it be the fabrication of a microprocessor or the building of a concrete bridge. From an understanding of the operations involved, they must be able to arrive at a realistic schedule of time, materials, and manpower. Therefore, emphasis should be placed on courses in engineering design, economics, business law, and psychology.

[5-12] The construction industry is the largest industry in the United States.

Observe due measure, for right timing is in all things the most important factor.

Hesiod, ca. 700 B.C.
Works and Days

OPERATIONS

In modern industrial plants, the number and complexity of machines, the equipment and buildings to be cared for, and the planning needed for expansion have brought forward the need for specialized engineers to perform services in these areas. If a new manufacturing facility is to be constructed, or an addition made to an existing facility, it will be the duty of a plant engineer to perform the basic design, prepare the proposed layout of space and location of equipment, and to specify the fixed equipment such as illumination, communication, and air conditioning. In some cases, the work of construction will be contracted to outside firms, but it will be the general responsibility of the plant engineer to see that the construction is carried on as it has been planned.

After a building or facility has been built, the plant engineer and an appropriate staff are responsible for maintenance of the building, equipment, grounds, and utilities. This work varies from performing routine tasks to setting up and regulating the most complex and automated machinery in the plant.

To perform these functions, the plant engineer must have a wide knowledge of several branches of engineering. For land acquisition and building construction, civil engineering courses will be needed.

[5-13] The plant engineer is a critical element in the safe and efficient operation of the modern industrial plant.

For work with power generation equipment, mechanical and electrical backgrounds are essential. For work in specialized parts of the plant, knowledge may be needed in such fields as chemical, metallurgical, nuclear, petroleum, or textile engineering. Work activities will be in one or more of the following areas: plant layout and design, construction and installation, maintenance–repairs–replacement, operation of utilities, or plant protection and safety.

Plant engineers should be able to compare costs of operating under various conditions and set schedules for machines so that the best use will be made of them. In the case of chemical plants, they will also attempt to regulate the flows and temperatures at levels that will produce the greatest amount of desired product at the end of the line.

In the dual role as a plant and operations engineer, it is important to evaluate new equipment as it becomes available to see whether additional operating economies can be secured by retiring old equipment and installing new types. In this function the engineer must frequently assume a salesperson's role in order to convince management that it should discard equipment that, apparently, is operating perfectly and spend money for newer models. Here the ability to combine facts of engineering and economics is invaluable.

Plant and operations engineers must be able to work with people and machines and to know what results to expect from them. In this part of their work, a knowledge of industrial engineering principles is valuable. In addition, it is desirable to have a basic understanding and knowledge of economics and business law.

SALES

An important and sometimes unrecognized function in engineering is the realm of applications and sales. As is well known, the best designed and fabricated product is of little use unless a demand either exists or will be created for it. Since many new processes and products have been developed within the past few years, a field of work has opened up for engineers in presenting the use of new products to prospective customers.

[5-14] Engineers must be articulate and able to present their ideas with clear, concise presentations.

Discoveries and their subsequent application have occurred so rapidly that a product may be available about which even a recent graduate may not know. In this case, it is the responsibility of the engineer in sales who has intimate knowledge of the principles involved to educate possible users so that a demand can be created. In this work the engineer must assume the role of a teacher. In many instances the product must be presented primarily from an engineering standpoint. If the audience is composed of engineers, the sales engineer must "talk their language" and answer their technical questions. But if the audience includes nonengineers, the sales engineer must present the features of the product in terms that can be easily understood.

In addition to acquiring a knowledge of the engineering features of a particular product, sales and application engineers must also be familiar with the operations of the customer's plant. This is important from two standpoints. First, they should be able to show how their product will fit into the plant, and also they must show the economics involved to convince customers that they should buy it. At the same time, they must point out the limitations of their product and the possible changes necessary to incorporate it into a new situation. For example, a new bonding material may be available, but in order for a customer to use it in an assembly of parts, a special refrigerator for storage may be necessary. Also, the customer would need to be informed of the necessity for proper cleaning and surface preparation of the parts to be bonded.

A second reason that the sales and application engineer must be familiar with a customer's plant operations is that it is here that many times new requirements are generated. By finding an application area in which no apparatus is available to do the work, the sales engineer is able to report back to the company that a need exists and that a development operation should be undertaken to produce a device or process to meet the need.

Almost all equipment of any complexity will need to be accompanied by introductory instructions when it is placed in a customer's plant. Here the application engineer can create goodwill by conducting an instruction program outlining the capabilities and limitations of the equipment. Also, after the equipment is in service, maintenance and repair capabilities by competent technical personnel will serve to maintain the confidence of customers.

[5-15] Sales should not be made unless the product will serve the needs of the client.

Sales and applications engineers should have a basic knowledge of engineering principles and should, of course, have detailed knowledge in the area of their own products. Here the ability to perform detailed work on abstract principles is of less importance than the ability to present one's ideas clearly. A genuine appreciation of people and a friendly personality are desirable personal attributes. In addition to basic technical subjects, courses in psychology, sociology, public speaking, and human relations will prove valuable to the sales and applications engineer.

Usually, an engineer will spend several years in a plant learning the processes and the details of the plant's operation and management policies before starting out to be a member of the sales staff. As a sales engineer you represent your company in the mind of the customer. Therefore, you must present a pleasing appearance and give the customer a feeling of confidence in your engineering ability.

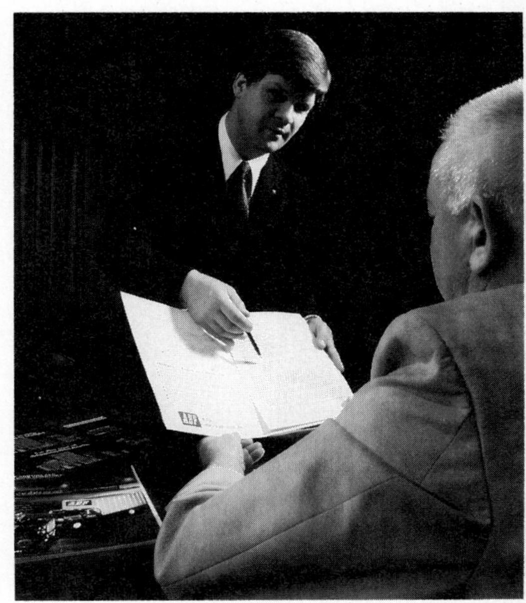

[5-16] Sales engineers must have a basic knowledge of engineering principles and detailed technical knowledge in the area of their own products.

MANAGEMENT

Results of recent surveys show that the trend today is for corporate leaders in the United States to have backgrounds in engineering and science. It has been predicted that within five years, the *majority* of corporation executives will be persons who are trained in engineering and science as well as in business and the humanities, and who can bridge the gap between these disciplines.

Since the trend is toward more engineering graduates moving into management positions, let us examine the functions of an engineer in management.

The basic principles used in the management of a company are generally similar whether the company objective is dredging for oys-

The first principle of management is that the driving force for the development of new products is not technology, not money, but the imagination of people.
David Packard
Quoted in Industrial Engineering, *August 1984, p. 61*

Man's greatest discovery is not fire, nor the wheel, nor the combustion engine, nor atomic energy, nor anything in the material world. It is the world of ideas. Man's greatest discovery is teamwork by agreement.
J. Brewster Jennings

The question "Who ought to be boss?" is like asking "Who ought to be the tenor in the quartet?" Obviously, the man who can sing tenor.
Henry Ford

ters, building diesel locomotives, or producing microcomputers. These basic functions involve using the capabilities of the company to the best advantage to produce a desirable product in a competitive economy. The use of the capabilities, of course, will vary widely depending on the enterprise involved.

The executive of a company, large or small, has the equipment in the plant, the labor force, and the financial assets of the organization to use in conducting the plant's operations. In management, the engineer must make decisions involving all three of these items.

A generation ago it was assumed that only persons trained and educated in business administration should aspire to management positions. However, now it has been recognized that the education and other abilities which make a good engineer also provide the background to make a good management executive. The training for correlating facts and evaluating courses of action in making engineering decisions can be carried over to management decisions on machinery, personnel, and money. In some cases, the engineer is technically strong but may be less experienced in the realm of business practicability. Therefore, it is in the business side of an operation that the engineer usually must work harder to develop the necessary skills.

The engineer in management is concerned more intimately with the long-range effects of policy decisions. Where the design engineer considers first the technical phases of a project, the engineer in management must consider how a particular decision will affect the employees who work to produce a product and how the decision will affect the people who provide the financing of the operation. It is for this reason that the management engineer is concerned less with the technical aspects of the profession and relatively more with the financial, legal, and labor aspects.

This does not imply that engineering fundamentals should be minimized or deleted. Rather the growing need for engineers in management shows that the type and complexity of the machines and processes used in today's plants require a blending of technical and business training in order to carry forward effectively. This trend is particularly strong in certain industries, such as aerospace and electronics, where the vast majority of executive managerial positions are occupied by engineers and scientists. As other industries become computer intensive and automated, a similar trend in those fields also will become apparent.

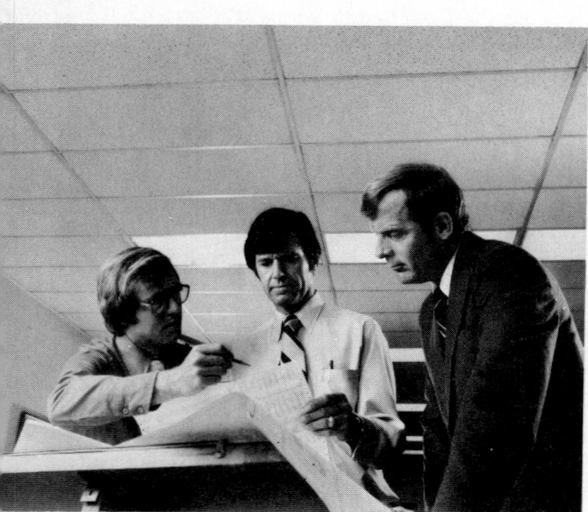

[5-17] It is important for the engineer in management to listen to the advice and counsel of subordinate engineers.

The education that an engineer in management receives should be identical to the basic engineering education received in other engineering functions. However, young engineers usually can recognize early in their careers whether or not they have an aptitude for working with people and directing their activities. If you have the ability to "sell your ideas" and to get others to work with you, probably you can channel your activities into managerial functions. You may start out as a research engineer, a design engineer, or a sales engineer, but the ability to influence others to your way of thinking, a genuine liking for people, and a consideration for their responses will indicate that you probably have capabilities as a manager.

Of course, management positions are not always executive positions, but the ability to apply engineering principles in supervisory work involving large numbers of people and large amounts of money is a prerequisite in management engineering.

OTHER ENGINEERING FUNCTIONS

A number of other engineering functions can be considered that do not fall into the categories previously described. Some of these functions are testing, teaching, and consulting.

As in the other functions, there are no specific curricula leading directly toward these types of work. Rather, a broad background of engineering fundamentals is the best guide to follow in preparing for work in these fields.

In testing, the work resembles design and development functions most closely. Most plants maintain a laboratory section that is responsible for conducting engineering tests of proposed products or for quality control on existing products. The test engineer must be qualified to follow the intricacies of a design and to build suitable test machinery to give an accelerated test of the product. For example, in the automotive industry, not only are the completed cars tested, but also components, such as engines, brakes, and tires, are tested individually to provide data to be used in improving their performance. The test engineer must be able also to set up quality control procedures for production lines to ensure that production meets certain standards. In this work, mathematics training in statistical theory is helpful.

A career in teaching is rewarding for many persons. A desire to help others in their learning processes, a concern for some of their personal problems, and a thorough grounding in engineering and mathematics are desirable for those considering teaching engineering subjects. In the teaching profession, the trend today is toward the more theoretical aspects of engineering, and a person will usually find that teaching is more closely allied with research and development functions than with others. Almost all colleges now require the faculty to obtain advanced degrees, and a person desiring to be an engineering teacher should consider seriously the desirability of obtaining a doctorate in his or her chosen field.

More and more engineers are going into consulting work. Work as an engineering consultant can be either part time or full time. Usually, a consulting engineer is a person who possesses specific skills in addition to several years of experience, and may offer services to advise and work on engineering projects either part time or full time.

Frequently, two or more engineers will form an engineering consulting firm that employs other engineers, technicians, and computer-aided design specialists, and will contract for full engineering services on a project. The firm may restrict engineering work to rather narrow categories, such as the design of irrigation projects, power plants, or aerospace facilities, or a staff may be available that is capable of working on a complete spectrum of engineering problems.

On the other hand, as a consulting engineer you may prefer to operate alone. Your firm may consist of only your own skills such

[5-18] Teaching is a rewarding activity of engineering. Frequently, the engineering professor is the first person to introduce the student to the ethics and responsibilities of the profession.

He who knows not, and knows not that he knows not, is a fool. Shun him.
He who knows not, and knows that he knows not, is simple. Teach him.
He who knows, and knows not that he knows, is asleep. Waken him.
He who knows, and knows that he knows, is wise. Follow him.

Arabic Apothegm

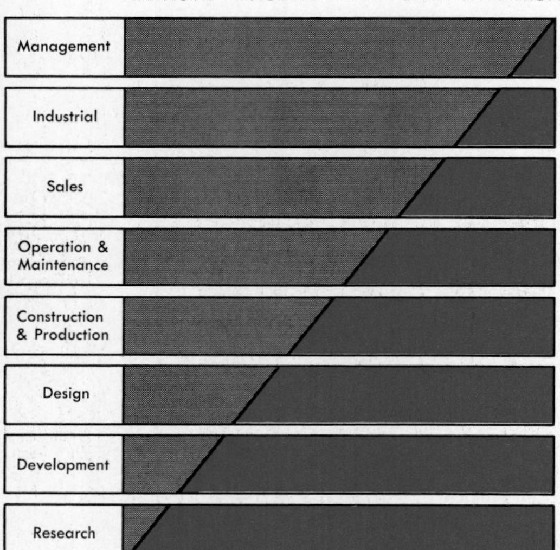

[5-19] Application of principles in various engineering functions.

that, in a minimum time, you may be able to advise and direct an operation to overcome a given problem. For instance, you may be employed by an industrial plant. In this way the plant may be able to solve a given problem more economically, particularly if the required specialization is only occasionally needed by the plant.

As may be inferred, a consulting engineer must have specific expertise to offer, and must be able to use his or her creative ability to apply individual skills to unfamiliar situations. Usually, these skills and abilities are acquired only after several years of practice and post-graduate study.

Consulting work is an inviting part of the engineering profession for a person who desires self-employment and is willing to accept its business risks to gain an opportunity for financial reward.

ENGINEERING FUNCTIONS IN GENERAL

As described in previous paragraphs, training and skills in all functions are basically the same, that is, fundamental scientific knowledge of physical principles and mathematics. However, it can be seen that research on one hand and management on the other require different educational preparations.

For work in research, emphasis is on theoretical principles and creativity, with little emphasis on economic and personnel considerations. On the other hand, in management, primary attention is given to financial and labor problems and relatively little to abstract scientific principles. Between these two extremes, we find the other functions with varying degrees of emphasis on research- or managerial-oriented concepts.

Figure [5-19] shows an idealized image of this distribution. Bear in mind that this diagram merely depicts a trend and does not necessarily apply to specific instances.

To summarize the functions of the engineer, we can say that in all cases the engineer is a problem identifier and solver. Whether it be a mathematical abstraction that may have an application to the design of a space station or a meeting with a bargaining group at a conference table, it is a problem that must lie identified and reduced to its essentials and the alternatives explored to reach a solution. The engineer then must apply specialized knowledge and inventiveness to select a reasonable method to achieve a result, even in the face of vague and sometimes contradictory data. The engineer has been able, in general, to accomplish this as evidenced by a long record of successful industrial management and productivity.

PROBLEMS

5-1. Discuss an important scientific breakthrough of the past year that was brought about by an engineering research effort.

5-2. Discuss the differences between engineering research and engineering development.

5-3. Interview an engineer and estimate the percentage of his or her work that is devoted to research, development, and design.

5-4. Discuss the importance of the engineer's design capability in modern industry.

5-5. Investigate the work functions of the engineer and write a brief essay describing the function that most appeals to you.

5-6. Discuss the importance of the sales engineer in the total engineering effort.

5-7. Interview an engineer in management. Discuss the reasons that many engineers rise to positions of leadership as managers.

5-8. Compare the engineering opportunities in teaching with those in industry.

5-9. Investigate the opportunities for employment in a consulting engineering firm. Discuss your findings.

5-10. Discuss the special capabilities required of the engineer in construction.

5-11. Perpetual motion powered by electricity was often favored as an idea by nineteenth-century inventors. Figure [5-P11] illustrates one of these designs. Discuss how it was supposed to work. Why did it fail to work?

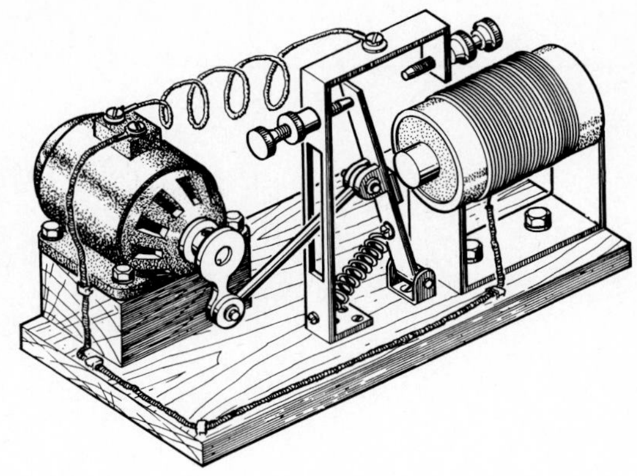

[5-P11] A proposed perpetual motion machine.

Chapter 6

Professional Responsibilities of the Engineer

[6-1] By virtue of education and experience, the engineer is better equipped than most people to foresee and appreciate problems as well as to identify and assess alternatives.

Profession—The pursuit of a learned art in the spirit of public service.

American Society of Civil Engineers

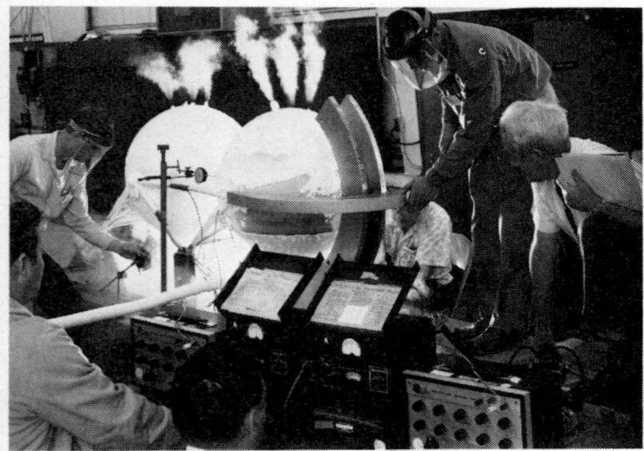

[6-2] Professionalism includes the establishment of performance standards and safety criteria.

THE ENGINEER AS A PROFESSIONAL PERSON

Who is a professional? As generally used in the sense of the learned professions, a professional person is one who applies certain knowledge and skill, usually obtained by college education, for the service of people. In addition, a professional person observes an acceptable code of conduct, uses discretion and judgment in dealing with people, and respects their confidences. Also, professional persons usually have legal status, use professional titles, and participate in a professional organization.

Knowledge and skill above that of the average person is a characteristic of the professional. Where a worker will have specific skills in operating a particular machine, a professional person is considered able to apply fundamental principles that are usually beyond the range of the average worker. The knowledge of these principles as well as the skills necessary to apply them distinguishes the professional. The engineer, because of an education in the basic sciences, mathematics, and engineering sciences, is capable of applying basic principles for such diverse things as improving the construction features of buildings, developing processes that will provide new chemical compounds, or designing canals to bring water to arid areas.

An important concept in the minds of most persons is that a professional person will perform a service for people. This means that service must be considered ahead of any monetary reward that a professional person may receive. In this respect the professional, individually, should recognize that a need for personal services exists and seek ways to provide a solution to these needs.

Engineering may be considered to be a profession insofar as it meets these characteristics of a learned professional group:

- Knowledge and skill in specialized fields above that of the general public.
- A desire for public service and a willingness to share discoveries for the benefit of others.
- Exercise of discretion and judgment.
- Establishment of a relation of confidence between the engineer and client or the engineer and employer.

- Self-imposed standards for qualifications (such as accredited schools, registration laws, and the formulation and conduction of licensing examinations.
- Acceptance of overall and specific codes of conduct.
- Formation of professional groups and participation in advancing professional ideals and knowledge.
- Recognition by law as an identifiable body of knowledge.

With these as objectives, students should pursue their college studies and training in employment so as to meet these characteristics within their full meaning and take their places as professional engineers in our society.

Members of a profession—A group of people who have dedicated themselves to unselfish services to humanity through the application of knowledge and skills possessed by the group.

E. C. Easton
"An American Engineering Profession,"
Journal of Engineering Education, *April 1962*

Technology can have no legitimacy unless it inflicts no harm.

Admiral H. G. Rickover
"Humanistic Technology," Mechanical Engineering, *November 1982*

SOCIAL RESPONSIBILITY: THE PIVOTAL ROLE OF ENGINEERS

Society has a special challenge for the engineering profession: use acquired knowledge for the benefit of humankind without endangering the surrounding environment or adding risks to the lives and safety of individuals. The response of engineers to this challenge will be on continuous public display because they design, produce, and operate many physical systems and products which have a direct impact on both individuals and society as a whole. When flawed, large-scale projects such as dams, bridges, and jet aircraft are apt to lead to serious accidents. Flawed designs of industrial process equipment may cause illness or injury to thousands of workers who must use the machines and ancillary chemicals. Apparent beneficial engineering operations such as power generating plants can also lead to environmental damage, depletion of resources, and other unique problems. Computers and robots used in factories to increase production and improve quality may also require a redistribution of specific worker manual skills. Another major concern of society today is the development of nuclear weapons and space warfare techniques which could place our entire civilization in danger. All of these resulting conditions must be dealt with by society.

The engineering profession plays a pivotal role in this important function of controlling technology for the benefit of society. Because of their education and experience, engineers are especially equipped to anticipate problems and to evaluate alternatives. As technology becomes more complex, society makes engineering increasingly responsible for this control.

Leonardo da Vinci writes in his autobiographical notes that he has discovered how to build a submarine which has a special application for naval warfare. This underwater boat would allow the user to sneak into a busy harbor without being seen and drill holes in the bottom of ships. History records, however, that during his lifetime he withheld disclosure of this invention because he believed that it would be an abomination to mankind.

Victor Paschkis
Conference on Engineering Ethics, *American Society of Civil Engineers, New York, 1975.*

Development of the Role of Engineering

For centuries society seemed assured that technologists could (and would) regulate themselves to assure a high level of competence and responsible behavior. However, as technology became more complex, society called for more safeguards and regulation. Design catastrophes such as boiler explosions, mine cave-ins, and building collapses began to change the image of engineering from one where its members are always considered to be a part of a respected guild, to one where its participants needed more than self-regulation.

[6-3] Yeah . . . I know it leans a little to the starboard. Actually, that's intentional . . . to compensate for the spongy land I bought on sale. According to a new method of calculating that I've worked out, it will right itself in about two months.

[6-4] The Tacoma Narrows Bridge was designed for static loading . . . but not for dynamic loading.

There is a difference, however, between "state of the art" design (which ultimately may prove to be unsafe) and risk design. There is much about the universe and the laws of nature that we do not yet understand. As engineers, we are expected to execute designs using principles that are based on the most current understanding of natural physical laws. We would expect no less of our family physician, whom we would expect to treat a body ailment using the most modern understanding of medicine and its effect on the human body. However, even the best of designs *under these circumstances* may fail. The collapse of the Tacoma Narrows Bridge at Puget Sound, Washington, in 1940, only four months after its completion, is an example. The bridge had two spans of 2800 feet with a width of only 39 feet, and the deck was stiffened throughout its length by means of 8-foot-deep plate girders. The design seemed to be adequate because similar designs had not failed in over 50 years of use. Such was not the case, however. The location of the bridge and its particular configuration made it susceptible to severe torsional vibration and aerodynamic instability. Ultimately, in winds of only 42 mph, the vibrations became so violent that the deck was torn away and crashed into the water [6-4]. Since that time a number of design methods have been developed to circumvent this problem, among them to shape the deck like a shallow airfoil. But, in 1940, who would have thought that a bridge should be designed like an airplane?

Risk design involves allowing sufficient tolerance above the point of failure, even under the most unusual circumstances, so that failure will not occur. This involves knowledge and judgment as well as conjecture as to what unexpected and unforeseen events might occur. The tolerance above the expected failure level is called the *factor of safety*. However, if the design is to work at all, it must also be compatible with the natural laws of nature. For example, a highway bridge might have a factor of safety of 2, meaning that the bridge is designed to hold up to two times the load that is anticipated to cause failure. This reasoning would not work in the case of airplane design. If the component parts of the airplane were all oversized by two times, the airplane would be so heavy that it would not fly. It is more customary, then, for airplane designers to use factors of safety of 1.02, or even less for many of the structural components. The bridge is overdesigned by 200 percent, and some parts of the airplane by only 2 percent. Now, knowing this, will you be more nervous when you fly to Hawaii for a vacation? Actually, there is little cause for concern. Many of the components of the space shuttle were designed to equivalent factors of safety of 1.00, and this design has proven to be safe. More and more, emphasis is being directed today to overestimating expected design loads rather than using a true factor of safety.

How can society be assured that appropriate engineering decisions are being made? Because of people's concerns about the consequences of these decisions, almost a century ago engineers in the United States began to impose design regulations through technical societies. These activities, which first became significant near the end of the nineteenth century, were aimed at standardizing engineering methods, and at creating a controlled professional environment for

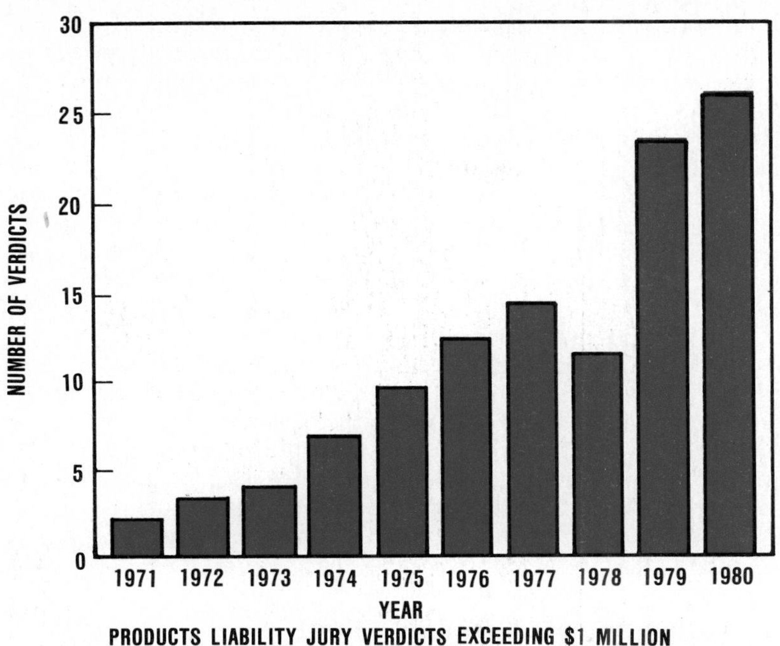

PRODUCTS LIABILITY JURY VERDICTS EXCEEDING $1 MILLION

[6-5] In the early 1960s, product liability lawsuits comprised only a small portion of the litigation in this country. Then came the deluge, exemplified by a nearly sixfold increase between 1971 and 1976. Between 1974 and 1981, the number of suits grew at an average annual rate of 28 percent, more than three times faster than other civil suits.[1]

engineers. These early efforts led to more formalized regulation through licensing requirements and the formation of ethical codes.

This formalized regulation within the profession was the major method for control of technology until after World War II (1950), when society, becoming increasingly concerned about the impact of technology, began to demand improved performance from those in the engineering profession. This demand took the form of new public laws in the areas of environmental protection, occupational safety and health, and consumer protection. For example, the Occupational Safety and Health Act (OSHA) of 1970 was a significant piece of legislation aimed at providing every working person in the nation safe and healthful working conditions and preserving our human resources. Among the functions included was the authority to prescribe regulations that would protect workers from potentially toxic substances or harmful physical agents that might endanger their safety and health.

By imposing legal controls, society is sending a message that engineers must be held accountable for the consequences of their designs. The degree of this accountability depends on the particular duties of the engineer and the nature of the technical problems being solved. However, there is little doubt that the intent is to hold engineers legally responsible for their actions—or inactions.

One important outcome of this imposition of legal control is the increased emphasis on technical competency and on having a high

Responsibility n.—state of being responsible; that for which any one is responsible; a duty; a charge; an obligation.

Webster's Dictionary

[1]Data from *Personal Injury Valuation Handbooks: Injury Valuation Reports, Current Award Trends,* Vol. 1, No. 258, (1982). Jury Verdict Research, Inc., Solon, Ohio

The study of ethics is the study of human action and its moral adequacy.

Kenneth E. Goodpaster, Ethics in Management, *Boston: Harvard Business School, 1984*

It is not enough that you should understand about applied science in order that your work may increase man's blessings. Concern for man himself and his fate must always form the chief interest of all technical endeavors, concern for the great unsolved problems of the organization of labor and the distribution of goods—in order that the creations of our mind shall be a blessing and not a curse to mankind. Never forget this in the midst of your diagrams and equations.

Albert Einstein, 1879–1955
Address, California Institute of Technology, 1931.

How can a person talk of pursuing the good before knowing what the good is?

Socrates,[2] *469–399 B.C.*

level of current knowledge related to specific assignments. For example, recent court decisions indicate that the engineer is expected to seek out information on all known hazards associated with a particular project *before beginning any design*. The level of responsibility is determined by what a prudent and knowledgeable engineer would have done.

Working in complex organizations, engineers may believe that their decisions have no impact, but the necessary control of technology starts with knowledgeable individuals taking responsible actions. At all levels, judgments are made which have widespread effects.

In many cases the engineer is the first to realize that there may be ecological considerations or perhaps a problem with user safety, and that a design change is needed. Such an early discovery would be very important to reveal because, as the project develops, corrective measures become more difficult to make and costs for changes increase rapidly. Engineers are also frequently asked to give their expert opinions to those who make policy decisions.

ETHICS: A DECISION-MAKING PROCESS

The satisfactory completion of a design project requires that the project engineers work together as a team, each making decisions related to their own design responsibilities but which may also affect the efforts of other team members. Effective team operation also requires good communications, clear assignments, and a recognition of the feelings and needs of others.

Making decisions that have interpersonal, moral, and legal implications can be more complex and difficult for engineers than might be the case for other professionals (such as physicians), because most engineers are not self employed. In fact, over 95 percent of graduate engineers are employees of consulting firms, industrial corporations, or governmental agencies. The result is that in some cases the engineer is faced with a dilemma—individual ethics versus job security. As an engineer, you will have an obligation to protect the well-being of society, act in accordance with a strong moral code, and at the same time be a reliable employee.

To operate effectively in such a system of interrelated decisions, engineers need to base their actions on a code or pattern of behavior that is built around a moral point of view which can be used to decide between good and bad, right or wrong, virtue or vice. Professional engineering societies have developed and are continuing to develop codes of ethics as they make decisions that affect their employers, society in general, and their fellow engineer members. These developed codes provide valuable guidelines, but ethical conduct must always be an individual attribute and a matter of personal conscience [6-6]. It requires a lifelong process of learning and examination to set values that determine responsible and ethical action.

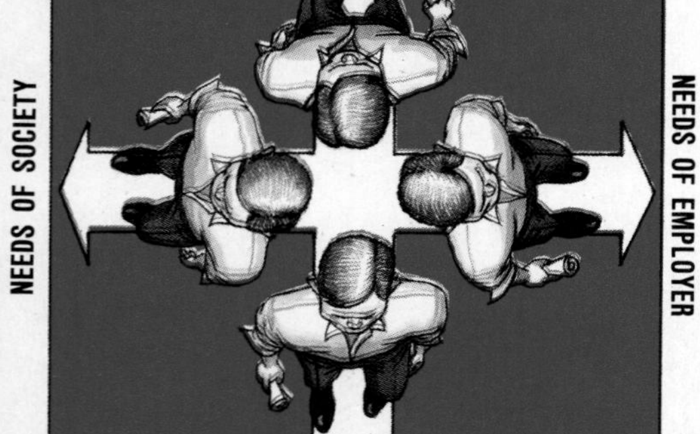

[6-6] The engineer is pulled in different directions.

[2]Quoted by L. L. Nash, "Ethics Without the Sermon," *Harvard Business Review,* November–December 1981, p. 79.

Because ethical decisions usually involve relations with other people and affect peoples' lives, they cannot be reduced to an analytical process such as solving a physics problem. However, there is a place for a reasoning pattern in ethics which is built around a moral point of view, and it starts with an ideal as basic as the Golden Rule. Using this moral guide, an engineer can make judgments and decisions with a well-defined set of ethical values. This set of ethical values might include such personal actions as:[3]

Golden Rule

Do unto others as you would have others do unto you.

All people have a special dignity or worth.

- Obey the law.
- Keep promises, contracts, and employment agreements.
- Respect the rights of others.
- Be fair, do not lie or cheat.
- Avoid harming others.
- Prevent harm to others.
- Help others in need.
- Help others in the application of these values.

It takes a wise person to give the answer to a technical question that involves the conflicting rights and desires of a number of people. Yet the engineer is often required to give such an answer and on very short notice.

Philip L. Alger et al., Ethical Problems in Engineering, *New York: Wiley, 1984*

For many decision situations, a personalized set of values such as the above will be sufficient. Such a value system functions as a pattern of behavior which demands much from the engineer, but it can also work as a base to provide inner satisfaction and be an effective guide as the engineer works with others.

Sometimes, however, there may be conflicts between certain items within the individual set of values, or interpersonal conflicts may arise with people who do not have the same set of values or may attach different weights to the various values.

For example, suppose that an engineer on an automobile design team is responsible for the catalytic converter. The present design frequently fails to meet governmental standards. The engineer has been developing and testing a new technique which shows promise of making the emission control system even more effective than is needed to meet the minimum government emissions standards. The engineer strongly recommends to the project engineer that the improved design be selected. However, the engineer estimates that an additional two weeks would be required to adequately test the new technique for verification of the anticipated results. The supervisor in charge of this phase of the design is anxious to meet the overall design schedule and does not want to delay the project. The supervisor therefore approves the old system design with an emission control device which has proven to be only marginally satisfactory in past applications. The action taken results from a disagreement of values between the two individuals and can create a conflict for the engineer as a course of action is being considered.

In this case, the catalytic converter design engineer may be forced into a second level of ethical decision making where additional judgmental thinking is involved [6-7]. Here the search will be for additional criteria that can be used to select the applicable ethical values, clarify their applicability in this particular circumstance, and

[3]Kenneth E. Goodpaster, *Ethics in Management* (Boston: Harvard Business School, 1984).

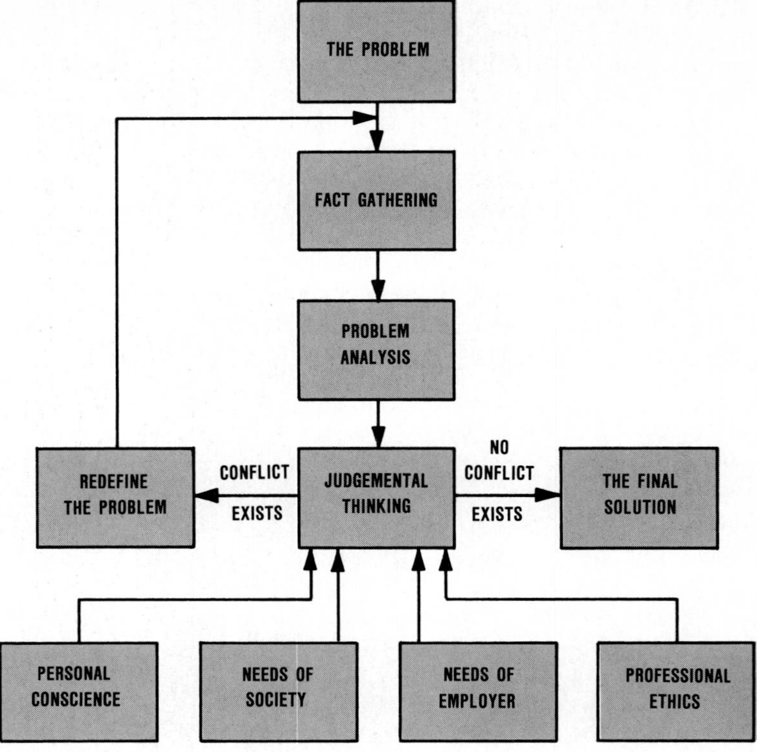

[6-7] A block diagram representing the steps involved in the process of making ethical decisions.

resolve any conflicts between the values. Such judgmental thinking requires looking at *three key areas:*

1. What action represents the strongest duty or responsibility?
2. What action most fairly respects the rights of all people involved?
3. What action maximizes all benefits versus all costs?

[6-8] Three key areas . . .

In struggling to find the proper ethical position in this situation, the engineer should start with the detailed problem analysis and include the extensive fact gathering. To establish neutrality and reduce defensive emotions, an attempt should be made to look at the other side of the problem. (In this example, how does the supervisor see the problem?) Before making a final decision the engineer needs to determine how everybody will be affected by his or her decision. What would you have done?

Studies indicate that most people will accept voluntary risks in the range of one fatality risk in 1000 (equivalent to driving a car 300,000 miles or smoking 1400 cigarettes in a lifetime.) On the other hand, many people believe and several government agencies require that risk imposed involuntarily must not exceed one fatality in 1,000,000.
Dennis J. Paustenbach, "Risk Assessment and Engineering in the 80's," Mechanical Engineering, *November 1984, p. 55*

USING RISK ANALYSIS TO AID IN MAKING ETHICAL DECISIONS

All people in their daily lives are regularly involved in activities that require weighing benefits against possible risks. They analyze the consequences and then proceed with risk-taking actions such as driving a car, sky diving, surfing, smoking, skiing, or even overeating.

Similarly, engineers in their professional capacity make risk/benefit decisions. Although these two types of decisions require similar thinking processes to arrive at logically selected actions, there is a major difference in the decision outcomes. In one case, people voluntarily accept actions involving a self-determined degree of risk. In the other case, engineers make judgments about the degree of risk to which someone else will be involuntarily or unknowingly exposed. Engineering decisions such as determining safety factors in a bridge design, or deciding on the location of a waste disposal facility, directly expose people to risk.

The far-reaching effects of engineering decisions involving involuntary risks and the complexity of today's risk-related problems emphasize the need for carefully making reasonable and ethical judgments. This requires a systematic risk analysis process which includes anticipating hazards, determining the degree of acceptable risk, and then selecting the best risk/benefit alternative. The various steps in the risk analysis process can be grouped into two major areas: *risk assessment,* which includes the steps required to determine the numeric probability of an adverse effect, and *risk management,* which includes the steps required to analyze risk consequences and implement selected action.[4]

[6-9] Surfing, like manned exploration of space, involves risk.

Risk Assessment

The end result of risk assessment is an expression of risk that defines the probability of harm which could result from a given action or situation. This could range from an expression that states the risk associated with placing wastes in a selected disposal site, to a number that defines the reliability of equipment design. The expression should combine all elements included in the risk assessment, such as extent or concentration of the hazard, effect or consequence of exposure to the hazard, and possibility of exposure. All of these factors are combined to describe the nature and magnitude of the risk along with all uncertainties associated with the entire process. Uncertainties can result from the manner in which the process is implemented, or basing results on unreliable or insufficient data.

When involved in assessing risks, engineers must carefully implement the process and allow for possible inadequacy of information and test data. They must recognize that assessment cannot be based on blind dependence on what is assumed to be good information or objective data. Judgments must be made concerning evaluation and application of the data. There may be an inner tension created by the gap between what is known and what must be assumed.

One risk element that often falls in the gap between what is known and what must be assumed is the possibility of human failure. When assessing risks, engineers need to consider carefully how possible human errors could cause failures or compound the consequences of design failures. An example of how an improper human reaction to an equipment failure can create serious consequences is

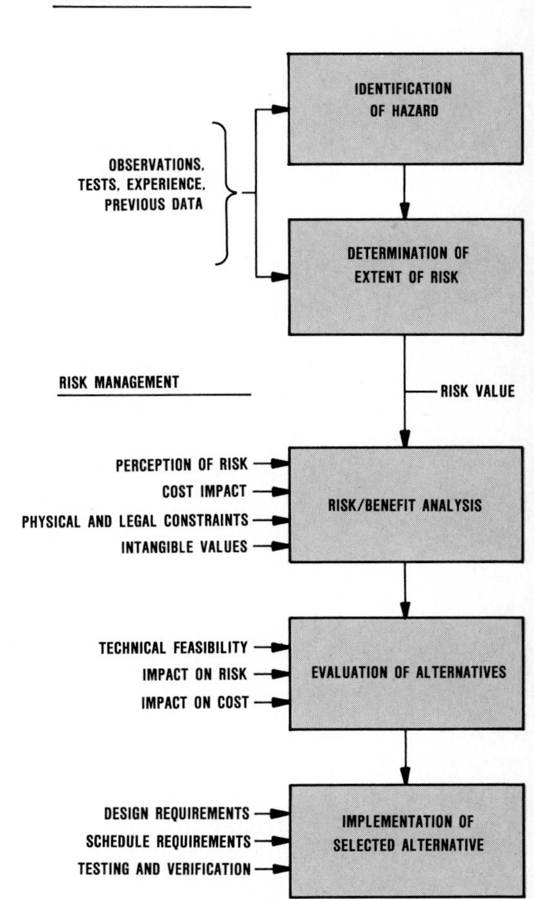

[6-10] Risk analysis.

[4]A. Alan Moghissi, "Risk Management—Practices and Prospects," *Mechanical Engineering,* November 1984.

Two Key Questions for Engineers in Risk Analysis

- Does a significant risk exist?
- Are there feasible alternatives for avoiding, preventing, or decreasing risk?

Engineers must be trained in risk assessment and must be equipped to confront ethical dilemmas as they appear.

Lay Wilkinson, President of Union Carbide Corp.
Engineering Times, *December 1984*

Typical Expressions of Risk

- Certain radiation emission from a nuclear power plant equals specified health risk.
- Exposure to specified concentration of benzene over a 70-year lifetime equals specified added fatality risk.

As far as we know, our computer has never had an undetected error.

Conrad H. Weisert, Union Carbide Corporation
Quoted by OMNI, *September 1979*

All substances are poisons. The right dose differentiates a poison and a remedy.

Philippus Paracelsus, 1493–1541

illustrated in the poison-gas leak that occurred at a Union Carbide Corp. pesticide plant in Bhopal, India, during the night of December 4, 1984, causing the world's worst industrial accident.[5] Deadly methyl isocyanate gas diffused undetected through this city of over 600,000 people for over 40 minutes. Over 1700 people were killed and 300,000 others were left with long-term serious injuries. What went wrong, the equipment or the people?

Subsequent investigations of this accident revealed that the gas leak was caused by a large increase in pressure inside one of three storage tanks that contained the liquefied gas which was used to manufacture insecticides. When the pressure became too intense, a valve opened, allowing gas to escape into a "scrubber" mechanism. This was a safety feature designed to neutralize any leaking gas and vent it as harmless gas into the atmosphere. Apparently, one of the scrubbers was not operable, due to a maintenance problem. The pressure of the operating scrubber was so high and the flow of gas so fast that the "scrubber" could not neutralize the gas *before* it escaped into the atmosphere. Backup safety procedures called for two workers to be assigned to monitor the tank gas pressure and to cool the tank by spraying it with water whenever the escape valve opened. However, the workers panicked, ran away, and failed to cool the tank. This allowed the deadly fumes to escape for over 45 minutes before supervisors with protective gas masks could arrive and cool the tank. The government of India has charged that the plant's designers are at fault because "the plant's design and manufacturing controls were not foolproof to ensure absolute avoidance of violent chemical reactions and leakage of toxic gases." The company contends that "the workers at the Bhopal plant were poorly trained and failed to prevent the accident." Who do you think was responsible for this disaster? Do you believe there were elements of risk which were not properly considered in the risk assessment process?

Risk Management

Using the risk value obtained from the risk assessment procedure to decide what action to take is called *risk management*. It involves weighing the calculated risk against benefits, desired results, or requirements and then selecting a cost-effective alternative to achieve the best risk/benefit ratio. Safety, physical, and chemical risks are all present to some degree in the work environment. It is the designer's responsibility to see that these risks are reduced to the lowest level possible.

In some cases, risk management may involve weighing a calculated reliability or risk value against a criterion of acceptability or standards established by industry, government (OSHA), or a certifying agency. The risk management process is based on the premise that no action is totally risk free; there are no rewards without risks, no operating systems without potential failures.

[5]This unfortunate catastrophe is described by the authors using the most current Associated Press accounts available at the time of this text's publication. It is likely that the matter will be in litigation for several years. Also, the long-term effects of the atmospheric pollutants could significantly increase the reported casualties.

The improper perception of risk by the people affected by the risk is a serious obstacle to effective risk management. Society tends to perceive risks in two broad categories: high probability/low consequences and low probability/high consequences. A typical high probability/low consequence action is cigarette smoking, which has been proven to result in a high percentage of lung cancer. However, because cancer appears many years after the smoker starts to smoke, and the smoker is the prime victim, society has been slow to impose even minimum restrictions. A typical low probability/high consequence action is the generation of energy with nuclear reactors. Society reacts to this action with extensive and costly regulations. NASA's loss of the space shuttle Challenger has been viewed as a national disaster because the previous 24 successful flights had given the public a false impression that little risk was involved.

Another major factor in the risk management process is the impact of financial costs. This can involve determining costs related to taking required actions, costs involved in various trade-offs, and costs involved if no actions are taken to reduce the risks. Lost along with the $1.2 billion space craft Challenger were a $100 million communications satellite and a smaller $10 million payload that was to have studied Halley's comet.

In disasters such as that which occurred in Bhopal, India, management may be considered to be personally liable and be required to pay the consequences. Immediately after the Bhopal disaster, five plant officials were arrested on negligence charges (they were eventually released) and everyone involved will have to live with the uneasy concern that a more realistic assessment of risk could have averted the disaster. Corporate management makes key economic cost/benefit decisions and then expects the engineer to use his or her technical expertise to make certain that these decisions, along with other risk/benefit decisions, are properly implemented.

LOCAL INJURY

erythema depilation vesiculation
necrosis gangrene

ACUTE RADIATION SYNDROME

nausea vomiting diarrhea
anemia leukopenia

CHRONIC INJURY

anemia leukemia cataracts
neoplasia
shortening of life span (?)
genetic mutation (?)

HOW TO HANDLE A RADIATION VICTIM

BEGIN EMERGENCY PROCEDURES IMMEDIATELY

resuscitate and stabilize patient
get detailed history
give symptomatic-treatment of systemic and skin reactions

DETERMINE INTENSITY OF IRRADIATION AS SOON AS POSSIBLE

map the involved anatomic area

DECONTAMINATE PATIENT

remove clothing and store for analysis
remove penetrating missiles
clean wounds surgically and seal with plastic
wash—do not shower—patient
shampoo and cut—do not shave—hair

PREPARE PATIENT FOR EVACUATION TO RADIATION MEDICAL CENTER

dress in hospital gown
wrap in blankets
shield with plastic sheet

[6-11] What exposure to radiation can do.[6]

MAJOR TECHNOLOGICAL AREAS REQUIRING ETHICAL DECISIONS

Atmospheric Contamination

A major challenge for technology today is to provide for society's increasing demand for energy without polluting the air that we must have to live. Engineers functioning in their jobs are faced with ethical decisions related to this technological challenge.

Air pollution poses a special challenge for engineers because it is a complex problem resulting from an interaction of individual pollutants that originate from a combination of sources. Coal-burning power plants, smelters, and motor vehicles all produce pollutants which can individually cause severe damage or can interact with other pollutants or substances in the atmosphere to form still another damaging agent. For example, hydrocarbons and nitrogen oxides produced mainly by gasoline-powered vehicles combine in the presence of sunlight to produce ozone, a known cause of damage to crops, forests, and human health. As discussed in Chapter 2, carbon dioxide, a direct by-product of all fossil-fuel combustion, is possibly

[6] *Emergency Medicine,* March 1970, p. 61

Recent studies warn that within 60–80 years the carbon dioxide concentration will be twice existing levels. This higher concentration is expected to cause a global temperature rise sufficient to raise sea levels, diminish water supplies, and alter rainfall patterns.
Sandra Pastel
Air Pollution, Acid Rain and the Future of Forests
Washington, D.C.: Worldwatch Institute, 1984

TABLE 6-1 SOURCES OF SULPHUR AND NITROGEN OXIDE EMISSIONS IN THE UNITED STATES

	Percent of total	
Source	***Sulphur dioxide***	***Nitrogen oxides***
Electric utilities	66	29
General industries	22	22
Smelters	6	1
Motor vehicles	3	44
Homes, businesses	3	4

Source: Adapted from Sandra Pastel, *Air Pollution, Acid Rain and the Future of Forests* (Washington, D.C.: Worldwatch Institute, 1984), pp. 15–16.

Because acid rain, ozone and the build-up of carbon dioxide have common origins, they can have common solutions.
Sandra Pastel
Air Pollution, Acid Rain and the Future of Forests
Washington, D.C.: Worldwatch Institute, 1984

[6-12] ". . .quite frankly Johnston when I asked you to solve the emission problem from that stack, I expected a somewhat more sophisticated solution. . ."

a long-term serious pollutant because of its heavy concentration and potential for causing extensive worldwide climate change.

Sulphur and nitrogen oxides emitted during the burning of fossil fuels and the smelting of metallic ores, unlike carbon dioxide, do not stay in the atmosphere, but eventually return to the earth in some form to produce their own damaging effects (Table 6-1). Some stay as gases in the region of the pollutant source, where they are visible in the haze that hangs persistently over many cities. Some are deposited in dry form on surfaces such as leaves or needles, where they react with moisture to form damaging acids. Some of the sulfur and nitric oxides combine with water vapor existing in the air to form sulfuric and nitric acids, which are considered to be the major constituents of acid rain, a newly recognized pollutant now being intensely studied and evaluated because of its potential for worldwide environmental damage.

Environmental specialists, industrialists, and legislators continue to debate about the causes of acid rain and the need for corrective action, while the pollution effects continue to accumulate. It is estimated that at least 180 lakes in the Adirondack Mountains in New York have become so acidic in content that fish and almost all other forms of aquatic life can no longer survive. Over 100 miles of streams in the Monongahela National Forest of West Virginia are in a similar condition. Killing fish and other aquatic life may not be the only effect of acid rain; cattle and game animals that eat waterside vegetation and birds that feed on fish and aquatic insects may also be affected by acidic food or even experience a total loss of food sources. Damage to trees in the United States is still confined to a few species at high altitudes in localized areas, but the Black Forest in Germany has sustained damage in many species, in various types of soils, and in a variety of altitudes. Scientists in both Germany and the United States are continuing to study both the extent of direct damage and tree rings to determine the short-term and long-term effects of acid rain on forests.[7]

Although we do not fully understand all of the reasons why forest destruction is occurring, the effects of acid rain and other

[7]Bob Thomas, "Acid Rain," *Arizona Republic,* June 1983.

pollutants are apparently stressing sensitive forests beyond their ability to survive. Weakened in this way, many trees lose their resistance to natural calamities such as drought, insect attacks, and frost. In some cases the pollutants alone cause direct injury or decline in growth. The mechanisms are complex and may take decades of additional research to provide us with a complete understanding.[8]

In 1963, the U.S. government passed the Clean Air Act, which recognized that since the largest portion of air pollution is caused by urban centers that cross state borders, federal financial assistance and leadership is essential.

At various times since 1967 the law has been amended to emphasize research efforts, establish national emission standards for motor vehicles, establish economic approaches for controlling pollution, and allow citizen action suits against the Environmental Protection Agency (EPA, the agency designated to implement and enforce the law).[9]

Although this law provides a basis for national action, its effectiveness depends on responsible action by engineers and other technical professions. This responsible action must start with research to discover or verify the complex mechanisms producing the damage. This is especially important in the case of acid rain, where so many possible sources and pollutants are involved. A related area of research is the investigation of technologies that could be used to reduce pollution produced by combustion.

Another important responsibility of engineers related to solving problems of this nature is to clarify and translate technical information into a form that is understandable to the decision makers in order that it can be used as a basis for judgment. Decisions related to the control of air pollution are particularly difficult to make because they require a close cooperation between states, regions, and even nations. For example, erecting a high smokestack to reduce localized pollution in one area may result in gases being carried downwind causing acid rain pollution many hundreds of miles away. Also, some regions or nations may be reluctant to clean up their own area if pollution continues to come in from neighbors who have taken no action. For example, pollutants originating from electric utilities and smelters in the coal-using states of Ohio, Indiana and Illinois are thought to contribute to the extensive damage throughout Ontario, Canada, and into Nova Scotia.[10] The smelters at Copper Cliffs, Ontario, Canada, also contribute greatly to this problem.

In addition to the complexity of causation and wide-ranging effects of air pollution, economic factors also bring about their own problems for the decision makers. On one side they might hear, "Implementing this change to reduce pollution will increase the cost of the end product." On the other side they might also hear, "Environmental damage is a cost which should be included in the price

[6-13] One consequence of the drive to purify urban air over the last two decades has been the construction of tall smokestacks, to better disperse pollutants into the atmosphere. These smokestacks, along with high levels of emissions, sent pollutants traveling hundreds of kilometers before returning to the earth's land and waters.

[6-14] How much is clean air worth?

[8]Sandra Pastel, *Air Pollution, Acid Rain and the Future of Forests* (Washington, D.C.: Worldwide Institute, 1984), p. 7.

[9]*Environment Reporter (Federal Laws)* (Washington, D.C.: The Bureau of National Affairs, Inc., 1984).

[10]Sandra Pastel, "Protecting Forests from Air Pollution and Acid Rain," in *State of the World, 1985* (New York: W. W. Norton, 1985).

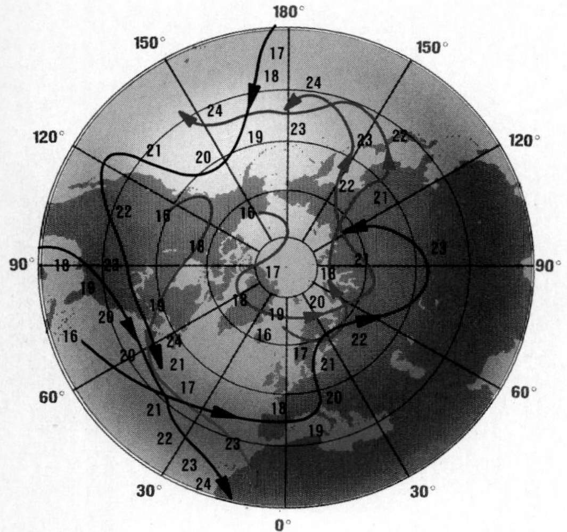

[6-15] Movement of air around the world can be determined by identifying specific air parcels and tracking them. The tracks shown above were identified by Edwin Danielsen, National Center for Atmospheric Research. The parcels in the troposphere are shown in black and those in the stratosphere are shown in color. The numbers represent successive days in April, 1964.

that consumers pay. Those creating the pollution must pay for it." Which advice is correct? An amendment to the Clean Air Act passed in 1984 provides for economic approaches to controlling air pollution, such as incentives for positive actions.

However, the engineer designing systems involving any aspect of pollution control or who may be placed in a position of evaluating various solutions, will be faced with using economic factors in the decision-making process.

In addition to making technical decisions related to their job responsibilities, engineers have an opportunity to respond to pollution problems as interested, responsible, and knowledgeable citizens.

The Clean Air Act states that any person may commence a civil action against the EPA where there is alleged to be a failure to perform any act or duty required under the act. The "What would you do?" case below is taken from an actual EPA case. In the actual situation, EPA allowed the manufacturer to proceed with their proposed alternative plan. However, several individuals have sued to force the EPA to order a recall of the inferior equipment.

Chemical Contamination of Underground Water Supplies

Pollution of groundwater caused by toxic chemicals leaking from waste storage sites is also rapidly becoming a major environmental concern in this country [6-16 and 6-17]. It affects every state in the

An Example of a Decision in Engineering Ethics: What Would You Do?

When engines exceed pollution control standards, *recall* is the only remedy outlined in the Clean Air Act.

U.S. Appeals Court Ruling, October 26, 1984.

Court mandates recall of vehicles for emission-standards violations.

Headline
Arizona Republic, *October 27, 1984.*

Assume that you are a chemical engineer working for the Environmental Protection Agency responsible for investigating motor vehicle engine pollution control standards. You have found that one group of engines produced by a major automobile manufacturer for the current model of automobiles now being sold exceeds pollution control standards. You report your findings to your superior, who notifies the manufacturer to determine the direction of corrective recall action that will be implemented.

The manufacturer submits a plan to you proposing to leave the current engines alone, but to make sure that next year's models would be especially designed for lower emissions to make up for the excessive levels emitted by the current models. It is argued that their plan would save them $11.8 million and in addition would save the owners $25.8 million in fuel costs because the repairs necessary to improve emission performance would increase fuel consumption. They also argue that their proposal would be better for the environment because emissions from all of next year's models would be below the standard, while not all of the current recalled models would be brought in for repair.[11]

The proposal from the manufacturer comes to you for your recommendation. What do you recommend: an acceptance of the manufacturer's proposal, or initiation of immediate recall action? Include your reasons in your recommendation.

[11] Adapted from "Court Mandates Recall of Vehicles for Emission-Standard Violations," *Arizona Republic,* October 27, 1984.

nation and threatens the water supply of half of the nation's population. It is especially dangerous because its effects are gradual and less apparent than the effects of other pollutants, such as smog or the sight of dead trees caused by acid rain.

Pollution of underground water supplies can confront an engineer with ethical considerations similar to those related to the acid rain problem discussed above. Like acid rain, water pollution is a complex problem resulting from many types of pollutants emanating from a wide range of sources. Water-contaminating chemicals, which can cause cancer and other less traumatic illnesses, vary from pesticides originating in agricultural areas to volatile organic compounds that often are used in high-tech industries. Also, like acid rain, these contaminants can affect wide areas many miles distant from the pollutant source. The resources often affected are huge underground water reservoirs, called *aquifers,* that underlie much of the country. These aquifers now provide water for half of the nation, including more than 80 percent of the rural population which live in areas previously considered free from possible pollution sources.

Similar to the problems associated with acid rain, laws and regulations alone have proven to be inadequate for solving the problem. There is a need for action by responsible and knowledgeable people at key points in the decision making process. For example, a federal law[12] passed in 1976 aimed at controlling toxic wastes requires operators of the thousands of waste facilities around the country to monitor leakage from their installations. The law allows for issuance of an interim permit that will be valid until a monitoring system can be established. However, the final permit that would be issued after monitoring is established would require more stringent and usually more expensive anticontamination efforts. Therefore, many operators postpone establishing monitoring equipment until forced to do so by a court order. The U.S. Government Environmental Protection Agency has delegated responsibility for administering and enforcing this program to the state agencies, which in many cases have been unable or unwilling to force the operators to obey the law.

In addition to making decisions concerning the disposal of wastes that cause water pollution, engineers can play a key role in making decisions at the point where the wastes are generated. Organizations generating the waste know more than anyone else about their wastes and the processes that generate it. Therefore, they can be most effective in investigating possibilities for reducing the toxicity and quantity of their wastes. One way to begin is by investigating possibilities for in-plant process changes. An example of a possible process change is in the steel-finishing industry, where before steel products are painted, they must be treated to remove rust. The most commonly used method is to dip the steel in an acid bath. This generates large amounts of waste acid. Most of the rust, however, could be removed mechanically, thereby reducing the volume of acid required.[13] Another area for investigation at the source could be the possibility of recycling or reuse of materials to reduce wastes.

The solution to a coming water crisis may be more elusive and expensive than the energy crisis.
Representative Mike Synar, Oklahama, Chairman of House Government Operations Environment Subcommittee
Quoted in Arizona Republic, *October 26, 1984*

[6-16] In the past, municipal water supplies were subject to contamination because of improper disposition of animal wastes . . .

[6-17] . . . today the culprit is more likely to be improperly discarded manmade chemicals.

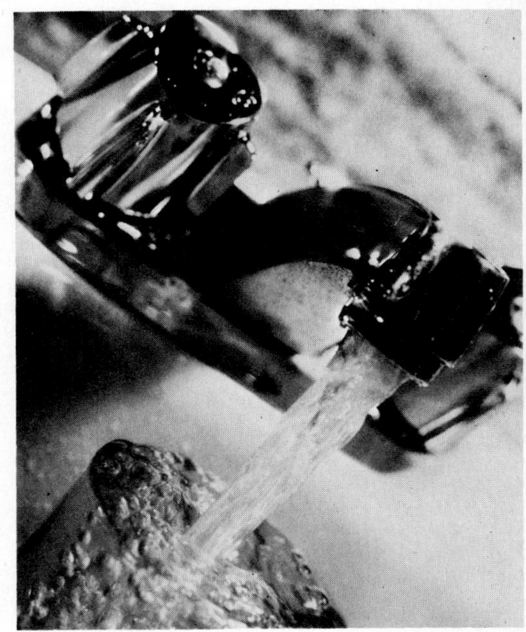

[12]Toxic Substances Control Act (15 USC 2601-2629), October 11, 1976.

[13]David Anderson and Beth Fentrup, "How Should We Dispose of Hazardous Wastes?" *Civil Engineering,* April 1984, p. 42.

Contamination of underground water supplies is bad and getting worse because state and federal laws and programs do not provide sufficient protection.

Congressional Office of Technology Assessment Quoted in Arizona Republic, *October 26, 1984*

Another decision area for engineers is in the selection of the most effective disposal method for the remaining waste. Such a decision may often involve research and special technical investigation, and in general keeping current on new techniques and procedures. It may also involve resisting pressure to take the path of least resistance or of lowest cost instead of the ethical path.

WATER-TAINT MONITORING 'HAS FAILED'
TOXIC-DUMP CHECKS FLAWED, EPA ADMITS

By PHILIP SHABECOFF

New York Times

WASHINGTON—The key government program to monitor toxic waste contamination of underground water supplies is not working, according to a report drafted by Environmental Protection Agency officials.

The report says that a large majority of site operators are not doing the job as required, that many of the states have proved unable or unwilling to make the operators obey the law and that the agency itself had been deficient in overseeing the states and in ensuring that standards for protecting the water supplies are met.

Arizona Republic, *October 16, 1984*

CONTROLLING OCCUPATIONAL HEALTH AND SAFETY HAZARDS

A demanding challenge for engineers today is helping society control the effects of toxic agents on people working in various industrial environments. The effects of industrial toxic agents maybe more apparent and intense than toxic agents in the general environment because workers may be exposed to higher concentrations over long and continuous periods. Unless controlled, this continuous exposure could result in occupational illnesses ranging from skin disorders to possibly cancer, depending on the extent of exposure to the toxic agent. The number of agents that could harm people working in various industrial environments is large and growing. The National Institute for Occupational Safety and Health has developed a registry that lists over 60,000 toxic substances which are manufactured or used in some workplaces. Within this group, they have identified approximately 1500 suspected cancer-causing agents.[14]

Increased awareness of the number of toxic agents and their effects has resulted in an increasing demand for protection. Since 1965, the U.S. government has enacted over 30 pieces of legislation aimed at controlling the use of toxic substances [6-18]. Similar legislation has been introduced in other industrialized countries.[15]

The cheapest way of disposing of waste is to dig a pit and throw the waste in it. The rationale for such facilities is that dilution is the solution to pollution. You mix a little waste with a lot of clean air, rainwater or groundwater and your waste problems are solved. But dilution only gives the illusion of being a solution.

David Anderson and Beth Fentrup, "How Should We Dispose of Hazardous Wastes?" Civil Engineering, *April 1984, p. 43*

[14]Mary Jane Bolle, *Effectiveness of the Occupational Safety and Health Act: Data and Measurement Problems* (Washington, D.C.: Congressional Research Service, Library of Congress, 1984).

[15]Dennis J. Paustenback, "Occupational Safety—Are You Professional?" *Mechanical Engineering,* March 1984, p. 80.

What Would You Do?

Assume that you are an engineer working for a high-tech company that makes extensive use of a trichloroethylene (TCE), an industrial solvent that causes cancer and affects the liver, kidneys, and nervous system. Your company has established a storage site and you are responsible for designing a monitoring system for detecting possible leakage. You submitted your design to your supervisor for approval two months ago, but to date have not received any indication that the corrective design solution has been reviewed. Two weeks ago when you questioned your supervisor about the status, he said it was still in the "approval cycle." What will you do now? (If you decide to wait another two weeks and again find that no decision has been made, then what will you do?)

By enacting this legislation, society is saying that the health and safety aspects of any project should concern the engineer as much as its efficiency, timeliness, overall quality, or cost. To implement the legislation, industrial management looks to engineering to recommend the most cost-effective alternative for meeting the requirements for minimizing health risks.

Typical recommendations could involve the substitution of a less hazardous substance, changing a processing technique, selection of a control technique, or selection of various types of personal protection equipment. Engineers would also be involved in the creation of safety and monitoring procedures.

A major difficulty in controlling occupational health hazards is the uncertainty of information concerning both the hazards and the control techniques. Most occupational illnesses are clinically indistinguishable from general chronic-type diseases, and diseases such as cancer may develop long after the worker leaves the hazardous work environment. Also, because most of the data concerning effects of toxic substances is based primarily on animal tests, it may not be applicable for controlling human diseases. This uncertainty concerning hazards results in an uncertainty concerning the relative effects of various control techniques and procedures.

Today's engineers, however, cannot hide behind this possible lack of reliable information. Design engineers must know occupational and environmental exposure limits and how to design their equipment to meet or exceed these limits. This requires also that the engineer work closely with health and other engineering personnel and to be familiar with current technical literature on the subject.

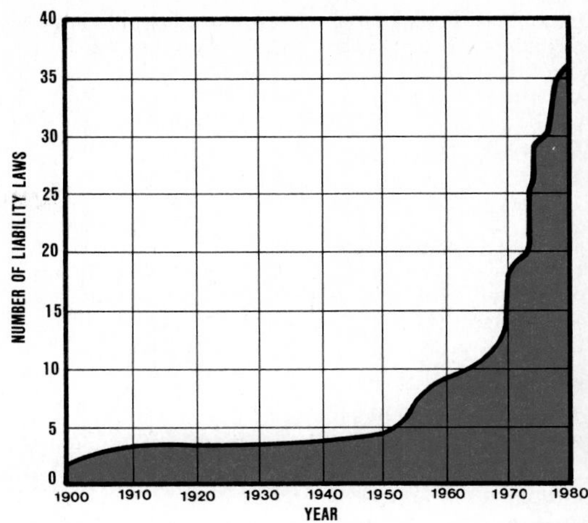

[6-18] The increase in the number of federal laws regulating hazardous materials in the United States over the past 20 years has been dramatic.

The science of preventive medicine involves a study of its causes and prevention. The knowledge and application of preventive measures takes us into regions that are engineering rather than medical.

E. B. Phelps et al.

Public Health Engineering, *Vol. 1, (New York: Wiley, 1984)*

What Would You Do?

Assume that you are an engineer working for the Environmental Protection Agency and are responsible for cleaning out a site that has been cited as a major pollution source in the immediate area. Cleaning the site requires moving contaminated materials to a new site in an adjoining state. People living in the area around the new site are registering major objections to this move and local authorities there claim they have test data which show that the new site is already a pollution problem. Authorities in your agency claim that the new site is now safe and that the added contaminants will not create a problem. What will you do?

- Assume the information from your agency is correct and proceed with the cleanup as planned?
- Obtain information available from the new site to determine if moving new material into the site will create new problems or add to an already existing problem?

If you do investigate and determine that moving contaminants have the potential of creating additional pollution problems at the new site, but you also calculate that the overall potential dangers would be much less at the new site than those existing at the site being cleaned, what action would you take?

Caveat venditor (let the seller beware) is replacing the old adage *caveat emptor* (let the buyer beware).

Kenneth E. Goodpaster,
Ethics in Management,
(Boston: Harvard Business School, 1984)

Today, it is reasonable to expect that if a manufacturer sells his product, he will be strictly liable in almost every circumstance should a user be injured and elect to sue to recover damages.

Verne L. Roberts
1984 Institute for Product Safety, Durham, N.C.

Intentional Misuse???

In 1979, a 14-year-old boy died after intentionally inhaling the freon propellant from a can of Pam (an aerosol sprayed on cooking pots to prevent sticking) in order to experience a tingling sensation in the lungs. The can carried a warning: "Avoid direct inhalation of concentrated vapors. Keep out of reach of children."

The boy's mother in suing the manufacturer of Pam, maintained that this warning was inadequate, since the company knew that 45 other teenagers had previously died from inhaling the fumes.

The jury, ignoring the fact that the product had been deliberately misused, awarded her $585,000.

"At issue: Product Liability," Shell News,
Vol. 52, No. 4 (1984)

Two 19th century reports from the Caribbean describe the ready availability of an axe in every sugar mill to amputate a slave's arm should it be caught in the in-running nip point of the rollers used to crush the sugar cane.

C. G. A. Oldendorp, Der Geschicht der Mission der Evangelische Bruder von Inseln S. Thomas, S. Croix, *S. Jan, Barby Johann Jacob Bossart, 2 vols., 1777*

At the heart of this problem is a changing focus on the liability statutes. At one time, the issue was one of moral responsibility—had the maker or seller of a product been negligent in its construction or sale? Now the changing focus has turned the spotlight on the product itself—is there a condition associated with the product that creates an unnecessary hazard or danger? This switch in emphasis led to the concept of "strict liability," which allows courts and juries to assign liability even though there has been no *fault* or lapse of moral responsibility. Liability can now result if the court deems that a product's design, construction, operating instructions or safety warnings make its use unreasonably hazardous.

Shell News, *Vol. 52, No. 4 (1984), p. 18*

Society's Demand for Safety

Engineers today work within a society which expects technology to provide all that is needed for an improved life-style and at the same time demands a high degree of protection to accompany this luxury. This demand for safety in everything from consumer products to airplane travel takes the form of regulations, codes, and laws aimed at controlling all of the design, production, construction, and service functions required to provide the good life.

The emphasis on product safety has increased rapidly during the past 30 years. Consumerism, a social movement emphasizing the rights of buyers and users, has resulted in the passage of a Consumer Product Safety Act which holds the manufacturer liable if the product contains a flaw (arising from production or design) that makes it unreasonably dangerous. Working with this legislation, courts have moved toward strict liability, holding manufacturers responsible for any product defect that results in an injury.

Concurrently with increasing emphasis on safety for consumer products, there is also increasing emphasis with regard to safety for products and machinery used in an industrial setting. Pressure for improved machinery design and safety procedures has come from various sources, including the worker, safety and health groups, and employers who recognize that a safe working environment has many benefits, including improved production. This recognition of the value of safety is a major change from the approach used in early industrial plants, where workers were considered expendable and the major objective was to make certain that injuries did not interfere with production.

Machine safety design requirements are based on codes and regulations that emphasize identification and elimination of hazards rather than specifying safety requirements for the use of each specific machine. The codes emphasize designing to make certain that machine operators cannot proceed with their work unless the required safety conditions prevail [6-19].

Another area of safety emphasis is public transportation, especially automobiles and airplanes. Consumer advocate organizations, such as the privately funded Center For Auto Safety, have been very active in pressuring for required safety legislation and monitoring performance. The major government agency involved in regulating automobile safety is the National Highway Traffic Safety Administration (NHTSA) of the Department of Transportation. The major emphasis in airplane safety today is for all accidents to be "survivable." Airplane accidents are termed survivable if the fuselage is not severely damaged on impact or if there is at least one survivor or if exit from the plane is possible. Because most of the fatalities with this type of accident are fire related, safety research is being aimed at techniques or materials which could prevent fires or reduce their effects. This involves fuel additives, protective face hoods, interior materials, and seat construction. It is a challenging area with many performance, cost, and safety trade-offs.

Meeting the requirements for safety in areas such as consumer products, industrial machinery, transportation, and construction continues to be a challenge, even with the existing increased empha-

TABLE 6-2 INJURY RATES BY INDUSTRY FOR 1982 (RATE PER 100 FULL-TIME WORKERS)

Construction	14.5
Agriculture	11.3
Manufacturing	9.9
Transportation	8.4
Trade (wholesale and retail)	7.1
Services	4.8

Source: Mary Jane Bolle, *Effectiveness of the Occupational Safety and Health Act: Data and Measurement Problems* (Washington, D.C.: Library of Congress, 1984).

[6-19] The safety of the workplace must be a concern for the engineer.

sis and regulations. During 1983, according to the National Transportation Safety Board, 43,000 people were killed on U.S. highways. Government statistics indicate that each year approximately 30,000 Americans are killed as a result of accidents related to faulty consumer products,[16] and injuries continue at unsatisfactory rates in most areas of industry (Table 6-2).

Safety problems will continue and could even become more severe as technology becomes more complex and the increase in population results in added demands for products and services.

Society expects engineers to play a major role in meeting the requirements for safety. Along with an ethical responsibility, meeting safety requirements is also a serious legal responsibility. As suggested above, safety legislation and governmental regulations are tending to hold engineers legally liable for the consequences of their designs. Any design flaw that contributes to an injury could lead to legal claims against the engineers involved with designing or developing the product [6-20].

DEVELOPING AN ENGINEERING RESPONSE TO THE DEMAND FOR SAFETY

Use Risk Analysis to Select the Best Alternative

To provide a safe product or project, engineering teams must use a systematic risk analysis procedure which starts by identifying all potential hazards [6-10]. This requires determining where the product will be used, how it will be used, and who will use it. The identified hazards can then be used along with applicable standards, regulations, and codes to determine the basic safety requirements. These are weighed against other factors, such as cost, schedule, and technical feasibility, to determine the most feasible alternative. How this selection was made is a critical consideration in product liability cases. If there is a safety problem, the outcome of any lawsuit could depend on whether another safer alternative was available to the designers.

[6-20] Along with the ethical responsibility, meeting safety requirements is also a serious legal responsibility.

[16]Kenneth E. Goodpaster, *Ethics in Management* (Boston: Harvard Business School, 1984), p. 107.

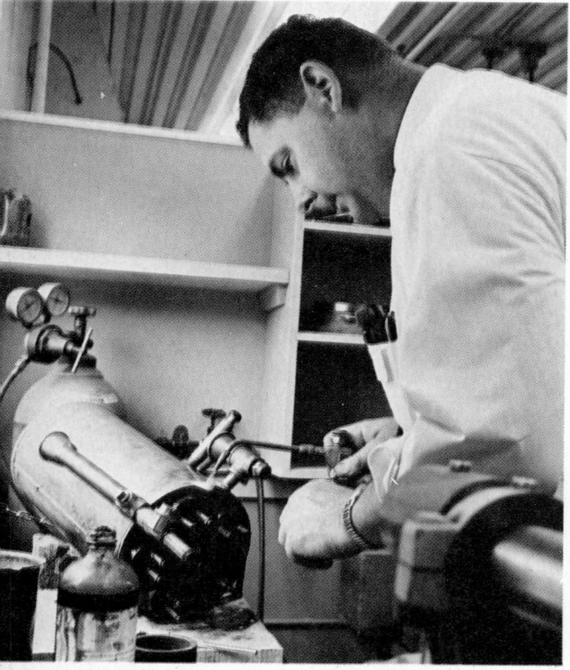

[6-21] The worker's eyes and hands should be protected.

[6-22] Protective eye shields and protective clothing can do much to reduce injury.

Example 6-1

An example of how the selection of a design alternative affects the safety record of a product is the Pinto, a subcompact automobile manufactured by Ford for six years starting in 1970. Pintos were sold in spite of recognized safety problems which ultimately resulted in a costly recall program.

An early design decision concerned the selection of the location for the gas tank to achieve maximum safety. The main hazard to be considered was the possibility of fires caused by rear-end collisions of the vehicle. The design team selected a location under the rear floor and behind the rear axle, which was the normal location in most other automobiles at that time. A location above the rear axle was considered but not selected, because it would increase the threat of fire in the passenger compartment. Also, it required installation of a filler pipe which was more apt to be damaged during a collision. In addition, an over-the-axle location would not be suitable for station wagons and hatchback models which were to be considered for production after the Pinto two-door sedan (the first model to be designed) was in production.

A major problem in selecting this location and designing for maximum fuel-system safety was the lack of adequate and consistent safety standards concerning rear-end collisions. Ford originally designed the tank in accordance with an internal company standard, referred to as "a 20-mph fixed barrier regulation." The National Highway Traffic Safety Administration (NHTSA) proposed an alternate rear-end fuel system integrity standard approximately 18 months after the Pinto design program had been started. Although not strictly required to do so, Ford tested against this standard and altered the fuel tank design before production began. The NHTSA standard was not fully implemented until six years after the first Pinto was produced.

The final enactment of the NHTSA standard resulted from pressure by consumer groups and Congress who were becoming increasingly concerned about safety problems with the Pinto. Rear-end collisions were causing fires which resulted in serious injuries and, in several cases, fatalities. In the six years from proposing the new standard to enacting it, NHTSA was hampered by the lack of relevant and meaningful statistical information which could be used to specifically identify hazards and serious consequential results.

After years of defending the safety of the Pinto, Ford finally conceded that NHTSA had identified specific design features which needed to be changed to reduce the risk of gas leakage caused by rear-end collisions. They agreed to recall over 1.5 million automobiles for replacement of the fuel filler pipe and installation of a shield across the front of the fuel tank.

Ford's estimate for the cost of this recall program was over $20 million. Table 6-3 and Figure [6-23] summarize this example.

Remember the User

When designing for safety, engineers should be aware of errors the user can make as well as errors the designers can make. In addition to being an ethical responsibility, preventing user errors or reducing their effects is becoming a very serious legal requirement.

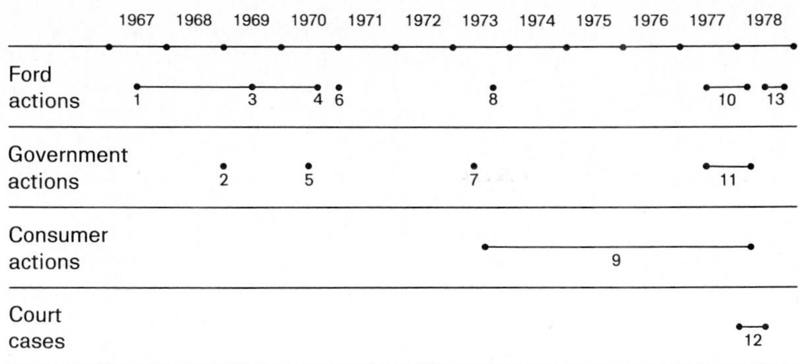

[6-23] Story of the Ford Pinto.

Procedural Problems: Rolls-Royce, the luxury car manufacturer, was forced to recall 2000 Silver Shadows in 1978 because one owner reported the brakes had failed.

The *Almanac of Investments* reports the company claimed its autos never broke down, they merely "failed to proceed."

It is an easy task to formulate a plan of accident preventing devices after the harm is done, but the wise engineer foresees the possible dangers ahead and embodies all the necessary means of safety in the original design.

J. H. Cooper, "Accident-Preventive Devices Applied to Machines," ASME Transactions, *Vol. 12 (1891), p. 249*

TABLE 6-3 STORY OF THE FORD PINTO

1. *June 1967:* Ford starts design and development of the Pinto.
   - Assembles a special team of engineers dedicated to development of a Pinto model within 40 months.
   - Establishes management design constraints: must not weigh an ounce over 2000 pounds; not cost a cent over $2000.
   - Makes key decision after trade-off studies: locate gas tank under rear floor, behind the rear axle. (No government standards on gas tank design existed during the early design stage.)
2. *January 1969:* A government agency (National Highway Traffic Safety Administration—NHTSA) proposes the first rear-end integrity standard.
   - Requires that a stationary vehicle should leak less than 1 ounce of fuel per minute after being hit by a 4000-pound barrier moving at 20 mph (called the 20-mph moving barrier).
3. *June 1969:* Ford tests and alters the Pinto design to meet the proposed government standard.
4. *August 1970:* First Pinto rolls off the assembly line and sales begin.
   - The vehicle meets the schedule, weight, and cost goals.
5. *August 1970 (after production of the first Pinto):* NHTSA proposes new requirements.
   - Calls for changes from the existing 20-mph moving-barrier standard to a more severe fixed-barrier standard.
   - Indicates a possibility of a long-term requirement for a 30-mph fixed-barrier standard.
6. *August 1970–January 1971:* Ford considers whether to test vehicles against the existing or the proposed governmental standard.
   - Decides to continue designing to meet the original 20-mph moving-barrier standard.
   - Decides against making gas tank modifications to meet a possible future 30-mph fixed-barrier standard.
7. *August 1973:* NHTSA proposes a new 30-mph moving-barrier standard effective 1976 for all 1977 models.
   - Also adopts a fuel system standard applicable to rollover accidents.
8. *Late 1973:* Ford agrees to modify rollover standards after testing and making controversial cost/benefit claims.
9. *Late 1973–May 1978:* Consumer actions place pressure on NHSTA to force Pinto recall actions.
10. *September 1977–May 1978:* Ford publicly refutes consumer claims and defends safety record.
11. *September 1977–May 1978:* NHTSA investigates Pinto fuel tank systems and determines that safety problems require recall actions.
12. *February 1978–March 1978:* Series of court actions focuses attention on Pinto safety problems.
    - *February 1978:* California jury assesses $125 million in punitive damages in a case involving rupture and explosion of the fuel tank on a 1972 Pinto.
    - *March 1978:* Pinto owners in Alabama and California file class action suits demanding that Ford recall all Pintos built from 1971 through 1976 and requiring modification of their fuel system.
13. *March 1978–June 1978:* Ford initiates a recall and modification program.
    - *March 1978:* Recalls 300,000 1976 Pintos.
    - *June 1978:* Recalls 1.5 million Pintos.
    - Agrees to replace fuel filler pipe and install a polyethylene shield across the front of the fuel tank. Total estimated cost for modification program $20 million after taxes.

Source: Adapted from Kenneth E. Goodpaster, *Ethics in Management* (Boston: Harvard Business School, 1984).

In 1974, a Pennsauken, New Jersey, police officer was severely injured when his patrol car—a Dodge Monaco—spun off a rain-slicked highway and slammed sideways into a steel pole 15 inches in diameter. The policeman sued the Chrysler Corporation, contending that the car's design was unsafe because it did not have a rigid steel body.

Chrysler argued that the car's flexible body design provided maximum passenger protection in front- or rear-end collisions, by far the most numerous types of accidents, and that rigid construction would add enough weight to appreciably reduce fuel efficiency and increase operating costs. Therefore, to meet federal regulations for both fuel economy standards and front-end collision survivability, Chrysler maintained, the design was optimal.

The jury awarded the plaintiff $2 million.

"At Issue: Product Liability," Shell News, *Vol. 52, No. 4 (1984)*

[6-24] The DC-10 jet aircraft has experienced some design defects, particularly with regard to the operation of the cargo door.

[6-25] Passenger-restraint systems were tested recently in a full-scale crash test of a Boeing 720 aircraft. In the 20-year period since the jet age began in 1959, 933 jetliner accidents occurred worldwide. Less than 30 percent of the 12,668 passengers aboard were killed. For comparison, in 1983 more than 43,000 people were killed on American highways.

User experience with cordless telephones, a new consumer convenience product, illustrates the need to be aware of safety problems that can occur when possible user errors are not adequately considered. For convenience and portability, the "ringer" in most cordless telephones is located in the earpiece, where it must be loud enough to attract attention to incoming calls. In most cases a "standby/talk" switch is provided to allow the user to disable the "ringer" when making a call. There is the possibility, however, of the user holding the telephone to an ear without switching to "talk." In this case, the ring of an incoming call can cause damage to nerve endings in the ear.

The Federal Consumer Product Commission became aware of this problem and ran tests which they said showed that sound levels produced by the ringer present the potential for unacceptable levels of injury as well as the probability of occurrence resulting in an unacceptable risk. They issued a consumer warning to remind users to switch from "standby" to "talk" when placing a call.[17]

A possible outcome from this safety problem could be lawsuits on a national level because of the extensive use of cordless telephones. After these safety problems became apparent, manufacturers of cordless telephones placed warnings directly on the phones or provided instruction sheets to accompany them.

There is No Such Thing as an Insignificant Detail

Engineers working on teams designing large complex products or structures need to remember that overall safety is determined by the safety of each individual component. The selection and application of each component should be the result of a risk analysis made to identify possible hazards and to select the best alternatives for eliminating or minimizing these hazards. In most cases this requires a thorough analysis to determine how the components interact and how user operation of one component can affect overall safety.

Example 6-2

An airplane is an example of a complex mechanism where the design of what might otherwise be considered to be an insignificant item could affect overall safety. The failure of one switch, a break in an oil line, or a ruptured tire could result in a serious accident. On March 3, 1974, a Turkish Airline DC-10 jet aircraft crashed after taking off from the Paris, France, airport, killing 346 people [6-24]. Safety investigations made after the crash determined that the accident was caused by a faulty locking mechanism on a cargo door. At an altitude of approximately 10,000 feet, the cargo door burst open, creating a pressure differential within the aircraft which caused the floor of the passenger compartment to collapse. Hydraulic control lines running under the floor panel to the rear control surfaces were crushed, causing the aircraft to be uncontrollable.

There were several interrelated safety factors involved in this accident. The DC-10 cargo doors open outward and are closed be-

[17]Chuck Hawley, "Harmful Earful," *Arizona Republic,* October 15, 1984.

fore takeoff with a latching mechanism that is operated from the outside by the baggage handler. In the design used on the crashed aircraft, proper operation of the latch was very dependent on adjustments made during maintenance. Also, it was possible for the baggage handler and flight crew to be convinced that the door was properly latched when in actuality certain of the latching pins were not fully engaged. Other critical factors included strength of the floor panels and the location and construction of the control lines. All of these conditions were considered in the design changes that have subsequently been implemented by the manufacturer to prevent similar accidents from occurring.

[6-26] and [6-27] A Boeing 720 aircraft skids across the dry lake bed (December 1, 1984) after being guided to a remote-controlled crash. The purpose was to test the effectiveness of a new, specially treated kerosene fire-retardant fuel as well as improved aircraft safety features.

Make Certain That Design Changes Do Not Reduce the Safety of the Original Design

Engineers working on design teams often need to make design changes that are required to meet new design requirements, incorporate new technology, or to reduce costs. After the product has been in use, field experiences related to performance or safety (such as the DC-10 accident) may indicate that changes in the design are required. These changes must be made with the same care and control used to complete the original design.

Example 6-3

An example of how an improperly implemented change can affect safety is the catastrophe that occurred in Kansas City, Missouri, on July 17, 1981. A Hyatt Regency Hotel fourth-floor skywalk collapsed and caused a second-story skywalk to fall with it to the hotel lobby, killing 114 persons and injuring more than 200 others [6-28].

City records, visual examinations by experts, and photographic evidence indicate that sometime during the construction process a design change was made that altered the weight distribution on the skywalk supporting structures and doubled the resulting stress on beams supporting the fourth-floor skywalk[18] [6-29].

This accident points out that when changes are made, all safety factors need to be considered. Also, it is essential to check to make certain that all details of the design are properly implemented.

Maintain Design Integrity

Designing for safety must be more than just good intentions. A project team dedicated to meeting safety requirements must carefully implement design decisions through the assembly or construction processes. Inspections and tests are completed at critical stages and any detected discrepancies are corrected. Changes in construction procedures or in the design are made whenever test results or field experiences show that a basic problem exists. This type of program requires a management/engineering team that is dedicated to product integrity.

You can go a long way towards checking structural viability when you review the plans. Basically the success of determining the structural integrity is in the plan.

William Bullard, Planning Director, Independence, Missouri, Kansas City Star, *July 21, 1981*

[18] Rick Alm and Thomas G. Watts, "Critical Design Change Is Linked to Collapse of Hyatt's Skywalk," *Kansas City Star,* July 21, 1981.

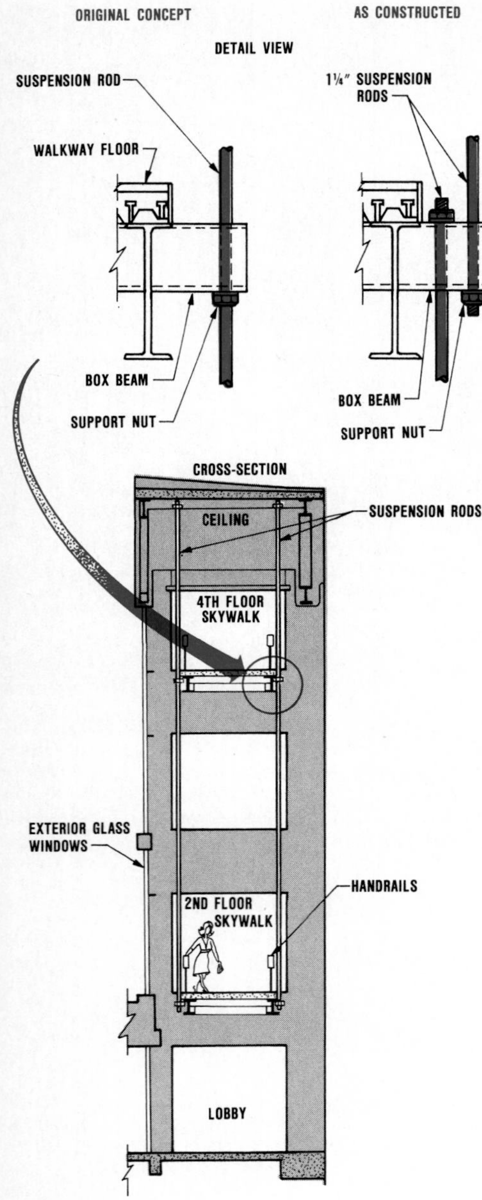

[6-28] The Kansas City Hyatt Regency skywalk.

Improper implementation of the steps required for product integrity can lead to serious contract problems, economic sanctions, and lawsuits. National Semiconductor Corporation, a major supplier of microchips to the U.S. government, paid over $1.7 million in civil and criminal penalties in March 1984. The firm had been indicted on charges which claimed that microchips sold to the government between 1978 and 1981 had been inadequately tested. The indictment included criminal charges of mail fraud and making false statements regarding the testing program. The government originally proposed to ban National Semiconductor from doing any more business with the Defense Department. However, this serious economic threat was dropped after the firm took several corrective actions: creation of an independent quality auditing group, reassignment of the several managers who were responsible for quality control during the problem years, institution of a company policy requiring dismissal of any employee who fails to follow government regulations in the future, and the creation of an internal company hotline which allows any employee to report anonymously any wrong doing.[19]

POSSIBLE SOURCES OF SUPPORT FOR ENGINEERS

Organizational Support

Engineers need not believe that they stand alone as they make critical ethical decisions related to their job responsibilities. Because society emphasizes the need for correct decisions and recognizes the difficulty of decision making, there are many sources of support for the engineer. The first of these is the company organization for which the engineer works. In addition to ethical considerations, the possi-

[19] *Electronic Buyers News,* August 13, 1984.

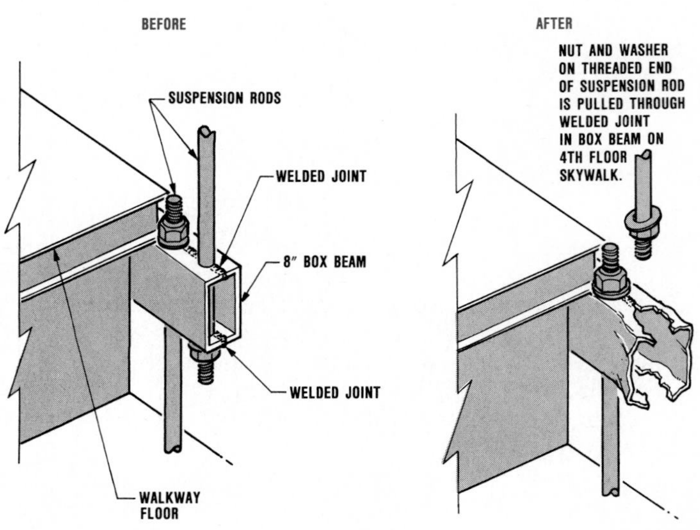

[6-29] The before failure and after failure drawings show the design change that allowed the entire walkway to collapse. The original design called for the walkway to be supported by continuous rods [6-28].

bility of costly lawsuits and damage to reputation make today's business firms especially interested in making correct decisions.

In most cases disagreements with company management over ethical approaches are settled in an amicable manner since all concerned want to find the best solution to the problem. However, occasionally a case involving several issues, priorities, and conflicting values may result in a dispute between an engineer and project supervision. In this case, the first choice for the engineer is to try to solve the dispute using established organizational channels and procedures for resolving conflicts.

If an impasse develops with organizational management, the engineer is faced with choosing between three alternatives: give up, appeal within the organization, or report the problem to persons outside the organization for possible resolution of the problem. The easiest alternative is the first one—total surrender. Before taking this easy way out, an engineer needs to ask the question, "Can I live with myself if I stand by and let something go on that could result in serious injury to others or perhaps have other serious consequences?" The last alternative, called "whistle blowing," is a last-resort alternative because of the obvious negative effects it will have on present working conditions and on your future career possibilities within the company. You may wish to seek outside counsel. Before resorting to outside help, however, you should carefully review the validity of the position you have taken and the importance of the issue. Writing a position statement can help to focus your thinking on the facts. In any case, take the position that is ethically proper.

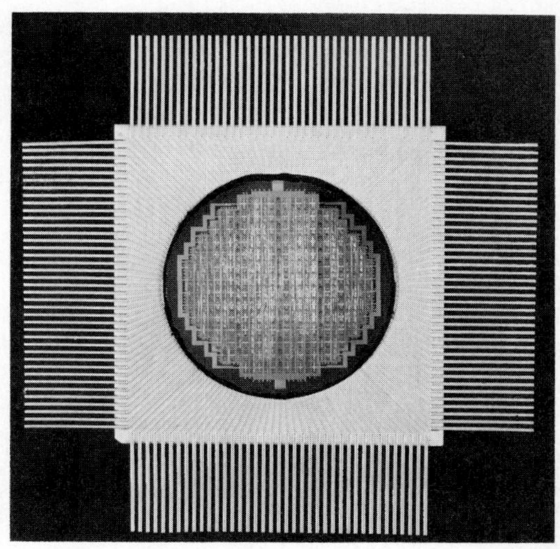

[6-30] Valid testing of semiconductor components has become a concern of the government. If a critical chip fails in a satellite or undersea detection device, the security of the United States might be at risk.

What to Do in Case of an Impasse

1. Prepare a position statement which focuses on the facts and considers all facets of the issue.
2. Use all available appeal procedures within the organization.
3. Seek help and advice from professional friends, colleagues, and technical societies.
4. Go outside if the conflict has significant ethical or safety implications and cannot be resolved inside the organization.

Technical Societies

Engineers and scientists have organized a number of technical societies in various fields of specialization (Table 6-4). The first societies were originally formed over 100 years ago to allow engineers to band together to exchange ideas, improve their technical knowledge, initiate design regulation, and to learn new skills and techniques

What Would You Do?

Assume that you are an engineer in charge of the design of a microwave oven that has been designed for general consumer use. Quality control tests in your organization show that radiation leakage from the ovens is in excess of U.S. government standards. Although your research shows that the level of leakage from your ovens is substantially below the true hazard level identified by health professionals, you have made a design change to make certain that all units now in production and all future units will meet the government standard. However, 10,000 units which could violate the government standard are now in the stores for Christmas sales. Should you recommend that your firm initiate a recall program, keep quiet, or take some other action?

TABLE 6-4 ENGINEERING AND TECHNICAL SOCIETIES

Code	*Name*	*Address*	*Year organized*	*Total membership*
AcSoc	Acoustical Society of America	335 East 45th St. New York, NY 10017	1929	5,750
APCA	Air Pollution Control Association	P.O. Box 2861 Pittsburgh, PA 15230	1907	7,921
AAAS	American Association for the Advancement of Science	1776 Massachusetts Ave., N.W. Washington, DC 20036	1848	139,000
AAEE	American Academy of Environmental Engineers	P.O. Box 269 Annapolis, MD 21404	1955	2,400
AAES	American Association of Engineering Societies	345 East 47th St. New York, NY 10017	1979	n.a.
AACE	American Association of Cost Engineers	308 Monongahela Bldg. Morgantown, WV 26505	1956	6,100
AAPM	American Association of Physicists in Medicine	335 East 45th St. New York, NY 10017	1958	2,200
ACI	American Concrete Institute	22400 West 7 Mile Road Detroit, MI 48219	1906	14,716
ACM	Association for Computing Machinery	11 West 42nd St., 3rd Floor New York, NY 10036	1947	58,000
ACS	American Ceramic Society, Inc.	65 Ceramic Dr. Columbus, OH 43214	1898	8,152
ACS	American Chemical Society	1155 16th St., N.W. Washington, DC 20036	1876	131,764
AES	Audio Engineering Society, Inc.	60 East 42nd St. New York, NY 10165	1948	9,600
AIAA	American Institute of Aeronautics and Astronautics	1633 Broadway New York, NY 10019	1932	35,448
AIA	American Institute of Architects	1735 New York Ave., N.W. Washington, DC 20006	1857	42,132
AIChE	American Institute of Chemical Engineers	345 East 47th St. New York, NY 10017	1908	62,000
AICE	American Institute of Consulting Engineers	345 East 47th St. New York, NY 10017	1910	420
AIME	American Institute of Mining, Metallurgical, and Petroleum Engineers, Inc.	345 East 47th St. New York, NY 10017	1871	99,734
AIP	American Institute of Physics	335 East 45th St. New York, NY 10017	1931	61,000
AIPE	American Institute of Plant Engineers	3975 Erie Ave. Cincinnati, OH 45208	1954	8,675
AMS	American Mathematical Society	201 Charles St. Providence, RI 02940	1888	20,392
ANS	American Nuclear Society	555 N. Kensington Ave. LaGrange Park, IL 60525	1954	14,650
APS	American Physical Society	335 East 45th St. New York, NY 10017	1899	32,781

continues next page

TABLE 6-4 ENGINEERING AND TECHNICAL SOCIETIES (CONTINUED)

Code	*Name*	*Address*	*Year organized*	*Total membership*
APHA	American Public Health Association	1015 15th St., N.W. Washington, DC 20005	1872	28,268
ASAE	American Society of Agricultural Engineers	2950 Niles Ave. St. Joseph, MI 49085	1907	11,300
ASCE	American Society of Civil Engineers	345 East 47th St. New York, NY 10017	1852	92,747
ASEE	American Society For Engineering Education	Eleven Dupont Circle Washington, DC 20036	1893	10,060
ASEM	American Society for Engineering Management	301 Harris Hall, University of Missouri, Rolla, MO 65401	1979	1,011
ASM	American Society for Metals	Metals Park, OH 44073	1913	45,104
ASQC	American Society for Quality Control	230 W. Wells St. Milwaukee, WI 53203	1946	43,345
ASHRAE	American Society of Heating, Refrigerating and Air-Conditioning Engineers, Inc.	345 East 47th St. New York, NY 10017	1894	49,000
ASLE	American Society of Lubrication Engineers	838 Busse Highway Park Ridge, IL 60068	1944	3,850
ASME	American Society of Mechanical Engineers	345 East 47th St. New York, NY 10017	1880	111,645
ASNE	American Society of Naval Engineers, Inc.	1452 Duke St. Alexandria, VA 22314	1888	6,800
ASSE	American Society of Safety Engineers	850 Busse Highway Park Ridge, IL 60068	1911	19,000
ASTM	American Society for Testing and Materials	1916 Race St. Philadelphia, PA 19103	1898	28,692
AWRA	American Water Resources Association	St. Anthony Falls Hydraulic Laboratory Miss. River at 3rd Ave. S.E. Minneapolis, MN 55414	1964	2,485
AWWA	American Water Works Association, Inc.	6666 W. Quincy Ave. Denver, CO 80235	1881	29,475
CEC	Consulting Engineers Council of the United States of America	1155 15th St., N.W. Washington, DC 20005	1959	2,300
IES	Illuminating Engineering Society of North America	345 East 47th St. New York, NY 10017	1906	7,665
IEEE	Institute of Electrical & Electronics Engineers, Inc.	345 East 47th St. New York, NY 10017	1884	240,068
IES	Institute of Environmental Sciences	940 East Northwest Highway Mt. Prospect, IL 60056	1956	2,122
IIE	Institute of Industrial Engineers	25 Technology Park/Atlanta Norcross, GA 30092	1948	39,000
ITE	Institute of Transportation Engineers	525 School Street, S.W. Suite 410 Washington, DC 20024	1930	6,911

continues next page

TABLE 6-4 ENGINEERING AND TECHNICAL SOCIETIES (CONTINUED)

Code	Name	Address	Year organized	Total membership
ISA	Instrument Society of America	67 Alexander Dr. P. O. Box 12277 Research Triangle Park, NC 27709	1945	36,000
NACE	National Association of Corrosion Engineers	P. O. Box 218340 Houston, TX 77218	1945	14,000
NAPE	National Association of Power Engineers, Inc.	176 West Adams St., Suite 1914 Chicago, IL 60603	1882	8,900
NICE	National Institute of Ceramic Engineers	65 Ceramic Dr. Columbus, OH 43214	1938	1,904
NSPE	National Society of Professional Engineers	2029 K Street, N.W. Washington, DC 20006	1934	79,370
ORSA	Operations Research Society of America	Mount Royal and Guilford Aves. Baltimore, MD 21202	1952	6,700
SESA	Society for Experimental Stress Analysis	14 Fairfield Drive Brookfield Center, CT 06805	1943	2,900
SIAM	Society for Industrial and Applied Mathematics	117 S. 17th St., 14th Floor Philadelphia, PA 19103	1952	6,000
SAE	Society of Automotive Engineers	400 Commonwealth Dr. Warrendale, PA 15096	1905	42,060
SES	Society of Engineering Science	c/o Dept of Engineering Science Virginia Tech, Blacksburg, VA 24061	1963	400
SME	Society of Manufacturing Engineers	P.O. Box 930 Dearborn, MI 48121	1932	71,714
SNAME	Society of Naval Architects and Marine Engineers	One World Trade Center Suite 1369 New York, NY 10048	1893	13,400
SPE-AIME	Society of Petroleum Engineers of AIME	6200 N. Central Expressway Drawer 64706 Dallas, TX 75206	1913	54,413
SPE	Society of Plastics Engineers, Inc.	14 Fairfield Dr., Brookfield Center, CT, 06805	1941	23,500
SWE	Society of Women Engineers	345 East 47th St. New York, NY 10017	1950	14,000

[6-31]. In addition, the National Society of Professional Engineers is concerned primarily with the legal and registration aspects of the entire field of engineering.

These technical societies can provide various degrees of support for engineers as they are faced with making critical ethical decisions in the profession. Advice from sympathetic, objective, and experienced colleagues can be very valuable. Some societies may provide direct support for engineers involved in organizational disputes resulting from ethical issues. This support may include an investigation of the circumstances surrounding the situation and preparation of a report which can be used to help resolve the conflict. In some cases legal support may be provided.[20]

Another major area of support provided by technical societies is the formation of codes of ethics. These provide guidelines for individual members by indicating how general principles of ethical behavior can be applied in specific circumstances, and establishing publicly approved standards of behavior. Present society codes place heavy emphasis on the responsibilities of professional integrity and give less attention to the rights of engineers or to the protection of engineers who may take job risks instead of standing firm for ethical principles.

[6-31] Participation in technical society activities are important in the professional development of the engineer.

ENGINEERING REGISTRATION

An important element for any professional is to gain respect and trust of the public. One way to do this is through state or national registration, which provides a license to practice. It is also a commitment to public service and to conducting a trustworthy career. Licensing implies that the professional will be guided by ethical codes and imposes standards of competence and integrity. Registration encourages continued efforts to remain up to date, thus upgrading the profession.

Registration could become even more important if legal liability and consumer protection groups require engineers to go through a licensing procedure which periodically documents their competence and good judgment.

Three things are to be looked to in a building: that it stand on the right spot; that it be securely founded; that it be successfully executed.
Johann Wolfgang von Goethe, 1749–1832
Elective Affinities

In an examination those who do not wish to know, ask questions of those who cannot tell.
Sir Walter Raleigh, 1552–1618
Some Thoughts on Examinations

The first step in obtaining professional engineering (PE) registration can be accomplished at or near the end of your senior year in college. This is passing the Engineer-In-Training (EIT) examination on engineering fundamentals. It includes problem solving from a number of basic subjects, beginning with mathematics, physics, and chemistry and extending to mechanics, engineering economics, thermodynamics, and electrical network theory. If you have mastered the principles of engineering economy in Chapter 8 of this text, for example, you should be able to pass that portion of the EIT examination. Other steps in the registration process include four or more years of pertinent professional work experience and passing a professional examination.

PROFESSIONAL RESPONSIBILITY AND THE ENGINEERING STUDENT

Engineers do not automatically become responsible professionals the day they receive their baccalaureate degrees, start work on their first professional assignment, join a professional society, or meet the minimum requirements for professional registration. Becoming a responsible professional engineer requires continuous acquisition of

[6-32] Fundamental tenets of professionalism are developed prior to graduation from college.

[20]Stephen H. Unger, *Controlling Technology: Ethics and the Responsible Engineer* (New York: Holt, Rinehart and Winston, 1982).

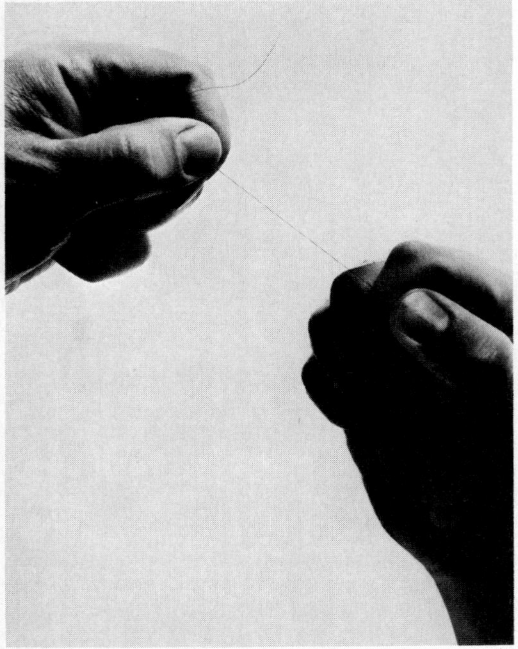

[6-33] Designing to the limits of the system is analogous to the stretching of a thread. On the one hand, the engineer strives for the most efficient and least expensive design . . . while on the other hand society insists on a tolerance to insure safety and minimum risk.

the technical knowledge and background required to help solve specific problems, and development of a professional attitude that is needed to work effectively with others in making critical ethical decisions.

Engineering students preparing to become responsible professionals in their chosen field need to emphasize development and application of a professional attitude concurrently with completing the course work required to earn their degree. This means developing and applying the same set of ethical values that they will someday apply as they work in industry as professionals. Engineering students need to complete course requirements with the same integrity, diligence, and persistence that will someday be required to meet on-the-job responsibilities. The same degree of honesty should be used in completing course assignments and in supplying answers to test questions. Begin to develop desirable work habits and personal discipline in completing project assignments. It is difficult to picture someone who cheats on examinations being totally trustworthy when required to make difficult on-the-job ethical decisions.

Preparing student research papers requires special attention to honesty. Students need to be certain that their papers represent their personal research effort and that the work supplied by others is properly documented. Developing a proper work ethic in preparing student papers will carry over after the student graduates.

Developing the ability to work with others is another important area in preparing to be a responsible professional. Working on team projects associated with some courses, engineering students have an opportunity to develop their ability to assume personal responsibility for their share of the team effort and to interact with others. This will involve developing communicating skills, sharing technical knowledge, and making certain that each team member gets his or her earned acknowledgment for work done. Participation in student technical organizations and extracurricular activities can also give you an opportunity to develop teamwork skills.

PROBLEMS

6-1. Discuss the similarities and dissimilarities of engineering and other professions. What do you think is the most significant common factor? Why is this so?

6-2. Discuss the factors that make engineering a unique profession. What are the reasons for this uniqueness?

6-3. How should these common and unique factors (see Problems 6-1 and 6-2) affect the approach you use as you prepare to enter the profession of engineering?

6-4. Interview three graduate engineers, each practicing in a different field of engineering. Obtain their answers to these questions:

- Why do you consider yourself a professional?
- What do you consider to be the major challenges you encounter as you practice your profession?
- What do you consider to be the most important factors in preparing to be a professional?

Prepare a summary of the results of the interviews. In what ways were the answers received from the three professionals similar? What major differences did you see? How do you explain these differences?

6-5. Discuss the reasons why a professional person such as a registered engineer, an attorney, or a physician will not bid competitively on the performance of a service.

6-6. In interviewing for permanent employment, a senior student in a California engineering school agreed to visit on two successive days a company in Chicago and a company in Detroit. Upon her return home both companies sent her checks to cover her expenses, including round-trip airfare. Discuss the appropriate actions that should have been taken by the student.

6-7. The majority of all engineering designs require some extension of the engineer's repertoire of scientific knowledge and analytical skills. How can the engineer determine whether or not this extension lies beyond the "areas of his or her competence"?

6-8. Assume that you are working in your first professional position for Sillwell Co., a firm that has developed an inexpensive household specialty that they hope will find a huge market among housewives. They want to package this product in 1-gallon and ½-gallon sizes. A number of container materials appear to be practical—glass, aluminum, treated paper, steel, and various types of plastics. As an engineer assigned to the manufacturing department, you have completed a container-disposal study which shows that the disposal cost for 1-gallon containers can vary by a factor of 3, depending on the weight of the container, whether it can be recycled, whether it is easy to incinerate, and whether it has good landfill characteristics.

The company's marketing specialist believes that the container material with the highest consumer appeal is the one to use, although it presents the most severe disposal problem. He states that the sales potential would be at least 10 percent less if the easiest-to-dispose-of container were used because it would be less attractive and distinctive.

The results of your study have been forwarded in a report to company management, but you are concerned that your study is not being properly considered and that management will follow the recommendations of the marketing expert. Do you think you have an ethical responsibility to present a stronger case before management and make specific recommendations as to which container should be used? If you think you should make a stronger case, how would you proceed? What action would you take if after considering your recommendation, the company implements the marketing expert's recommendation?

6-9. Jerry Williams is a chemical engineer working for a large diversified company on the east coast. For the past two years he has been a member—the only technically trained member—of a citizen's pollution-control group working in his city.

As a chemical engineer, Williams has been able to advise the group regarding what can reasonably be done about abating various kinds of pollution, and he has even helped some smaller companies to design and buy control equipment. (His own plant has air and water pollution under good control.) As a result of Williams's activity, he has built himself considerable prestige on the pollution-control committee.

Recently, some other committee members started a drive to pressure the city administration into banning the sale of phosphate-containing detergents. They have been impressed by reports in their newspapers and magazines on the harmfulness of phosphates.

Williams believes that banning phosphates would be misdirected effort. He tries to explain that although phosphates have been attacked in regard to the pollution of the Great Lakes, his city's sewage flows from the sewage-treatment plant directly into the ocean. And he feels that nobody has shown any detrimental effect of phosphate on the ocean. Also, he is aware that there are conflicting theories on the effect of phosphates, even on the Great Lakes (e.g., some theories put the blame on nitrogen or carbon rather than phosphates, and suggest that some phosphate substitutes may do more harm than the phosphates).

In addition, he points out that the major quantity of phosphate in the city's sewage comes from human wastes rather than detergent.

Somehow, all this reasoning makes little impression on the backers of the "ban phosphates" measure. During an increasingly emotional meeting, some of the committeemen even accuse Williams of using stalling tactics to protect his employer, who, they point out, has a subsidiary that makes detergent chemicals.

Williams is in a dilemma. He sincerely believes that his viewpoint makes sense and that it has nothing to do with his employer's involvement with detergents (which is relatively small, anyway, and does not involve Williams's plant). Which step should he now take?

A. Go along with the "ban phosphates" clique on the grounds that the ban won't do any harm, even if it doesn't do much good. Besides, by giving the group at least passive support, Williams can preserve his influence for future items that really matter more.

B. Fight the phosphate foes to the end on the grounds that their attitude is unscientific and unfair, and that lending it his support would be unethical. (Possible outcomes: his ouster from the committee, or its breakup as an effective body.)

C. Resign from the committee, giving his side of the story to the local press.

D. Other action (explain).

6-10. Assume that you are a chemical engineer working for a small chemical company. You have been assigned the task of taking periodic samples of the effluent in a river resulting from drainage from an overflow pipe. The sampling location was selected by a representative of the state health department.

The sampling program consistently indicates a pollution rate well within the allowable limits. You notice, however, that the sampling site has been incorrectly chosen and does not detect pollution resulting from a discharge that flows through a deep pipe not visible from the surface.

Revealing the existence of the overlooked pipes could expose your company to major expense required to lower the actual pollution rate to within the limits of the discharge permit.

What do you think is the proper action for you to take? Do you think you would be legally liable if you remained silent? What would you do if you notified your company but the sampling location was not changed?[21]

6-11. Smith and Jones worked together for three years on a major research project and had nearly completed a paper for joint presentation to the national meeting of their engineering society. Smith was fired by their company for poor work habits and insubordination. Since Smith no longer works for the company, the management wants Jones to complete the paper and present it at the national meeting with no credit to Smith. What action should Jones take:

A. Complete the paper and present it as the only author.

[21] Adapted from Stephen H. Unger, *Controlling Technology: Ethics and the Responsible Engineer* (New York: Holt, Rinehart and Winston, 1982).

B. Complete the paper and present it with acknowledgment to Smith (risking management displeasure).
C. Stop preparation of the paper.
D. Take some other action.
Discuss your choice.

6-12. Select one of the student engineering societies at your school for investigation. Attend at least one meeting; talk to several officers and members. Prepare a short report answering these questions: What are the major objectives of this society? Why do the members you talked to belong to the society? What do they expect to obtain from the meetings? How are they working to improve their society? Would you like to belong to the society? Why or why not?

6-13. Determine the requirements in your state for obtaining a license as a professional engineer. What engineering fields of specialization are recognized by your state registration board for licensing?

6-14. Review a copy of an Engineering Code of Ethics and discuss the following questions:

- What items are included which help to provide a practical standard of behavior for practicing engineers?
- Are there any irrelevant items which could be eliminated? Why do you think they are irrelevant?
- Are there items which are confusing and that might be misinterpreted? Suggest ways for rewriting which could improve clarity.
- Could you use this code as a standard for yourself as you study to be an engineer? What areas would be the most difficult to comply with?

6-15. Interview some OSHA personnel in your area. What are the most prevalent violations? What are the most serious violations in terms of the number of people that are affected? In your opinion are any of these violations caused by poor engineering design?

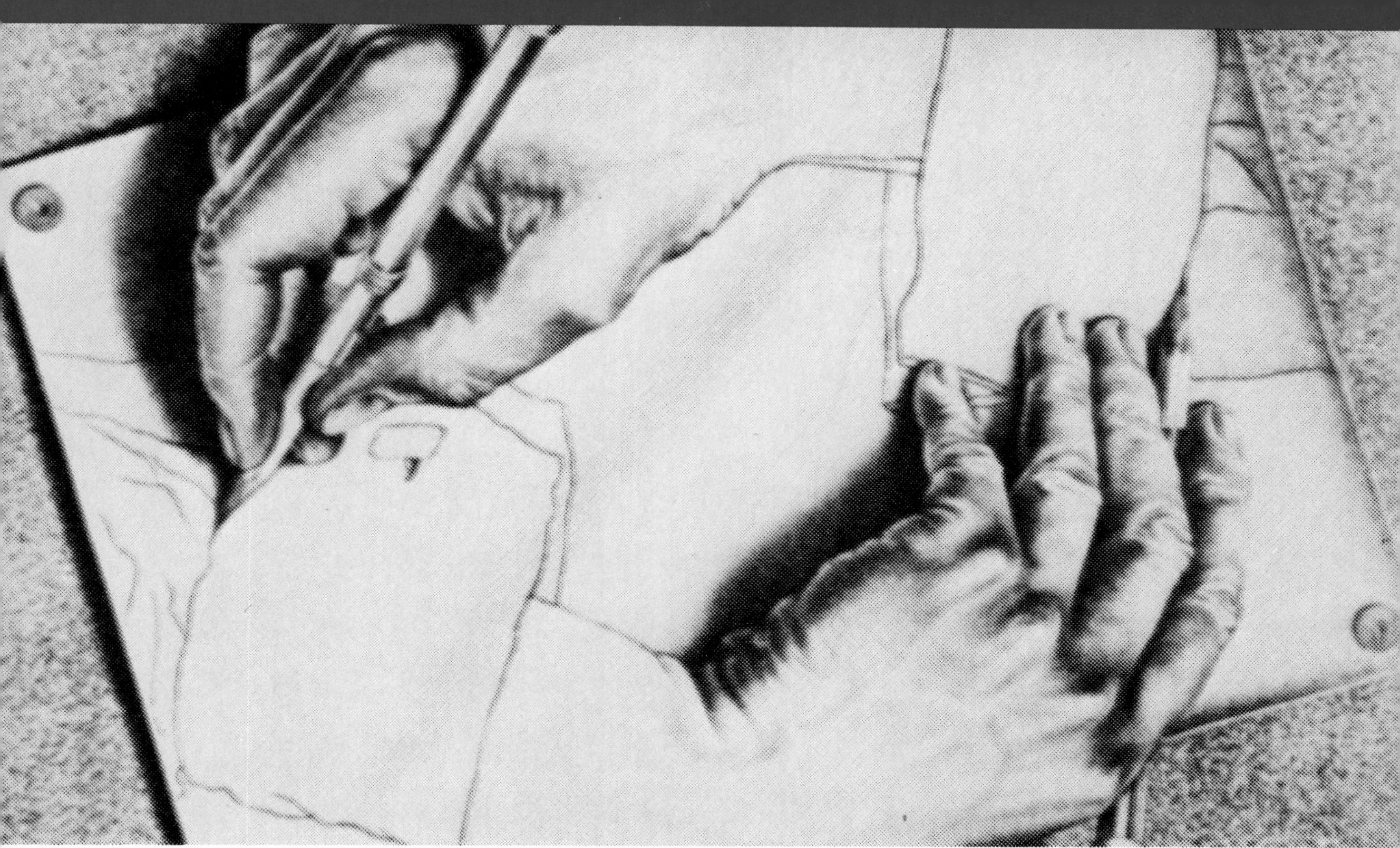

Section Two

The Engineer's Toolkit for Problem Solving

Chapter 7

Freehand Drawing and Visualization

Drawing is the ability to translate a mental image into a visually recognizable form. Simply, it requires the use of an instrument to mark or stain a surface. Today there are three general methods of producing drawings: freehand drawing, projection drawing, and computer-aided drawing (CAD). There are many variations of these three types, such as artistic drawing, design drawing, engineering drawing, architectural drawing, technical illustration, and cartography. Each drawing type has its own inventory of fundamental tenets that must be mastered by the beginner if competency is desired.

Of the general methods of drawing, the freely drawn artform, *freehand drawing,* is the oldest. Prehistoric man, wishing to record the results of a triumphant hunt, drew pictures of his experiences on rocks or on the walls of caves [7-1]. Until after the Middle Ages this was the only type of drawing. Drawing became more formalized when *projection drawing* was developed in France in the eighteenth century by Gaspard Monge (1746–1818). It was first used to simplify the design and construction of military fortifications. Descriptive geometry and engineering drawing were developed from these principles [7-2].

Engineering drawing, together with the principal of *tolerances,* unlocked man's ability to produce interchangeable parts in the manufacturing process. Engineering drawing is a precise discipline based on a thorough understanding of the principles of orthographic projection. Since this form of drawing values the accuracy that results from the application of projection theory, it has been greatly enhanced by the development of *computer-aided drawing*. Along with the graphical description of an object, computer-aided drawing creates an extensive data base detailing the attributes of the object. A computer is not only very adept at producing and manufacturing a graphical image, but also very efficient at data recording and manipulating for use in design and manufacturing. Where product drawings are generated by computer-aided drawing, the production machines that manufacture the products can receive their operating instructions directly from the data base in the computer. This results in fewer errors and a considerable savings in man-hours of work.

Today, using the power and diversification of the computer, it is possible to transfer freehand "idea drawings" into computer drawings, where they can be scrutinized simultaneously by many people. Objects can be easily rotated, sectioned, and viewed from myriad positions. Revisions can be readily made, if necessary, and then the final design can be permanently recorded in the computer's memory

[7-1] A prehistoric drawing of hunters.

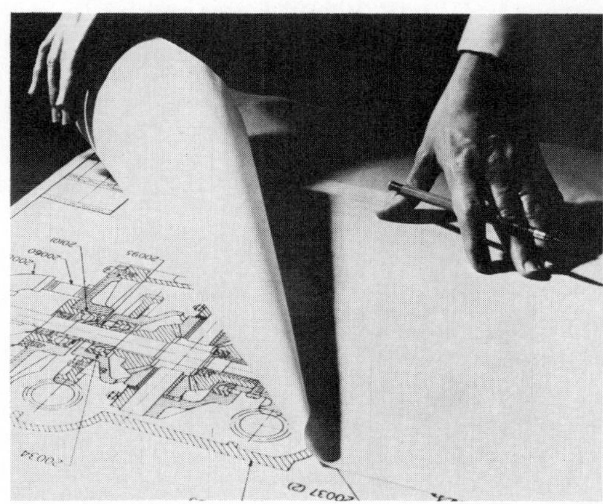

[7-2] Engineering drawing is a mechanically precise form of drawing compared to freehand drawing.

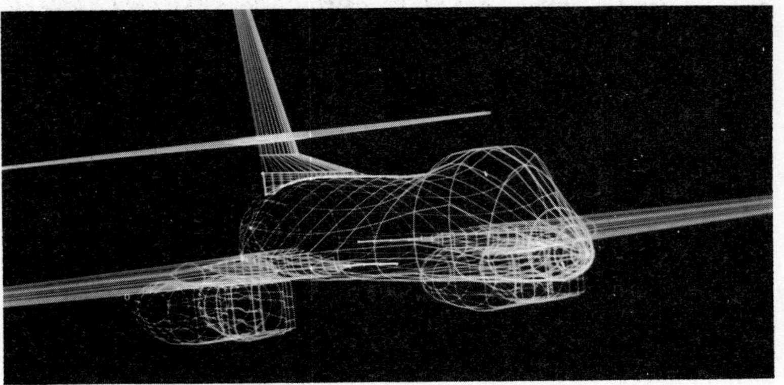

[7-3] Computer drawings are used extensively in industry.

[7-3]. Consequently, in industry more and more drawings of various types are being generated on computers rather than by engineers and drafters using pencil and paper. Engineering drawings can best be developed if engineers know how to select and use the appropriate computer software. This saves both time and money, and the result is much less susceptible to error.

Engineering drawing is not the best medium to use when a creative design engineer wants to convey an idea to nontechnical people. Even a superior knowledge of engineering drawing is not effective in this environment. A drawing form is needed that can be understood by all members of today's management teams, which frequently include both technical and nontechnical personnel. For most people, freehand pictorial drawing is the most easily and universally understood when it is necessary to realistically represent ideas that are intended to show "how things work."

One of the earliest "engineers" who used this medium to record his design ideas was Leonardo da Vinci (1452–1519). This man was a genius of great versatility. Although he is still acclaimed today as one of the greatest of all artists, his contributions as an architect, engineer, and inventor are perhaps even more impressive. In retrospect, it is interesting to note that the common thread that ran through his work in these seemingly diverse fields was his excellence in freehand drawing and visualization. In many cases, the only keys we have today to his brilliance are the idea drawings that he made [7-4]. His ability to produce realistic sketches makes the visualization of his thoughts possible [7-5]. Although most of us do not possess the mental talents of da Vinci, we can achieve skill in freehand drawing through proper training and experience. We can develop our ability to *draw with our pencil* those things that exist only in our imagination. As with da Vinci's sketches, our sketches can be used by others to see and understand the images that exist only in our minds. The objective of the authors is to instill within each engineering student the desire, confidence, and some of the proficiency demonstrated in the freehand drawing of Leonardo da Vinci.

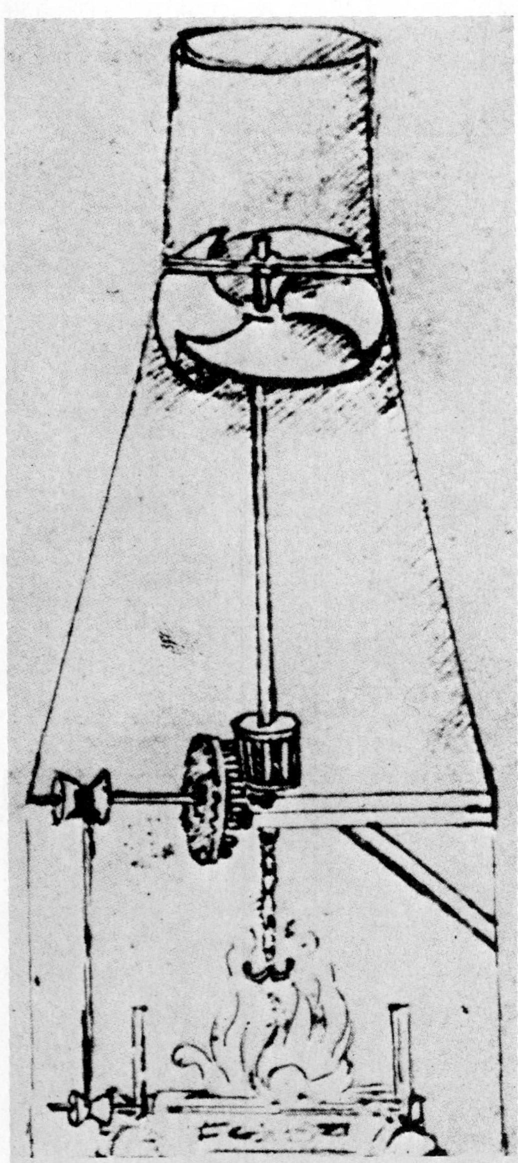

[7-4] Today, 400 years after Leonardo da Vinci drew them, his imaginative design drawings are still easy for the average person to understand.

This chapter has been written for the student who has had little or no formal instruction in freehand drawing. Its purpose is to develop a person's proficiency in freehand drawing and visualization. This will greatly improve your ability to record quickly the essence of your ideas as you follow the sequential steps in the design process (see Chapter 12).

The authors make no claim that the methods used here to teach freehand drawing are unique or that they are more effective than the time-honored techniques that have been developed throughout the ages by artists from every culture.[1] Indeed, we have drawn heavily on the experience of both traditionalists and advocates of the right-brain methodology in developing our teaching methods. This chapter is not intended to air pedagogical arguments or to produce artists. Of one thing we are certain, however; the methods described here *will work for you* if you:

1. Put aside any negative experience that you may have had with drawing. Be willing to learn.
2. Believe that you can learn to draw with skill.
3. Acquire the ability to perceive images accurately with your eyes.
4. Master a few basic skills in using a pencil to create images.
5. Follow each recommended drawing exercise in sequence. Don't leave any out. Do more if you have time or experience difficulty in improving your drawing skill.

Several of the illustrations used in this chapter are examples of student work. In each instance the students have consented for their work to be shown to serve as realistic examples of freehand drawing skills.

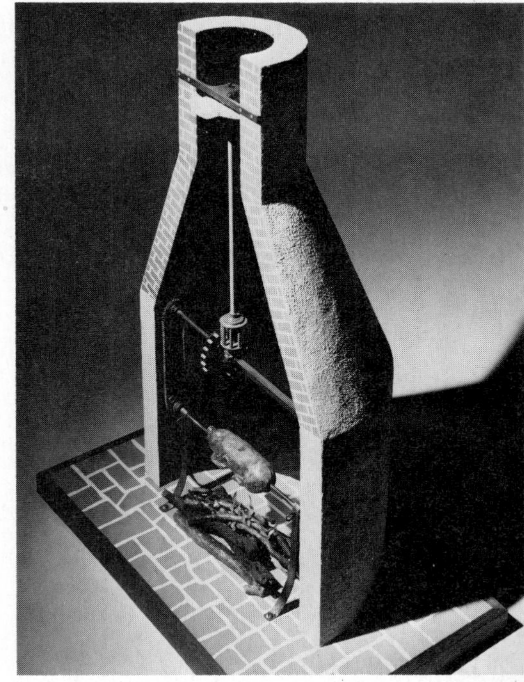

[7-5] It is easy today to build a model of one of Leonardo's conceptual drawings made over 500 years ago.

If you are an average engineering student, you will have emphasized in your prior schooling those subjects that were said to be good preparation for an engineering career: mathematics, physics, chemistry, biology, foreign language, and perhaps typing or computer programming. These are certainly all appropriate subjects, but primarily they are subjects that require intense stimulation and interaction with the left hemisphere of your brain [7-6]. The development of the right hemisphere of your brain (which is more concerned with music and art awareness, insight, imagination, and three-dimensional forms) has probably lagged behind. Engineering drawing, which is a subject that in former years was a requirement for most engineering students, is also a subject that draws heavily on interaction with the left hemisphere of the brain. On the other hand, freehand drawing, being free of technical symbols, is dominated by the right hemisphere of the brain. Because of the nature of your prior schooling, it is understandable that your proficiency in drawing and visualization may not equal your skills in mathematics and science. Since some of the exercises developed in this chapter are based on the conclusions above, let us briefly examine the scientific basis for the two-hemisphere brain theory.

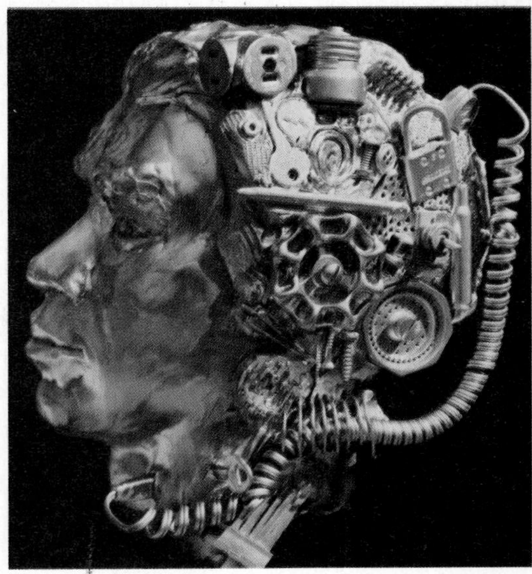

[7-6] The left hemisphere of the brain processes the verbal, mathematical, and scientific information used extensively in engineering. The right hemisphere processes information related to artistic awareness, imagination, and intuitive insight.

THE BRAIN

The human brain is a 1500-gram information processing organ of miraculous configuration. For protection, it is housed in the skull. For efficiency, it is located near the eyes, nose, ears, and mouth. It

[1]Kimon Nicolaides, *The Natural Way to Draw* (Boston: Houghton Mifflin, 1941).

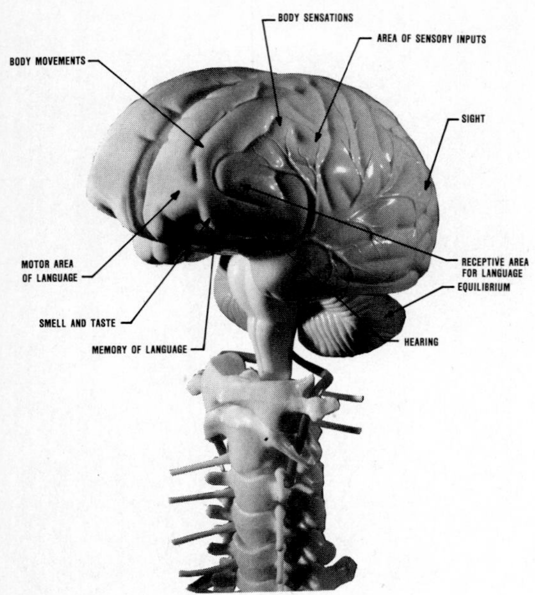

[7-7] The brain is the command center of the body.

The soul never thinks without a mental picture.
Aristotle, 384–322 B.C.

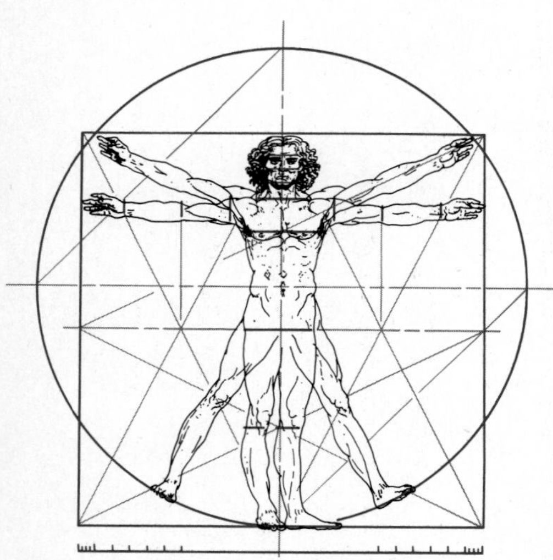

[7-8] The right hemisphere of the brain controls the body actions of the left side. Conversely, the left hemisphere of the brain controls the body's right-side actions.

also sits atop the remaining part of the central nervous system located within the spinal cord, which collects information from the muscles and skin [7-7]. The brain's fundamental working unit is the *neuron*. Each neuron is different from all the others. It is estimated that there are at least 10 billion neurons in each person's brain. Their function is to communicate with each other through electrical and chemical messages in a vast neuronal system network whose total number of connections exceeds 10^{15}.[2]

For reasons that no psychologist or neuroscientist fully understands, our brain is divided into two apparently symmetrical halves or hemispheres. In appearance, each hemisphere seems to be a mirror reflection of the other, but they are not precisely alike. In terms of capability, they are also not equally proficient in processing different types of information. Midway below the top of the brain, the two hemispheres are connected by a web of nerve fibers called the *corpus callosum*. This connection allows the two hemispheres to process information between them and even for one side to take over (temporarily or permanently) the normal duties of the other side.

The human body is a remarkable machine. Although it does not have interchangeable parts, it does have some back up (pinch hitter) capabilities. If one part fails or weakens, another similar part can carry some of the original responsibility. For example, we have two arms, legs, eyes, ears, lungs, kidneys, and cerebral hemispheres. Over time, however, the brain's hemispheres, unlike other organs, develop specializations of their own. As one grows older, each hemisphere is less able to carry out the functions of the other.

The design of the nervous system is such that each cerebral hemisphere receives and transmits information primarily from the opposite half of the body [7-8]. Thus the left hemisphere of the brain controls the actions of the right side of the body—the right hand, the right leg, and so on. The right hemisphere specializes in control of the left-side body components. However, there are other differences between the two hemispheres of the brain. In general, the right hemisphere is more proficient with visualizing and remembering events and faces and with other spatial and emotional functions such as visual construction tasks, artistic awareness, musical appreciation, and intuitive insight. The left hemisphere is more dominant with regard to written and spoken language and motor skills, analytical reasoning, and sequenced goal-oriented activities [7-9].

These conclusions about the specialized information-processing capabilities of the two cerebral hemispheres were made possible by extensive research by a number of neurosurgeons and experimental psychologists over the past 40 years. In the late 1940s a neurosurgeon, William Van Wagenen, performed the world's first split-brain surgery, severing an epileptic patient's corpus callosum. As soon as the patient's seizure charges were unable to cross the corpus callosum, the formerly untreatable epileptic spasms stopped. More important, there was apparently no detectable impairment in the patient's mental functions. In 1981, Roger Sperry was named a recipient of the Nobel Prize in Psychology and Medicine for his

[2] Richard M. Restak, *The Brain* (New York: Bantam Books, 1984).

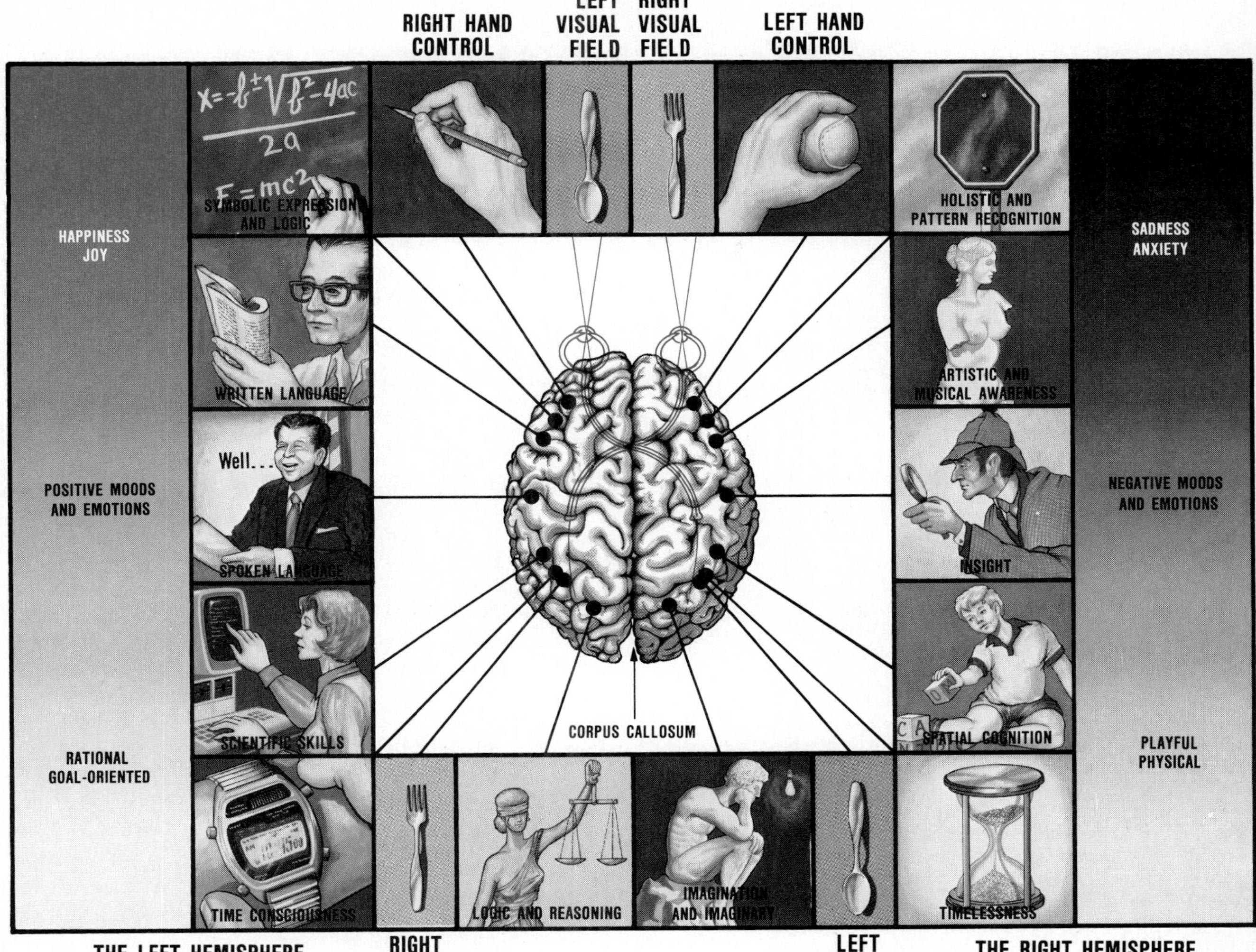

[7-9] The right and left hemispheres of the brain are assigned different primary functions.

split-brain research on epileptic patients.[3] This research further clarified the two modes of thinking in a split-brain person. In contrast, however, another English psychologist and Nobel Prize recipient, Sir John Eccles, disputes Sperry's belief that split-brain patients have two separate minds and separate spheres of consciousness.[4]

The purpose of this little discussion is to show that even the most learned of men still disagree on specific details concerning how the brain works. However, there is now general agreement that for

The main theme to emerge . . . is that there appear to be two modes of thinking, verbal and nonverbal, represented rather separately in left and right hemispheres, respectively, and that our educational system, as well as science in general, tends to neglect the nonverbal form of intellect. What it comes down to is that modern society discriminates against the right hemisphere.

Roger W. Sperry
"Lateral Specialization of Cerebral Function in the Surgically Separated Hemispheres," 1973

[3]R. W. Sperry, "Hemisphere Disconnection and Unity in Conscious Awareness," *American Psychologist* Vol. 23, (1968).

[4]J. Eccles, *The Brain and Unity of Conscious Experience: The 19th Arthur Stanley Eddington Memorial Lecture* (Cambridge: Cambridge University Press, 1965).

Right-hemisphere vs. Left-hemisphere Mental Processes

Over twenty years ago an anthropologist, Thomas Gladwin, contrasted the nonverbal right-hemisphere check and balance mode of thinking of the South Pacific Trukese navigators with the more rigidly structured left-hemisphere mode of thinking of navigators in the European tradition.

As an example, a native navigator of a multi-man sailing canoe will begin a 100-mile voyage across open ocean with a clear mental image of the relative positions of his home island, other islands in the area, and his destination island. This is often a tiny dot of land less than a mile across, and visible from any distance only because of the height of the coconut trees growing on its sandy soil. He has no navigational instruments of any type. Rather, he relies on dead reckoning and he continually adjusts his direction to compensate for changing wind and wave conditions. He sets his course by the rising and setting of stars, having memorized for this purpose the knowledge gleaned from generations of observations of the directions in which stars rise and fall through the seasons. A good navigator can tell by observing wave patterns, for example, when the wind is shifting its direction or speed, and by how much. Although the entire process is mental and involves an incredible number of complex decisions, the Truk navigator cannot describe in words how it all was accomplished. For the most part he has used right-hemisphere of the brain mental processes.

On the other hand, Western navigators carefully plan their voyage in advance. A course is plotted on a chart and this, in turn, provides the criteria for decision. Unless the navigator is sailing a direct point-to-point course, he does not carry in his mind a physical sense of where he is going. Once the plan of the trip is conceived and put into operation, the navigator need only carry out each step consecutively to arrive at the preplanned designation. If asked, he can describe precisely how each navigation instrument (map, compass, sextant, etc.) was used to assist in helping achieve his goal. The Western navigator uses primarily left-hemisphere of the brain mental processes.

Thomas Gladwin, "Culture and Logical Process," in Explorations in Cultural Anthropology, pp. 167–77; New York: McGraw-Hill, 1964.

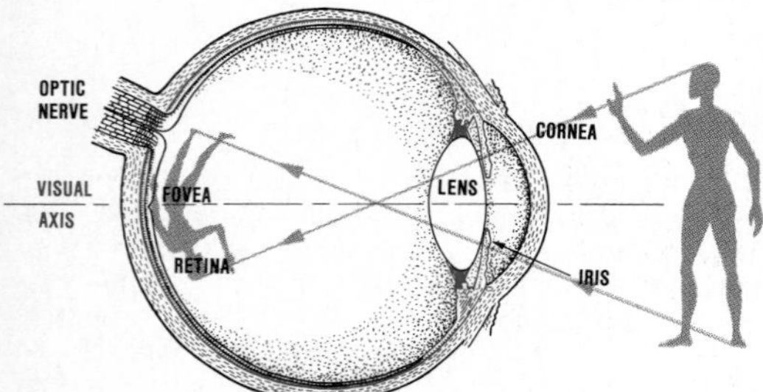

[7-10] Sight is achieved when the eye conveys a sense of vision to the brain. Coded information in the form of electrical impulses is arranged in a way that the objects being seen are represented to the brain.

healthy individuals the verbal, analytic left-hemisphere part of the brain is dominant for most individuals in the Western world. Similarly, the nonverbal, global hemisphere is known to perceive reality in its own way and to experience and process information independently. It is this underdeveloped hemisphere that we must work with if we want to develop our visualization and freehand drawing skills. It is the authors' belief that the ultimate objective is to arrive at a state where both hemispheres work together with equal intensity and proficiency. We must be whole-brain engineers—not entirely left brain or entirely right brain.

So much for the scientific reasons why *most of us* need some help in strengthening right-brain responses in learning to visualize, to understand unusual combinations and arrangements, and to draw with our pencil what we see with our eyes, or what we perceive in our imagination. The first step is to learn to *see* in this way. Let us begin.

SEEING AND ILLUSION

We learn about the external world through interaction of our brain and our senses—sight, hearing, smell, taste, and feeling. Of these, sight is the most used by the average person, and it is also the one that gives us the greatest confidence. At one time or another most of us have made the statement, "I won't believe it until I see it." Sight is achieved when the eye conveys a sense of vision to the brain. We use sight to detect light intensity, recognize images and patterns, estimate distances (depth perception), distinguish colors, perceive motion, and to aid in controlling our bodily actions.

The eye is a remarkable organ of the body. Each of the two eyes accepts light rays through its lens and projects them onto the retina. The two visual pathways cross behind the retinas [7-10], each response passing to the opposite hemisphere of the brain—a shift that makes unified, three-dimensional (stereoscopic) vision possible. The pictorial image has been inverted by the lens. The resulting upside-down picture is then encoded within the neuron lacework of the brain for storage and deciphering. What we see, then, is the result of both the visual stimulus that reaches the brain and the brain's interpretation of the stimulus.

We perceive scenes differently depending on our sociological and psychological conditioning. Several people observing the same street corner accident will frequently swear under oath that conflicting events occurred. Perceiving what we see improves with practice. How perceptive are you? One measure of a person's ability in this area is how successfully and quickly you can identify embedded figures in a more complex figure. Figure [7-11] is an example of such a problem. See how long it takes you to identify the location of the small hexagonal area within the larger pyramidal figure. It is interesting that when this type of test is taken by members of different cultures, some of the fastest solutions were provided by Eskimo hunters. They had an easy time with this type of problem because they had been solving similar and more complex tasks by distinguishing polar bear shapes on distant ice packs.

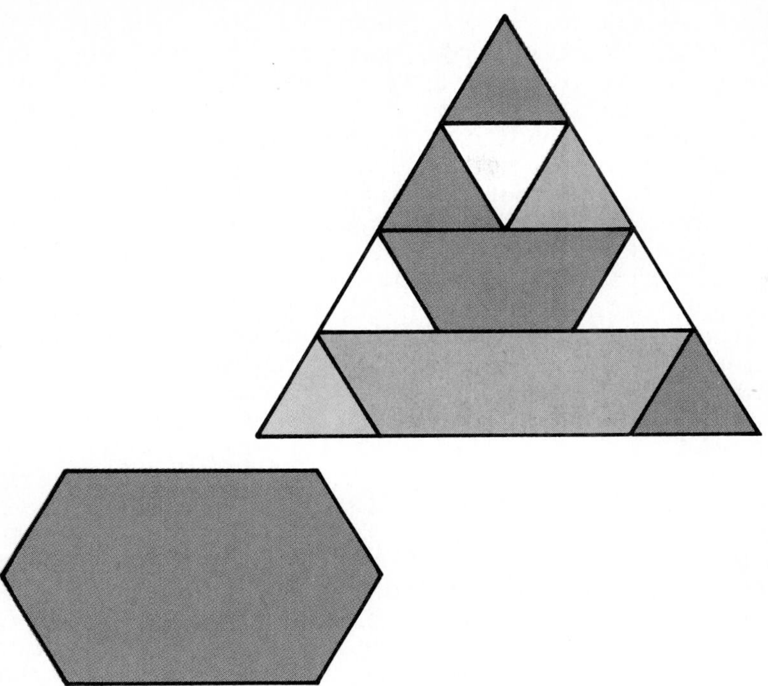

[7-11] Find the hidden hexagon.

[7-12] M. C. Escher's "Waterfall."

There are occasions where the eye's visual signals and the brain's interpretative messages are in conflict. Because of this, it is not unusual for us to see what our brain tells us we are seeing—not what our eyes actually envisage. Such visual conflicts are called *illusions*. The drawing "Waterfall" by M. C. Escher [7-12] is an example of such a conflict. The water falling onto the millwheel seems to be the source of perpetual motion. On examination we see that all of the water that flows onto the waterwheel eventually runs uphill until it can drop again and provide the energy necessary to grind the grain. Our brain warns us that such a condition is a violation of the natural laws of nature. Even knowing this, our eyes continue to deceive us.

In 1970, M. Gardner[5] created an illusion by incorporating an impossible figure into a picture [7-13]. (A more modern version of Gardner's illusion appeared recently as a three-legged blue-jeans advertisement.) In 1960, C. F. Cochran[6] used his "freemish" crate to demonstrate effectively that *seeing is not necessarily believing* [7-14 and 7-15].

If we are to become proficient in drawing shapes that represent real or imagined images, we must understand something of illusionary representation. The drawing of the duck and rabbit [7-16], first used in 1900 by the psychologist Joseph Jastrow, and the 1915 car-

[7-13] An impossible situation. What did the artist do to make this illusion?

[5]M. Gardner, "Of Optical Illusions, from Figures That Are Undecidable to Hot Dogs That Float," *Scientific American,* Vol. 272 (May, 1970), p. 124.

[6]C. F. Cochran, Letter to *Scientific American,* Vol. 214 (June, 1966), p. 8.

[7-14] How was this crate constructed?

[7-15] . . . by illusion.

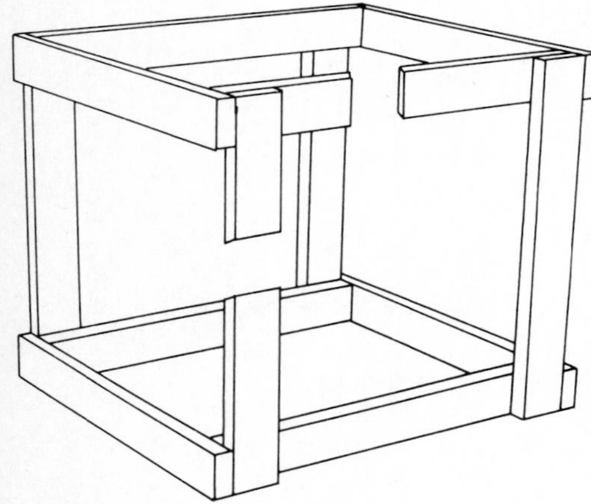

toon drawing of the "Wife and Mother-in-law" by W. E. Hill [7-17] are other examples of ambiguous illusions that are difficult for the eye and brain to assimilate agreeably.

In viewing, the eye reports to the brain both familiar and unfamiliar states of being. The brain measures this information against a data bank (memory) of past experience. It then communicates the comparison to you so that you can act accordingly. For example, if you are driving a car on the highway and you suddenly see a blinking red light in your rear view mirror, your first split-second reaction would likely be to put your foot on the car's brake pedal. If you have been exceeding the speed limit, your overall feeling will probably be one of despair. On the contrary, if you are in the process of being kidnapped by a bank robber, your feeling will more likely be one of apprehension, if not ecstatic relief. Similarly, what appears to be attractive or beautiful to one person may be seen as repulsive or ugly to another.

The brain is responsible for representing a person's perception of the outside world at any specific point in time. For this reason, even though the eyes of two individuals may seem the same scene or action, the resulting messages from the two brains may be quite different [7-18]. The competition between foreground and background relationships was one of the earliest conflicts to be identified. Generally, the eye will see in the foreground an outline or object that we will call the *figure*. It has structure and is viewed as being separate from its boundless, shapeless surroundings that we will call the *ground*. Knowing the orientation of the figure with respect to the ground is also important. Figure [7-19] is an example of this rela-

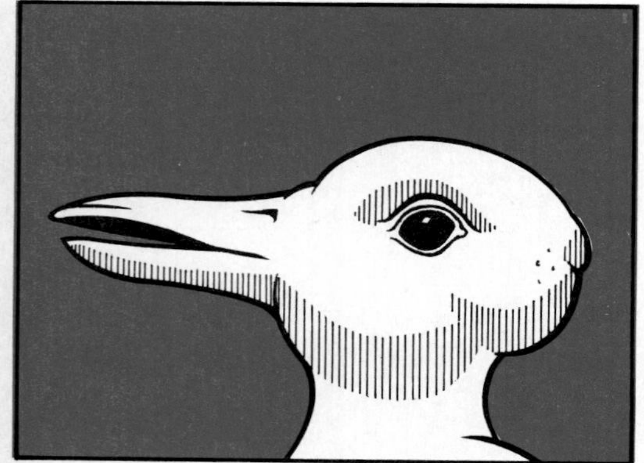

[7-16] Duck or hare?

[7-17] Wife or mother-in-law?

tionship. The white area representing the vase will be recognized by most people as being the figure and the surrounding black area as being the ground. Now turn the figure upside down. The two facing black faces, which were considered to be the ground before, now peer at each other in front of a lighted doorway. Your brain will allow you to alternately see the vase or the two faces but not both at the same time. If you hide one-half of the drawing by covering it with a sheet of paper, the remaining face is still very easy to identify, but the visible white area has no meaning. Your brain is not accustomed to identifying a one-sided vase.

There are other optical illusions that should be introduced to those who want to learn how to draw. In the late nineteenth century, psychologist Wilhelm Wundt described the simplest of visual illusions—that a vertical image looks longer than a horizontal image of equal length. Figure [7-20] illustrates this. Note that the vertical height of the top hat appears to be greater than the width of the brim, yet they are equal in length. The Gateway Arch in St. Louis is another example of this illusionary distortion. It is 630 feet tall and 630 feet wide [7-21]. Artists of ages past have known that a tree looks much shorter when it has been felled than when it is standing. Figure [7-22] is actually square but would look more like one if the top and bottom lines were covered.

In 1889, Franz Müller-Lyer presented a configuration of two lines of equal length that produced a false impression [7-23]. The two horizontal lines are of equal length, although they do not appear to be so. See how many variations of this phenomenon you can contrive. As a variation, the minute hand and hour hand in [7-24] are the same length.

The mind is susceptible to error by drawing conclusions too soon and using too few data. This is particularly true when we see a few members of an apparent series and conclude that all the other

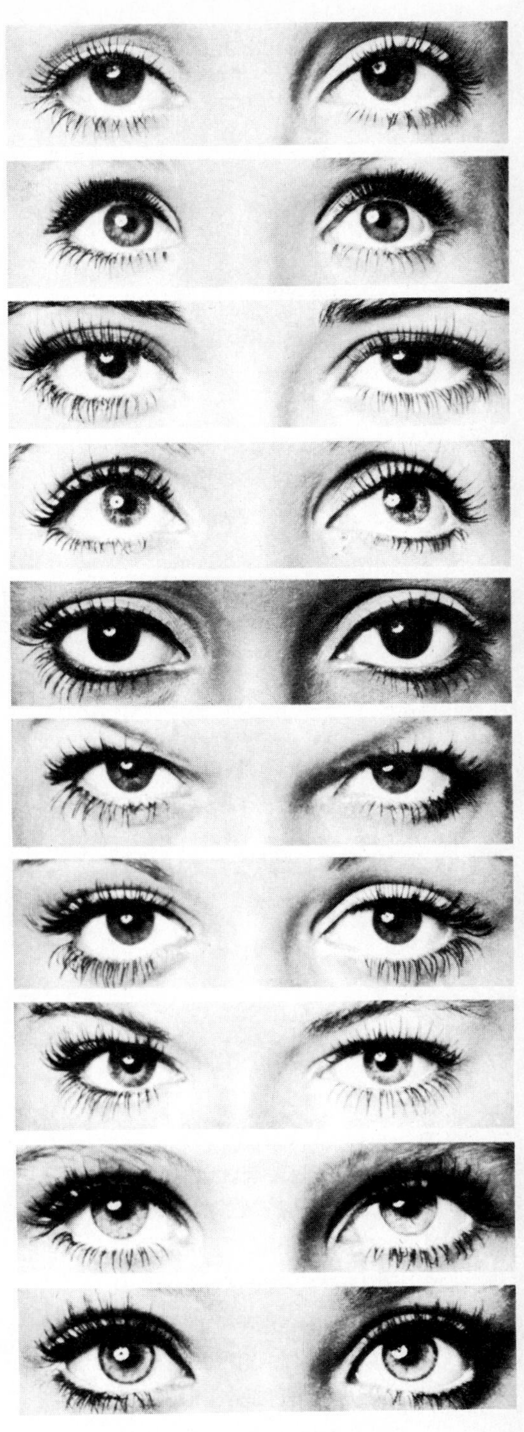

[7-18] Ten pairs of eyes viewing the same scene or action will convey 10 different messages to their respective brains.

[7-19] Vase or twins?

[7-20] Square top hat.

[7-21] The St. Louis Gateway Arch.

[7-22] How many pancakes should you remove to have a square stack?

members will follow in turn. J. Fraser's spiral [7-25] presents us with a dramatic example of this problem.[7] Only a tracing pencil will convince us that we are not looking at a spiral at all. We are seeing a series of concentric circles.

This is not a chapter on illusions, but it is important to understand that we must learn to draw what our eyes actually see—not what our brain tells our eyes they are seeing. Later in the chapter where appropriate, other illusions—and how they affect our perception of what we are seeing—will be discussed. We are now ready, however, to try our hand at drawing.

FREEHAND DRAWING

Anyone can learn to draw! The reason some people can pick up a pencil and draw anything with realism is because their hand is able to duplicate what their eye is seeing. They do not have to think about telling the hand what the eye is seeing. It just seems to be a natural process. Fortunately for most of us, however, this is an acquired ability that can be developed and enhanced through a series of exercises that will strengthen the coordination of the eye–brain–hand team. Experience has shown that it is much easier to draw well than it is to play expertly on a musical instrument. Neither skill is acquired by inheritance. Like learning to master music, learning to draw requires both instruction in the fundamentals and directed practice. Learning to draw is basically learning to see. It is necessary to practice until the eye can accurately perceive images that the brain will correspondingly translate into hand movement.

[7]J. Fraser, "A New Visual Illusion of Direction," *British Journal of Psychology,* Vol. 2 (1980), p. 307.

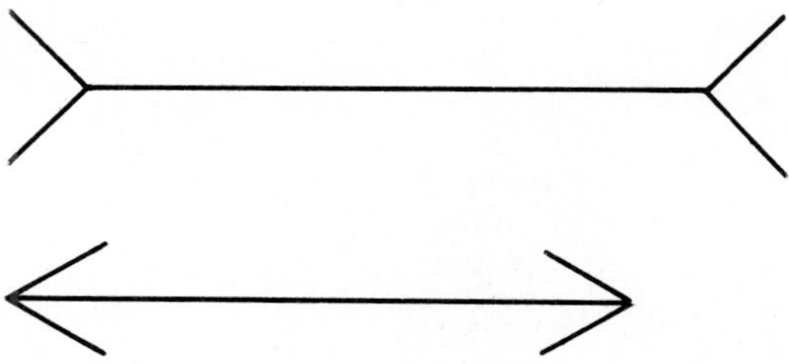

[7-23] Which horizontal line is longer?

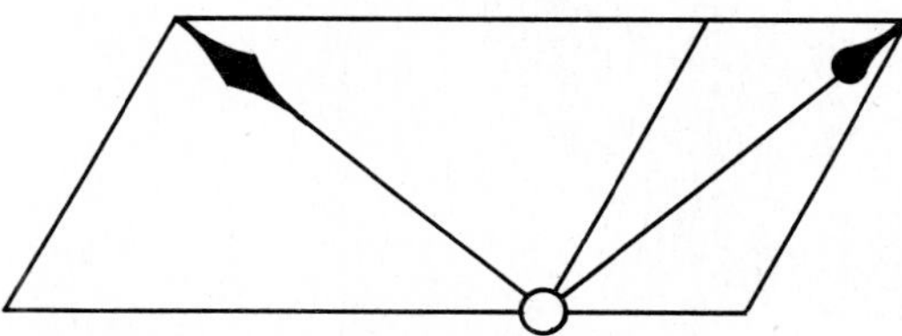

[7-24] Is the hour hand as long as the minute hand?

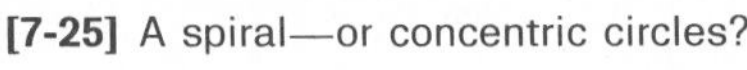

[7-25] A spiral—or concentric circles?

The drawing exercises in this chapter will help you to represent things as they *appear to the eye*—not how they actually exist in reality. You must learn to make the transition from a condition where your brain controls your eye and tells it what it *should be seeing* to a preferable one where your eye reports to your brain what it is *actually seeing*. Remember, there is a vast difference in what we know about our surroundings and that which we actually see.

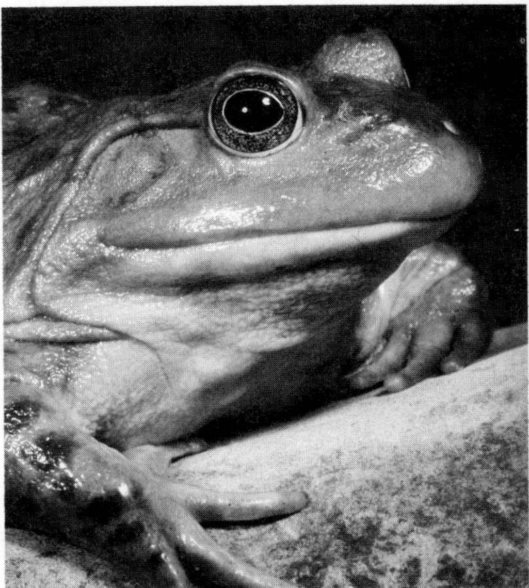

[7-26] Frogs are practically blind, and their eyes are stationary in their sockets. However, they have other remarkable abilities. They receive, process, and relay information *important to the frog* and filter out everything else. For example, a juicy bug flying toward him is important; one flying away is not. The frog's eye signals his brain if:
(a) An object is flying toward him,
(b) The object is "bug size,"
(c) The object is flying at "bug speed,"
(d) The object is within range.
Everything else is ignored. The frog's eye also makes other life and death decisions for him without bothering his feeble brain. A sudden shadow, for example, will trigger a danger signal causing the reflexive jumping mechanism to function. His eye–brain dominance would make him a good candidate to learn to draw—if he could only hold a pencil.

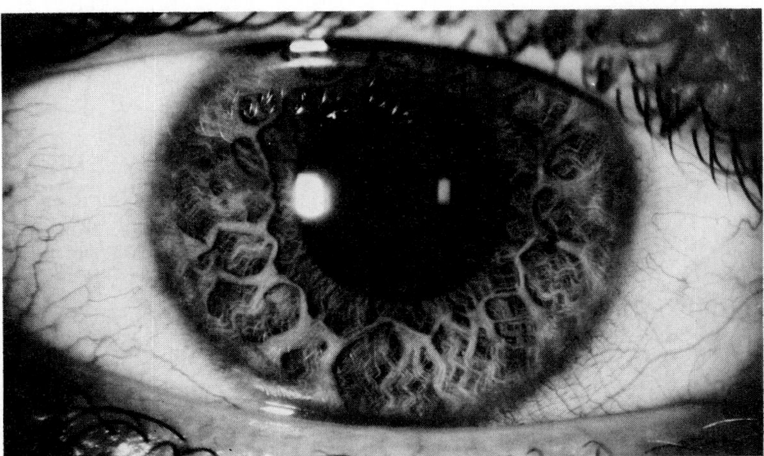

[7-27] The human eye, unlike that of the frog, moves. It also negotiates with the brain to interpret the meaning of transmitted light. Your proficiency in drawing depends on how well your hand can replicate the image that the eye is seeing, not the image that the brain thinks the eye should be seeing.

Once we can master the skill of seeing, we can develop proficiency in translating this into a two-dimensional image. Since we see in three dimensions, the task of accurately transferring an image to a two-dimensional form on paper may seem overwhelming. Do not despair. There are several techniques that we can use to help make the transition easier. Here, illusion becomes our friend—not our enemy. Before beginning the first drawing exercise, we must obtain appropriate drawing materials.

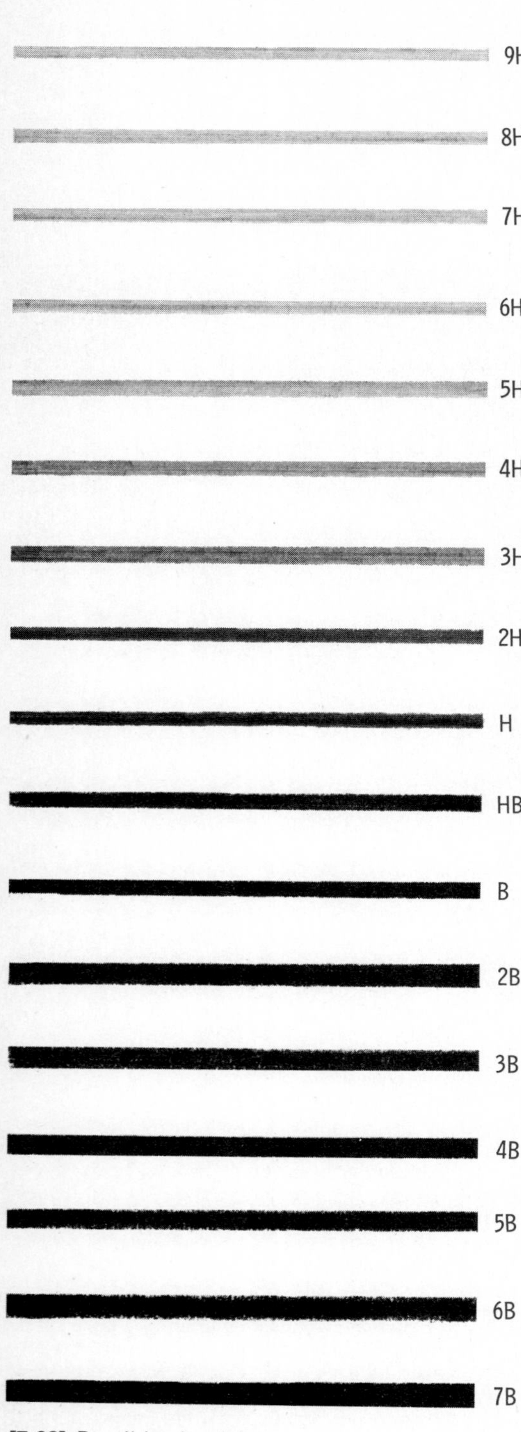

[7-28] Pencil lead weights.

MATERIALS FOR DRAWING

It takes very little to begin drawing. Basically you need only two items: a pencil and a piece of paper. This sounds simple enough, but it is useful to understand something about your materials before you begin. First, the *pencil.* Graphite pencil leads are divided into two categories, soft (the B series) and hard (the H series). An HB pencil falls midway between the two types [7-28]. The most commonly used weight for freehand drawing is 2B. We will use this pencil to begin our drawing exercises. Rather than changing pencils to obtain gradations in line intensity, we will learn to vary the pressure on the pencil. Applying light pressure will produce fine, lightly textured lines. Heavy pressure will produce thicker, denser strokes in the areas you desire them. However, as you progress you will find that a variety of pencil weights can add interest to your drawings. A soft pencil will make a darker, thicker line, and a harder pencil will make a faint but sharper line. You will find that the softer pencils are much easier to draw with. This is why it is recommended that you begin with a 2B pencil. Do not use a lead holder or an ultrathin mechanical pencil. If you use a wooden pencil, sharpen it so that it has a conical-shaped point. Always have three or four sharpened pencils available before you begin to draw.

You will quickly discover that there are a number of different types of drawing paper to choose from. Drawing papers may be purchased in pads (spiral or glue bound) or as loose sheets. A good size is 11 by 14 inches. The beginner should start with a drawing paper that has a little "tooth" to it. This means that it has some surface texture that will let the graphite from the pencil adhere to it easily. The "slicker" the paper, the more difficult it will be to draw on it with a pencil.

You may want to purchase some standard, inexpensive newsprint paper to practice drawing lines and to do some of the quick gesture drawing exercises. However, the majority of your drawings should be done on good-quality, white, bond drawing paper. Remember, all of these materials are relatively inexpensive, and unwanted drawings can be discarded without emotion. Always assume, however, that you will keep every drawing you begin. Every drawing should be a potential "keeper."

For the beginner, *erasers* offer a temptation for indecisive and sloppy work. They should not be used! Do not even buy one. If a line is drawn in the wrong place, redraw a better one. Do not concern yourself with the original line. It will fade into insignificance as you proceed. If not, start over. After you have mastered the fundamentals in this chapter and have gained considerable proficiency, you will find that a *kneaded eraser* can serve as an invaluable drawing tool. By pressing the paper with a kneaded eraser, unwanted lines and tones can be removed without destroying the crispness of the other linework. At that point you may also want to acquire an *erasing shield*. Now, let's get back to the pencil.

[7-29] Hold the pencil comfortably.

HOLDING THE PENCIL

The manner in which you hold a pencil is a personal matter. By now you probably have acquired a consistent and comfortable way to hold a pencil. The most effective method is illustrated in Figure [7-29]. You may wish to try this position if you do not now use it. Use the bottom of the little finger as a steadying point on which to slide the hand.

KEEPING A SKETCHBOOK

The artist's sketchbook is an important tool used in learning to see. It helps to make drawing more than just a classroom activity and brings it into your everyday life. It is a method for recording your ideas, thoughts, and experiences, as well as providing a convenient method to practice your drawing skills. You will be interested to note your improvement from the onset of your sketchbook to its final pages.

For maximum effectiveness your sketchbook should be used daily. This may seem difficult at first, but you will notice that in time you will place a gradual dependency on it for practicing new methods and freely expressing your ideas. It can be considered to be a journal, and you may want to make comments on new techniques or elaborate on ideas for later development. In selecting a sketchbook, keep in mind that it should be an easy size to carry, but not so small that it is difficult to make full-sized drawings. Make sure that its construction is sturdy and able to withstand the punishment of being carried with your other books. A variety of styles are available at your bookstore, ranging from a hardback, bound volume to a spiral bound, or cardboard cover style. A good size to purchase would be 9 by 12 inches or 11 by 14 inches. Be sure to date each page of your sketchbook, not only to note the progress of your improvement, but for future reference to keep in mind when and where your ideas and drawings were made.

Now we are ready for our first exercise.

MAKING A LINE

A line drawing is the most direct means of expression, and lines are the basis for all pencil drawings. They define the shapes of objects and give meaning to form. Lines alone can be interesting and expressive, and variation in line character can add relevance and feeling to any drawing. As you draw lines, a variation in line quality can be achieved by varying the pressure used on your pencil. In Figure

[7-30] Line can add spatial dimension. The accent and variation of line used here help to give it a plastic quality.

[7-30] notice the expressive quality that is achieved by the variation of line weights. After practice, this will become second nature to you. You will find that your hand will automatically vary the pressure on the pencil as you are drawing. This will not be a deliberate attempt on your part to change the line widths, but rather it will happen instinctively as you move the pencil across the paper. When this happens it will be encouraging, because it is a spontaneous response of the hand to what the eye is seeing.

Exercise 7-1

(a) Holding your pencil in a normal drawing position, make a series of horizontal lines running all the way across the page. The pencil's movement should not be the result of the movement of your fingers. Rather, the pencil will be moved by the simultaneous motion of the hand and forearm. Your wrist will be locked and your fingers will move very little. Vary the pressure on your pencil as it glides across the page. This should cause the line to have varying width and *value* (relative lightness or darkness). Make several pages of these horizontal lines. Experiment with making the lines vary in value in the same places on the page so that an optical illusion of motion is created. The lines may even begin to look like waves even though they are actually straight lines, with the only variation being in line width and value [7-31].

(b) Now make a series of vertical lines using the same technique. Try to keep the spacing between lines equal to develop control over your pencil and the paper. Once again, vary the line pressure to create an optical illusion of lines that are in motion running across the page. The vertical lines may be more difficult to draw because you are used to making and seeing lines running left to right across a page. This unfamiliar arm movement may also require more practice to achieve a page filled with straight vertical lines.

(c) The last series of lines are made in the shape of circles and ellipses. First, make a series of circles in horizontal rows running across the page. Try varying the pressure on the pencil in different positions on each circle. Some will have a dark area at the top of the circle, some at the sides and some at the bottom. Now, continue by making rows of ellipses, varying the angle of each. Make several rows of ellipses that slant toward the left side of the paper as well as the more familiar form, to the right side.

Do not get discouraged if your pages of lines are not neat. These are exercises to help you develop a *mastery* over the pencil and paper. Beauty is not our main concern. Let the pencil and paper know that you are in charge! They will soon become eager servants waiting to perform whatever you ask.

This exercise is designed to help train your eyes to see rather than to rely on what your mind believes to be true. Many familiar objects do not look the same when they are taken out of a familiar context. Salvador Dalí was a master at being able to take familiar objects and place them in an unfamiliar setting to cause a disturbed feeling [7-116]. Notice how much longer you spend looking at one of his paintings as you try to recognize all the objects. In many cases our eyes are telling us one thing while our mind is telling us another. It is important to train your mind to rely on your eyes and to transfer what the eyes see to the hand, and in turn to the pencil.

[7-31] Learning to vary the pressure on your pencil is an essential skill that is learned through repetition of line exercises.

Exercise 7-2

(a) Find a line drawing of a familiar person, such as the one of Albert Einstein [7-32]. If you have seen his picture before, he is easily recognized and named because your mind identifies the familiar features of his face and tells you it is Albert Einstein. Place the line drawing in front of you *in the upside-down position* and begin to copy it on your paper. Your drawing will also be upside down since it will match the drawing in front of you. Start at the top of the drawing and progress downward. Try to copy each line on your drawing as it appears on the original drawing. Keep in mind the length and shape of each line and its relationship to the other lines and to the edges of the paper. It may help you to define your drawing space. Sometimes it is difficult for beginners to visualize that the edge of the paper can also be the edge of the drawing. Once you get started with your drawing you will find that you are more interested in copying the lines and shapes rather than in identifying the familiar features of Albert Einstein. Do not turn your drawing upright until you are completely finished.

[7-32] Turn me upside down.

(b) Figure [7-33] is an upside-down photograph of a well-known American actor. In its unnatural orientation you may have difficulty recognizing the person as John Wayne. However, do not concern yourself with identification. Again your task is to *copy the upside-down image* on the photograph onto your drawing paper. Proceed just as you did in copying the drawing of Albert Einstein. Copy each line, reproducing line connections, angles, and contour shapes. You will find that it will be more difficult to copy the wrinkles and contours on the photograph than it was to copy those on the Albert Einstein drawing. You will have to make your eyes work harder in transferring information to the brain, and the brain must work harder to sort out the characteristics for duplication. Of course, the photograph has subtitles of shading that were not present in the line drawing. Finish the drawing in one sitting. When you have completed the drawing, turn it right-side up. You will be amazed at the quality of your effort.

(c) Once you have completed parts (a) and (b) you may want to try another by first tracing a photograph to get it into line form. Then turn it upside down and try to redraw it as in part (a). You will be surprised at how proficient you can become at recreating a drawing, because you are taking more time to see what is actually there.

(d) This exercise should be done in groups of four or more. Acquire an 8- by 10-inch photograph of a familiar person. Cut the photo up into as many equal-size pieces as there are people in your group. Mix the pieces up and let each person in the group pick a piece. Each person will now make a line drawing of that section of the photo. Make sure that each person is making the line drawing on the same-size piece of paper so that the proportions of each drawing will be compatible. Your piece of the photograph may show only one eye and a piece of the nose, but try to be as accurate as possible in recreating your piece. After each person is finished, place all the line drawings together to recreate the photograph. See how accurately your familiar person is recreated in the drawings. It will surprise you to see what details some people emphasize and others ignore. This points out that we all see differently and recognize subtleties in expression to different degrees, but we all recognize the importance of seeing.

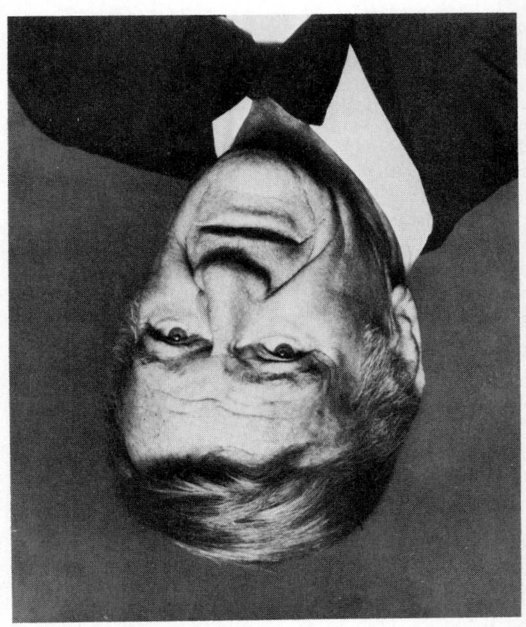

[7-33] Draw what you see.

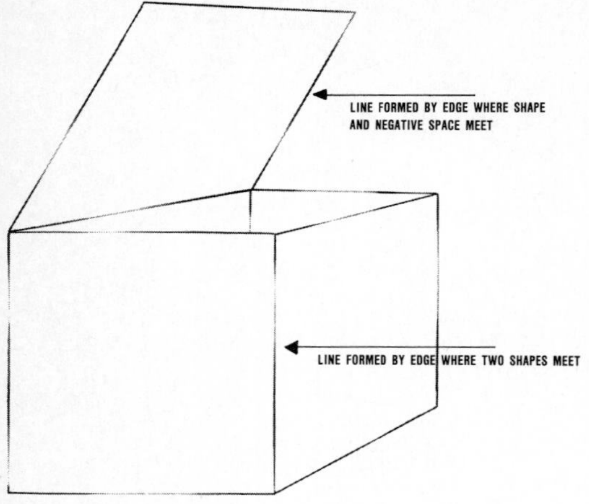

[7-34] Draw the visible edges.

CONTOUR DRAWING

The simplest line has length, direction, width, and value. Often the line is used to describe the bounds of contours, separating each area or volume from adjacent ones. In this role, it is its most expressive self. The preceding exercises have helped us to develop a sensitivity for drawing the line as an isolated entity. Now we will use the line to create contours.

A contour line is merely a line that describes an *edge*. When two shapes meet, they form an edge, which we also call a *contour*. For instance, the front face and side face of a box meet and form an edge. When we draw all the edges that we can see, we have a contour of the box [7-34]. A contour drawing then consists of all the edges where shapes meet, including the edge created where air (space) meets an object. If you open the box, you will notice that there is an edge made by the top of the lid as it meets the open air as well as other edges that are made by the sides of the box meeting each other. Contour drawings never show value, texture, or shading. They are pure line drawings and consist only of the edges of an object.

Contour drawing will do more to improve your drawing ability than any other exercise. This is because it emphasizes seeing more than any other drawing technique. Done properly, contour drawing will give you the confidence and understanding to draw anything. In this chapter we will progress through three stages of contour drawing: pure contour drawing, modified contour drawing, and cross contour drawing.

First, we want to develop a distinct kinship between the object you are drawing, your eyes, and your hand. Your eyes must learn to focus on and follow along the features of the object as if they were touching it. In this sense you may consider your eyes as a substitute for your hands. This technique will take time and practice, but it will open a whole new world for you once you have become accomplished at contour drawing [7-35].

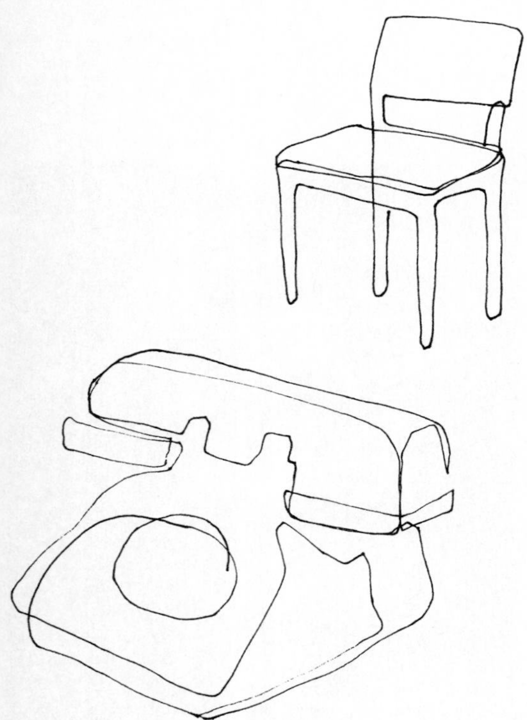

[7-35] Contour drawings project the essence of character of the objects seen.

Pure Contour Drawing

Pure contour drawing develops the eye–hand coordination that is invaluable in being able to draw. Some of the following techniques may seem frustrating at first, but they are designed to enhance your coordination and make it a natural response in later drawings. All contour drawings are to be done very slowly. At first you may feel like it is taking forever just to complete one small drawing. As you become more accomplished and really begin to learn to see, the elapsed time of drawing will not seem so significant. There are a few important points or "rules" to remember when working on contour drawings. These are:

1. *Do not look at your paper.* Since you will not be making any lines that your eye does not see, and you are keeping your eye on the object at all times, there is no need to look at your paper until you have completed the entire drawing. This will become your greatest temptation, *but resist it!*

2. *Keep drawing at the same rate of speed.* Your eye should be moving consistently along the edges of the object and your hand (with pencil) should be following at the same rate. Your hand should not move faster or slower than your eye.
3. *Place the object you are drawing very close to you.* You need to be able to see details easily, and it is also important that your eye is not distracted by other objects that may be between you and the object being drawn.
4. *Make sure you will not be interrupted for at least 30 minutes.* This will allow you to create an environment for total concentration. People talking, music, or television can be a major distraction to the concentration necessary for pure contour drawing.

Keeping the guidelines above in mind, we are ready to move on to your first pure contour drawing.

There is only one right way to draw and that is a perfectly natural way. It has nothing to do with artifice or technique. It has nothing to do with aesthetics or conception. It has only to do with the act of correct observation, and by that I mean a physical contact with all sorts of objects through all the senses.
Kimon Nicolaides, The Natural Way to Draw, 1941

Exercise 7-3

Your first contour drawing will be of your hand. You may think this will be easy, since your hand is a very familiar object to you. We are still going to make your mind work extra hard in seeing every detail in that hand. *We will do this by having you draw with the opposite hand to the one that you use to write.* If you are normally right-handed, in this exercise you will use your left hand to hold the pencil, and vice versa [7-36]. Because it will take tremendous concentration for you to instruct your "uncoordinated" hand to make each line the way your eye tells it to, you will find that you will be forced to draw at a much slower pace, but with surprising accuracy. Keep in mind all the previously discussed guidelines—do not look at your paper, do not let your eyes get ahead of your hand, and keep the pencil moving at the same rate. If you can not seem to resist looking at your paper during your first contour drawings, turn your body around so that you are forced to look away from your paper if you look at the hand you are drawing. Since this drawing position is very awkward and the paper may have a tendency to move, you may wish to secure the edges of your paper to the drawing table with tape. Now let's begin the first drawing.

Hold your nondrawing hand slightly below eye level. Spread your fingers slightly. Pick a central point on that hand for your eye to rest on. Pretend that your eye is actually touching the hand. Position your pencil somewhere in the center of your paper. Do not look back down at your pencil until the drawing is completed. Slowly begin to trace the edge of your hand with your eye. At the same time your drawing hand will be moving the pencil on the paper as if it were your eye. Try to observe very closely all the little details of wrinkles and bends along the edge of your hand. Perceive in your mind that your pencil is recording everything exactly as you are seeing it. Do not pause or pick up your pencil during the entire drawing. Keep concentrating on making the pencil follow what the eye is actually seeing. This will be especially difficult since you are drawing with the opposite hand that you are used to writing with and coordination skills will be lacking. It is possible, however, to concentrate hard enough to make a teamwork relationship develop between the eye, mind, and hand. Once you have completed your drawing,

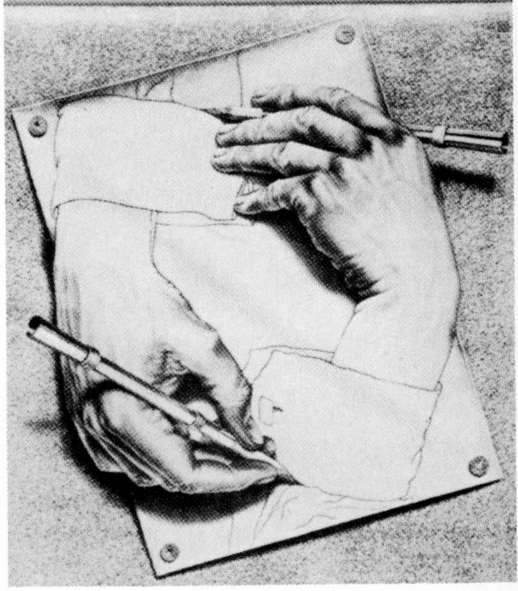

[7-36] Practice drawing with either hand.

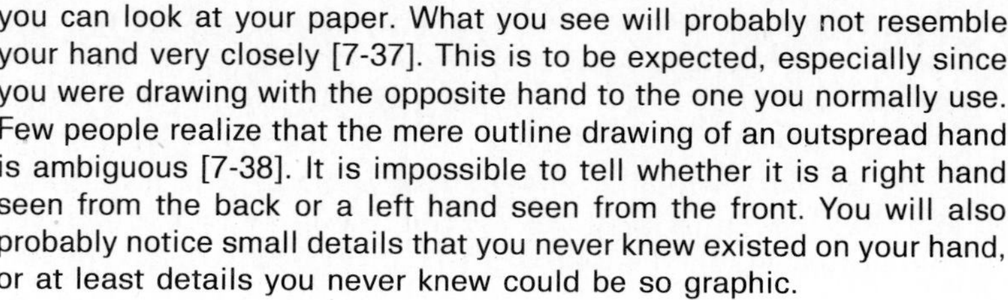

you can look at your paper. What you see will probably not resemble your hand very closely [7-37]. This is to be expected, especially since you were drawing with the opposite hand to the one you normally use. Few people realize that the mere outline drawing of an outspread hand is ambiguous [7-38]. It is impossible to tell whether it is a right hand seen from the back or a left hand seen from the front. You will also probably notice small details that you never knew existed on your hand, or at least details you never knew could be so graphic.

Exercise 7-4

Now we will repeat Exercise 7-3. This time switch and use the hand for drawing that you are normally used to using. If you are right-handed, you can go back to using your right hand, and if you are left-handed, you may once again use your left hand. Now, complete another contour drawing of your opposite hand. (This time it will be the opposite hand that you will be drawing.) Hold your hand slightly below eye level [7-39]. You may wish to pose your hand differently for this drawing, such as a clenched fist or in a pointing position [7-40]. Once again, pick a point on the outside of your hand for your eye to rest on. Slowly follow along that edge with your eye, matching each eye movement with a corresponding pencil movement on the paper. Remember, you are still not allowed to look at your paper, and you should not let the eye move any faster than the hand can move or the hand faster than the eye. Keep the pace slow and consistent, and do not pause at any point during the drawing. This drawing should take anywhere from 15 to 30 minutes if done correctly. You may find that this process becomes tedious as you try to move slowly around the hand, including each little bump and wrinkle. But, as time passes, you will find that you have become enamored with seeing,

[7-37] Drawings made with your nondrawing hand will be crude and should not be judged as to accuracy.

[7-38] The silhouette of a hand (a) may represent either a right hand (b) or a left hand (c).

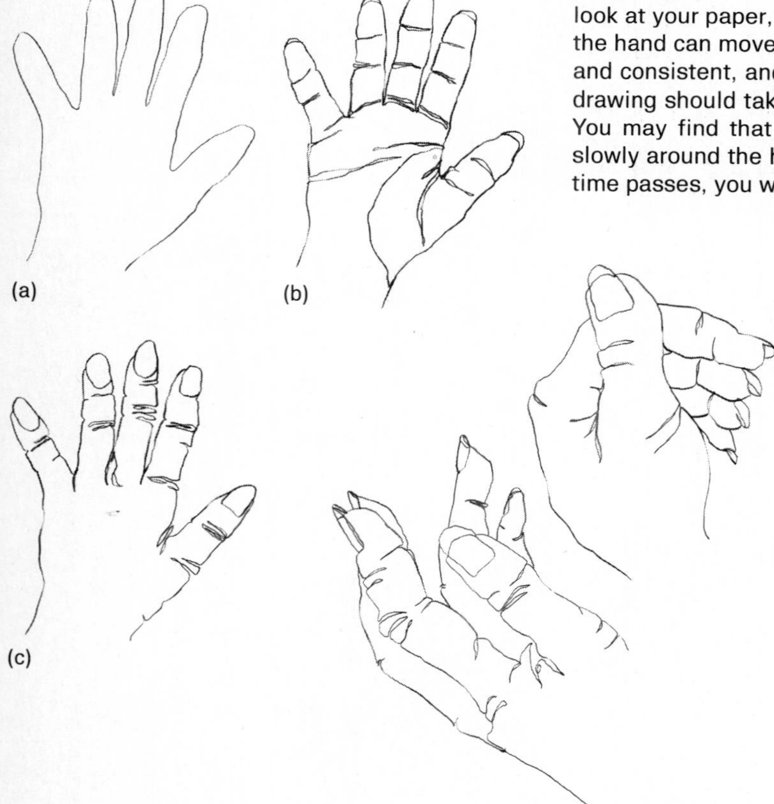

[7-39] Practice drawing your hand in various positions.

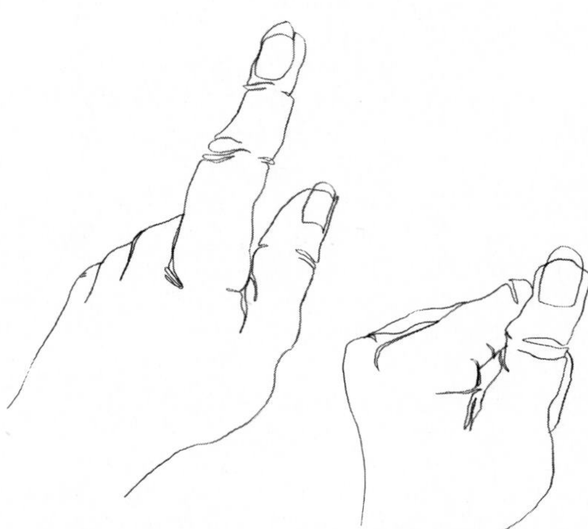

[7-40] Clinch your fist or point your finger and make a contour drawing using your normal hand to draw.

perhaps for the first time. Most students have found that this form of drawing takes more energy and concentration than any of the other types of drawing. It is for this reason that it is vital that you emphasize pure contour drawing in your daily practice drawings. Many professional artists begin each session with pure contour drawings just to increase their level of awareness and concentration.

Modified Contour Drawing

Modified contour drawing is done exactly the same way as pure contour drawing with the exception of one guideline. Now you may look at your paper occasionally, but *only* for the purpose of establishing proper lengths and proportions of lines and making sure that angles are correct. You will also need to glance at your paper from time to time if you need to reposition your pencil to begin a new line on the object. You will still make your pencil follow what your eye is following and, although you are allowed to look at your paper, *you should never be looking at your paper while your pencil is moving*. Since your pencil should only be following what your eye is seeing on the object, it is impossible to make a mark on the paper while looking at the paper and not the object. Be sure you do not use the freedom of being able to look occasionally at your paper as a crutch. It is only intended to be a guide in establishing more realistic relationships.

[7-41] Contour drawings are made up of lines described by an edge.

Exercise 7-5

Take off one of your shoes, or even better, trade shoes with another person in your class. Place the shoe on the table or desk immediately in front of your paper. Pick a point on the shoe to place your eye. Slowly start along the edge of the shoe with your eye, and just as in pure contour drawing, follow that same edge with your pencil. When you get to the end of the edge, you may want to check the length of the line you just made and then reposition your pencil for the next contour. To do this you may glance down at your paper briefly. Do not make any marks on your paper while you are looking at it. Once the pencil is repositioned, begin again. Repeat the same procedure until the entire shoe has been drawn [7-41]. This drawing may take as long as 1 hour if done properly. Remember that you are still only drawing contours—no shading or texture should be indicated on this drawing. Concentrate on what you are actually seeing and its relationship to the whole object. When you have completed your drawing, you should note a dramatic difference in the accuracy of this drawing when compared to your previous pure contour drawings. There will still be inaccuracies in proportion and angles, but these will be improved through additional techniques and exercises that we will do later. Notice that your drawing is actually beginning to look like the object you viewed [7-42]. Now you are learning to see, and in turn you are learning to draw.

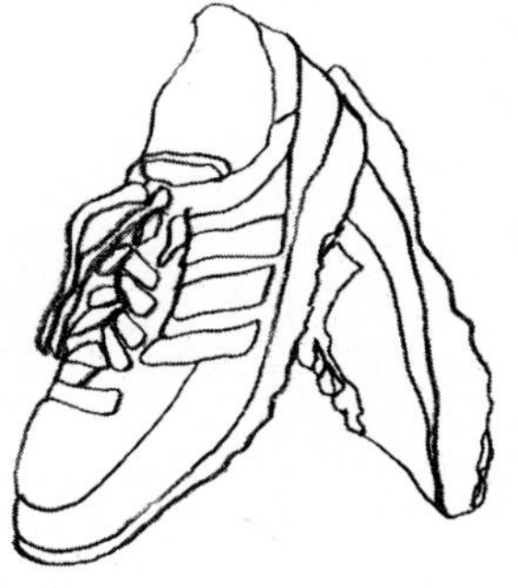

[7-42] Contour drawing is a good method to simplify and at the same time add a certain emotional quality.

Exercise 7-6

Take a sheet of paper, preferably torn from a spiral notebook. Crumple this paper in a wad and then let it slowly open out. Now place it in front of your drawing paper. This will be the subject for our second

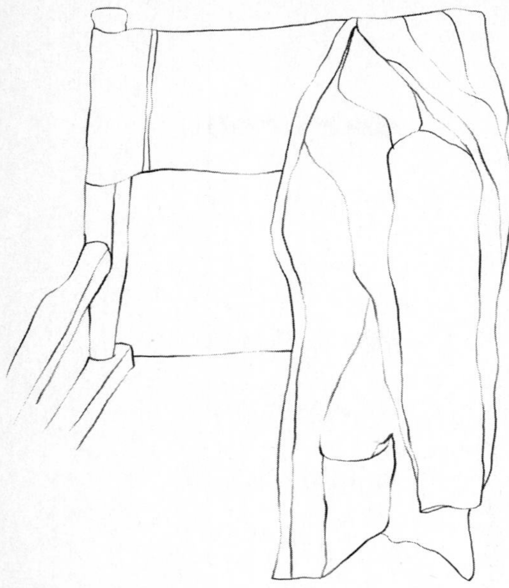

[7-43] Notice that varying the line pressure on the pencil will add character to the continuous-line drawings.

[7-44] Crumpled paper provides a variety of edges for contour drawing.

modified contour drawing. You may find the piece of paper more challenging than the shoe, since it is unique unto itself. No two papers crumple exactly the same, and it will not be as familiar as the shoe might have been. Also, if you have torn the paper from a spiral notebook, you may find all the detail around the torn edge to be quite time consuming to draw if done accurately. Remember, only look at your drawing to reposition your pencil or to check on proportion and angles. Be sure to include all the contours of the different surfaces that have been created when you crumpled the paper.

[7-45] Plants make good subjects for contour drawing.

Exercise 7-7

Now we need to really challenge our ability to see. Take an object that has a lot of detail in it. Some good examples are: an ear of shucked corn, a woven basket, a hand egg beater, a pine cone, or any complex plant or flower. If you browse through the vegetable section of a grocery store, you can find numerous suitable items (artichokes, ginger root, pineapple, and so on) that would be perfect subjects for a challenging modified contour drawing. Spend at least 1 hour making your drawing. Try to see every detail in the object. Remember, you are not to include any shading or shadows, but you should define all the edges (contours) of the details. You will be amazed at the accuracy of your drawing. Notice also that your drawings are becoming more and more expressive. Try to keep in mind your very first exercise of varying the hand pressure placed on the pencil. You should now be automatically creating interesting lines without even thinking about it. Make several other contour drawings using complex objects. When you feel proficient at contour drawing, continue to challenge yourself by combining several objects into entire still-life settings. The more contour drawings you can complete, the more accomplished you will become.

[7-46] Even the simplest of objects make interesting contour drawings.

Cross-Contour Drawing

Cross contours are drawings that help to explain the volume or third dimension of an object. Many students mistake a contour drawing for an "outline" drawing. The two are not synonymous. An outline drawing is very "flat" and shows only the outside lines of an object.

It lacks character. On the other hand, a contour drawing defines the edges of *all* the surfaces, inside the object as well as outside [7-47]. The cross-contour drawing helps to make this more obvious. Cross-contour drawing can define an object using *no* outside edge contour lines. Rather, the object will be defined by showing inside contours only.

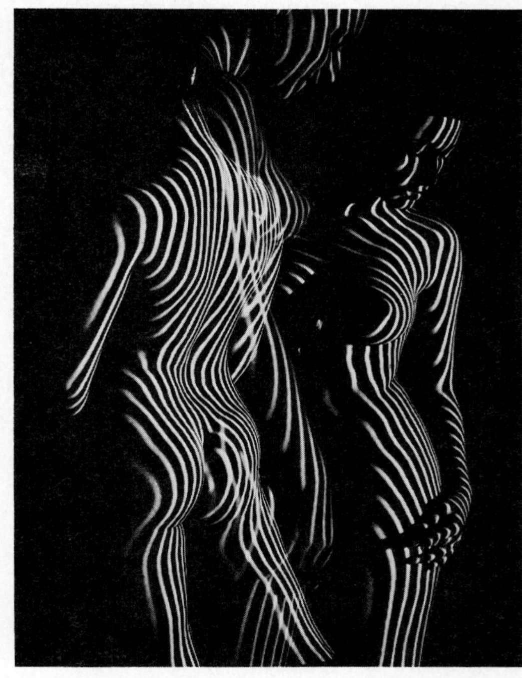

[7-47] Contours add definition and depth to shape.

Exercise 7-8

Once again you will use your hand for a subject. Hold it out in front of you so that you can still follow all the contours with your eye. You may need to study it by placing your elbow on the desk top. Fix your eye on any outside point, just as you did in the first contour drawing. This time, instead of moving your eye along the edge of the hand, move it across the hand at a right-angle path to the outside edge. At the same time your pencil will just be recording on the paper what your eye is actually seeing. Move your eye in this manner until it reaches the other side of the hand. At this point, move your eye back across the hand to the side where you first began. There are no visible contour lines on your hand to guide you, but there is a contour path along which your eye followed. A cross-contour line will show all the individual dips into the hollow places and rise over all of the little bumps and raised veins [7-48]. Students sometimes benefit from actually making drawn black ink lines across their hands to show the path of the cross contour and to get the feel for the shape of the hand. Continue crossing back and forth over your hand until the shape of the hand appears. You will have made no *outside* contour lines to define the hand. Instead, all the cross-contour lines will have defined the extent of the hand by their stopping and starting points. This is also a good time to practice line quality since you should feel a definite variation on the pressure placed on your pencil. More pressure should be applied as you push into hollow places and less pressure as you climb over bumps.

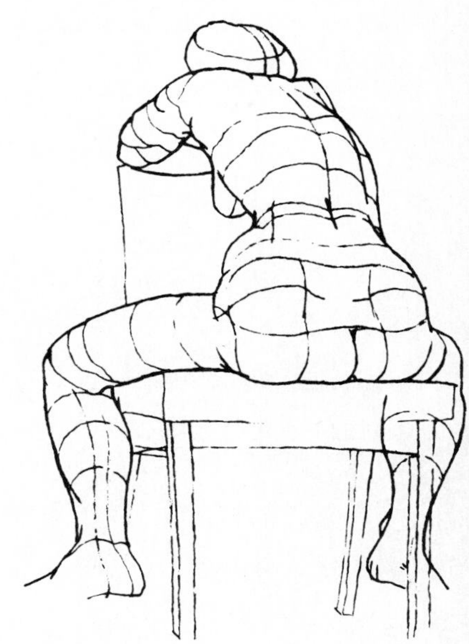

[7-48] A cross-contour line defines on the drawing the image of depth as seen by the eye.

Exercise 7-9

Now use an inanimate object as a subject for a cross-contour drawing [7-49]. You probably have several that you used for the modified contour drawings. Make a cross-contour drawing of one of the same objects. Now compare the two drawings. Your cross-contour drawing should define the outside shape of the object just as accurately as the original contour drawing. If you are having trouble with cross-contour

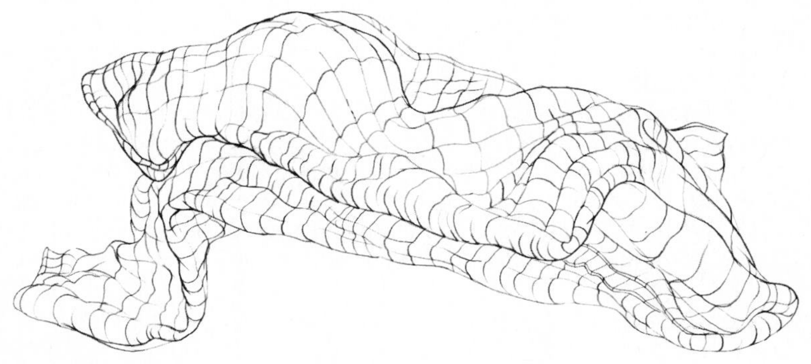

[7-49] A drapery is a good subject for cross contour study. The sea-like roll made of curving surfaces is defined in its up-and-down movement.

[7-50] Do you sense motion or a third-dimensional quality by looking intently at this drawing?

[7-51] Your drawing can produce a sense of action and drama.

drawing, select an object on which you can easily draw across or around (a crumpled box, a detergent bottle, an old shoe) and can actually draw black lines across or around the object. Now make your eyes slowly follow those lines. As your eyes follow those lines, your pencil should be transferring the corresponding image onto paper. A quick way to get an impression of cross-contour drawing is to crumple a piece of lined notebook paper. See what happens to the lines once they are made to go around all the folds and bends in the paper. Practice cross-contour drawing until you are confident that you are able to perceive the volume and mass of an object as well as the outside shape.

LEARNING TO SEE MOVEMENT: GESTURE DRAWING

Gesture drawing is another approach to *learning to see*. The hand continues to duplicate what the eye is actually seeing. A gesture drawing records not what an object looks like or what the object is, but the movement or action of the object. Figure [7-50] is an optical illusion of motion that was produced by A. Michael Noll of Bell Telephone Laboratories. He used computer-generated line patterns to produce wavelike effects that appear to have a convincing three-dimensional reality.[8] In gesture drawing you are going to be relating to what the object is *doing,* its motion and action [7-51]. The practice of gesture drawing will not only help to enhance your ability to see, but just as important, will add feeling and energy to your drawings.

Gesture drawings, just like contour drawings, still include only what is actually seen by the eye. The details of the subject will be ignored and only the action or movement of the object will be included in your drawing. Do not get caught up with making the object actually look like what it is but try to make the movement recognizable. In Figure [7-52] the action of hitting the tennis ball is more important than the identification of the athlete.

If a gesture drawing appears to be a meaningless scribble after your first effort, it does not matter. The main concern is that it should capture the essence of what was happening at the moment that you were making the drawing. Courtroom drawings are an excellent example of how gesture drawing is put to use. They seem to be alive and take on strong personality just by virtue of their gestural quality. Courtroom artists have become very skilled at being able to see quickly and record the action taking place at the moment.

Exercise 7-10

Just as contour drawing makes you believe that you are actually touching the object with your eyes, gesture drawing makes you believe you are actually doing whatever the object is doing. Your pencil will follow the movement exactly, as if it were making the movement happen. You will place your pencil on the paper and not pick it up until the gesture drawing is completed. All linework should be one continuous

[8] John McCarthy, "Information," in *Scientific Technology and Social Change* (San Francisco: W. H. Freeman, 1974), p. 229.

line. There is no time to pick up your pencil and place it back down on the paper. In most cases the action would be completed by the time you could look at your paper and back up at the model.

Before beginning a gesture drawing, place your pencil at a starting point on the paper that will allow enough room to show all the action taking place. A good way to begin gesture drawing is to draw from a model. This can be a classmate or a roommate making quick poses of some action. Start with simple movements such as walking, jogging in place, raising arms, bending down, or clapping hands. These are very simple and not strenuous to the person doing the modeling. Time your drawings and gradually work into longer periods of time. Start with 15-second drawings [7-53]. At first this may seem impossible, and on your first few drawings you will probably hardly have your pencil on the paper by the time the 15 seconds goes by.

It does not matter where you begin your drawing as long as you have picked one point for your eye to start and then to move it along the lines of movement. Remember that you are recording movement only, not detail. Your first drawings will be unrecognizable scribbles [7-54].

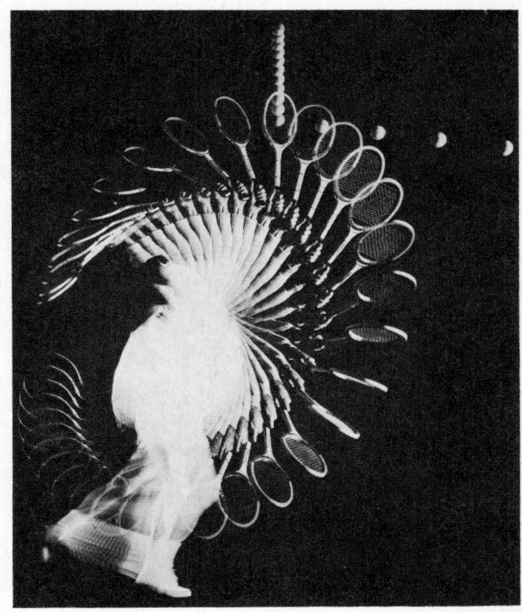

[7-52] Capture the motion . . . .

[7-53] Gesture drawing . . . capture the mood or action in 15 seconds.

[7-54] Your first gesture drawings may appear to be meaningless stick figure scribbles.

[7-55] A gesture drawing that was completed in 3 to 5 minutes.

However, be sure that your "scribbles" resemble the movement that actually took place. As you increase your time on each drawing (progressively add time in 10-second intervals) you will be surprised at how your drawings will begin to take on the characteristics of a person. Continue making pages of gesture drawings until you have completed a series of drawings varying in time from 15 seconds to 2 minutes. It will be necessary to have someone time each drawing, since you cannot be looking at a clock and the model simultaneously.

When starting a gesture drawing, it may be helpful to look quickly for the longest line in the object. In the case of a human subject, you would look for the spine. Then decide what kind of line is being created—vertical, horizontal, diagonal, or a curve. Where is the center of weight being transferred? Just remember to keep your eye and hand moving together, never letting your pencil leave the paper or letting your eye leave the subject.

Exercise 7-11

Now that you have warmed up with the quick "scribble" gestures, you can move to gesture drawings that are done over a slightly longer period of time. Increase the time allowed for each action to 3 to 5 minutes. Your model may have difficulty holding one pose for 5 minutes, so slow down the model's action and perhaps repeat it several times. Also, pick actions that involve more than one type of motion (walking several steps and then kicking the foot, or perhaps clapping the hands). This will cause your final drawing to show a series of gestures drawn one on top of another to show all the action that has taken place. Be sure to start your drawing at an appropriate point on the paper so that the complete range of the action can be recorded. You now have a drawing that

[7-56] The sequence of actions is the most important thing to capture with your eye.

includes a series of actions. Make sure that you do not use the extra time to worry about details of the subject, only record the essence of the movement. We only want to see what the subject is *doing,* not if the subject has hair or is wearing shoes. After completing several drawings, go back and recall each of the actions that took place in each particular drawing. You will be amazed at how your drawings are beginning to become recognizable as records of particular action. Try letting your classmates or friends guess at the action that took place in each of your gesture drawings.

[7-57] Gesture drawings capture the character of the action.

Exercise 7-12

In this exercise we will be combining contour drawing and gesture drawing. Gesture drawing includes more than drawing objects that are moving. It also includes describing the *character* of an action [7-57]. In this sense a plant growing up the side of a wall has gesture, and a crumpled rug on the floor has gesture, even though you do not see the action as it happens. It is important to capture this *implied action* in your drawings. By combining contour drawing for accuracy of detail and gesture drawing to capture the "gesture" or character of the object, we can actually make the subject "come alive." This type of drawing may be identified as a "quick contour" drawing.

Pick a point of focus on a stationary human model just as you did in contour drawing. You will still be looking only at the model and not at your paper. You will be moving your eye much faster and your pencil will be moving much faster than in contour drawing. As you move faster over the contours, your eyes will eliminate much of the detail and you will retain only the "gesture" or pose of the model, not the detail [7-58]. Make sure that you are still making a conscious effort to draw only what you see. Try to move at the same pace throughout the drawing. Draw the entire figure, but do not worry about personal characteristics (hair, eyes, long or short nose, and so on) unless they have a direct relationship to the "feel" of the pose (such as long hair blowing back and forth). Now look back at your original scribble gesture drawings and compare them to your recently drawn quick contour drawings. You should be able to

[7-58] Gesture drawings may be made in a variety of styles.

[7-59] Inanimate objects can have a gestural quality which gives them an implied action.

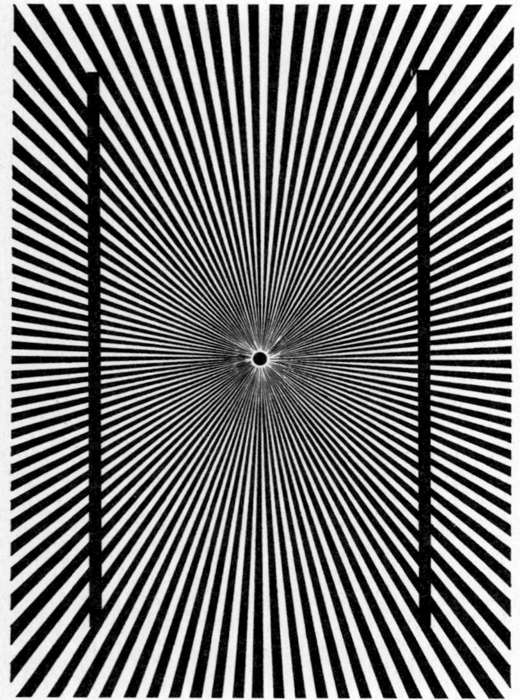

[7-60] Tilt away from you to restore parallelism.

detect the same feeling of action in your quick contour drawing as you have recorded in your moving gesture drawings.

For additional practice in gesture drawing, take a pad of paper and pencil to a playground or park. Watch the children at play or people enjoying a picnic. Make some quick gesture drawings as well as several quick contour drawings. More and more your drawings should begin to clearly define the characteristics of each of the different subjects used in the drawings.

LEARNING TO SEE IN SPACE

One of the more difficult transitions in drawing is to take a three-dimensional subject and then transfer it to a two-dimensional medium and have it still appear to be a three-dimensional subject. What we want to do is create a three-dimensional illusion. We are still going to draw exactly what we see, but we are going to use various illusionary techniques to represent what we see. This will help us to overcome what our mind is telling us that we perceive, rather than what our eyes are actually seeing.

Several methods have been developed to represent three-dimensional space accurately on a two-dimensional surface. A commonly used method is called *linear perspective*. European artists developed this technique during the Renaissance period, and various other forms of perspective have been developed from this method. It is only an approximation of the view seen by the eye. Although perspective can be studied in depth and broken down into a precise technique, you, as a beginning drawing student, will only need to become aware of the various types of perspective and how to apply them to your drawings.

Linear Perspective

In attempting to depict a three-dimensional image on a two-dimensional piece of paper, illusions again play a role in guiding our brain in interpreting what our eye is seeing. In Figure [7-60] the ends of the two parallel vertical lines appear to bend inward. In 1861, Edward Hering discovered that when parallel lines are superimposed on a radial field of lines, they will appear to lose their parallelism. This is because the radial lines are seen both as a two-dimensional pattern on a flat background, and as vanishing lines to a one-point perspective (three-dimensional) drawing. The brain knows that no actual object can be simultaneously two-dimensional and three-dimensional. If you tip the book away from you so that the three-dimensional aspect disappears, the two lines will again appear straight and parallel. This is rather convincing proof that the convergence of lines within our field of vision are important factors in influencing the brain's interpretation of what is seen.

Most artists do not use formal projection system procedures to establish linear perspective in their drawings. Because they draw what their eyes actually see, artists can create a drawing by just looking. However, it is important for us to become generally aware of some of the formal principles involved in perspective drawing so that we can develop the ability to create proper perspective in our beginning drawings.

If you were to stand in the middle of a set of railroad tracks and look down the tracks into the distance, you would notice that the two rails appear to meet in the distance [7-61]. You know for a fact that this condition could not possibly be true since the train continues to run on parallel tracks to its final destination. If this scene were to be drawn relying on what our mind tells us rather than what our eyes actually see, we would draw railroad tracks as two parallel lines running the length of the paper. Such a representation, however, would destroy the illusion of space and take away the "perspective" feeling of the drawing. Instead, let's consider what our eyes are actually seeing.

In perspective drawing it is again possible to use illusionary tactics to deceive the brain [7-62]. Equal-sized figures do not appear to be equal [7-63]. In Figure [7-63] the naturally appearing perspective lines of the background are drawn to suggest a three-dimensional scene. The people are actually the same size individually, but the person on the extreme right appears to be considerably larger. The converging lines of the background confuse the brain as it attempts to make a definitive judgment. This is possible because the brain already has a large data base that offers convincing evidence that equal-sized figures *must* appear to diminish as their images recede into the distance.

If you were to look directly down a road such as shown in Figure [7-64] at the Dallas–Fort Worth Regional Airport, you would see that the light poles seem to be getting progressively shorter as they recede into the distance. At the farthest extreme, the last pole would appear as a point. You will also notice that this point rests on a line that separates the sky and the earth. This is what the eye actually sees, although reality does not exist in this form. In linear perspective, the line where the sky and earth meet is called the

[7-61] Parallel tracks appear to meet on the horizon.

[7-62] The brain has stored the information that from experience the railroad rails are parallel. It will not allow the eye to conclude that the two railroad ties are identical in size . . . which they are.

[7-63] Which guard is tallest?

[7-64] The last pole vanishes as it rests on the horizon.

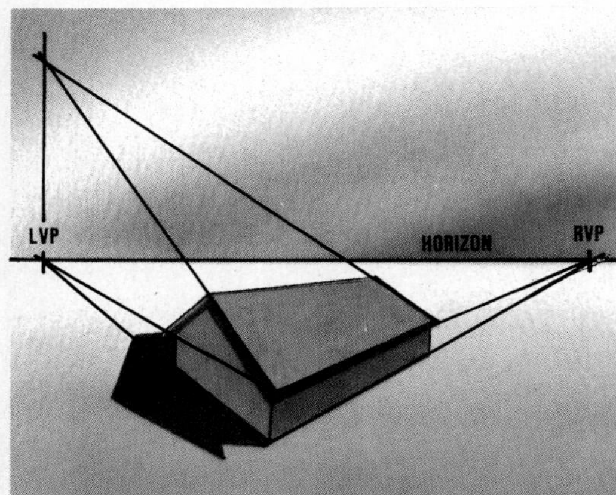

[7-65] Different planes vanish to different vanishing points—on, above, or below the horizon.

[7-66] Bird's-eye view.

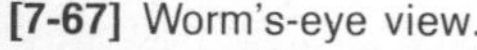

[7-67] Worm's-eye view.

horizon line. The point at which the light poles appear to disappear is called the *vanishing point*. Because the illusion of objects disappearing as they recede into the distance appears with absolute regularity, the sizes and spaces between objects can be mathematically determined, if necessary. However, for simple drawing purposes in this chapter, we will be using "sight" perspective and not relying on any mathematical or projective systems. It will be helpful, however, to understand something about how the system works.

Parallel edge lines of horizontal and vertical surface planes vanish to right and left vanishing points lying on the horizon line [7-65]. The parallel edge lines of angled surfaces (such as the pitched roof of a house) also vanish to points on the horizon or to points directly above or below these points. In Figure [7-65] we can see the horizontal edge lines vanishing to points on the right and left ends of the horizon line. The angled lines of the roof plane vanish to a point directly above the left vanishing point. So much for theory, now let's get back to drawing.

The position of the viewpoint in a perspective drawing is determined by where the viewer is located in relationship to the horizon line. If the viewer is positioned above the horizon line, as if looking out of an airplane, the view is considered to be a "bird's-eye" view [7-66]. If the viewer is positioned below the horizon line, as if standing in a deep depression, the view is considered to be a "worm's-eye" view [7-67]. If the viewer is positioned at eye level with the horizon line, the view is said to be an "eye-level" view. All parallel lines will disappear to the same vanishing point.

Perspective drawings also vary according to whether the objects have one, two, or three vanishing points. The most common case is two vanishing points. For example, if you look at the near corner edge of a rectangular box, you will be able to see two of its vertical sides [7-68]. The top and bottom horizontal edge lines of each side will seem to be disappearing in different directions, thus creating two vanishing points. A drawing of this condition would be considered to be a *two-point perspective* [7-69]. *One-point perspectives* have only one vanishing point [7-70], and *three-point perspectives* have three vanishing points [7-71].

Although the vanishing points and the horizon line may not be located within the boundaries of your drawing paper, you must remember that they are always present. To avoid distortion it may be necessary to locate the vanishing points of your drawing some distance off your paper. This results in a more subtle use of linear perspective and therefore makes the drawing more believable.

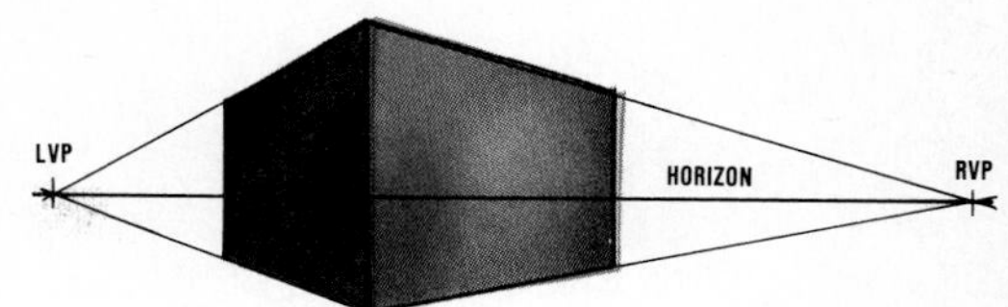

[7-68] Draw a box, using a left and right vanishing point.

[7-69] Two-point perspective.

[7-70] One-point perspective.

[7-71] Three-point perspective.

[7-72] Dürer's wire grid.

Albrecht Dürer (1471–1528) was one of the first artists to develop a device for helping to cope with perspective. In drawing the human figure, perspective is usually referred to as *foreshortening*. Dürer developed a windowlike wire grid that enabled him to foreshorten his visual image as he peered through it at the model [7-72]. The model's head, which is much farther away from the artist, appears to be smaller than it is in reality, and in turn, the feet and legs appear to be much larger than they really are because they are closer to the artist. Dürer then drew his drawing on a piece of paper that matched the wire grid exactly. He copied what he saw through the wire grid, line for line, angle for angle. When Dürer completed his drawing, he had created a foreshortened drawing of the model. Again, this was a tool used to help the artist see since he could not rely on what his mind told him. This is why the Renaissance artists developed the system of linear perspective, to help them overcome what their minds were telling them about what their eyes were actually seeing [7-73]. Through various exercises you can develop a sense

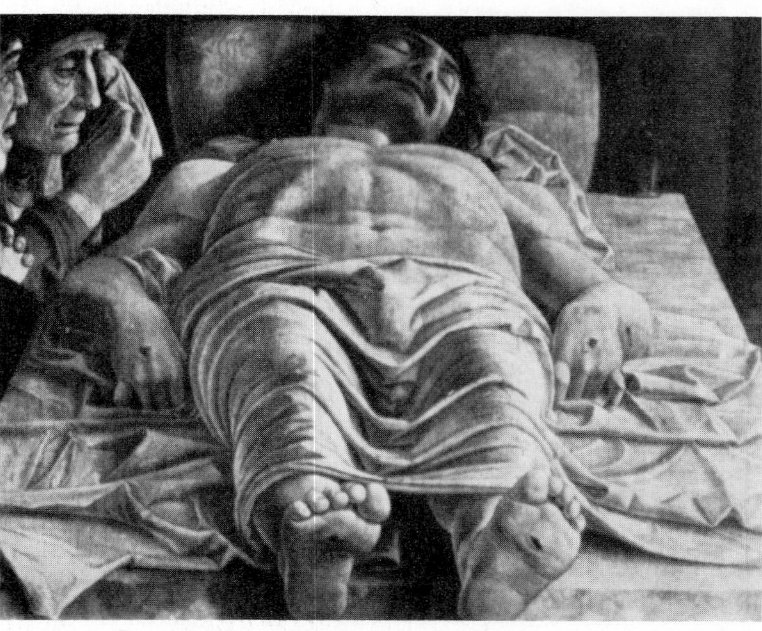

[7-73] The Crucified Christ . . . a famous painting showing the use of foreshortening.

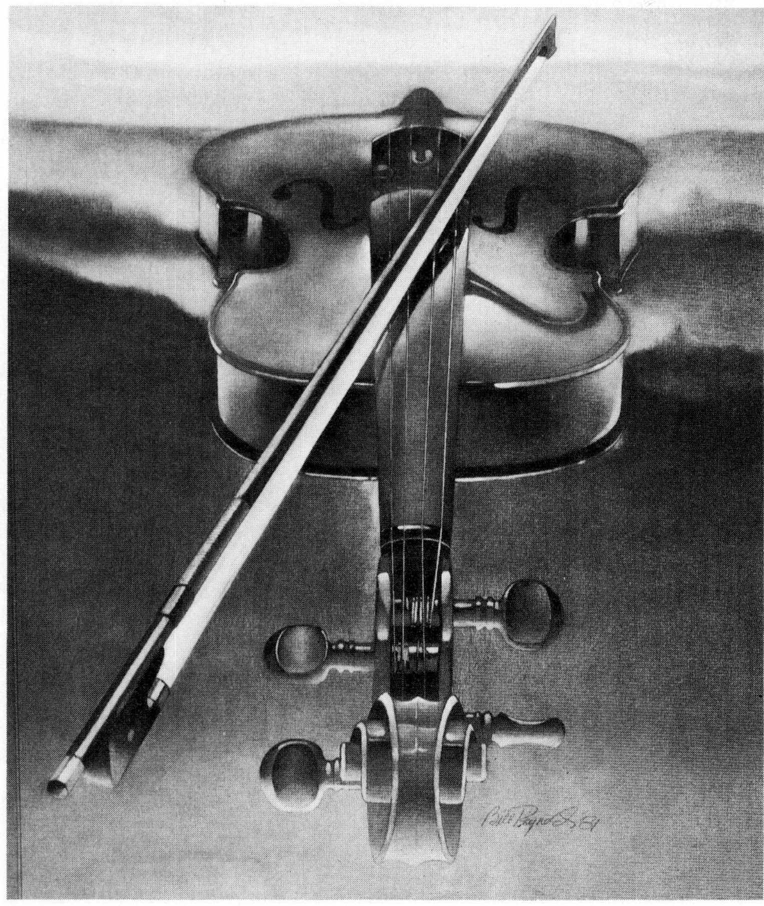

[7-74] Foreshortening adds a sense of realism.

of perspective as a type of second nature. Just as you learned to see detail, you will begin to see perspective in the objects that you draw [7-74] and [7-75].

Exercise 7-13

The horizon line is the foundation reference used for perspective drawing. Remember, it is assumed to be straight and horizontal. Therefore, at a point located approximately in the middle of your paper, draw a horizon line. This is where we will assume that the earth and sky meet. To make a simple one-point perspective drawing, you will only need one varnishing point. Locate it near the center of the horizon line by making a small dot. Now we are going to draw a simple cube. Draw the front face of the cube (this will appear to you as a square) as if you were looking at it directly. Place it either above, directly on, or below the horizon line, but place it to one side of the vanishing point. For this drawing we will want to see at least one additional side of the cube, so make sure that the vanishing point is not covered by any part of the cube. Since we know that all horizontal parallel lines disappear together to the vanishing point, we can now draw lines from each of the corners

[7-75] By use of foreshortening, even common objects can be given a dynamic and interesting quality.

[7-76] View several different box locations.

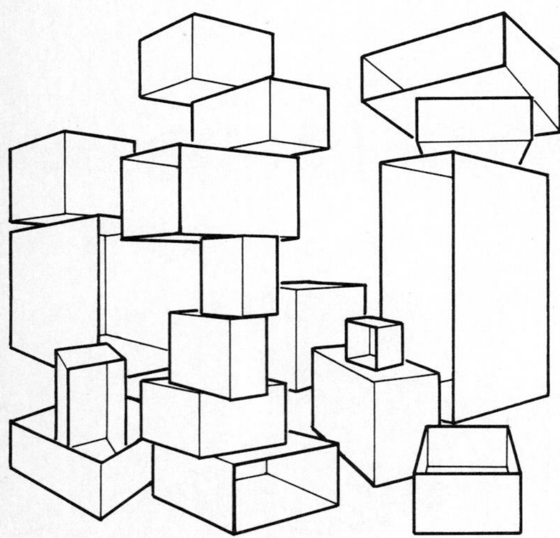

[7-77] Group of boxes.

of the cube's front face to the vanishing point. This will allow us to create the sides of the cube. In our drawing we will assume that the vertical lines still remain vertical. We can now complete the cube by drawing the remaining edges. We have created the illusion in drawing form of the image that our eye is describing. The cube is three-dimensional. Our mind still tells us that all the sides of a cube when measured should be equal, but in reality, if we drew the cube with its receding sides equal in length to the front edges, we would not have created the illusion of a cube at all.

Now repeat this same exercise using two-point perspective. You should place an additional vanishing point on your horizon line. This time, rather than looking at the front face of the cube, you will be looking at the vertical edge created by the two sides of the cube. Therefore, you will begin your drawing using a line representing the point of intersection of the two sides, rather than a square. Connect the top and bottom points of the line with the two vanishing points to create the two sides to your cube. Once again, add the vertical lines for the sides and the parallel lines to create all the remaining edges of your cube.

Try making several cube drawings using a worm's-eye view, a bird's-eye view, and an eye-level view [7-76]. Remember that an eye-level view will be obtained when the cube overlaps the horizon, a worm's-eye view when the cube is above the horizon, and a bird's-eye view when the cube is below the horizon. Next challenge yourself by drawing perspective drawings of groupings of boxes of several different sizes that overlap one another. You may also want to try several groupings of boxes positioned at many different angles, necessitating many different vanishing points [7-77].

Exercise 7-14

In these next drawings you will be able to tap your creative imagination. Using a visible horizon line and two vanishing points, you will draw an imaginary city street scene. In this case we will place the vanishing points off the edges of the drawing paper to avoid distortion of the drawing. The viewpoint can be either worm's-eye, bird's-eye, or eye-level. The more dramatic scenes are created by using either a worm's-eye or bird's-eye view. Try to imagine that you are on a street corner looking down two different streets. You will begin by drawing the vertical corner edge of a building. Remember, each side of the building will then disappear to the vanishing points. Once the first building is drawn, you can continue to draw other buildings along each street. Let your imagination run wild. The city scene may be one of the future, or perhaps it may be a scene from the old Wild West. Add all the little special details that would help describe your city, such as store-front signs, windows, doors, parking meters, hitching rails, cars, sidewalks, and bricks or other surface textures. Keep in mind that all these objects will also shrink to the same vanishing points as your buildings. If you have even one item that you forget to draw in perspective, your drawing will suddenly become less believable. After you have completed your drawing, you may want to lightly erase (did you really throw your eraser away?) your horizon line and vanishing points, since these were merely used originally as guidelines to help you draw in perspective. You will be amazed at how dramatic your drawing looks.

[7-78] This drawing of an ancient city street uses one-point perspective at eye level.

Exercise 7-15

This exercise emphasizes the idea of foreshortening in the use of perspective when drawing the human figure. You will be practicing learning to use perspective as a "second nature" rather than as a formal method of determining distance. Adjust your paper and pencil as if you were going to make a contour drawing. Now, stretch your foot out in front of you so that it is easily visible. If you are seated, you will not have to place your foot out far to be able to see the top of your shoe. You will be applying the principle of foreshortening as you make this drawing. Start by drawing the top of your leg, which is much closer to your eye than your shoe. Now complete the drawing of your shoe. If you are careful about drawing only what you see, your shoe should appear to be much smaller than the top of your leg and knee. Another easy variation of this drawing would be to use your hand and arm instead of your foot. In this case stretch your arm out as far in front of you as possible. Begin to draw what you see, from the top of your arm all the way out to your hand and fingers. Remember, draw only that which you are actually seeing. Your upper arm will appear to be much wider and larger than your hand and fingers, which are farther away. Both of these drawings will give you a rather dramatic demonstration of foreshortening.

Exercise 7-16

Another excellent method for discovering how to capture a feeling of perspective is to draw several "corner" drawings. All we need for a model is the corner of a room. If you are in a classroom, you may want to choose a corner that has either windows or perhaps a chalkboard attached to the wall close to the corner. Place a piece of furniture in the corner (an extra desk and two or three chairs). Give yourself several interesting subjects to draw by adding small details such as a stack of books, coats, pencils, trash can, and so on. Now, sitting back away from the corner, begin to draw all these items as you see them. You will begin by first drawing the corner, including the edges of the walls. Keep in

[7-79] The Egyptians were one of the first to use the principle of overlapping shapes to represent depth.

mind that you will need a horizon line and vanishing points. (Remember, these may not be visible on the paper.) Draw *only* what you are actually seeing. The items closer to you will appear larger than the items in the back that are directly connected to the wall. Later your may want to produce some drawings of corners in rooms at home, or you may prefer to stand out on a street corner and draw a perspective drawing looking down the street. The world about you is full of exciting possibilities that through the use of dramatic perspective can be represented with dynamic and impressive drawings.

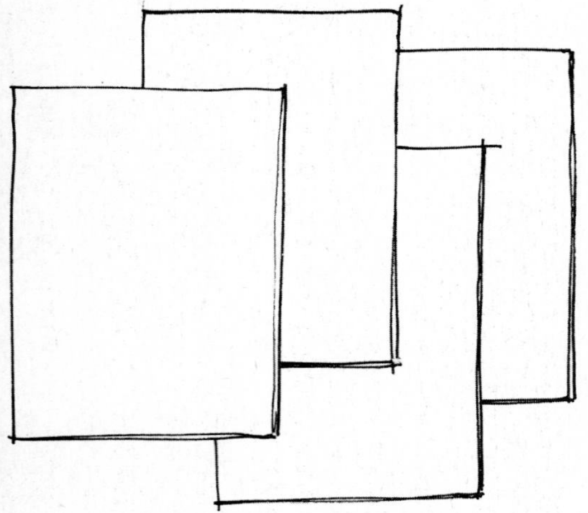

[7-80] Overlapping shapes create the illusion of depth . . . but how far apart are they?

Perspective Illusions

In addition to linear perspective, there are several other visual methods that are used to create the illusion of three-dimensional space. These methods are not as formal as linear perspective and are more directly related to drawing "what you see" rather than what actually exists. The first method involves the use of *overlapping shapes.* When an object or part of an object is hidden behind another object, it is interpreted by your mind as being located farther away than the shielding object [7-79 and 7-80].

However, it is not always evident just how far away the farthest object is located. You might not know, for example, whether a few millimeters or several meters separate the shapes. Sometimes this arrangement can cause a deceptive illusion. There are instances when artists use this type of illusion as a part of their art.

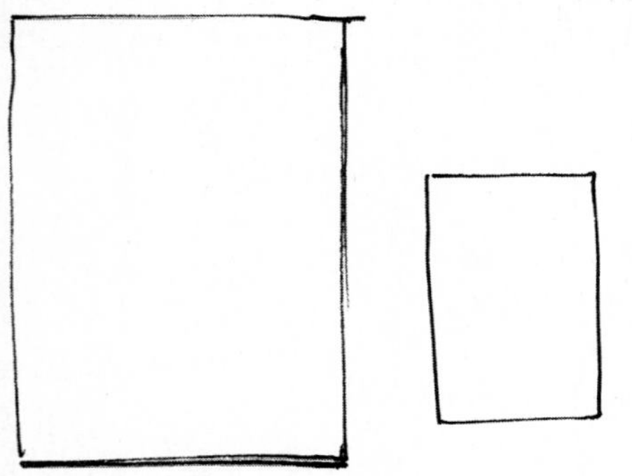

[7-81] If the sizes are identical . . . the larger one must be closer.

Another method of creating perspective illusions is that of *relative size.* If you see two objects that appear to be identical and one appears to be larger than the other, it is natural to assume that the objects are actually the same size and that the one appearing larger is closer [7-81]. This illustrates the power of the brain to convince the eye that the larger object is closer. You are not always aware of your brain's work since it functions automatically and frequently beyond the level of consciousness. Your mind stores a large amount of information concerning the relative size of objects, and it is particularly adept in making instantaneous judgments about the size of the objects that your eye sees. For example, if you see a traffic sign and an airplane at the same time, and the sign appears to be larger, you ordinarily will assume that it is closer to you [7-82]. The cumulative effect of overlapping shapes and relative size increases the illusion of depth [7-83 and 7-84].

Another condition affecting the illusion of depth is *aerial positioning*. This method is based on the fact that the human eye is located about 1.6 meters above the ground. This gives the human an advantage over most animals by providing an increased ability to see over objects and into the distance. Closer small objects therefore appear lower in your field of vision and objects of the same size that are farther away appear higher [7-85]. Look across the room. The chairs that are close appear lower in your field of vision and those farther away appear higher [7-86]. Elevation in the field of view is one of the most firmly established of the illusionary depth cues used by artists working in the European tradition.

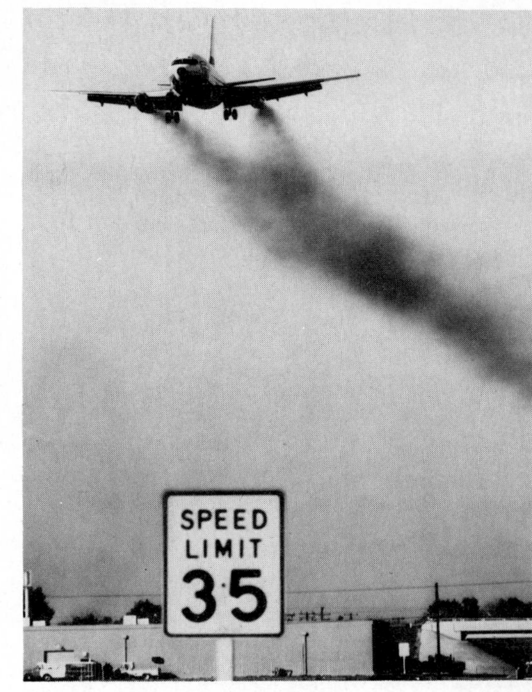

[7-82] Where would the sign be located if it were the same actual size as the airplane?

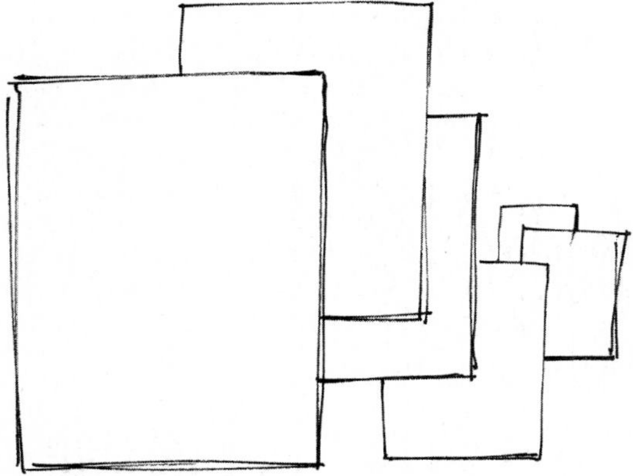

[7-83] Combining overlapping shapes and relative sizes . . . increases the illusion of depth.

[7-84] The brain interprets the combination of overlapping shapes and relative sizes as a function of distance from the observer.

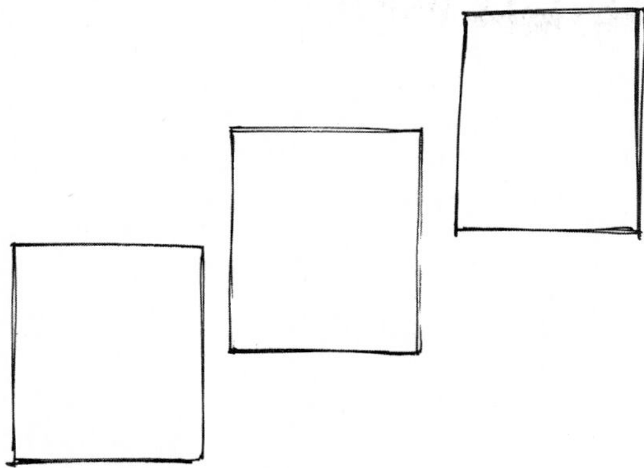

[7-85] Aerial positioning: closer objects appear lower in the field of vision.

[7-86] The nearest chair will appear lowest in the visual field.

[7-87] More distant objects appear less distinct.

[7-88] Distant shapes are less distinct. This increases the illusion of depth.

Another method of illusion involves the clarity of the atmosphere. This includes the effects that air, smoke, fog, and haze have on the appearance of objects. As an object recedes into the distance, detail becomes less defined and more obscure [7-87]. If the light level remains constant, more distant surfaces tend to appear as middle-tone grays. Careful visualization of grass on a golf course will help us to understand this principle. In the near foreground, you can see individual blades of grass with contrasting dark shadows and brightly lit surfaces. Some 10 meters away, the grass surface appears as a uniform fine texture with suggestions of darks and lights. As the fairway recedes, continues on to the next hill, and approaches the horizon, the individual grass blades are no longer identifiable and only shades of green-gray appear. The saturation, or intensity, of the green hue also recedes with distance. Taking these atmospheric effects into account will help to add increased illusion of depth in a drawing [7-88]. It is most effective, however, to combine all the methods discussed of creating illusions—linear perspective, overlapping shapes, relative size, aerial positioning, and atmospheric effects [7-89].

[7-89] Which depth illusion techniques are used here?

Exercise 7-17

Place several simple objects at varying distances from you. Put the smallest item closest to you and the largest items much farther away. An example would be to place a pencil very close to your drawing paper. Place your notebook some distance away. Now draw these two objects. You know that in reality the pencil is a much smaller object than the notebook, but because of its proximity to your eyes, it will appear to be much larger. This is the way these objects should appear on your drawing. Also note that in your vision field the notebook may appear to be above the pencil. Therefore, you will place it above the pencil in your drawing. Try several drawings like this using various objects. In each instance you should still be drawing using the continuous-line drawing method. Look only at the item while you are drawing, and leave out any shading or shadows.

Exercise 7-18

To gain experience with atmospheric perspective, we will draw a landscape or at least an outside drawing [7-90]. (The principle of atmospheric perspective is also applicable for indoor scenes.) Find a landscape in which you can see for several miles. Practice drawing objects very close to you in the scene that you are drawing. Include as much

[7-90] Atmospheric perspective gives the illusion of distance.

detail as possible. Then add additional items in the background. These will not include very much detail, and you may wish to use aerial positioning as well. Of course, it is still important to remember that you should use linear perspective in all these drawings. The other methods of creating depth illusion are also important to enhance the feeling that you are viewing a real three-dimensional scene.

Exercise 7-19

Overlapping shapes is a very easy illusion to use. A simple still-life arrangement will provide you an excellent example of overlapping shapes. Select "everyday" items that you are familiar with, such as a bowl of fruit, a stack of books, a closet full of clothes, or dishes stacked on the kitchen counter. All of these scenes should be drawn to give the illusion of three-dimensional space by using the method of overlapping shapes. Try to find as many items as you can that naturally shows overlapping, and make several drawings.

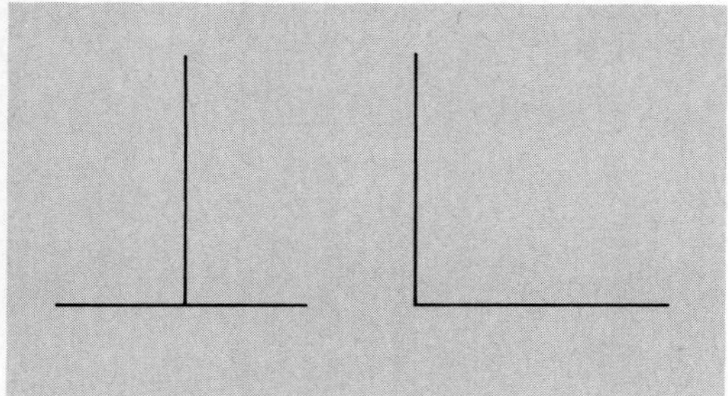

[7-91] Which horizontal line is the longer?

SEEING IN PROPORTION

You may notice in your contour drawings that you have difficulty determining sizes of objects or parts of objects as they appear in relationship to other objects. This is caused by your inability to see "in proportion." Proportion is the size of the parts when measured in relationship to the size of the whole. When you first made a drawing of your hand, you may have drawn a thumbnail that ended up being half as large as the drawn size of your thumb. You did this because it was difficult to determine how large to make the thumbnail in relationship to the rest of the thumb and in turn in relationship to the rest of the hand. Once again your mind has dictated a different arrangement than your eye is actually seeing.

In looking at a doorway, you know the height is supposed to be much longer than the width. *Because your mind knows this,* you may have difficulty in drawing a doorway any other way. Until you learn to draw what your eye is seeing, you may draw doorways that are "out of proportion." There are several illusions that point this out clearly. In Figure [7-91] which horizontal line is the longer? They appear to be different lengths, but if you make a quick measurement, you will see that they are in reality the same length. At this point it is important to convince your eyes to see what you know to be true. In the case of the Müller-Lyer illusion [7-23], you established a sense

[7-92] King Tut's School of Freehand Drawing.

of proportion by measuring the lines to assure yourself that they were actually the same length. This is the same method that an artist uses to establish proportion in drawings.

It would be much too time consuming and impractical to use a ruler and protractor to measure the accuracy of every line and angle drawn. Therefore, a simple method of measurement estimation has been developed. This method has commonly been referred to as *sighting* [7-92]. It is not a precise form of measurement and would never be used in making a scaled drawing. However, it is a quick form of measurement that improves the accuracy of the drawing. Sighting will help you determine relationships between height and width as well as to establish comparative relationships between angles. Once again, let's use the example of the door.

Take your pencil and hold it out directly in front of your face, using a straight, locked arm. Align it with the vertical edge of the door. Sight past the pencil with one eye closed and position the pencil in line with the side of the door that you wish to measure. Slide your thumb along the pencil to adjust its length, matching the apparent distance on the door [7-93]. Now that you have established this vertical measurement, turn your pencil horizontally and align it in a similar way with the width of the door. (Do not move your thumb from its original measured vertical distance on the pencil.) Now continue to "sight across" your pencil and establish the relationship between the height and width of the door [7-94]. Does the door appear to be twice as tall as it is wide, or perhaps does the width measurement appear to be slightly more than one-third the door height? Using this method of "sighting" over your pencil, you should be able to establish comparative relationships between any types of objects or even parts of objects. For instance, hold your thumb directly out in front of you. Now, using the method of sighting over a pencil, establish the size relationship between your thumbnail and your thumb.

It is also possible to establish comparative relationships between angles using the sighting method. Hold your pencil straight out in front of your eyes. Find an angle on the object you are drawing and adjust the pencil so that it lays parallel to or on top of one of the edges [7-95]. Now look at your pencil. At what angle from the horizontal is it tilted? As an alternative, you can hold your pencil horizontally and use it as a base reference [7-96]. Continue to check angles in this manner to establish the relationship of all the angles in the drawing. Remember that a false angle will raise a controversy between your eye and brain, causing an illusion.

Be sure to check the angle and proportions of all of your lines using the sighting method. Once you have developed skill in using this method, many drawings that you previously may have perceived to be "too difficult" are now suddenly easily drawn. You now have a method to make all the objects "fit" together harmoniously in a drawing.

[7-93] Sight over a marker to estimate distances and proportions . . . hold your arm straight with elbow locked.

[7-94] Compare the width of the door with its previously measured height.

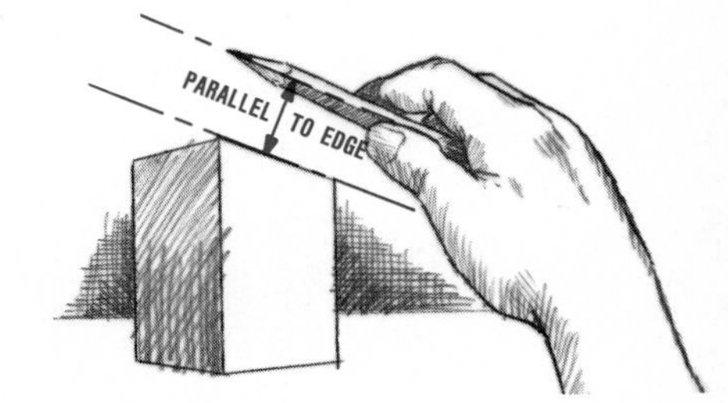

[7-95] A marker held parallel to an inclined edge helps to estimate the angle of decline or incline.

[7-96] A marker held horizontally helps estimate angles.

Exercise 7-20

This exercise will help you use sighting to establish proportions and angles. Tape several rectangular pieces of paper of different sizes at various heights on the wall in front of the classroom. You may want to tape them reasonably close together. Also, tape them in such a way that the top and bottom of each sheet is parallel to the floor and ceiling. Now using the sighting method, make a composite drawing of all the sheets of paper as they hang on the wall. Establish the various sizes in relationship to each other and to their locations on the wall. Include any windows, chalkboards, or bulletin boards that may be on the wall. Once you have completed this drawing, reposition all the papers at various angles to one another. Again use the sighting method to establish the edge angles as well as the paper sizes. Repeat this exercise until you feel confident using the sighting method in establishing sizes as well as angles.

Exercise 7-21

Align several books of various sizes in front of you. Adjust them so that several are standing and slightly opened and several are lying flat on the table. Now draw the set of books as you see them. Use sighting to establish the varying sizes and angles of the books in relationship to one another. This exercise should be repeated several times until you are confident that your drawings are accurate enough to be believable.

SEEING POSITIVE AND NEGATIVE SPACES

Everything we look at has both positive and negative space. Sometimes, as seen in several illusion examples, we have difficulty establishing which is the positive and which is the negative space [7-97]. Generally, the eye will search out the positive space, and the brain will take charge and interpret the result. The illusion of vases and faces dramatically points this out [7-19]. When viewed one way, with the white area as a solid background, the drawing appears to be two faces looking at each other. When your eye concentrates on the black area as being the center of attention, the image of a vase appears. However, either way, there is still both positive and negative space. It is this combination of positive and negative space that makes up the composition of every drawing.

[7-97] Which way are the ducks flying?

[7-98] The judicious use of positive and negative spaces can add interest to your drawing.

The composition of a drawing may be viewed as the arrangement that you have chosen for all the elements of your drawing (both positive and negative). Therefore, you are placing all the positive and negative spaces on the drawing surface. Although we are conditioned through what our mind has learned (to search first for the positive spaces), readjusting our eye's priorities in the simple illusion of the vases and faces shows us that proficiency in drawing can change this ingrained habit. This illusion also illustrates the importance of viewing both positive and negative space images with equal importance.

[7-99] Negative space can be used to create the existence of an "implied" image. Here you *assume* the lady is in bed because of the white space that suggests the existence of bed covers.

[7-100] A few lines suggest the existence of a chair. The absence of detailing of the chair actually adds interest to the drawing and concentrates your attention on the old man . . . which is the intent of the person who made the drawing.

Many times the artist uses negative space to create the existence of an "implied" positive image. Because of this we can see that it is just as important to consider the spaces *that you are not drawing* as it is to give attention to the solid object (spaces) *that you are drawing* [7-98 to 7-100]. It is important that you carefully fit all the pieces of each drawing together, including both the positive and negative spaces. Because we are readjusting your learned behavior, the exercises used here to help you learn to view negative space may seem a bit frustrating at first. Practice will help you to start to perceive many different types of negative spaces, rather than just the traditional view of seeing only solid objects. Suddenly you may find yourself looking at the spaces created by the stadium bleachers rather than the bleachers themselves. Photographers have a finely developed sense for perceiving negative space since they spend a lot of time viewing both shadows and streams of light. The area created by a small ray of light or even a reflection can help to make a very dramatic photograph.

Exercise 7-22

In this exercise we are going to use the example of the faces–vase illusion. You are going to recreate such an illusion *from your imagination.* If you are right-handed, draw a subject's profile on the left side of the paper. If you are left-handed, draw a profile on the right side of your paper. Make up your own version of a profile. Let your imagination run free. You can make a strange looking face or even put warts on the end of the person's nose. Next make horizontal lines at the top and bottom of the profile. Once this is done, repeat the profile in reverse on the opposite side of the paper. Your second profile should match exactly your first one in order to complete the drawing. Now take your pencil and blacken in all the area made inside the vase. What is the result? Have you drawn a vase or two faces? What you originally drew as two profiles has become a vase.

Exercise 7-23

On a sheet of paper draw a square, a circle, a triangle, and a rectangle. Do not let them touch. Now make the negative spaces created by these shapes stand out by drawing around each shape. This will create a shape for the negative space. Using your pencil, darken in the negative space. This should help you begin to visualize that a negative space forms one or more shapes. It should also be clear to you now that positive and negative shapes actually share the same edges. When you draw the edge for one, you have in effect created the same edge for the other. Make several more of these shape drawings to help you become aware of the negative shapes that you are creating.

Exercise 7-24

You are now going to *create something by drawing nothing.* Find yourself a chair or desk and place it at some distance in front of you but positioned so that you have an unblocked view of it. Now in your imagination place the image of the chair on your paper in such a way that several edges of the chair touch the outside edges of the paper. The result will be a rather large drawing since this one chair will fill the whole page. Begin to determine *only the negative spaces* created by the chair. You will have to look at the chair for quite some time before your mind begins to relate *only* to the negative spaces. It may help you to cut a small rectangular window in the center of a piece of cardboard and hold it in such a way that the chair is framed within that rectangle. This will help you to limit the outside edges that correspond with the outside edges of your drawing paper. The window also acts as a "frame" for the negative shapes. Now draw all the negative shapes that you see. Once you have drawn all the negative spaces, you should be able to see an accurate outline drawing of the chair. Remember, all edges made by the negative spaces are also edges of the positive spaces. Once you have completed the drawing, take your pencil and darken all the negative shapes. The result should be a solid white chair on a dark background.

Exercise 7-25

Repeat several negative space drawings using arrangements of selected items. An example might be a grouping of books, pencils, crumpled paper, and a decorative plant. Household items may also be used, such as a coffee pot, silverware, soda bottles, and so forth. Just make sure that you are arranging them to create a number of negative space images between the items. Remember, if you have trouble relating the arrangement of the items to the outside edges of the paper, use the cut rectangular window to help establish the outside edge relationships. Continue to darken the negative shapes to help you visualize them as being just as important as the positive shapes.

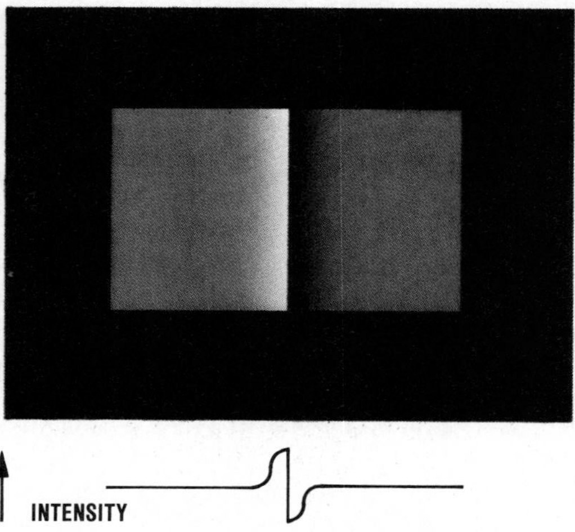

[7-101] These two patterns look much the same. But the distributions of light, plotted below each one, show that the right one is an illusory edge created by a sudden local change in intensity in the middle of an area of equal intensity.

LEARNING TO SEE VALUE

Value is a term used to describe how light or how dark each surface of an object appears to a viewer. The relationship between the values of two areas is called *value contrast*. For most everyone, our world is made up of colors—some light, some dark. It is not often that we get the opportunity to see objects in strictly black and white or even only in shades of gray. Although the world is very colorful, each color takes on a specific value depending on the intensity of that particular color and the light being reflected from it. A yellow color has a much lighter value than a deep burgundy or even a black, but it has a darker value than pure white. The value of this specific yellow would change however, depending on the amount of light being reflected from it. A yellow lemon sitting in a dark corner would appear to be a much darker value than would a deep red apple positioned in front of it that had a bright light shining directly on it. This gives us a good example of value contrast.

Again the illusionary capacity of our eye–brain team must be taken into account. Some 100 years ago, an Austrian physicist, philosopher, and psychologist, Ernst Mach, discovered that if two juxtaposed uniform surfaces are seen reflecting different amounts of light, the border between them appears to be lighter than the light area on the side of the latter, and vice versa. The left and right areas in Figure [7-101] appear to be much the same, except in brightness.[9] The distribution of light plotted below the figure shows that the outer edge is illusory, created by a sudden change in intensity in the center of an area of equal intensity. Cover this apparent edge with your pencil and see how the two areas suddenly reflect the same brightness.

Value can add dramatic impact to a drawing. We use differences in value to help us interpret and recognize objects. A strong contrast of value on a person's face can give it a very dramatic, strong impact,

[7-102] This drawing by Alfred Leslie illustrates the dramatic effect that can be achieved by use of strong contrast.

[9] R. L. Gregory and E. H. Gombrich, eds., *Illusion in Nature and Art* (New York: Scribner, 1973), p. 22.

[7-103] A chiaroscuro by Rembrandt.

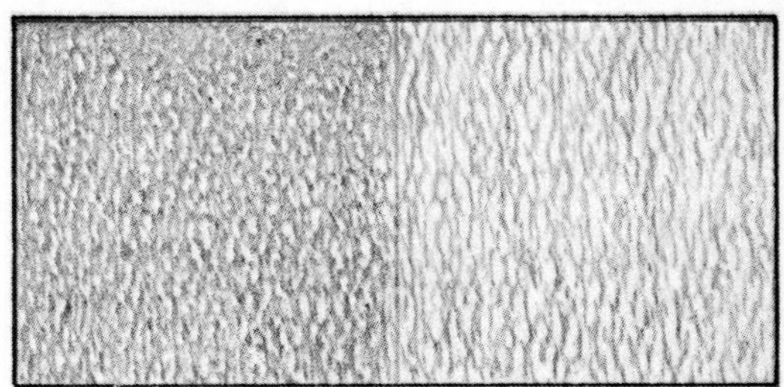

Low value contrast . . . light values adjacent.

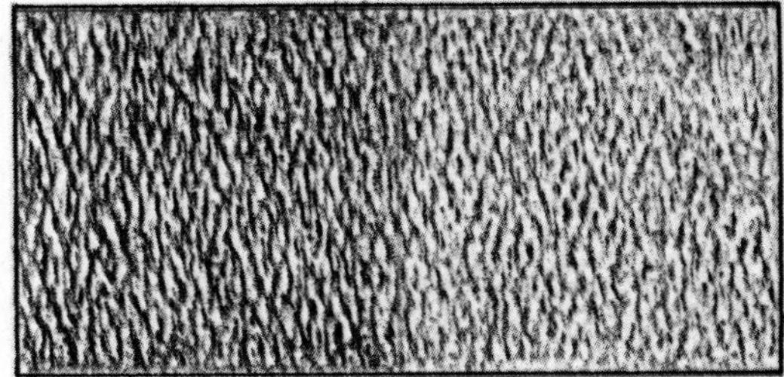

Low value contrast . . . heavy values adjacent.

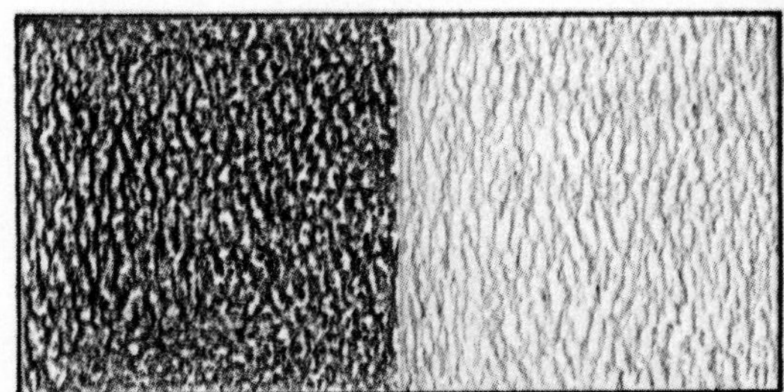

High value contrast . . . light and heavy values adjacent.

[7-104]

perhaps emphasizing rough features [7-102]. Softer values with minimum contrast can make a face look sweet or delicate. One of the first artists to make use of strong value contrast and dramatic lighting was Rembrandt (1606–1669). A strong use of contrast between the light and dark values is called *chiaroscuro* [7-103]. This method of painting became very popular during the Renaissance period.

Value can also help us to imply lines. If you place a very dark value next to a very light value, creating a strong value contrast, there is no need to draw a line to indicate an edge. The eye will automatically see a line where the two values meet [7-104 and 7-105]. Value contrasts can also define textures and patterns. For example, a burlap bag does not have the look and feel of burlap until its texture is defined through the use of value contrast. Values will actually define texture by indicating how lights and darks are created by the surface of an object. This creates a simulated tactile quality [7-106]. Several French painters and early Americans of the nineteenth century such as Harnett perfected a method of using values in what are called *trompe-l'-oeil,* or "fool-the-eye" paintings. Their ability to "see values" is so acute that it is hard to distinguish between real and simulated textures [7-107]. Although you may never be able to develop such a keen sense of value perception, it is important to be able to see values in drawing.

[7-105] Value contrast decreasing with distance provides the illusion of deep space.

Exercise 7-26

The first step in learning to see value is to make a value scale. This will also help you to learn the subtleties that may be achieved by using only one density of pencil lead. Begin by drawing nine vertically connected boxes on your paper. They do not need to be perfect, but they should be aligned adjacent to each other. Leave the top box blank. This represents the lightest value possible, the blank paper. Next go to the

[7-106] Values are used to simulate a texture, such as a wooden surface.

[7-107] William M. Harnett (1848–1892) was a master at producing drawings that had a photographic quality of realism.

bottom box and using your pencil, darken it as much as you possibly can. Make sure that you get maximum coverage from your pencil. This represents the darkest possible value that you can achieve in your drawing. Now progressively darken each box in value, moving down the value scale (from light to dark) from the first box. Progressively darken the value in each box from the one immediately above it. Try to be consistent in the amount that each box is darkened to give you a smooth, uniform scale [7-108].

You can now move on to experimenting with various pencils ranging from a soft pencil (6B), which will give you a very black final value, to a very hard pencil (6H), which will give you a very light, subtle final value. If you want to play with values, there are many ways to experiment, such as making a dark, black value with soft lead and then erase in gradations to obtain the various shades. You could also take a dark pencil and smear the lines with your fingers.

Sometimes artists will use white pencils along with their graphite pencil to increase the values in a drawing. In such a case it might even be possible to obtain a value that would be viewed as being lighter than the plain white paper. Colored paper, such as cream or buff, may also be tried to allow the white pencil to be noticeably lighter in value than the paper. You may wish to refer to the value scale when drawing. This will help you determine appropriate values as you proceed with your drawing.

There are several methods of creating value with a pencil. Smudging a drawn line with the finger has already been mentioned. Rubbing a line with a kneaded eraser is another. Two other common methods are crosshatching and stippling. *Crosshatching* uses a series of short parallel lines to create a pattern and then going back on top of them with another series of parallel lines drawn at approximately a right angle to the first set. Multiple layers of parallel lines can be used to intensify the value [7-109]. You will develop your own style of

[7-108] A value scale . . . the value contrast between each of the pairs of shaded blocks is equal.

[7-109] Crosshatching uses a series of short parallel lines to create a pattern.

[7-110] Stippling uses dots to create value. The value desired can be achieved by varying the number and distribution of dots.

[7-111] A range of values can be achieved by stippling.

crosshatching as you progress with your practice drawings. When a light value is desired, you will draw the lines farther apart and make fewer layers of them. For a darker value, draw the lines closer together and layer them heavily one on top of the other.

Stippling uses dots to create value [7-110 and 7-111]. With the point of your pencil make a series of dots very close to one another until the area becomes gray or even black. By varying the number and distribution of dots, you will determine the value that you desire. Both crosshatching and stippling are valuable tools when you are trying to show textures by using values.

Exercise 7-27

Gather as many white items as possible for this drawing. Examples might include an egg, a baseball, a pillowcase, a cleansing tissue, a piece of paper, a plate, a sock, and a bar of soap. Use your imagination: you will be surprised at how many white items you will be able to find. Place all these items together. Now draw the assembled collection using only subtle values to indicate the various shapes. Your drawing will end up as a very soft drawing, and it will express subtle changes in value [7-112]. This type of drawing is sometimes called "high key." Take your time and really study the subtleties of each value. A variation of this exercise would be to find as many black items as possible and arrange them together for a similar value drawing. Look around and you can probably find other items that have very similar values. These can be used to make another subtle value drawing, this time in "low key."

Exercise 7-28

Take a piece of fruit such as an apple, lemon, or orange, and place it in front of you [7-113]. Now using values only, draw the object. Do not use any lines to define the existing edges seen by your eye, only values. You might want to continue making value drawings using other objects. A glass jar, for example, may reflect a lot of light and therefore it may be easier to define the edges using values. If you have a lamp or strong light, you can also achieve some dramatic effects through the use of strong value contrast.

Exercise 7-29

For centuries artists have used draperies to study value. The shadows and changes in values that are created by the folds and curves in a soft piece of cloth provide interesting subject matter. Take a plain sheet or white towel and drape it over a chair in front of you. If you have a lamp or strong light, this will help to create additional value or at least stronger values. If you do not have extra lighting, don't worry, the values still exist. Just look for them. Now draw the piece of cloth using the values that have been created by the folds and bends. One piece of cloth should provide subject matter for several hours of drawing. Just keep changing the way the cloth is draped over the chair.

DRAWING FROM YOUR IMAGINATION

The first sentence of this chapter states that "drawing is the ability to translate a mental image into a visually recognizable form." The exercises that followed were designed to help train your mind to block out preconceived images and to relate to your hand what your eye was actually *seeing*. By now you will have discovered that you can draw, if you learn to *see*. We are now going to go a step further and involve your imagination. Everyone has an imagination and possesses the ability to create mental images. This creative imagination is put to use in a variety of ways. Some people sing creatively, tell jokes creatively, cook creatively, dance creatively, or even solve engineering problems creatively.

Drawing from your imagination can involve several different aspects of visualization. You may draw exactly what you are seeing in your mind (daydreaming or fantasizing), you may add emotion or character to what your eye is actually seeing, or you may make a drawing entirely from your imagination that serves as your statement (belief) or opinion about something [7-114]. Whatever the purpose of this type of drawing, it will involve the use of your imagination.

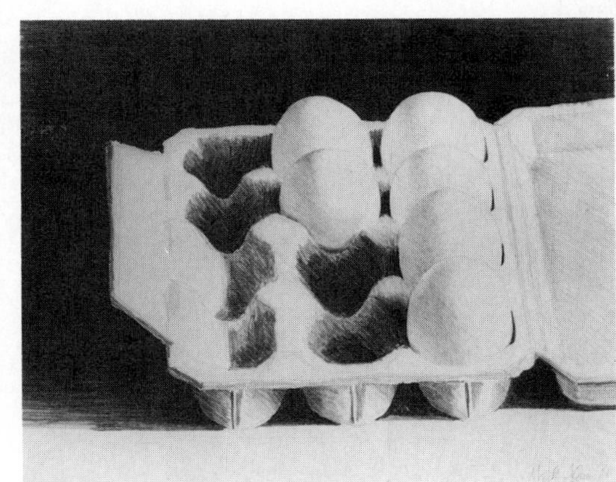

[7-112] Eggs make good models.

DRAWING FROM A DREAM

One of the most common examples of drawing from a dream, or daydream, is the simple "doodle." Most everyone at one time or another has practiced the art of doodling. In many homes a notepad is kept next to the telephone for message taking. If you examine these notes you will find various forms of doodling: around the telephone numbers; a person's name will be embellished with flowers; geometric designs will run up or across the page; or even simple lines and circles will be drawn around the message for emphasis. These are all forms of imagination drawing.

Exercise 7-30

Make a geometric shape in the middle of your drawing paper (square, triangle, circle, and so on). Now make the shape become mechanical. Add gears, wheels, chains, cranks, or whatever it would take to

[7-113] Vegetables make good subjects for drawing.

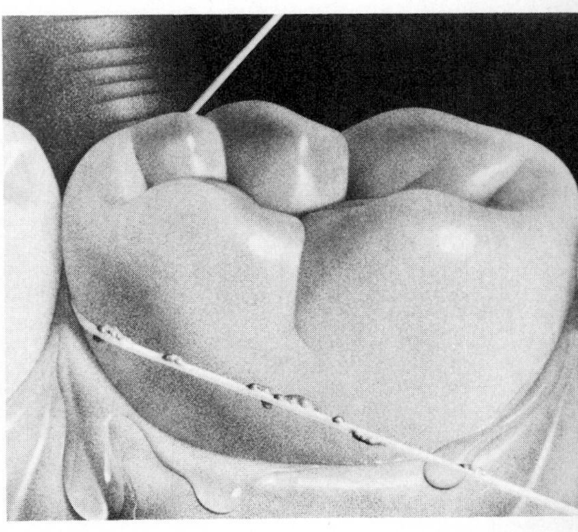

[7-114] If you were a piece of dental floss, imagine how you would spend Thanksgiving afternoon.

[7-115] Ordinary household objects can be altered in their normal functions to accomplish strange or unusual tasks.

make this shape become a mechanical object. Let your imagination run wild. Do not let yourself be inhibited by thoughts of what function your little mechanical object will provide. Just make it mechanical. Now make it interact with another shape. Is it pushing a square, or pulling a circle? Now cause your second shape to interact with another of the shapes. Continue drawing in this manner until you have portrayed a chain reaction "happening" around your paper. Draw all the little shapes so that they are different, and make them perform different actions with the other shapes. You may want to give some of your shapes human characteristics, such as eyes, mouths, or teeth.

Exercise 7-31

Draw a familiar, ordinary household object such as a radio, telephone, water glass, or chair. Make a simple line drawing that does not portray any emotional qualities. Now place your object in an unfamiliar environment that it would not normally be found in [7-115]. You can also make the object's size deviate from that which would seem realistic. For example, your telephone may be sitting in the middle of the Gobi Desert, or the radio could be rolling down the freeway. Salvador Dalí was an expert at placing objects in unfamiliar places or distorting them in unusual ways, such as the melting clock [7-116]. Many times this type of drawing becomes a catalyst for new inventions or improved ways to use an old familiar object.

Exercise 7-32

Draw the essence of a dream. Close your eyes and let your mind wander. You may want to sit back in your chair and spend a few minutes letting your mind fantasize [7-117]. Now start to draw. This form of fantasy should eliminate any preconceived ideas concerning size, proportion, function, and convention. Dreams let us depart entirely from the world around us and can represent a spectrum of emotions from playful to horrifying. Let your pencil translate onto the paper what your mind is fantasizing.

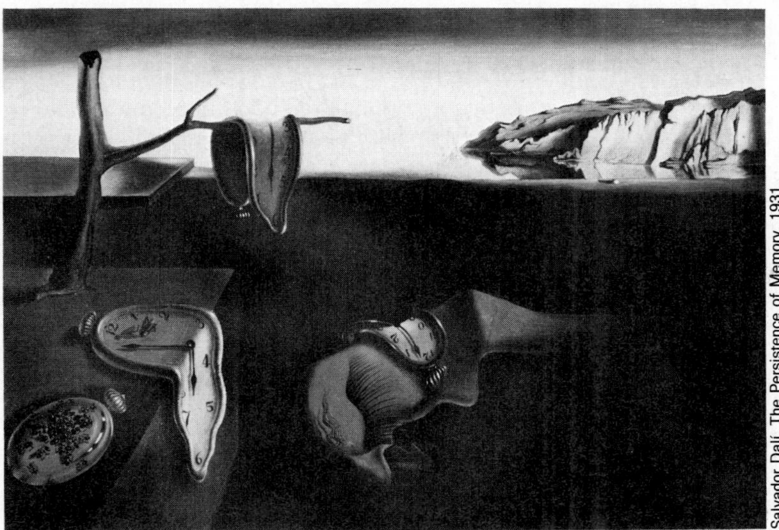

Salvador Dalí, The Persistence of Memory, 1931.

[7-116] Salvador Dalí was a master at giving authenticity to the improbable.

[7-117] Dreams can provide subjects for imaginary drawings.

DRAWING FROM EMOTION

Transferring emotion into a drawing adds character and life to the drawing. Emotions expressed in the drawing can give the added emphasis needed to create a responsive emotion from the person viewing the drawing. It is important that your drawings reflect your emotions; otherwise, they will become static and boring [7-118].

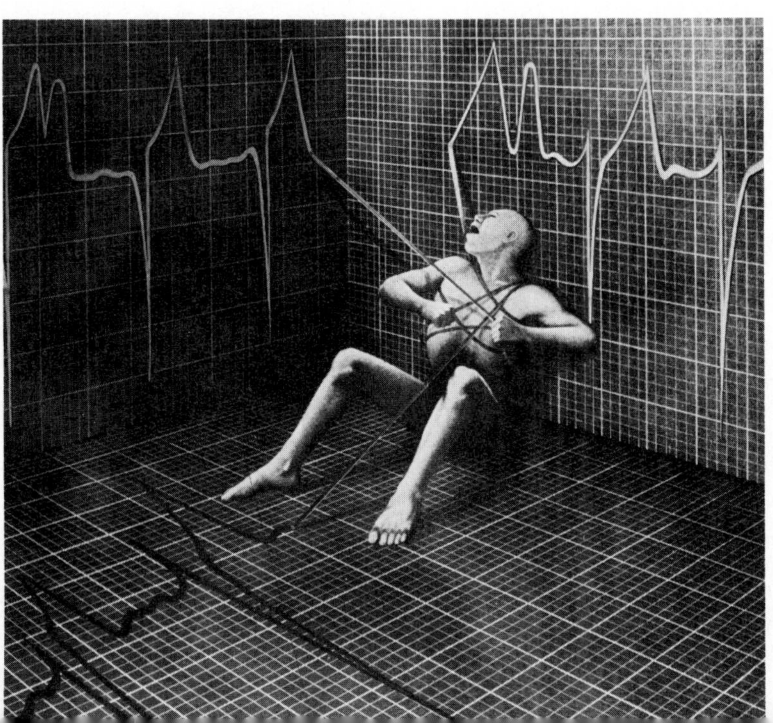

[7-118] Drawings can be a good outlet for expressing your emotions.

[7-119] Walk softly . . . carry a big stick.

Exercise 7-33

Make three drawings of a tree that is growing in three different weather conditions. Draw what you would perceive the tree to look like on a sunny afternoon, on a stormy night, and during an early morning sunrise in winter. Put into the drawing what you would feel like in each of these three situations. Try to evoke those same feelings into that of the tree. Ask someone to decide which tree drawing belongs to each one of the three conditions. Use values in your drawing to emphasize the "mood" of the weather.

Exercise 7-34

After all the contour drawings, your hand should now be a very familiar object. Let's draw your hand again, but try to draw it to represent three states of emotion. How would your hand look when you are angry (clenched fist)? How would you hold your hand in a very relaxed or even a sleeping position? Now, how would you hold your hand(s) if you are very nervous? There are many other emotions that can be expressed by your hand. Draw your hand in other unusual emotional situations.

DRAWING TO MAKE A STATEMENT

Political cartoonists are asked daily to make statements or to give an opinion through their drawing. It is usually very easy for the reader to see at a glance the point the cartoonist is trying to convey [7-119]. Muralists for years have used their art form to make statements and express opinions concerning governments, poverty, and famine. Mexico City is a prime example, where artists have expressed their opinions to the government and to the people of Mexico.

Exaggeration is a method often used by artists of all types to emphasize their opinion. To use exaggeration effectively, however, you must rely heavily on your imagination.

Exercise 7-35

Take an ordinary object with which you may be exasperated or frustrated (a ringing telephone, an inoperative electric can opener, a car that always breaks down, a door that always gets stuck). Now draw this object trying to convey the opinion that you have about it.

In contrast, draw something that gives you a lot of pleasure (a soft pillow, your favorite blanket, your favorite baseball glove). Just by looking at the drawing an observer should be able to tell the difference in your opinion of each of these objects.

DRAWING WITH THE COMPUTER

Once you have developed the ability to see, and in turn, the ability to draw, you can begin to go beyond the limitations of the traditional pencil and paper as a medium and move into a world where the computer is becoming commonplace. So far, you have used your hand and a pencil to transfer to paper what your eye sees. Today, advanced technology has provided us with a new tool, the microcomputer. With it we can produce these same images that our mind perceives and our eye sees. In this respect the computer acts as an extension of the human hand, just as did the pencil. Through electronics these images are traced in light on a phosphor-coated screen, making this technique similar to the pencil-and-paper method.

There are many advantages to using a computer to assist us in drawing. The computer has other almost magical capabilities. Once the image is stored in the computer's memory, it can be duplicated, enhanced, and modified in many ways. An image in the mind can be transformed from a simple concept sketch into a more precise drawing that can be used for analysis and to make manufacturing working drawings. All of this information can be integrated into a data base capable of driving machine tools that will manufacture the part. The engineering workplace of today is not complete without the computer.

While computation and data processing have been the major contributions of computers in the past, interactive computer graphics is a rapidly developing branch of computer applications. For many years computers were too costly and cumbersome to be readily available to engineering designers. Today desktop, portable microcomputers are readily available to almost all designers, allowing for increased use of their graphic capabilities. Without exception these small machines have comparable or even greater computing power than many of the large mainframe computers of just a few years ago.

In addition to miniaturization, equipment and software costs have steadily declined, making interactive computer graphics systems more attractive. Technology in the microcomputer world is rapidly transforming the personal computer into a faster and more powerful machine with expanded memory and higher-resolution graphics.

It is important to understand some commonly used computer terms. The term "software" refers to the prewritten instructions or programs that cause the computer to execute certain operations. "Interactive" refers to the mode of operation where the user interacts

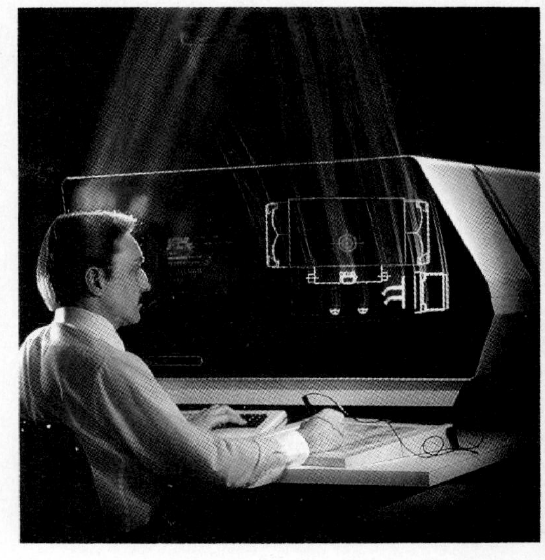

[7-120] Interactive computer graphics workstation.

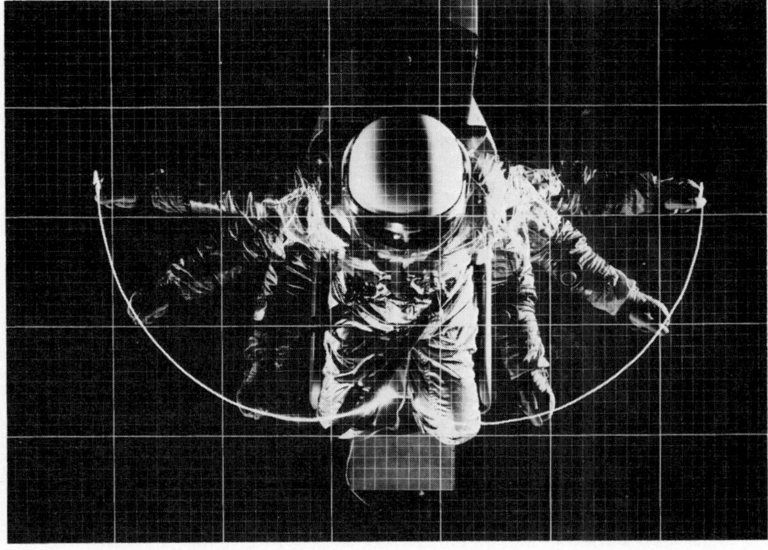

[7-121] Photographic motion study of an astronaut.

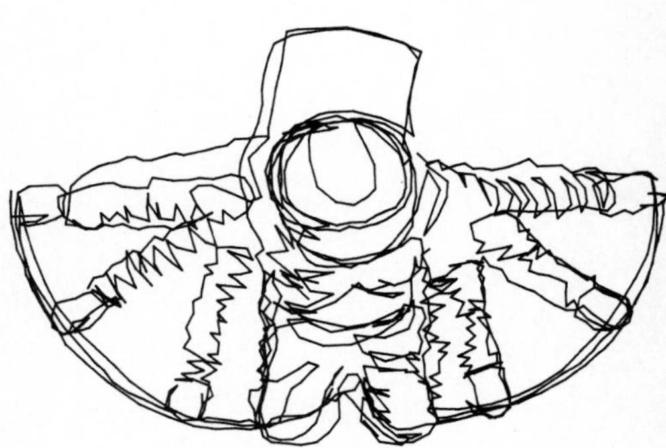

[7-122] Computer-aided gesture drawing of an astronaut.

with the computer by responding to prompts or questions posed by the computer, making choices, and entering data at appropriate times during the execution of a program.

Applications software is designed to be understandable to the average user. This is referred to as "user friendly." Computer programming skills, although quite useful in advanced applications of computer graphics, are not necessary for our use here.

Most interactive graphics systems share the following characteristics.

1. They are menu driven. This means that a list of commands called the *menu* is presented to the user, who selects commands in response to queries from the computer.
2. They are interactive. The user supplies data, makes decisions, and selects operations for the computer to execute, usually in response to prompts or queries from the computer. The execution is in real time. That is, the response of the computer takes place immediately, rather than being processed at a later time.
3. They display graphic and textual material on a phosphor-coated screen, referred to as a cathode ray tube or "CRT," similar to a television screen. Some systems use a single screen, whereas others use two screens: one for textual material and another for graphic images. Some systems use a monochromatic display; others use multiple colors.
4. They may produce plots or "hard copies" of the graphic material (as well as textual material) on paper or plastic film such as Mylar, when desired, or at appropriate points in the design development process.
5. They allow the user to enter, display, and manipulate the geometry and textual material that constitute a model of some real-world entity, such as a manufactured part or a structure. The model, or data base, is stored digitally by the computer in its memory and can be retrieved by the user when desired. Complex mathematical computations are undertaken by the computer on the user's commands to modify the image, causing dynamic changes to the model display. Typical commands that are available cause the image to change location on the screen (translate), change size (scale and stretch), and rotate. Images may also be mirrored, duplicated, merged with others, and easily modified in a number of other ways. Using well-designed applications software should become second nature to the user after a brief familiarization period. Manipulations of the graphic model should be as easy as using a pencil or eraser.

There are many varieties of computer software to support drawing available on the market today. They usually support a particular class or manufacturer of microcomputers. They have a variety of features and capabilities, and you will want to find the particular features that best fit your needs. We will now examine two popular graphic application programs, and compare their features and the similarities which they have to our now familiar pencil-and-paper method of drawing.

PC Paintbrush and MacPaint

Two of the most popular graphics programs for microcomputers have similar features. PC Paintbrush[10] was developed for the IBM PC[11] and IBM compatibles and the MacPaint[12] operates on the Apple Macintosh[12] computer. Although neither of these systems were designed specifically for drafting or design applications, they are extremely user friendly, allowing them to generate freehand graphic images easily and quickly.

These systems are almost self-tutoring and after an hour or so at the controls you should be familiar with their major capabilities. In the margin of the screen you will find the main menu, which consists of graphic symbols (or icons) for commands and functions. These symbols are universal enough to communicate their function quickly and easily to the new user. These graphic symbols (we can think of them as tools in our toolbox) consist of a paintbrush, a spray can, a paint roller, and an eraser. Each of these symbols function just as their name implies. The paint brush makes lines, the spray can makes dots like a spray paint pattern, the paint roller covers areas with solid color, and the eraser moves back and forth to erase images on the screen.

Editing (making changes on a drawing) is easy, and drawings may be cleaned up and minor adjustments made to lines and patterns. A zoom command enlarges the drawing so that individual

[10]PC Paintbrush is a product of Z-SOFT.

[11]IBM PC is a product of International Business Machines Corporation.

[12]McPaint and Macintosh are products of Apple Computer, Inc.

[7-123] PC Paintbrush sketch of a pine cone.

[7-124] Stop-motion photograph of a golf swing.

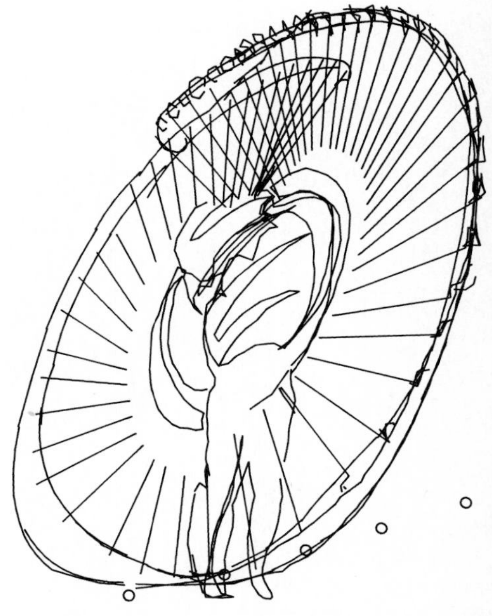

[7-125] Computer-aided gesture drawing of a golf swing.

picture elements (pixels) may be seen and changed. Various patterns are available to fill areas, and in color versions a color pallet is displayed for easy selection.

Text may be added to the drawing in a number of different sizes and styles. Once placed on the screen, text elements may be edited like any other graphic material. The entire drawing, or any portions of it, may also be moved to new locations.

Menu choices are made by moving a cursor arrow with a mouse, and pressing the select button when the arrow points to the icon of interest. The selected command is highlighted to show you the current operating mode. Submenus may be entered with cursor selection. These submenus are temporarily displayed on the screen, to allow you a choice, and then disappear again, allowing unobstructed access to the drawing. These systems are fun to operate, and this feature is a great advantage to the new user, who can quickly learn to explore the graphic solutions to problems without worrying about computer controls, protocol, programming, or other information processing that interferes with spatial perception. There are a number of software packages available for personal computers which operate in a similar way to PC Paintbrush and MacPaint. They are known generically as "paint systems."

AutoCAD

AutoCAD, a product of Autodesk, Inc., is one of the most widely used computer-aided design software systems. It is designed to run on the IBM PC and compatibles. AutoCAD's design and drafting features are beyond the scope of our discussion here, but this program also makes available a sketching mode. In this mode, computer sketching compares favorably with the freehand techniques we have learned in this chapter.

AutoCAD can be used with a variety of input devices, or pointers, such as a mouse, light pen, stylus, or puck with a digitizing table [7-126]. All of these devices may be used to input positional data in the sketch mode. For visualization, however, the stylus or the mouse

[7-126] The user can communicate with the computer using an input device, such as a stylus.

are probably the best choices for simulating the normal motor control of a marking device, such as a pencil or pen.

In other types of AutoCAD drawing, data for locations must be explicitly entered by pointing to geometric features such as points, lines, and arcs, or by entering coordinates for each location. In the sketch mode, the motion of the pointer is captured as a series of linked lines. However, the individual lines may be consolidated into a single entity by using the block command, inserted into other drawings, scaled, stretched, rotated, and mirrored. The drawing may be plotted in your choice of colors, line types, and scales.

[7-127] Computer-aided drawing of Albert Einstein.

Exercise 7-36 Computer-Aided Upside-Down Drawing

Refer to Exercise 7-2 and use the computer in sketch mode to produce a drawing from a photograph, such as the John Wayne photo [7-33]. You may want to turn the photo on its side to better fit the computer screen proportions. As in Exercise 7-2, concentrate on drawing lines, angles, and proportions. When finished, plot your drawing. Figure [7-127] is an example of such a computer drawing that has been made using a mouse.

Exercise 7-37 Computer-Aided Modified Contour Drawing

Refer to the discussion of contour drawing in Exercises 7-3 to 7-5. Using a computer graphics system, produce a modified contour drawing. Use your hand or some other familiar object as a model. While the cursor is moving remember to look at the model and not at the screen. This may be an even more difficult task than it was when you were drawing with a pencil. Think of your eye as moving over the contours of the model. Remember, the computer is acting as a recording device for your perceptions. When you are finished, plot your drawing at several different sizes. Figure [7-128] is an example of a computer-aided contour drawing.

Exercise 7-38 Computer-Aided Modified Cross-Contour Drawing

Again refer to the discussion of contour drawing in Exercises 7-3 to 7-9. Choose a complex subject such as a pine cone, crumpled paper, shoe, or flower. Now, using a mouse, produce a cross-contour drawing, following the discussion of Exercise 7-9.

When you are finished, plot your drawing at half-size or smaller. Notice how shrinking the scale of the drawing during plotting actually enhances its appearance. Graphic artists speak of the effect of scale reduction as "tightening up" a drawing.

Exercise 7-39 Computer-Aided Foreshortened Drawing

Refer to Exercise 7-15. Using the computer as a drawing device, make a drawing of your leg and foot, or your hand and arm, as described in Exercise 7-15.

Exercise 7-40 Computer-Aided Perspective Drawing

Refer to Exercise 7-16 and produce a drawing of the view that you see looking toward the corner of your computer laboratory. Use your knowledge of both sighting and linear perspective. Remember to use the principles of overlap and size reduction as you work.

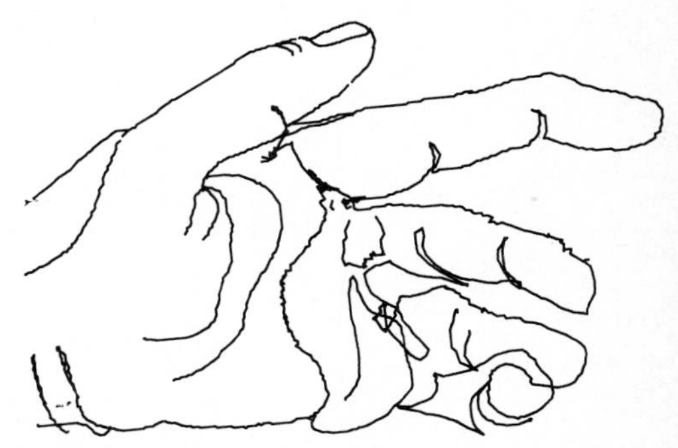

[7-128] Computer-aided modified contour drawing

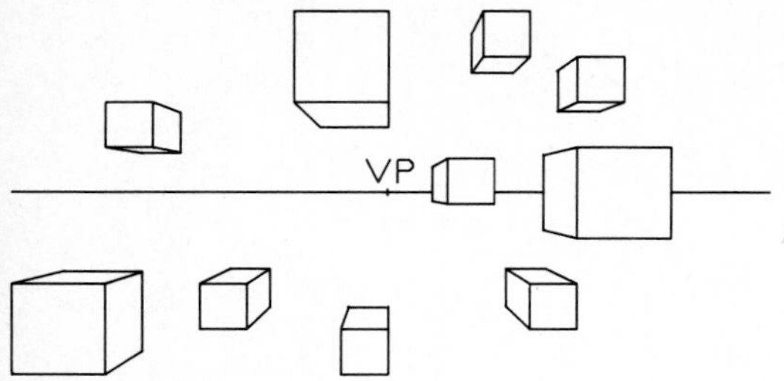

[7-129] One-point perspective cubes drawn on AutoCAD.

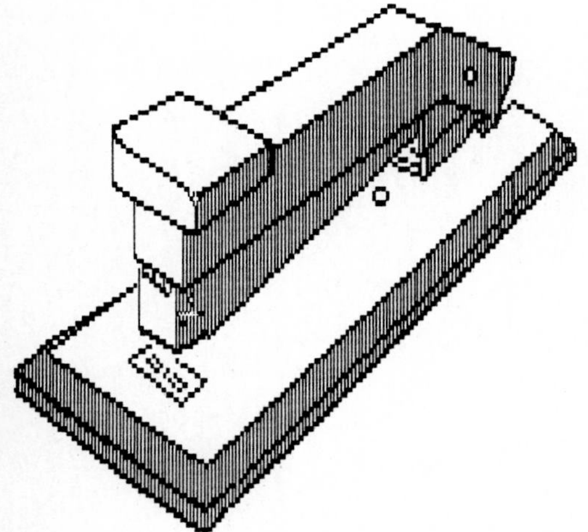

[7-130] Perspective sketch made with PC Paintbrush.

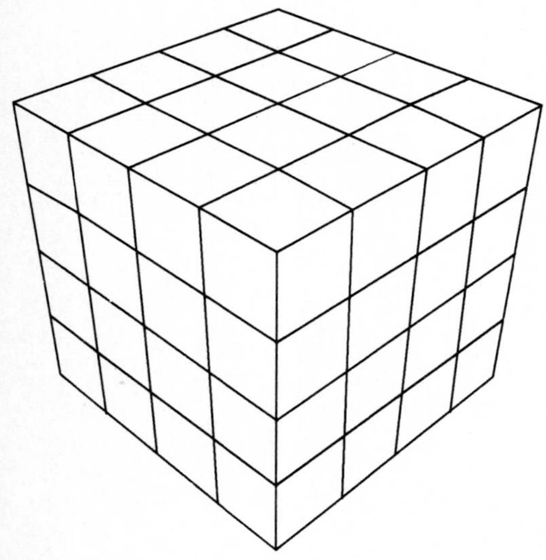

[7-131] Computer-generated three-point perspective grid.

Exercise 7-41 Computer-Aided Linear Perspective Drawing

Refer to Exercise 7-13. This time try to reproduce the exercise using the computer. First, draw a horizontal line through the middle of the screen. This will serve as the horizon line. Next, draw eight or ten perspective cubes following the instruction of Exercise 7-13. Place the cubes above, below, and across the horizon line. Start each cube by drawing a square representing its front face. Next, draw lines connecting the corners of each cube with the vanishing point, a point placed on the horizon line at about the center of the screen. Estimate the length of the receding sides of the cube and make the vertical and horizontal edges parallel to those of the square. Clean up the drawing by breaking and erasing the vanishing lines. Figure [7-129] is an example of a one-point perspective drawing of a series of cubes drawn using AutoCAD.

Now repeat the exercise, using two vanishing points. Experiment with the location of the vanishing points (they should be located on the horizon). You will find that placing vanishing points relatively close together produces strong perspective convergence (parallel lines appear to converge noticeably toward the vanishing point). This convergence may be used for dramatic effect or to create a sense of size. Very large objects such as buildings or ships will seem to have noticeable perspective convergence.

On the other hand, small objects or objects farther away from us may display very little perspective convergence. Figure [7-130] is an example of a drawing produced with the PC Paintbrush System.

Exercise 7-42

Computer-aided drawing can also make use of grids. Figure [7-131] is a user-generated three-point perspective grid useful for sketching a variety of subjects. Better control of proportions and parallelism may be achieved in computer-aided drawing by using grids that are composed of dots or lines. These grids can be changed in scale to suit the need and are provided as a common feature of the software. You can also generate special applications. Orthogonal grids are the most common type. Some systems provide layering, so that the grid may be stored separately from the drawing in the computer's memory. Isometric grids are also available in some systems for use in pictorial drawing.

If you have access to software package featuring grids, try using them for drawing. Lines drawn parallel to the grid lines will have a realistic appearance in the resulting drawing.

Layering, a computer graphics technique that allows you to redraw over a quick sketch to improve proportions and clarity, may be used to prepare a design development drawing in a series of iterations while retaining the option of plotting each state of the development separately. Figures [7-132] and [7-133] show an example of this technique.

If your computer graphics system allows layering, try to develop a drawing in several phases, using freehand drawing to start the process [7-134]. Progressively improve the drawing by working on successive layers [7-135 and 7-136] or in different colors if you have a color system. This use of layering is a technique used by many designers to develop a finished drawing through the use of successive tracing paper overlays.

Circular elements of sketching subjects are usually represented in the drawing by ellipses. Stretching and rotation, two operations made easy with computer graphics, may be performed on a circle to create the correct appearance in pictorial drawings. Figure [7-137] is a drawing produced with the three-point grid.

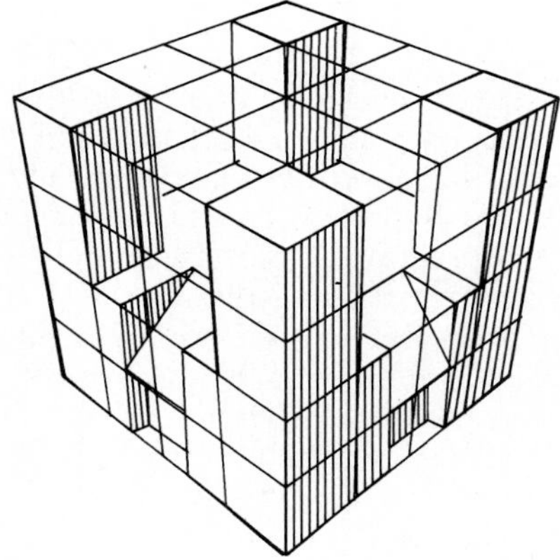

[7-132] Three-point perspective drawing superimposed on the grid.

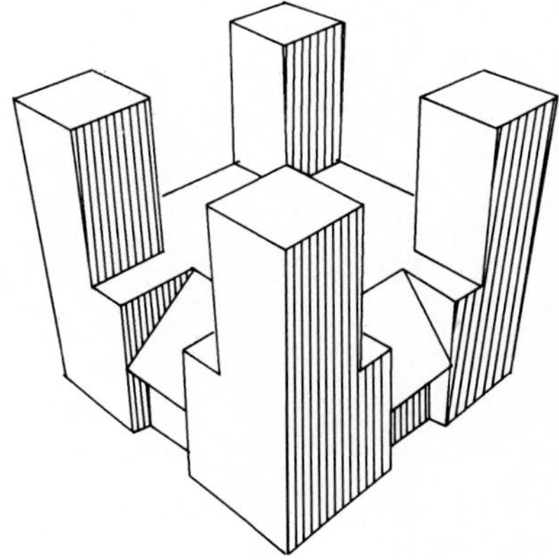

[7-133] Three-point perspective drawing plotted alone.

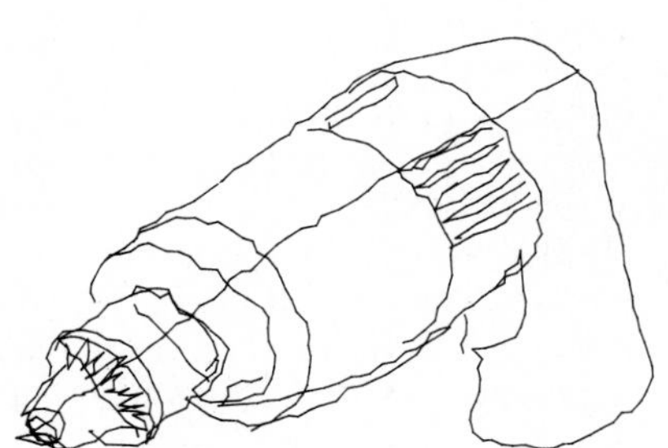

[7-134] Preliminary design using AutoCAD.

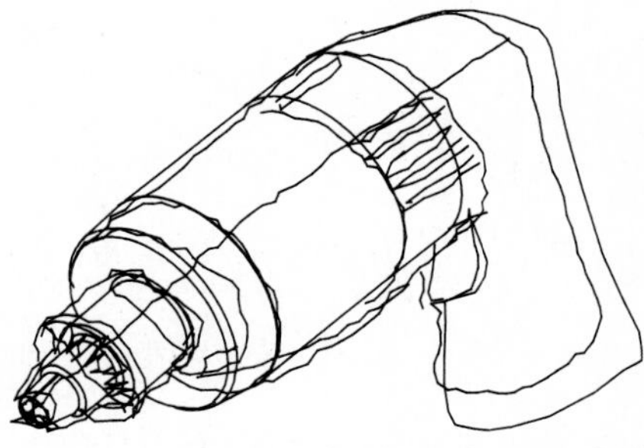

[7-135] Preliminary sketch with refinements superimposed.

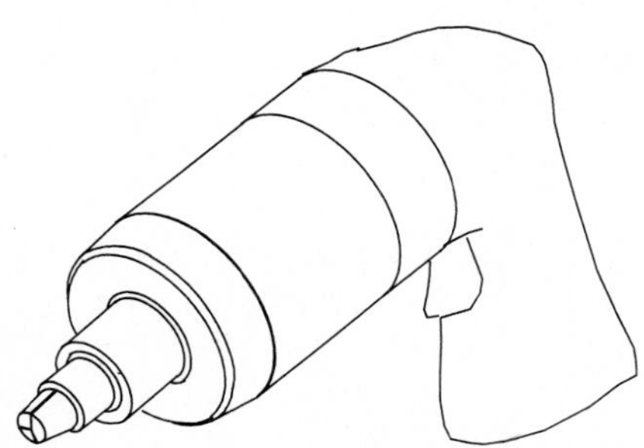

[7-136] Refined design sketch.

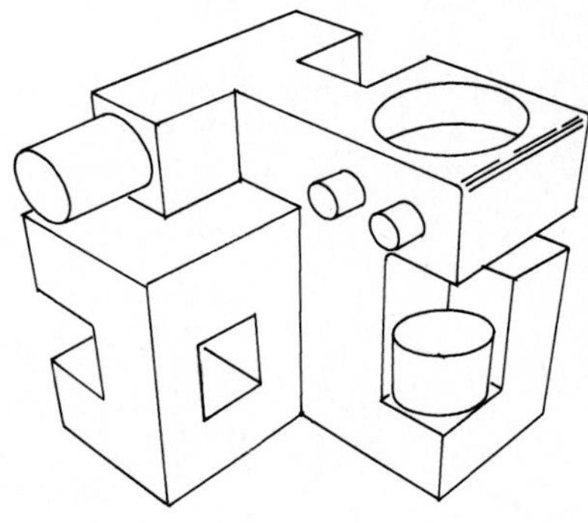

[7-137] Computer-aided three-point perspective drawing.

Chapter 8

Engineering Economy

When a man says money can do anything, that settles it: he hasn't any.

Ed. Howe

Time is money.

Benjamin Franklin

Money is a stupid measurement of achievement but unfortunately it is the only universal measure we have.

Charles P. Steinmetz

[8-1]

In earlier chapters we learned that engineering is a profession that exists only to serve the needs of society. It is a people-oriented activity purposely directed toward the satisfaction of human wants and needs. These needs range from the physical, cultural, and economic to the spiritual. We also learn from our study of physics, and from the natural laws of thermodynamics in particular, that energy can neither be created nor destroyed and that *perfection is impossible* (from this it has been concluded that perpetual motion machines cannot be constructed). As an example, friction and heat losses are commonly found to be nemeses of perfection in the design of machines. These losses are the natural costs of operation. This means that every natural system in our universe operates in a manner such that its energy dissipates, and its behavior is irreversible. More simply, *there is a cost associated with the use of everything.*

Engineering economics is a study of various methods used to evaluate the worth of physical objects and services in relation to their cost. A mastery of these methods is particularly valuable, because they can be used to evaluate the costs before they are incurred. In this way the specific costs associated with the various design alternatives can be evaluated *before* investments of time and money are made. Engineering economy then becomes a very important tool of the engineer, because the same end result can often be attained by several different methods, each with its unique costs. However, this consideration is not unusual. Such an evaluation process occurs almost daily with most people.

There are a number of options available to a lending agency or person who possesses a sum of money. Some of these options are:

1. Hoard it in a secure place (i.e., bury it in the ground).
2. Loan it to someone for love or goodwill with expectancy of its return at a future date (i.e., loan it to your brother or to the local girl scouts organization).
3. Exchange it for products and services to enhance personal satisfaction (i.e., buy food, clothing, jewelry, or medical care).
4. Exchange it for potentially productive goods or properties (i.e., real estate, a taxi, or machine shop equipment).
5. Lend it on condition that the borrower will repay the principal sum with accrued interest at a future date (i.e., U.S. Treasury bonds, or a savings and loan company).

In any case, money is considered to be a valuable asset in all cultures, and the ability to use it wisely is a highly regarded attribute.

TABLE 8-1 PREDICTED CASH FLOWS FOR THE FORD AND CHEVROLET

	Initial cost		*Operating cost*		*Maintenance cost*		*Salvage cost*	
Year	*Ford*	*Chevy*	*Ford*	*Chevy*	*Ford*	*Chevy*	*Ford*	*Chevy*
0	$5000	$4000	—	—	—	—	—	—
1	—	—	$1100	$1000	$200	$250	—	—
2	—	—	1100	1000	300	400	—	—
3	—	—	1100	1000	400	550	—	—
4	—	—	1100	1000	500	700	—	—
5	—	—	1100	1000	600	850	−$1000	−$500
Totals	$5000	$4000	$5500	$5000	$2000	$2750	−$1000	−$500

However, the options that are available concerning the strategies of using money are not well known by most people.

For example, let us examine Mary Brown's dilemma. She is considering purchasing either a Ford or a Chevrolet. Both are used cars, but the Chevrolet is a year older than the Ford. Using data supplied by friends who own similar models, Mary made an evaluation of the investment worth of the two automobiles. She assumed a five-year life for the two vehicles. Her estimated requirements for money to initiate and maintain these possible acquisitions over the years are shown in Table 8-1. Economists refer to such cash outflows or needs (and cash inflows) as *cash flows*. The salvage costs shown in year 5 are negative since they represent money coming back to Mary.

As shown in Table 8-2, Mary then tabulated the total cost for each automobile. On the basis of these calculations, Mary concluded that the Chevrolet was the better buy.

However, because of an ommision on Mary's part, she should not infer from these calculations that it is more cost-effective to purchase the Chevrolet. She has ignored the cost of interest, often referred to as the time value of money. That is, in summing the cash flows over the projected years of life, she has implicitly assumed that $1 spent in year 1, for example, is the same as $1 spent in year 4.

This assumption is not valid. Remember, everything costs something . . . including the use of money. For example, Mary found that the dealership who offered the Ford for sale was willing to finance

TABLE 8-2 SUMMATION OF ESTIMATED COSTS

Cost	*Ford*	*Chevrolet*
Initial	$ 5000	$ 4000
Operating	5500	5000
Maintenance	2000	2750
Salvage	−1000	− 500
Total five-year cost	$11,500	$11,250

The propensity to truck, barter, and exchange . . . is common to all men, and to be found in no other race of animals.

Adam Smith, 1723–1790
The Wealth of Nations

It is a socialist idea that making profits is a vice. I consider the real vice is making losses.

Winston Churchill

There are two things needed in these days; first, for rich men to find out how poor men live; and, second, for poor men to know how rich men work.

E. Atkinson

In modern business it is not the crook who is to be feared most, it is the honest man who doesn't know what he is doing.

Owen D. Young

. . . No business, no matter what its size, can be called safe until it has been forced to learn economy and to rigidly measure values of men and materials.

Harvey S. Firestone

The successful producer of an article sells it for more than it cost him to make, and that's his profit. But the customer buys it only because it is worth *more* to him than he pays for it, and that's his profit. No one can long make a profit *producing* anything unless the customer makes a profit *using* it.

Samuel B. Pettengill

I don't like to lose, and that isn't so much because it is just a football game, but because defeat means the failure to reach your objective. I don't want a football player who doesn't take defeat to heart, who laughs it off with the thought, "Oh, well, there's another Saturday." The trouble in American life today, in business as well as in sports, is that too many people are afraid of competition. The result is that in some circles people have come to sneer at success if it costs hard work and training and sacrifice.

Knute Rockne

[8-2]

Money, which represents the prose of life, and which is hardly spoken of in parlors without an apology, is, in its effects and laws, as beautiful as roses.
Ralph Waldo Emerson
Nominalist and Realist, *1848*

Where profit is, loss is hidden nearby.
Japanese Proverb

Most men believe that it would benefit them if they could get a little from those who *have* more. How much more would it benefit them if they would learn a little from those who *know* more.
W. J. H. Boetcker

You're worth what you saved, not the million you made.
John Boyle O'Reilly
Rules of the Road

the purchase cost at 10 percent. For the Chevrolet she had to resort to a loan from her bank, where the interest rate was 15 percent.

The question arises as to how we should enter these interest rates into Mary's calculations. In order to explain this, let us review the basic methods of handling cash flows over time.

THE NATURE OF INTEREST

If we were to ask the question "would you rather have $1.00 now or $1.00 next year?", most people would choose to take $1.00 now. The principal reasons for this choice are (1) $1.00 now may be invested and earn interest so that it will be worth more than $1.00 after one year; and (2) due to inflation, $1.00 now will purchase more goods and services than $1.00 will a year from now. For the time being, we will ignore the effects of inflation and direct our attention toward the study of interest and its effects on cash flow.

In borrowing and lending situations, *interest* may be thought of as money paid (if you are borrowing) or earned (if you are lending) for the time use of money. It is the *time value of money*. The longer the money is held, the more the interest that is paid or earned. The magnitude of the interest rate is a function of the risk of loss of the borrowed sum, administrative expenses, and desired magnitude of profit or gain. The money being held by the borrower is called the *principal*. It will be held for one or more *periods of time* (weeks, months, quarters, years, etc.). The interest will be earned at a specified rate per period. This is the *interest rate*. For the most part, the rate of interest charged is a function of conditions experienced or specified by the lender. We will develop most of the concepts regarding interest from a borrower–lender viewpoint since this condition is familiar to most people. Later, we will modify the concept of interest so that it is more appropriate for engineering design decisions.

A loan transaction and the interest associated with it are viewed differently by the lender and the borrower. For example, the lender will consider the sum loaned as a negative cash flow and the payments received as positive cash flows. On the other hand, the borrower will consider the principal sum received as a positive cash flow and the repayments as negative cash flows.

To simplify understanding of the principles of engineering economy, there are several terms that need definition as follows:

i = interest rate for a given interest period (often given as a percent, but always used as a decimal fraction)

n = number of compounding interest periods

P = principal sum that exists at the beginning of an interest period at a time regarded as being the present

F = sum of money that exists in the future at the nth interest period measured from a time regarded as being the present

A = single amount in a series of n equal payments made at the end of each interest period

Since there are a variety of ways in which loans can be repaid, we are often faced with the problem of comparing different alternatives that are possible. Thus we need a way to determine the equivalence of the alternatives. Two alternatives are said to be equivalent if they produce the same effect on the system. The concept of equivalence is very important in engineering economy studies. Computational techniques that provide methods of verifying the equivalence of alternatives are the backbone of engineering economic studies. They allow the analyst to evaluate the economic effect of a single project or proposal in many different ways which will result in the same conclusion. Equivalence also permits two or more alternatives to be compared on the basis of either present worth, annual worth, future worth, and so on, with the assurance that whichever basis is used for comparison, the same alternative will be superior. More will be mentioned about equivalent methods of comparing alternatives later in the chapter.

Money is the seed of money, and the first guinea is sometimes more difficult to acquire than the second millions.

Jean Jacques Rousseau
Discours sur l'origine et le fondement de l'inégalité parmi les hommes, *1754*

90 percent of new products fail, in the sense that they are pulled off the market within four years of launch; in the more specialized areas, such as selling to the Original Equipment Market rather than to the consumer, possibly as many as two-thirds of the new products and processes lose money.

Design News
April 27, 1970

SIMPLE INTEREST

In simple-interest situations, the interest earned is directly proportional to the capital involved in the loan, and the interest rate is applied each period only to the principal amount.

Example 8-1

$100 is borrowed for five years at a simple interest rate of 10 percent per year. How much is owed at the end of five years?

Solution:

In general, if a sum of money or principal (P) is borrowed for (n) time periods at a simple rate (i) per period (note that i is used as a decimal fraction), the amount owed (F), called a *future amount,* at the end of n periods is given by

$$F = P + nPi = P(1 + ni) \tag{8-1}$$

$$= \$100(1 + 5 \times 0.10)$$

$$= \$150$$

From the following tabulation we can see that this relationship is consistent throughout the lending period.

End of year	*Interest during year*	*Amount owed at end of year*
1	$10	$110
2	10	120
3	10	130
4	10	140
5	10	150

There is one rule for industrialists and that is: Make the best quality of goods possible at the lowest cost possible, paying the highest wages possible.

Henry Ford

Success is that old ABC—ability, breaks, and courage.

Charles Luckman
Quoted in the New York Mirror, *September 19, 1955*

An idea, in the highest sense of that word, cannot be conveyed but by a symbol [of factor label (ed.)]
Samuel Taylor Coleridge, 1772–1834
Biographia Literaria

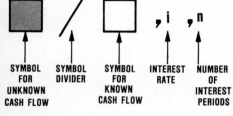

COMPOUND INTEREST

Most economic situations in business, government, and industry are governed by compound rather than simple interest. Therefore, for the remainder of this chapter all calculations will assume that the interest rate is compound rather than simple.

There are six common compound-interest factors (mathematical relationships) that we will use in comparing alternatives. These are shown in Table 8-3. To solve problems they are used in conjunction with cash flow quantities P, F, and A (see definition of terms, p. 208) as follows:

$$\text{desired amount} = (\text{given amount}) \times \begin{pmatrix}\text{discrete compound-}\\ \text{interest factor}\end{pmatrix}$$

Examples:

$$F = (P) \times (\text{single-payment compound amount factor})$$

$$A = (F) \times (\text{sinking fund factor})$$

$$P = (A) \times (\text{uniform series present worth factor})$$

In the material that follows, the compound interest factors are explained and example problems are used to illustrate their application. In each case, the formulas used are for lump-sum cash flows and for interest that is compounded at the end of finite-length periods (such as a month or a year). To simplify the calculation process, tables of interest factors are included in Appendix III. Also, a type of shorthand code will be used to refer to these six types of problems. The code consists of four parts and is referred to as the interest *factor label*. In each case one part of the label will be unknown and the other three parts will be known. Placement of information within the label is invariable, and the parts are ordered as shown in Figure [8-3].

, i , n

SYMBOL FOR UNKNOWN CASH FLOW | SYMBOL DIVIDER | SYMBOL FOR KNOWN CASH FLOW | INTEREST RATE | NUMBER OF INTEREST PERIODS

[8-3] The factor label.

Examples of factor use:

1. Solve for F, the equivalent future worth. Given: P, i, n; use the label $(F/P, i, n)$.
2. Solve for P, the equivalent present worth. Given: A, i, n; use the label $(P/A, i, n)$.
3. Solve for the uniform payment, A. Given: F, i, n; use the label $(A/F, i, n)$.

When used as a part of a cash flow diagram it is convenient to use the circular form of the factor label in the manner shown in Figure [8-4]. Use of the factor label in cash flow diagrams is discussed in more detail in the following section.

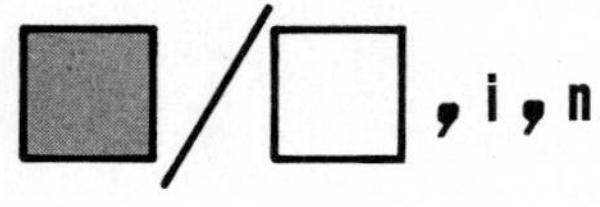

BECOMES

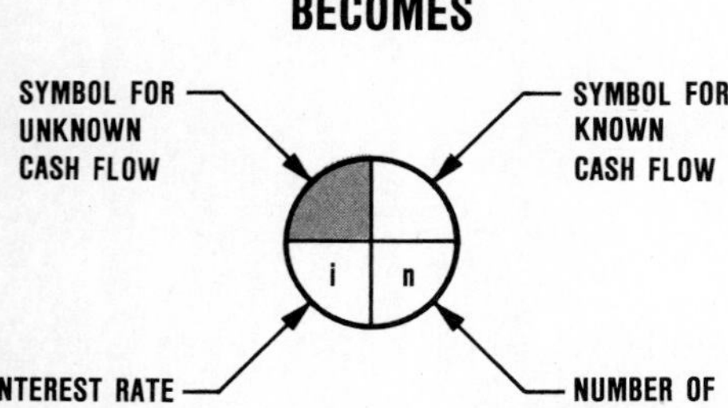

[8-4] The circular factor label.

CASH FLOW DIAGRAMS

In engineering economy studies, it is often helpful to draw diagrams that represent the flow of cash over time. These diagrams are similar to the free-body diagrams to be introduced in Chapter 11. Vectors whose lengths are proportional to the magnitude of the cash flows

TABLE 8-3 DISCRETE COMPOUND-INTEREST FACTORS AND LABELS

Desired	Given	Factor label	Factor[a b]	Name of factor
			Single payment	
F	P	F/P, i, n	$(1+i)^n$	Single-payment future worth
P	F	P/F, i, n	$\frac{1}{(1+i)^n}$	Single-payment present worth
			Equal payments	
F	A	F/A, i, n	$\frac{(1+i)^n - 1}{i}$	Uniform series future worth
A	F	A/F, i, n	$\frac{i}{(1+i)^n - 1}$	Sinking fund
P	A	P/A, i, n	$\frac{(1+i)^n - 1}{i(1+i)^n}$	Uniform series present worth
A	P	A/P, i, n	$\frac{i(1+i)^n}{(1+i)^n - 1}$	Capital recovery

[a] Note that three of these factors are reciprocals of the other three factors.
[b] Calculated values of the discrete compound interest factor are given in Appendix III for various i's and n's.

Symbols
i = Interest rate for a given interest period (often given as a percent, but always used as a decimal fraction).
n = Number of compounding interest periods.
P = A principal sum that exists at the beginning of an interest period at a time regarded as being the present.
F = A sum of money that exists in the future at the nth interest period measured from a time regarded as being the present.
A = A single amount in a series of n equal payments made at the end of each interest period.

are plotted proportionally in a vertical direction. Upward vectors represent positive cash flows or money earned. Downward vectors denote negative cash flows, or money paid. The horizontal axis represents time. Figure [8-5] illustrates the relationship between a present sum P and a future sum F using cash flows over time. Note that in Figure [8-5] the cash flow at time 0 is downward, whereas at time n, it is in an upward direction. This represents the situation as viewed by the lender. From the borrower's viewpoint, the directions of the vectors would be reversed. Without specifying the nature of the situation, the direction of the cash flow vectors is completely arbitrary.

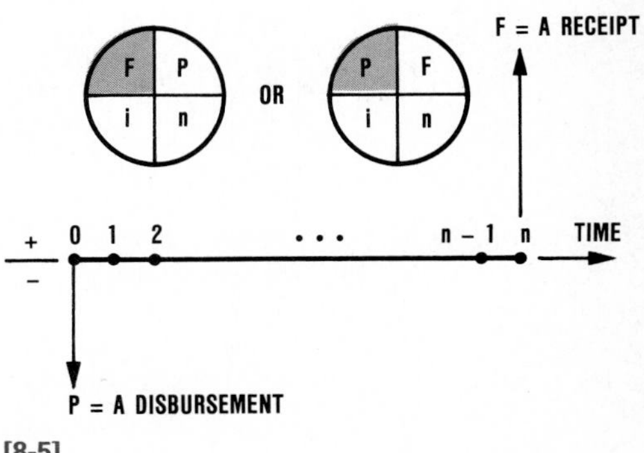

[8-5]

COMPOUND INTEREST—SINGLE PAYMENTS

In compound-interest situations, the interest rate is applied each period to the principal amount plus all previous interest charges (or earnings). This means that for each time period the borrower pays

The most dangerous thing in the world is to try to leap a chasm in two jumps.

David Lloyd-George

Mere parsimony is not economy. . . . Expense, and great expense, may be an essential part of true economy.

Edmund Burke
Letter to a Noble Lord, *1796*

interest *on the interest previously paid* plus the interest on the principal that remains. This is advantageous to the lender.

Example 8-2

Suppose that the 10 percent per year interest rate of Example 8-1 is changed to a compound rate (i.e., $i = 10$ percent per year compounded yearly). How much is owed at the end of five years?

Solution:

Let us first consider this problem from a year-by-year viewpoint. We could tabulate the interest owed as follows:

End of year	***Interest during year***	***Amount owed at end of year***
1	\$10.00	\$110.00
2	11.00	121.00
3	12.10	133.10
4	13.31	146.41
5	14.64	161.05

Note that the interest charged during any given year is the result of applying the interest rate (i) to the amount owed at the end of the preceding year. Table 8-4 generalizes the result for any principal (present value) (P), interest rate (i) per period, and accumulated amount (future value) (F).

We have seen that the relationship between a single amount P (present value) and future amount F is given by $F = P(1 + i)^n$. The factor $(1 + i)^n$ is called the *single payment future worth factor* (see Table 8-3) and its label is

$(F/P, i, n)$ or [F P / i n]

The term F/P is read as "F given P." For example, this is the factor to be used to find F, given a value for $P = \$100$ (and also given the interest rate per period, 10 percent, and the number of periods, five).

TABLE 8-4 FUTURE VALUE, *F*, OF SINGLE AMOUNT—COMPOUND INTEREST

Period	***Interest during period***	***Accumulated amount at the end of period (F)***
1	Pi	$P + Pi = P(1 + i)$
2	$P(1 + i)i$	$P(1 + i) + P(1 + i)i = P(1 + i)^2$
3	$P(1 + i)^2 i$	$P(1 + i)^2 + P(1 + i)^2 i = P(1 + i)^3$
.	.	.
.	.	.
.	.	.
n	$P(1 + i)^{n-1} i$	$P(1 + i)^{n-1} + P(1 + i)^{n-1} i = P(1 + i)^n$

Thus:

$$F = P(F/P, i, n) \quad (8\text{-}2)$$
$$= P(1 + i)^n$$
$$= \$100(1 + 0.10)^5$$
$$= \$100(1.6105) \quad \text{(obtain this value from Appendix III.)}$$
$$= \$161.05$$

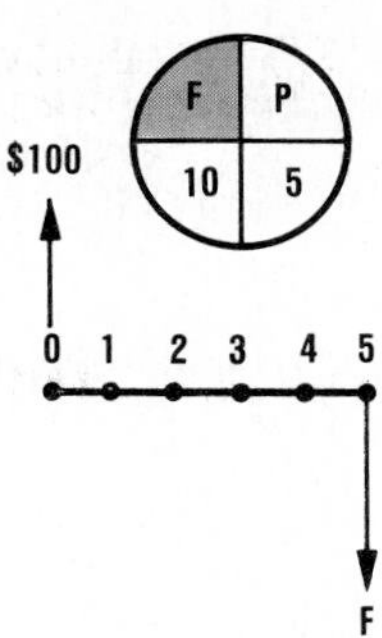

[8-6]

Suppose that the future amount F is known and i and n have been specified, and it is desired to identify the present value P that will accumulate to F at interest rate i per period after n periods. The cash flow diagram for this situation is represented by Figure [8-6]. Equation (8-2) is solved for P. The result is

$$P = F\left[\frac{1}{(1 + i)^n}\right] = F(P/F, i, n) \quad (8\text{-}3)$$
$$= 161.05\left[\frac{1}{(1 + 0.10)^5}\right]$$
$$= \$100$$

The factor $1/(1 + i)^n$ is called the *single-payment present worth factor* (see Table 8-3) and is abbreviated as $(P/F, i, n)$ or "P given F, i, n." Remember that

$$(P/F, i, n) = \frac{1}{(F/P, i, n)}$$

Example 8-3

How much money must be deposited in an account that earns 2 percent per quarter compounded quarterly so that $1000 accumulates after two years?

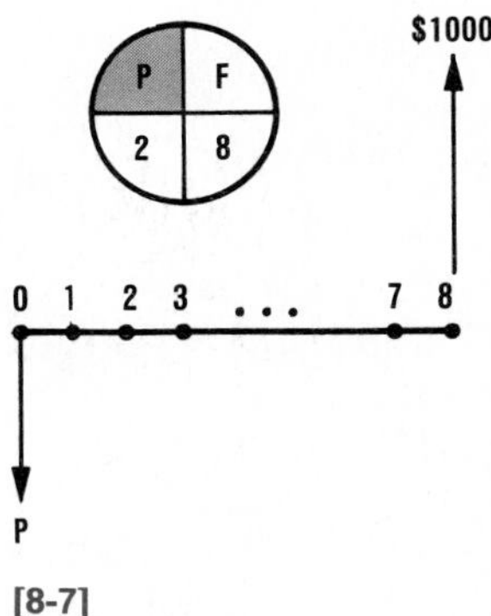

[8-7]

Solution:

The first step in solving the problem is to draw a cash flow diagram that describes the transaction [8-7]. Use the label $(P/F, i, n)$ in Table 8-3.

$$P = F(P/F, i, n) = F\left[\frac{1}{(1 + i)^n}\right]$$
$$= \$1000\left[\frac{1}{(1 + .02)^8}\right] = \$1000\left[\frac{1}{1.1717}\right]$$
$$= \$1000(0.8535) = \$853.49$$

Notice that the value of n used in Example 8-3 is 8, since two years = eight quarters (a quarter is a three-month period). It is important to note that in all engineering economy calculations there must be agreement between i and n with respect to their labels. For example, if i is expressed as "percent per month compounded monthly," then n must also be expressed in months. The label associated with i is always "percent per time period compounded every time period." Usually, n is converted to time periods specified by i.

The justification of private profit is private risk.
Franklin D. Roosevelt

The highest use of capital is not to make more money, but to make money do more for the betterment of life.
Henry Ford

He who will not reason, is a bigot; he who cannot is a fool; and he who dares not, is a slave.
William Drummond

Tables for the P/F, F/P, and other factors soon to be developed are provided in Appendix III for selected values of i and a large range of values for n. The appropriate table to be used in a particular calculation is identified by the appropriate value of i. The value of n is found along the first column and then the numerical value of the factor is read below the appropriate factor symbols in the row associated with n.

For example, with i at 2 percent, the table headed "Decimal Interest Rate = .02" with $n = 8$ yields a P/F value of 0.8535 as calculated in Example 8-3. The 2 percent table is appropriate for all compounding periods where the interest rate is 2 percent per period, for example, 2 percent per month compounded monthly (in this case, n represents months), 2 percent per quarter, compounded quarterly (n represents quarters), 2 percent per year compounded yearly (n represents years), and so on.

COMPOUND INTEREST—SERIES OF EQUAL PAYMENTS

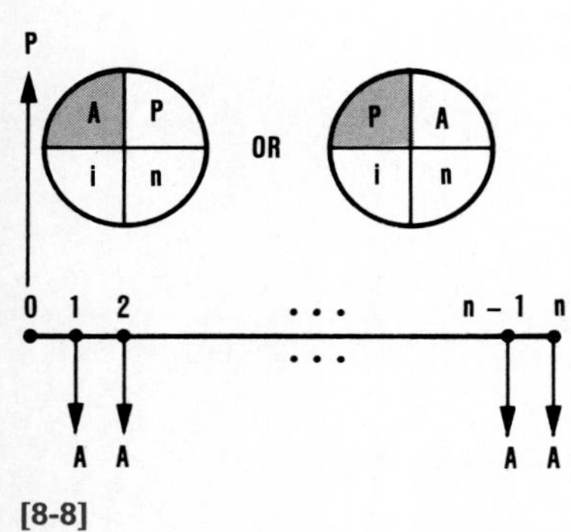

[8-8]

Suppose that you wish to borrow P dollars and plan to repay it at i percent compound interest in a series of equal payments of A dollars each over n periods. In considering such a loan you will want to know corresponding loan payments, A, for varying loan amounts (P's). The cash flow is shown in Figure [8-8]. Note that the first A value will occur one time unit after the P has been received.

It can be shown that the value of A is given by

$$A = P\left[\frac{i(1+i)^n}{(1+i)^n - 1}\right] = P(A/P, i, n) \qquad (8\text{-}4)$$

The factor $(A/P, i, n)$ is known as the *capital recovery factor* (See Table 8-3).

If A is known and the value of P is desired, then

$$P = A\left[\frac{(1+i)^n - 1}{i(1+i)^n}\right] = A(P/A, i, n) \qquad (8\text{-}5)$$

The factor $(P/A, i, n)$ is known as the *uniform series present worth factor* (see Table 8-3).

Example 8-4

A bank has offered to loan \$4000 toward the purchase of a microcomputer. This loan will be financed at a rate of $1\frac{1}{2}$ percent per month compounded monthly over 48 months. What is the amount of each monthly payment?

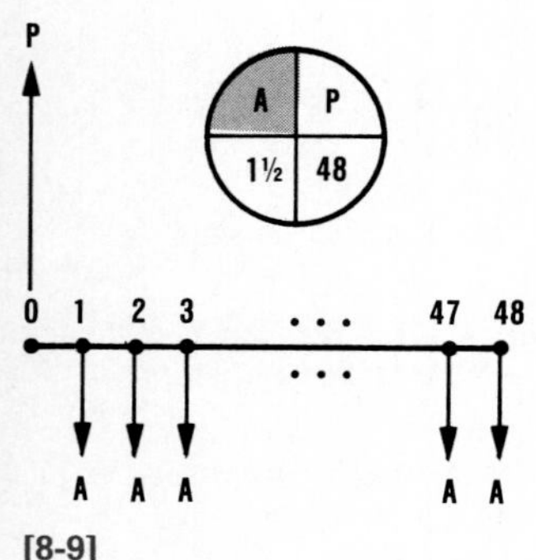

[8-9]

Solution:

First draw the cash flow diagram [8-9]. Using Appendix III, we have

$$A = \$4000(A/P, 1\tfrac{1}{2}, 48)$$
$$= 4000(0.0294)$$
$$= 117.60$$

Example 8-5

How much money must Henry Jones deposit now which will earn a rate of interest of 10 percent per year compounded annually for him to be able to withdraw \$3000 for each of four years beginning 10 years from now so that he can provide for his son's college education?

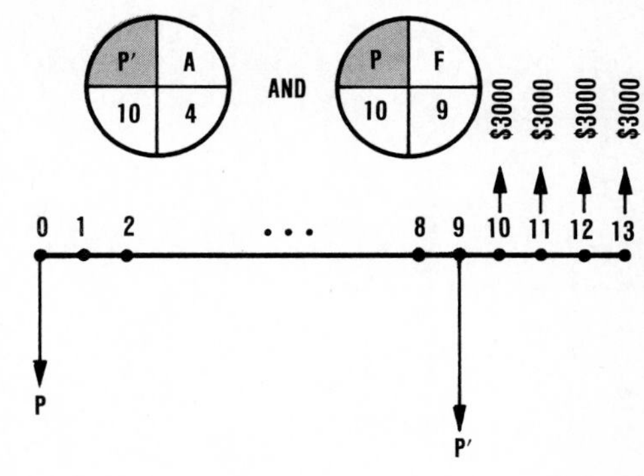

[8-10]

Solution:

In this problem we are attempting to find the value of P at time 0 [8-10]. However, it is first necessary to convert the four \$3000 withdrawals to an equivalent deposit P' at time 9. (Time 9 is one time unit before the first A value.) Then the P' value will be treated as a future amount (an F) and its value will be converted to an equivalent value at $t = 0$. Use the label $P/A, i, n$.

$$\begin{aligned} P' &= \$3000(P/A, 10, 4) \\ &= \$3000(3.1699) \qquad (\textit{Appendix III}) \\ &= \$9509.70 \end{aligned}$$

then use the label $P/F, i, n$.

$$\begin{aligned} P &= \$9509.70(P/F, 10, 9) \\ &= \$9509.70(0.4241) \\ &= \$4033.06 \end{aligned}$$

Another situation occurs frequently in evaluating the worth of money. This is when it is desirable to determine the future value of a series of equal payments. The cash flow diagram for this situation is shown in Figure [8-11]. As we can see, equal payments of (A) are deposited every period at i percent interest per period (compounded every period) and the accumulated amount (including interest), F, results after n periods.

Note that in this case the accumulated value F occurs at the same time as the last deposit, A. This means that the last deposit earns no interest. The value of F is given by the relationship

$$F = A\left[\frac{(1+i)^n - 1}{i}\right] = A(F/A, i, n) \tag{8-6}$$

The factor $(F/A, i, n)$ is called the *uniform series future worth factor* (see Table 8-3). Solving Equation (8-6) for A yields

$$A = F\left[\frac{i}{(1+i)^n - 1}\right] = F(A/F, i, n) \tag{8-7}$$

The factor $(A/F, i, n)$ is called the *sinking fund factor* (see Table 8-3).

Let us examine several examples that illustrate the use of the $(F/A, i, n)$ and $(A/F, i, n)$ factors.

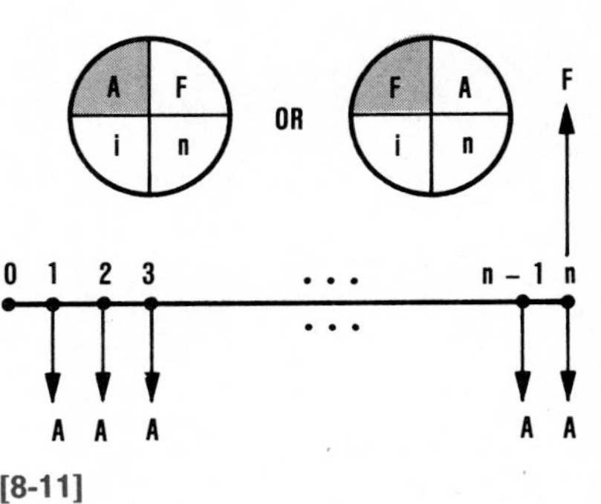

[8-11]

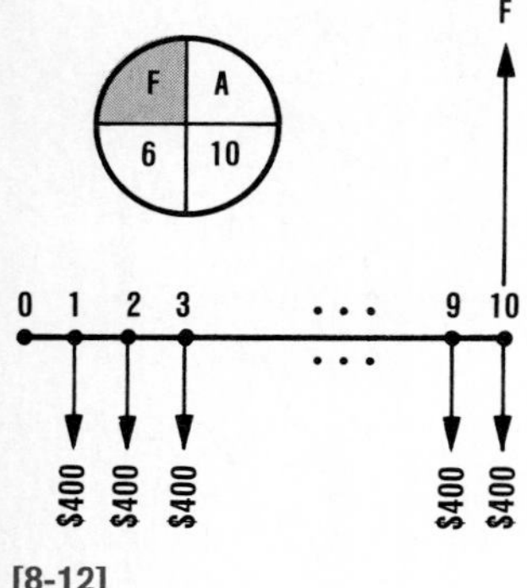

[8-12]

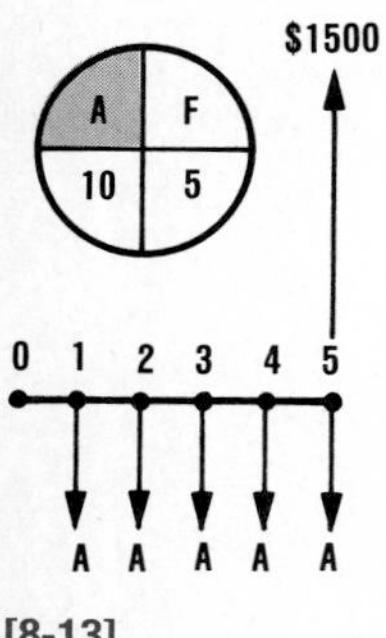

[8-13]

Example 8-6

$400 is deposited every six months into an account that earns 6 percent interest every six months, compounded semiannually. How much money has accumulated after five years?

Solution:

The cash flow diagram is shown in Figure [8-12]. Use the label (F/A, i, n).

$$F = \$400(F/A, 6, 10)$$
$$= \$400(13.1808)$$
$$= \$5232.72$$

Example 8-7

How much money must be deposited annually into an account that earns 10 percent interest per year compounded annually so that $1500 is accumulated after five years?

Solution:

The cash flow diagram is shown in Figure [8-13]. Use the label (A/F, i, n).

$$A = \$1500(A/F, 10, 5)$$
$$= \$1500(0.1638)$$
$$= \$245.70$$

Example 8-8

$50 is deposited every month for two years. Beginning one year after the last deposit, the entire accumulated amount will be withdrawn in four equal monthly payments. What is the amount of these payments? Interest is at 2 percent per month compounded monthly.

Solution:

The cash flow diagram is shown in Figure [8-14]. From the cash flow diagram we can see that there are three tasks to be performed. First, we must determine the future value F of the 24 equal payments of $50 each. Second, we must determine the value of F eleven months hence, X. Finally, we will disburse X in equal payments A over the next

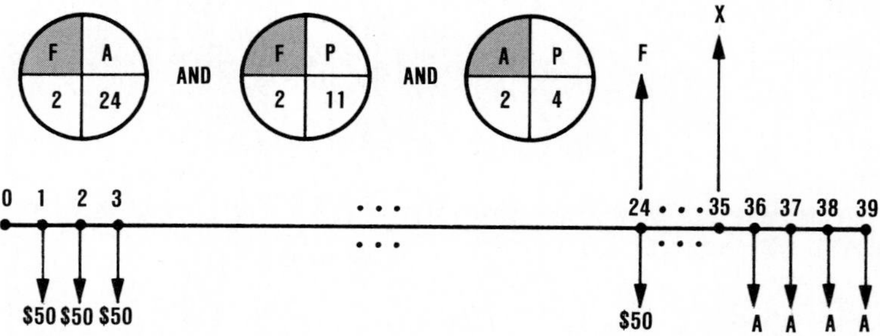

[8-14]

four-month period. The calculations for this transaction would be as follows:

Use the label $(F/A, i, n)$.

$$F = \$50(F/A, 2, 24)$$
$$= \$50(30.4218)$$
$$= \$1521.09$$

Hint: X is actually another future sum F.

Use the label $(F/P, i, n)$.

$$X = \$1521.09(F/P, 2, 11)$$
$$= \$1521.09(1.2434)$$
$$= \$1891.32$$

Use the label $(A/P, i, n)$.

$$A = \$1891.32(A/P, 2, 4)$$
$$= \$1891.32(0.2626)$$
$$= \$496.66$$

GRADIENT SERIES

There are many situations where the cash flow from one period to the next differs by a constant amount, or gradient. Such a gradient cash flow is illustrated in Figure [8-15]. In examining Figure [8-15] we see that this particular situation involves both a uniform cash flow, U, and a cash flow of multiples of the gradient amount, G. It is often desired to convert a gradient series to an equivalent present amount (P) or an equivalent annual amount (A).

Gradient cash flow problems can be solved more easily if they are divided into their component parts. (This is called decomposition.) In the case of Figure [8-15], the cash flow diagram is the result of adding two other cash flow diagrams, one representing the cash flow of several equal payments, and one representing the cash flow of several sequential payments that increase by the constant amount G. This process of decomposition is shown in Figure [8-16].

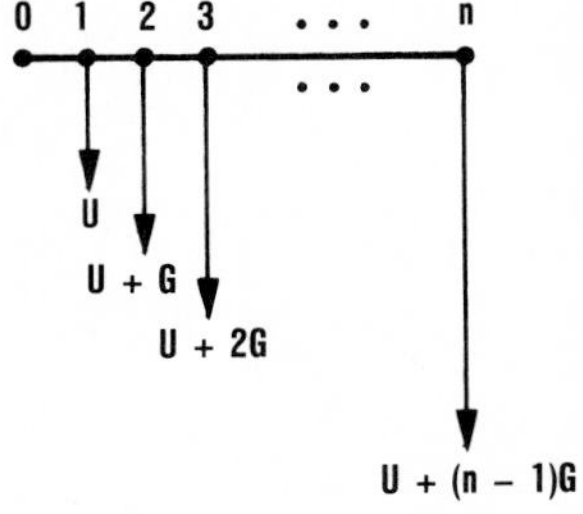

[8-15]

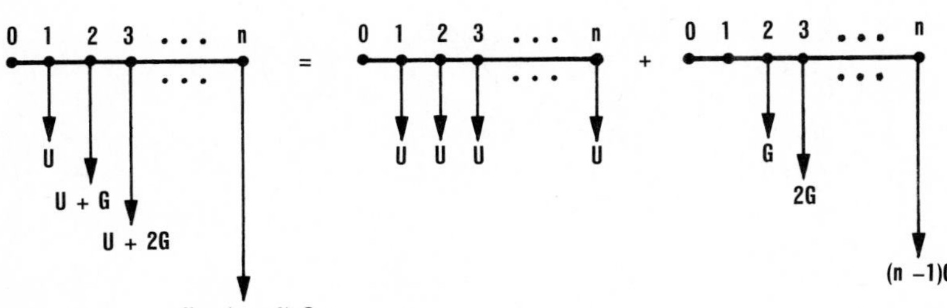

[8-16]

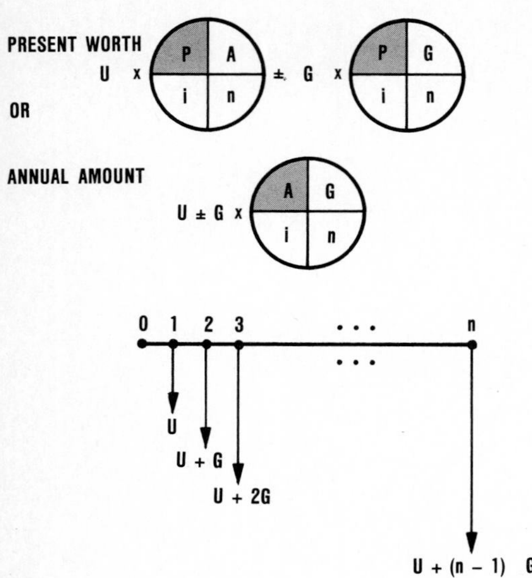

[8-17]

Decomposition of Gradient Cash Flow

Every gradient cash flow has a uniform portion (U), which is always the amount shown in period 1, and a gradient portion (G), which begins in period 2. Special factors have been developed to convert a gradient cash flow to either a uniform series or a present worth. To convert a gradient cash flow to a uniform series, we use

$$A = U \pm G(A/G, i, n) \qquad (8\text{-}8)$$

Note from Figure [8-16] that U is already a uniform amount and the $(A/G, i, n)$ factor converts the gradient amounts into a uniform series. The factor $(A/G, i, n)$ is called the *gradient uniform series factor*. To convert a gradient cash flow to a present worth, the appropriate formula is

$$P = U(P/A, i, n) \pm G(P/G, i, n) \qquad (8\text{-}9)$$

We see from Equation (8-9) that multiplying U by $(P/A, i, n)$ results in a present worth, while $(P/G, i, n)$ converts the gradient amounts into a present worth. In Figure [8-17], the notation differs slightly from that used with the first six interest factors. To illustrate conversion to present worth, U has been written as a multiplier of $(P/A, i, n)$ since it is actually a uniform series value (or A). G has been written as a multiplier of $(P/G, i, n)$ for consistency. To illustrate conversion to a uniform amount, U (already a uniform amount), is added to G multiplied by $(A/G, i, n)$.

The factor $(P/G, i, n)$ is called the *gradient present worth factor*. We can see from the cash flow diagram of Figure [8-16] that the first G value does not appear until period 2. It is important to recognize that this condition has already been considered in the development of both the $(A/G, i, n)$ and $(P/G, i, n)$ factors. Values for the P/G and A/G factors for selected values of i are given in Appendix III. Let us consider a cash flow situation involving gradient payments.

Table 8-5 summarizes the basic interest factors. Computer programs for generating interest tables are shown in Figures [8-18] and [8-19]. Interest tables for selected values of i that have been developed from the BASIC programs of this chapter are given in Appendix III. The use of wide paper is required for the tables to print properly.

Example 8-9

To settle a debt, the lender has requested that \$500 be deposited the first year into an account that will earn 12 percent per year compounded annually. In each subsequent year for a 10-year period \$50 less than the preceding year will be deposited. The borrower wishes to consider an alternative payment plan using uniform annual payments for the same rate of interest for the 10-year period. What would be the equivalent annual deposit over a 10-year period?

Solution:

First, we will draw a cash flow diagram representing these transactions [8-20]. Then [using the label $(A/G, i, n)$]

TABLE 8-5 SUMMARY OF THE BASIC INTEREST FACTORS

Cash flow diagram	*To find*	*Given*	*Use*
	F	P	F = P(F/P,i,n)
	P	F	P = F(P/F,i,n)
	P	A	P = A(P/A,i,n)
	A	P	A = P(A/P,i,n)
	F	A	F = A(F/A,i,n)
	A	F	A = F(A/F,i,n)
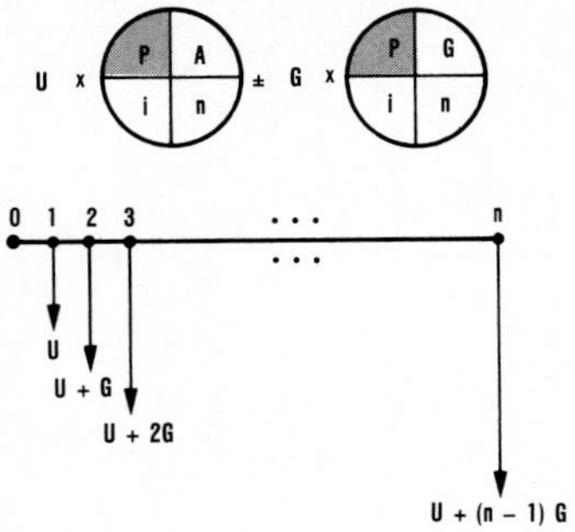	P	Gradient series	P = U(P/A,i,n) ± G(P/G,i,n)
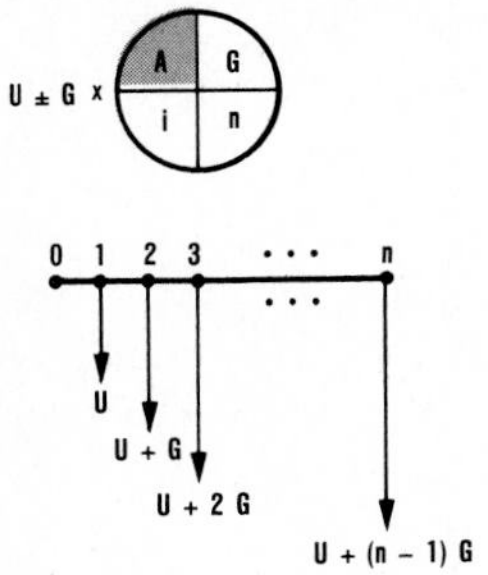	A	Gradient series	A = U ± G(A/G,i,n)

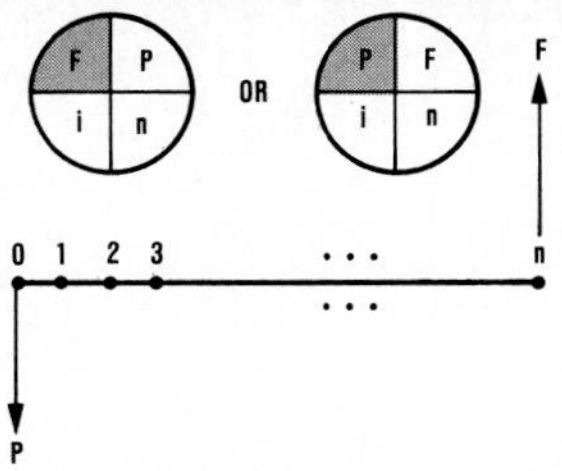

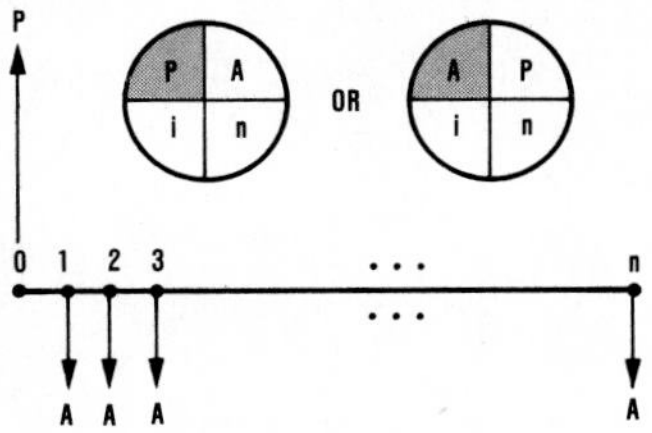

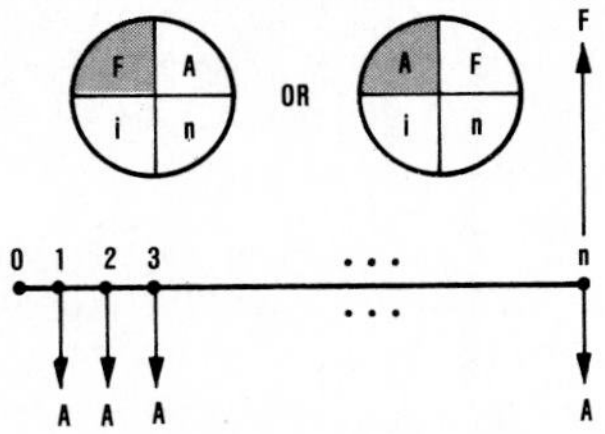

[8-18] BASIC language program for generating large interest tables (see Chapter 19).

```
10 CLS
20 INPUT "ENTER INTEREST RATE";I
30 CLS
40 LPRINT "                         DECIMAL INTEREST RATE =",I
50 LPRINT "  "
60 LPRINT "   PERIOD       F/P          P/F          F/A          A/F          P/A
  A/P          P/G          A/G  "
70 LPRINT " "
80 FOR J=1 TO 40
90 FP=(1+I)^J
100 PF=1/FP
110 FA=(FP-1)/I
120 AF=1/FA
130 PA=FA/FP
140 AP=1/PA
150 PG=(PA-J*PF)/I
160 AG=1/I-J/I*AF
170 LPRINT USING "####.####    "; J,FP,PF,FA,AF,PA,AP,PG,AG
180 NEXT J
190 END
```

[8-19] BASIC language program for generating small interest tables (see Chapter 19).

```
10 CLS
20 INPUT "ENTER INTEREST RATE";I
30 CLS
40 LPRINT "                         DECIMAL INTEREST RATE =",I
50 LPRINT "  "
60 LPRINT "   PERIOD       F/P          P/F          F/A          A/F          P/A
  A/P          P/G          A/G "
70 LPRINT " "
80 FOR J=1 TO 120
90 FP=(1+I)^J
100 PF=1/FP
110 FA=(FP-1)/I
120 AF=1/FA
130 PA=FA/FP
140 AP=1/PA
150 PG=(PA-J*PF)/I
160 AG=1/I-J/I*AF
170 IF (J>36) AND (J MOD 6 >< 0) THEN 190
180 LPRINT USING "####.####    "; J,FP,PF,FA,AF,PA,AP,PG,AG
190 NEXT J
200 END
```

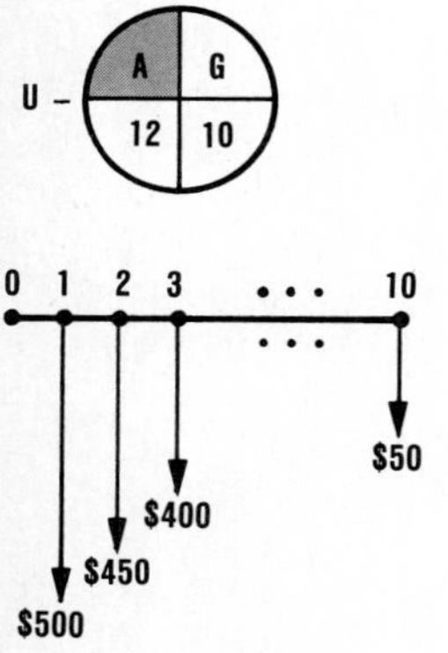

[8-20]

$$A = \$500 - \$50(A/G, 12, 10)$$
$$= \$500 - \$50(3.5847)$$
$$= \$500 - \$179.24$$
$$= \$320.76$$

Note that since the deposits *decreased* by a constant amount from year to year, a negative gradient was used. Also, note that the solution procedure did not require use of the yearly cash flow values. Only the first cash flow (*U*) and the gradient (*G*) were used to solve this problem.

Let us examine another example of a gradient series.

Example 8-10

It is proposed to open a savings account with a $30 deposit. It will earn interest at a rate of 2 percent per quarter compounded quarterly. During each quarter $10 more than the amount of the previous quarter will be deposited. What will be the accumulated amount of savings after five years of deposits?

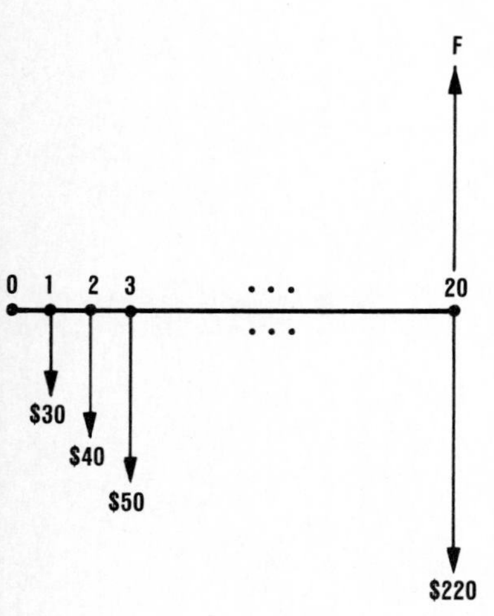

[8-21]

Solution:

Let us draw a cash flow diagram representing these transactions [8-21]. This problem can be solved by either of two methods as follows:

Method 1: Obtain the equivalent present worth first.

Step 1: Convert to a present worth [8-22].

$$P = \$30(P/A, 2, 20) + \$10(P/G, 2, 20)$$

$= \$30(16.3514) + \$10(144.5976)$

$= \$490.54 + \$1445.98$

$= \$1936.52$

Step 2: Calculate the future worth [8-23].

$F = \$1936.52(F/P, 2, 20)$

$= \$1936.52(1.4859)$

$= \$2877.48$

Method 2: Obtain the equivalent annual amount first.

Step 1: Convert to an annual amount [8-24].

$A = \$30 + \$10(A/G, 2, 20)$

$= \$30 + \$10(8.8431)$

$= \$30 + \$88.43$

$= \$118.43$

Step 2: Calculate the future worth [8-25].

$F = \$118.43(F/A, 2, 20)$

$= \$118.43(24.2973)$

$= \$2877.53$

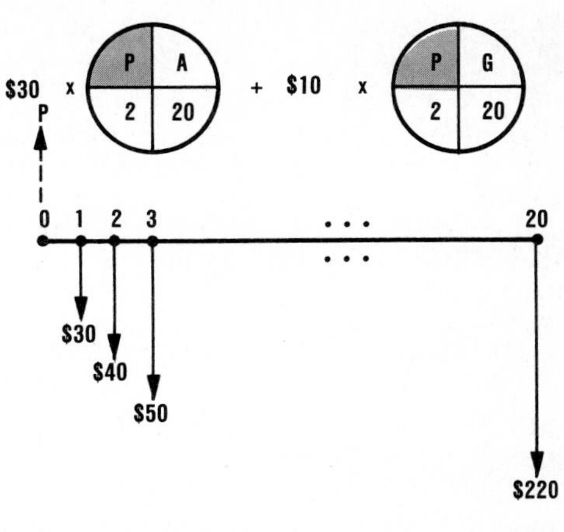

[8-22]

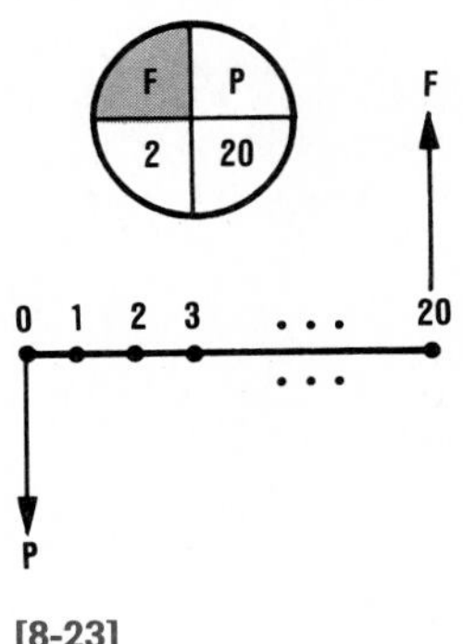

[8-23]

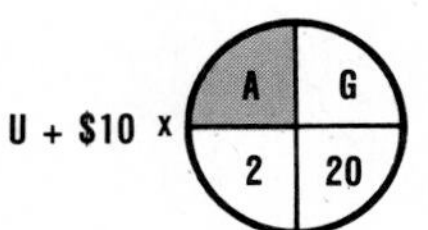

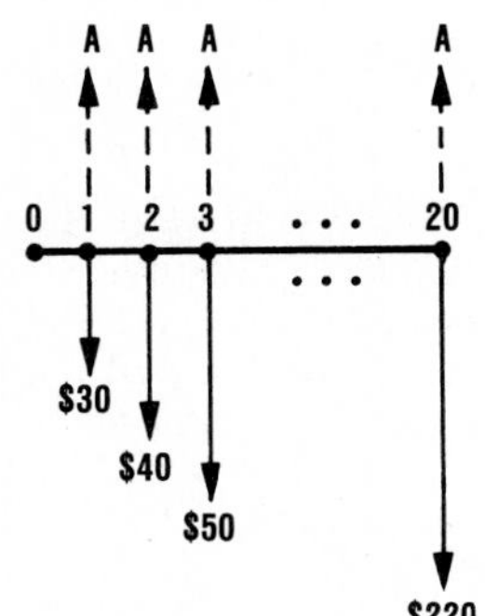

[8-24]

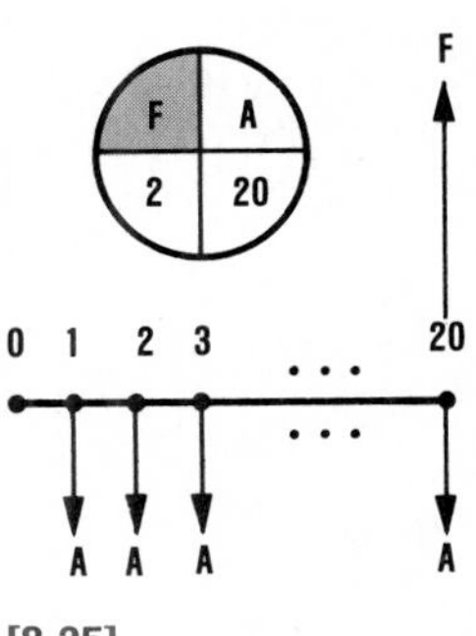

[8-25]

Problems

8-1. \$1000 is borrowed for a period of four years. How much must be paid by the borrower at the end of four years if:

a. 15 percent simple interest is available?

b. Interest is at 15 percent per year compounded yearly?

8-2. What amount must you now deposit at 8 percent per year compounded yearly to accumulate \$2000 after five years?

8-3. With interest at 7 percent per year compounded yearly, what equal annual amount would have to be deposited at the end of each of the next six years to accumulate \$5000?

8-4. How much should you now deposit in an account that

pays 10 percent per year compounded yearly so that you may withdraw $1000 each year for eight consecutive years beginning one year from now?

8-5. A wealthy relative has offered you $10,000 now or $4000 next year, $5000 after two years, and $2000 after three years. You conclude that you could earn 8 percent per year compounded yearly with any amounts you receive. Which of the two offers should you take? *Hint:* Compare the present worth of the three installments with $10,000.

8-6. Solve Problem 8-2 if interest is at 8½ percent per year compounded yearly. *Hint:* Use the mathematical expression in Table 8-3 for the appropriate factor.

8-7. Solve Problem 8-3 if interest is at 7¼ percent per year compounded yearly. *Hint:* Use the mathematical expression in Table 8-3 for the appropriate factor.

8-8. Resolve Mary Brown's dilemma (see Table 8-1) by comparing the present worth of ownership of the Ford at 10 percent per year compounded yearly and the Chevrolet at 15 percent per year compounded yearly.

8-9. Convert the following cash flow to an equivalent uniform amount if the periods are years and the interest rate is 20 percent per year compounded yearly. *Hint:* First find a present worth (P) or future worth (F) value and then convert that value to a uniform amount (A).

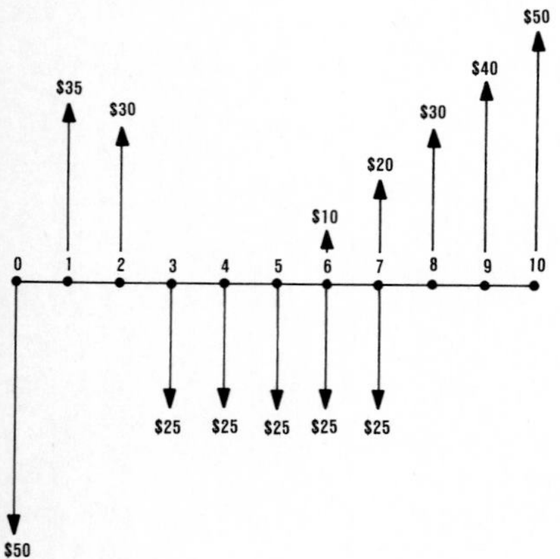

[8-P9]

8-10. The annual maintenance cost of a machine is $500 the first year, $700 the second year, $900 the third year, and so on (i.e., $200 increase each year). If the machine is to be used for an eight-year period, compute the equivalent uniform annual cost of maintenance as well as the present worth. Interest is at 15 percent per year compounded yearly.

8-11. Suppose that a cost reduction project is expected to save $1000 during the first year, $900 during the second, $800 during the third year, and so on (i.e., a $100 decrease each year). Treating these savings as end of the year values, compute the equivalent annual savings over eight years if interest is 18 percent per year compounded yearly.

8-12. Suppose that a person is to receive $1000 at the end of year 1, $1500 at the end of year 2, $2000 at the end of year 3, and so on (i.e., $500 increase each year) for 10 years. What is the accumulated amount of this cash flow after 10 years if interest is at 7 percent per year compounded yearly? *Hint:* First calculate either the present worth (P) or the equivalent annual worth (A) and then convert to a F value.

8-13. $1000 is borrowed at $t = 0$. It was to have been repaid in 10 equal installments beginning at $t = 1$. How much money is required to pay off the loan just after the fourth payment is made? Interest is at 15 percent per period. *Hint:* Find the amount, A, of each of the payments required over the 10 periods. Then convert the last six of these A values to a single amount at period 4.

8-14. In Figure [8-P14], if the downward arrows are deposits and the upward arrows are withdrawals, what value of X will close the account if $i = 6$ percent per period? *Hint:* Using the appropriate factors, move the six $100 deposits to time 16 and then move the three $50 deposits to time 16. X is the algebraic difference of these two amounts.

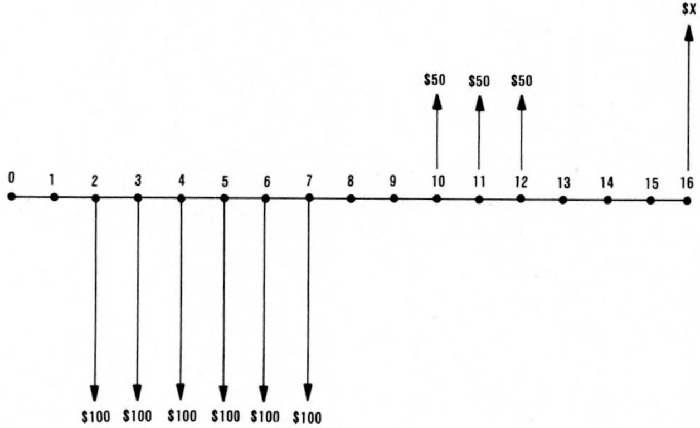

[8-P14]

8-15. In Figure [8-P15], what is X if the downward arrows are deposits and the seven withdrawals will close the account? $i = 6$ percent per period. *Hint:* Using the appropriate factors, convert the gradient series to a single amount at time 0. Then multiply X by (P/A, 6, 4); equate this value to the time 0 equivalent gradient amount, and solve for X.

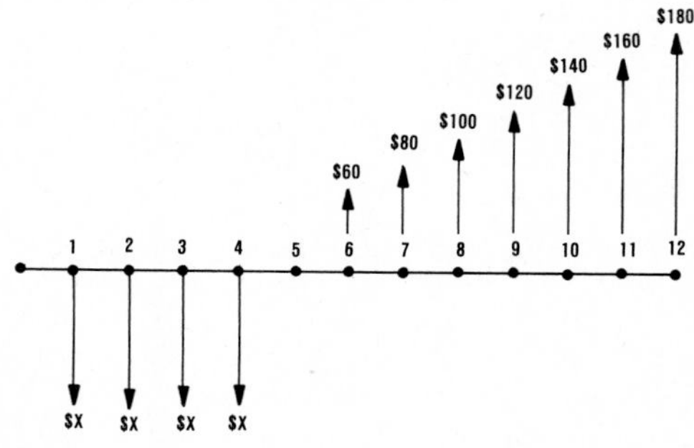

[8-P15]

NOMINAL AND EFFECTIVE INTEREST RATES

An annual interest rate with compounding occurring more than once a year is referred to as a *nominal interest rate* and is given the symbol r. For example, 10 percent per year compounded semiannually is a nominal rate. Similarly, 12 percent per year compounded quarterly is a nominal rate. Most often, borrowing and lending agencies will quote nominal rates to customers. For example, you might be offered an automobile loan with "monthly payments at 18 percent." This quotation is a nominal rate, since a more accurate description of this interest rate would be 18 percent per year compounded monthly. To determine the amount of each monthly payment of a loan, the annual interest rate compounded monthly must be converted to a monthly rate compounded monthly. To do this, simply divide the nominal rate by the number of compounding periods per year, in this case 12. That is, 18 percent per year compounded monthly = 18/12 = $1\frac{1}{2}$ percent per month compounded monthly. Now we have the correct equivalent interest value (per period compounded every period) to use in the interest factors.

Industry prospers when it offers people articles which they want more than they want anything they now have. The fact is that people never buy what they need. They buy what they want.

Charles F. Kettering

Example 8-11

Bill Smith has just agreed to purchase a new fiberglass fishing boat. The boat costs \$7872.86, including sales tax. A down payment of \$2872.86 is required. The \$5000 balance will be financed. Find the amount of equal monthly payments that will be required if the repayment period is four years and the interest is 18 percent per year compounded monthly.

Solution:

The cash flow diagram is shown in Figure [8-26]. Use the label (A/P, i, n).

$$A = \$5000(A/P, 1\tfrac{1}{2}, 48)$$
$$= \$5000(0.0294)$$
$$= \$147.00$$

In general, if a nominal rate r is specified with m compounding periods per year, r must be divided by m to obtain an equivalent rate that has the label "per period compounded each period." This must be done so that the appropriate interest factor may be used in the calculations. In documents that accompany the lending agreement, the nominal rate, r, is usually expressed as a percent and is called the "annual percentage rate." When funds are solicited for deposit in an interest-bearing account, the advertisement often presents r as an "annual yield" rate. Accompanying this rate will be a disclosure specifying how often interest will be compounded. For example, a depositor may be offered a 9 percent annual rate with quarterly compounding. For this quotation, an accumulated amount after k quarters, the potential investor would use the following formula:

$$F = P\left(1 + \frac{0.09}{4}\right)^k$$

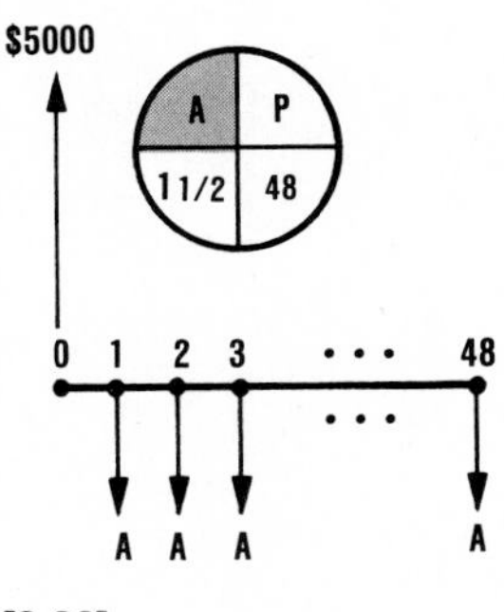

[8-26]

Note that $(1 + 0.09/4)^k$ is a F/P factor $(F/P, r/m, k)$. It is expressed here in formula form since tables may not exist for a $\left(\frac{0.09}{4}\right)$ value of i.

In general, if you want to accumulate a sum of money, F, by investing a principal amount, P, with a nominal interest rate, r, and m compounding periods per year, the accumulated amount after k compounding periods is found using

$$F = P\left(1 + \frac{r}{m}\right)^k$$

On the other hand, if a known present amount, P, is to be converted to an equivalent series of uniform amounts, A, over k compounding periods with a nominal interest rate, r, and m compounding periods per year, A is given by

$$A = P\left(A/P, \frac{r}{m}, k\right)$$

which becomes

$$A = P\left[\frac{(r/m)(1 + r/m)^k}{(1 + r/m)^k - 1}\right] \tag{8-10}$$

Obviously if r/m results in a value, i, for which tables exist, we may use

$$A = P(A/P, i, k)$$

The use of a nominal interest rate is illustrated in Example 8-12. These calculations are typical of the manner in which repayment schedules for consumer loans are developed.

Example 8-12

$8000 has been borrowed toward the purchase of an outdoor swimming pool. The annual interest rate is 20 percent compounded monthly. There is a four-year repayment schedule, using monthly payments. What is the amount of each monthly payment?

Solution:

In solving this problem [8-27] we can use Equation (8-10).

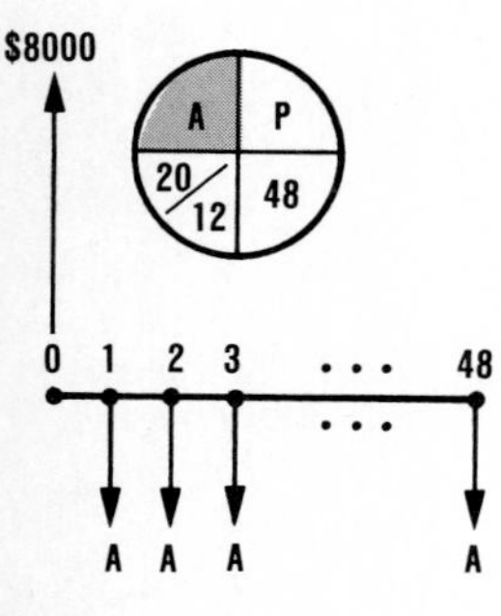

[8-27]

$$A = \$8000\left(A/P, \frac{r}{m}, k\right)$$

$$= 8000\left[\frac{(0.20/12)(1 + 0.20/12)^{48}}{(1 + 0.20/12)^{48} - 1}\right]$$

$$= \$8000(0.03043)$$

$$= \$243.44$$

There is a simple means of converting any nominal rate to a rate that is expressed as "per year compounded annually." Any rate that carries the label "per year compounded annually" is called an *effective*

TABLE 8-6 NOMINAL AND EFFECTIVE RATE COMPARISON

Nominal rate	*Calculation*	*Effective rate (percent)*
18% *per year compounded semiannually*	$\left(1+\frac{0.18}{2}\right)^2-1$	18.81
18% *per year compounded quarterly*	$\left(1+\frac{0.18}{4}\right)^4-1$	19.25
18% *per year compounded monthly*	$\left(1+\frac{0.18}{12}\right)^{12}-1$	19.56
18% *per year compounded daily*	$\left(1+\frac{0.18}{365}\right)^{365}-1$	19.71
18% *per year* compounded continuously	$e^{0.18}-1$[a]	19.72

[a] $\lim m \rightarrow \infty\ (1 + 0.18/m)^m = e^{0.18}$

rate. Effective rates provide a convenient yardstick for comparing two or more nominal rates. The conversion is as follows:

$$i_{\text{effective}} = \left(1 + \frac{r}{m}\right)^m - 1 \qquad (8\text{-}11)$$

where, as before, r is the nominal rate expressed as a decimal and m is the number of compounding periods per year. Table 8-6 illustrates the use of the effective rate equation.

Note in Table 8-6 that as the number of compounding periods per year increases, the effective rate also increases and approaches a limit of 19.72 percent. It can be shown that $\lim m \rightarrow \infty\ (1 + r/m)^m = e^r$. It is interesting to note that some savings institutions promote daily and even continuous compounding of interest ("heart beat" interest rates). But at a nominal rate of 8 percent, the effective rates for quarterly, monthly, daily, and continuous compounding are only 8.24, 8.30, 8.32, and 8.33 percent, respectively. For the small investor, these differences are minimal.

Problems

8-16. A loan of $500 is to be repaid in 12 equal monthly installments. Interest is at 18 percent per year compounded monthly. What is the amount of each installment payment. *Hint:* First convert 18 percent per year compounded monthly to a rate per month compounded monthly. Then use the appropriate factor with this rate and 12 periods.

8-17. How much will accumulate in an interest-bearing account after five years if $50 is deposited every three months and the interest rate is 8 percent per year compounded quarterly? *Hint:* Convert the 8 percent per year compounded quarterly to a rate per quarter compounded quarterly and then use the appropriate factor with this rate and 20 quarters.

8-18. How much must you deposit now at 10 percent per year compounded yearly to accumulate $6000 after four years?

8-19. Solve Problem 8-18 if the interest rate is 12 percent per year compounded (a) semiannually; (b) quarterly; (c) monthly.

8-20. Suppose that in Problem 8-18, identical deposits will be made at the end of years 1, 2, 3, and 4 rather than a single deposit at time 0. What is the amount of these deposits?

8-21. You have decided to withhold a portion of your monthly take-home pay for deposit in an account that pays 6 percent per year compounded monthly. $100 is withheld from the first month's pay and in each subsequent month, an additional $10 is withheld (i.e., $110 per month 2, $120 for month 3, etc.). How much is in your account after two years?

8-22. What effective rate corresponds to (a) 12 percent per year compounded semiannually; (b) 12 percent per year compounded quarterly; (c) 12 percent per year compounded monthly; (d) 12 percent per year compounded instantaneously?

8-23. You are financing a $5000 automobile loan with 48 equal monthly payments at a rate of 18 percent per year compounded monthly. You have just made your fifteenth payment and now wish to pay off the entire loan. What amount is necessary to do this? *Hint:* See the hint for Problem 8-13.

8-24. What nominal rate r, expressed as "percent per year compounded monthly" corresponds to an effective rate of 14 percent per year compounded yearly? *Hint:* Use Equation (8-11) to solve for r with $m = 12$ by taking the logarithm of both sides of the equation.

8-25. Modify Figure [8-25] by changing line 80 to "FOR J = 1 TO 360." Then use Figure [8-25] with a decimal interest rate of 0.0108333. This is the rate per month corresponding to 13 percent per year compounded monthly (i.e., 0.13/12 = 0.0108333), a typical mortgage rate. Use the tables to compute the monthly payments necessary to purchase a home if $110,000 is being financed at 13 percent per year compounded monthly and the mortgage is for (a) 20 years; (b) 25 years; (c) 30 years.

THE NATURE OF INTEREST IN ENGINEERING ECONOMY STUDIES

Within an organization, there will be many projects that will compete for use of the organization's money. However, not all projects will be funded. Only those that make a compound return on the money invested in them (i.e., a rate of return) at least as great as the company's minimum attractive rate of return (MARR) have a chance of being funded. Among these, the likelihood of being funded depends on how large the estimated rate of return will be and how much of the company's money at this particular time is available for investment in engineering projects.

The company's MARR is usually based historically on what the company has been able to earn on previous engineering projects. Or it is based on the "cost of capital," which is the company's cost of borrowing money. Hence interest as we have used the term in the borrowing/lending situation developed in this chapter becomes "rate of return on investment."

METHODS OF ANALYSIS IN ENGINEERING ECONOMY STUDIES

The economic feasibility of launching engineering designs and other engineering projects are usually evaluated using one of the following methods:

An engineer is an unordinary person who can do for one dollar what any ordinary person can do for two dollars.

1. Present worth (PW).
2. Annual worth (AW).
3. Future worth (FW).
4. Rate of return (RR).
5. Ratio of earnings to investment (RI).
6. Payback period (PP).

These methods are nothing more than applications of the concepts and factors which have already been discussed. The first five methods are equivalent treatments of the same cash flow. The result of any engineering economy study will be a recommendation; that is, when a single project is evaluated, the recommendation will be either to undertake the project or not to undertake the project. When several competing projects are compared, either a single project will be recommended or perhaps more than one will be recommended. If a project is recommended, it may or may not be undertaken by the company. An economic evaluation is only one of several considerations that must be made before a project is funded. Other considerations include environmental impact of the project, availability of funds for projects of this type, safety and health considerations, effect of the project on employees of the company, compliance with government regulations, and the judgment of the manager or managers within the company who must ultimately approve or disapprove such projects. In most instances, projects must be "sold" to the decision makers within the company, on the basis of both economic and noneconomic considerations. It has been said that a good idea (or project) will sell itself, but this belief does not seem to be valid in the competitive arenas of business, government, and industry.

Certainly there are lots of things in life that money
won't buy, but it's very funny—
Have you ever tried to buy them without money?
Ogden Nash
The Terrible People

It should be recognized that cash flows that will occur throughout the life of a design or other engineering project must be predicted during the developmental stages of the project in order to estimate the final total costs. Although the investment at time zero (the present time) is usually known with some certainty, most of the cash flows used in engineering economy evaluations are estimates of future returns, costs, or savings. The accuracy of the projected final costs will be as good or bad as the individual estimates that make up the total. For this reason, great care must be exercised in forecasting the future cash flows.

Each of the first five methods always seems to result in the same recommendation (i.e., do or do not undertake the project) when applied to the same cash flow situation. However, method 6, the payback period method will sometimes result in a different recommendation than the other methods. For each of the six methods listed above, incomes or savings are treated as a positive cash flow and are given a "plus" sign. Investments, payments, and other costs are treated as negative cash flow and assigned a "minus" sign.

Present Worth Method

The term *present worth* (PW) method refers to a sum of money whose value at the present time (time zero) is equivalent to a series of cash receipts (+) and disbursements (−), each occurring at the end of a successive interest period following time zero. In determining the present worth of these inflows and outflows, their values must be discounted to time zero at an interest rate equal to MARR.

The algebraic sum of the inflow and outflow present worths is then computed. The resulting sum is called the *net present worth* (NPW). For a given project, if the net present worth is positive, it can be concluded that the project will earn more than the MARR. If the net present worth is negative, the project will probably earn less than the MARR. If the net present worth is zero, the project is predicted to earn at a rate exactly equal to the MARR.

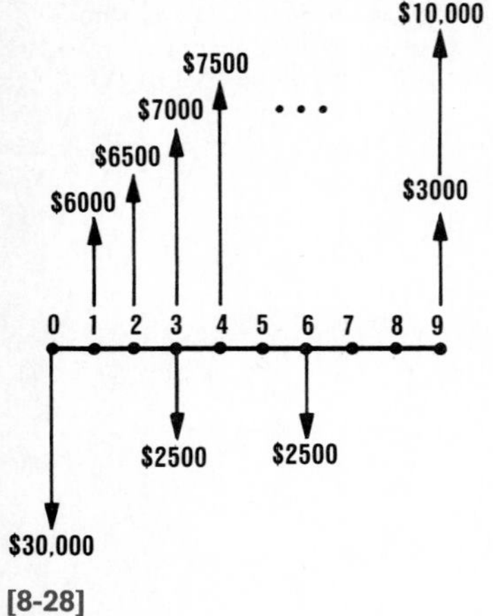

[8-28]

In general, if a single project is being evaluated, the net present worth must be greater than zero in order for the project to be recommended. If two or more competing projects are being compared, the project having the largest *positive* net present worth will be recommended. However, there are cases where a project having a negative net present worth may be recommended. For example, if a problem must be solved and all projects directed at solving the problem result in a negative present worth, that project having the algebraically largest (although negative) net present worth would probably be selected.

Let us examine a situation involving net present worth.

Example 8-13

A small fabricating plant has purchased a metal-forming machine to perform several operations which previously had been done manually. The cost of the machine, including delivery and installation, is $30,000. The machine is expected to have a useful life of nine years and a salvage value of $3000 at that time. Estimated savings due to reduced labor costs and increased productivity are $6000 in year 1 with an increased savings of $500 for each successive year. The machine will need to be overhauled in years 3 and 6, with estimated costs of $2500 for each overhaul. The company's MARR is 15 percent. Compute the net present worth (NPW) of the investment.

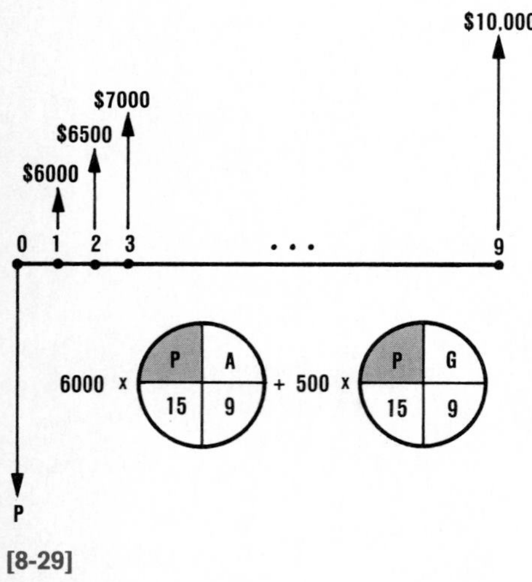

[8-29]

Solution:

The cash flow diagram is shown in Figure [8-28]. This problem will be solved in four steps.

Step 1: Convert the gradient series to a present worth [8-29].

$$\begin{aligned} \text{PW}_G &= \$6000(P/A, 15, 9) + \$500(P/G, 15, 9) \\ &= \$6000(4.7716) + \$500(14.7548) \\ &= \$28{,}629.60 + \$7377.40 \\ &= \$36{,}007.00 \end{aligned}$$

Step 2: Convert each overhaul cost to a present worth [8-30].

$$\begin{aligned} \text{PW}_O &= -\$2500(P/F, 15, 3) - \$2500(P/F, 15, 6) \\ &= -\$2500(0.6575 + 0.4323) \\ &= -\$2500(1.0898) \\ &= -\$2724.50 \end{aligned}$$

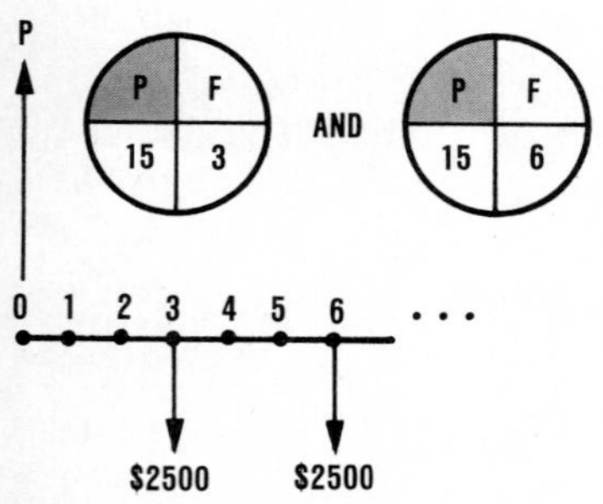

[8-30]

Step 3: Convert the salvage value to a present worth [8-31].

$$\begin{aligned} \text{PW}_S &= \$3000(P/F, 15, 9) \\ &= \$3000(0.2843) \\ &= \$852.90 \end{aligned}$$

Step 4: To obtain the net present worth (NPW), add all present worths, including the machine cost.

$$\text{NPW} = -\$30{,}000 + \$36{,}007 - \$2724.50 + 852.90$$
$$= \$4135.40$$

Since the NPW is positive, the investment in the new machine earns more than 15 percent. Thus the investment is recommended.

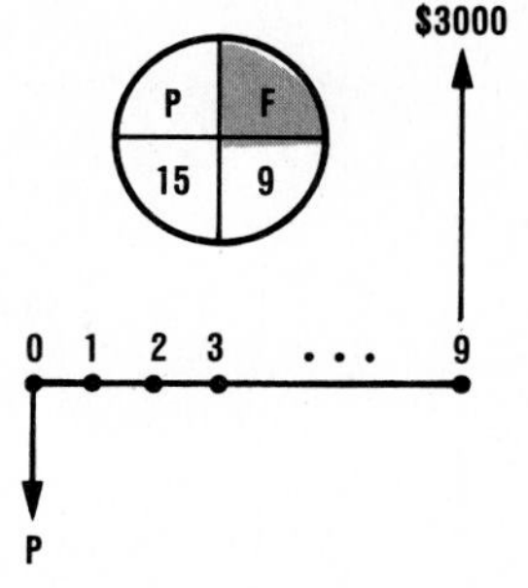

[8-31]

Annual Worth Method

The term *annual worth* (AW) refers to a uniform annual sum of money that is equivalent to a particular schedule of cash receipts and disbursements, minus an amount that represents the equivalent uniform annual cost of the capital invested.

In this method, the cash flows and capital recovery cost equivalency are converted to an equivalent net annual worth in a manner similar to the present worth method just discussed. If the AW is greater than zero, the project earns at a rate greater than MARR. If the AW is less than zero, then the project earns at a rate less than MARR. If the AW is zero, the project earns at a rate exactly equal to MARR. The comments made earlier concerning criteria and exceptions for recommending projects after a PW analysis are also valid for AW.

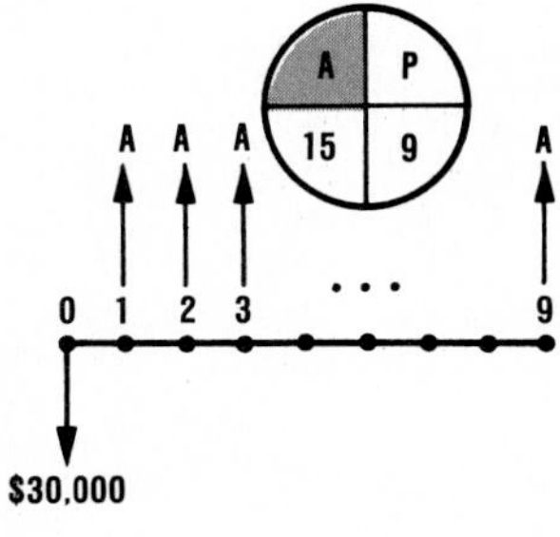

[8-32]

Example 8-14

Compute the AW for the cash flow of Example 8-13.

Solution:

This problem is also solved in five steps.

Step 1: Convert the machine cost to an annual amount [8-32].

$$\text{AW}_m = -\$30{,}000(A/P, 15, 9)$$
$$= -\$30{,}000(0.2096)$$
$$= -\$6288.00$$

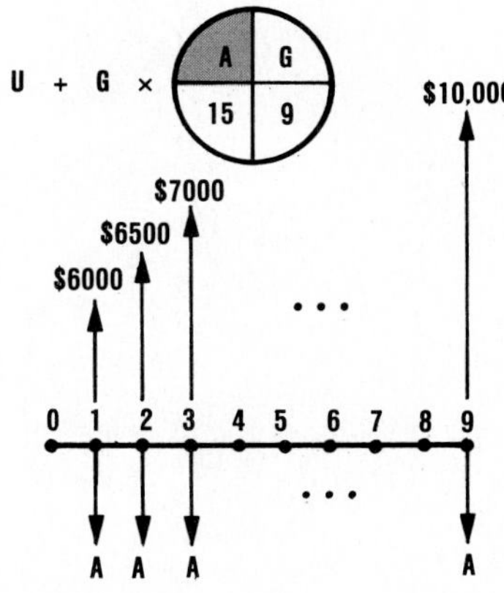

[8-33]

Step 2: Convert the gradient series to an annual amount [8-33].

$$\text{AW}_G = \$6000 + \$500\ (\text{A/G}, 15, \text{a})$$
$$\text{AW}_G = \$6000 + \$500(3.0922)$$
$$= \$7546.10$$

Step 3: Convert the overhaul costs to a future amount [8-34] at year 9.

$$\text{FW}_O = -\$2500(F/P, 15, 6) + \$2500(F/P, 15, 3)$$
$$= -\$2500(2.3131 + 1.5209)$$
$$= -\$2500(3.8340)$$
$$= -\$9585.00$$

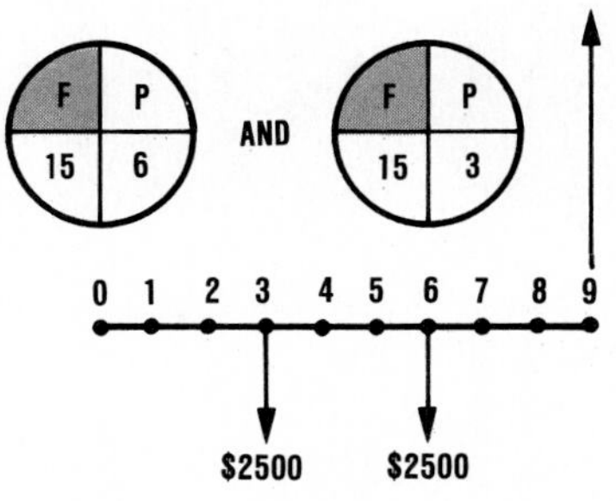

[8-34]

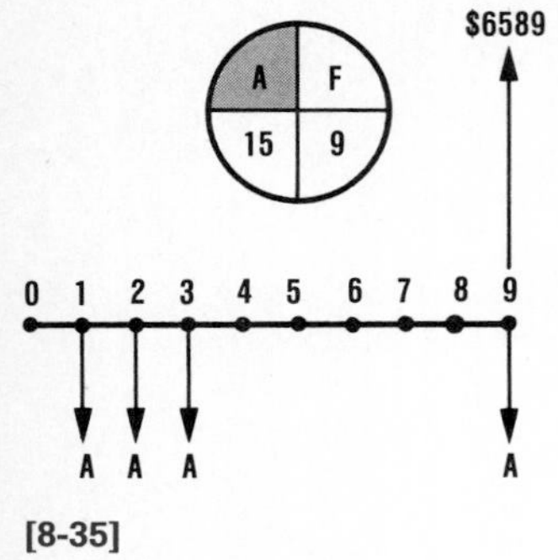

[8-35]

Step 4: Convert the salvage value and the FW_O value to an annual amount [8-35].

$$AW_{O+S} = (-\$9585 + 3000)(A/F, 15, 9)$$
$$= -\$6589(0.0596)$$
$$= -\$392.47$$

Step 5: Sum the annual worths of steps 1, 2, and 4.

$$AW = -\$6288.00 + \$7546.10 - \$392.47$$
$$= \$865.63$$

The net annual worth is positive and the project might be recommended. Note that the PW solution to Example 8-13 could have been used to obtain the AW. This computation would yield the following result:

$$AW = \$4135.40(A/P, 15, 9)$$
$$= \$4135.40(0.2096) = \$866.78$$

The $1.15 difference in the two solutions is due the rounding off of numbers in using the factors.

[8-36]

When looms weave by themselves man's slavery will end.

Aristotle

Other Methods of Evaluation

The management of a company is often more interested in determining the future worth of costs and returns on investment than in knowing their equivalent present worth. This is particularly true if there is some future profit goal or debt limitation that influences current activities. Such an evaluation is called the *future worth* (FW) method. In the FW method all cash flows are "moved" to the end of the last period. Again, the MARR is used with the appropriate interest factors. The algebraic sum, called the net future worth, is computed at this point. Projects are judged competitively as to their economic promise as with the PW and AW methods.

The rate of return on an investment is a universal measure of economic success. A popular evaluation comparison, called the *rate-of-return* (RR) *method,* determines an effective interest rate i for which the present worth of receipts and disbursements will be equal. The RR method is most often referred to as the *internal rate-of-return method.* It gives the return on the unrecovered portion of the investment. This value is then compared with the MARR. If a project's rate of return meets or exceeds the MARR, it will probably be recommended. When two or more competing projects are evaluated using this method, the one with the largest rate of return is recommended, provided that it meets or exceeds the MARR. There are special cases (as mentioned earlier) where a project with a rate of return less than the MARR may be selected.

In the *ratio of earnings* (or savings) to *investment* (RI) *method,* the present worth (using the MARR) of the positive cash flows are divided by the present worth (using the MARR) of the negative cash flows, treated as positives. Any project earning more than the MARR will have a ratio greater than 1.00. A project earning less than the MARR will have a ratio less than 1.00. When two or more projects are compared, the project with the largest earnings or savings to investment ratio which is greater than 1.00 will be recommended. As before, there are exceptions that may result in a project with a ratio less than 1 being recommended.

Payback Period Method

Liquidity is a measure of how quickly an investment can be recovered. In some situations it is advantageous to determine a project's liquidity rather than its profitability. An evaluation technique used in this situation is the payback period (PP) method. This method avoids the need to calculate the cost of capital while still recognizing the financial concern for limited resources. It reflects caution against unexpected cost increases while employing a specified time period in which the original investment must be recouped.

In this method, the initial investment is divided by the difference of the annual receipts and the annual disbursements that have occurred as a result of the investment. Each company will usually have a minimum payback period which each project must meet. When two or more projects are compared, the one with the smallest payback period is usually recommended. This method ignores the concept of the time value of money and for this reason it does not enjoy widespread use. The payback period method will often select an alternative that is different from the one selected by the other five methods. It is difficult to apply when the cash flows vary significantly from year to year.

Design Example

Part of the life-cycle cost analysis which occurs during the design stage of product development is the estimation of cash flows throughout each year of product life. These estimates may be made for a single design or for two or more competing designs. Consider the projected cash flows, estimated during the design phase for each of two competing designs A and B, as shown in Table 8-7. Each design has an estimated five-year useful life. The initial costs include materials and labor to manufacture each design plus transportation and setup costs, operating costs including energy costs, direct and indirect labor, and consumable materials used. Maintenance costs are

TABLE 8-7 CASH FLOWS: DESIGNS A AND B

Design	*Initial cost*		*Operating cost*		*Maintenance cost*		*Development cost*		*Salvage cost*	
Year	*A*	*B*	*A*	*B*	*A*	*B*	*A*	*B*	*A*	*B*
0	$12,000	$10,000	—	—	—	—	—	—	—	—
1	—	—	$600	$500	$500	$300	$2000	$3000	—	—
2	—	—	600	500	600	500	1000	2000	—	—
3	—	—	600	500	700	700	—	—	—	—
4	—	—	600	500	800	900	—	—	—	—
5	—	—	600	500	900	1100	—	—	−$3000	−$2000
Totals	$12,000	$10,000	$3000	$2500	$3500	$3500	$3000	$5000	−$3000	−$2000

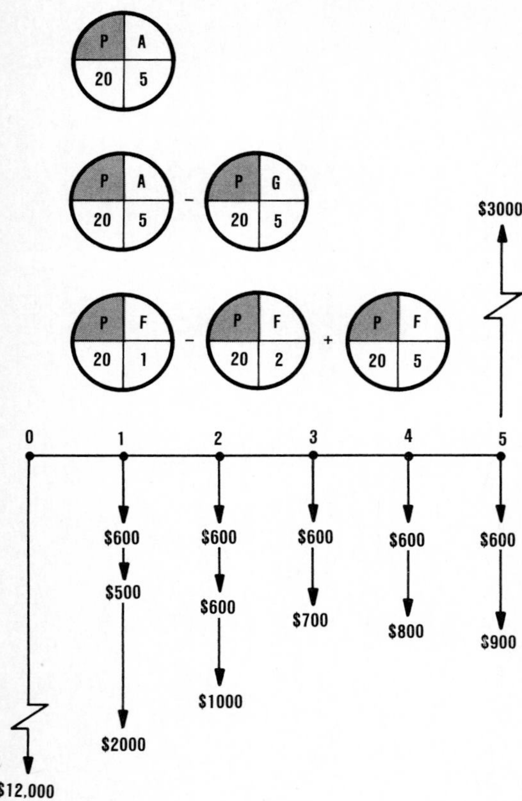

[8-37] Design A cash flow diagram.

for service and repair. The development costs include the costs of anticipated improvements to each design, which will be made only during years 1 and 2. The salvage value is a return (hence a negative cost) to the owners as a result of trade-in or sale of the device at the end of its useful life.

Although costs may actually occur throughout the year on daily, weekly, monthly, or some other basis, the rule-of-thumb convention used in engineering economy evaluations treats these costs as if they will occur at an instant in time at the end of the year in which they actually occurred.

Once the cash flows have been tabulated, the two alternatives may be compared in terms of their net costs. From Table 8-8 we see that design B has a higher overall cost in spite of the fact that its purchase price was less.

Since all cash flows were costs except for the salvage values, the costs were given positive signs and the salvage values were assigned negative signs. This situation is an example of the special cases mentioned earlier where a design will expend more than it earns. The designs will be compared using the present worth method with a MARR of 20 percent. In this analysis we will treat costs as negative worth and salvage values as positive worth.

Solution:

The step-by-step analysis will not be given in this example. However, you will recognize that the solution procedure is quite similar to that of Example 8-13.

For design A [8-37]:

$$\begin{aligned} PW_A &= -12{,}000 - 600(P/A, 20, 5) \\ &\quad -\$500(P/A, 20, 5) - \$100(P/G, 20, 5) \\ &\quad -\$2000(P/F, 20, 1) - \$1000(P/F, 20, 2) + 3000(P/F, 20, 5) \\ &= -\$12{,}000 - \$600(2.9906) - \$500(2.9906) \\ &\quad -\$100(4.9061) - \$2000(0.8333) - \$1000(0.6944) \\ &\quad +\$3000(0.4019) \\ &= -\$12{,}000 - \$1794.36 - \$1495.30 \\ &\quad -\$490.61 - \$1666.60 - \$694.40 + 1205.7 \\ &= -\$16{,}935.71 \end{aligned}$$

TABLE 8-8 COST COMPARISON: DESIGNS A AND B

Cost	*Design A*	*Design B*
Initial	$12,000	$10,000
Operating	3000	2500
Maintenance	3500	3500
Development	3000	5000
Salvage value	−3000	−2000
Totals	$18,500	$19,000

One of the cheapest ways to design something is not to design it at all.

Gordon L. Glegg
The Design of Design, *1969*

For design B [8-38]:

$$\begin{aligned} PW_B &= -\$10{,}000 - \$500(P/A, 20, 5) \\ &\quad -\$300(P/A, 20, 5) - \$200(P/G, 20, 5) \\ &\quad -\$3000(P/F, 20, 1) - \$2000(P/F, 20, 2) \\ &\quad +\$2000(P/F, 20, 5) \\ &= -\$10{,}000 - \$500(2.9906) - \$300(2.9906) \\ &\quad -\$200(4.9061) - \$3000(0.8333) - \$2000(0.6944) \\ &\quad +\$2000(0.4019) \\ &= -\$10{,}000 - \$1495.30 - \$897.18 - \$992.02 \\ &\quad -\$2499.90 - \$1388.80 + \$803.80 \\ &= -\$16{,}469.40 \end{aligned}$$

We see that design B has the highest PW (although negative), and this design is recommended over design A. From Table 8-8 we observe that design B expended more dollars than design A. However, design B was selected because of the pattern by which the money was spent (fewer dollars spent earlier and more dollars spent later, relative to design A).

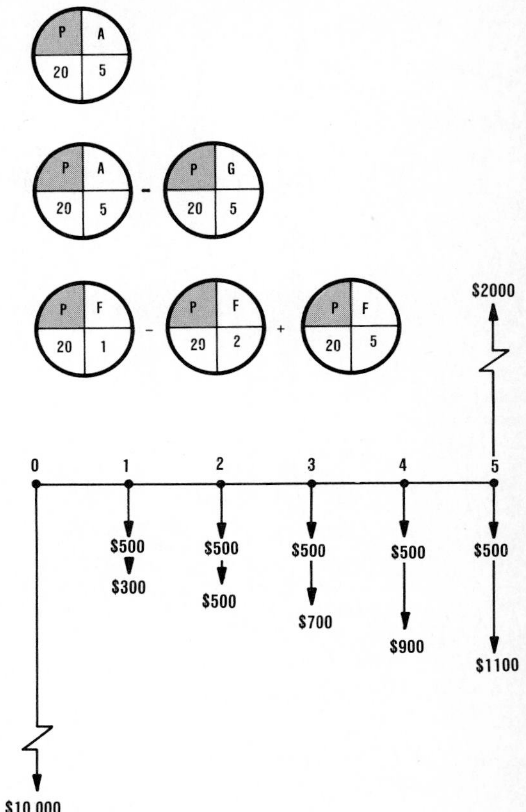

[8-38] Design B cash flow diagram.

UNEQUAL LIVES

In the examples comparing two projects presented in this chapter, each project has had the same estimated life so that a uniform comparison could be made by using any of the methods previously discussed. Realistically, situations will often arise where among two or more projects, at least one will have a life endpoint that is different from the others. In these cases, a common period of time, called a *planning horizon,* must be used in comparing the alternatives. The most obvious choices to be selected for the planning horizon are either the life of the shortest-lived alternative or the life of the longest-lived alternative.

If the life of the shortest-lived alternative is chosen as the planning horizon, the cash flow of each of the other alternatives must be truncated and new salvage values must be estimated. If the life of the longest-lived alternative is chosen, cash flows must be extrapolated for each of the other alternatives. In this case, cost patterns are sometimes assumed to follow the same pattern as estimated to occur during the original life, unless more realistic estimates are made.

Some studies may use the least common multiple of the lives of the alternatives. For example if three alternatives have lives of 6, 8, and 12 years, the estimated planning horizon would be 24 years and the cash flow of the six-year alternative would repeat three more times, the cash flow of the eight-year alternative would repeat two more times, and the cash flow of the 12-year alternative would repeat once. Most engineering economists regard the use of common multiples as the planning horizon as a very poor and unrealistic approach. Consider, for example, two alternatives with estimated lives of 7 and 13 years, respectively. The use of 91 years as a planning horizon is undoubtedly much too distant for any planning purposes whatsoever. Other studies may elect to use a planning horizon that does not coincide with either the longest- or shortest-lived alternative, nor does it represent a multiple of the longest-lived alternative. In these studies, combinations of the approaches already described are used.

CAPITALIZED COST

There are three faithful friends—an old wife, an old dog, and ready money.

Benjamin Franklin, 1738

In some engineering economy studies, a method of evaluation called capitalized cost (CC) is used. With this method, the planning horizon is assumed to be an indefinitely long period of time, or at least a large number of years. This is especially true in considering the economic feasibility of large construction projects such as bridges, dams, or stadiums that may have estimated lives of 50 years or more. In many of these cases, the life of the project is difficult to estimate, although it is known to be an indefinitely long length of time. The method of analysis for such projects assumes an infinite planning horizon. As long as the number of periods (planning horizon) is large, the error introduced by such an assumption is small since the limit of any compound interest factor as n approaches infinity is quite similar to any corresponding value for large n.

For most of the capitalized cost analyses, the time period is expressed in years. In a capitalized cost analysis, the only factor that differs from an analysis with a finite planning horizon is the uniform series present worth factor, $P/A, i, n$.

If n is assumed to be ∞, then

$$\lim_{n\to\infty} (P/A, i, n) = \lim_{n\to\infty} \frac{(1+i)^n - 1}{i(1+i)^n} \tag{8-12}$$

$$= \lim_{n\to\infty} \frac{1 - \dfrac{1}{(1+i)^n}}{i}$$

$$= \frac{1}{i}$$

Therefore, in a capitalized cost study, all uniform annual costs are multipled by $1/i$ to obtain the present worth, and periodic costs that occur every k years must first be multiplied by $(A/R, i, k)$ and then multiplied by $1/i$ to obtain the present worth.

Example 8-15

A new superhighway will be constructed between two cities. The initial cost is \$1,200,000 per mile with expected annual maintenance costs per mile of \$75,000. The highway will be resurfaced every five years at an estimated per mile cost of \$300,000. What is the capitalized cost per mile? The MARR is 20 percent per year.

Solution:

The cash flow diagram is shown in Figure [8-39].

$$CC = \$1{,}200{,}000 + \$75{,}000\left(\frac{1}{0.2}\right) + \$300{,}000(A/F, 20, 5)\left(\frac{1}{0.2}\right)$$

$$= \$1{,}200{,}000 + \$75{,}000\left(\frac{1}{0.2}\right) + \$3000(0.1344)\left(\frac{1}{0.2}\right)$$

$$= \$1{,}200{,}000 + \$375{,}000 + \$201{,}600$$

$$= \$1{,}776{,}600$$

Note that the costs of \$300,000 every five years were first converted to annual costs by multiplying by (A/F, 20, 5) so that the annual costs could be converted to a present worth by multiplying by 1/0.2.

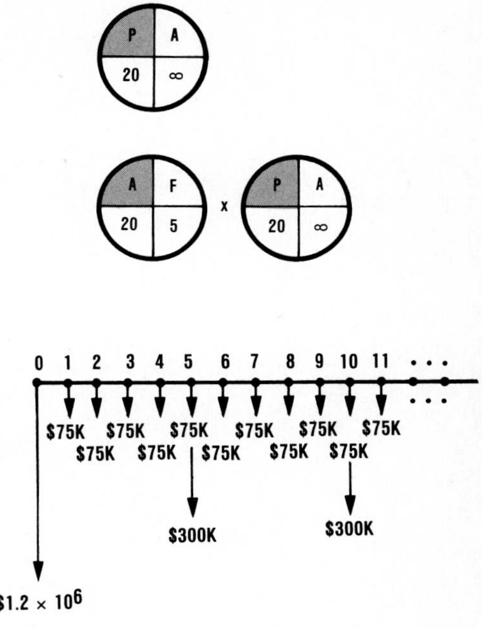

[8-39]

INFLATION IN ECONOMY STUDIES

Inflation refers to a general increase in price levels which brings about a reduction in the purchasing power of a unit of money. Inflation is usually ignored in economy studies since it is assumed to have the same effect on all projects being considered. However, there are occasions when the engineer may be interested in the present or annual worth of cash flows that have been predicted considering the effects of inflation. Inflation, like interest, is a compound rate. For example, if fuel costs are expected to be \$100 during the first year[1] and will increase, due to inflation, by 10 percent in each succeeding year, the projected fuel costs for the first five years will be:

Year 1: \$100

Year 2: \$110 = \$100 + 0.10(100)

Year 3: \$121 = \$110 + 0.10(110)

Year 4: \$133.10 = \$121 + 0.10(121)

Year 5: \$146.41 = \$133.10 + 0.10(133.10)

If it is desired to obtain the present worth of a future cash flow that is subjected to inflation at a compound rate, j, and the rate of return is i, the factor $(1 + j)$ may be used repeatedly to obtain the inflated value in each year of the study and the factor $(1 + i)^{-n}$ may be used with each of these cash flows and the appropriate value of n to determine the present worth.

Example 8-16

The cost of replacement parts for a new robot design is expected to be \$900 during the first year. Due to inflation their cost will increase by 12 percent per year in subsequent years. The MARR is

[1] Note that there is no inflation for the first year; this is an assumption for this problem only—indeed, for many other problems, inflation may be present in the very first year.

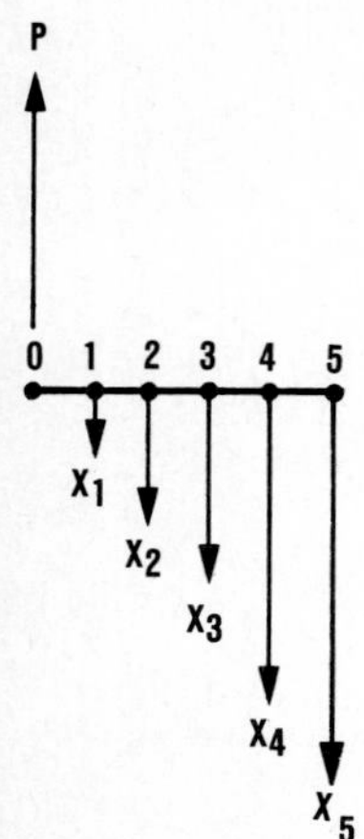

[8-40]

20 percent. What are the equivalent present worth and annual worth values of this inflated cash flow over a five-year period [8-40]?

Solution:

Year	Inflated cash flows		Present worth value	
1		\$ 900.00	\$ 900.00$(1 + 0.2)^{-1}$ =	\$ 750.00
2	\$ 900.00(1 + 0.12) =	1008.00	1008.00$(1 + 0.2)^{-2}$ =	700.00
3	\$1008.00(1 + 0.12) =	1128.96	1128.96$(1 + 0.2)^{-3}$ =	653.33
4	\$1128.96(1 + 0.12) =	1264.44	1264.44$(1 + 0.2)^{-4}$ =	609.78
5	\$1264.44(1 + 0.12) =	1416.17	1416.17$(1 + 0.2)^{-5}$ =	569.12
			Net present worth	\$3282.23

Note that the factor $(P/F, 20, n)$ could have been used in place of $(1 + 0.2)^{-n}$. \$3282.23 represents the present worth of the costs for robot replacement parts, which have increased from year to year due to inflation. This type of calculation is most common in problems involving inflation, as the resulting present worth may be thought of as the amount of money that must be deposited in an interest-bearing account (20 percent for this problem) to be withdrawn as needed to pay for future inflated costs.

Sometimes it is desired to obtain a present worth value that removes the inflationary effects and yields an amount which reflects only the costs in today's (year 0) economy. In this case, the effects of inflation are removed by multiplying the future inflated amount by $(1 + j)^{-n}$ and this value is converted to a time 0 cost by multiplying by $(1 + i)^{-n}$. Thus

$$\begin{aligned}\text{present worth} &= \text{future worth}(1 + j)^{-n}(1 + i)^{-n} \\ &= \text{future worth}(1 + i + j + ij)^{-n} \\ &= \text{future worth}(1 + k)^{-n}\end{aligned}$$

where $k = i + j + ij$. In Example 8-16,

$$k = 0.20 + 0.12 + (0.20)(0.12) = 0.344$$

and

$$\begin{aligned}\text{PW} = \$\ & 900.00(1 + 0.2)^{-1} \\ &+ 1008.00(1 + 0.344)^{-1}(1 + 0.2)^{-1} \\ &+ 1128.96(1 + 0.344)^{-2}(1 + 0.2)^{-1} \\ &+ 1264.44(1 + 0.344)^{-3}(1 + 0.2)^{-1} \\ &+ 1416.17(1 + 0.344)^{-4}(1 + 0.2)^{-1} = \$2691.55\end{aligned}$$

The same result could have been obtained from

$$\begin{aligned}\text{PW} &= \$\ 900.00(P/A, 20, 5) \\ &= 900.00(2.9906) \\ &= \$2691.55\end{aligned}$$

INCOME TAXES AND ECONOMY STUDIES

Nothing is certain but death and taxes.
Benjamin Franklin
Letter to M. Leroy, *1789*

The effects of state and federal income tax rates on the merit of individual engineering projects or on the selection of a single project from among two or more projects can be quite pronounced. It is particularly important to consider the effects of income taxes whenever machines or equipment are being replaced. In these instances, a depreciation schedule is used [the schedule must be approved by the Internal Revenue Service (IRS)] to obtain the book value at the time of replacement. Book value is the original worth of a property less the accumulated depreciation. The *book value* is compared with the salvage or resale value to obtain the capital gain or capital loss. It is beyond the scope of this chapter to explain after-tax engineering economy studies, but you should recognize that there are conditions (such as replacement studies) where an after-tax analysis should be made. Some companies have both a before- and after-tax MARR for use in engineering economy studies. For example, the before tax MARR may be 20 percent, and the after-tax MARR may be 12 percent.

Depreciation in Economy Studies

The assets of the company include property, buildings, machines, equipment, goods in inventory, as well as accounts receivable and operating capital. Some of these assets, such as the machines that are used to manufacture the company's products and the equipment used for material handling, testing, and inspection, will need to be replaced from time to time due to wear-out or functional obsolescence (the machine or equipment is considered to be outdated). The expected time that a machine or equipment will be used by a company until it "wears out" or becomes obsolete is called the *economic life* of the asset.

Example 8-17

A machine might be purchased for \$100,000 and is expected to be used for five years. At that time it can be sold as a trade-in on another similar machine for \$20,000. From this example, we can see that this machine has lost \$80,000 of its value over its five-year economic life. The loss in value over time is called the *depreciation* of the asset. The average loss per year for this asset is

$$\frac{\$100{,}000 - \$20{,}000}{5} = \frac{\$80{,}000}{5} = \$16{,}000 \text{ per year}$$

Value depreciation and tax depreciation may differ. The Internal Revenue Service (IRS) of the U.S. Treasury Department recognizes loss of value over time and allows companies to set up a *tax depreciation schedule* which lists how much of the asset's value will be assumed lost each year. Often IRS regulations allow a tax depreciation schedule that depreciates the book value faster than the actual resale value declines. If the depreciation schedule is approved by the IRS, then during each year of use of the machine or equipment, the depre-

Profits decline more during (an economic) crisis than wages. It is because of the fall in profits that unemployment occurs.
Jean Fourastié

ciation, or loss of value during that year, may be deducted from corporate profits before the corporate income tax rate is applied to the corporate profits. There are two types of tangible assets recognized by the IRS for assets placed into service after 1980. Section 1245 class property refers to manufacturing equipment, tools, machinery, office equipment, and vehicles. Section 1250 class property is represented by buildings, parking lots, and other exterior fixtures of the company. Section 1245 class properties are assigned a three- or five-year economic life. Small vehicles and certain manufacturing tools may be depreciated over a three-year period and the balance of the other Section 1245 class properties, such as machinery, equipment, and other manufacturing tools, are assigned a five-year economic life. Section 1250 class property is assigned either a 10- or a 15-year economic life. There are other assignments for public utility properties which are not discussed here. A specified percentage of the first cost may be depreciated each year according to the accelerated cost recovery system (ACRS). The ACRS allowable percentages for Section 1245 class properties are shown in Table 8-9.

The depreciation allowances for the machine of Example 8-17 are as follows:

Year	*Depreciation allowance during the year*	*Book value at end of year*
1	\$15,000	\$85,000
2	22,000	63,000
3	21,000	42,000
4	21,000	21,000
5	21,000	0

The company using this schedule may deduct \$15,000 from corporate profits during the first year of the machine's use before those profits are subjected to income taxes. If the company's income tax rate is 46 percent, this means that \$15,000 of the company's profits during this year will not be taxed, since these profits are offset by depreciation. The company will pay 0.46(\$15,000) = \$6900 less in taxes during the first year of the machine's use than it would if the machine were not depreciable. Similarly, the depreciation allowances

Labor can do nothing without capital, capital nothing without labor, and neither labor nor capital can do anything without the guiding genius of management; and management, however wise its genius may be, can do nothing without the privileges which the community affords.

W. L. MacKenzie King
Canadian Club Speech, *1919*

TABLE 8-9 ACRS PERCENTAGES FOR SECTION 1245 CLASS PROPERTY

Year	*Three-year property*	*Five-year property*
1	25	15
2	38	22
3	37	21
4	—	21
5	—	21

in years 2 through 5 may be used to offset taxes in those years. Thus depreciation schedules are used by companies to shelter part (approximately one-half of the depreciation allowance each year) of corporate profits.

Note that the ACRS depreciation percentages sum to 100 percent and thus the entire value of the asset is depreciated over a five-year period. If, indeed, the asset is sold for its estimated salvage value ($20,000 in Example 8-17) at the end of the fifth year, the company would have to pay taxes on this gain (excess of the resale or exchange price over book value). Similarly, if the machine was sold before the end of the fifth year, the selling price would be compared with the book value at the time of the sale and the gain or loss would be computed so that a tax rate or tax credit could be applied.

It should be emphasized that the selection of depreciation schedules is not arbitrary, and in any event all schedules must be approved by the IRS.

There is this difference between those two temporal blessings, health and money: Money is the most envied, but the least enjoyed; Health is the most enjoyed, but the least envied; and this superiority of the latter is still more obvious when we reflect, that the poorest man would not part with health for money, but that the richest would gladly part with all their money for health.

Charles C. Colton

PROBLEMS

8-26. A company must decide between purchasing and leasing a large computing system. A five-year planning horizon will be used in the study. A new computing system will cost $600,000 and operating expenses are expected to be $25,000 per year. Maintenance charges will not be incurred by the company until the second year as the warranty extends over the first year. Maintenance costs in years 2 through 5 are expected to be $35,000. $40,000, $45,000, and $50,000, respectively. The system can be resold by the company for $150,000 after five years. On the other hand, a comparable system can be leased for $100,000 per year, with the payments being made at the *beginning* of each year. This is common practice in leasing arrangements. Operating expenses will be the same as those for ownership—$25,000 per year. In the case of a lease arrangement, the company will not have to pay any maintenance charges. The company's MARR is 20 percent per year compounded yearly. Compare the two alternatives by calculating a present worth for each. *Hint:* To convert the beginning of the year lease costs to end of the year values, multiply by $(1 + 0.2)$. You now have five end-of-year costs in years 1 through 5 which can be converted to a present worth by multiplying by $(P/A, 20, 5)$.

8-27. The owner of an automatic car wash has been having difficulties in keeping the equipment operational. A decision must be made to either overhaul the present equipment or to replace it with new equipment. An overhaul will cost $10,000 and the equipment is expected to experience operating costs of $3000 per year over the next 10 years. Maintenance costs are expected to be $1000 in year 1 and will increase by $600 each year thereafter (i.e., $1600 in year 2, $2200 in year 3, etc.). If overhauled, the equipment will have no salvage value after 10 years.

On the other hand, new equipment costs $25,000 and is expected to have operating expenses of $1000 per year over the next 10 years. Maintenance charges are expected to be $500 in year 1 and will increase by $100 each year thereafter. This equipment can be resold after 10 years of use for $5000. Furthermore, if new equipment is purchased, the present equipment can be resold now for $5000.

Compare these two alternatives on the basis of annual worth (i.e., compute the net annual costs of each) if the MARR is 18 percent per year compounded yearly.

8-28. Which of the following three proposals is recommended if a MARR of 15 percent per year is required? Compare on the basis of net future worth (i.e., convert each cash flow to a value at the end of year 5).

YEAR-BY-YEAR CASH FLOWS

Year	Proposal A	Proposal B	Proposal C
0	−$100,000	−$50,000	−$20,000
1	20,000	20,000	5,000
2	30,000	20,000	5,000
3	40,000	20,000	5,000
4	50,000	20,000	5,000
5	60,000	20,000	15,000

8-29. A bridge has a first cost of $75,000, annual maintenance costs of $5000, and must be repainted every five years at a cost of $8000. If $i = 10$ percent per year compounded yearly, what is the capitalized cost?

8-30. A stadium has a first cost of $4,000,000 and annual maintenance costs of $300,000. Seats are replaced every 15 years at a cost of $400,000. Concrete supporting columns are replaced every 20 years at a cost of $1,000,000. What is the capitalized cost if $i = 15$ percent per year compounded yearly?

8-31. Two possible types of road surface are being considered with cost estimates per mile as follows:

	Type I	*Type II*
First cost	$80,000	$120,000
Resurfacing period	12 years	18 years
Resurfacing cost	$50,000	$ 50,000
Average annual cost	$ 4,000	$ 3,000

Compare these two types on the basis of capitalized cost using an interest rate of 8 percent per year compounded yearly.

8-32. Suppose that a product costs $1000 today. How much will be required to purchase this product three years from now if the rate of inflation is expected to be 8 percent each year? *Hint:* Simply multiply $1000 by the (*F*/*P*, 8, 3) factor.

8-33. With respect to Problem 8-32, we will now consider the time value of money. With the same inflation rate of 8 percent, how much should be deposited now at a rate of return of 7 percent per year compounded yearly to have enough money to buy the product at the end of three years? *Hint:* Multiply the answer to Problem 8-32 by (*P*/*F*, 7, 3).

8-34. Reconsider Problems 8-32 and 8-33. Suppose that the inflation rate is expected to be 8 percent in year 1, 9 percent in year 2, and 10 percent in year 3. The rate of return is still 7 percent per year compounded yearly. Under these conditions, how much should now be deposited to have enough money to buy the product at the end of three years?

8-35. A machine has a purchase price of $35,000 and an estimated salvage value of $19,000 after a three-year life. Compute the depreciation allowance during each of the three years of economic life and the book value at the end of each of the three years using the ACRS allowable percentages for Section 1245 class properties.

8-36. Suppose that the machine of Problem 8-35 is sold after one year for $27,000. As a result of this transaction only, would the owners of the machine pay a gains tax or receive a tax credit?

8-37. A piece of equipment has an estimated five-year life and has a purchase price of $95,000 and an estimated salvage value of $10,000 after the fifth year of use. Compute the depreciation allowance during each of the five years of economic life and the book value at the end of each of the five years using the ACRS allowable percentages for Section 1245 class properties.

8-38. Suppose that the equipment of Problem 8-37 is sold after three years for $40,000. As a result of this transaction only, would the owners of the equipment pay a gains tax or receive a tax credit?

Chapter 9

The Presentation of Engineering Analyses

THE NEED TO DOCUMENT ANALYSES

There is a subset of the infamous Murphy's "laws"[1] that seemingly governs bureaucracies. Two members of this subset are cited here in order to give you an idea of their "tongue-in-cheek" flavor: (1) massive expenditures obscure the evidence of bad judgments (known as the First Bureaucratic Bylaw), and (2) a system that performs a certain function or operates in a certain way will continue to operate in that way, regardless of need or changed conditions (you may recognize this as Newton's law of systems inertia).

Another important one of these "laws" that relates more particularly to the material in this chapter states: "A memorandum is written not to inform the reader, but to *protect* the writer." Called Acheson's rule of the bureaucracy, this statement implies that effective communication is less important than protecting the job of the bureaucrat.

Although this may seem to violate all engineering codes of ethics, engineers occasionally do need to obey this "law." They often write for protection reasons. This statement is made without hesitation and without tongue-in-cheek. We will provide some clarification below.

The documentation in writing of ideas, analyses, and results has a number of purposes other than personal protection. Foremost among these purposes is communication—informing others of the author's thoughts, decisions, progress, and conclusions. In our rapidly changing and complex technology, seldom will an entire project be carried out from conception to completion by a single person. The team approach is crucial for much of the nation's industries. There may be bankers to be persuaded, fellow workers (including,

[9-1] Why me?

[1]There is some evidence to suggest that Murphy was a development engineer at Wright Field (Ohio) Aircraft Lab. In 1949, being frustrated with a measuring device that was not working because it had been incorrectly wired by a technician, he was reportedly heard to say, "If there is any way to do it wrong, he will." Over the years, this rule has been generalized to: "If anything can go wrong, it will." It has also been expanded into a large set of corollaries. As Dickson has written, "It has been suggested that Murphy . . . helped more people get through crises, deadlines, bad days, the final phases of projects, and attacks by inanimate objects than either pep talks, uplifting epigrams, or the invocation of traditional rules. It is true that if your paperboy throws your paper into the bushes for five straight days it can be explained by Newton's law of gravity. But it takes Murphy to explain why it is happening to you" (see Paul Dickson, "The Official Rules of Engineering," *New Engineer,* December 1978, p. 56).

perhaps, your boss) to be informed, patent attorneys to be educated, and perhaps even licensing boards to be convinced. Lack of communication among any of the team members who are working on the project is almost always catastrophic.

More and more industrial communication is being handled through computerized electronic mail or through the use of digital data bases (rather than by exchanging "hardcopy"). However, the form of the documentation is not the important feature. The content and ease of assimilation (reading and understanding) are the important features. There must be an *effective* transfer of information!

Documentation is also necessary for protection, as has been pointed out previously. However, unlike the bureaucrat, who may write to protect his job, engineers write mainly to protect their credibility and their ideas. Patents, for example, are never granted for ideas that exist only in an inventor's mind. Ideas must be documented in hard copy; otherwise, legal protection for original ideas cannot be guaranteed. This does not mean that formal patent applications must be filed on every idea, but at the very least, the inventor's personal written log should describe each idea, its date of conception, and the identities of witnesses.

For internal documentation and communication, many companies (and engineering professors) require engineers to keep a written log or engineering notebook. When this log is properly used, all ideas, notes, derivations, progress, and meeting minutes are recorded in it. Neatness and conciseness are not required, but completeness, in the sense that the record conveys enough information so that a knowledgeable reader can understand the intent, is paramount.

An occasional review of this information often restimulates old ideas and or rejuvenates a stagnant project. Companies often require that each filled workbook be placed on file within the company archives.

Often, engineers must be able to document particular information in order to establish and/or protect their professional credibility. For example, if you were requested to serve as an expert technical witness in a court case, you would generally be asked to document your credentials. Your testimony, either given in the courtroom or through depositions, would be scrutinized as much for its ability to establish your competence and credibility, as for its ability to inform the court on matters pertinent to the case.

Engineering students are no different than practicing engineers in their need to communicate effectively. Their success or failure in their educational pursuit depends on their ability to express themselves through the written word. Faculty, like clients of practicing engineers, have great difficulty separating those students who do not understand a subject from those who understand it but cannot properly express themselves.

THE DETAILS OF DOCUMENTATION: PRESENTATION, UNCERTAINTIES, RESOLUTION, AND MISTAKES

Presentation

When shopping in your favorite grocery store or drugstore, did you ever notice the emphasis placed on product packaging? It is a fact that marketing success often depends on how attractive the product appears, rather than on just functionality alone. The failure of the Edsel and the success of the Mustang automobiles [9-2 and 9-3] cannot be explained by the fact that the Edsel was a more comfortable car and offered much more interior space than the Mustang. The difference in marketing appeal was focused on the Mustang's exterior "packaging" and sporty appeal, not on the relative merits of the two interiors.

[9-2] The Edsel—a marketing failure.

Unfortunately, packaging (i.e., the method of presentation) plays a strong role in the acceptance of technical documentation. But fortunately, this desirable attribute of attractiveness in packaging usually manifests itself somewhat differently in the case of technical analyses than for commercial products. The readers of technical literature are not looking for the racing stripe down the side or the color that just matches their stockings. If they were looking for fantasy and excitement they would be reading Ian Fleming's books about James Bond.

[9-3] The Mustang—a marketing success.

The astute technical reader is looking for *conciseness, clarity, and accuracy*. These attractive features "sell" technical literature. It should not always be necessary for the reader to rederive the formulae, reperform the calculations, or replot the data in order to fill in missing pieces and to understand the work. On the other hand, the reader should not have to be carried through every step of the problem in order to comprehend the concept.

There must be a middle ground. The key to finding it is for engineers to understand their purpose in writing and to know the level of their readers.

There are many documentation formats that can provide the structure for a technical document. Most large companies require a standard format for their reports; many smaller companies do not. However, all successful formats have some similar features, some of which are:

1. An abstract (a very short, less than one page, description of the full report).
2. A statement of the problem or subject being addressed.
3. An explanation of the problem solution or conclusions reached.
4. A description of the significant features of the solution or findings.
5. The supporting data.

Later in this chapter we provide examples of problem solutions that meet the conditions of conciseness, clarity, and accuracy. But first we wish to explore some of the problems that often arise in the manipulation of numbers and the performance of calculations.

A single photographic reproduction is said to have the same intrinsic value as one thousand eloquent articulations.
Translation: A picture is worth a thousand words.

A rotating fragment of mineral collects no bryophytic plants.
Translation: A rolling stone gathers no moss.

If it is necessary to undertake to complete a task, one should strive to complete the job in an exemplary manner.
Translation?

It is impossible to form an authoritative opinion concerning a hardbound compilation of literary information by mere observation of its protective bindings.
Translation?

Douglas' Law of Practical Aeronautics: When the weight of the paperwork equals the weight of the plane, the plane will fly.

Attributed to airplane builder Donald Douglas

Tables. A compact and efficient method of presenting data is in the form of *tables* of numbers. As with equations, tabular data have independent and dependent variables. In Table 9-1 gauge number is the *independent variable* since the other quantities listed in the table are commonly determined (calculated or measured) for each gauge. The other quantities (cross-sectional area and ohms per kilometer) are known as *dependent variables.*

Tables with only one independent variable (such as Table 9-1 below) are sometimes referred to as *single-parameter* tables. Similarly, tables with two independent variables (such as are found in Appendix III) are referred to as *two-parameter* tables. Tables in which there are three independent variables are usually represented as a series of two-parameter tables. Each table in such a series might give data for two of the variables, but each table would represent a different value of the third variable. Such a collection of tables is given in Appendix III. Tables for data with four or more independent variables are rarely encountered.

Tables are particularly useful for data in which no interpolation is required (where only discrete values of the independent variable are possible) or for data in which the precision of the numbers would be compromised by plotting in graphical form (discussed in the next

TABLE 9-1 WIRE TABLE FOR STANDARD ANNEALED COPPER [AMERICAN WIRE GAUGE (BROWN & SHARPE) SOLID CONDUCTOR; TEMPERATURE = 20°C]

Wire gauge	*Diameter (mm)*	*Cross-sectional area (mm^2)*	*Resistance (ohms/km)*
0000	11.6840	107.219	0.1608
000	10.4038	85.011	0.2028
00	9.2659	67.432	0.2557
0	8.2525	53.488	0.3224
1	7.3482	42.409	0.4065
2	6.5430	33.624	0.5128
3	5.8268	26.665	0.6463
4	5.1892	21.149	0.8153
5	4.6203	16.766	1.0279
6	4.1148	13.298	1.2963
7	3.6652	10.551	1.6345
8	3.2639	8.367	2.0610
9	2.9058	6.631	2.5988
10	2.5883	5.261	3.2772
12	2.0526	3.309	5.2100
14	1.6276	2.081	8.2841
16	1.2908	1.309	13.1759
18	1.0236	0.823	20.9482
20	0.8118	0.518	33.3005

section). Table 9-1 contains physical and electrical data for copper wire in the standard wire sizes in the range from 0000 to 20 gauge. Since, for economic reasons, wire of standard size is almost always used, there is usually no interpolation required when using this table.

Tables should always be organized for the ease of use of the reader. This means, above all, that only fundamental data should be presented—data that can be useful in many different ways, not just in an apparatus similar to the one that you have chosen for your experiment. For example, if you were measuring the viscosity of a lubricating oil as a function of temperature by measuring the time it takes for the oil to flow out of a container (the Saybolt technique for viscosity determination), the measurements you might make are temperature and time intervals. However, no one except you is really interested in the time intervals you measure; what is much more useful are viscosities that are implicit in the time intervals measured. You need to use a "data reduction" scheme to deduce viscosity from your measurements so that you could tabulate the fundamental quantity viscosity versus temperature.

Other rules that aid the user are:

1. Align the decimal points in each column of similar numbers.
2. Use scientific notation for large numbers.
3. If users will commonly interpolate from the table, choose convenient intervals for tabulating the independent variables. Intervals of 1, 2, 5, 10, or multiples thereof, are preferred.
4. Label all data with the names of the quantities being tabulated and their units.
5. Title the table with enough information to make clear its contents and limitations.
6. Place extended information about such things as limitations and sources of nonoriginal data in footnotes to the table.

Graphs. The use of tables to display data is only one way of presenting information to prospective users. Often, graphical presentation is a superior method of showing data that might otherwise be placed in a table. Sometimes, graphical and tabular methods complement each other and the inclusion of both is useful.

Graphs are presentation forms that allow the data to be expressed as "pictures" in which lengths or areas are used to depict the size of the numbers. When displayed in this way, lengths and areas are much more quickly assimilated than are collections of numbers. The old adage "A picture is worth a thousand words" is certainly true in this case.

Types of Graphs. A graphical display of information may take any of several forms, depending on the type and use of the information to be presented. Often, *pictographs* provide for the most rapid interpretation to the widest possible audience. They are used routinely in newspapers and on television. An example is shown in Figure [9-4].

Pie charts are ideal for presenting data that represent various parts of a whole or total entity. Such charts allow the user to understand quickly the contribution of each part to the total. In Figure

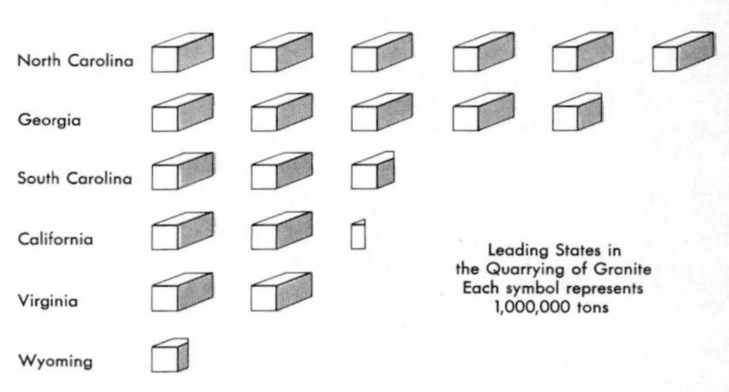

[9-4] Pictograph.

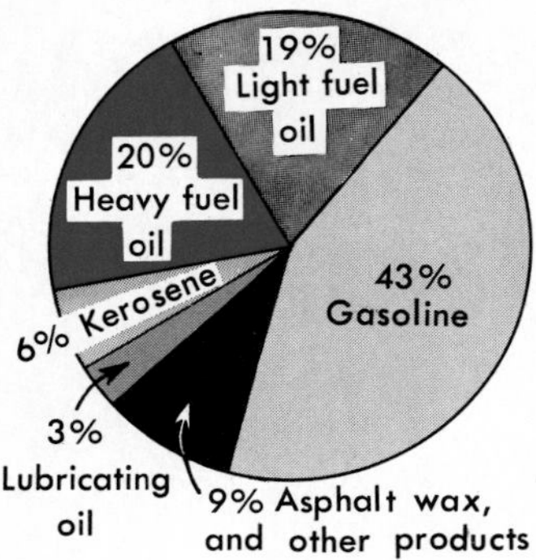

[9-5] Pie chart (data expressed as parts of a whole).

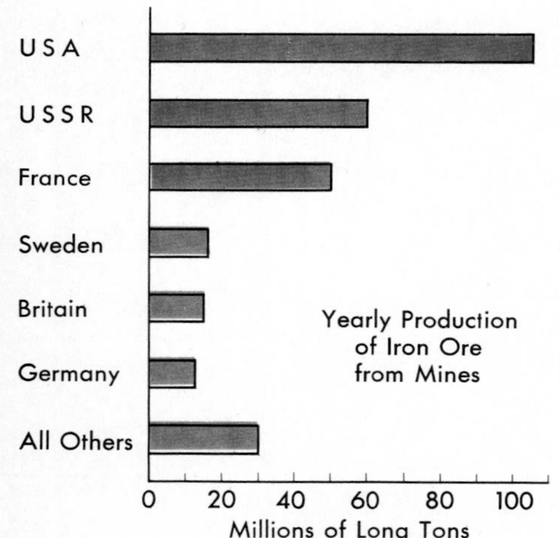

[9-6] Horizontal bar graph.

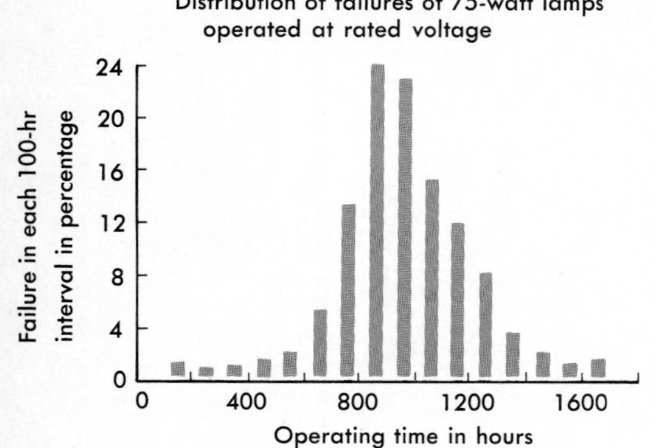

[9-7] Vertical bar graph.

[9-5], you can readily see from the sizes of the slices of the "pie" that gasoline is by far the largest product in the refining of crude oil.

Bar charts are graphs in which data for the dependent variable are shown as bars whose lengths are proportional to the size of the data. In Figure [9-6], for example, *country* is the independent variable while the *amount of iron ore* is the dependent variable. We can see that the United States produces almost twice as much iron ore as does its next closest competitor.

A *histogram* is a special type of bar chart in which the dependent variable is usually the *number of "things"* that fall into "ranges" (often called *bins* or *cells*) shown on the independent variable axis. In the histogram of Figure [9-7] the vertical bars represent the *number of light bulbs* (expressed as a percentage of those tested) that failed in various ranges of operating hours. The largest fraction, here 24 percent, failed within 800 to 900 hours of operation although some, 1 to 2 percent, lasted more than 1600 hours. Figure [9-7] readily gives you information on how long you might expect a light bulb from the particular group tested, to last. You can see that histograms are convenient for showing statistical data.

The bulk of engineering data are presented using *line graphs* or *x-y* plots. Such graphs are usually more exact and better suited for use in interpolation, extrapolation, and investigation of data trends. In *linear* axis or *linear* scale line graphs, the distance along each axis is proportional to the numbers representing the data [9-8 and 9-9]. In *semilogarithmic* or *semilog* plots, one axis of the plot (the linear one) is *proportional to the numbers* representing one of the variables, while the other axis (the logarithmic one) is *proportional to the logarithm of the numbers* representing the other variable [9-10]. *Log-log* plots are graphs in which both axes are *proportional to the logarithms of the numbers* representing the respective variables [9-11]. On any logarithmic axis, it is *not* conventional to label the axis with the value of the logarithm of the numbers plotted, but rather to label the axis in terms of the data numbers being plotted (which are the antilogs of the logarithm numbers). This fact always makes the logarithmic axes appear nonlinear [9-10 and 9-11] and more difficult to read; that is, equal increments on logarithmic axes do not correspond to equal intervals for the variables being plotted. For example, Figure [9-10] shows that the first major division along the horizontal logarithmic axis represents a change from 10^4 to 10^5 loading cycles while the next major horizontal division represents a change from 10^5 to 10^6, considerably more.

Semilog plotting, where only one variable is given a log scale, is convenient in cases where

1. the variation of the data is such that it may be desirable to compress the larger values of one variable (for example when one variable spans a large range of numbers), or
2. there is an exponential behavior of the data.

To illustrate the latter phenomenon, let us consider data that satisfy the following form of equation:[2]

$$y = e^x$$

[2]Similar analyses apply to the equation $y = 10^x$.

Taking logarithms (to the base 10) of both sides yields

$$\log_{10} y = \log_{10} (e^x)$$

or

$$\log_{10} y = x \log_{10} e = 0.4343x$$

A plot of (log y) versus (x) would be a straight line. *Thus exponential relationships plot as straight lines on semilog graphs and, conversely, straight lines on semilog graphs signify exponential relationships.*

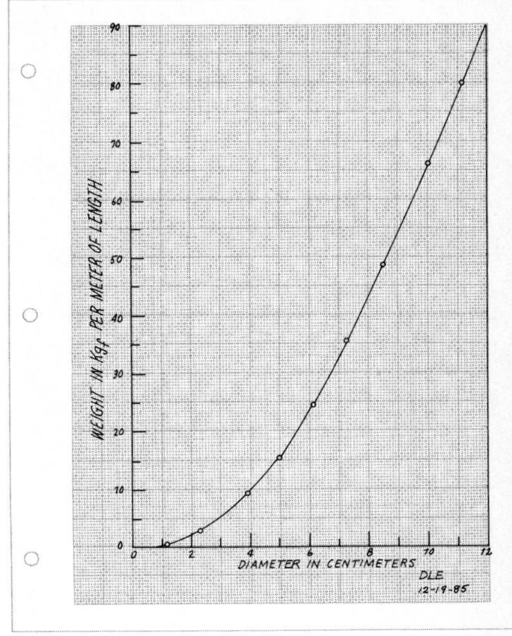

[9-8] Linear scale graph.

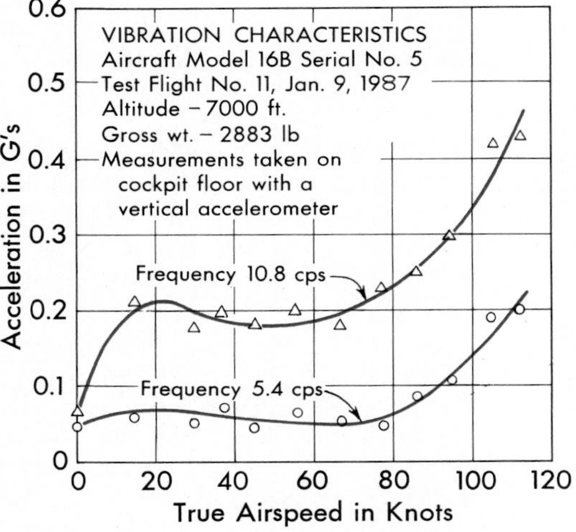

[9-9] Linear scale graph.

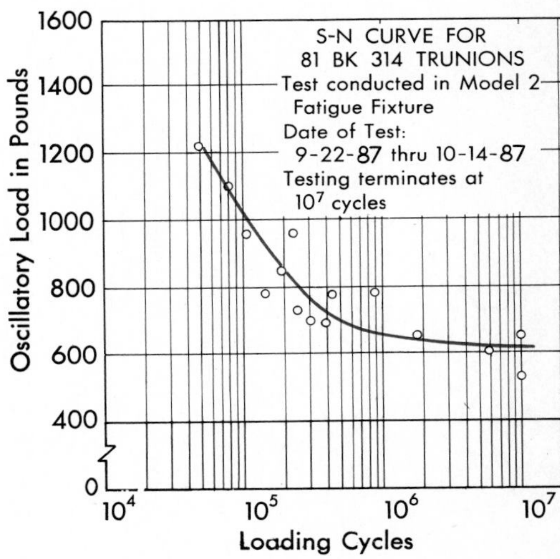

[9-10] Semilog plot.

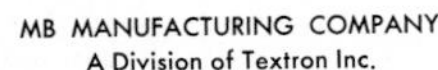

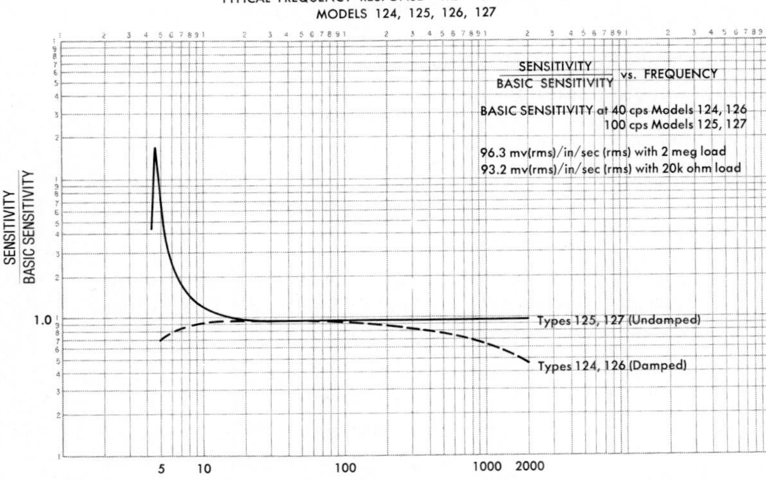

[9-11] Log-log plot.

Log-log plotting, where both variables are assigned log scales, is convenient in the following types of cases:

1. The variation of the data is such that it may be desirable to compress values of both variables (e.g., when both variables span large ranges of numbers).
2. The data satisfy algebraic equations containing power relationships or roots.

To illustrate the latter phenomenon, consider data that satisfy the following form of equation:

$$y = \alpha\, x^n$$

where α and n are constants. Taking logarithms of both sides yields

$$\log y = \log \alpha + n \log x$$

The latter equation represents a straight-line relationship between ($\log y$) and ($\log x$). The line would have a slope of (n) and a y intercept of ($\log \alpha$) in the ($\log y$) versus ($\log x$) plane. *Thus power relationships between variables plot as straight lines on log-log graphs and, conversely, straight lines on log-log graphs demonstrate that there are power relationships between the variables.*

Each power of 10 that can be plotted on a logarithmic axis is said to be a *cycle*. For example, Figure [9-10] is a three-cycle semilog plot since three powers of 10 (10^4 through 10^7 loading cycles) or three orders of magnitude in the number of loading cycles can be accommodated on the horizontal axis.

Polar graphs are often used where a variable quantity is to be examined with respect to various angular positions. Examples are the light output of luminous sources, the response of microphone pickups, and the behavior of rotating objects at various angular positions. An example of a graph plotted on polar coordinate paper is shown in Figure [9-12].

Graphing Practice. Much graphing is now done using application programs on computers. For example, most spreadsheets (discussed in Chapter 18) have graphing capabilities. In addition, there are many special applications graphing programs for micros, minis, and mainframes. Depending on the software package being used, the user may or may not be able to exercise much control over the plots. However, there are some important rules that need to be followed to promote ease of use. In some cases, this may mean that a new set of axis labels will have to be added to a machine-made graph. The important rules are as follows:

1. It is customary to plot the independent variable along the abscissa (the horizontal axis) and the dependent variable along the ordinate (the vertical axis).
2. The scale for each axis must be suitable for the paper and the data being used. On linear plots, scale divisions of 1, 2, 5, 10, or a multiple of these numbers, are *mandatory* for ease of use. Never use a scale that requires awkward interpolation. For example, labeling points along an axis as 21.07, 23.27, 25.47, and so on, does not promote ease of use. For logarithmic axes, label

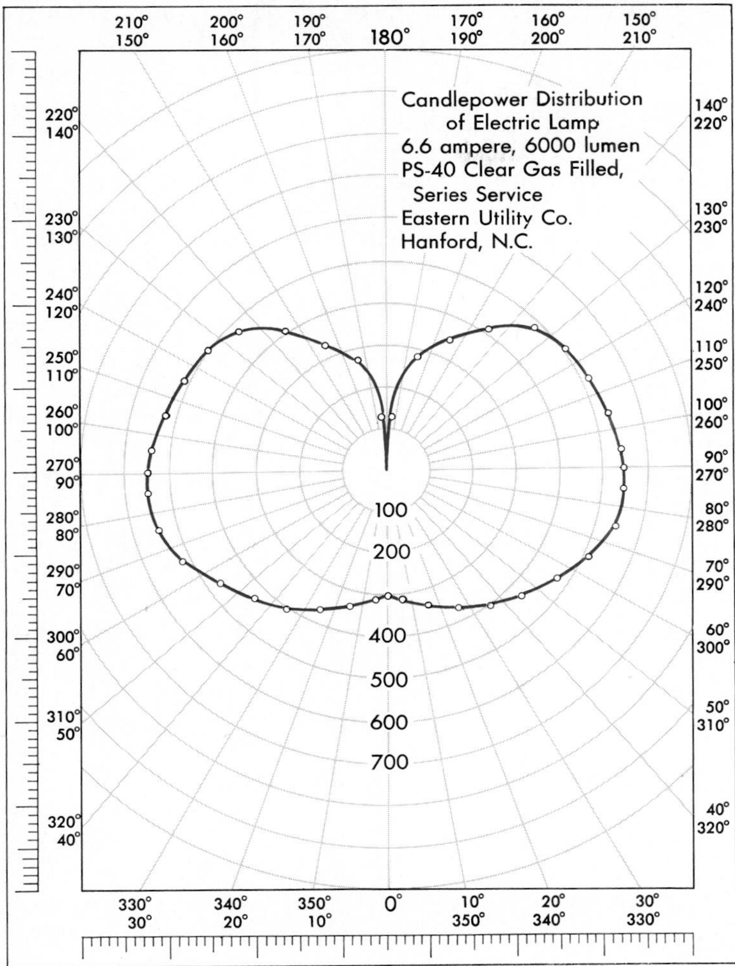

[9-12] Polar graph.

the powers of 10 [9-10 and 9-11], and perhaps some intermediate values if space permits.

3. On linear axes, it is desirable to show zero as the beginning of the ordinate and abscissa axes unless this would compress the data unnecessarily. If compression is severe, show zero at the origin and use a broken axis line to alert the reader that the data are compressed (see the ordinate on Figure [9-10]). Logarithmic axes never go to zero since log(0) is undefined. Therefore, choose the origin to be the lowest power of 10 that will permit plotting of the data.
4. The points for each data set being plotted should be identified by using distinctive identification symbols such as small squares, triangles, diamonds, or other simple geometric figures [9-9]. Distinctive linework such as solid line, dashed line, long dash–short dash line also may be used to aid identification of independent data sets.

5. Graphs may be drawn for theoretical relationships, empirical relationships, or measured relationships. Curves of theoretical relationships will not normally have point designations [9-8]. Empirical relationships should form smooth curves or straight lines, depending on the form of the mathematical expression used [9-9]. Measured data points that can*not* be expected to have an underlying mathematical or empirical relationship, should be connected by straight lines drawn from point to point. Otherwise, data obtained from measured relationships should be drawn to average the plotted points (called curve fitting) [9-9 and 9-10]. In this situation, curves representing the measured data do not necessarily go from center to center of the points.
6. Curves used to approximate the behavior of the data points *should not* be drawn through the symbols that enclose the plotted points, but rather stop at the perimeter of the symbols [9-10].
7. Exponential notation should be used when labeling large numbers on the abscissa and ordinate.
8. The abscissa and ordinate variables should be labeled together with their respective units of measure: for example, *Weight in Newtons*.
9. The title of the graph should refer to the ordinate values first and the abscissa values second. The title should also include other descriptive information, such as sizes, weights, names of equipment, date that the data were obtained, location where the data were obtained, serial numbers of the apparatus, name of manufacturer of apparatus, and any other information that would help describe the graph.
10. The title should be placed on the sheet where it will not interfere with the curve.
11. The name (or the initials) of the person preparing the graph and the date the graph was plotted should be placed on the sheet [9-8].

Uncertainties

It should be obvious that the diameter of a saucepan and the diameter of a diesel engine piston, although each may measure about 15 cm, will usually be measured with different accuracies. Similarly, a measurement of the area of a large ranch that is valued at $0.05 per square meter would not be made as accurately as a measurement of a piece of commercial property that is valued at $10,000 per square meter. Given enough resources (equipment and money), we could increase the accuracy of almost any measurement, but there are many situations that simply do not merit such effort. There is almost always, however, a need to know the approximate accuracy in any measurement we make. The concepts of uncertainty and significant figures are often used to express the accuracy of single measurements.

Uncertainties are the differences between observed measurements (or calculated numbers) and the true values of those measurements (or numbers). Since physical quantities cannot be measured exactly, the true values cannot be known. The engineer's idea of uncertainty, then, is that of an inaccuracy, where complete accuracy is impossible. The word "error" is often used as a synonym for uncertainty in science and engineering literature, but since it is often used as a synonym for mistake or blunder in layperson's conversation, it is preferable not to use the term at all.

Realizing that inaccuracies are a fact of life, we attempt to assess the amount of uncertainty we might have in any measurement or in any calculation that makes use of measurements. We try to establish *bounds* on the uncertainty, a *lower bound* and an *upper bound* between which the true value is presumed to lie. For example, we might say that the measurement is 4.173 ± 0.002 meters to indicate an uncertainty of 0.002 meter. In this case, the true value is presumed to be bracketed between the upper bound of 4.175 meters ($4.173 + 0.002$) and a lower bound of 4.171 meters ($4.173 - 0.002$). Bounds are sometimes given as a percentage of the quantity: $57 \pm 3\%$ liters. Here the bounds are said to be *relative bounds*. The uncertainty in this case is 3 per cent of 57, or about 1.7 liters and implies an upper bound of about 58.7 and a lower bound of about 55.3.

More commonly in engineering calculations, the uncertainty is implied by the manner in which the value itself is stated (with no $\pm$ quantities attached). The usual practice is to retain no more than one questionable digit for any measured or calculated quantity. The uncertainty then is taken to be one of the units which the questionable digit represents (for example, 1, 10, or 100), or plus or minus one-half of the unit. If the 6 in 4026 is the questionable digit, we normally would say that the uncertainty is 1, or ± 0.5, since the 6 represents the number of 1's in 4026. We will amplify on this technique in the discussion of significant figures below.

The uncertainties or inaccuracies that appear in results arise from three causes. One cause is the uncertainties that exist in the original data used in the calculation. These always cause uncertainties in the final output and, for most scientific and engineering work, are the most troublesome.

Another cause of uncertainties in the output of any calculation is the numerical technique used. Finally, the third cause is related to the limit on the number of digits that can be retained during any computation. We discuss the latter two sources of error briefly in a later section on numerical and rounding errors.

Significant Figures. A *significant figure*, or *significant digit*, is any *digit* within a number that may be considered to be reliable as a result of measurements or mathematical computations. For example, if the digit 3 within the number 143,580 representing some measurement, truly represents the number of thousands in that measurement, then 3 is said to be a significant figure. Owing to the way in which measurements are made, if the digit 3 in the number 143,580 is a significant figure, then the 1 and the 4 are also significant figures. As you proceed from left to right in a number (for example, from the 3 to the 5 to the 8, etc.), you will eventually encounter a digit which is doubtful due to the way that it was measured or determined (due to the accuracy of an apparatus or an instrument or of human precision).

All the digits in a number, from left to right, up to and including the doubtful one, are said to be significant figures. Therefore, if the 8 in 143,580 is the doubtful digit, then this number is said to contain, or be accurate to, five significant figures. What this means is that the actual value, if it could be determined, should be somewhere between 143,575 and 143,585. Note that the 8 is really in doubt. Note also that once we are certain of the doubtful digit, we will have just found another way of expressing the uncertainty in the number. That is, 143,580, with the knowledge that 8 is the doubtful digit, is equivalent to writing the absolute bounds as 143,580 ± 5, or the relative bounds as 143,580 ± 0.0035%.

In the number 12.43, there are four significant figures if 3 is the doubtful digit, while in 0.123456, there are six significant figures if 6 is the doubtful digit.

For numbers in which the doubtful digit is located to the right of the decimal point, it is misleading to add further digits (including zeros) to the right of the doubtful digit. To do so violates a standing tradition of using the doubtful digit as the last digit and hides the true accuracy of the number. Hence 12.43, by tradition, implies that the number is accurate up to the doubtful digit of 3, whereas 12.430 implies that the 3 is a reliable number and the 0 is the doubtful digit.

For numbers in which the doubtful digit is located to the left of the decimal point, we cannot omit the zeros that may lie between the doubtful digit and the decimal point. Thus the number 987,600. does not make clear which is the doubtful digit. It could be the 6 or it could be either of the two zeros. In such a case we can write the number in scientific notation with the doubtful digit shown to the right of the decimal point as the last digit in the number (as in 9.876×10^5). This same condition can be represented in the computerized world, where E-formats (in FORTRAN) and ^^^^-formats (in BASIC) are simple to request. However, in writing code, we often do not know the uncertainty of the numbers that will eventually be entered as data.

To avoid decreasing the accuracy of a calculated set of numbers, a programmer usually assigns the outputs more digits than are actually significant based on the uncertainties of the inputs. Thus the outputs often look more accurate than they really are. We shall return to this problem in our discussion on resolution.

Resolution

Calculators and computers have the capability of retaining many digits of *resolution* in the numbers they handle. Most calculators will accept the entry of numbers with up to 10 digits. The resolution of numbers accepted by computers depends on the word size and on whether the number is integer or real.[3] For real numbers, the smallest number of digits accepted by computers is usually 7 and the largest, 15.

---

[3]INTEGER and REAL classifications of numbers relate to how numbers are stored in a computer's memory. They are explained in detail in Chapter 17.

It is often very easy to be lulled into believing that all the digits that can be displayed are always significant. But there is a very important distinction between resolution (the ability to display many digits) and accuracy (the requirement that all digits used are accurate or meaningful).

For example, if we were to multiply 2.2 × 1.1 on most handheld calculators, we might observe results that vary from 2.42 to 2.42000000. Using the discussion above as a reference in determining the doubtful digit, these results would imply anything from three significant figures to nine significant figures. Let us explore the accuracy of this calculation for a moment.

Based on the lower bounds or the uncertainty implied in the numbers 2.2 and 1.1, the smallest product that we could expect would be 2.15 × 1.05 or 2.2575. Similarly, using the upper bounds, the largest product that we might expect would be 2.25 × 1.15 or 2.5875. From this it is apparent that the doubtful digit in the product is the one immediately following the decimal point. Thus the original product of 2.4200 . . . has a doubtful digit of 4, giving the number only two significant figures. The product written as 2.4 implies the correct uncertainty, since the 2 is accurate and the 4 is doubtful.

Products and Quotients. Products and quotients have no more significant figures than those contained in the least accurate number used as an input. For example, in the multiplication of two numbers, one having five significant figures and one having only two, the product will have only two significant figures.

Sums and Differences. Sums and differences are no more accurate than the doubtful digit among the numbers being added or subtracted that represents the largest unit (1's, 10's, 100's, etc).

Example 9-1

If 301×10^3 and 4028 are being added, the number representing the thousands (1000's) in the final result will be doubtful since the number of thousands in the first number is in doubt. Therefore, the sum, 305,028, has 5 as its first doubtful digit going from left to right. This sum should, by convention, be written as 305×10^3 in order to imply no more than its true accuracy.

Example 9-2

If you owned 5.01×10^3 acres of land and purchase 0.1 acre more to obtain frontage on a highway, you are implying a false accuracy to say that your total acreage is now 5010.1. Your holdings after the acquisition are still *somewhere* between 5005 and 5015 acres, just as they were previously. Of course, you could pay to have your land surveyed more accurately so that you can make legitimate use of more resolution in the numbers.

So it is in engineering measurements, there is an economic trade-off between increased accuracy (which may require more expensive instruments, better test rigs, and so on) and the *need* for the accuracy.

Subtraction of Nearly Equal Numbers. Great care must be exercised when taking the difference of two numbers that are close to each other in size.

Example 9-3

Subtracting 1110 ± 5 from 1160 ± 5 gives 50 ± 10, which shows that the 5 is doubtful; the difference in this case has only one significant figure. The uncertainty, especially the relative uncertainty has greatly increased, because the true answer lies somewhere between 40 (1155 − 1115) and 60 (1165 − 1105). In the notation introduced in this text, this would be represented by 50 ± 10 percent or 50 ± 20 percent. The original numbers were represented by relative bounds of 1110 ± 0.45 percent and 1160 ± 0.43 percent.

The point of the examples above is that you should never let all the numbers that a calculator or computer can display lull you (or your client or your instructor) into a false sense of accuracy. Calculating machines do not increase the accuracy of the input. If imprecise data are entered, then imprecise numbers will be generated, regardless of the number of digits of readout possible.

A similar problem exists in making laboratory measurements. Improved designs in digital equipment have made possible data acquisition systems capable of high resolution (displaying many digits). Connected to low-accuracy measuring instruments, these devices are fantastic in providing many insignificant digits in their output. If these outputs are connected directly to a computer for data reduction and manipulation, it is exceedingly easy (but ***naive***) to believe the final displayed result of, say, 30.0034801°C in a situation where the thermocouple (a temperature-measuring device) being used is accurate to only ±0.1°C.

Numerical and Rounding Errors. Nearly all numerical calculations involve some degree of approximation. This is due mainly to the fact that we have to deal with finite limits on the numbers of digits storable and on the numbers of operations performable. For example, representing the reciprocal of three (1/3) with a finite number of terms is never truly accurate. Similarly, including only a finite number of terms in an infinite series for a trigonometric function or an exponentiation always involves some error.

Errors that we encounter or introduce because of our finite mathematical procedures are called *numerical errors*. These errors are introduced when, for example, we use a finite number of terms in an infinite series or when we stop an iterative calculation before we reach the truly correct answer.

The fact that all numbers must be represented on a computer or calculator in a finite number of digits can cause *rounding errors*. Combining (adding or subtracting) large and small real numbers, or subtracting two nearly equal numbers, often requires very large resolution (many digits) to maintain numerical accuracy.[4] When more

[4]We use the term *numerical accuracy* here to represent the accuracy in computing with numbers in which all the digits are assumed to be significant. As we discussed previously, all the digits may or may not be significant or meaningful based on the accuracy of the data entered.

digits are required to represent a number than a machine can physically handle, the numbers must be truncated or rounded. In *truncation,* all the digits beyond the maximum number permitted by the machine are discarded. In *rounding,* all the digits beyond the maximum number are discarded, but the last retained digit is increased by one unit if the closest dropped digit is greater than 5.

Numerical and rounding errors are closely related. Once generated, these errors may propagate and increase significantly under certain conditions. This is particularly true in large "number-crunching" programs, where 10^6 or even 10^9 calculations are often performed. The study of numerical and rounding errors and their interplay forms a large part of the important field of *numerical analysis.* The subject is beyond the scope of this text.

Fortunately, the resolution and capacity of most calculators and computers keep these numerical and rounding errors to negligible proportions in many cases.

Mistakes

When computations are performed by calculators or computers, there is a great tendency to believe the results. The attitude is: "If the calculator did it, it must be correct." The results are often accepted without question, even when a little reflection would indicate that something is seriously wrong with the answer. Indeed, if computations do turn out to be wrong but cause no harm to life or property, the computer usually gets the blame. Nearly all of us have seen it happen. However, when incorrect calculations lead to serious problems, such as a building collapse or a chemical plant explosion, the situation ceases to be a humorous inconvenience. In these serious circumstances you should note that computer reputations do not suffer, calculators are not sued, and computing machines are never incarcerated. It is the people responsible for the calculations who are held responsible.

A *mistake* can be thought of as an error in action, calculation, or judgment that has been caused by poor reasoning, inadvertence, or carelessness. Even a seemingly minor mistake, such as leaving out one keystroke in programming, can produce results that are totally invalid. Let us examine three common sources of mistakes.

1. *Poor reasoning that results from using an inaccurate mathematical description.* The mathematical relationships used in engineering calculations must be valid or none of the calculations that result from their use will be valid. Calculations that result from the use of incorrect mathematical descriptions may appear to proceed logically and even give reproducible results. But this is of no consequence. If the mathematical model is inaccurate or unsuitable, the results obtained will be equally unsuitable.
2. *Poor reasoning that results from using invalid computational procedures.* Fortunately, many improper computational procedures are discovered by error trapping routines in calculators and computers. Some of these mistakes are mathematical, such as trying to take roots of negative numbers, performing division by zero, or attempting to find the arccosine of numbers with magnitudes

Identification and Correction of Mistakes

How might you recognize that you have a questionable result, and, if so, how do you clear up the problem so that you can rely on the results obtained? There are several things you can do, but all methods are not applicable to all situations. Several of the more common procedures are listed below:

- *Evaluate the reasonableness of results.* In most real applications of computation you will be dealing with physical quantities that have some meaning to you. In these cases your intuition is often a guide to the numerical range of results that might be appropriate.
- *Make mental estimates of the range of the expected result before you begin the calculation process.* In this way, you will avoid letting the outcome of the calculation influence your judgment of the expected results. If the result varies significantly from your expectations, check your procedures as well as your calculation. Example: In solving a problem involving parallel resistors, your understanding of the problem should tell you two things: first, that the equivalent resistance will be less than that of the smallest resistor in the circuit, and second, that the outcome cannot be negative. Any results that violate these bounds will be incorrect. Try to apply this type of reasoning in advance for every problem that you solve.
- *Check dimensional consistency.* Verify dimensional consistency for all of your proposed calculations before beginning the calculations. Recheck the dimensions and units if the results are widely different from your expectations.
- *Step-by-step checking.* Another technique is to look closely at intermediate results. If computer calculations are being performed, print out the intermediate results in which you might be otherwise uninterested. If the detailed result for any step differs significantly from an estimate you might make, or better still, from a hand calculation you can perform, stop immediately and identify the reason.
- *Double checking.* Repeat the entire calculation. If possible, use another sequential problem solving

procedure for the second attempt. Forcing yourself to use a different technique may keep you from repeating a simple mistake.

- *Printing.* Examine printouts of the programs to ensure that you have entered the coding correctly.
- *Test data.* Run test or fictitious data through your procedure. This should be data for which you know the answer. If it works correctly for the test data, then you can treat your data with greater, but not absolute, confidence.
- *Practice.* Build your proficiency in devising and executing computational procedures by working many problems and exercises. Work problems found in other books. Those with answers given can be particularly helpful, but avoid looking at the answers until you have made your own mental estimate of the result and have actually calculated it. If your result does not match the answer given, practice the troubleshooting techniques above to identify and correct the difficulty. By following this procedure you will develop a facility and flexibility in using your calculator and give yourself confidence in your ability to organize and execute computations.

You must also devote attention to guarding against computational blunders. All of us make a mistake from time to time. However, it is important that you are able to recognize mistakes and correct them when they occur.

larger than 1. Others result from improper or incorrect coding. Although it is annoying to have the computer call these improper procedures to your attention, it is a far better condition than the alternative: invalid results masquerading as valid ones. Many of these self-identifying mistakes can be corrected as soon as they are encountered. Some may be more difficult to trace to their unique cause. But at least you have been made aware that a mistake exists.

Much more serious are computational procedures that are accepted by the computer as being valid, but are not really the proper steps for the solution of the problem at hand. Calculators and computers give you no warning that such logical errors exist. The machines cannot decide whether the sequence is appropriate for your problem.

It is possible, with totally invalid procedures, to get answers that are close enough to the true or expected results that they look valid *purely by accident*. Even a nonworking clock is correct twice a day! At these two times of day you might not even recognize that something is amiss.

3. *Inadvertence and carelessness.* Avoiding the two sources of mistakes described above will not ensure valid results. The problem-solving procedure must be executed properly, and the results must be recorded accurately before you are finished. Carelessness and lack of attention to detail are probably the biggest obstacles on the path to achieving correct results. Thus, consistent vigilance is the price to obtain correct calculations. Remember, you must execute the computations as carefully as you create them.

THE METHODOLOGY OF ENGINEERING ANALYSES AND PROBLEM SOLVING

One characteristic of engineers that sets them apart from other technical professionals is their ability to solve applied problems. Within each engineer this characteristic is exemplified by an inquisitive mind that has been forged by the rigors of an educational environment that requires repeated problem solving and analysis. Finally, it is fine-tuned by the practice of the profession.

At some time within the educational process, each student usually realizes that there is a general methodology that is useful in solving a large class of engineering problems, whether these problems are mechanical, electrical, or chemical in nature. Some students recognize this early in their educational careers, whereas others discover it much later. Once you have become aware of this general methodology and can put it to use, your capability in problem solving will greatly increase.

The overall scheme of this methodology embodies the following steps:

1. Define the system (that portion of the universe) to which you are going to devote your attention.

2. Conceptualize the problem in your mind (i.e., try to visualize it in your mind; use simple sketches and think about it).
3. Identify the knowns and unknowns in the problem.
4. Model the problem in some appropriate manner.
5. Make the necessary analyses or conduct the required experiments that will provide needed information.
6. Evaluate the effect that assumptions, made in step 4, have on the final solution. Remodel the problem if necessary.

Each of these steps is discussed and explained in more detail in the sections that follow.

Define the System

This first step is important because it allows you to concentrate your attention on the subject at hand. You do not want to concern yourself with all the problems of the universe; you can solve many of those later. It is important to realize that the physical laws of nature all deal with *something*. That is, they apply to some physical object: a box sliding across the floor, a resistor in an electrical circuit, the temperature of a piece of metal, and so on.

Therefore, in solving technical problems that are amenable to solution through an application of the fundamental physical laws, you will need a clear definition of your system. If you have difficulty defining your system, you will have trouble applying the fundamental laws. This is an extremely important point and, experience tells us, one that can never be overemphasized.

Conceptualize and Sketch the System

During your college studies you will find that various disciplines assign different names to these conceptualized system sketches. In mechanics they are called *free-body diagrams;* in the study of fluid mechanics and thermodynamics (energy) they are often called *control volumes;* in electrical sciences they may be termed *circuit diagrams*. However, the purpose of the idealized representation is always the same: to focus your attention on the key elements of the problem as you start your solution.

Most engineering instructors will require a system sketch as a part of each problem solution. Whether or not it is required, you should make it an automatic part of your solution.

Identify the Knowns and Unknowns

This may seem to be an obvious step in any problem-solving methodology, but it is listed here for additional emphasis. Unless you make this a conscientious part of your overall "problem-solving toolkit," it is likely to be overlooked. The best way to approach this step is to make two lists: one list for the knowns and one for the unknowns.

This step helps you to "size up" the problem. By coming to grips with the specific information that is known and determining what questions are to be answered, the choice of problem solution techniques is usually made much clearer.

Students often wonder why engineers must study calculus and even higher-level mathematics. The answer is that much engineering modeling is accomplished by using mathematics. If some new process is to be developed or a new device built, it is usually much more economical to do mathematical modeling to refine the design than to set up the process in the laboratory or build the actual device in the model shop. That is, fewer physical models and prototypes have to be built and tested when mathematics is used first to narrow the final design choices. Hardware models are very expensive to build because of their labor, space, and equipment requirements. Perhaps more important, mathematical modeling can cut the time necessary to bring a product or process to market. This places a company in a more visible and competitive position.

Model the Problem

Analytical and/or experimental modeling, whichever may be appropriate for the problem at hand, plays an important function in engineering problem solving. The role of modeling in engineering is discussed in more general terms in Chapter 11.

In general, junior- and senior-level engineering courses will teach engineering modeling. The degree of sophistication of these models will grow as you delve deeper into your studies.

Computer-aided engineering (CAE), which is just another form of mathematical modeling, is developing rapidly because it has the potential to simplify modeling, thereby reducing development costs and bringing products to market quicker.

Conduct Analyses and/or Experiments

The execution of this step depends on the level of sophistication of the modeling required to solve the problem. If, for example, only algebraic equations are involved in the mathematical modeling, you must have the same number of independent algebraic equations as there are unknowns in the problem.

Remember Peers' Law

The solution to a problem changes the problem.
John Peers, President, Logical Machine Corp.

Evaluate the Accuracy of the Final Result

There are many factors that control the accuracy of the final solution to an engineering problem. For example, assumptions must always be made in the modeling step and these can greatly influence the result. The correctness of the assumptions you have made must always be confirmed.

Also needing confirmation here are the units of each term. To talk about a temperature of 3 or an area of 406 is nonsense. You must always ask, "3 *what*?" or "406 *what*?" As will be pointed out in Chapter 10, a thorough units analysis should be a standard part of your problem-solving technique.

EXAMPLES OF ENGINEERING ANALYSES

The computation paper used for most calculations is $8\frac{1}{2}$ by 11 inches in size, with lines ruled both vertically and horizontally on the sheet. Usually, these lines divide the paper into five squares per inch, and the paper is commonly known as cross-section or engineering calculation paper. Many bookstores stock paper that has the lines ruled on the reverse side of the paper so that erasures will not remove them. A fundamental principle to be followed in the use of the paper is that the problem work shown should not be crowded and that all steps of the solution should be included.

Several styles of model problem sheets are shown in Figures [9-14] and [9-15]. Notice in each sample that an orderly sequence is followed in which the known data are given first. The data are followed by a brief statement of the requirements and then the engineer's solution.

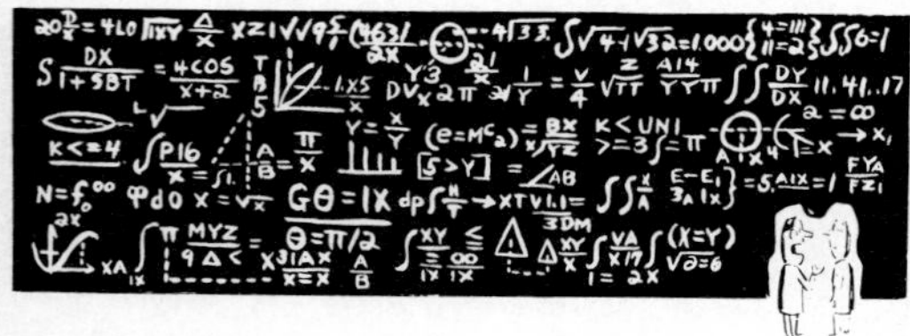

[9-13] *"I FOUND THE ANSWER BUT I FORGOT THE PROBLEM."*

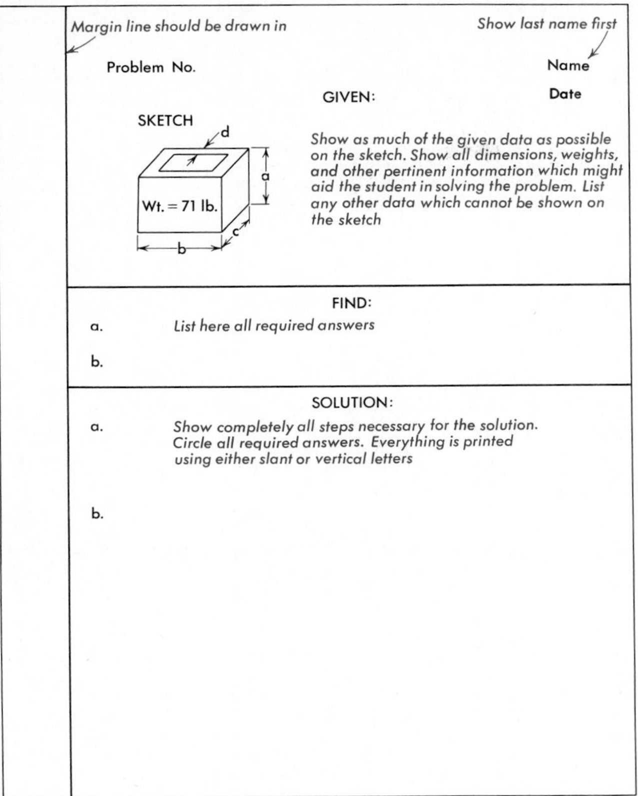

[9-14] Model problem sheet, style A. This style shows a general form that is useful in presenting the solution of mensuration problems.

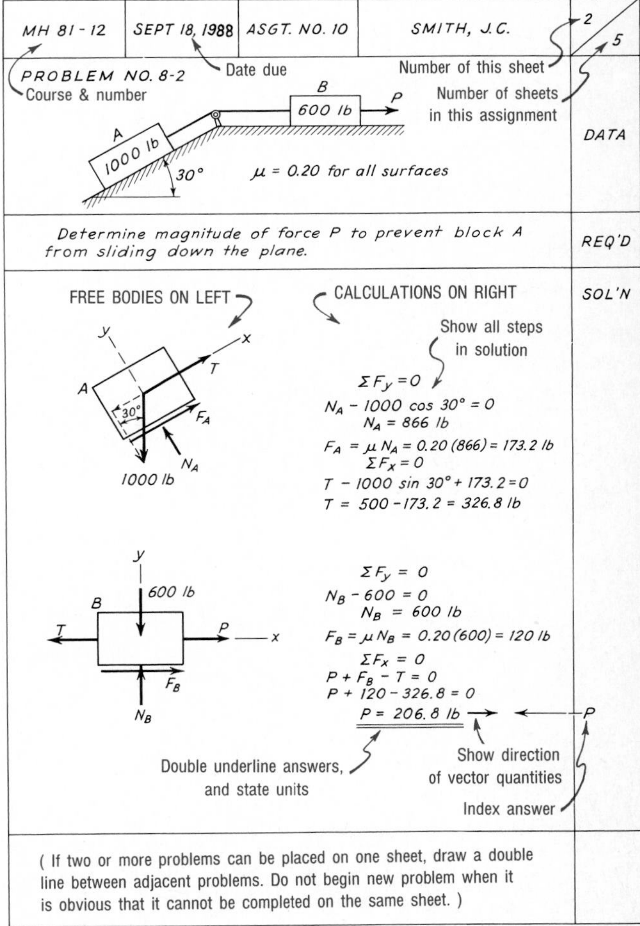

[9-15] Model problem sheet, style B. This style shows a method of presenting stated problems. Notice that all calculations are shown on the sheet and that no scratch calculations on other sheets are used.

PROBLEMS

9-1. Plot a graph showing the relation of weight as the dependent variable to diameter as the independent variable for round steel rods. Plot values for every quarter-inch up to and including $3\frac{1}{2}$ in. in diameter. (See Figure [9-8].)

WEIGHT OF ROUND STEEL RODS IN POUNDS PER LINEAL FOOT (BASED ON 489.6 LB/FT³)

Size (in.)	*Weight (lb/ft)*	*Size (in.)*	*Weight (lb/ft)*
$\frac{1}{4}$	0.167	2	10.66
$\frac{1}{2}$	0.668	$2\frac{1}{4}$	13.50
$\frac{3}{4}$	1.50	$2\frac{1}{2}$	16.64
1	2.68	$2\frac{3}{4}$	20.20
$1\frac{1}{4}$	4.17	3	24.00
$1\frac{1}{2}$	6.00	$3\frac{1}{4}$	28.30
$1\frac{3}{4}$	8.18	$3\frac{1}{2}$	32.70

9-2. Plot a graph showing the relation of normal barometric pressure of air vs. altitude. Plot values up to and including 15,000 ft.

Altitude (feet above sea level)	*Normal barometric pressure (in. Hg)*	*Altitude (feet above sea level)*	*Normal barometric pressure (in. Hg)*
0	29.95	5,000	24.9
500	29.39	6,000	24.0
1000	28.86	7,000	23.1
1500	28.34	8,000	22.2
2000	27.82	9,000	21.4
2500	27.32	10,000	20.6
3000	26.82	15,000	16.9
4000	25.84		

9-3. Plot a graph of maximum horsepower transmittable by cold-drawn steel shafting vs. diameter for a speed of 72 rpm based on the formula

$$\text{hp} = \frac{D^3R}{50}$$

where hp = horsepower

D = diameter of shaft, in.

R = rpm of shaft

Calculate and plot values for every inch diameter up to and including 8 in.

9-4. *a.* Plot a graph for the following experimental data of the period in seconds vs. the mass of a vibrating spiral spring.

Period (sec)	*Mass (g)*	*Period (sec)*	*Mass (g)*
0.246	10	0.650	70
0.348	20	0.740	90
0.430	30	0.810	110
0.495	40	0.900	130
0.570	50	0.950	150

b. Using the same data, plot a graph of period squared vs. the mass of a vibrating spring.

9-5. Plot a graph on the variation of the boiling point of water vs. pressure.

Boiling point (°C)	*Pressure (mm Hg)*	*Boiling point (°C)*	*Pressure (mm Hg)*
33	38	98	707
44	68	102	816
63	171	105	907
79	341	107	971
87	469	110	1075
94	611		

9-6. *a.* Plot a graph of the following measured values of sliding force vs. the normal force for a wood block on a horizontal wood surface.

Sliding force (g)	*Normal force (g)*
100	359
130	462
155	555
185	659
210	765
240	859

b. Determine the slope of the line plotted and compare with the average value of the coefficient of sliding friction obtained from individual readings of normal force and sliding force.

Slope = $\tan\theta$ (where θ is the angle that the line makes with the abscissa) $\tan\theta = \frac{y_2 - y_1}{x_2 - x_1}$

9-7. Plot the variation of pressure vs. volume, using data as obtained from a Boyle's law apparatus.

Pressure (cm Hg)	***Volume (cm^3)***	***Pressure (cm Hg)***	***Volume (cm^3)***
50.3	23.2	76.8	15.1
52.5	22.4	79.7	14.7
54.5	21.5	82.7	14.1
56.9	20.9	84.2	13.6
59.4	19.6	87.9	13.2
63.0	18.5	90.6	12.8
65.3	17.8	93.5	12.5
67.2	17.3	95.7	12.3
72.6	16.1	101.9	11.4
74.5	15.6		

9-8. Using data in Problem 9-7, plot a graph pressure vs. the reciprocal of the volume.

9-9. Plot the relation between magnetic flux density (B, tesla) and magnetizing force (H, ampere-turns/meter) for a specimen of tool steel. This graph will form what is customarily called a B–H curve.

***H* (ampere-turns/m)**	***B* (tesla)**	***H* (ampere-turns/m)**	***B* (tesla)**
0	0	18,970	1.466
2,710	0.900	21,680	1.486
5,420	1.180	24,390	1.498
8,130	1.302	27,100	1.523
10,840	1.375	29,810	1.535
13,550	1.409	35,220	1.557
16,260	1.422		

9-10. The formula for converting temperatures in degrees Fahrenheit to the equivalent reading in degrees Celsius is

$$°C = \frac{5}{9}(°F - 32°)$$

Plot a graph so that by taking any given Fahrenheit reading between 0 and 220° and using the graph, the corresponding Celsius reading can be determined.

9-11. The equation which expresses the variations of electric current with time in an inductive circuit is

$$i = I_0 e^{\frac{-Rt}{L}}$$

where i = current, amperes
I_0 = initial steady-state value of current and is a constant
e = base of the natural system of logarithms and is approximately 2.7183
R = resistance, ohms, in the circuit and is constant
t = time, seconds, measured as the current i varies
L = inductance, henrys, and is constant

Let

$$I_0 = 0.16 \text{ ampere}$$
$$R = 1.2 \text{ ohms}$$
$$L = 0.5 \text{ henry}$$

Calculate and plot values of i as t varies from 0 to 0.5 sec.

9-12. Plot the variations of efficiency with load for a $\frac{1}{4}$-hp, 110-V, direct-current electric motor, using the following data taken in the laboratory.

Load output (hp)	***Efficiency (percent)***	***Load output (hp)***	***Efficiency (percent)***
0	0	0.175	56.5
0.019	24.0	0.195	58.0
0.050	42.0	0.248	59.1
0.084	44.9	0.306	58.0
0.135	50.7	0.326	56.2

9-13. Plot the values given in Problem 9-1 on semilog paper [log(weight) versus size].

9-14. Plot the values given in Problem 9-2 on semilog paper [log(pressure) versus altitude].

9-15. Plot the values given in Problem 9-7 on semilog paper [log(pressure) versus volume].

9-16. Plot the values given in Problem 9-11 on semilog paper [log(i) versus t].

9-17. Plot the values given in Problem 9-1 on log-log paper.

9-18. Plot the values given in Problem 9-3 on log-log paper.

9-19. Plot the values given in Problem 9-7 on log-log paper.

9-20. The following data were taken from an acoustical and electrical calibration curve for a type 1126 microphone. The test was run with an incident sound level of 85 dB perpendicular to the face of the microphone.

Frequency (Hz)	***Relative response (dB)***
20	−40
50	−29
100	−19
400	−5
1,000	+1
2,000	+1
3,000	0
6,000	−4
10,000	−11

Plot a graph on semilog paper showing the decibel response with frequency.

9-21. The electrical frequency response of a type X501 microphone is given below.

Frequency (Hz)	*Relative response (dB)*
20	−40
40	−33
80	−22
100	−18
200	−11
400	−5
600	−2
1,000	+1
2,000	+2
4,000	−1
6,000	−4
10,000	−10

Plot a graph on semilog paper showing the decibel response with frequency.

9-22. A Weather Bureau report gives the following data on the temperature over a 24-hr period for October 12.

Midnight 47°	10 am 55°	6 am 63°
2 am 46°	Noon 68°	8 pm 58°
4 am 44°	2 pm 73°	10 pm 57°
6 am 43°	4 pm 75°	Midnight 57°
8 am 49°		

Plot the data.

9-23. A test on an acorn-type street lighting unit shows the mean vertical luminous intensity distribution to be as shown below.

Midzone angle (deg)	*Candlepower at 10 ft*	*Midzone angle (deg)*	*Candlepower at 10 ft*
180	0	85	156
175	0	75	1110
165	0	65	1050
155	1.5	55	710
145	3.5	45	575
135	5.5	35	500
125	8.5	25	520
115	13.5	15	470
105	22.0	5	370
95	40.0	0	370

Plot the data. (While data for only half the plot are given, the other half of the plot can be made from symmetry of the light pattern.)

9-24. The luminous intensity distribution of a 400-W, type J-H 1 fluorescent lamp used for street light service was measured with a photometer, and the following data were obtained:

Midzone angle (deg)	*Candlepower at 10 ft*	*Midzone angle (deg)*	*Candlepower at 10 ft*
180	0	75	7700
165	0	72	8600
145	0	65	7100
135	3	55	5300
125	20	45	4300
115	100	35	3500
105	700	25	2700
95	1200	15	2300
85	3000	5	2100
		0	2000

Plot the data. (While data for only half the plot are given, the other half of the plot can be made from symmetry of the light pattern.)

9-25. From data determined by you, draw a pie chart to show one of the following.

a. Consumption of sulfur by various industries in the United States.
b. Budget allocation of the tax dollar in your state.
c. Chemical composition of bituminous coal.
d. Production of aluminum ingots by various countries.

9-26. Make a bar chart showing the number of male students registered in your school for each of the past 10 years.

9-27. Plot a graph on rectangular coordinate paper of $N = 1.296^x$ for values of x from −9.0 to +9.0 in 0.5 increments.

9-28. Plot a graph on rectangular coordinate paper of the equation $N = 0.813^x$ for values of x from −9.0 to +9.0 in 0.5 increments.

9-29. Prepare a line chart that will permit converting readings from grams to ounces up to 64 oz and a line chart that will convert readings from pounds to kilograms up to 10 lb.

9-30. A series of test specimens of a crank arm, part 466-1, was tested for the number of cycles needed to produce fatigue failure at various loadings. The results of the tests are shown below. Plot a graph of load against operating cycles (S–N curve) on semilog paper [log(cycles) versus load].

Specimen number	*Oscillatory load (lb)*	*Operating cycles to produce failure*
1	960	1.1×10^5
2	960	2.2×10^5
3	850	2.4×10^5
4	800	4.2×10^5
5	800	6.0×10^5

Specimen number	Oscillatory load (lb)	Operating cycles to produce failure
6	700	2.4×10^5
7	700	5.1×10^5
8	650	1.8×10^6
9	600	7.7×10^6
10	550	1.0×10^7

(Note that the proper number of significant figures may not be given in the reading.)

9-31. Compute the percent error.
a. Reading of 9.306 ± 0.003
b. Reading of 19165 ± 2.
c. Reading of 756.3 ± 0.7
d. Reading of 2.596 ± 0.006
e. Reading of 13.750 ± 0.009
f. Reading of 0.0036 ± 0.0006
g. Reading of 0.7515 ± 0.02
h. Reading of 12,835 ± 20
i. Reading of 382.5 ± 5
j. Reading of 0.03 ± 0.03

9-32. Compute the numerical error.
a. Reading of 35.219 ± 0.03 percent
b. Reading of 651.79 ± 0.01 percent
c. Reading of 11.391 ± 0.05 percent
d. Reading of 0.00365 ± 2 percent
e. Reading of 0.03917 ± 0.6 percent
f. Reading of 152 ± 4.0 percent
g. Reading of 0.0575 ± 10 percent
h. Reading of 7.65×10^7 ± 7 percent
i. Reading of 3.080×10^{-4} ± 2.5 percent
j. Reading of 32.5×10^{-2} ± 30 percent

9-33. A surveyor measures a property line and records it as being 3207.7 ft long. The distance is probably correct to the nearest 0.3 ft. What is the percent error in the distance?

9-34. The thickness of a spur gear is specified as 0.875 in., with an allowable variation of 0.3 percent. Several gears that have been received in an inspection room are gaged, and the thickness measurements are as follows: 0.877, 0.881, 0.874, 0.871, 0.880. Which ones should be rejected as not meeting dimensional specifications?

9-35. A rectangular aluminum pattern is laid out using a steel scale which is thought to be exactly 3 ft long. The pattern was laid out to be 7.42 ft by 1.88 ft, but it was subsequently found that the scale was incorrect and was actually 3.02 ft long. What were the actual pattern dimensions and by what percent were they in error?

9-36. A resident of a city feels that his bill for water is considerably too high, probably because of a defective water meter. He proposes to check the meter on a do-it-yourself basis by using a 1-gallon milk bottle to measure a volume of water. He believes that the volume of the bottle is substantially correct and that the error of filling should not exceed plus or minus 2 tablespoons.
a. What would be the probable maximum error in gallons per 1000 gallons of water using this measurement?
b. Using the milk bottle, he draws 10 full bottles of water and observes that the meter indicates a usage of 1.345 ft^3 of water. If the average rate for water is $1.05 per 1000 ft^3, by how much could his water bill be too high?

Addition of Laboratory Data

9-37. Add and then express the answer to the proper number of significant figures.

a.	*b.*	*c.*
11.565	858.7	1.39395
4.900	404.3	8.7755
226.55	54.42	10.6050
82.824	19.8	49.201
17.668	8.775	88.870
108.77	12.04	108.887

d.	*e.*	*f.*
757.1	16.59	0.32
54.540	0.0531	6171.0
11.5	11.72	255.5
1.0375	285.5	80.60
378.64	4.41	715.55
4372.1	0.0748	3707.

g.	*h.*
6282.6	38.808
545.81	11.955
122.55	35.306
334.75	67.332
98.88	105.65
28.77	575.75
1.059	

Subtraction of Laboratory Data

9-38. Subtract and then express the answer to the proper number of significant figures.

a.	*b.*	*c.*
6508.	8.104	0.04642
3379.	7.891	0.0199

d.	*e.*	*f.*
731.16	7.114	10276.
189.28	16.075	61581.

g.	*h.*	*i.*
118.72	0.016	766.
366.	0.1513	−516.16

j.	*k.*	*l.*
0.8280	−933.0	−156.2
−0.023	77.12	0.0663

Multiplication of Laboratory Data

9-39. Multiply and then express the answer to the proper number of significant figures.

a. 5167. 238.	*b.* 32105. 5.28	*c.* 535.58 0.2759
d. 84.636 30869.	*e.* 1.03975 54682.	*f.* 0.0548 0.00376
g. 14.7410 0.7868	*h.* 47.738 0.065	*i.* 15903. 0.00469
j. −9757 0.05478	*k.* 7.5427 −542.16	*l.* −0.0989 −11.6507

Division of Laboratory Data

9-40. Divide and then express the answer to the proper number of significant figures.

a. $\frac{3928.}{5636.}$	*b.* $\frac{216.75}{53.83}$	*c.* $\frac{7.549}{3.069}$
d. $\frac{539.77}{1.6303}$	*e.* $\frac{0.5322}{0.343}$	*f.* $\frac{8831.}{128.75}$
g. $\frac{73.65}{127.1}$	*h.* $\frac{4.91}{1598.}$	*i.* $\frac{0.2816}{5383.}$

Chapter 10

Dimension and Unit Systems

THE LAWS OF NATURE

Perceptions. We interpret the universe by evaluating our perceptions. More than any other group, those individuals who are generally called *scientists* try to make sense out of their perceptions. They look for reoccurrences of observed phenomena and develop concepts to predict the happenings around them. The *physical scientists* who have preceded us have distilled the predictive schemes and inventions of man down to what we now call the *physical laws of nature*. Some examples are:

Conservation of mass.

Newton's second law.

Conservation of energy.

Law of universal gravitation.

Second law of thermodynamics.

Maxwell's equations of electrodynamics.

We will consider some of these in more detail later in this text; others, along with more advanced examples, must be left for a more in-depth study in science and engineering.

The fact that general "laws" are available and have such wide applicability has allowed scientists and engineers to make giant technological strides. Fields such as the social sciences, in which the governing "laws" are not yet well known or where the governing equations cannot yet be written down, have not advanced nearly as rapidly as the fields of science and engineering. It is one thing for an observer, after having studied a particular cylinder rolling down a given incline, to postulate a theory for that cylinder on that incline. It is another to be able to predict the behavior of any cylinder on any incline. The physical laws of nature allow us the luxury of the latter type of prediction.

> Nature will tell you a direct lie if she can.
> *Charles Darwin, 1809–1882*

Although it may sound as if we fully understand nature and can accurately predict all physical happenings, such is not always the case. It must be remembered that what we call the physical laws of nature do not come from nature; they were not handed to us emblazoned on stone tablets. The "laws" are the inventions of man. They simply represent the sum total of our experiences to date expressed in ways that make possible more than just educated guesses as to how nature might behave in the future. In many areas we can predict with

Lo, for your gaze, the pattern of the skies!
What balance of the mass, what reckonings
Devine!
Here ponder too the Laws which God,
Framing the universe, set not aside
But made the fixed foundations of his work.

Edmund Halley
The Ode Dedicated to Newton

much confidence and precision; indeed, in some fields we can calculate with almost absolute certainty.

There are times and places when the physical laws, as we understand them, appear to break down. Fortunately, history tells us that most of these apparent breakdowns are the result of a misapplication of the laws rather than errors in the laws themselves.

However, there have been instances where the "so-called" basic laws simply did not fully explain observed phenomena. In these cases, scientists had to go to work to unravel the mysteries, uncover new facts, and help establish new theories. So it has been over the ages; this pursuit of knowledge for knowledge's sake continues today and indeed, must go on.

Sir Isaac Newton wrote in the Preface to *Principia:* "I heartily beg that what I have here done may be read with forbearance; and that my labors in a subject so difficult may be examined, not so much with the view to censure, as to remedy their defects." Do you think that Newton really anticipated that others would find limitations to what has become known as "Newtonian" mechanics? What are some of the limitations?

DIMENSIONS[1]

The physical laws discussed above relate observable or measurable quantities. Examples of common quantities are distance (L), time (t), force (F), mass (M), electrical charge (Q), area (A), volume (V), speed (S), work (W), power (P), temperature (T); the list goes on. Notice here that we are referring to the measurable *phenomena* themselves and *not* to the numbers that represent the sizes or magnitudes of the measurements. These measurable phenomena are called *dimensions*.

You may realize that some of these dimensions can be expressed in terms of other dimensions. For example, speed, a dimension, is defined to be distance divided by time, both of which are dimensions. In referring to speed, we may argue about whether we actually mean distance traveled divided by total elapsed time (usually called average speed) or speed at some instant of time (usually called instantaneous speed, a quantity that you will become more familiar with as you study calculus). However, such discussions are not important here. What is important is to understand the concept of measurable quantities or dimensions.

Dimensions can arbitrarily be assigned scales of measure. However, since dimensions are usually related to one another through the fundamental laws of nature and the definitions of man, we usually do not assign an arbitrary measuring scale to every dimension but only to a select set. We arbitrarily select some dimensions and then let the physical laws tell us what the resulting dimensions of the other measurable quantities will be. The selected dimensions are called *primary* or *fundamental dimensions*. The remaining dimensions, those that can be expressed as combinations of the primary ones, are called *secondary* or *derived dimensions*.

[1]Letters set in blue type signify dimensions, whereas quantities set in black type represent variables that have magnitudes.

For example, if we were to choose length and time as primary dimensions, we could derive speed as a secondary dimension since it is defined to be distance per unit time. On the other hand, we could just as easily take length and speed as primary dimensions and use time as a secondary dimension derived from speed and distance.

If we were to assume that length, time, *and* speed are all primary dimensions, we would have to alter our relationship between the three to get meaningful results. In the latter case our definition is over specified. We will have trouble if we want to assign values arbitrarily to all three variables in an equation that involves only three variables. We have to introduce a dimensional constant (an invariant number, often experimentally determined, that has dimensions associated with it) into the equation to make it correct and useful. This process should become clearer as we discuss k_n and k_g, constants in Newton's second law and the classical law of gravity, respectively.

The quantities that are used in the equations of science and engineering nearly always have some physical meaning or interpretation: a force, a distance, a voltage, and so on. Thus, essentially all of the equations we use should be dimensionally homogeneous, since they relate observed quantities. "Dimensionally homogeneous" means that when the dimensions of each term in the equation are substituted for the term and algebraic reduction is carried out, both sides of the equation will contain the same collection of dimensions ordered in the same arrangements.

At first glance, some equations may not appear to be dimensionally homogeneous, but that is probably only because the dimensions of a constant within the equation have been neglected.

[10-1] Length, force, and time are measurable quantities.

Example 10-1

A, an area, and S, a speed, are involved in the following equation relating physical quantities or variables:

$$Q = S(A - P)$$

Determine the dimensions on both P and Q.

Solution:

Since the equation relates physical phenomena, it should be dimensionally homogeneous. P must have the same dimensions as A since they are added together (you cannot add apples and oranges). Also, Q must have the same dimensions as the product of SP. We know[2]

$$A \overset{d}{=} L^2 \qquad \text{and} \qquad S \overset{d}{=} L/t$$

since areas, being the products of two lengths, have dimensions of L^2, and speeds have traditional dimensions of L/t. Using this information

$$P \overset{d}{=} L^2$$

and

$$Q \overset{d}{=} \left\{\frac{L}{t}\right\}L^2 \overset{d}{=} L^3/t^2$$

[2]The symbol d above the = denotes "dimensionally equivalent to," meaning that the left-hand side of the equation has the same dimensions as the right-hand side of the equation.

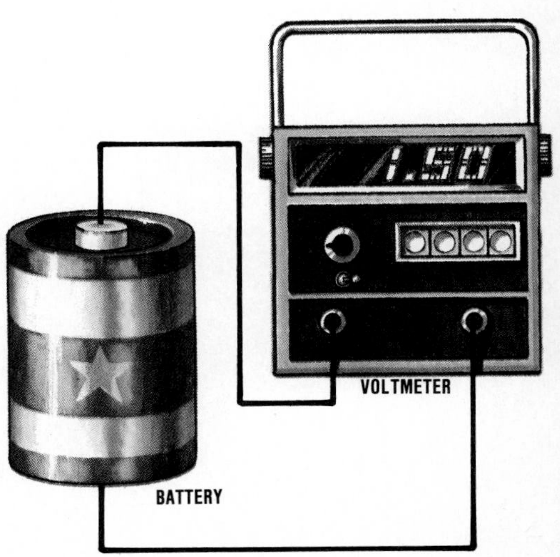

[10-2] Voltage is a measurable quantity.

Example 10-2

Solve for the dimensions of the conversion factor k in the expression

$$\frac{L^2t^3T}{F^4} \stackrel{d}{=} k\left(\frac{L^5tF^2}{Q^2}\right)$$

Solution:

$$k \stackrel{d}{=} \frac{t^2TQ^2}{F^6L^3}$$

Checking:

$$\frac{L^2t^3T}{F^4} \stackrel{d}{=} \overbrace{\frac{t^2TQ^2}{F^6L^3}}^{k}\left(\frac{L^5tF^2}{Q^2}\right)$$

$$\frac{L^2t^3T}{F^4} \stackrel{d}{=} \frac{L^2t^3T}{F^4} \stackrel{d}{=} L^2t^3TF^{-4}$$

Problems

Solve for the dimensions of the conversion factor k.

10-1. $\frac{FTL^2}{tM^3} \stackrel{d}{=} k\frac{t^5M}{T^2}$

10-2. $k\left(\frac{QM}{TF^2}\right) \stackrel{d}{=} \sqrt{L^4ItQ^8}$

10-3. $t^2\sqrt{LM^5} \stackrel{d}{=} k\left(\frac{FT^2}{M^3}\right)$

10-4. $k(Ft^2TL^{-2}M^{-3}) \stackrel{d}{=} M^5LtF^{-3}$

10-5. $M^2FT^{-5}L^{-2} \stackrel{d}{=} k\sqrt{MTt}$

10-6. $\sqrt{LT^3F^{-2}M} \stackrel{d}{=} k\sqrt{TF^3M^6}$

10-7. $k\frac{\sqrt{T^3Q}}{L^2F^{-2}} \stackrel{d}{=} MTLF$

10-8. $k(F^2T\sqrt{Lt^{-2}}) \stackrel{d}{=} t^{-3}T^{-2}$

10-9. $FL^3Q^{-1}M^{-3} \stackrel{d}{=} k\sqrt{L^2Q^{-1}}$

UNITS

The scales or size subdivisions that we use in expressing the magnitudes of the various dimensions are referred to as *units*. Dimensions are measured in terms of units. Micrometers (μm), millimeters (mm), feet, yards, miles, and light-years are all examples of units used to express magnitudes of the dimension of length (L). Table 10-1 gives some typical units used to express the magnitude of common dimensions.

TABLE 10-1 TYPICAL UNITS FOR COMMON DIMENSIONS

Common fundamental dimensions	*Typical units*
Length, L	Micrometers, millimeters, feet, yards, meters, rods, chains, miles, light-years
Time, t	Microseconds, seconds, minutes, hours, days, months, years, decades, centuries
Force, F	Dynes, newtons, ounces, pound-force, tons
Mass, M	Grams, pound-mass, kilograms, slugs

We pointed out in the preceding section that the equations of science and engineering are nearly always dimensionally homogeneous. Dimensionally homogeneous equations must also be homogeneous or balanced in terms of the units. Thus, substitution of the applicable units for each term in the equations, and reduction of the algebraic expressions that result, should yield left- and right-side values that are identical. This salient fact can be quite useful in solving problems, although it is seldom exercised by beginning students.

If the left-hand side of a dimensionally homogeneous equation has the dimensions of FL/T (force · length/time), then the right-hand side must reduce to dimensions of FL/T. If the left-hand side of a dimensionally homogeneous equation has *units* of (pound · meters)/year, then the right-hand side must reduce to the same units of (pound · meters)/year. In the latter case if the right-hand side reduced to (newton · meters)/second while the left-hand side was still in units of (pound · meters)/year, then the equation would contain a units error, although it would still be dimensionally correct. Results derived in the presence of such an error should never be trusted; however, such equations may yield to correction, barring a more complicated error, by the introduction of simple unit conversion factors. If the equation cannot be balanced dimensionally, this most often indicates that more serious problems exist.

Example 10-3

The stress in a certain column may be calculated by the following relationship:

$$\sigma = \frac{F}{A}\left[1 + \left\{\frac{\ell}{k}\right\}\frac{R}{\pi^2 nE}\right]$$

where[3] σ = induced stress, lb-force/in.^{2}

F = applied force, lb-force

A = cross-sectional area of member, in.^{2}

ℓ = length of bar, in.

k = radius of gyration, in.

R = elastic limit, lb-force/in.^{2}

E = modulus of elasticity, lb-force/in.^{2}

n = dimensionless coefficient for different end conditions

(a) Is this equation dimensionally homogeneous?
(b) Do the units balance?

Solution:

Collecting all the dimensions on both sides of the equation results in

$$\frac{F}{L^2} \overset{?}{=} \frac{F}{L^2}\left[1 + \left\{\frac{L}{L}\right\}^2 \frac{F/L^2}{F/L^2}\right] \overset{d}{=} \frac{F}{L^2}$$

[3]lb-force (lb_f) means pounds of force. We discuss this expression and its meaning later in this chapter.

which shows dimensional homogeneity. Collecting all of the units on both sides of the equation:

$$\frac{\text{lb-force}}{\text{in.}^2} \overset{?}{=} \frac{\text{lb-force}}{\text{in.}^2}\left[1 + \left\{\frac{\text{in.}}{\text{in.}}\right\}^2 \frac{\text{lb-force/in.}^2}{\text{lb-force/in.}^2}\right]$$

$$= \frac{\text{lb-force}}{\text{in.}^2}$$

This demonstrates that the equation is homogeneous in units.

Problems

10-10. Is the equation $a = (2S/t^2) - (2V_1/t)$ dimensionally homogeneous if a is an acceleration, V_1 is a velocity, t is a time, and S is a distance? Prove your answer by writing the equation with fundamental dimensions.

10-11. Is the equation $V_2^2 = V_1^2 + 2as$ dimensionally correct if V_1 and V_2 are velocities, a is an acceleration, and s is a distance? Prove your answer by rewriting the equation in fundamental dimensions.

10-12. In the homogeneous equation $R = B + \frac{1}{2}CX$, what are the fundamental dimensions of R and B if C is an acceleration and X is a time?

10-13. Determine the fundamental dimensions of the expression $B/g\sqrt{D} - m^2$, where B is a force, m is a length, D is an area, and g is the acceleration of gravity at a particular location.

10-14. The relationship $M = \sigma I/c$ pertains to the bending moment for a beam under compressive stress. σ is a stress in F/L^2, C is a length L, and I is a moment of inertia L^4. What are the fundamental dimensions of M?

10-15. The expression $V/K = (B - \frac{7}{3}A)A^{5/3}$ is dimensionally homogeneous. A is a length and V is a volume of flow per unit of time. Solve for the fundamental dimensions of K and B.

10-16. Is the expression $S = 0.031V^2/fB$ dimensionally homogeneous if S is a distance, V is a velocity, f is the coefficient of friction, and B is a ratio of two weights? Is it possible that the numerical value 0.031 has fundamental dimensions? Verify your solution.

10-17. If the following heat transfer equation is dimensionally homogeneous, what are the units of k?

$$Q = \frac{-kA(T_1 - T_2)}{L}$$

A is a cross-sectional area in square feet, L is a length in feet, T_1 and T_2 are temperatures (°F), and Q is the amount of heat (energy) conducted in Btu per hour.

10-18. In a swimming pool manufacturer's design handbook, for a pool whose surface area is triangular, you find the following formula: $V = 3.74Rt\theta$, where V = volume of pool in gallons, R = length of base of triangular-shaped pool in feet, t = altitude of triangular-shaped pool in feet if t is measured perpendicular to R, and θ = average depth of pool in feet. Prove that the equation is valid or invalid.

10-19. You are asked to check the engineering design calculations for a sphere-shaped satellite. At one place in the engineer's calculations you find the expression $A = 0.0872\Delta^2$, where A is the surface area of the satellite measured in square feet, and Δ is the diameter of the satellite measured in inches. Prove that the equation is valid or invalid.

10-20. The U.S. Navy is interested in your torus-shaped lifebelt design and you have been asked to supply some additional calculations. Among these is the request to supply the formula for the volume of the belt in cubic feet if the average diameter of the belt is measured in feet and the diameter of a typical cross-sectional area of the belt is measured in inches. Develop the formula.

10-21. From the window of their spacecraft two astronauts see a satellite with foreign insignia markings. They maneuver for a closer examination. Apparently the satellite has been designed in the shape of an ellipsoid. One of the astronauts quickly estimates its volume in gallons from the relationship $V = 33.8ACE$, where the major radius (A) is measured in meters, the minor radius (E) is measured in feet, and the end-view depth to the center of the ellipsoid (C) is measured in centimeters. Verify the correctness of the mathematical relationship used for the calculations.

10-22. An engineer and his family are visiting in Egypt. The tour guide describes in great detail the preciseness of the mathematical relationships used by the early Egyptians in their construction projects. As an example he points out some peculiar indentations in a large stone block. He explains that these particular markings are the resultant calculations of "early day" Egyptians pertaining to the volume of the pyramids. He says that the mathematical relationship used by these engineers was $\odot = \square\uparrow$, where $\odot$ was the volume of pyramid in cubic furlongs, $\square$ was the area of the pyramid base in square leagues, and $\uparrow$ was the height of the pyramid in hectometers. The product of the area and the height equals the volume. The engineer argued that the guide was incorrect in his interpretation. Prove which was correct.

10-23. Develop the mathematical relationship for finding the weight in drams of a truncated cylinder of gold if the diameter of the circular base is measured in centimeters and the height of the piece of precious metal is measured in decimeters.

10-24. Is the equation $F = WV^2/2g$ a homogeneous expression if W is a weight, V is a velocity, F is a force, and g is the acceleration of gravity? Prove your answer.

THE PHYSICAL LAWS OF NATURE AND DIMENSION/UNIT SYSTEMS

We concentrate first on "mechanical" dimensions (those playing an important role in mechanics: distance, speed, acceleration, force, mass, time) and their relationship to the physical laws. Later we discuss "electrical and magnetic" and "thermal" dimensions.

Newton's Second Law

As an introduction to *dimension/unit systems* consider the dimensions involved in Newton's second law. This law relates the forces acting on an object to both its mass and its acceleration (how fast the object acquires or loses velocity). In translated form, Newton's own statement says: "The change of motion is proportional to the motive force impressed; and is made in the direction of the right line[4] in which that force is impressed."[5] The "change of motion" here refers to the change of motion of a defined object. Newton defined "motion" to be "velocity and quantity of matter conjointly," or mass × velocity (the quantity we now call momentum).

Stated as an equation, Newton's second law for a constant-mass object is given by[6]

$$\mathbf{F} = k_n M \mathbf{a} \tag{10-1}$$

where we have expressed the "change of motion" as the mass, M, multiplied by the acceleration, **a** [the latter being the rate at which the velocity is changing,[7] dimensionally (L/t^2)]. In the equation we have also introduced k_n as a constant of proportionality (Newton said "the change of motion is proportional to the motive force").

One of the confusing things to students in the application of this law has always been the distinction between mass and weight. Mass is a property of all matter that does not depend on the environment

[10-3] America's "number one" sport depends on the athletes' ability to use a force to strike a mass and impart to it an acceleration. The force from the bat acts over a very short time period. It produces an acceleration that takes the ball velocity from about 90 mph toward home plate, to perhaps 140 mph toward the outfield fence.

[4]The word "right" in the phrase "of the right line" is redundant and should be deleted if it causes difficulty in interpreting the meaning.

[5]From Motte's Translated Revised, Newton's *Principia,* now published by the University of California Press, Berkeley (1962) (the *Principia* was first published July 5, 1686).

[6]The boldface symbols that are used here and in some other equations in this chapter represent *vector* quantities . . . quantities that have both direction and magnitude. Vectors are explained in more detail in Chapter 11.

[7]The word "rate" implies "the time rate." Another way of expressing "the rate at which the velocity is changing" is to say, "the amount by which the velocity has changed in a unit of time." If the magnitude of the velocity of an object was 3 m/s at some earlier instant of time but was 3.4 m/s, 2 s later, then "the rate at which the velocity changed," or the magnitude of the acceleration, was $(3.4 - 3)/2 = 0.2$ (m/s)/s. Velocity and acceleration are both vector quantities having direction as well as magnitude. The direction of the acceleration is the same as the direction of the change of the velocity vector (i.e., the vector difference of the velocity vector at the end of the time period and the velocity vector at the beginning of the time period). This short discussion should be a hint that vectors and their properties are very important to engineers. You will learn more about them as you get further into your academic program.

[10-4] With the same effort, a hurdler on the moon could jump higher than if he were on the earth.

or on the surroundings in which the matter finds itself.[8] This book has a mass, you have a mass, and each star has a mass. Each of these masses is independent of whether it is in the vicinity of some other object, is isolated in outer space, or is darting through the universe.

However, masses attract each other according to the law of universal gravitation (discussed in the next section). That is, two objects exert forces on each other, forces that attempt to pull the objects together. In many situations we call this force of attraction the *weight*. For example, the earth applies a gravitation attraction to you (and you to it). The force on you, or your weight, though, varies depending on how physically separated you are from the earth and how close you are to other objects that attract you. Thus your *weight is a force* and that force depends on your environment or surroundings.

Weight changes as the surroundings change. In the vicinity of the moon, for example, your weight is considerably reduced. This is because the moon, when you are close to it, does not attract you with the same force as does the earth, when you are near it. Remember, however, that your mass, which does not depend on your surroundings, would be independent of whether you are on the moon or on the earth.

If we consider Newton's second law to be dimensionally homogeneous (experience tells us it must be), we have some flexibility in choosing what we will consider to be the primary dimensions. If you apply the principles of algebra, you will note that we cannot arbitrarily choose primary dimensions for all the quantities in the equation ($\mathbf{F}$, k_n, M, and $\mathbf{a}$). If we did, how can we expect both the left- and right-hand sides to reduce to the same set of dimensions?

Also, if we choose any two of the quantities to have arbitrary primary dimensions, only the *combination* of the remaining two is determined. That is, the secondary dimensions of either of the remaining two quantities are not uniquely specified.

A physicist named Eotvos, over a 25-year period starting in about 1890, showed that the masses that appear in this equation are the same masses that appear in Newton's second law. His work is just one more example of the detailed devotion and effort that has made possible our present understanding of nature and our advanced state of technology.

We conclude from the discussion above that the dimensions of three of the four quantities in Equation (10-1) can be specified arbitrarily. The dimensions of the fourth and last quantity are then uniquely determined by the mathematical form of the equation. Table 10-2 shows all of the primary and secondary possibilities available to us (we have listed the dimensions of $\mathbf{a}$ as L/t^2, using the definition of $\mathbf{a}$).

The Law of Universal Gravitation

Newton was also the first person to formulate what is now called the (classical) law of universal gravitation. This principle expresses the force ($\mathbf{F}$) with which any two bodies in the universe attract each other as

$$\mathbf{F} = \frac{k_g M_1 M_2}{r^2} i \qquad (10\text{-}2)$$

[8]The mass that we are discussing here is known in the world of physics as the "nonrelativistic or rest mass." This label applies to objects that travel at speeds considerably less than the speed of light (3×10^{10} m/s). The mass of an object that travels at speeds approaching the speed of light is different from the mass the object has at much slower speeds. The theory of relativity deals with very fast moving objects.

TABLE 10-2 SOME POSSIBLE CHOICES FOR PRIMARY DIMENSIONS IN NEWTON'S SECOND LAW

Primary quantity	Primary dimension	Secondary quantity	Secondary dimension
Option 1			
F	F		
a	L/t^2	M	Ft^2/L
k_n	(1)[a]		
Option 2			
M	M		
a	L/t^2	F	ML/t^2
k_n	(1)[a]		
Option 3			
F	F		
M	M	k_n	Ft^2/ML
a	L/t^2		
Option 4			
F	F		
M	M	a	F/M
k_n	(1)[a]		

[a](1) is intended to imply unity or no dimensions.

[10-5] The law of universal gravitation expresses the force with which any two bodies in the universe attract each other.

where M_1 and M_2 are the masses of two gravitating bodies 1 and 2, respectively, r is the distance of separation, i is a unit vector (magnitude 1, no dimensions, and a direction along the line between bodies 1 and 2), and k_g is a constant of proportionality.

In establishing dimension/unit systems, we have to make sure that our choices are consistent with this equation, as well as with Newton's second law. The dimensions and units of any three of the four dimensional quantities in Equation (10-2) (the four are **F**, M, r, and k_g) can be arbitrarily specified; however, when we do this, the fourth is then determined automatically. In nearly all common dimension/unit systems, the dimensions of **F**, M, and r are determined before any consideration is given to the law of gravitation. This means that k_g will then be uniquely determined; it cannot be arbitrarily specified.

Example 10-4

Suppose that you carried out an experiment to measure the gravitational attraction between two objects. Object A has a mass of 4.4×10^{20} heavies (abbreviated hev), while object B has a mass of 7.4×10^5 heavies (you have chosen these nonstandard units because you have not yet read the next section). When you separate A and B by 2.01×10^{-2} strides (st), the magnitude of the force of attraction you measure is 4.68×10^{-23} pulls (pu). Determine the magnitude and units of k_g in the law of universal gravitation. (Although Newton formulated the law of universal gravitation, he could never determine the value of k_g.)

Solution:

Solving Equation (10-2) for k_g yields (here F means the magnitude of **F**)

$$k_g = \frac{Fr^2}{M_1M_2}$$

Substituting the appropriate values yields

$$k_g = \underbrace{4.68 \times 10^{-23}}_{\text{pu}} \cdot (\underbrace{2.01 \times 10^{-2}}_{\text{st}})^2 / (\underbrace{4.4 \times 10^{20}}_{\text{hevs}} \cdot \underbrace{7.4 \times 10^5}_{\text{hevs}})$$

or

$$k_g = 5.8 \times 10^{-53} \text{ pu} \cdot \text{st}^2/\text{hev}^2$$

Now in these units, we can always use the relationship given by

$$F = 5.8 \times 10^{-53} (\text{pu} \cdot \text{st}^2/\text{hev}^2) M_1M_2/r^2$$

to relate force, masses, and distance of separation. That is, we have a general law that holds good in these nonstandard units.

THE NEED FOR STANDARDS

As the world continues to grow effectively smaller through transportation and communications marvels, the advantages of having all people use the same dimensional systems and units should be obvious. Many of the unique dimension/unit systems used by earlier cultures have disappeared under attempts to standardize these important features from culture to culture.

Unfortunately, we still do not have one standard dimension/unit system in use throughout the world. The common ones that persist today can be classified into three types based on what are chosen as primary dimensions and what result as secondary dimensions. These are exhibited in Table 10-3. Notice that length (L) and time (t) are primary dimensions in all three categories, but mass (M) and force (F) are not.

Electrical quantities are more involved than those listed in Table 10-3 and will be discussed later.

The SI System

The SI system (from Le Système International d'Unités or International System of Units) is the most internationally accepted dimension/unit system. Of all the industrial nations of the world, only the United States has not converted to the SI system in an extensive way. After several years of investigation by Congress, legislation was enacted in 1974 to implement a changeover to the SI system. However, two factors have worked against this implementation. First, no specific date was set for a mandatory changeover, and second, the conversion process is extremely expensive because it requires much physical retooling of machinery and equipment. The mental conversion that is required to allow people to begin to "think" in terms of SI quantities (kilograms, meters, etc.) is also a deterrent.

TABLE 10-3 MOST COMMON CHOICES FOR DIMENSIONING SYSTEMS

Category 1	*Category 2*	*Category 3*
	Measurable quantities having primary dimensions	
—	Force	Force
Length	Length	Length
Time	Time	Time
Mass	—	Mass
k_n[a]	k_n[a]	—
	Measurable quantities having secondary or derived dimensions	
Force	—	—
—	Mass	—
—	—	k_n[a]
k_g[a]	k_g[a]	k_g[a]
Velocity	Velocity	Velocity
Acceleration	Acceleration	Acceleration
Area	Area	Area
Volume	Volume	Volume
Energy	Energy	Energy
Power	Power	Power
Pressure	Pressure	Pressure
Density	Density	Density
Specific weight	Specific weight	Specific weight

[a]The dimensions of k_n and k_g are discussed on pages 278, 281, and 285.

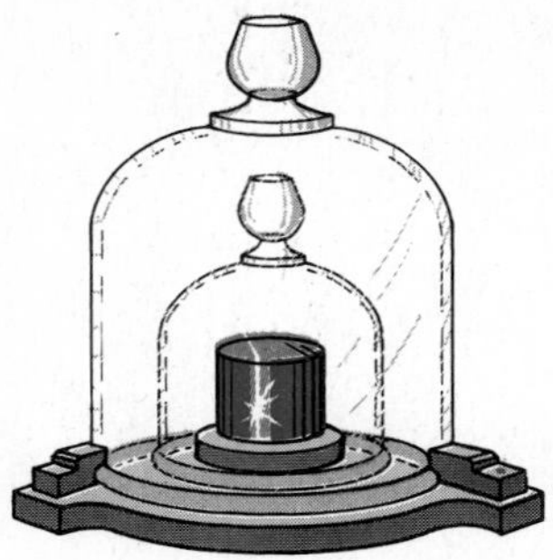

[10-6] The standard for the unit of mass, the *kilogram,* is a cylinder of platinum-iridium alloy kept by the International Bureau of Weights and Measures at Paris. A duplicate in the custody of the National Bureau of Standards serves as the mass standard for the United States. This is the only base unit still defined by an artifact.

TABLE 10-4 METRIC PREFIXES

Prefix	*Multiplication factor*	*SI symbol*
exa	10^{18}	E
peta	10^{15}	P
tera	10^{12}	T
giga	10^{9}	G
mega	10^{6}	M
kilo	10^{3}	k
hecto	10^{2}	h
deca	10^{1}	da
deci	10^{-1}	d
centi	10^{-2}	c
milli	10^{-3}	m
micro	10^{-6}	μ
nano	10^{-9}	n
pico	10^{-12}	p
femto	10^{-15}	f
atto	10^{-18}	a

Of course one of the attractive features of the metric system, on which the SI system draws, is that prefixes such as *milli, centi,* and *kilo* that are commonly added to unit names are all related by powers of 10 (see Table 10-4); this makes it a decimal-based system.

Mass. In the SI system, the standard unit for the fundamental dimension of mass is the kilogram (kg) [10-6]. The standard for this unit has been internationally accepted since 1889. It is defined to be the mass (not the volume, weight, or composition) of a certain platinum–iridium cylinder that is maintained under carefully controlled conditions at the International Bureau of Weights and Measures in Sèvres, France.

The gram (g), another unit that is often used for mass, is, by definition, 1/1000 of a kilogram.

Length. The standard unit for length is the meter (m). As our technology has advanced over the years, the definition of the meter has been refined to the point where it now personifies, perhaps more than the other dimensions, the high-tech world in which we live. A meter, as redefined in 1960, is that distance in space that is equivalent to 1,650,763.73 wavelengths (in a vacuum) of the orange-red light emitted by krypton 86 (a specific isotope of that element) [10-8]. For most of us this is an abstract definition.

It is, of course, much easier to develop an intuitive feeling for this unit after seeing a "meter" stick in a high school or college freshman physics lab (a meter stick is a few inches longer than a yardstick). But because all materials expand and contract with temperature and pressure, any definition based on the length of a certain material has its limitations. (Similarly, the wavelength of light varies with the index of refraction of the material through which it passes; thus, the standard krypton 86 radiation must be measured in a vacuum.)

[10-7] A balance can be used to compare the mass of one object with another (usually known) mass. However, the balance really compares the weight of one object with the weight of the other. Since both masses are located in the same gravitational field, the masses are equal when the weights are equal.

Length—*METER*—m The meter is defined as 1,650,763.73 wavelengths in vacuum of the orange-red line of the spectrum of Krypton-86.
1
SPECTRAL LAMP
PRODUCES ARC IN KRYPTON-86 GAS, EMITTING LIGHT WAVE RADIATION
2
SPECTROSCOPE
SEPARATES LIGHT AND THE DISTINCTIVE ORANGE-RED BAND IS FOCUSED INTO THE INTERFEROMETER
R
V
INTERFEROMETER
TRANSLUCENT MIRROR SPLITS LIGHT BEAM: HALF TO 4a, HALF TO 4b
3
4b
MOVABLE MIRROR
b2
b1
1,650,763.73 wavelengths
ONE STANDARD METER
4a
FIXED MIRROR
5
LIGHT WAVES FROM 4a INTERFERE WITH LIGHT WAVES FROM 4b, FORMING LIGHT AND DARK FRINGES
METER BAR
6
OBSERVING TELESCOPE
DISPLAYS FRINGE PROGRESSION ACROSS FIELD OF VIEW AS MIRROR 4b IS MOVED. FROM STARTING POSITION b1, DARK FRINGES ARE COUNTED AS MIRROR 4b IS MOVED BACKWARDS SLOWLY. EACH FRINGE EQUALS EXACTLY ONE-HALF WAVELENGTH PROGRESSION. BLOCKING OR PHOTOELECTRIC DETECTION METHODS SIMPLIFY COUNTING.

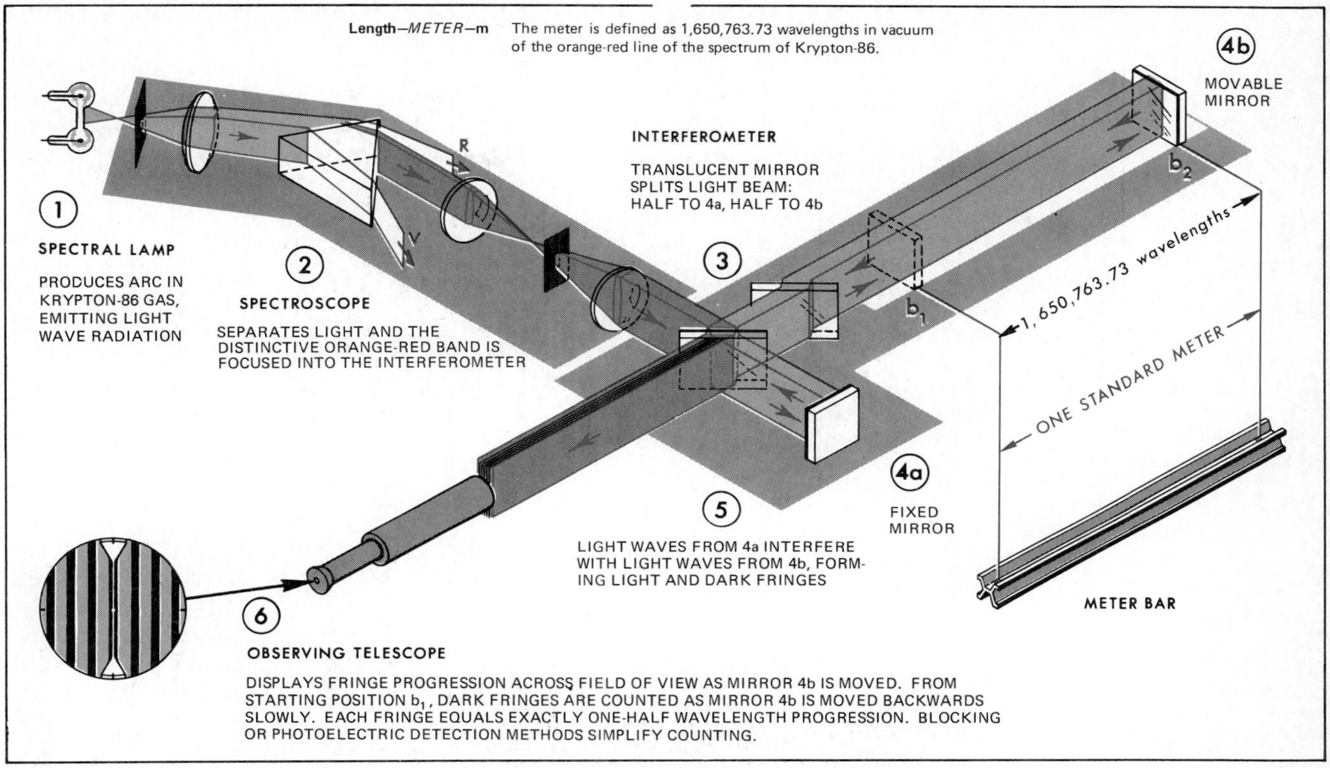

[10-8] Implementation of the standard meter.

Although the meter is the standard length in the SI system, the units of millimeter (mm), centimeter (cm), and kilometer (km) are often used to simplify the writing of digits. If a device is only 1 mm long, it is often inconvenient to list it at 0.001 m.

Time. Once you get the wavelengths of krypton 86 counted, you might want to get started counting the vibrations of a cesium 133 atom, for it holds the key to the definition of the standard time unit, the second. In the time required to count 9,192,631,770 vibrations of this atom, 1 standard second will have elapsed [10-9]. Obvi-

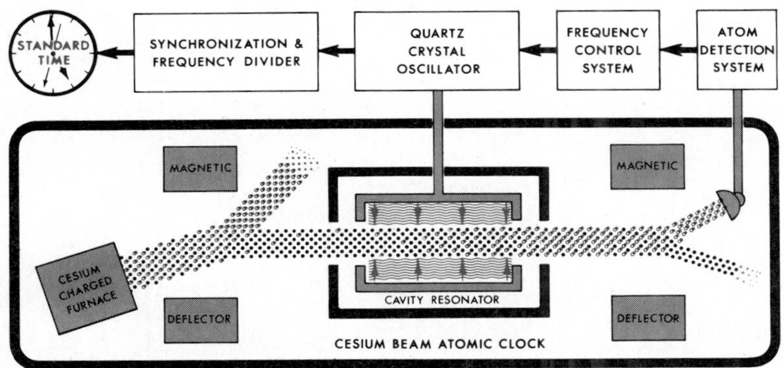
STANDARD TIME
SYNCHRONIZATION & FREQUENCY DIVIDER
QUARTZ CRYSTAL OSCILLATOR
FREQUENCY CONTROL SYSTEM
ATOM DETECTION SYSTEM
MAGNETIC
MAGNETIC
CESIUM CHARGED FURNACE
CAVITY RESONATOR
DEFLECTOR
DEFLECTOR
CESIUM BEAM ATOMIC CLOCK

[10-9] Implementation of the standard second.

The adoption of a standard based on some very fundamental phenomenon, such as the orange-red light of krypton, allows a unit to be precisely reproduced anywhere in the world, indeed the universe. For example, if scientists and engineers throughout the world need to compare lengths accurately, each could obtain some krypton 86 and set up his or her own local standards lab. This would ensure that person A is talking about the same length as person B. If the standard meter were based on the length of some agreed-upon bar (which it was prior to 1960), kept in a single place at specified conditions, each worker would have to produce copies of that bar, keep them at the same conditions, and at some time make direct comparisons against the standard by traveling to the home of the original.

You may have noticed that the accepted standard for the kilogram mass does not yet have this same type of fundamental definition that leads to independent reproducibility. In the case of kilograms, "traceability of a measurement to the standard," a term often used in standards work, implies going to France and making a direct comparison against the original.

[10-10]

ously, this international standard, set in 1967, also reflects our modern era.

However, the technique of counting atomic vibrations is quite analogous to counting the number of swings of the pendulum in a wall clock and equating a given number of these swings to a certain period of time. Only the accuracy, reproducibility, and preciseness are different.

k_n. In the SI system, the constant k_n in Newton's second law [Equation (10-1)] is arbitrarily taken to be unity (a magnitude of 1 with no dimensions). We can make this assumption as long as the dimensions and units of force (the only variable left in Newton's law that is not yet defined) are set by the equation and not by us. We will ensure this in the next paragraph.

Force. With the choice of 1 for k_n, Newton's second law [Equation (10-1)] reduces to

$$\mathbf{F} = M\mathbf{a} \tag{10-3}$$

in the SI system. It is through this law that the (secondary) dimensions of force are established. Thus force must have the dimensions of the product of $M\mathbf{a}$ or ML/t^2.

When we begin to explore forces in the SI system, Newton's law tells us that a force of 1 kg · m/s^2 is required to accelerate a mass of 1 kg at 1 m/s^2. This comes from direct substitution

$$\mathbf{F} = 1\ \text{kg} \cdot 1\ \text{m/s}^2 = 1\ \text{kg} \cdot \text{m/s}^2 \tag{10-4}$$

This gives us a meaningful way of assigning numbers to forces. If a force produces an acceleration of 2 m/s^2 when applied to 1 kg of mass, the force must have a magnitude of 2 kg · m/s^2. Two important features should be noted from this exercise. The first is that we can now compare one force with another; the second is that we have ensured that Newton's law works in this dimension/unit system.

Since the strange set of units for the dimension of force seems awkward (have you ever seen a spring scale that was graduated in units of kg · m/s^2?), it is customary to rename any force of 1 kg · m/s^2 to be a *newton*. This is the standard, but *derived,* unit in the SI system. Note that it is intimately related to having a standard kilogram, a standard meter, and a standard second. Note also that the term *newton* applied to forces is an alias [10-10].

Example 10-5

Consider a mass of 68.4 kg acted on by a force of 667.2 N. What is the acceleration of this mass under these conditions?

Solution:

From Newton's second law, the magnitude of the acceleration **a** is

$$a = \frac{F}{M} = \frac{667.2\ \text{N}}{68.4\ \text{kg}} = 9.75\ \text{N/kg}$$

Substituting the collection of units for which newton is an alias results in

$$a = 9.75 \frac{\text{kg} \cdot \text{m}}{\text{s}^2} \cdot \frac{1}{\text{kg}} = 9.75 \text{ m/s}^2$$

k_g. In the SI system, we have seen that M and L are primary dimensions, whereas F is a secondary dimension. We can also see that Equation (10-2) demonstrates that if F, M, and L are all chosen, then k_g cannot be arbitrarily specified. It must have dimensions of[9]

$$k_g \stackrel{\text{d}}{=} \frac{F \cdot L^2}{M^2} \stackrel{\text{d}}{=} \overbrace{\frac{M \cdot L}{t^2}}^{F} \frac{L^2}{M^2} \stackrel{\text{d}}{=} \frac{L^3}{M \cdot t^2} \tag{10-5}$$

In terms of the standard units in the SI system (kg, m, N), k_g is an experimentally determined constant. The accepted value is

$$k_g = 6.673(\pm 0.003) \times 10^{-11} \text{ m}^3/(\text{kg} \cdot \text{s}^2) \tag{10-6}$$

The English Engineering System

The English Engineering system has been used in the English-speaking countries of the world for decades. However, most of these countries have now effectively changed to the SI system in order to better participate and compete in the international marketplace and to take advantage of the multiples of 10 that separate various subunits within the SI system. Ironically, even England now shuns the English Engineering system in favor of the SI system.

At the present time, the English Engineering system's chief user, but not necessarily a proponent, is the United States, where there are two common forms in use. These are the *FMLt* system and the *FLt* system. The latter is sometimes called the British Gravitational system.[10] The two systems differ in the choices of primary dimensions.

The English Engineering system is so deeply embedded in the U.S. engineering profession that it will probably be at least a decade or two before the adoption of the SI system can be considered successful. Most of the engineering journals where important work can be published and disseminated now require that SI units be used in all papers. American journals, however, will also accept the English Engineering equivalents if they are entered in parentheses alongside the SI values.

Most machinery that is being shipped abroad from the United States today is dimensioned in SI units. Also, most computer-aided drafting packages available today allow the user to work in either the SI or the English Engineering system, while the machine does the

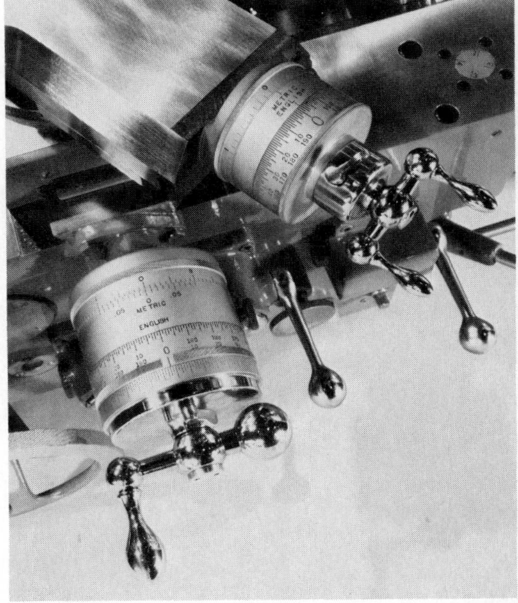

[10-11] These machine tools are calibrated in both SI and English units.

[9]Remember, the symbol d above the equal sign is intended to imply "dimensional equivalence," whereby the terms on the left and right sides of the equation have the same dimensions. Numerical equality is not intended.

[10]The dimension/unit systems that define force to be one of the primary dimensions at one time had to rely on the gravitational attraction between the earth and a standard mass in order to define a standard force unit. Some authors often use the word *gravitational* as a part of the system name. Examples of such systems are the English Engineering system *FLt* and the metric mks system, the latter of which defines a kg_f as a primary force unit. Other systems, which define M as a primary dimension, such as the SI system, are sometimes referred as *absolute* systems. Since the English *FMLt* system chooses both F and M as primary, it is your choice to call it a gravitational or an absolute system.

The only problem with tradition is that it sometimes lasts too long.

Source unknown

conversion internally if required. The conversion to the SI system is continuing, but the pace has slowed considerably from what it was in the mid-1970s.

The English Engineering *FMLt* System

Force, mass, length, and time are all considered to be primary dimensions in the *FMLt* system. Thus this system necessarily falls into category 3 in Table 10-3; it is one of the very few that does. The *FMLt* system is most deeply embedded in the disciplines of engineering concerned with fluid flow, thermodynamics (the study of energy), and heat transfer (these subjects are sometimes grouped under the term *thermosciences*).

This system is one of the most confusing of all because of the use of the word "pound" as a measurement of both force and mass. Students should always speak carefully and with preciseness of meaning when using this word.

Mass. The standard unit for the primary dimension of mass in the English Engineering *FMLt* system is the pound-mass, abbreviated as lb_{mass} or lb_m. Although there is an interesting history behind its evolution and adoption, the pound mass is now simply defined in terms of the standard kilogram-mass that was discussed under the SI system. This definition is

$$1 \text{ lb}_m = 0.453\ 592\ 37 \text{ kg} \tag{10-7}$$

or the more familiar relationship

$$1 \text{ kg} = 2.204\ 622\ 6 \text{ lb}_m \tag{10-8}$$

Distance. The standard and most often used unit for the dimension of length is the foot (ft), defined to be

$$1 \text{ ft} = 0.3048 \text{ m (exactly)} \tag{10-9}$$

Other length units often used are the inch, the yard, the statute mile, and the nautical mile.

Time. The standard unit or scale for the dimension time is the second, defined the same way as in the SI system.

Force. Force is considered to be a primary (not derived) dimension in the English Engineering *FMLt* system. The standard unit for measuring the dimension of force in this system is the pound-force, abbreviated lb_{force} or lb_f. It is paramount that the reader know and understand that there is a fundamental difference between lb_m and lb_f, just as there is a fundamental difference between a kilogram (a mass) and a newton (a force) in the SI system.

We arbitrarily say that 1 lb_f is that force, when applied to a standard 1 lb_m (which is now well defined), which produces an acceleration of 32.1740 ft/s^2 (both foot and second are also now well defined). There are some details that need to be accommodated, such as how we measure the acceleration, but the experiment itself is well defined.

Now we have a unit that allows us to put meaningful numbers on forces just as we did in the SI system. (However, in the SI system, we did not arbitrarily choose the dimensions and units of force, but let Newton's law do it for us.) All that remains is to ensure that when we use these arbitrarily chosen units of F, M, L, and t in our physical laws of nature, we will get meaningful (i.e., correct) results. The discussions on k_n and k_g below show how this is done for Newton's second law and the universal law of gravitation.

You may wonder how the number 32.1740 ft/sec^2 was chosen to be a part of the definition of lb$_f$. Why wasn't 1 or 10 or perhaps, 136.3 ft/sec^2 chosen? In the evolution of the English system, people talked of taking the standard mass (1 lb$_m$) to a given place on earth (approximately sea level and 45° latitude) and measuring the gravitational force applied to that mass by the earth. This pull or force was called (calibrated as) 1 lb$_f$. In English units, the acceleration of an unrestrained object under just the influence of gravity at the subject location was 32.174 ft/s^2 (9.807 m/s^2). An object weighing 1 lb$_m$ at the subject location would have a mass of 1 lb$_m$. Thus the definition is appropriate.

The acceleration produced by only gravitational attraction acting on an unrestrained object is called the *local acceleration of gravity*. As was found by experiments conducted from the leaning tower of Pisa in Italy, such an acceleration is independent of the shape (neglecting air resistance) or mass of the object. However, this acceleration does depend on location with respect to the mutually attracting body. *At sea level on earth,* the local acceleration of gravity g, at any latitude ϕ, may be approximated from the following relationship:

$$g = 32.09(1 + 0.0053 \sin^2\phi) \text{ ft/s}^2$$

or

$$g = 9.78(1 + 0.0053 \sin^2\phi) \text{ m/s}^2$$

k_n. As pointed out earlier, there must be some flexibility in the equations which relate measurable quantities so that arbitrary choices on the dimensions of quantities that appear in the equations do not overspecify the problem. In the English Engineering *FMLt* system, since F, L, M, and t are all arbitrarily defined, we cannot arbitrarily assign a value to k_n. If Newton's second law is to be satisfied for a 1 lb$_m$ accelerating at 32.1740 ft/s^2 under the action of a force of magnitude of 1 lb$_f$, then we must have

$$1 \text{ lb}_f = k_n \cdot 1 \text{ lb}_m \cdot 32.1740 \text{ ft/s}^2 \tag{10-10}$$

or

$$k_n = \frac{1}{32.1740} \frac{\text{lb}_f \cdot \text{s}^2}{\text{ft} \cdot \text{lb}_m} \tag{10-11}$$

Notice that the dimensions of k_n are $F \cdot t^2/(L \cdot M)$, not t^2/L. Notice also that the units are not (s^2/ft). The lb$_f$ and the lb$_m$ do not cancel mathematically because they are not the same thing. A 1 lb$_m$, being a unit of mass, is independent of its environment. No matter where you take it, as long as you retain the original collection of matter that made up the 1 lb$_m$, it will remain a mass of 1 lb$_m$. At a location on earth where the local acceleration of gravity is 32.174 ft/s^2, the 1 lb_m will weigh, or be attracted by the earth, with a force of 1 lb_f. At any location where the local acceleration of gravity is different from 32.174 ft/s^2, the weight or force of attraction of a 1 lb_m will be different than 1 lb_f.

Forces can act on masses and cause the masses to accelerate, but forces are not masses. Mass is a property that a collection of matter possesses. Force is something that a collection of matter can experience. There is a very fundamental, but sometimes subtle, difference in the two.

It is common in engineering work that uses the *FMLt* system, particularly the thermosciences, to use a term called g_c, which is simply $1/k_n$. In these cases Newton's law is written as

$$F = \frac{1}{g_c} Ma \tag{10-12}$$

where for the *FMLt* system, $g_c = 32.1740 \text{ ft} \cdot \text{lb}_m/(\text{lb}_f \cdot \text{s}^2)$. You will find the inclusion of g_c (or k_n, for that matter) to be somewhat controversial. Some engineering literature and many textbooks use it, some do not; some instructors require it, some do not.

Unfortunately, the tendency on the part of many students, when using Equation (10-12), is to associate g_c with the acceleration due to gravity. However, it should be clear by now that the term does not even possess the dimensions, not to mention the units, of an acceleration.

If g_c is used, it must be viewed as a fundamental constant inherent in the dimension/units system being used. Its value depends only on the dimension/units system chosen and not on the strength of the local gravitational field or other physical phenomena. In the English Engineering system being discussed here, the value is $32.1740 \text{ ft} \cdot \text{lb}_m/(\text{lb}_f \cdot \text{s}^2)$; in the SI system, it is unity.

The personal preference of the authors is to write the basic equations without k_n or g_c when using any of the dimension/unit systems, including the English Engineering *FMLt* system. Instead, we prefer a very careful units analysis of every equation used. Inconsistencies will appear during units reduction (the algebraic reduction of units involved in the equation) if g_c should have been included but was not. For example, if you are calculating a force and the units analysis produces a $\text{ft} \cdot \text{lb}_m/(\text{lb}_f \cdot \text{s}^2)$ instead of a lb_f, you will know that something is wrong. A little experience will tell you that all you need do is divide by g_c. We reiterate this point in the section on units conversion later in this chapter.

With a workable Newton's second law in both the SI and English Engineering *FMLt* systems, we can now express 1 lb_f in terms of the force unit in the SI system, the newton [remember, however, that the dimensions of force in the SI system are derived, whereas those in the English Engineering *FMLt* system are not]. The result is

$$1 \text{ lb}_f = 4.448\,221\,615\,260\,5 \text{ N} \tag{10-13}$$

Example 10-6

Establish the foregoing relationship between lb_f and newtons.

Solution:

To determine force from Newton's second law, we need mass and acceleration. Therefore, let us assume that we are dealing with 1 kg of mass and that we wish to accelerate it at 1 m/s^2 (we may as well start off with some "nice" numbers). In the English system, the 1-kg mass is equivalent to 2.205 lb_m, while the acceleration of 1 m/s^2 is equivalent to

$$1\frac{\text{m}}{\text{s}^2}\cdot\frac{1\ \text{ft}}{0.3048\ \text{m}} = 3.281\ \text{ft/s}^2$$

In the metric system, a force of

$$F = 1\ \text{kg}\cdot 1\ \text{m/s}^2 = 1\ \text{N}$$

is required to produce the assumed acceleration. In the English system, a force of

$$F = 2.205\ \text{lb}_\text{m}\cdot 3.280\ \text{ft/s}^2 = 7.233\ \text{lb}_\text{m}\cdot\text{ft/s}^2$$

The units do not look familiar for a force. But since they involve lb_m, we should sense that we need g_c here (the constant we discussed above and not the local acceleration of gravity)

$$F = 7.234\ \frac{\text{lb}_\text{m}\cdot\text{f}_\text{t}}{\text{s}^2}\cdot\frac{1}{32.174}\ \frac{\text{lb}_\text{f}\cdot\text{s}^2}{\text{ft}\cdot\text{lb}_\text{m}}$$

$$= 0.225\ \text{lb}_\text{f}$$

Thus 1 N = 0.225 lb_f, or 1 lb_f = 4.45 N.

k_g. As in the SI system, all the quantities in the universal law of gravitation have now been determined in the Engineering *FMLt* system, except k_g. The dimensions and units of k_g are found from the form of the equation. The numerical value must come from experimental data. It is

$$k_g = 1.068\,91\times 10^{-9}\ \text{lb}_\text{f}\cdot\text{ft}^2/\text{lb}_\text{m}^2 \qquad (10\text{-}14)$$

Example 10-7

Suppose that your bathroom scales indicated that your weight (the force with which the earth attracts you) was 150.0 lb_f (if you are familiar with the accuracy of bathroom scales, you should question the resolution of this number). Knowing that the local acceleration of gravity was 32.00 ft/s^2, calculate your mass in kg.

Solution:

Since metric units were requested, perhaps we should convert the given quantities (here your weight and the local acceleration of gravity) to metric values. Since we found earlier that 1 lb_f = 4.45 N, we easily calculate that

$$\text{Wt} = 150.0\ \text{lb}_\text{f}\cdot 4.448\ \text{N/lb}_\text{f} = 667.2\ \text{N}$$

The local acceleration of gravity can also be converted since we know the conversion between ft and m:

$$g = 32.0\ \text{ft/s}^2\cdot 0.3048\ \text{m/ft} = 9.75\ \text{m/s}^2$$

Newton's law then tells us that

$$M = \frac{F}{a} = \frac{667.2\ \text{N}}{9.754\ \text{m/s}^2} = 68.40\ \frac{\text{N}\cdot\text{s}^2}{\text{m}} = 68.40\ \text{kg}$$

Before we leave this example, it may be constructive to convert this to mass in lb_m.

$$M = 68.40\ \text{kg} \cdot 2.205\ \text{lb}_m/\text{kg} = 150.8\ \text{lb}_m$$

Note that the weight and mass in English units are not numerically equal, since the local acceleration of gravity differs slightly from the value of 32.174 ft/s^2 used to standardize the pound-mass.

The English Engineering *FLt* System

Force, length, and time are commonly said to be the primary dimensions in the *FLt* system. Thus this system necessarily falls into category 2 in Table 10-2. This system is used more in the engineering disciplines that deal primarily with mechanics (the study of the relation between applied forces, the motion of objects, and the stresses transmitted through the objects) than those that study thermoscience subjects.

This system is just as confusing as the *FMLt* system because the definitions within it seem to go in circles. This is more a result of the evolution of the system than anything else. If you enjoy the history of technology, you would find the study of this evolution quite interesting.

What we wish to stress here is that all the standards for the units used in this system, as well as those we discussed previously, are well established. Some, like the kg and its relative, the lb_m, require traceability to a single piece of matter, unique in the world. Others are based on physical phenomena that are highly reproducible, allowing anyone with the appropriate apparatus to duplicate the standard.

The word "pound" is used in the Engineering *FLt* system only in connection with forces. To avoid confusion, however, it is good practice to retain the additional word "force" (e.g., pound-force or lb_f) when specifying this unit.

Distance. The standard and most often used unit for the dimension of length is the foot (ft), just as in the *FMLt* system.

Time. The most common units of the dimension time are seconds, which are defined the same way as in both the English Engineering *FMLt* and SI systems.

Force. Force is considered to be a primary, not derived, dimension in the English Engineering *FLt* system, just as it is in the *FMLt* system. However, its definition in the *FLt* system seems to go in circles, because although mass is said to be a derived quantity, the definition of the standard force unit depends on having a defined mass.

The standard unit for force is the pound-force. Basically, it is defined the same way as the lb_f is defined for the *FMLt* system. Note, however, that it depends on the definition of the lb_m. We have

$$1\ lb_f\Big|_{FLt} = 1\ lb_f\Big|_{FMLt} = 4.448\ 221\ N\Big|_{SI} \qquad (10\text{-}15)$$

k_n. k_n is arbitrarily defined to be unity.

Mass. To satisfy Newton's second law with arbitrarily chosen units on F, L, t, and k_n, we must now let the equation tell us the appropriate size or unit for mass. We have come nearly full circle; what we must do now is redefine the magnitude of the unit of mass which we will need for satisfying Newton's second law. That is, we have defined 1 lb_m in order to define 1 lb_f. Now we are defining a new size or unit for mass so that we can use Newton's second law correctly with $k_n = 1$. Solving that equation for mass, we find

$$M = \frac{F}{k_n a} \stackrel{d}{=} \frac{F}{1 \cdot L/t^2} \stackrel{d}{=} \frac{Ft^2}{L} \qquad (10\text{-}16)$$

If we were to calculate the magnitude of the mass that would accelerate at 1 ft/s^2 under the action of a force of 1 lb_f, Newton's law tells us that the mass would be

$$M = \frac{1\ lb_f}{1\ ft/s^2} = 1\frac{lb \cdot s^2}{ft} \qquad (10\text{-}17)$$

A mass of 1 $lb_f \cdot s^2/ft$ has traditionally been given the name *slug*. This is an alias in the same way that N is an alias for a force of 1 $kg \cdot m/s^2$ [10-12]. Newton's second law in the English Engineering *FLt* system becomes

$$\mathbf{F} = M\mathbf{a} \qquad (10\text{-}18)$$

as long as forces are measured in lb_f, lengths in feet, times in seconds and masses in slugs. Note here for the *FLt* system that $k_n = 1/g_c = 1$, which is considerably different from k_n or g_c in the English Engineering *FMLt* system. It may be obvious now that g_c in the *FMLt* system, combined with the definition of the slug, reduces to

$$\begin{aligned} g_c &= 32.1740\ ft \cdot lb_m/(lb_f \cdot sec^2) \\ &= 32.1740\ lb_m/slug \end{aligned} \qquad (10\text{-}19)$$

This means that g_c just plays the role of a mass conversion constant in Newton's second law in the English Engineering *FMLt* system. In fact, this shows that

$$1\ slug = 32.1740\ lb_m = 14.593\ 879\ kg \qquad (10\text{-}20)$$

If size were proportional to mass (be careful here, for in the physical world, size and mass are not necessarily related), Figure [10-13] demonstrates the relative magnitudes of the lb_m, kg, and slug.

k_g. All the quantities in the universal law of gravitation have now been determined in the Engineering *FLt* system, except k_g. The dimensions and units can be determined from the form of the equa-

[10-12]

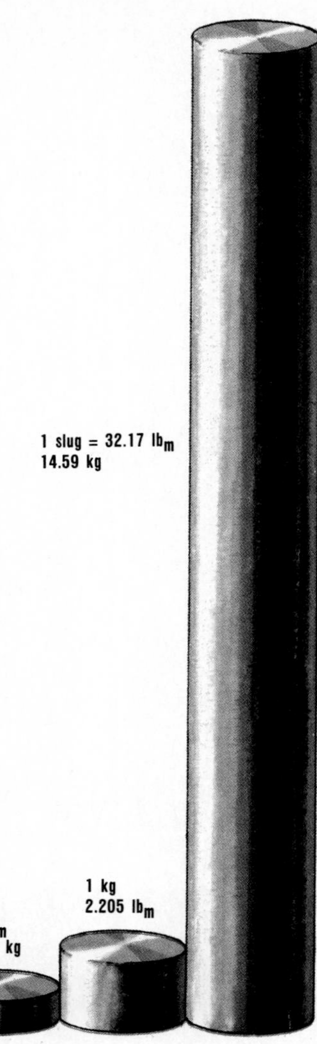

[10-13] If these figures represent pieces of the same material, then the size of the figures denote the relative sizes of the lb_m, kg, and slug.

tion; the numerical value must come from experimental data. It is

$$k_g = 3.439\,1 \times 10^{-8}\ \text{ft}^4/(\text{lb}_f \cdot \text{s}^4)$$

or

$$k_g = 3.439\,1 \times 10^{-8}\ \text{ft}^3/(\text{slug} \cdot \text{s}^2) \qquad (10\text{-}21)$$

The latter set of units is obtained from the former by using the alias slug for the collection of terms, $\text{lb}_f \cdot \text{s}^2/\text{ft}$.

Example 10-8

Solve for the mass (in lb_m) which is being accelerated at 6.14 ft/s^2 by a force of 196 lb_f.

Solution:

$$F = Ma \quad \text{or} \quad M = \frac{F}{a}$$

$$M = \frac{196\ \text{lb}_f}{6.14\ \text{ft/s}^2} = 31.9 \frac{\text{lb}_f \cdot \text{s}^2}{\text{ft}}$$

Notice that the units do not involve lb_m. However, the collection of units that we do get is known by the alias, the slug. Hence

$$M = 31.9\ \text{slugs}$$

Since the conversion between slugs and lb_m is 1 slug = 32.147 lb_m, then

$$M = 31.9\ \text{slugs}(32.174\ \text{lb}_m/\text{slug})$$
$$= 1.03 \times 10^3\ \text{lb}_m$$

The multiplicative factor 32.174 lb_m/slug is nothing more than unity since the numerator and the denominator are equal to one another. Therefore, use of it anywhere in an equation does not change the equality, but serves only to convert the units.

TRYING YOUR HAND AT A NEW DIMENSION/UNIT SYSTEM

Now it is your turn. Let's see how well you understand dimension/unit systems. To allow you to try your hand at relating them to the physical laws, we will pose the following possibility.

The Problem Posed

A UFO lands at the United Nations building and the emerging UFOnauts (Pronounce it ū · fon · auts. Do you have a better word?) declare that the Earth is now the property of their home planet, Gamma 7. Their initial list of orders contains the demand that their home dimension/unit system supplant that (or those) presently in use on Earth. The Secretary-General of the UN tells the visitors that things have become so computerized here on earth that everyone but you has forgotten how all these units and dimensions are related. The UFOnauts then seek you out and order you to prepare the conversion factors.

From discussions with the UFOnauts, you learn that they also use what we refer to as Newton's second law and the law of universal gravitation (reassuring, isn't it?). They dislike extraneous numbers and have arbitrarily chosen k_n and k_g to be unity (1). They have chosen the mass of a captured klingon cruiser, called the *heavy* (hev), as their standard unit of mass and the total length of their high priest's two noses to be the standard unit of distance, the *long* (ℓ) (it was either that or count millions of wavelengths of orange-red light). The standard force unit, called a *pull* (pu), is derived from the law of gravity, and the standard time unit, called *time warps,* is the time required to move a mass of 1 heavy, 1 long under the constant application of 1 pull, starting from rest.

Assuming a conversion factor between heavies and kilograms and between longs and meters, determine the dimensions of their pulls and their time warps. Establish the conversion factors between

Seconds and time warps.

N, lb_f, and pulls.

What is your height and weight in Gamma 7 units? How would you establish the validity of the assumptions you made on the relationship between klingon cruisers and kilograms and between longs and meters?

The Solution

In the Gamma 7 system, all the dimensional quantities in the law of gravitation are specified, except one. Perhaps we should begin by analyzing that equation. Dimensionally, we have

$$F = k_g M_1 M_2/r^2 \overset{d}{=} (1) \cdot M \cdot M/L^2 \overset{d}{=} M^2/L^2 \qquad (10\text{-}22)$$

(i.e., force will have the dimensions of M^2/L^2). This should not be too disturbing to you, since, if you recall, force in the SI system has the strange dimensions of $M \cdot L/t^2$, and you have become familiar with that. Now, since the standard units of mass and length are set in the Gamma 7 system, we will consider a gravitating system of two masses, each one of magnitude 1 heavy, separated by a distance of 1 long. The sketch of the system is shown in Figure [10-14].

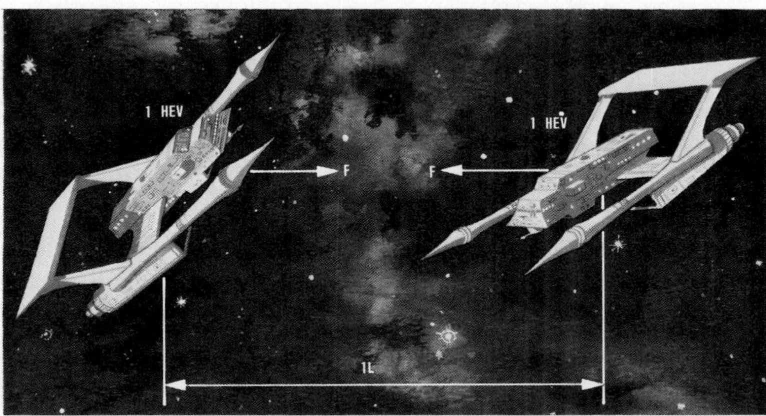

[10-14] Two klingon cruisers attracting each other at a distance of 1 long.

Computing the magnitude of F from the law of gravitation, we find

$$F = (1) \cdot \frac{(1)\ \text{hev} \cdot (1)\ \text{hev}}{(1)\ \ell \cdot (1)\ \ell} = 1\ \text{hev}^2/\ell^2 \qquad (10\text{-}23)$$

In follow-up meetings with the UFOnauts you would verify that this is indeed the size of force that they have labeled as a pull. Any unknown force can be calibrated by comparing it to the force of gravitational attraction between two 1-hev masses placed 1 ℓ apart. Now we have established that

$$1\ \text{pull} = 1\ \text{hev}^2/\ell^2 \qquad (10\text{-}24)$$

In Newton's second law, there is only the acceleration remaining to be defined. Since our concept of acceleration involves distance and time, we need to be just a little careful here in order to separate out time dimensions. That is, we already have a standard unit for distance (the long, ℓ), and rather than place a standard on acceleration, we want to use distance and acceleration to specify time. This can be done in the following way. Newton's law, solved for the acceleration, yields

$$a = \frac{F}{M} \stackrel{d}{=} \frac{M^2/L^2}{M} \stackrel{d}{=} M/L^2 \qquad (10\text{-}25)$$

In terms of units and magnitude, a 1-hev$^2/\ell^2$ force applied to a 1-hev mass would yield an acceleration of

$$a = \frac{(1)\ \text{hev}^2/\ell^2}{(1)\ \text{hev}} = 1\ \text{hev}/\ell^2 \qquad (10\text{-}26)$$

Our earthly concept of acceleration says that this quantity has the dimensions of L/t^2, that is,

$$a \stackrel{d}{=} L/t^2 \qquad (10\text{-}27)$$

and we have no reason to doubt that such a concept would apply to Gamma 7. This means that the dimensions in Equation (10-26) must be equivalent to the dimensions in Equation (10-27), or

$$L/t^2 \stackrel{d}{=} M/L^2 \qquad (10\text{-}28)$$

Thus time, in the Gamma 7 system, must have the *dimensions* of

$$t \stackrel{d}{=} \sqrt{L^3/M} \qquad (10\text{-}29)$$

Time *units* or scales of measure are a little more involved. First, knowing that a mass of 1 hev under the action of a 1 pull of force (1 hev$^2/\ell^2$) will accelerate at 1 hev/ℓ^2, we consider such a mass with this acceleration. From the physics involved, we obtain the equation that relates acceleration, initial velocity, and time to distance traveled (this is rectilinear motion with constant acceleration). The equation is[11]

[11]This equation should be familiar to those who are taking or have taken physics. It links displacement, velocity, and acceleration (all in the same direction, here the x direction), with time for an object under constant acceleration. x_1 and v_1 are values of the location (or displacement) and velocity, respectively, at some time $t = t_1$, while x_2 is the displacement at some later time t_2. a_x is the constant acceleration.

$$x_2 - x_1 = 0.5a_x(t_2 - t_1)^2 + v_1(t_2 - t_1) \qquad (10\text{-}30)$$

We imagine our object starting at time $t_1 = 0$ with position $x_1 = 0$ and initial velocity $v_1 = 0$, and solve for the time required for the object to travel a distance of $x_2 - x_1 = 1\ \ell$. We obtain

$$t = \sqrt{2}(\ell^2/\text{hev})^{1/2} \qquad (10\text{-}31)$$

This is what the UFOnauts have called a time warp (tw). That is,

$$1\ \text{tw} = \sqrt{2}(\ell^2/\text{hev})^{1/2} \qquad (10\text{-}32)$$

Now for conversion of some of our units into their units. Assuming that 1 hev is equivalent to 5.0×10^3 kg, and 1 ℓ is the same as 0.50 m (remember, the ℓ includes the length of *both* noses), the gravitating system used above will result in a force of attraction in SI units of

$$F = 6.67 \times 10^{-11}\ m^3/(\text{kg} \cdot \text{s}^2)\frac{(5.0 \times 10^3\ \text{kg})^2}{(0.5\ m)^2}$$

$$= 6.67 \times 10^{-3}\ \text{kg} \cdot m/\text{s}^2 = 6.67 \times 10^{-3}\ N \qquad (10\text{-}33)$$

Therefore,

$$1\ \text{pull} = 6.67 \times 10^{-3}\ \text{N} = 1.50 \times 10^{-3}\ \text{lb}_\text{f} \qquad (10\text{-}34)$$

The amount of earthly time necessary for an accelerating mass of hev, starting from rest, under the action of 1 pull, to travel 1 ℓ would be found from

$$0.50\ \text{m} = \frac{6.67 \times 10^{-3}\ \text{N}}{5.0 \times 10^{30}\ \text{kg}}t^2 \qquad (10\text{-}35)$$

from which

$$t = 6.1 \times 10^2\left\{\frac{\text{m} \cdot \text{kg}}{\text{N}}\right\}^{1/2} \qquad (10\text{-}36)$$

Reverting back the collection of units for which Newton is an alias, we find

$$t = 6.12 \times 10^2\left\{\frac{\text{m} \cdot \text{kg}}{\text{kg} \cdot \text{m/s}^2}\right\}^{1/2}$$

or

$$t = 6.12 \times 10^2\ \text{s} \qquad (10\text{-}37)$$

which is also 1 tw.

Now with that help, what is your height and weight in the Gamma 7 system? What is the time period for your classes?

Problems

10-25. Change 100 N of force to lb_f.

10-26. In the English *FLt* system, what mass in slugs is necessary to produce a weight of 15.6 lb_f at standard conditions?

10-27. In the English *FMLt* system, what mass in lb_m is necessary to produce a weight of 195.3 lb_f at standard conditions?

10-28. An interstellar explorer is accelerating uniformly at 58.6 ft/sec^2 in a spherical spaceship which has a total mass of 100,000 slugs. What is the force acting on the ship?

10-29. At a certain instant in time a space vehicle is being acted on by a vertically upward thrust of 497,000 lb_f. The mass of the space vehicle is 400,000 lb_m, and the acceleration of gravity is 32.1 ft/sec^2. Is the vehicle rising or descending? What is its acceleration? (Assume that "up" means radially outward from the center of the earth.)

10-30. Some interstellar adventurers land their spacecraft on a certain celestial body. Explain how they could calculate the acceleration of gravity at the point where they landed.

10-31. Using the relationship for g on page 281 and the fps gravitational system of units, determine the weight, at the latitude 0°, of a stainless steel sphere whose mass is defined as 150 $lb_f \cdot sec^2/ft$.

10-32. The mass of solid propellant in a certain container is 5 kg. What is the weight of this material in newtons at a location in Greenland where the acceleration of gravity is 9.83 m/sec^2? What is the weight in newtons?

10-33. If a gold sphere has a mass of 89.3 lb_m on earth, what would be its weight in lb_f on the moon, where the acceleration of gravity is 5.31 ft/sec^2? What is the weight in SI units?

10-34. Assuming that the acceleration due to gravitation is 5.31 ft/sec^2 on the moon, what is the mass in slugs of 100 lb_m located on the moon? In SI units?

10-35. A silver bar weighs 382 lb_f at a point on the earth where the acceleration of gravity is measured to be 32.1 ft/sec^2. Calculate the mass of the bar in lb_m and slug units.

10-36. The acceleration of gravity can be approximated by the following relationship:

$$g = 980.6 - 3.086 \times 10^{-6} A$$

where g is expressed in cm/sec^2 and A is an altitude in centimeters. If a rocket weighs 10,370 lb_f at sea level and standard conditions, what will be its weight in dynes at an elevation of 50,000 ft? In SI units?

10-37. At a certain point on the moon the acceleration due to gravitation is 5.35 ft/sec^2. A rocket resting on the moon's surface at this point weighs 23,500 lb_f. What is its mass in slugs? In lb_m? In SI units?

10-38. If a 10-lb weight on the moon (where g = 5.33 ft/sec^2) is returned to the earth and deposited at a latitude of 90° (see p. 281), how much would it weigh in the new location?

10-39. A 4.37-slug mass is taken from the earth to the moon and located at a point where g = 5.33 ft/sec^2. What is the magnitude of its mass in the new location?

10-40. The inertia force due to the acceleration of a rocket can be expressed as follows:

$$F = Ma$$

where F = unbalanced force

a = acceleration of the body

M = mass of the body

a. Given: a = 439 ft/s^2; M = 89.6 $lb_f \cdot s^2/ft$
Find: F in lb_f; in SI units.

b. Given: F = 1500 lb_f; M = 26.4 $lb_f \cdot s^2/ft$
Find: a in ft/s^2; in SI units

c. Given: $F = 49.3 \times 10^5$ lb_f; a = 32.2 ft/s^2
Find: M in $lb_f \cdot s^2/ft$; in SI units.

d. Given: M = 9650 $lb_f \cdot s^2/ft$; a = 980 cm/s^2
Find: F in lb_f; in SI units.

10-41. Choose k_g to be unity, and assume that F, t, and M are primary dimensions. Use the law of universal gravitation to define the secondary dimension of length. What are the dimensions of k_N?

ELECTRICAL, MAGNETIC, AND THERMAL QUANTITIES

Electrical and Magnetic Quantities

There are several observable or measurable quantities that are important in the fields of electricity and magnetism. In addition to length (L), force (F), work (W), power (P), and charge (Q), mentioned in the section entitled "Dimensions,"[12] these include electrical current (I), electrostatic potential (ϕ), electric field [or voltage] (E), resistance (R), capacitance (C), inductance (L), magnetic flux [or magnetic induction] (B), auxiliary field [or magnetic field strength] (H), plus a few others. Each of these can be considered to be a dimension.

From the list above, it appears that there are a large number of new and different dimensions with which we must learn to deal. However, due to the many fundamental laws that have been found to govern these fields, nearly all of the new dimensions are secondary

[12]You should not be surprised to see force and length (and their product, work) play intimate roles in electromagnetic theory. It is this interrelationship which makes, among other things, electromagnets attract ferrous metals and motor shafts turn.

ones. That is, these new dimensions we have listed can be expressed in terms of a much smaller subset of primary dimensions.

In discussing the lack of standards in dimensions and units encountered in mechanics (quantities involved in Newton's second law and the law of universal gravitation), we mentioned SI units and the two sets of commonly used English engineering units. Although the fields of electricity and magnetism do not embrace English engineering units, they do use several different sets of metric units. Among these are the rationalized mks system, the absolute esu system, and the absolute emu system.

The chief difference between the various dimension/unit systems used in electricity and magnetism is the set of primary dimensions that each system uses. Since the derived or secondary sets are chosen from the primary ones, the derived dimensions are also different. A more minor difference among the various systems is the actual units commonly used to quantify each dimension.

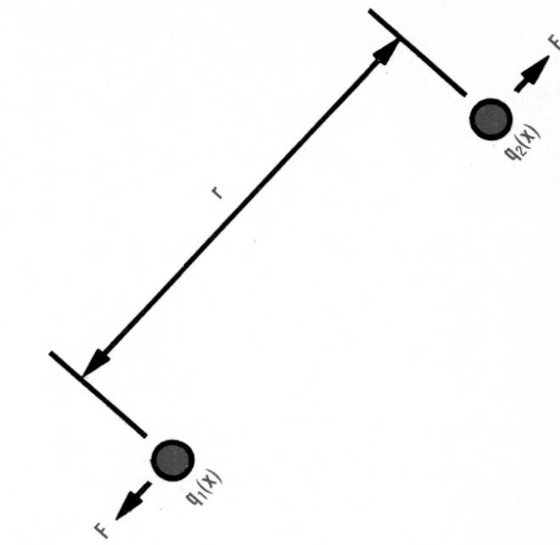

[10-15] Two like charges (i.e., both +, or both −) repelling each other.

We will cover only one example of dimension/unit systems in electricity and magnetism because to address all of the different dimensions would require more space than we can devote here. We will cover this single example to reinforce the understanding of dimensions and units that you have gained from the earlier sections in this chapter. The important concept is that in establishing dimension/unit systems, we should let fundamental laws form as many of the relationships between quantities as we can. We should avoid overspecifying the quantities that are to be considered as primary.

The example shown below uses Coulomb's law, which relates the forces that two charged particles exert on one another. In equation form this law is expressed as

$$F = k_c \frac{q_1 q_2}{r^2} \tag{10-38}$$

where F is the magnitude of the force, q_1 and q_2 are the electrostatic charges on particles 1 and 2, respectively, r is the distance of separation between the two particles, and k_c is a constant of proportionality. You may remember this law being demonstrated in high school physics experiments, where it was shown that like charges repel and unlike charges attract each other.

Two of the quantities, force and distance, contained in Equation (10-38) were encountered earlier in our discussions on dimensions in mechanics. The other two, k_c and the q's, are new. Were we to establish arbitrarily that charge, q, and k_c are going to be primary dimensions, we would have to accept the fact that either force or length (or the combination of both) would be derived dimensions. This, of course, is going to lead to incompatibility with our concepts of force and length as being the same dimensions that we established earlier in this chapter. That would be both unfortunate and unwise.

A better alternative is to consider that force and length have already been established by our work in mechanics, and then to let Equation (10-38) establish either the dimensions of q or k_c (or the combination). The rationalized mks system and the absolute esu system do just that, although they differ on whether q or k_c is to be considered a primary dimension.

The Rationalized MKS System

Dimensions. In the rationalized mks system, the dimension of charge (Q) is declared to be a primary dimension. The dimensions of force and length are taken from the SI system, where they have dimensions of ML/t^2 and L, respectively. With these choices, then, only the dimensions of k_c remain to be determined from Equation (10-38). Solving this equation for k_c yields

$$k_c = \frac{Fr^2}{q_1 q_2} \overset{d}{=} \frac{F \cdot L^2}{Q^2} \overset{d}{=} \frac{M \cdot L}{t^2}\frac{L^2}{Q^2} \overset{d}{=} \frac{M \cdot L^3}{t^2 \cdot Q^2} \tag{10-39}$$

It is more traditional to write Equation (10-38) as

$$F = \frac{1}{4\pi\epsilon_0}\frac{q_1 q_2}{r^2} \tag{10-40}$$

where $\epsilon_0 = 1/(4\pi k_c) \overset{d}{=} Q^2 \cdot t^2/(M \cdot L^3)$. The constant ϵ_0 is usually referred to as the dielectric constant or "permittivity" of free space.

Units. In the rationalized mks system, the common unit of mass is the kilogram, the unit of length is the meter, the unit of time is the second, and the unit of charge is the coulomb. One coulomb (C) is the charge found on 6.24196×10^{18} electrons.

For a given pair of charged particles held a known distance apart, the force of attraction or repulsion is predetermined by nature and cannot be set arbitrarily. This means that for Equation (10-40) to give the correct answer, the value of ϵ_0 (or k_c) cannot be set arbitrarily. It must be determined experimentally. The common value used is $\epsilon_0 = 8.854 \times 10^{-12}\ \mathrm{C^2 \cdot s^2/(kg \cdot m^3)}$. Since a newton of force is $1\ \mathrm{kg \cdot m/s^2}$, the collection of units on this ϵ_0 is identical to $\mathrm{C^2/(N \cdot m^2)}$. With this substitution, ϵ_0 is also equal to $8.854 \times 10^{-12}\ \mathrm{C^2/(N \cdot m^2)}$.

The Absolute ESU System

Dimensions. In the absolute esu system, k_c is arbitrarily taken as unity (a value of 1, no dimensions). The dimensions of force and length are assumed to be the same as the respective dimensions in the SI unit systems (but a different set of units is commonly used, as we will note in the next section). This leaves only the dimensions of charge to be determined. Solving Equation (10-40) for the product of the q's, we find

$$q_1 q_2 = \frac{F \cdot r^2}{k_c} \overset{d}{=} F \cdot L^2 \overset{d}{=} \frac{M \cdot L}{t^2} \cdot L^2 \overset{d}{=} \frac{M \cdot L^3}{t^2} \tag{10-41}$$

Thus the dimensions of charge are $(M \cdot L^3/t^2)^{1/2}$ in the absolute esu system.

Units. The traditional units in the absolute esu system are the centimeter for length, the gram for mass, and the second for time. We have already noted that k_c has been chosen equal to 1. The (secondary) unit for force in Coulomb's law is the dyne, which is an alias for $1\ \mathrm{g \cdot cm/s^2}$. A dyne is also the same as 10^{-5} N.

Two identical charged particles that repel each other with a force of 1 dyne (i.e., 1 g · cm/s^2) when placed 1 cm apart must have a charge of 1 (dyne · cm^2)$^{1/2}$ or 1 (g · cm/s^2)$^{1/2}$, according to the absolute esu version of Coulomb's law. Since this looks like an appropriate place for an alias, we say that a charge of 1 (dyne · cm^2)$^{1/2}$ will be called a *statcoulomb*. Thus the statcoulomb is an alias in the same way that the newton is an alias for 1 kg · m/s^2 of force. Now, in the following form of Coulomb's law,

$$F = \frac{q_1 q_2}{r^2} \tag{10-42}$$

we will get meaningful results for forces (in dynes) when we use r's in cm and q's in statcoulombs.

Thermal Quantities

Dimensions. In the subjects of energy and thermodynamics, there is another set of dimensions, or measurable quantities, that become important. Fortunately, like the fields of electricity and magnetism, nearly all of these can be considered to be derived or secondary dimensions. Through basic laws and fundamental definitions, they can be related to the dimensions already discussed in this chapter (mass, length, time, force, etc.). Essentially the only exception to this is the fundamental quantity temperature (T). It must be considered as a primary dimension in problems involving the storage of energy within the atoms, molecules, and crystal structure that make up matter, the so-called internal energy.

Units. There are four sets of units in current use that allow us to put quantitative values or units on the primary dimension of temperature. Two of these four are said to be "absolute" temperature scales and two are said to be "relative" scales.

The absolute scales have their origin in the theory and principles of the science of thermodynamics. In this field we describe absolute zero as that temperature where nearly all molecular motion ceases.

The two relative scales are the Celsius (C) and the Fahrenheit (F) scales. The extent of the Celsius scale was defined by assigning the freezing temperature of water an arbitrary value of 0 "degrees," and the boiling temperature (at 1 atmosphere of pressure) an arbitrary value of 100 "degrees." The Fahrenheit scale was similarly established by assigning an arbitrary value of 32 degrees to the freezing point of water and a value of 212 degrees to the boiling point. Thus, the Celsius scale has 100 steps or "degrees" between the freezing and boiling points of water, while the Fahrenheit scale has 180 steps or degrees. From this we can see that a Celsius degree is not the same size as a Fahrenheit degree. The ratio of 100 to 180 is in the proportion of 5 to 9. This is the source of the $\frac{5}{9}$ and $\frac{9}{5}$ that appear in the conversion tables in Appendix I.

The two absolute scales are the Kelvin (K) scale and the Rankine (R) scale. Both start with 0 (zero) at absolute zero temperature, but each assigns different increments as the temperature increases. They are analogous to using two different measuring tapes, one calibrated

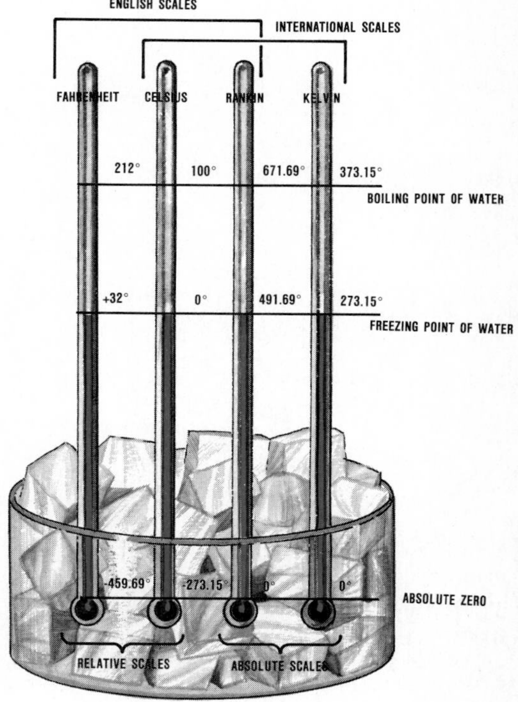

[10-16] Temperature scales.

in feet and one calibrated in meters, to measure distance from a specified point. Both measuring tapes might begin with their zero at the specified point but each progresses at different increments. When the tape marked in feet reads 10 ft, the tape marked in meters will show slightly over 3 m.

The size of the increments between temperatures is the same in the absolute Kelvin scale as in the relative Celsius scale. Similarly, the size of the increments between temperatures or the degrees is the same in the absolute Rankine scale as in the relative Fahrenheit scale. This means that the Kelvin degree is $\frac{9}{5}$ the size of a Rankine degree.

When you specify temperature it is imperative that you specify the scale that you are using. To say the temperature is 45 degrees or that the temperature has changed by 3 degrees is meaningless.

The Kelvin and Celsius scales are often referred to as the International scales since they, like the SI unit system, are almost uniformly adopted around the world. The Rankine and Fahrenheit scales, sometimes referred to as the English temperature scales, are now in use only in the United States.

UNITS CONVERSION

Conversion between units within the same dimension is not too difficult. That is, converting from one unit of mass to another unit of mass, or from one unit of length to another unit of length, requires

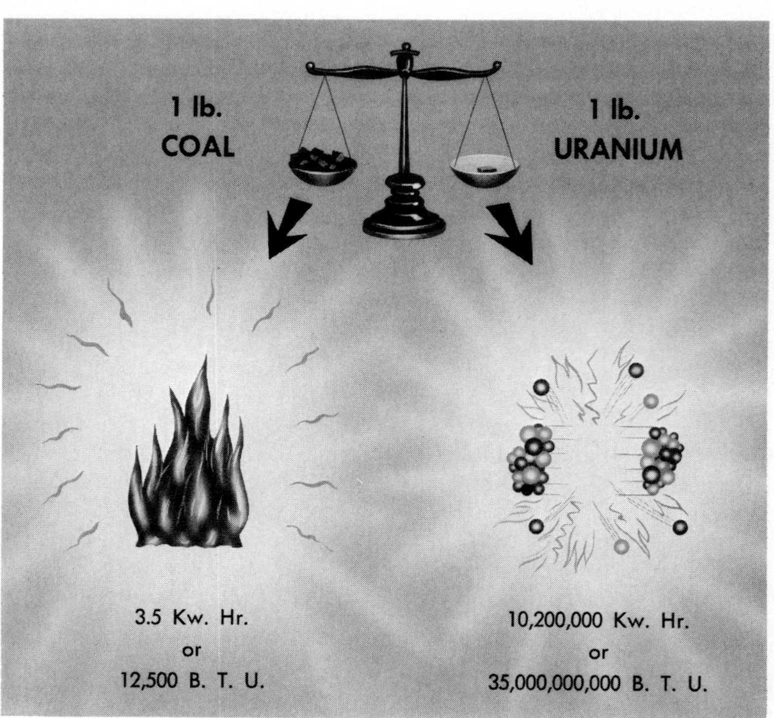

[10-17] Relative energy conversion from coal and nuclear fuels. In working with such energy sources, the engineer must be able to convert from one unit system to another.

the use of a simple conversion factor. A table of conversion factors is included in Appendix I to aid in this process.

Conversion between two or more different dimensions should not be attempted. It is wrong both conceptually and numerically. You can no more convert force units into mass units than you can length units into time units.

Engineers frequently work in several systems of units in the same calculation. In this case it is only necessary that each dimension be expressed in any valid set of units from any of the various unit systems. Numerical equality and unit homogeneity may be determined in any case by applying conversion factors to the individual terms of the expression.

It is recommended that engineers avoid using stereotyped conversion symbols such as g_c or J (the mechanical equivalent of heat) in writing mathematical expressions that represent some physical phenomenon. If one of these, or any other conversion factor, is needed in an equation to achieve unit balance, it can then be added. Since many different unit systems may be used from time to time, it is best to add conversion factors only as they are needed.

Unfortunately, in much of the engineering literature, the equations used in a particular situation include one or more conversion factors. Therefore, considerable care must be exercised in using these expressions since they represent a "special case" rather than a "general condition." The engineer should form a habit of always checking the unit balance of all equations.

[10-18]

Example 10-9

Change a speed of 3000 miles per hour (mi/hr) to m/s.

Solution:

Calling the speed *V*, we find

$$V = \left[3000\ \frac{\text{mi}}{\text{hr}}\right]\underbrace{\left[\frac{5280\ \text{ft}}{1\ \text{mi}}\right]}_{\text{Conversion factor 1}}\underbrace{\left[\frac{1\ \text{hr}}{3600\ \text{s}}\right]}_{\text{Conversion factor 2}}\underbrace{\left[\frac{0.3048\ \text{m}}{\text{ft}}\right]}_{\text{Conversion factor 3}}$$

$$= 1341\ \text{m/s}$$

Here again, the three conversion factors are each equivalent to unity since the numerator and denominators are equivalent. These factors do not change the validity of the equation, only the units.

Example 10-10

Using the weight and mass from Example 10-7, determine the mass of the earth assuming that a distance of 6379 km separates you from the center of the earth.

Solution:

Since this problem involves gravitational attraction, perhaps the law of universal gravitation, expressed in Equation (10-2), would be a good place to start. Since we know your mass and the force of attraction (your weight) and the distance of separation, we can calculate the mass of the earth from

$$M_e = \frac{Fr^2}{M_{you}k_g} = \frac{667.2\ \text{N} \cdot (6379 \times 10^3\ \text{m})^2}{68.40\ \text{kg} \cdot 6.673 \times 10^{-11}\ \text{m}^3/(\text{kg} \cdot \text{s}^2)}$$

$$= 5.95 \times 10^{24}\ \text{N} \cdot \text{s}^2/\text{m}$$

or, using the fact that the newton is an alias for a kg · m/s^2,

$$M_e = 5.95 \times 10^{24} \frac{\text{kg} \cdot \text{m}}{\text{s}^2} \cdot \frac{\text{s}^2}{\text{m}} = 5.95 \times 10^{24}\ \text{kg}$$

DERIVED DIMENSIONS/UNITS OF OTHER COMMONLY USED QUANTITIES

We end this chapter with a brief discussion of several common quantities defined and used by engineers. Conversion factors for various units are supplied in Appendix I.

Area and Volume

The dimensions of area and volume are L^2 and L^3, respectively. Commonly used units are:

Area: m^2 (SI preferred), cm^2, mm^2, in.^{2}, ft^2, yd^2, miles^2, acres, hectares.

Volume: m^3 (SI preferred), cm^3 (or cc), liter, in^3, ft^3, yd^3 (often

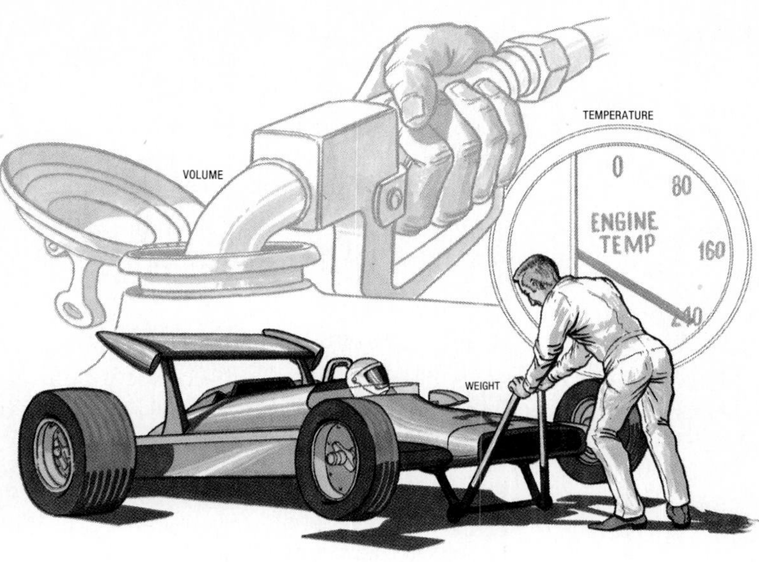

[10-19]

called simply a yard for concrete or sand), gallons, Imperial gallons, barrels, bushels, board feet (for lumber), cords (for firewood).

Work

Work is the product of a force and a distance through which that force acts. Energy is the ability or capacity for doing work. Although the two quantities are conceptually different, they have the same dimensions and can, therefore, be expressed in the same units. The dimensions are $F \cdot L$. Commonly used units are kilojoules (SI preferred), joules, calories, kilocalaries, electron-volts, in. $\cdot$ lb_f, ft $\cdot$ lb_f, therms, British thermal units (Btu), and horsepower-hours.

Power

Power is the time rate at which work is done. Hence its dimensions are $F \cdot L/t$. Common units are kilowatts (SI preferred), watts, calories/s, in. $\cdot$ lb_f/s, ft $\cdot$ lb_f/s, horsepower, and Btu/hr.

[10-20]

Pressure

Pressure is force per unit area, giving rise to dimensions of F/L^2. Common units are pascals (SI preferred), atmospheres, bars, $lb_f/in.^2$, lb_f/ft^2, millimeters or inches of Mercury, inches of water.

Mass Density

Mass density is mass per unit volume. Therefore, its dimensions are M/L^3. Common units are kg/m^3 (SI preferred), grams/liter, lb_m/ft^3, $slugs/ft^3$.

Specific Weight or Weight Density

Specific weight (or weight density) is a measure of the weight of a substance per unit volume. Its dimensions are F/L^3. Common units are N/m^3 (SI preferred), kg_f/m^3, $lb_f/in.^3$, lb_f/ft^3.

[10-21]

PROBLEMS

10-42. Let the mass of the earth (see Example 10-10) be 1 bigm, the distance 6379 kilometers be 1 long, and your weight be 1 ltwt. Express Newton's second law and the universal law of gravitation in forms that work for these new quantities (retain the standard second as the unit of time).

10-43. Referring to Example 10-10, calculate the mass of the sun if the earth (see Example 10-10) has an orbital diameter of 1.49×10^7 km and the force of attraction between the two celestial bodies is 1.44×10^{25} N.

10-44. Convert 76 N to dynes and lb_f.

10-45. Convert 2.67 in. to angstroms and miles.

10-46. Convert 26 knots to feet per second and meters per hour.

10-47. Convert 8.07×10^3 tons to newtons and ounces.

10-48. Convert 1.075 atmospheres to dynes per cm^2 and inches of mercury.

10-49. Convert 596 Btu to foot-pounds and joules.

10-50. Convert 26,059 watts to horsepower and ergs per second.

10-51. Convert 75 angstroms to feet.

10-52. Express 2903 ft^3 of sulfuric acid in gallons and cubic meters.

10-53. Change 1 Btu to horsepower-seconds.

10-54. A car is traveling 49 mi/hr. What is the speed in feet per second and meters per second?

10-55. A river has a flow of 3×10^6 gal per 24-hr day. Compute the flow in cubic feet per minute.

10-56. Convert 579 qt/sec to cubic feet per hour and cubic meters per second.

10-57. A copper wire is 0.0809 cm in diameter. What is the weight of 1000 m of the wire?

10-58. A cylindrical tank 2.96 ft high has a volume of 136 gal. What is its diameter?

10-59. A round iron rod is 0.125 in. in diameter. How long will a piece have to be to weigh 1 lb?

10-60. Find the weight of a common brick that is 2.6 in. by 4 in. by 8.75 in.

10-61. Convert 1 yd^2 to acres and square meters.

10-62. A container is 12 in. high, 10 in. in diameter at the top, and 6 in. in diameter at the bottom. What is the volume of this container in cubic inches? What is the weight of mercury that would fill this container?

10-63. How many gallons of water will be contained in a horizontal pipe 10 in. in internal diameter and 15 ft long, if the water is 6 in. deep in the pipe?

10-64. A hemispherical container 3 ft in diameter has half of its volume filled with lubricating oil. Neglecting the weight of the container, how much would the contents weigh if kerosene were added to fill the container to the brim?

10-65. What is the cross-sectional area of a railroad rail 33 m long that weighs 94 lb/yd?

10-66. A piece of cast iron has a very irregular shape and its volume is to be determined. It is submerged in water in a cylindrical tank having a diameter of 16 in. The water level is raised 3.4 in. above its original level. How many cubic feet are in the piece of cast iron? How much does it weigh?

10-67. A cylindrical tank is 22 ft in diameter and 8 ft high. How long will it take to fill the tank with water from a pipe which is flowing at 33.3 gal/min?

10-68. Two objects are made of the same material and have the same weights and diameters. One of the objects is a sphere 2 m in diameter. If the other object is a right cylinder, what is its length?

10-69. A hemisphere and cone are carved out of the same material and their weights are equal. The height of the cone is 3 ft. $10\frac{1}{2}$ in., while the radius of the hemisphere is 13 in. If a flat circular cover were to be made for the cone base, what would be its area in square inches?

10-70. An eight-sided wrought iron bar weighs 3.83 lb per linear foot. What will be its dimension across diagonally opposite corners?

10-71. The kinetic energy of a moving body in space can be expressed as follows:

$$KE = \frac{MV^2}{2}$$

where KE = kinetic energy of the moving body

M = mass of the moving body

V = velocity of the moving body

a. Given: $M = 539\ \text{lb}_f \cdot \text{s}^2/\text{ft}$; $V = 2900$ ft/s
Find: KE in ft · lb_f; in SI units.

b. Given: $M = 42.6\ \text{lb}_f \cdot \text{s}^2/\text{ft}$; $KE = 1.20 \times 10^{11}$ ft · lb_f
Find: V in ft/s; in SI units.

c. Given: $KE = 16{,}900$ in. · lb_f; $V = 3960$ in./min
Find: M in slugs; in SI units.

d. Given: $M = 143$ g; $KE = 2690$ in. · lb_f
Find: V in mi/hr; in SI units.

10-72. The force required to assemble a force-fit joint on a particular piece of machinery may be expressed by the following equation:

$$F = \frac{\pi dlfP}{2000}$$

where d = shaft diameter, in.

l = hub length, in.

f = coefficient of friction

P = radial pressure, psi

F = force of press required, tons

a. Given: $d = 9.05$ in.; $l = 15.1$ in.; $f = 0.10$; $P = 10{,}250$ psi
Find: F in lb_f; in SI units.

b. Given: $F = 4.21 \times 10^5\ \text{lb}_f$; $f = 0.162$; $P = 8.32 \times 10^8$ psf; $l = 1.62$ ft
Find: d in ft; in SI units.

c. Given: $d = 25$ cm; $l = 30.2$ cm; $f = 0.08$; $P = 9260$ psi
Find: F in tons; in SI units.

d. Given: $F = 206$ tons; $d = 6.23$ in.; $l = 20.4$ in.; $f = 0.153$
Find: P in lb_f/ft^2; in SI units.

10-73. The dynamic stress in the rim of a certain flywheel has been expressed by the following equation:

$$\sigma = 0.0000284\rho r^2 n^2$$

where σ = tensile stress, $\text{lb}_f/\text{in.}^2$
ρ = specific weight of material, $\text{lb}_f/\text{in.}^3$
r = radius of curvature, in.
n = number of rpm

a. Given: $\sigma = 200$ psi; $\rho = 0.282\ \text{lb}_f/\text{in.}^3$; $r = 9$ in.
Find: n in rpm; in SI units.

b. Given: $\rho = 0.332\ \text{lb}_f/\text{in.}^3$; $r = 23.1$ cm; $n = 200$ rpm
Find: σ in psi; in SI units.

c. Given: $\rho = 540\ \text{lb}_f/\text{ft}^3$; $n = 186$ rpm; $\sigma = 31.2 \times 10^3\ \text{lb}_f/\text{ft}^2$
Find: r in ft; in SI units.

d. Given: $\rho = 326\ \text{lb}_f/\text{ft}^3$; $n = 250$ rpm; $r = 0.632$ ft
Find: σ in lb/ft^2; in SI units.

10-74. The stress in a certain column may be calculated by the following relationship:

$$\sigma = \frac{F}{A}\left[1 + \left(\frac{l}{k}\right)^2 \frac{R}{\pi^2 nE}\right]$$

where σ = induced stress, psi
F = applied force, lb_f
A = cross-sectional area of member, in.^{2}
l = length of bar, in.
k = radius of gyration, in.
R = elastic limit, $lb_f/in.^2$
E = modulus of elasticity, $lb_f/in.^2$
n = coefficient for different end conditions

a. Given: $n = 1$; $E = 3 \times 10^7$ psi; $R = 4.2 \times 10^4$ psi; $k = 0.29$ in.; $l = 20.3$ in.; $A = 17.5$ in.^{2}; $F =$ 12,000 lb_f.
Find: σ in psi; in SI units.

b. Given: $\sigma = 11{,}500$ psi; $F = 6.3$ tons; $l = 2.11$ ft; $k =$ 0.41 in.; $R = 40{,}000$ psi; $E = 3.16 \times 10^7$ psi; $n = 2$
Find: A in ft^2; in SI units.

c. Given: $n = 1/4$; $E = 2.65 \times 10^7$ psi; $R = 3.21 \times 10^4$ psi; $k = 0.026$ ft; $A = 102$ cm^2; $F = 5.9$ tons; $\sigma =$ 10,000 psi
Find: l in ft; in SI units.

d. Given: $\sigma = 1.72 \times 10^6$ psf; $F = 1.33 \times 10^4$ lb_f; $l =$ 1.67 ft; $k = 0.331$ in.; $E = 7.87 \times 10^7$ psi; $n = 4$; $A = 14.2$ in.^{2}
Find: R in psi; in SI units.

10-75. Suppose that you chose k_c to be unity and desire to use electrostatic charge (q) as a fundamental quantity. What ramifications will there be in Newton's second law, the law of universal gravitation, and Coulomb's law?

Chapter 11

Problem Analysis

Engineering curricula contain more courses in physics and subjects derived from physics than from any of the other pure sciences. For example, all the topics studied in a freshman course in physics are covered again in more detail in courses that follow freshman physics. These topics include *motion, force, moments, energy, electricity, magnetism,* and *matter*. They are all of special concern to the engineer.

Our intent in this chapter and the next is to cover some of these beginning physics topics so that you may begin to use the problem-solving tools so useful to engineers. With proper practice, care, and maintenance, your tools will become natural extensions of your mental capabilities. With disuse and lack of care, your tools, like those of a mechanic, will become rusty and deteriorate.

This chapter reinforces these concepts of physics with problems whose solutions do not require the use of computers. The problems in this chapter require algebra, trigonometry, and at most, hand-held calculators. Chapter 19 then continues to reinforce physics topics, but stresses more involved or difficult problems that can be solved easily on the microcomputer.

MECHANICS

Since we can look at only a small selection of physics topics here, let us first consider the field of *mechanics*. *Mechanics is the physical science that describes and predicts the effects of forces acting on material objects.* There are three specialized branches into which the general field of mechanics may be divided for more specific studies. These are as follows:

1. Mechanics of rigid bodies.
   a. Statics.
   b. Dynamics.
2. Mechanics of deformable bodies.
3. Mechanics of fluids.
   a. Compressible flow.
   b. Incompressible flow.

Material objects may be in a state of rest or they may be undergoing some complicated motion. Our study of mechanics in this chapter will be concerned with an introduction to *statics,* that branch of mechanics that deals with the interaction of forces with rigid bodies that are at rest or moving with a constant motion. We will show how the engineering method is applied to problem solution in this field.

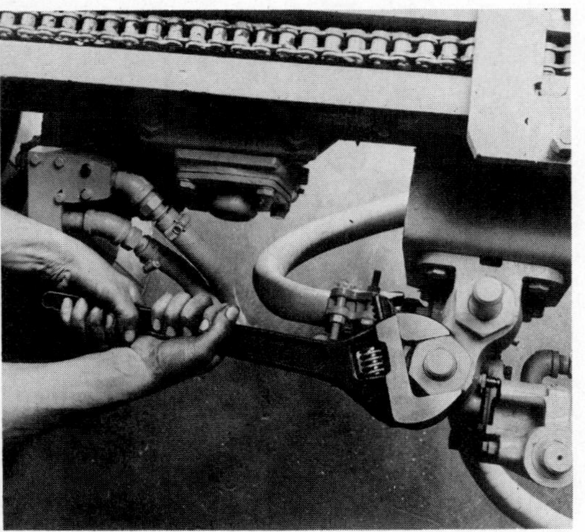

[11-1] The wrench applies a moment to the nut.

Fundamental Concepts and Definitions

Concepts used in our study of statics are *force, moments, space,* and *matter*. These concepts are basic and, for now, should be accepted on the basis of our general experience. A *force* is the result of the interaction of two or more bodies. It is through forces that objects communicate; one object can keep another object from moving, or it can cause another object to change its velocity and position. Forces have associated with them both a direction and a magnitude. That is, when you push on something in order to apply a force, you push in a certain direction with a certain intensity or magnitude.

Forces may result from physical contact between objects, or they may be applied without any physical contact. Examples of the latter no-contact forces are electrostatic attraction or repulsion of electrically charged particles, magnetic forces on charged particles moving in magnetic fields, and gravitational attraction of one object by another.

If a force is applied perpendicular to an object such as a wrench [11-1], there will be a tendency to cause the object to turn about some pivot point in either a clockwise or a counterclockwise direction. This tendency of a force to cause rotation about a given point is called a *moment* (in Figure [11-1] the point of interest is the center of the bolt that the wrench is gripping).

The direction of the tendency to cause turning will depend on the direction of the applied force. As almost anyone who has used a wrench knows that the magnitude of the turning tendency will depend on both the force that is applied perpendicular to the length of the wrench, and the point along the length of the wrench where the force is applied. The *length of the moment arm* is the perpendicular distance from the point of rotation to the line along which the applied force acts. In the case of the wrench, you can apply a larger moment either by applying a larger force or by increasing the length of the moment arm (that is why you often search for a wrench with a longer handle). The *magnitude of the moment* is calculated by multiplying the magnitude of the force by the length of the moment arm. Similar to the force, the moment has both a direction (e.g., clockwise or counterclockwise) and a magnitude.

Space is a region extending in all directions. Among other things, it contains points—locations or positions—that are important in mechanics problems. *Matter* is a substance that occupies space.

A *particle* may be said to be a highly concentrated amount of matter that occupies a single point in space. A *rigid* body is an object that is constructed entirely of particles that do not change their position in space relative to each other. The particles in rigid bodies are *always* located the same fixed distance from one another. Although no object is ever truly rigid, in many situations the deformation, or change in position of the particles with respect to one another, is very small. In these situations, neglecting the deformations generally has a nearly imperceptible effect on the analysis. Such is the assumption that is made in statics and it will be used in the remainder of this chapter. In the studies of the mechanics of deformable bodies and the mechanics of fluids this assumption is generally not true. The latter fields deal with the stresses that are caused by the internal forces that objects and substances must sustain, and the stretching or contraction—the deformation—that these stresses cause.

Forces and Moments as Vectors

Forces. In describing forces and moments, we have pointed out that both have an associated direction and magnitude. Both of these features, direction and magnitude, are properties of what are generally termed *vectors*. Quantities that have no directional properties but are specified only by a magnitude are called *scalar* quantities. Examples of vector quantities are force, moments, displacement, velocity, and acceleration. Examples of scalar quantities are temperature, volume, time, and energy. Vectors are said to be:

1. *Coplanar,* when all vectors lie in the same plane [11-2].
2. *Collinear,* when all vectors act along the same line [11-3].
3. *Concurrent,* when all vectors originate or intersect at a single point [11-4].

In this initial study of statics we shall deal mainly with concurrent, coplanar force systems.

It is sometimes advantageous to combine two forces into a single equivalent force, which we shall call a *resultant.* Often it is also advantageous to consider a force to consist of two or more *components* which, when combined, produce a resultant that is identical to the original force.

Combining and resolving (i.e., adding and subtracting) forces, indeed any vector quantities, must be handled according to certain strict rules of vector analysis. Vectors *cannot* be added together or subtracted in the same way that we do scalar quantities. Because of their directional dependence, vectors require special treatment in order for analyses to yield valid results. The general rules for vector addition and subtraction are presented in Topic 11-1. If you are not familiar with vector addition and subtraction, you will need to stop now and familiarize yourself with these rules before proceeding.

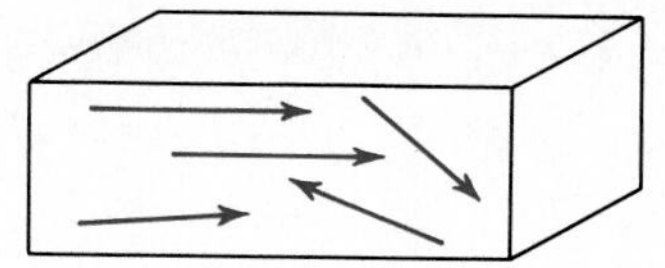

Coplanar Force System

[11-2] Coplanar force system.

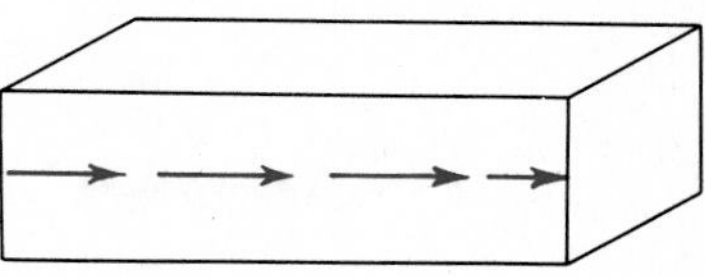

Collinear Force System

[11-3] Collinear force system.

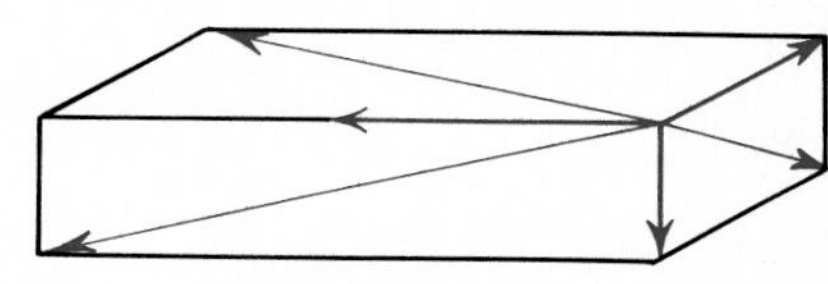

Concurrent Force System

[11-4] Concurrent force system.

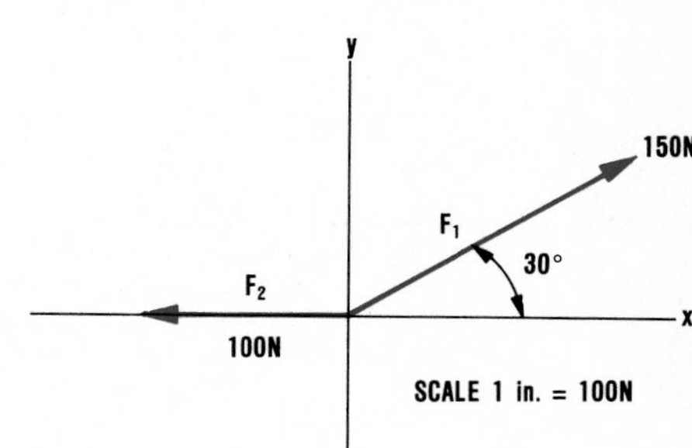

[11-5]

Example 11-1

In Figure [11-5], a force of 150 N is pulling upward from a point at an angle of 30° with the horizontal, while a force of 100 N is pulling to the left along the x axis. Use the graphical technique to find the resultant.

Solution:

In Figure [11-6], the lengths of the vectors were scaled (using an engineer's scale) to 1 in. equals 100 N. The 150 N force is 1.5 in. long acting upward at an angle of 30° with the horizontal. The 100 N force vector is 1 in. long and points directly to the left. In the graphic method the arrow tips should not extend completely to the end of the vector, since it is very easy to "overrun" the exact length of the measured line in the drawing of the arrowhead.

The resultant R of these two forces is 0.81 in. long, which, according to our scale, represents a force of 81 N. The resultant acts along a line that makes an angle of 68° with the positive x axis.

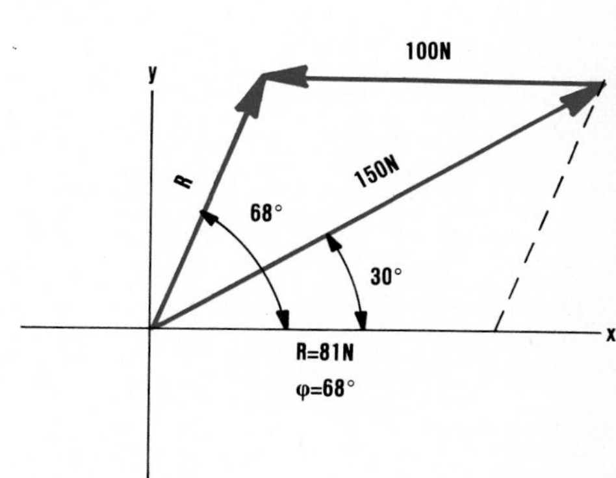

[11-6]

Topic 11-1

Addition and Subtraction of Vectors

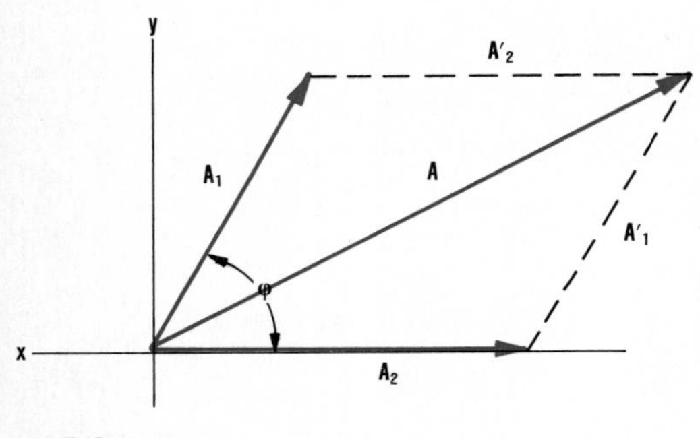

[11-T1]

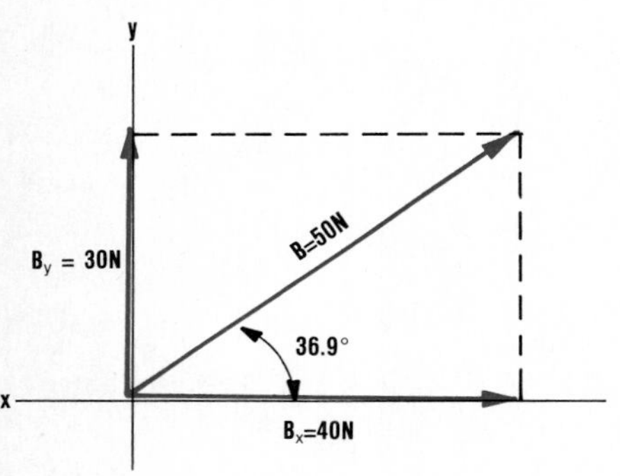

[11-T2]

THE METHOD OF RECTANGULAR COMPONENTS

The method most frequently used by engineers to find the resultant vector of a collection of vectors is the *rectangular component method.* In this method, the resultant is found from an analysis of the components of each vector in the collection. If there are two components for each member vector and, if the two components are perpendicular to each other, they are called *rectangular components.*

If each vector in the collection is resolved into rectangular components, then it is a simple matter to add the values of parallel components to obtain the values of the components of the resultant force. When the vectors being analyzed are forces, it is common, but by no means necessary, to choose component directions that are horizontal and vertical. However, any orientation of the directions or axes will produce equivalent results.

Figure [11-T1] shows a vector **A** having components A_1 and A_2 which are not perpendicular, and are not, therefore, rectangular components. In Figure [11-T2], we show a known vector **B** with its rectangular components B_x and B_y. Note that the lengths of the components B_x and B_y can be determined numerically by trigonometry from the direction and magnitude of **B**. Similarly, if B_x and B_y are known, then trigonometry can be used to determine the direction and magnitude of the vector **B**, the resultant of B_x and B_y. The components B_x and B_y may replace the vector **B** in any computation.

In using the rectangular component method, it is important to establish a set of coordinate directions and to consider the components of any vector positive if the components are in the positive coordinate direction or negative if the components are in the negative coordinate direction. We demonstrate the method in the following example.

Example 11-T1

Let us examine a concurrent coplanar vector system [11-T3] and resolve each vector into its rectangular components [11-T4].

Solution:

By trigonometry, F_x can be found using **F** and the cosine of the angle θ, or $F_x = F \cos \theta$. In the same manner $F_y = F \sin \theta$.

To keep the directions of the vectors better in mind, let us establish coordinate directions such that vectors acting to the right are positive

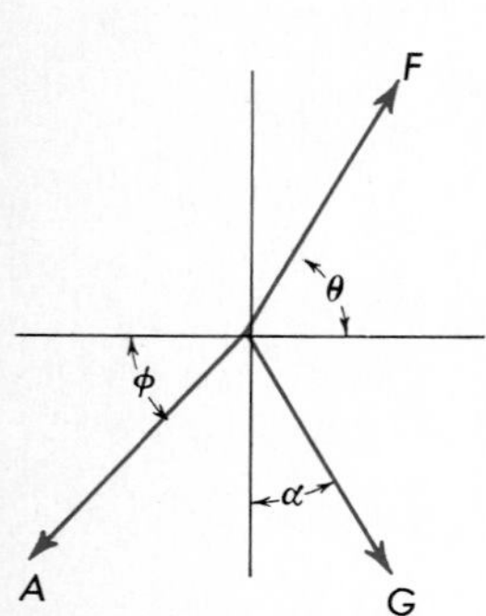

[11-T3] Three concurrent, coplanar vectors.

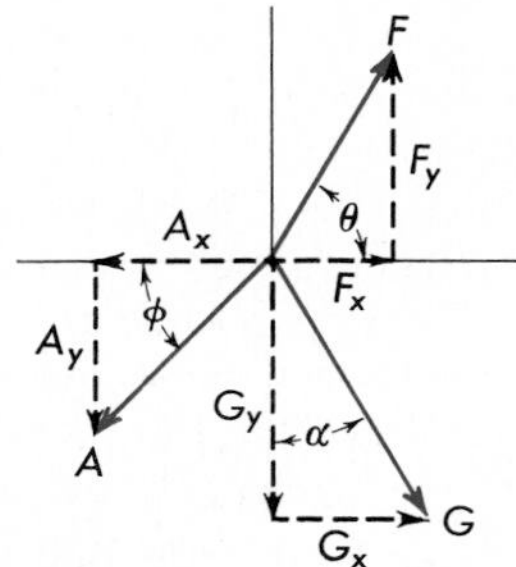

[11-T4] Components of the vectors in Figure [11-T3].

and those acting to the left are negative. Also, the vectors acting upward may be considered positive and those acting downward may be considered negative.

Force **G** has components $G_x = G \sin \alpha$ and $G_y = -G \cos \alpha$, while force **A** has components $A_x = -A \cos \phi$ and $A_y = -A \sin \phi$.

In adding vectors by the rectangular component method, it is often useful to construct a table of the components so that they easily can be added together. For example, the components of the resultant of the three vectors in Figure [11-T3] are $R_x = A_x + F_x + G_x$, and $R_y = A_y + F_y + G_y$. The magnitude of the resultant, R, is given by $(R_x^2 + R_y^2)^{1/2}$ and the angle it makes with the x axis is given by arctan (R_y/R_x).

The differences between two vectors can be found in a very similar manner. If it is desired to subtract **A** from **G** to get a vector **B**, then the components of **B** can be obtained from simply subtracting components, $B_x = G_x - A_x$ and $B_y = G_y - A_y$. The magnitude of **B** is determined from $\mathbf{B} = (B_x^2 + B_y^2)^{1/2}$; it makes an angle of arctan (B_y/B_x) with the x axis.

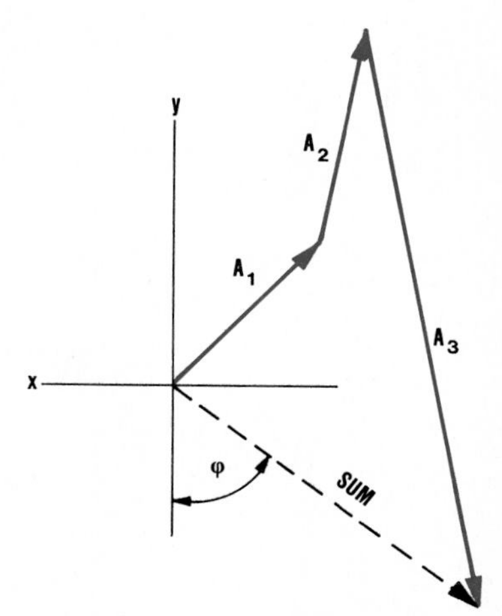

[11-T5] The dashed-line vector is the resultant (or sum) of vectors A_1, A_2, and A_3.

THE GRAPHICAL METHOD

The graphical method is not used as often as the method of rectangular components. It certainly does not lend itself to numerical computation in an algorithm as does the method of rectangular components. Also, it is generally not as accurate. However, a freehand drawn graphical diagram is often useful for visualizing a problem that may include many vectors.

In the graphical technique, two vectors are added by proportioning their lengths (i.e., scaling them) and drawing the vectors so that the head of one vector is just touching the tail of the other [11-T5]. Their sum (resultant) is then a vector that extends from the tail of the first vector to the tip of the last vector. If all the vectors are drawn to the same common scale, and their angles with respect to some set of axes, are correct, then the vector representing their sum can be scaled to obtain its magnitude. The angle that it makes with the chosen axes also gives its direction.

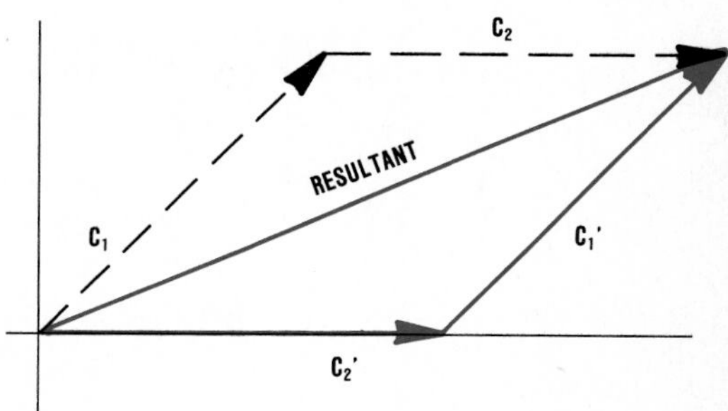

[11-T6] The parallelogram method of vector addition.

In vector addition, it does not matter which vector is used first in the problem-solving process. The same final vector representing the sum will be obtained regardless of the order in which you start. In Figure [11-T1], notice that the vector sum of $\mathbf{A}_1 + \mathbf{A}_2'$ is the same as the vector sum of $\mathbf{A}_2 + \mathbf{A}_1'$.

Figure [11-T6] shows that both sets of components C_1, C_2 and C_1', C_2' form a *parallelogram,* the resultant (diagonal) of which is the sum C_1, C_2 or the sum C_1', C_2'. The graphical technique is often referred to as the *parallelogram method.*

In the graphical technique, vector subtraction can be handled as a special case of vector addition. The vector difference $\mathbf{A} - \mathbf{B}$, is the same as the vector sum $\mathbf{A} + (-\mathbf{B})$. That is, subtracting vector **B** from vector **A** is the same as adding to vector **A** the vector $(-\mathbf{B})$. The vector $(-\mathbf{B})$ is just the vector **B** rotated by 180°.

In the graphical method, **A** is laid out to scale at the proper angle and the vector $(-\mathbf{B})$ is attached to it so that the tail of $(-\mathbf{B})$ is located at the tip of vector **A**. A vector drawn from the tail of **A** to the tip of $(-\mathbf{B})$ then represents the vector difference $\mathbf{A} - \mathbf{B}$ [11-T7].

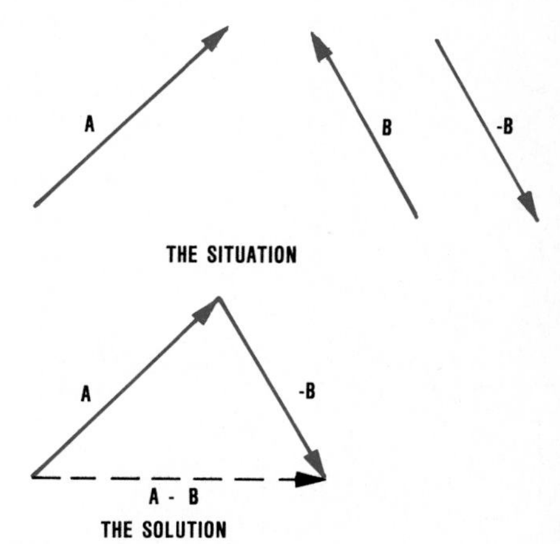

[11-T7] Subtracting vector **B** from vector **A**.

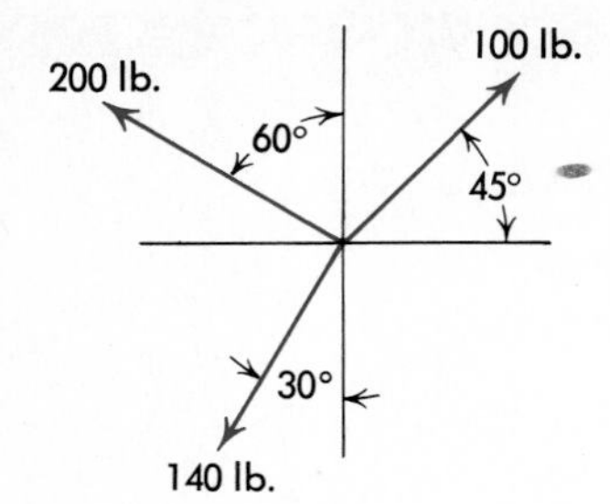

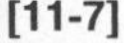
[11-7]

Example 11-2

Solve for the resultant, R of the three forces shown in Figure [11-7], using the method of rectangular components for the final resolution of the force system.

Solution:

We will demonstrate the use of a table for managing this problem.

Forces (lb_f)	***Horizontal component***	***Horizontal value (lb_f)***	***Vertical component***	***Vertical value (lb_f)***
100	100 cos 45° =	+70.7	100 sin 45° =	+70.7
200	−200 sin 60° =	−173.2	200 cos 60° =	+100
140	−140 sin 30° =	−70.0	−140 cos 30° =	−121.2
Total value	Positive	+70.7	Positive	+170.7
Total value	Negative	−243.2	Negative	−121.2
Sum	Horizontal	−172.5	Vertical	+49.5

The diagram for the solution is shown in Figure [11-8].

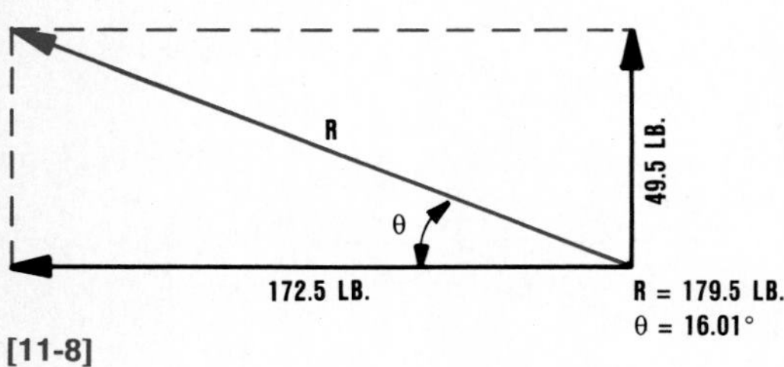

[11-8]

In mechanics we speak of the *line of action* of a force. This is simply the line in space which is collinear with the force being considered. The external effect of a force on a rigid body is independent of the point of application of the force along its line of action. Thus it would be immaterial whether you pushed or pulled an object as long as the push or the pull were applied along the same line in space. The effect on the object would be the same in either case. This principle is illustrated in Figure [11-9], where the application of two forces, one 18 lb_f and one 26 lb_f, are considered. All three possibilities yield the same effect on the object to which the forces are applied.

Problems

11-1. Find the resultant, in amount and direction, of the following concurrent coplanar force system: force A, 180 lb_f, acting S 60°W; and force B, 158 lb_f, acting S 80°W. Check your answer graphically, using a scale of 1 cm equals 50 lb_f.

11-2. Determine the amount and direction of the resultant of the concurrent coplanar forces system as follows: force A, 10 N, acting N 55°E; force B, 16 N, acting due east; force C, 12 N, acting S 22°W; force D, 15 N, acting due west; force E, 17 N, acting N 10°W.

11-3. Find the resultant and the angle the resultant makes with the vertical, using the following data: 10 lb_f, N 18°W; 5 lb_f, N 75°E; 3 lb_f, S 64°E; 7 lb_f, S 0°W; 10 lb_f, S 50°W.

11-4. a. In Figure [11-P4], using rectangular components, find the resultant of these four forces: A = 100 N, B = 130 N, C = 195 N, D = 138 N.

b. Find a resultant force that would replace forces A and B.

c. Separate force A into two components, one of which acts N 10°E and has a magnitude of 65 N. Give the magnitude and direction of the second component.

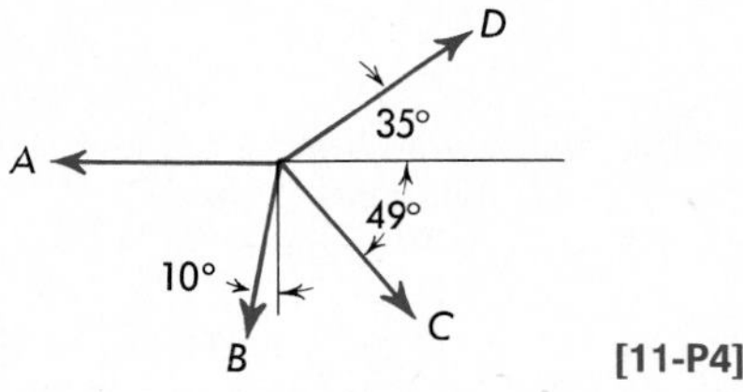

[11-P4]

11-5. A weight of 80 lb_f is suspended by two cords, the tension in AC being 70 lb_f and in BC being 25 lb_f, as shown in Figure [11-P5]. Find the angles α and θ.

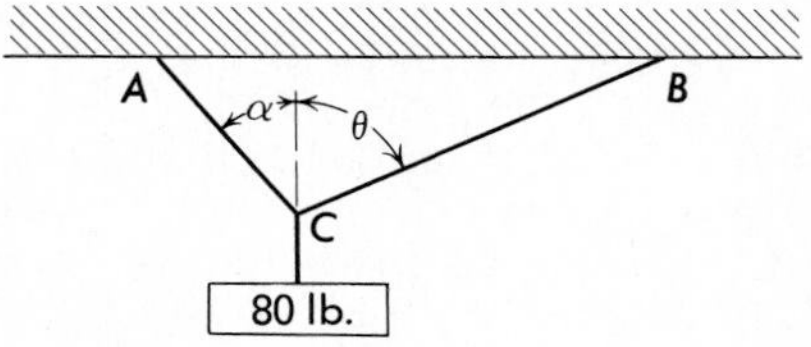

[11-P5]

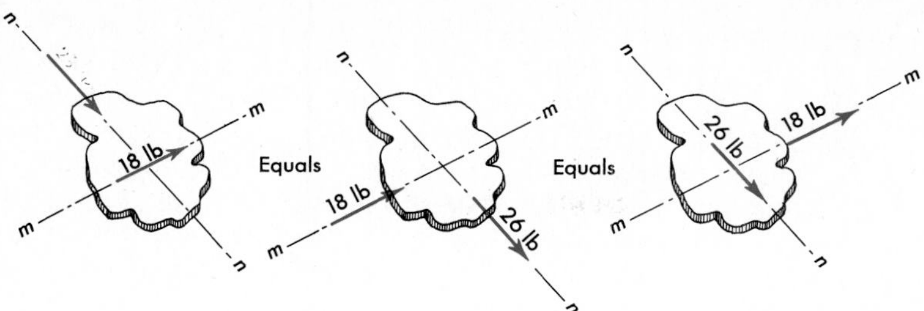

[11-9] Equivalent systems.

Moments. As stated previously, a moment is the product of a force and the length of a moment arm. Associated with the definition of the length of the moment arm is the point about which the moment acts. When speaking of moments, you must include the chosen point about which the moments are expressed. A given force will contribute different moments about different points on an object, since the length of the lever arm will be different for each point.

A given force will have a zero moment about any point along its line of action because the length of the lever arm will be zero for all such points. That is, the perpendicular distance between the point and the line of action is zero for any point along the line of action.

Give me a firm spot to stand on and a lever long enough, and I shall move the world.

Archimedes, ca. 250 B.C.

For nonzero moments, there are two conventional ways to convey information concerning the direction of a moment applied about a point on an object. One is to say the moments are either clockwise or counterclockwise. This works fine in problems involving coplanar forces, the type discussed in this chapter. In this situation, you need to view the plane containing the forces either from above or below and assign one direction (either clockwise or counterclockwise) as the positive direction and the other as the negative direction [11-10].

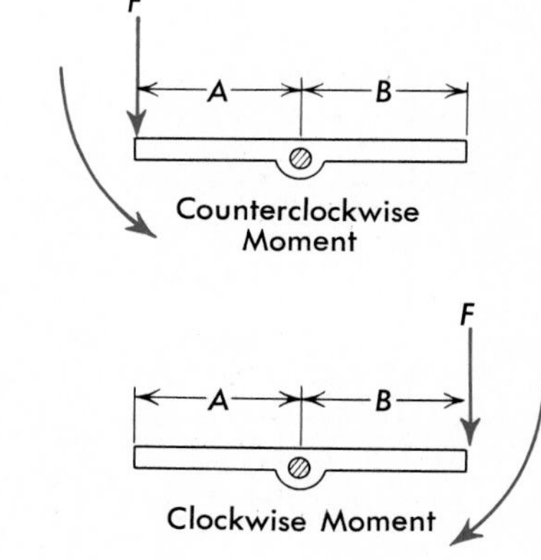

[11-10]

Since we will nearly always be concerned with both the forces and the moments they produce, we need to label all the assumed positive and negative directions that will be used in the solution of a problem. Figure [11-11] displays a convenient way of recording your choices for these assumed positive directions. This diagram immediately tells the author of the solution and the reader of the solution what directions have been assumed for positive forces and moments in the analysis that is being done. For each problem analysis, a small diagram showing this choice should be placed on the calculation sheet adjacent to the problem sketch. The diagram shown in Figure [11-11] will serve as a basis for problem analyses in this text.

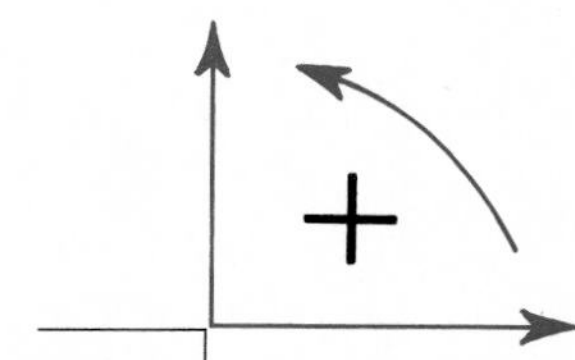

[11-11] Sign convention.

The other method for conveying information concerning the direction of moments makes use of vectors, not just directions such as clockwise or counterclockwise. By convention this is usually done by what is called the "right-hand rule." In this method, you extend the thumb of your right hand and form your hand into a loose fist. The thumb should be extended so as to be perpendicular to planes containing your curved fingers. When the planes containing your curved fingers are parallel with the plane containing both the force and the lever arm that produce a moment, then the right thumb extends in the direction of the vector representing the moment [11-12]. The length of the vector is proportional to the product of the

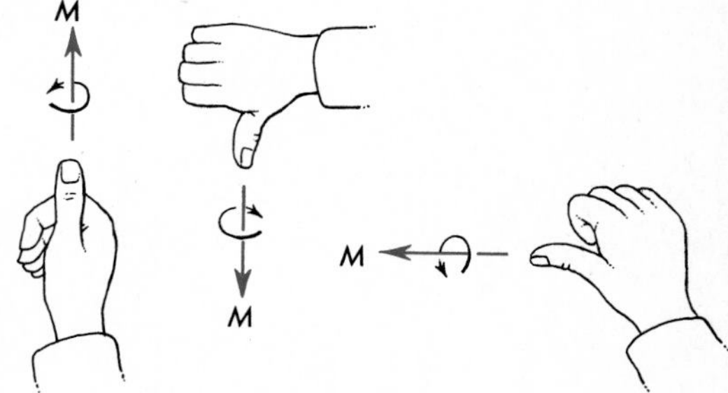

[11-12] Applying the right-hand rule, the thumb points in the direction of the vector representing the moment.

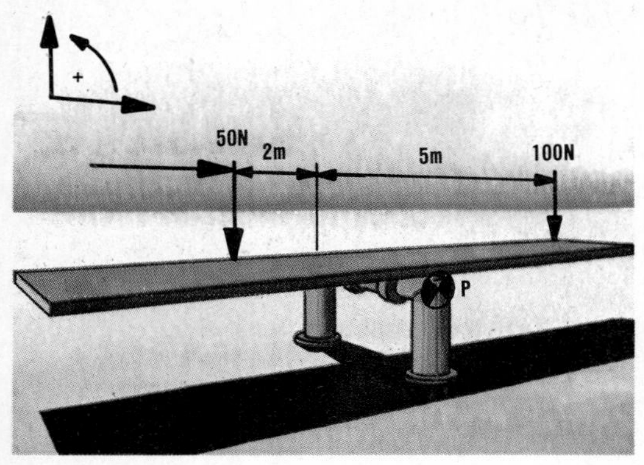

[11-13]

magnitude of the force and the length of the lever arm. This method is much more general than what we have chosen to use in this book. You will learn more about it in your future mechanics courses.

The moments about some common point created by any number of coplanar forces can all be treated as common components (really, components in the direction of your thumb using the right-hand rule). The resultant moment about some point is the sum of all the component moments about the same point. We demonstrate this in Example 11-3.

Example 11-3

Solve for the moments in Figure [11-13] that tend to cause turning of the beam about point P.

Solution:

Counterclockwise moment	=	(50 N) (2 m)	=	+100 N · m
Clockwise moment	=	−(100 N) (5 m)	=	−500 N · m
Resultant moment	=			−400 N · m

That is, the magnitude of the resultant is 400 N · m and the direction, determined from the sign, is clockwise.

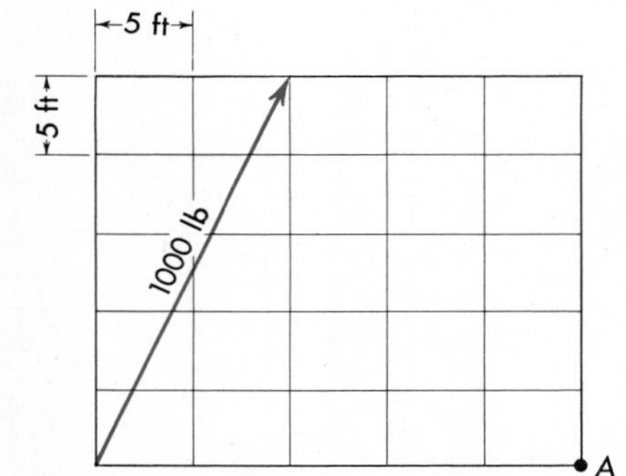

[11-14]

Moments are typically considered to have derived or secondary dimensions, rather than primary dimensions. Since a moment is the product of a force and a distance, its dimensions will be force × length or FL. By convention, moment units are usually expressed with the force unit being shown first, as N · m, lb_f · ft, lb_f · in., kip · ft (a kip is 1000 lb_f), and so on. This is done because *work* and *energy* also involve the product of distance and force (we shall consider work and energy shortly), and the units of length are usually placed before those of force (except for the N · m in the SI units), as in ft · lb_f and in. · lb_f.

The *moment of a force* about some given center is identical to the sum of the *moments of the components of that force* about the same center. This principle is commonly called *Varignon's theorem*. In problem analysis it is sometimes more convenient to solve for the sum of the moments of the components of a force rather than the moment of the force itself. However, the problem solutions will be identical.

Example 11-4

Solve for the total moment of the 1000-lb_f force about point A in Figure [11-14].

Solution A:

Moment of a force as shown in Figure [11-15]:

$$\theta = \arctan \frac{25}{10} = 68.2°$$

$$\text{Moment arm} = 25 \sin 68.2°$$

$$\text{Total moment} = (1000)\,(25 \sin 68.2°)$$

$$= 23{,}200 \text{ lb}_f \cdot \text{ft}$$

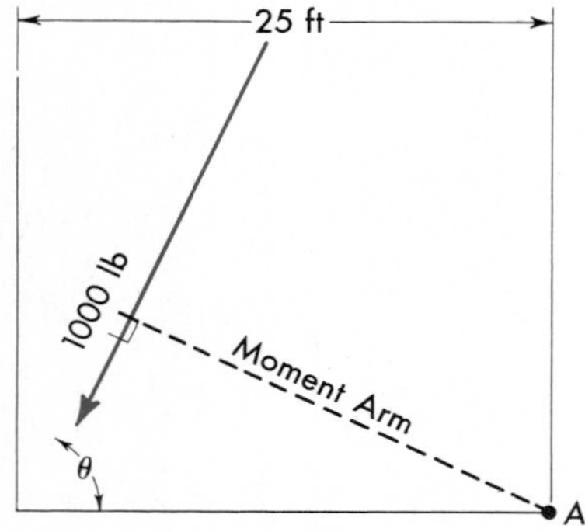

[11-15]

Solution B:

Moments of components of a force as shown in Figure [11-16]:

Vertical force component = 1000 sin 68.2°

Moment arm = 25 ft

Horizontal force component = 1000 cos 68.2°

Moment arm = 0

(Note that the horizontal component passes through the center *A*.)

Total moment = (1000 sin 68.2°) (25)

= 23,200 $lb_f \cdot ft$

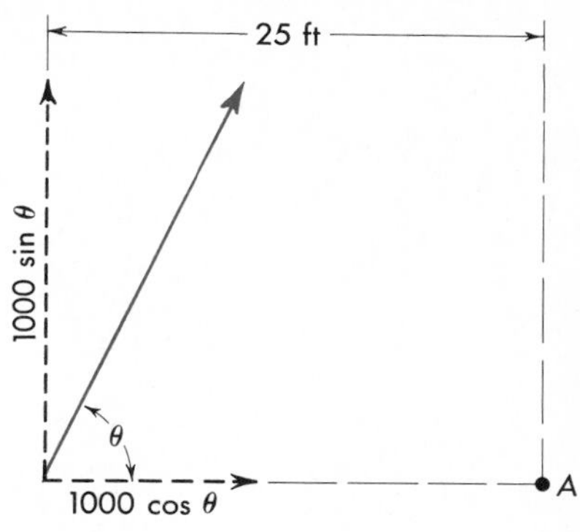

[11-16]

Free-Body Diagrams

The principles of statics are most easily put into practice by using *free-body diagrams*. Here we shall consider in more detail how to draw free-body diagrams and how to use them.

A free-body diagram is essentially a schematic drawing of the object on which you choose to direct your attention. All other objects in the universe other than the one you have chosen are removed from the free-body diagram and their actions are replaced by forces that give the same effect. Only external forces—forces that are applied to the chosen object by objects in the outside world[1]—are shown on the diagram. Forces that are internal to the subject—forces that are applied by one part of the chosen object on another part of the chosen object—are never shown on the free-body diagram.

Free-body diagrams allow you to concentrate on your chosen object without worrying about or being confused by all the other clutter in the universe. Their use should be part of your problem-solving methodology.

As an example, let us consider the free-body diagram of a ship moving in the water [11-17]. It is not necessary that the diagram of the ship be drawn to scale since the free-body diagram is only an imaginary concept.

[11-17] Ship moving through the ocean.

There are four external forces acting on the model: a forward thrust applied by the ocean at the ship's propeller; a friction drag, also applied by the ocean on the hull, that acts to retard the ship's motion; a buoyant force, also applied by the ocean to the hull, that keeps the ship afloat; and the ship's weight, which is due to the gravitational attraction of the earth and which may be considered as acting through the center of gravity of the ship [11-18].

The symbol ⊗ is used to denote the location of the center of gravity. Since free-body diagrams involve forces and moments, a coordinate system showing the assumed positive directions is very useful for purposes of orientation. The free-body diagram shown in Figure [11-18] would make possible an analysis of the relationships between the weight and buoyant force and between the thrust and

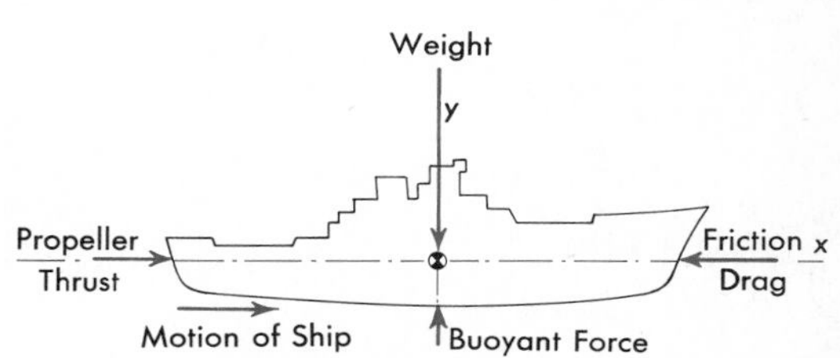

[11-18] Free-body diagram of the ship.

[1]These forces include both those that are applied by contact (such as forces applied by supports or cables that you do not wish to consider to be a part of your free-body diagram) and those forces applied without physical contact (such as the gravitational attraction or weight).

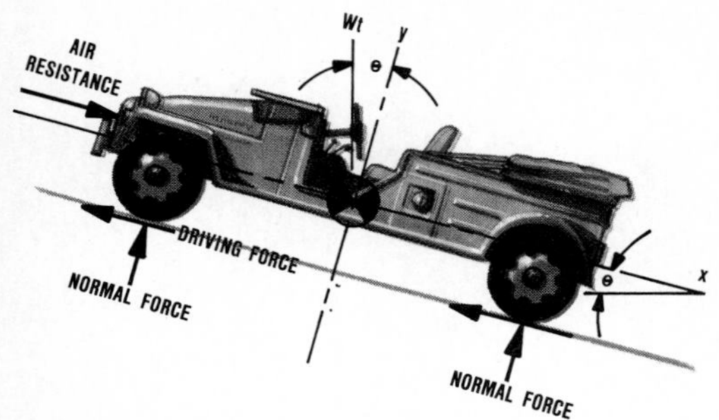

[11-19] Free-body diagram of a four-wheel-drive automobile moving up an incline.

drag. However, it would not, for example, be useful for determining the loads on the ship's engine mounts, for these are all internal to this free-body diagram. Another model (free-body diagram) of the engine alone would be required for this purpose.

Another example [11-19], shows the free-body diagram of a four-wheel-drive automobile driving up an incline. It is generally customary to show the object in a free-body diagram, in its true orientation with respect to common frames of reference. For example, here it would be awkward and confusing to draw the automobile in a horizontal position when it is actually going uphill.

General Suggestions for Drawing Free-Body Diagrams. To aid you in learning to draw free-body diagrams, we make the following suggestions:

1. *Free bodies.* Be certain that the body is *free* of all surrounding objects. Draw the body so that it is *free*. Do not show a supporting surface but rather show only the force vector which replaces that surface. Do not rotate the body from its original position, but rather, rotate the axes if necessary. Show all forces and label them. Show all pertinent dimensions and angles that you will need, but do not clutter the diagram with unnecessary features.
2. *Force components.* Forces are often best shown in their component forms. When replacing a force by its components, select the most convenient directions for the components. Never show both a force and its components by solid-line vectors; use broken-line vectors for one or the other, since the force *and* its components do not occur simultaneously.
3. *Weight vectors.* When the weight of the object is a significant force in the problem, always show it as a vertical vector with its tail or tip at the center of gravity. Place it so that it interferes least with the remainder of the drawing. It should always be drawn pointing vertically down.
4. *Direction of vectors.* The free-body diagram should represent the facts as nearly as possible. Always establish the line of action of each force. This will aid you in establishing moments for all forces, if you need to. Force vectors on free-body diagrams are not usually drawn to scale but may be drawn proportionate to their respective magnitudes.
5. *The beginning free-body diagram.* The first free-body examined in the solution of any problem should be one that involves the unknown quantity for which you wish to solve. However, many problems cannot be solved with just this single diagram. You may need to consider other free-body diagrams on your path to a solution.
6. *Two-force members.* When a two-force member is in equilibrium (we will discuss equilibrium in the next section), the forces are equal in magnitude, opposite in direction, and collinear. If the member is in compression (if it is being squashed or being compressed), the force vectors should point toward each other; if a member is in tension (if it is being pulled apart or stretched), the force vectors should point away from each other.

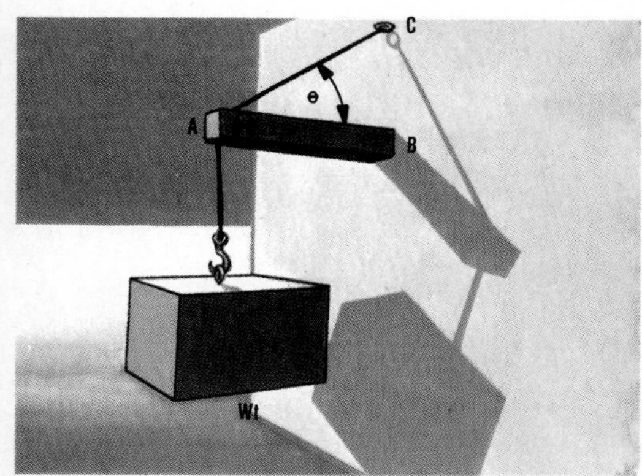

[11-20] The situation.

7. *Three-force members.* When a member is in equilibrium (we discuss equilibrium in the next section) and has only three forces acting on it, the three forces are always concurrent; that is, they go through the same point. In analyzing a problem involving a three-force member, one should recall that any set of concurrent forces may be replaced by a resultant force. Hence, if a member in equilibrium has forces acting at three points, it is a three-force member regardless of the fact that the force applied at one or more points may be replaced by two or more components.
8. *Concurrent force systems.* For a concurrent force system the size, shape, and dimensions of the body can be neglected, and the body can be considered to be a particle.

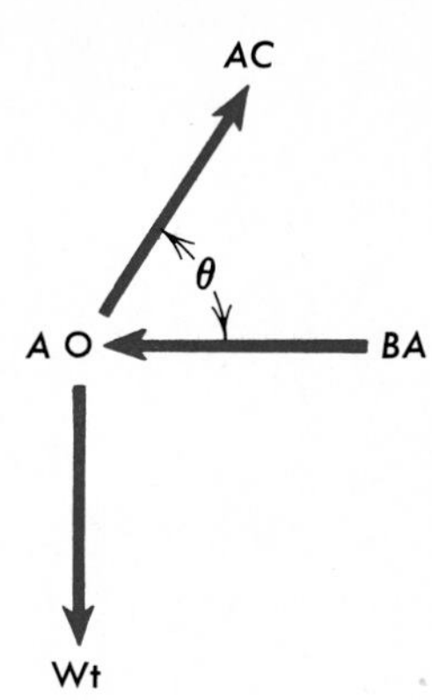

[11-21] Free-body diagram of point *A*.

Example 11-5

Draw a free-body of point *A*, shown in Figure [11-20].

Solution:

See Figure [11-21].

Table 11-1 provides examples of free-body diagrams for several common types of situations.

Problems

11-6. Solve for the algebraic sum of the moments in newton · meters about *A* when *h* is 20 ft as shown in Figure [11-P6].

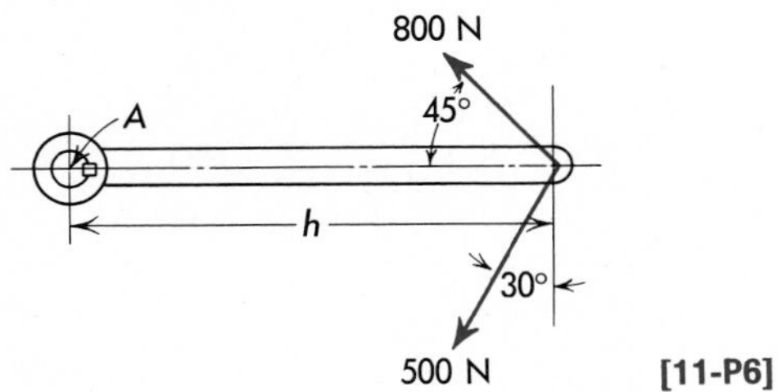

[11-P6]

11-7. a. Total the clockwise moments about *A*, *B*, *C*, *D*, and *E* in Figure [11-P7].
b. Total the counterclockwise moments about *A*, *B*, *C*, *D*, and *E*.
c. Solve for the algebraic sum of the moments about *A*, *B*, *C*, *D*, and *E*.

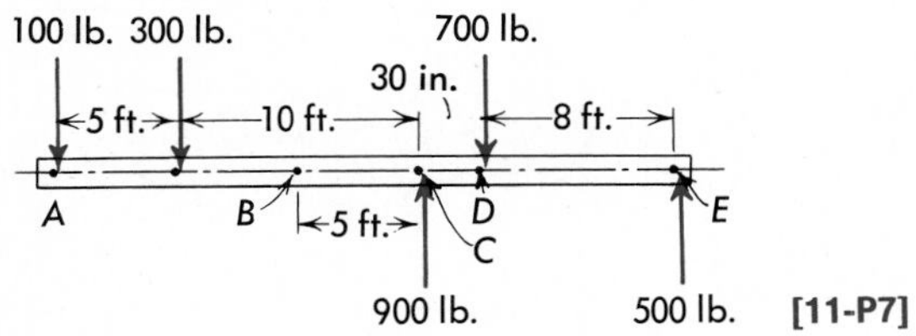

[11-P7]

11-8. Find the summation of the moments of the forces shown about *A* in Figure [11-P8]. Find the moment sum about *D*.

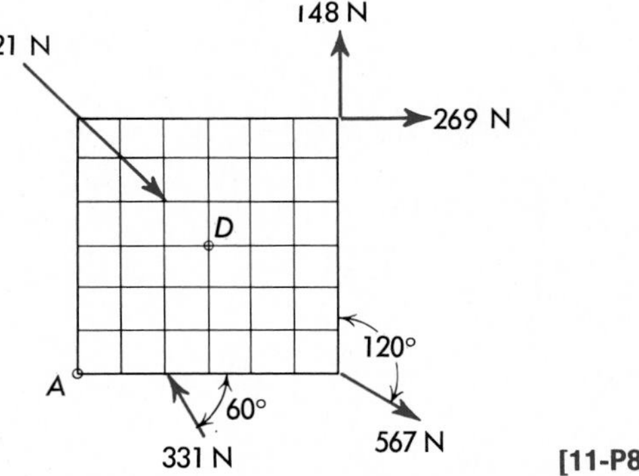

[11-P8]

11-9. What pull *P* is required on the handle of a claw hammer to exert a vertical force of 750 N on a nail? Dimensions in inches are shown in Figure [11-P9].

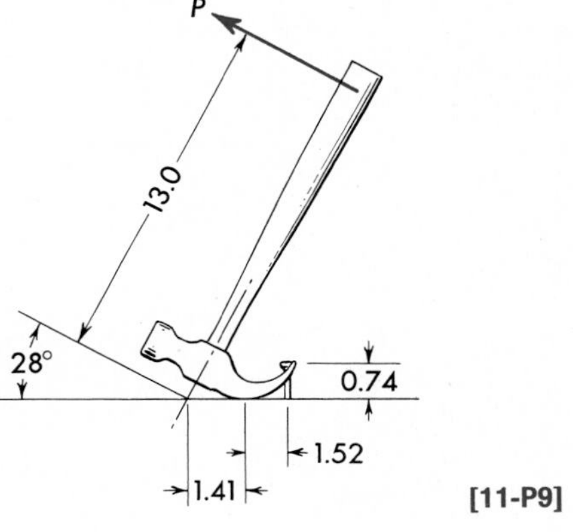

[11-P9]

TABLE 11-1

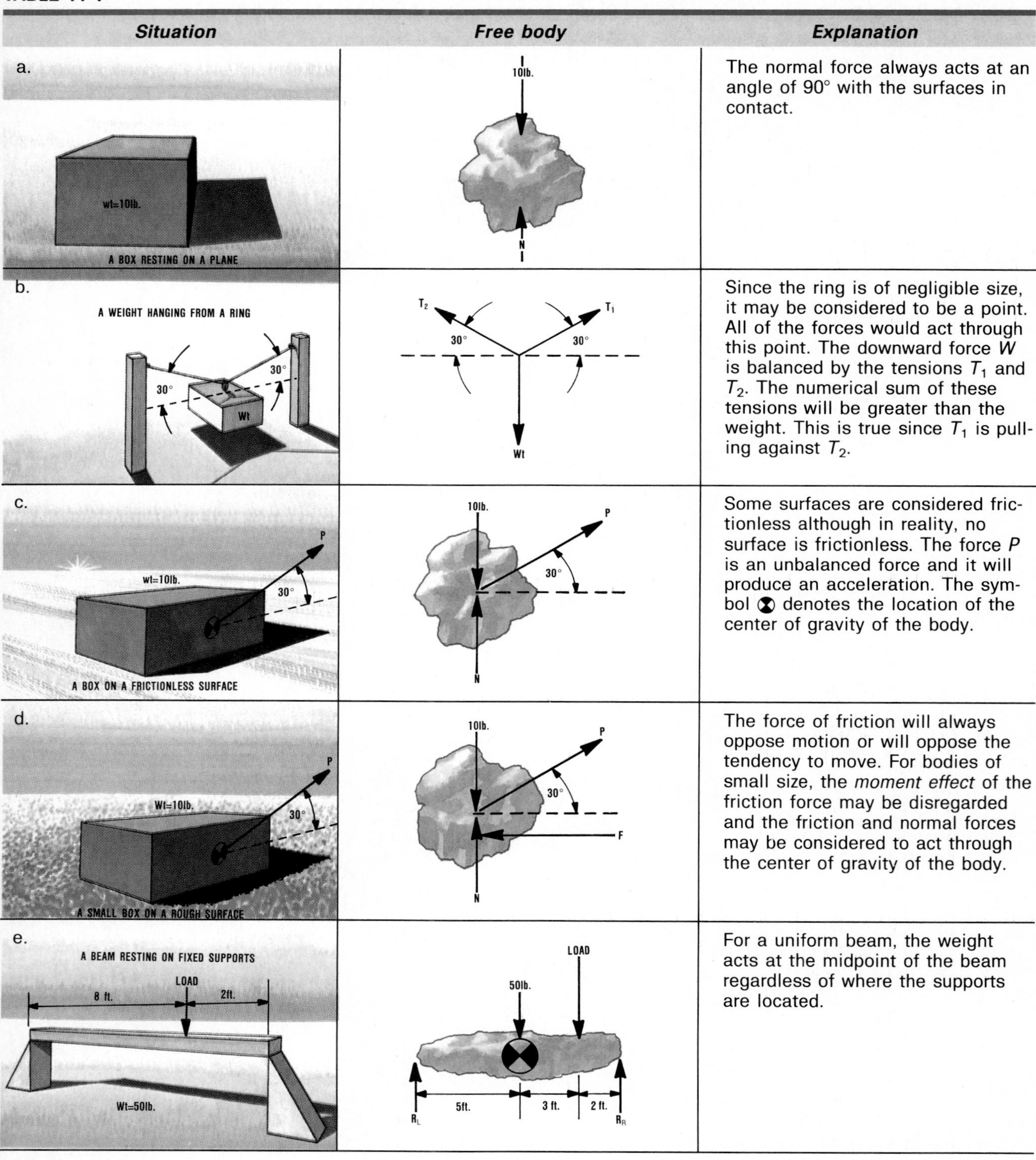

Situation	Free body	Explanation
a. wt=10lb. A BOX RESTING ON A PLANE	10lb. N	The normal force always acts at an angle of 90° with the surfaces in contact.
b. A WEIGHT HANGING FROM A RING 30° 30° Wt	T_2 T_1 30° 30° Wt	Since the ring is of negligible size, it may be considered to be a point. All of the forces would act through this point. The downward force *W* is balanced by the tensions T_1 and T_2. The numerical sum of these tensions will be greater than the weight. This is true since T_1 is pulling against T_2.
c. P wt=10lb. 30° A BOX ON A FRICTIONLESS SURFACE	10lb. P 30° N	Some surfaces are considered frictionless although in reality, no surface is frictionless. The force *P* is an unbalanced force and it will produce an acceleration. The symbol ⊗ denotes the location of the center of gravity of the body.
d. P Wt=10lb. 30° A SMALL BOX ON A ROUGH SURFACE	10lb. P 30° F N	The force of friction will always oppose motion or will oppose the tendency to move. For bodies of small size, the *moment effect* of the friction force may be disregarded and the friction and normal forces may be considered to act through the center of gravity of the body.
e. A BEAM RESTING ON FIXED SUPPORTS LOAD 8 ft. 2ft. Wt=50lb.	LOAD 50lb. 5ft. 3 ft. 2 ft. R_L R_R	For a uniform beam, the weight acts at the midpoint of the beam regardless of where the supports are located.

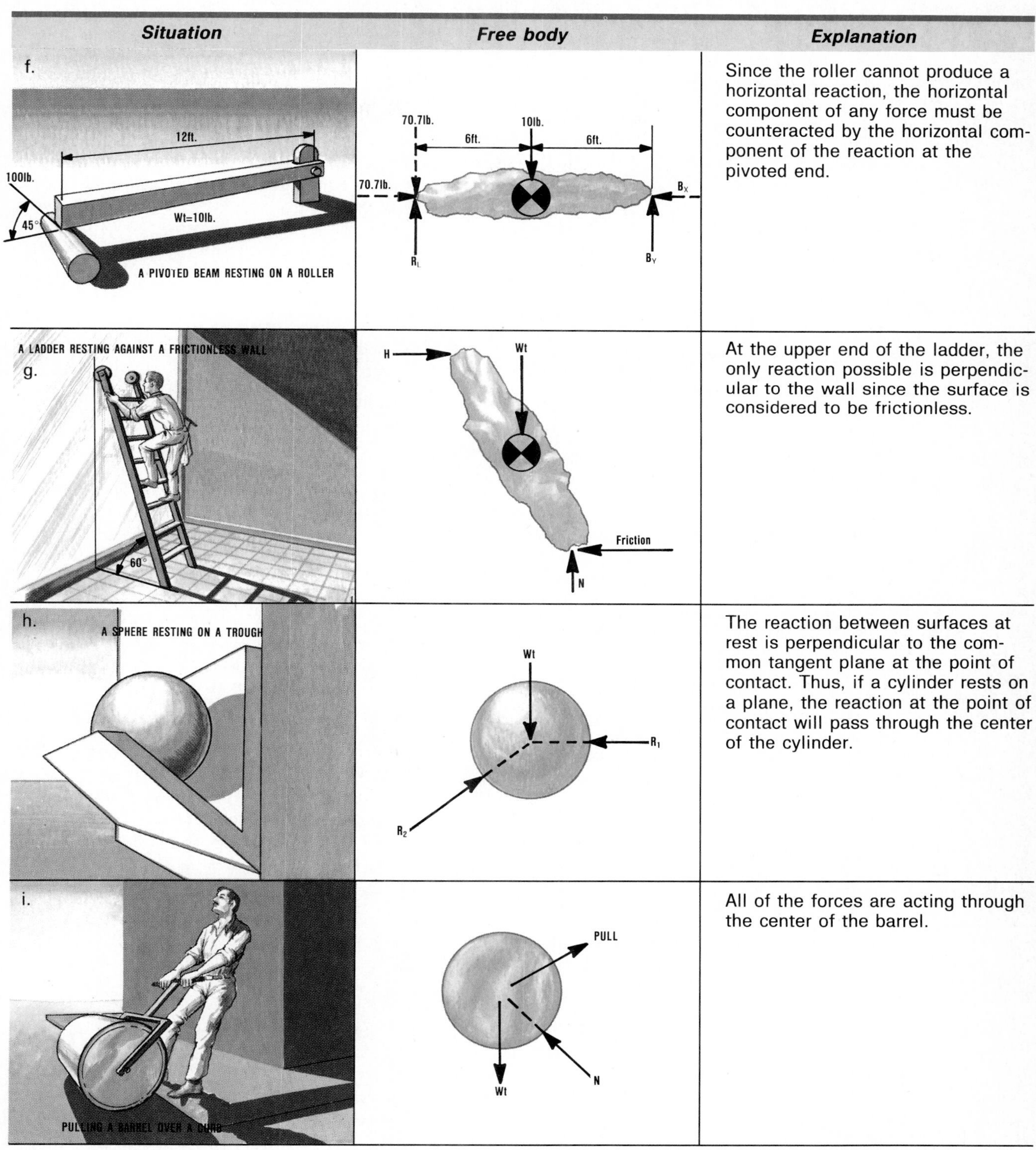

f. Since the roller cannot produce a horizontal reaction, the horizontal component of any force must be counteracted by the horizontal component of the reaction at the pivoted end.

g. At the upper end of the ladder, the only reaction possible is perpendicular to the wall since the surface is considered to be frictionless.

h. The reaction between surfaces at rest is perpendicular to the common tangent plane at the point of contact. Thus, if a cylinder rests on a plane, the reaction at the point of contact will pass through the center of the cylinder.

i. All of the forces are acting through the center of the barrel.

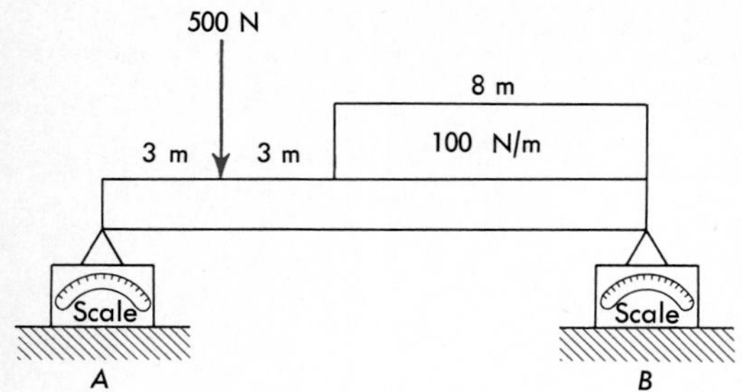

[11-22] The situation.

Equilibrium

The term *equilibrium* is used to describe the condition of any body when there is a balance of forces and moments such that the body has no tendency to change its position (or velocity if it is in motion). One requirement for equilibrium is that the resultant of all forces acting on the body equals zero. For example, the forces acting upward on a body in equilibrium must be balanced by other forces acting downward on the body. Also, the forces acting horizontally to the right are counteracted by equal forces acting horizontally to the left. Since no unbalance in moment or turning effect can be present when a body is in equilibrium, the sum of the moments of all forces acting on the body must also be zero. The moment center may be located at any convenient place on the body or at any place in space. We may sum up these conditions of equilibrium by the following equations:[2]

$\Sigma F_x = 0$ (the sum of all horizontal forces acting on the body equals zero)

$\Sigma F_y = 0$ (the sum of all vertical forces acting on the body equals zero)

$\Sigma M_0 = 0$ (the sum of the moments of all forces about any common point equals zero)

These equilibrium equations may be used to good advantage in working problems involving beams, trusses, and levers—indeed, any object that is either remaining motionless or is traveling at a constant velocity.

Example 11-6

A beam of negligible weight is supported at each end by a knife-edge. The beam carries a concentrated load of 500 N and one uniformly distributed load weighing 100 N/m, as shown in Figure [11-22]. Determine the scale readings under the knife-edges.

Solution:

The uniformly distributed load is equivalent to a resultant of $8 \text{ m} \times 100 \text{ N/m} = 800 \text{ N}$ acting at the center of gravity of the uniform-load diagram. Therefore, for determining support forces the entire distributed load can be replaced by a concentrated load of 800 N acting at a distance of 10 m from the left end, as shown in Figure [11-23].

Methodology:

1. Draw a free-body diagram of the beam.
2. Since there are no horizontal forces acting on the free body, $\Sigma F_x = 0$ is satisfied.
3. From $\Sigma F_y = 0$, we know that

$$A + B - 500 \text{ N} - 800 \text{ N} = 0$$

$$A + B = 1300 \text{ N}$$

[11-23] Free-body diagram for the beam in Figure [11-22].

[2] These equations are applicable for two-dimensional problems—problems involving coplanar force systems. ΣF_x is read as "the summation of all forces in the x direction."

4. From $\Sigma M_0 = 0$, we know that the moments about any point must equal zero. Let us take moments about point *A*.

$$\Sigma M_A = 0 \circlearrowright$$

$$(B)\ (14\text{ m}) - (500\text{ N})\ (3\text{ m}) - (800\text{ N})\ (10\text{ m}) = 0$$

$$B = \frac{1500\text{ N}\cdot\text{m} + 8000\text{ N}\cdot\text{m}}{14\text{ m}}$$

$$= \frac{9500\text{ N}\cdot\text{m}}{14\text{ m}}$$

$$= 679\text{ N} \uparrow$$

5. From the third step we saw that $A + B = 1300$ N. We can now subtract *B* from 1300 and obtain

$$A = 1300\text{ N} - 679\text{ N} = 621\text{ N} \uparrow$$

Note: The same answer for *A* could have been obtained by taking moments about *B* as a moment center.

In this book, problems involving trusses, cranes, linkages, bridges, and so on, should be considered to be *pin-connected,* which means that the member is free to rotate about the joint. That is, there will be no moments transmitted at this type of joint. Also, for simplicity, members are usually considered to be weightless.

By examining each member of the structure separately, forces transmitted at joints between members may be obtained by the conditions of equilibrium applied to the separate members. We demonstrate this in the following example.

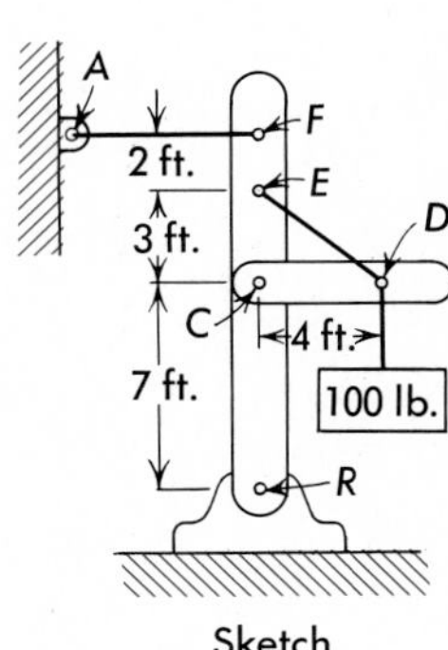

[11-24] The situation.

Example 11-7

Solve for the tensions in cables *AF* and *ED* and for the reactions at *C* and *R* in Figure [11-24].

Solution:

Equilibrium equations

$$\Sigma F_x = 0$$

$$\Sigma F_y = 0$$

$$\Sigma M_0 = 0$$

1. If we take moments about point *R* in free body 1 in Figure [11-25], we can obtain an equation that contains only one unknown, the tension in cable *AF*.

$$\Sigma M_R = 0$$

$$(12\text{ ft})\ (FA) - (100\text{ lb}_f)\ (4\text{ ft}) = 0$$

$$FA = \frac{400\text{ lb}_f\cdot\text{ft}}{12\text{ ft}} = 33.3\text{ lb}_f\ (\leftarrow)$$

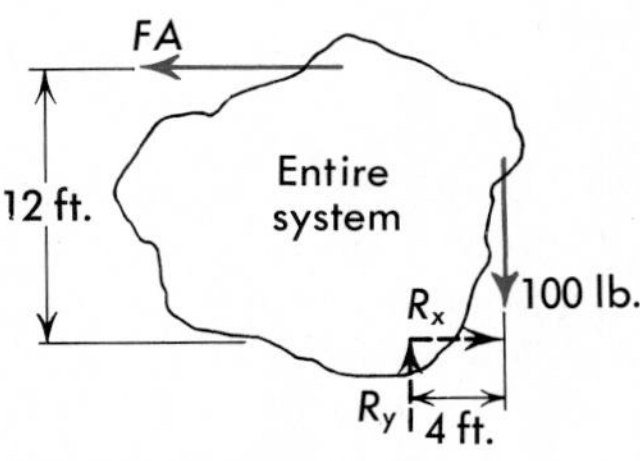

[11-25] Free body 1.

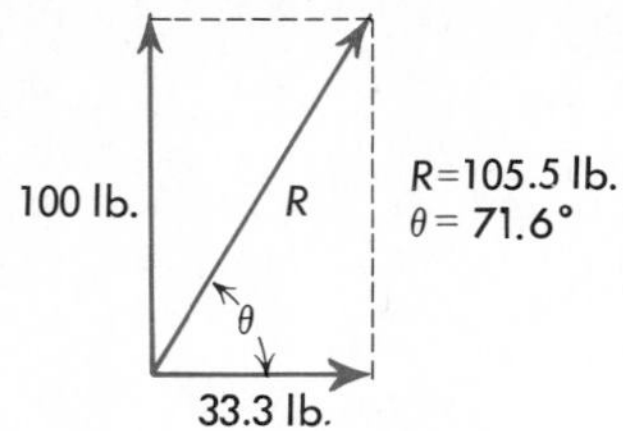

[11-26]

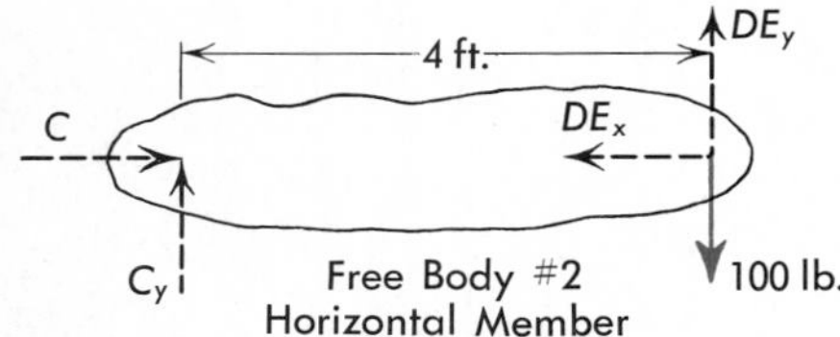

[11-27] Free body 2 horizontal member.

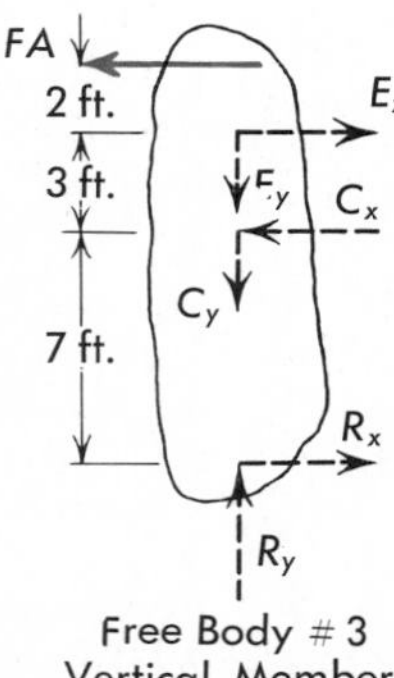

[11-28] Free body 3 vertical member.

2. R_x can be found from

$$\Sigma F_x = 0$$
$$R_x - FA = 0$$
$$R_x = FA = 33.3 \text{ lb}_f (\rightarrow)$$

3. Consider $\Sigma F_y = 0$ on Figure [11-25], to find R_y.

$$\Sigma F_y = 0$$
$$R_y - 100 = 0$$
$$R_y = 100 \text{ lb}_f \uparrow$$

Thus the resultant R is as given in Figure [11-26].

4. Now take moments about point C in free body 2 [11-27].

$$\Sigma M_c = DE_y \ (4) - 100\ (4) = 0$$
$$DE_y = 100 \text{ lb}_f (\leftarrow)$$

Then, since $DE_y = DE \cdot \frac{3}{5}$ (see [11-24]),

$$DE = \frac{100 \text{ lb}_f}{0.6} = 167 \text{ lb}_f$$

Also, from free body 2:

$$\Sigma F_y = 0$$
$$C_y = 100 \text{ lb}_f - 100 \text{ lb}_f$$
$$= 0$$

Further, from free body 2:

$$\Sigma F_x = 0$$
$$C_x = DE_x = 167 \cdot \frac{4}{5}$$
$$= 133 \text{ lb}_f \rightarrow$$

where the $\frac{4}{5}$ factor is from the geometry of the cable ED in Figure [11-24].

5. Free body 3 [11-28] can be used to double check the solution. Remember that in two-force members, such as cable DE, the reactions at each end will be equal in magnitude but opposite in direction; that is, E_x and E_y are equal to DE_x and DE_y, respectively.

Problems

11-10. A horizontal beam 20 m long weighs 1500 N. It is supported at the left end and 4 m from the right end. It has the following concentrated loads: at the left end, 2000 N; 8 m from the left end, 3000 N; at the right end, 4000 N. Calculate the reactions at the supports.

11-11. A 12-ft beam that weighs 10 lb_f per foot is resting horizontally. The left end of the beam is pinned to a vertical wall. The right end of the beam is supported by a cable that is attached to the vertical wall 6 ft above the left end of the beam. There is a 200-lb_f concentrated load acting vertically downward 3 ft from the right end of the beam. Determine the tension in the cable and the amount and direction of the reaction at the left end of the beam.

11-12. A horizontal rod 8 m long and weighing 120 N has a weight of 150 N hung from the right end and a weight of 40 N hung from the left end. Where should a single support be located so that the rod will balance?

11-13. An iron beam 12.7 ft long weighing 855 lb_f has a load of 229 lb_f at the right end. A support is located 7.2 ft from the load end.

a. How much force is required at the opposite end to balance it?

b. Disregarding the balancing force, calculate the reactions on the supports if one support is located 7 ft from the left end and the other support is located 4 ft from the right end.

11-14. A 2-ft-diameter sphere weighing 56 lb_f is suspended by a cable and rests against a vertical wall. If the cable *AB* is 2 ft long:

a. Calculate the angle the cable will make with the smooth wall.

b. Solve for the tension in the cable and the reaction at *C* in Figure [11-P14].

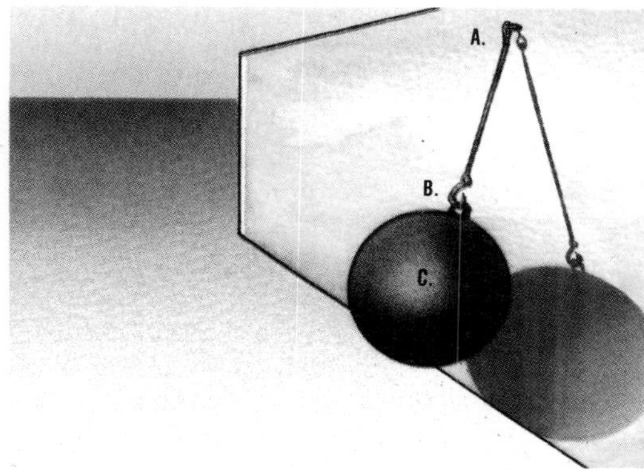

[11-P14]

11-15. What horizontal pull *P* will be necessary just to start the 1400 lb_f wheel over the 4-in. block in Figure [11-P15].

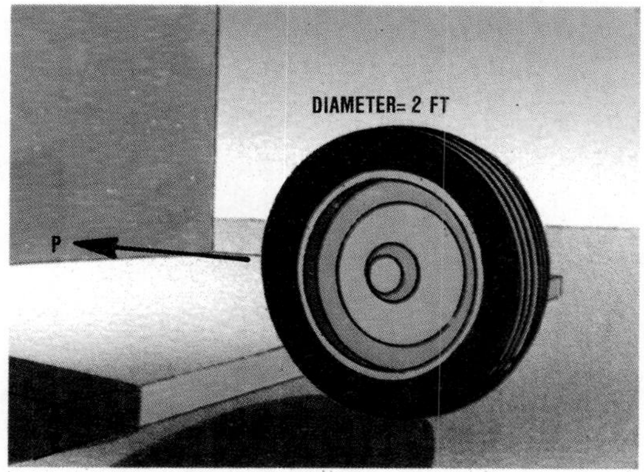

[11-P15]

11-16. A sphere weighing 206 N is placed in a smooth trough as shown in Figure [11-P16]. Find the two supporting forces.

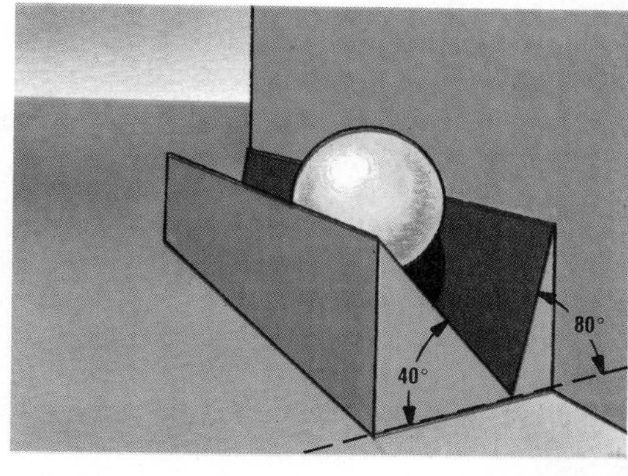

[11-P16]

11-17. Find the tension in *AB* and the compression in *BC* in Figure [11-P17].

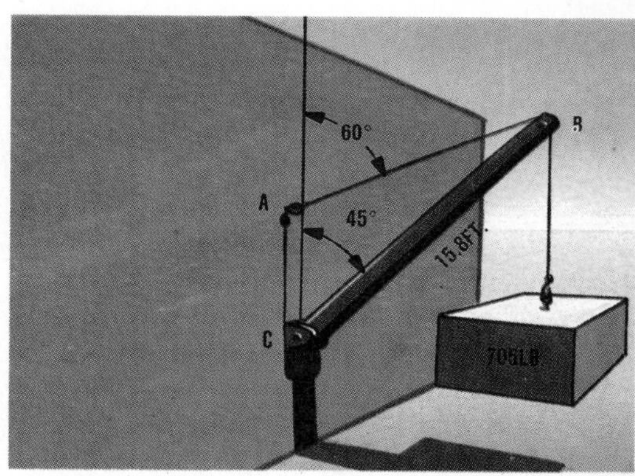

[11-P17]

11-18. A weight of 13,550 N is supported by two ropes making angles of 30° and 45° on opposite sides of the vertical. What is the tension in each rope?

11-19. (a) What is the tension in *BC* in Figure [11-P19]? (b) What is the amount and direction of the reaction at *A*?

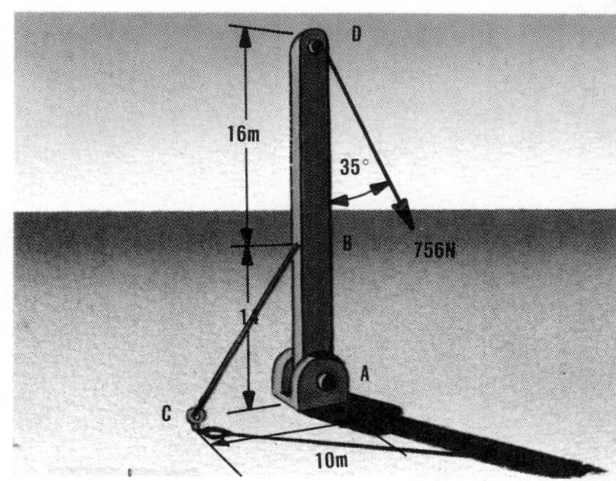

[11-P19]

Bodies in Uniform Linear Motion

Today, understanding motion and moving things is one of our most important tasks. In recent years even governments have become vitally interested in all types of motion—from the motion of atoms to that of satellites and celestial bodies. Moving people, animals, commuter trains, sleek automobiles, and jet airplanes are all a routine part of our daily life.

Motion exists when there is a *change of position* of an object with reference to some other object or point in space. For example, passengers in a jet airliner may be sitting still in the opinion of other passengers, but they would be in rapid motion relative to a farmer plowing in the field below. The airliner may be in *uniform motion* if it has balanced forces acting on it and is traveling at a constant velocity with respect to some truly stationary frame of reference. However, if the forces of jet thrust, air resistance, and gravity do not balance each other, the plane will be increasing or decreasing its velocity—it will be accelerating or decelerating.

Sometimes we speak of the motion of a body as *speed,* which refers to its rate of motion. The scientific term *velocity,* which refers to *rate of motion in a given direction,* is sometimes used incorrectly as a synonym for speed. Speed is the term used to designate the magnitude of velocity. Thus speed equals distance divided by time; its dimensions are L/t. Example units are m/s, ft/s, mi/hr, cm/s, yd/hr.

Velocity equals distance divided by time—expressed *in a given direction.* The dimensions are the same as those of speed. Example units include km/hr west, mi/hr north, and ft/s 30° east of north.

The Speed–Time Diagram

The study of motion becomes rather complex if we were to proceed directly to a full mathematical description. A simpler understanding can result if we can picture or visualize exactly what is taking place. For this reason extensive use will be made of the *speed–time* diagram as a means of pictorially representing the motions described. In addition, this treatment reduces the amount of memory work normally associated with the various relations.

In motion problems the total distance traveled is represented by the area that lies under the travel line of a speed–time diagram. For example, if an automobile travels at a uniform speed of 30 mi/hr for 30 minutes, it will cover a distance of 15 miles.

$$30\ \frac{\text{mi}}{\text{hr}} \times 30\ \text{min} \times \frac{1\ \text{hr}}{60\ \text{min}} = 15\ \text{mi}$$

The speed–time diagram for this situation is shown graphically as in Figure [11-29].

If the speed is constant, the distance traveled may be found by multiplying the ordinate value times the change in the abscissa value. In this case, the acceleration is zero, as indicated by the horizontal, straight line A-B.

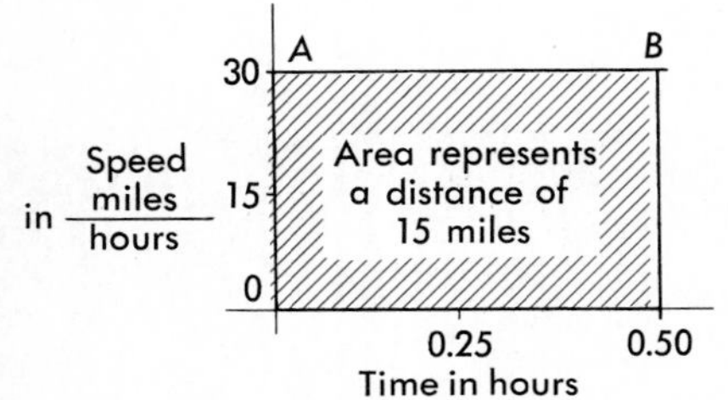

[11-29] Speed–time diagram.

Therefore, to work the problem above, the student need only draw the speed–time diagram and then find the area under the line *A-B* by simple mathematics.

Speed–time diagram principles may be summarized as follows:

1. The ordinate of the line at any instant will give the speed (V) at that instant.
2. Abscissa values give the time (t) consumed during travel.
3. The area under the travel line of the speed–time diagram gives the distance (S) traveled during the time interval under consideration.
4. The slope of the line at any point gives the acceleration (a) of the body at that point.

Slope may be defined as the *steepness* of a line and can be calculated by dividing the vertical rise by the corresponding horizontal distance.

Example 11-8

An automobile accelerates uniformly from a speed V_1 to a speed V_2 in time t [11-30].

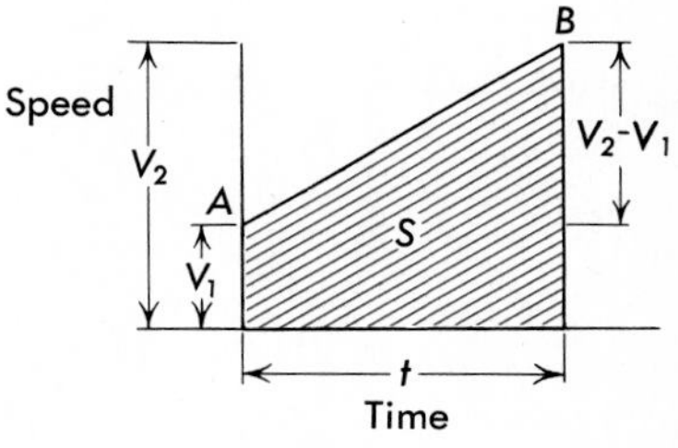

[11-30] Speed–time diagram.

Solution:

A speed–time diagram of the problem is drawn. The total distance traveled during time t can be calculated by solving for the area under the line *A-B*. This area is a trapezoid, and by simple geometry,

$$\text{Area} = \tfrac{1}{2}h(b_1 + b_2) \qquad \text{(see Appendix I)}$$

or

$$S = \tfrac{1}{2}t(V_1 + V_2) \qquad \text{(from speed–time diagram)}$$

The acceleration has been defined as the slope of the travel line. An examination of Figure [11-30] shows that this is also the change in speed ($V_2 - V_1$) divided by the elapsed time (t) that it took to make the change.

Stated algebraically, we have

$$a = \frac{V_2 - V_1}{t} \qquad \text{(slope of travel line)}$$

In some instances the term *average velocity* or *average speed* is used. Average speed is not generally an average of the initial and final speeds, although it may fortuitously turn out that way occasionally. Average speed may be expressed as

$$\text{Average speed} = \frac{\text{total distance traveled}}{\text{elapsed time during travel}}$$

Example 11-9

An automobile traveled a total distance of 100 mi at an average speed of 50 mi/hr. During the first 50 mi, the average speed of the automobile was 60 mi/hr. What was the average speed for the last 50 mi?

Solution:

$$\text{Average speed for trip} = \frac{\text{total distance}}{\text{total time}}$$

$$\text{Total time } (t) = \frac{100 \text{ mi}}{50 \text{ mi/hr}} = 2 \text{ hr}$$

For the first 50 mi:

$$\text{Time} = \frac{50 \text{ mi}}{60 \text{ mi/hr}} = 0.833 \text{ hr}$$

$$\text{Time remaining} = 2 \text{ hr} - 0.833 \text{ hr} = 1.167 \text{ hr}$$

For the last 50 mi:

$$\text{Average speed} = \frac{50 \text{ mi}}{1.167 \text{ hr}} = 42.8 \text{ mi/hr}$$

Many of the situations encountered in linear motion can be solved readily by the use of the speed–time diagram. Those problems that involve varying speeds and accelerations during any period under consideration should be handled with extra caution and should be clearly laid out on the speed–time diagram.

Example 11-10

A train travels 10 mi at a speed of 50 mi/hr and then uniformly increases its speed to 65 mi/hr during a 30-min period [11-31]. The train continues at this speed for 1 hr before being uniformly slowed to a stop with a deceleration of 650 mi/hr/hr. Find (a) the stopping time, (b) the distance traveled during acceleration, (c) the total time, and (d) the total distance traveled.

Solution:

(a) $$a = \frac{V_2 - V_1}{t} \quad \text{(slope of travel line as train stops)}$$

$$650 \text{ mi/(hr)}^2 = \frac{0 - 65 \text{ mi/hr}}{t_4 \text{ hr}}$$

$$t_4 = 0.10 \text{ hr}$$

(b) $$S = \frac{V_1 + V_2}{2} t \quad \text{(for a trapezoid)}$$

$$S_2 = \tfrac{1}{2} (50 \text{ mi/hr} + 65 \text{ mi/hr})\ 0.5 \text{ hr}$$
$$= 28.8 \text{ mi}$$

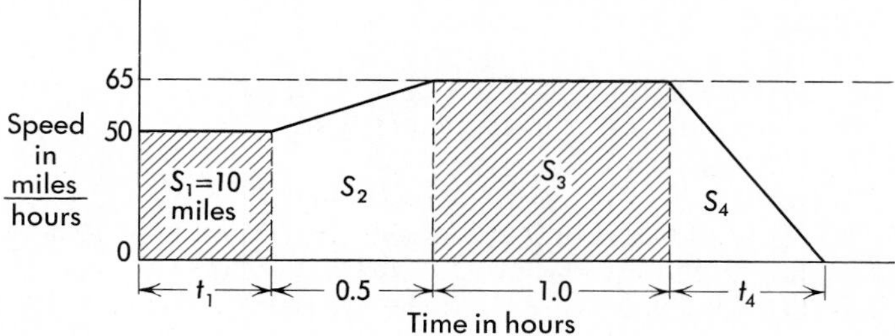

[11-31] Speed–time diagram.

(c)
$$t = \frac{S}{V_{av}}$$
$$t_1 = \frac{10 \text{ mi}}{50 \text{ mi/hr}} = 0.20 \text{ hr}$$
$$\text{Total time} = t_1 + t_1 + t_3 + t_4$$
$$= 0.20 + 0.50 + 1.0 + 0.10$$
$$= 1.80 \text{ hr}$$

(d)
$$S = (V_{av})\ (t)$$
$$S_3 = 65 \text{ mi/hr} \times 1 \text{ hr} = 65 \text{ mi}$$
$$S = \frac{V_1 + V_2}{2} t \qquad \text{(for a trapezoid)}$$
$$S_4 = \tfrac{1}{2}\ (65 \text{ mi/hr} + 0)\ 0.10 \text{ hr}$$
$$= \frac{6.5 \text{ mi}}{2} = 3.25 \text{ mi}$$
$$\text{Total distance} = S_1 + S_2 + S_3 + S_4$$
$$= 10 + 28.8 + 65 + 3.25$$
$$= 107 \text{ mi}$$

Problems

11-20. A ball is thrown vertically upward and, in time, falls back to the place of beginning. Starting from the time the ball leaves the hand, sketch a speed–time diagram that shows the motion involved. Add such explanation as you may deem necessary.

11-21. A man in a car travels a certain distance at an average speed of 8.49 m/s. After he arrives, he turns around and returns over the same route at an average speed of 5.81 m/s. What was the man's average speed both going and coming?

11-22. A car having an initial velocity of 30 km/h increases its speed uniformly at the rate of 10 m/s/s for a distance of 100 m.
a. What will be its final velocity?
b. How long will it require to cover this distance?

11-23. A 3300-lb_f automobile is traveling up a steep hill whose grade is 22 percent at a rate of 31 mi/hr when the power is shut off and the car is allowed to coast. Because of the loose gravel on the hill, the car comes to a stop in a distance of 125 ft. After traveling 75 ft, what will be the velocity of the car?

11-24. A car traveling at 20.12 m/s meets a train that is moving 14.75 m/s and the time required for the car to pass the train is 18 s. What is the length of the train in ft.?

11-25. A ball is dropped from the top of a tower 26.2 m high. Its acceleration is 9.81 m/s^2.
a. How long does it take it to reach the ground?
b. With what velocity does it strike the ground?

11-26. The speed of a ship traveling at the rate of 16 knots is uniformly retarded to 5 knots in a distance of 1 statute mile. If the rate of retardation continues constant:
a. What time in minutes will be required to bring the ship to rest?
b. How many meters will it have traveled from the point where the speed is 16 knots?

11-27. A train running on a straight, level track at 26.82 m/s suddenly detaches its caboose, which decelerates uniformly to a stop. After traveling 3218.7 m, the engineer notices the accident, and he stops the train uniformly in 50 s. At the instant the train stops, the caboose stops. What distance in meters did the engineer have to back up in order to hook onto the caboose?

11-28. A rock is dropped into a well. The sound of impact is heard 3 s after the rock is dropped. Sound travels 335.28 m/s. What was the depth of the well in meters?

Angular Motion

The motion we have just studied was linear motion and was concerned with the movement of a body or particle in straight-line travel. Many machines, however, have parts that do not travel in straight-line motion. For example, flywheels, airplane propellers, turbine rotors, motor armatures, and so on, all travel in curved paths or with angular motion. For purposes of study here, all bodies having angular motion will be considered to be rotating about a fixed center. While rotating about this fixed center there may be a *speeding up* or *slowing down* of the rotation rate.

Angular distance (usually designated by a Greek letter, such as θ, ϕ, β, etc.) may be measured in degrees, radians, or revolutions. A radian is defined as a central angle subtended by an arc whose length is equal to the radius of the circle.

$$1 \text{ revolution} = 360 \text{ degrees}$$

$$1 \text{ revolution} = 2\pi \text{ radians}$$

$$1 \text{ radian} = 57.3 \text{ degrees}$$

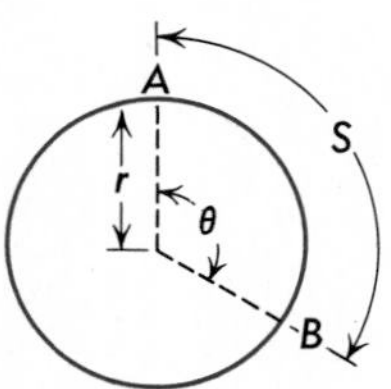

[11-32]

Example 11-11

Point A in Figure [11-32] travels through an angular distance $\theta = 120$ degrees, while moving to position B. Express this angular distance in revolutions and radians.

Solution:

$$\theta = 120 \text{ degrees} \times 1 \text{ revolution}/360 \text{ degrees}$$
$$= 0.333 \text{ revolution}$$
$$= 120 \text{ degrees} \times \pi \text{ radians}/180 \text{ degrees}$$
$$= 2.09 \text{ rad}$$

In angular motion, time is measured in the same units as for linear motion. Thus angular speed, which is an angular distance (θ) divided by time t, may have such units as rad/s, rev/min, deg/s, and so on. Angular speed, θ/t, is usually designated by the Greek letter ω (omega).

Angular acceleration can be found by solving for the slope of the travel line in the angular speed versus time graph, much like linear acceleration was obtained in linear motion. As in linear motion, we must divide the change in angular speed by the time it took to make the change. Angular acceleration, ω/t, is often represented by the Greek letter α (alpha).

There is a definite relation between angular motion and linear motion. Let us consider a point on the rim of a flywheel. In one revolution the point will travel through an angular distance of 2π radians or a linear distance of $2\pi r$ *linear units*. All points on a body will travel through the same angular distance during a period of time, but their linear distance moved will depend on the radii to the points under consideration. Therefore, linear distance is equal to angular distance in radians multiplied by the radius. Linear speed is found by multiplying the angular speed by the radius.

Length of arc (S):

$$S = r\theta$$

where θ is measured in radians.

Linear speed:

$$V = r\omega$$

where ω is the angular speed measured in radians per unit of time, and r is the radius measured from the point of rotation out to the point whose linear speed is desired.

Example 11-12

Point A in Figure [11-33] is located on the outside of a flywheel 6 ft in diameter. Point B is located on the inside of the rim 1 ft from point A. If the flywheel travels at 300 rev/min for 10 min, find (a) total angular distance traveled by point B in radians, (b) linear speed of point B in feet per minute, (c) linear distance traveled by point A in miles.

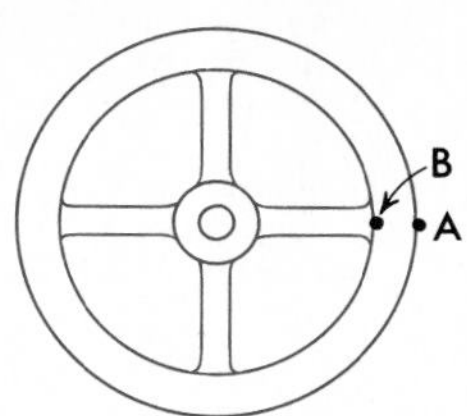

[11-33] The situation.

Solution:

Refer to Figure [11-34].

(a) $\theta = (300 \text{ rev/min})\ (10 \text{ min}) = 3000 \text{ rev}$
$= (3000 \text{ rev})\ (2\pi \text{ rad/rev}) = 18{,}900 \text{ rad}$

(b) $V = r\omega$
$= (2 \text{ ft/rad})\ [300 \text{ rev/min}(2\pi \text{ rad/rev})]$
$= 3770 \text{ ft/min}$

(c) $S = r\theta$
$= (3 \text{ ft/rad})\ (3000 \times 2\pi \text{ rad})$
$= 56{,}600 \text{ ft or } 10.7 \text{ mi}$

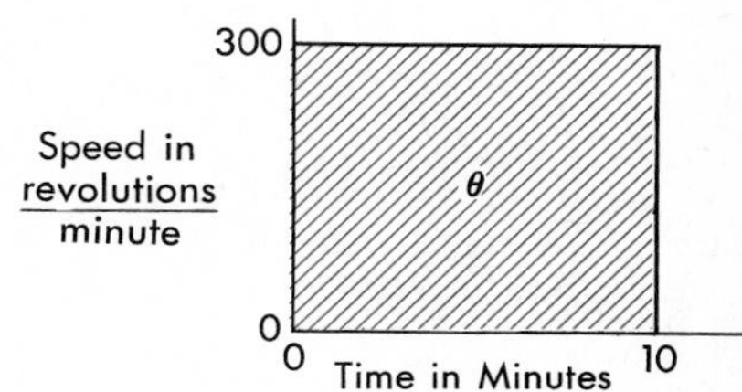

[11-34] Speed–time diagram.

Example 11-13

The flywheel of a gasoline engine [11-35] changes its angular velocity from 150 to 300 rev/min during a 5-min period. Solve for (a) the total distance traveled by a point on the rim of the flywheel, and (b) the angular acceleration of the point during this change.

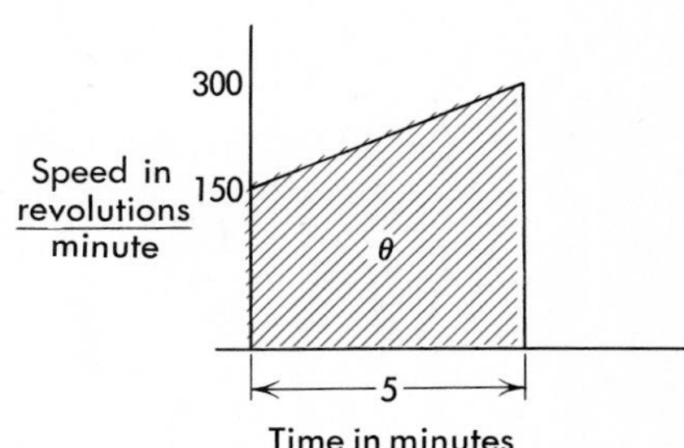

[11-35] Speed–time diagram.

Solution:

(a) θ = travel during the change
= area of the speed–time diagram under the travel line

$$= \frac{\omega_1 + \omega_2}{2} t$$
$$= 5 \text{ min} \cdot \frac{150 \text{ rev/min} + 300 \text{ rev/min}}{2}$$
$$= (5 \text{ min})\ (225 \text{ rev/min})$$
$$= 1125 \text{ rev}$$
$$= 7069 \text{ rad}$$

(b) α = angular acceleration
= slope of the travel line

$$= \frac{\omega_2 - \omega_1}{t}$$
$$= \frac{300 \text{ rev/min} - 150 \text{ rev/min}}{5 \text{ min}}$$
$$= 30 \text{ rev/min}^2$$
$$= 0.0524 \text{ rad/s}^2$$

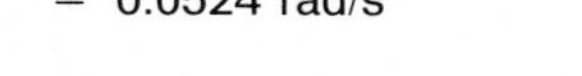

Problems

11-29. While going around a circular curve of 3000-ft radius, a train slows down from 36 to 13 mi/hr in a distance of 850 ft. Find the angular distance covered in degrees.

11-30. Two pulleys 8 and 17 in. in diameter are 10 ft apart on the same shaft and the 17-in. pulley runs 400 rev/min. How many radians per second does the small pulley turn?

11-31. Given $t = 5$ s; $\omega_1 = 50$ rev/s; and $\omega_2 = 15$ rad/s. Draw a speed–time diagram for the motion involved, calibrating both ordinate and abscissa. Is the angular acceleration positive or negative? Why?

11-32. Two wheels rolling together without slipping have a velocity ratio of 3 to 1. The driver (which is the smaller of the

two) is 0.3 m in diameter, and turns at 50 rev/min.
a. What is the speed in feet per second at a point on the surface of the wheels?
b. What is the angular velocity of the larger wheel?
c. What is the diameter of the larger wheel?

11-33. A locomotive, having drive wheels 2 m in diameter, is traveling at the rate of 90 km/h. What are the revolutions per minute of the drive wheels?

11-34. An elevator hoisting drum is decelerating at the rate of 15 rev/min/min. If the drum is brought to rest in 9 min, find (a) the total number of revolutions, and (b) the initial speed in radians per second.

11-35. A cylindrical drum $2\frac{1}{2}$ ft in diameter is rotated on its axis by pulling a rope wound around it. If the linear acceleration of a point on the rope is 36.9 ft/sec/sec, what will be the angular speed in (a) revolutions per minute at the end of 6 sec, and (b) radians per second? (c) How many turns will it have made during the 6 sec?

11-36. A wheel turns at an average speed of 50 rev/sec during a total angular distance of 15,900 rad. During the first 300 rev the average speed was 75 rev/sec. During the next 300 rev the average speed of the wheel was 53 rev/sec. Since the overall average speed was 50 rev/sec, what was the average speed during the remainder of the distance traveled?

Work

Many words used in physics and engineering have meanings that differ from their common, nontechnical meanings in everyday use. Once such word is *work,* as reference to any dictionary will show. In the common usage of this word, it may mean anything from merely a thinking process to the hardest sort of physical exertion. For over 200 years science has been trying to clear up the confusion regarding the use of the word *work* in concise scientific and technical writings.

Work is defined for our purposes as the product of a force F and a distance S through which the force acts; both force and distance must be measured along the same line. From this definition we can see that a force executes work on a body when it produces motion of the body—motion in the direction of the force. The force must be displaced through some distance. If there is no motion as a result of an applied force, there is no work done. You could push on an immovable object for hours and become quite exhausted, yet you would have done no work according to the technical definition.

Although the ideas advanced regarding work may not agree with the everyday usage of the word *work,* you must accept with an open mind the technical definition given above. It allows us to add precision to our technical communications.

The dimensions of work will be force × distance or FL. The units will be the product of a unit of force and a unit of length. For example, in English units, a common measure of work is the foot-$\text{pound}_{\text{force}}$ or ft · lb_f. One ft · lb_f of work is done on an object when a force of 1 lb_f is applied in moving an object through a distance of 1 ft in the direction of the force. In the event that force is not in the same direction that distance is measured, work could be calculated by using the component of force in the direction of the distance covered.

Example 11-14

A constant force of 50 N acting downward at an angle of 30° with the horizontal moves a box 10 m across a floor [11-36]. How much work is done on the box?

Solution:

In this example, only a portion of the 50-N force is effective in moving the box from position *A* to position *B*. This effective portion of the 50-N force is the horizontal component (50) (cos 30°) = 43.3 N.

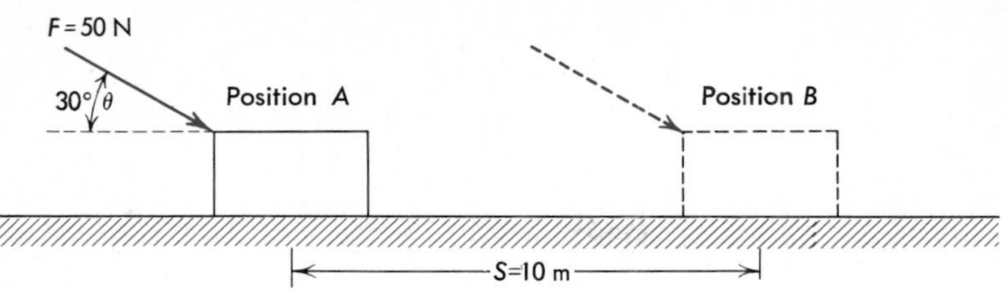

[11-36]

The vertical component (50) (sin 30°) does not produce any motion but serves only to press the box against the floor.

$$\text{Work} = (F \cos \theta)\ (S)$$
$$= (50\ \text{N})\ (\cos 30°)\ (10\ \text{m})$$
$$= 433\ \text{N} \cdot \text{m}$$

Example 11-15

A man carries a precision gauge weighing 38.5 lb_f up a flight of stairs that has a rise of 8 in. and a tread of 12 in. He climbs at the rate of two steps per second. How much work is done by the man on the gauge as it is carried up a stairway of 31 steps?

Solution:

Since we are attempting to find only the work done by the man on the gauge, we shall ignore other work. The work done, then, will be the weight of the gauge lifted[3] times the vertical height the gauge was lifted. The length of time to move the gauge does not enter into the computation for work.

$$\text{Vertical height} = \frac{(8)\ (31)}{12} = 20.7\ \text{ft}$$
$$\text{Work} = (F)\ (h)$$
$$= (38.5)\ (20.7)$$
$$= 796\ \text{ft} \cdot \text{lb}_f$$

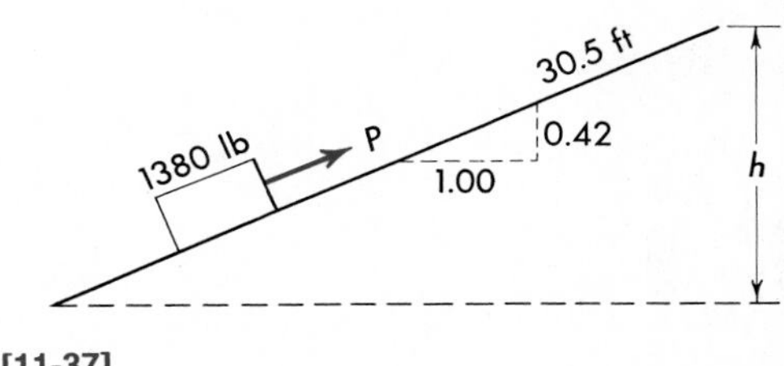

[11-37]

Example 11-16

A cable [11-37] is pulling a wooden crate of electronic computer parts, which weighs 1380 lb_f, up a frictionless ramp 30.5 ft long that rises to the second floor of a building at the rate of 0.42 ft vertically per foot horizontally. If the cable is pulling parallel to the ramp, what work is done on the crate by the cable?

Solution:

The work done by the cable is the force applied by the cable times the distance through which the cable moves parallel to the ramp. A free-body diagram of the crate [11-38] shows that P, the force applied to the crate by the cable, is given by

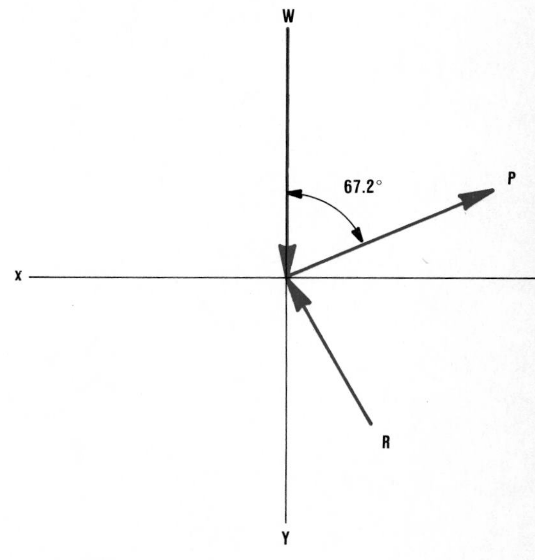

[11-38] Free body of block.

[3]We have assumed that the man moves at a uniform rate so that the gauge is in equilibrium at all times. Then the force that the man must apply to the gauge must equal the weight of the gauge.

$$\Sigma F_x = P - W \sin \theta = 0$$

or

$$P = W \sin \theta$$

where $\theta = \arctan 0.42 = 22.8°$. The work done is then

$$\begin{aligned} \text{Work} &= [W \sin(22.8)]30.5 \\ &= 1380 \sin(22.8)30.5 \\ &= 16{,}310 \text{ ft} \cdot \text{lb}_f \end{aligned}$$

Problems

11-37. It requires a constant horizontal force of 300 N to move a 4500 N table on casters. How much work would be done in moving the table 33 m over a level floor?

11-38. A skip hoist lifts a load of bricks to the third floor of a building under construction. The cable exerts an average pull of 2900 lb_f for a distance of 25.6 ft. How much work is done in lifting the loaded hoist?

11-39. A locomotive is pulling a string of 40 boxcars, each weighing 45 tons, at a constant speed of 35 mi/hr on a stretch of level track. The frictional resistance of the train is 8 lb per ton of weight. How much work is done by the drawbar pull of the engine in moving the train 1 mi?

11-40. A man carries a box weighing 200 N up a stairway of 17 steps. Each step is 0.25 m high and 0.3 m wide. How much work does he do in carrying the box up the stairway?

11-41. A belt passes over a pulley which is 1 m in diameter. If the difference in tension on the two sides of the belt is 300 N and the pulley is turning 530 rev/min, how much work is done per minute by the belt?

11-42. A rope is wrapped around a drum $6\frac{1}{2}$ in. in diameter. A crank handle 14 in. long is connected to the shaft carrying the drum. How much work would be done in lifting a weight of 75 lb_f a vertical distance of 55 ft? If the length of the crank handle is increased to 20 in., what will be the work done in lifting the weight as before?

11-43. Water is pumped against a constant head of 22 ft at the rate of 710 gal/min. How much work is done each minute in pumping the water?

11-44. A tractor is towing a loaded wagon weighing 1.4678×10^4 N over level ground and the average tension in the tow cable is 720.58 N. What work is done by the cable in moving the wagon 402.34 m?

Power

It is apparent that no interval of time was mentioned in our previous definition of work. In our modern civilization we frequently are as interested in the time of doing work as we are in getting the work done. For this reason the term *power* is introduced, which is the time rate of doing work. In symbol form:

$$\text{Power} = \frac{\text{work}}{\text{time}} \qquad P = \frac{W}{t} = \frac{F \cdot S}{t}$$

or it may be expressed as

$$\text{Power} = (\text{force})(\text{velocity}) \qquad P = FV$$

If a pile of bricks is to be moved from the ground to the third floor of a building, the job may be accomplished by moving one brick at a time, 10 bricks at a time, or the whole pile of bricks at once. The work done in any case is the same and is the product of the weight of the pile of bricks and the vertical distance through which the pile is moved. However, the time that will be taken will probably vary in each case, as will the capabilities of the lifting mechanism. To obtain an indication of the rate at which work can be done, we use the term *power,* which is a measure of how fast a force can move through a given distance.

The units of power in any system can be found by dividing work units by time units. In the English systems, power may be expressed as ft · lb_f/s, or ft · lbs/min. Since the days of James Watt and his steam engine, the horsepower (hp) has been a common unit of power and is numerically equal to 550 ft · lb_f/s or 33,000 ft · lb_f/min. The SI unit of power is the watt, which is 1 N · m/s. There are 746 watts (W) in 1 horsepower (hp).

$$\begin{aligned} 1 \text{ hp} &= 550 \text{ ft} \cdot \text{lb}_f/\text{s} \\ &= 33{,}000 \text{ ft} \cdot \text{lb}_f/\text{min} \\ &= 746 \text{ W} \\ &= 0.746 \text{ kW} \end{aligned}$$

Example 11-17

A box weighing 1100 lb_f [11-39] is lifted 15 ft in 3 sec. How much power is necessary?

Solution:

$$\text{Power} = \frac{\text{work}}{\text{time}}$$

$$= \frac{(1100)\ \text{lb}_f \cdot (15)\ \text{ft}}{3\ \text{s}}$$

$$\text{Horsepower} = \frac{\text{work in ft} \cdot \text{lb}_f}{(\text{time in s})(550)}$$

$$= \frac{(1100)(15)}{(3)(550)} = 10 \text{ hp}$$

$$\text{Power} = 746\ \frac{\text{W}}{\text{hp}}(10 \text{ hp}) = 7460 \text{ W}$$

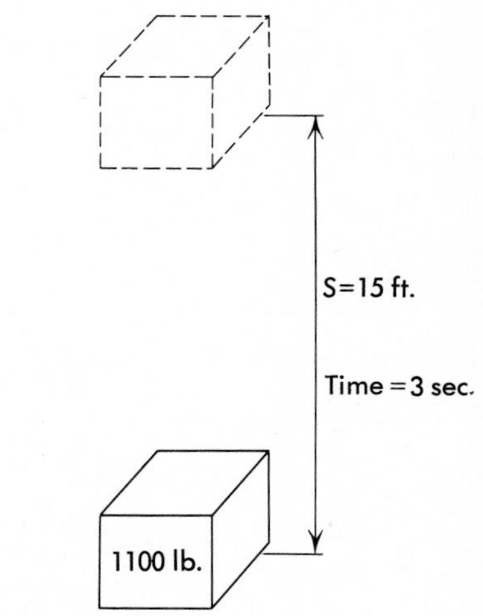

[11-39] The situation.

Electric power is usually expressed in watts or kilowatts. A kilowatt is 1000 W. When electric rates are prepared by utility companies, they customarily base their rates on the kilowatt-hour. Since the kilowatt-hour is the product of power and time, charges for electric services actually are charges for work or energy.[4] When you pay your electric utility bill, you are actually paying for work performed electrically or energy used rather than for how fast you use it. The kilowatt-hour is simply power consumed at the rate of 1 kW for 1 hr.

Example 11-18

How much will it cost to operate a 150-W electric light for 2.5 hr when the utility company charges are 8.5 cents per kilowatt-hour?

---

[4]Energy is discussed on page 331.

Solution:

$$\text{Work (or energy) in kWh} = (\text{power in kilowatts}) \times (\text{time in hours})$$

$$\text{Energy} = \frac{150}{1000}(2.5)$$

$$= 0.375 \text{ kWh}$$

$$\text{Cost of electric work (or energy)} = (\text{kWh}) \times (\text{cost per kWh})$$

$$= (0.375)(8.5)$$

$$= 3.19 \text{ cents}$$

Efficiency

The efficiency of any machine is expressed as the ratio of work output to the work input, or as a ratio of power output to power input. While efficiency has no units, it is usually expressed as a percentage.

$$\text{Efficiency of a machine} = \frac{\text{work output}}{\text{work input}} = \frac{\text{power output}}{\text{power input}}$$

$$\text{Percent efficiency} = \frac{\text{work output}}{\text{work input}} \times (100 \text{ percent})$$

$$= \frac{\text{power output}}{\text{power input}} \times (100 \text{ percent})$$

Example 11-19

What is the percent efficiency of a 12-hp electric motor that requires 9.95 kW of electric power when running at full load?

Solution:

The units of power input and power output must both be the same in order to calculate efficiency. We shall convert 12 hp to kilowatts and compute the ratio of power output to power input.

$$\text{Power output in kW} = (\text{power output in hp})\ (0.746)$$

$$= (12)(0.746)$$

$$= 8.95 \text{ kW output}$$

$$\text{Percent efficiency} = \frac{\text{power output}}{\text{power input}}\ (100 \text{ percent})$$

$$= \frac{8.95}{9.95}\ (100 \text{ percent})$$

$$= 90.0 \text{ percent efficiency}$$

The result in Example 11-19 would be the same if the power input in kilowatts had been converted to horsepower and efficiency

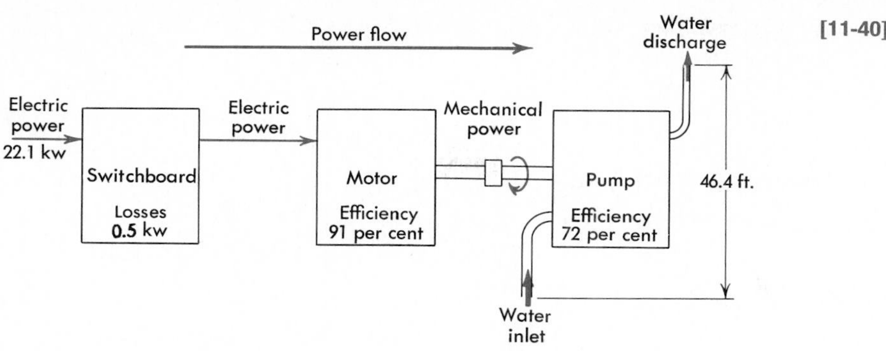

[11-40]

had been obtained as a ratio of horsepower output to horsepower input.

The power rating of motors and engines is the maximum output power that they are expected to deliver constantly, unless specifically stated otherwise. The input power will always be greater than the output power. For instance, a 100-hp electric motor can develop 100 hp at its pulley, but more than 100 hp will have to be supplied by the electric power line connected to the motor.

In some situations account must be taken of the efficiency of several machines as we trace the flow of power through them. As an example, let us consider a case in which an electric motor is connected to a pump that is pumping water. The motor obtains its power from electric lines that run through a switch. The data are given in Figure [11-40] in which small blocks are used to represent parts of the system.

Example 11-20

Power supplied to the system, as indicated by electric meters, is 22.1 kW. Find the amount of water delivered by the pump in cubic feet per second.

Solution:

Compute the output power of each part of the system in order, beginning with the switchboard.

$$\text{Power supplied to the motor} = 22.1 - 0.5$$
$$= 21.6 \text{ kW} \quad \text{(This is the power input to the motor)}$$

$$\text{Power output of the motor} = (21.6)(0.91)$$
$$= 19.7 \text{ kW}$$

If we assume no losses in the coupling between the motor and the pump, the power output of the motor is the same as the power input to the pump.

$$\text{Power output of the pump} = (19.7)(0.72)$$
$$= 14.2 \text{ kW}$$

Converting 14.2 kW to foot-pounds$_f$ per second:

$$\text{Power} = \frac{(14.2)(550)}{0.746}$$

$$= 10{,}550 \text{ ft} \cdot \text{lb}_f/\text{s}$$

The amount of water delivered may now be found if we remember that

$$\text{Power} = \frac{\text{work}}{\text{time}} = \frac{\text{weight} \times \text{height}}{\text{time}}$$

Then

$$\frac{\text{Weight (of water)}}{\text{time}} = \frac{10{,}500 \text{ ft} \cdot \text{lb}_f/\text{s}}{46.4 \text{ ft}}$$

$$= 22.6 \text{ lb}_f/\text{s}$$

Converting 225 lb$_f$/s to cubic feet per second:

$$\text{Volume of water per second} = \frac{22.6 \text{ lb}_f/\text{s}}{62.4 \text{ lb}_f/\text{ft}^3}$$

$$= 0.36 \text{ ft}^3/\text{s}$$

One additional item of information should be called to your attention. We notice that the efficiency of the motor is 91 percent and the efficiency of the pump is 72 percent. Considering the overall efficiency of both machines, the input to the motor is 21.6 kW, and the output of the pump is 14.16 kW. The overall efficiency of both machines can be found as follows:

$$\text{Overall efficiency} = \frac{\text{output}}{\text{input}} (100 \text{ percent})$$

$$= \frac{14.2}{21.6} (100 \text{ percent})$$

$$= 65.7 \text{ percent}$$

The overall efficiency of both machines could also be determined by finding the product of the individual efficiencies:

$$\text{Overall efficiency} = (0.91)(0.72) \times (100 \text{ percent})$$

$$= 65.5 \text{ percent}^5$$

Problems

11-45. An automobile requires 47 hp to maintain a speed of 55 mi/hr. What force in newtons is being exerted on it by the engine?

11-46. In a recent experiment, a student weighing 168 lb$_f$ ran from the first floor to the third floor of a building, a vertical distance of 26 ft, in 9 s. How much horsepower did he develop?

11-47. If a horse can actually develop 1 hp while pulling a loaded wagon at 3.5 mi/hr, what force does he exert on the wagon?

11-48. A car weighing 12,000 N is moving at constant speed up a hill having a slope of 17 percent. Neglecting friction, how fast will the car be moving when it is developing 20 kW?

11-49. An airplane engine that develops 2000 hp is driving the plane at a speed of 250 mi/hr. What thrust is developed by the propeller?

[5]The difference between 65.5% and 65.7% is due to round-off error in determining the 14.2 hp output.

11-50. A diesel engine runs a pump that pumps 18,000 gal of water per hour into a tank 65 ft above the supply. How many horsepower are required at the pump?

11-51. A car weighing 16,000 N is being towed by another car at a rate of 70 km/hr. The average force exerted by the tow cable is 1000 N. What horsepower is necessary to tow the car?

11-52. A tank holding 3500 gal of water is to be emptied by a small centrifugal pump. The water is 6 ft deep and is to be pumped to a height of 13.5 ft above the bottom of the tank. The pump is 65 percent efficient and is driven by a motor that develops $\frac{1}{4}$ hp. How long will it take to empty the tank?

11-53. A belt conveyor is used to carry crushed coal into a hopper. The belt carries 13 tons/hr up a 13 percent slope 45 ft long. The friction losses in the belt and rollers amount to 22 percent of the power supplied. How many kWs are needed to operate the belt?

11-54. A $\frac{1}{2}$-hp electric motor drives a pump that lifts 1200 gal of water each hour to a height of 15 m. What is the efficiency of the pump? While running, the motor requires 40 W of power. What is the efficiency of the motor? What is the overall efficiency of the motor–pump combination?

11-55. A pulley 10 in. in diameter is on the shaft of a motor that is running 1730 rev/min. If the motor is developing 20 kW, what is the difference in tension on the sides of the belt that passes over the pulley?

11-56. The piston of a steam engine is 12 in. in diameter and moves through a distance of 20 in. each stroke. The average pressure on the piston is 75 $\text{lb}_f/\text{in.}^2$ and the piston makes 250 power strokes per minute. How much horsepower is developed?

Energy

Another expression used extensively in mechanics is the term *energy*. Energy is a property or quality that an object possesses. The property energy is historically tied to the concept of work—it has been related either to the work that is done to put an object in its current energy state or to the work that could possibly be returned by the object. From this historical perspective, the energy possessed or stored in an object is a measure of its ability to do work.

Energy and work have the same dimensions and can be expressed in the same units. However, although work and energy are related and have the same dimensions, they involve slightly different, but important, concepts. Energy is something that is stored, while work is something that is being done. Our present understanding of the concept of work is that work is energy that is being transferred.[6]

The relationship between work and energy is much like the relationship between your banking transactions (deposits and withdrawals—money that is being transferred) and your bank balance (how much money is on deposit or in storage). Banking transactions are money in motion—money is either passing into or out of your account. Your bank balance is your reservoir of money—it is there in storage even if you go a long time without a transaction.

Work is the transaction in the energy business—when you do work you are transferring energy. Energy is equivalent to the bank balance—it is stored in matter.

In the banking world, transactions and balances have the same units (usually dollars). Similarly, in the energy world, work and energy have the same dimensions (FL) and may be expressed in the same units.

As every person learns soon after they open their first banking account, transactions *affect* the balance, but transactions and balances are not strictly one and the same. Obviously, there must be a rela-

[6]Later in your career, when you study thermodynamics, you will learn that there is another way of transferring energy, and that is by heat. The technical definition of heat (which is much different from the layperson's definition) is energy being transferred due to a temperature difference.

tionship between the two; in fact, you use that relationship each month to balance your account. If you can grasp this fundamental difference and relate it to work and energy, you will have less difficulty when you take your first thermodynamics course. (We suspect that your bank account will also benefit.)

Space does not permit an extended discussion here of all the various ways in which energy can be stored in matter. But we do wish to discuss the forms known as *potential* and *kinetic* energy.

Gravitational Potential Energy. The gravitational potential energy (often abbreviated PE) of a body is a property or quality that the body possesses *because of its elevation* or location in a gravitational field. This property is a measure of the body's ability or capacity to do work because of its elevation with respect to some vertical reference plane. Remember, energy is something that is stored or that an object possesses (a balance). Work is something that is performed or done on an object (a transaction).

Gravitational potential energy can be calculated from

PE = (weight) × (vertical height with respect to some vertical reference or datum)

Example 11-21

A 100-N box [11-41] is on a platform 10 m above the ground. What is its gravitational potential energy with respect to the ground?

Solution:

To analyze the problem, let us assume that the box is initially in position *B*. The work necessary to raise the box to position *A* is (10 m) (100 N) or 1000 n · m. Since energy and work done against gravitational forces are convertible,[7] the work of lifting the box evidently has gone to increasing its PE. The PE can then be found as the product of weight and the vertical distance above some reference plane. In this problem:

$$
\begin{aligned}
PE &= (W)(H) \\
&= (100\ N)(10\ m) \\
&= 1000\ N \cdot m
\end{aligned}
$$

Notice here that some work is done on the object (the box) and this increases its store of potential energy. Work is something that is done, energy is something that is possessed or stored. We shall return to this distinction again.

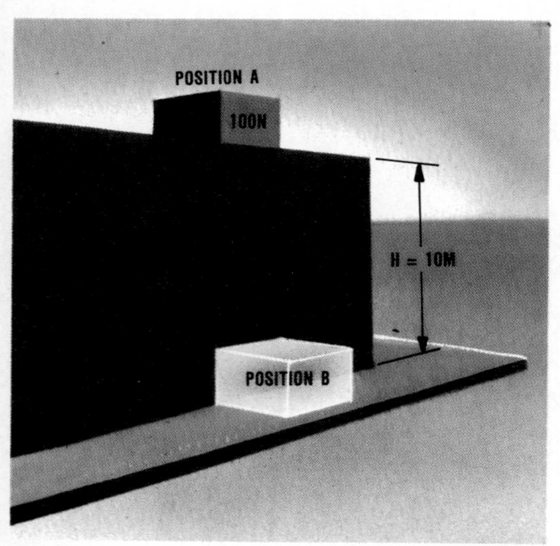

[11-41]

Kinetic Energy. The kinetic energy (KE) of a body is a property or quality that the body possesses *because of its speed* relative to some fixed reference system. This property is a measure of the object's ability to do work because of its motion. Although the dimensions of kinetic energy must be the same as the dimensions of work,

[7]Again, when you study thermodynamics, you will learn that not all work is convertible to potential and kinetic energy, and vice versa. In thermodynamics you will be introduced to internal energy. Some forms of work are converted directly into internal energy, not all of which can be converted back into useful work.

remember that the two concepts are not equivalent. The usual expression for determining KE is

$$KE = \frac{(\text{mass})(\text{speed})^2}{2}$$

The derivation of this expression is as follows. We first take an object of mass M that has no kinetic energy (consider it to have zero speed) and transfer some energy to it as work (remember that work is energy in transfer—a transaction). After we have done this work, the object will have some speed and thus some kinetic energy. We can then relate the amount of work done on the object with the amount of kinetic energy the object possesses when the work is finished.

From Newton's second law, a force (F) will produce an acceleration of our object of mass M, according to the relationship

$$F = (\text{mass})(\text{acceleration}) = Ma$$

If the force is constant and acts through a distance S while the body is accelerating, the work done is $(F)(S)$. Substituting the value of F above in the expression for work:

$$\text{Work} = (Ma)S$$

From the expression of accelerated motion,[8] the velocity acquired by a body starting from rest is $V^2 = 2aS$, or $S = V^2/2a$. Substituting this value of S in the expression for work, we get

$$\text{Work} = (Ma)\,\frac{V^2}{2a} = \frac{MV^2}{2}$$

Since this is the work (the transaction) to give the body a speed V, the work must have gone into increasing its kinetic energy (its balance), KE. If the balance started at zero, it must now have a balance of

$$KE = \frac{MV^2}{2}$$

which is the relationship we first postulated.

The relationship of units using the English FLt dimension/unit system, where the unit of mass is the slug (1 slug is 1 $\text{lb}_f \cdot \text{s}^2/\text{ft}$; see Chapter 10), is

$$KE = \overbrace{\left[\frac{\text{lb}_f \cdot \text{s}^2}{\text{ft}}\right]}^{\text{slug}} \left[\frac{\text{ft}}{\text{s}}\right]^2 = \text{ft} \cdot \text{lb}_f$$

As we pointed out previously, the dimensions and units of KE are identical with the dimensions and units of work. It should be remembered that in the English FLt dimension/unit system the mass of a body in slugs can be calculated by dividing the local weight of the body in pounds by the local acceleration due to gravity in feet per second per second. In the SI system the unit of energy is the joule (J), which is 1 newton · meter.

[8]Refer to Appendix II.

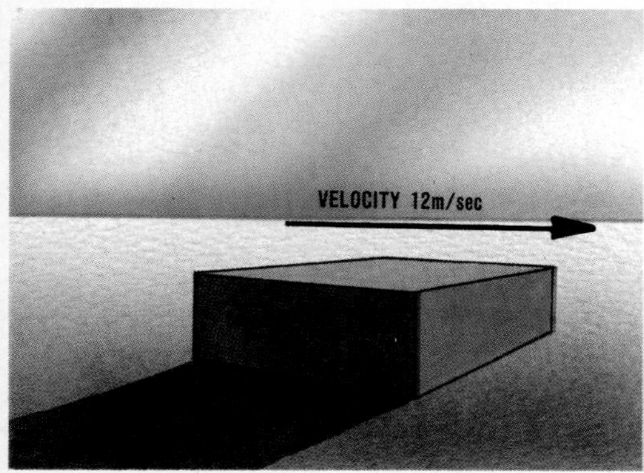

[11-42] The situation.

Example 11-22

A 10-N box [11-42] is moving with a velocity of 12 m/sec. What is its kinetic energy?

Solution:

$$\text{KE} = \frac{MV^2}{2}$$

$$= \frac{(10)(12)^2 \text{ N} \cdot \text{m}^2/\text{sec}^2}{(2)(9.8) \text{ m/sec}^2}$$

$$= 73.5 \text{ N} \cdot \text{m}$$

$$= 73.5 \text{ J}$$

The Law of Conservation of Energy. The relationship between work and energy can be summed up in what is known as the *law of conservation of energy.* This principle states that energy can be neither created nor destroyed but is either transformed from one form to another (neglecting mass–energy transformations) within an object or transferred to other objects or matter. In word form, this relationship is

Initial balance + deposits − withdrawals = final balance

just like your bank account. Note that there can be both positive (deposit) and negative (withdrawal) transactions.

It is extremely important to realize that this equation must be applied to something. If you cannot identify what that something is going to be, you will not be able to apply the equation. You will not be able to define the deposits and withdrawals, for example. If you want to apply it to your finances, you must define what account you are going to use. You cannot randomly apply it to part of one account along with part of some other account.

In analyzing energy problems you must define your system. Will it be the box on the floor, the lid on the box on the floor, or the box *and* the floor? Define your system by sketching it when you start your analysis.

In terms of potential and kinetic energies and work done, this word equation becomes

work done on the object (→ W_{to}) — work done by the object (→ W_{from})

Initial store ($PE_i + KE_i$) — Final store ($PE_f + KE_f$)

$$PE_i + KE_i \quad + \quad W_{\text{to}} \quad - \quad W_{\text{from}} \quad = \quad PE_f + KE_f$$

where the i and f subscripts imply initial and final, that is, before and after the works are accomplished, respectively. We should point out that PE is an alternate way of expressing the work done by gravity. Therefore, do not enter both the work done by gravity and gravitational potential energy into the equation above. Use only one or the other.

There is always the task of determining whether work is done *on* the object or *by* the object. Use the following steps for determining this:

1. Determine the direction of the force that acts *on* the object under study.
2. Determine the direction of the displacement of that force (i.e., which way did it move?).
3. Work is done *on* the object if the force and the displacement are in the same direction.
4. Work is done *by* the object if the force and the displacement are in opposite directions.

Example 11-23

The 100-N box in Figure [11-41], when in position *A*, has a PE of 1000 N · m relative to the lower level at position *B*. Its KE is zero because it is not moving. However, if we push the box to the edge of the platform so that it falls, we can see that, just as the box reaches position *B*, the height of the box above the ground is zero and its PE is zero. Calculate its KE and speed *V* as the box reaches position *B*.

Solution:

Evaluation of terms in the law of conservation of energy given above gives

$PE_i = 1000\ N \cdot m$

$KE_i = 0$

$PE_f = 0$

$KE_f = \text{unknown}$

Work = 0 (the only work done is by gravity, but since we have used PE, this already includes the gravity work)

Substituting yields

$$1000\ N \cdot m + 0 + 0 + 0 = 0 + KE_f$$

$$KE_f = 1000\ N \cdot m = 1000\ J$$

Since $KE_f = \frac{1}{2} MV_f^2$

and $M = \frac{W}{g}$

$$\text{Then } V_f = \sqrt{\frac{2\ KE_f}{M}} = \sqrt{\frac{2\ KE_f \cdot g}{W}}$$

$$= \sqrt{\frac{2\ (1000\ N \cdot m)\ 9.8\ m/s^2}{100\ N}}$$

$$= 14\ m/s$$

Example 11-24

A 1000-N pile-driving hammer falls 6 m onto a pile and drives the pile 6 cm. What is the average force exerted?

Solution:

Using the principles of energy and work, the energy of the moving hammer was transformed into work by moving the pile 0.06 m. We will choose the bottom-most point of travel of the top of the pile to establish our datum for potential energy. We will choose the hammer as our system and itemize the terms that appear in the conservation of energy equation.

$$\text{PE} = 1000\ (6 + 0.06)\ \text{N} \cdot \text{m}$$

$$\text{KE} = 0$$

$$\text{PE} = 0$$

$$\text{KE} = 0$$

$$W_{\text{to}} = 0$$

$W_{\text{from}} = 0.06(F)$ (the force on the hammer, which is up, and the displacement, which is down, are in opposite directions)

$$1000(6 + 0.06)\ \text{N} \cdot \text{m} + 0 - 0.06\ \text{m}\ (F) = 0 + 0$$

$$F = 1000(6.06)/0.06 = 1.01 \times 10^5\ \text{N}$$

Energy Stored in Springs. Another example of an energy–work conversion is in the use of springs. Using a coil spring as an example, if we compress one in our hands, we exert a force in order to shorten the spring. This means that we have exerted force through a distance and have done work or transferred energy to the spring. We have changed the stored energy in the spring since we have transferred energy to it. Traditionally, we call this stored energy *elastic potential energy*. We must not confuse it with gravitational potential energy which we have called PE earlier, but it is used in the same way.[9]

We know from experience that as the spring is compressed more and more, an increasing amount of force is required. In part (a) of Figure [11-43] there is no force on the 6-in. spring. As we slowly add weight to the spring, it will shorten. In part (b) the weight has been increased to 12 lb_f and the spring has been compressed until it is only 4 in. long. The applied force, which initially was zero, has been increased to 12 lb_f, which leads to an average force of 6 lb_f during compression.

We may take the average force, 6 lb_f, times the 2-in. movement of the spring as the work done, rather than take the small change of length due to each increase of force from zero pounds to 12 lb_f, and then add all the small increments of work. It can be shown by advanced mathematics that the increment method may be used, but for our purpose we shall use the average force multiplied by the distance the average force will act.

We use PE_s for the elastic potential energy stored in the spring and write the conservation of energy equation as

$$\text{PE}_i + \text{PE}_{si} + \text{KE}_i + \text{W}_{\text{to}} - \text{W}_{\text{from}} = \text{PE}_f + \text{PE}_{sf} + \text{KE}_f$$

Term by term, we have

$$\text{PE}_i = 0$$

$$\text{PE}_{si} = 0$$

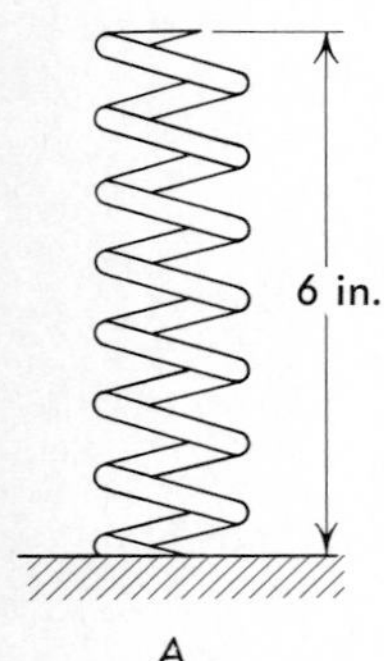

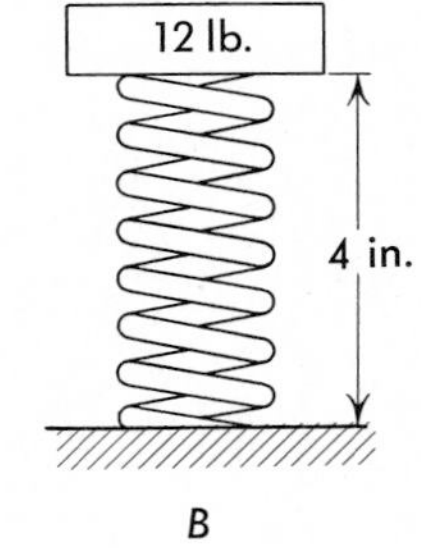

[11-43] The situation.

[9]As pointed out previously, there are many forms of stored energy. Some, like the energy stored by charged particles in an electrostatic field, are also called potential energy (electrostatic potential energy).

$\text{KE}_i = 0$

$W_{\text{to}} = (F_{\text{av}})(S)$ (the force on the spring and the displacement are in the same direction)

$= (F_1 + F_2)\dfrac{(S)}{2}$

$W_{\text{from}} = 0$

$\text{PE}_f = 0$

$\text{PE}_{sf} = \text{unknown}$

$\text{KE}_f = 0$

Assembling these in the law of conservation of energy gives

$$0 + 0 + \frac{(F_1 + F_2)(S)}{2} + 0 = 0 + \text{PE}_{sf} + 0$$

For springs, the *spring constant* or *spring rate,* K, is the ratio of the force applied on the spring to the change in length of the spring (i.e., $K = F/S$). Using this definition, $F = K \cdot S$ and the equation above becomes

$$0 + 0 + \frac{(KS + KS)(S)}{2} + 0 = 0 + \text{PE}_{sf} + 0$$

from which we find

$$\text{PE}_{sf} = \frac{KS^2}{2}$$

Example 11-25

A spring has a spring constant of 600 lb_f/ft. How much work is done by a force that stretches it 3 in.? What force was acting to stretch the spring 3 in.?

Solution:

The work done is

$$\begin{aligned} \text{Work} &= \frac{KS^2}{2} \\ &= \frac{600\ (\text{lb}_f/\text{ft})}{2}(0.25\ \text{ft})^2 \\ &= 300\ \text{lb/ft}(0.0625\ \text{ft}^2) \\ &= 18.8\ \text{ft-lb}_f \end{aligned}$$

The force to stretch the spring 3 in., or 0.25 ft, is found as follows:

$$\begin{aligned} F &= KS \\ &= (600\ \text{lb}_f/\text{ft})\ (0.25\ \text{ft}) \\ &= 150\ \text{lb}_f \end{aligned}$$

In SI units,

$$F = 667\ \text{N}$$

Problems

11-57. A car weighing 17 kN is moving 9 m/s. What is its kinetic energy? If the speed is doubled, by how much will the kinetic energy be increased?

11-58. How much potential energy is lost when a cake of ice weighing 1300 N slides down an incline 30 m long that makes an angle of 25° with the horizontal?

11-59. A train weighing 1100 tons is moving fast enough to possess 1.5×10^8 ft · lb_f of kinetic energy. What is its speed in miles per hour?

11-60. A car weighing 12 kN is moving with a speed of 13 m/s. What average force is needed to stop it in 19 m?

11-61. A hammer weighing 1 lb and moving 30 ft/s strikes a nail and drives it $\frac{3}{4}$ in. into a block of wood. What was the average force exerted on the nail?

11-62. A 22-caliber rifle fires a bullet weighing $\frac{1}{15}$ oz with a muzzle velocity of 1020 ft/s. The barrel is 26 in. long. What is the kinetic energy of the bullet as it leaves the muzzle? Assuming that the force on the bullet is constant while it moves down the barrel, what force was exerted on the bullet?

11-63. A ball weighing 11.12 N is dropped from the top of a building 38.1 m above the ground. After the ball is dropped, at what height will the kinetic energy and potential energy be equal?

11-64. It requires a force of 12 N to stretch a spring 3 cm. How much work is done in stretching the spring 9 cm?

11-65. A coil spring has a scale of 70 lb_f/in. A weight on it has shortened it 2.5 in., and when more weight is added it is shortened by an additional 0.75 in. What work was done by the added weight?

11-66. The floor of a car is 13.6 in. from level ground when no one is in the car. When several people whose combined weight is 573 lb_f get in the car, the floor is 11.9 in. from the ground. Assuming that the load was equally distributed to the front and rear wheels, what would be the force constant of the front spring system?

11-67. A weight of 130 N stretches a spring 1.6 cm. What energy is stored in the spring? What is the scale of the spring?

11-68. An iron ball weighing 7.5 lb_f is dropped on a spring from a height of 10 ft. The spring has a force constant of 70 lb_f/in. How far is the end of the spring deflected?

11-69. A bullet weighing 0.289 N and traveling with a velocity of 335 m/s strikes a large tree. Assuming that the bullet meets a constant resistance to motion of 18 kN, how far will the bullet go into the tree?

11-70. What horsepower motor is necessary to raise a 1200-lb_f elevator at a constant velocity of 12 ft/sec? (Assume no loss of power in the hoisting cables.) If the motor is 85 percent efficient, what is the kilowatt input?

11-71. Water flows into a mine that is 100 m deep at the rate of 0.472 m^3/s. What power should be supplied to a pump that is 60 percent efficient if it is to keep the water pumped out?

11-72. An electric motor is driving a pump that is delivering 750 gal of water per minute to a height of 83 ft. The motor has an efficiency of 81 percent and the pump has an efficiency of 73 percent. What power in kilowatts is supplied to the motor?

11-73. A freight train consisting of 60 cars, each weighing 50 tons, starts up a 1.5 percent grade with an initial speed of 15 mi/hr. The drawbar pull is 90 tons and the train resistance, including rolling resistance and air resistance, is 15 lb per ton of weight. At the top of the grade the speed is 30 mi/hr.
a. How long is the grade?
b. How much is the work of the drawbar pull?
c. How much work is done against gravity?

11-74. It is desired to install a hydroelectric station on a certain stream. The cross-sectional area of the stream is 800 ft^2. There is a fall of 48 ft obtainable and the velocity of the stream is 5 mi/hr. What would be the horsepower output assuming an overall efficiency of 75 percent?

11-75. A 1.75-ton car coasting at 15 mi/hr comes to the foot of a 2 percent slope. If it meets a resistance of 12 lb_f/ton on the slope caused by friction and windage, how far up the slope will it go before it stops?

11-76. A 3400-lb_f automobile is traveling 63 mi/hr up a 3 percent grade. The brakes are suddenly applied and the car is brought to a standstill. If the average air resistance is 54 lb and the rolling resistance is 20 lb_f/ton, what must the braking force be to stop the car in 300 ft?

ELECTRICAL THEORY

The Atom

The basis for explaining the behavior of electricity depends on our concept of the atomic structure of matter. Our present concept of the atom is one in which a system of electrons orbits a central nucleus. Some of these electrons in the outer orbit can be transferred to other atoms under the influence of such phenomena as electrical fields, heat, friction, and so on.

Materials differ widely in their tendency to transfer electrons, and all materials can be classified broadly into insulators or conductors as a measure of the ease with which electrons are transferred. For example, if hard rubber is stroked with a woolen cloth, friction will transfer electrons from the cloth to the hard rubber, but since the hard rubber atoms cling tightly to the electrons, little or no movement of the charges can then occur on the surface of the hard rubber.

On the other hand, if a piece of copper is charged, the charges will move readily through the copper, and unless insulating structures are provided, the charges usually will dissipate rapidly to other conducting media.

The concept of conductors or insulators then deals not with the production of electrical charges but rather with the relative ease with which charges are transferred.

Since the electrons appear to be moving in orbits, each electron will tend to produce a magnetic field due to its own motion. In almost every material, the orientation of the spins is such that the magnetic effects cancel and the resultant field is substantially zero. However, in the case of iron, nickel, and cobalt, and some of their alloys, the magnetic fields due to the electron spins do not cancel and the atoms or molecules do have definite magnetic patterns, and the material is magnetic.

In a classical experiment conducted by Robert A. Millikan early in the century, the numerical value of the charge on an electron was measured. As a result of this measurement, we find that approximately 6×10^{18} electrons flow through the filament of an ordinary 100-W 110-V electric light bulb per second.

Electric Currents

If charged particles, usually electrons, move in a conductor, the movement of the charges constitutes what is known as an electric current. Obviously, the charges will not move unless there is an excess of charges at one point and a deficiency at another. In the case of a simple electric cell, the tendency of one of the electrode materials to be chemically changed results in an ionization process that will produce a difference in charges on the electrodes. As long as an external path of conducting material exists, the charges flow from one electrode to another in an attempt to equalize the charges. A coulomb (c) is approximately 6.06×10^{18} electrons and a flow of 1 C/s past a given point in an electrical circuit is defined as a current of 1 ampere (A).

Voltage is basically a potential energy (electrostatic potential energy) per unit charge or a measure of the amount of work necessary to move a unit of charge (e.g., electrons) from one place to another in an electrostatic field. It is analogous to gravitational potential energy being related to the work required to move a mass vertically within a gravitational field. A voltage can be present even though the charges actually are not moving. For example, in a storage battery, a voltage, representing a state of separation of charges within the battery, exists regardless of whether the circuit is completed so that current can flow. This can be compared to having a pile of rocks on a raised platform. Potential energy due to the rock's elevated position is present even though the rocks are not moving.

The usual unit of voltage is the volt. This is the potential necessary to cause a current flow of 1 A through a resistance of 1 ohm Ω (an ohm is defined at the end of this section).

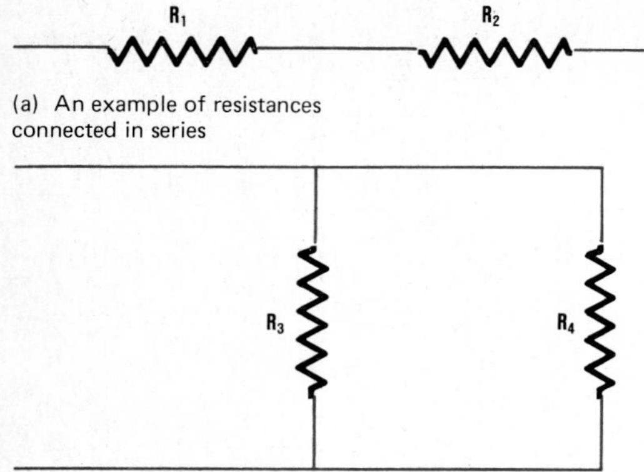

(a) An example of resistances connected in series

(b) An example of resistances connected in parallel

[11-44] Two basic ways in which resistances can be connected.

Resistance to flow of an electric current exists because of the difficulty of moving electrons from one atom to another. All materials have some resistance to current flow except that certain metals at temperatures near absolute zero temperature (approximately −459°F), appear to have negligible resistance. Commonly used materials having quite low resistances at ordinary temperatures are silver, copper, and aluminum. All metals are good conductors; however, the three mentioned are among the best-conducting materials. Other substances having relatively low resistance are carbon and solutions containing ions. Almost without exception, all other materials are insulators having resistances from thousands to millions of times that of the metals. In some cases, insulators at ordinary temperatures will become fairly good conductors at the temperatures of several hundred degrees and upward. Glass and some plastics possess this property of having a markedly lower resistance at elevated temperatures. The unit of resistance is the ohm and it is defined as the resistance of a column of mercury 1 mm^2 in cross section and 106.3 cm long held at a temperature of 0°C.

Laws and Principles

A well-known relation of electrical quantities in a circuit is Ohm's law. Stated briefly, it says that in a conductor, the ratio of the voltage to the current is a constant. Of course, like many laws, it has some limiting conditions, the major one being that the temperature of the conductor must remain constant. In symbol form:

$$\frac{V \text{ (voltage)}}{I \text{ (current)}} = R \text{ (resistance)}$$

This means that in a circuit of fixed resistance, if the voltage of the circuit is doubled, the current (flow) will also double.

There are two basic ways in which circuit elements can be connected. These are series and parallel connections. Examples are given in Figures [11-44 and 11-45].

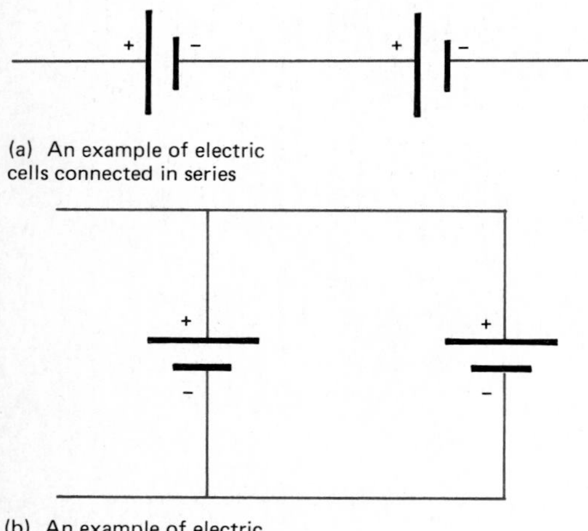

(a) An example of electric cells connected in series

(b) An example of electric cells connected in parallel

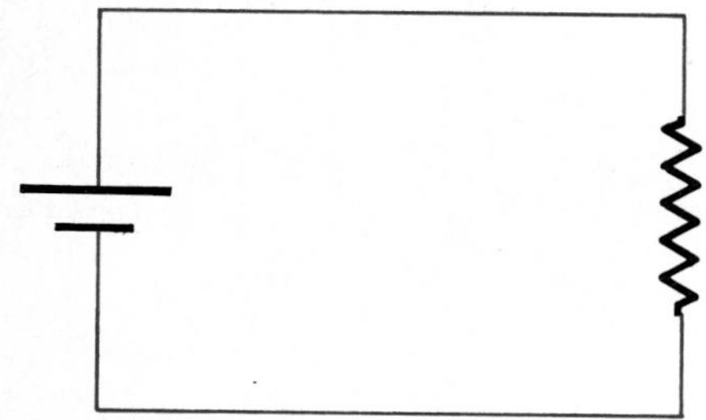

(c) An example of an electric cell and a resistance connected in series

[11-45] Series and parallel arrangements of circuit elements.

Series Circuits

As an example of an application of Ohm's law, if a simple series circuit is sketched showing a voltage source [sometimes referred to as an electromotive force (emf)] with a voltage of 28.3 V in series with a resistance of 2.10 Ω, the current can be computed readily. First, draw a simple sketch using conventional symbols and label the known quantities [11-46]. Second, solve for the unknown quantities. Applying Ohm's law yields

$$\frac{V}{I} = R$$

$$I = \frac{V \text{ (volts)}}{R \text{ (ohms)}}$$

$$= \frac{28.3}{2.10} = 13.48 \text{ A}$$

This assumes that the resistance of the source of the voltage or emf and of the connecting wires is negligible.

For another example, let us take a circuit where several resistances are connected in series as shown in Figure [11-47]. For this type of circuit, first add all the resistances to get a sum which is the equivalent of all the resistances together. This sum is 8.81 Ω. Then apply Ohm's law:

$$I = \frac{V\text{ (volts)}}{R\text{ (ohms)}}$$

$$= \frac{31.8}{8.81} = 3.61\text{ A}$$

Since the circuit elements are all in series, the same current flows through each element. We can compute the voltage across each resistance since a part of the total available voltage is used for each.

As a check, the sum of the individual voltages across the resistances, frequently called the total voltage drop across the resistance, can be obtained and should be the same as the voltage source full voltage, within round-off accuracy.

$$V_1 + V_2 + V_3 = \mathrm{IR}_1 + \mathrm{IR}_2 + \mathrm{IR}_3 = 31.8\text{ V}$$

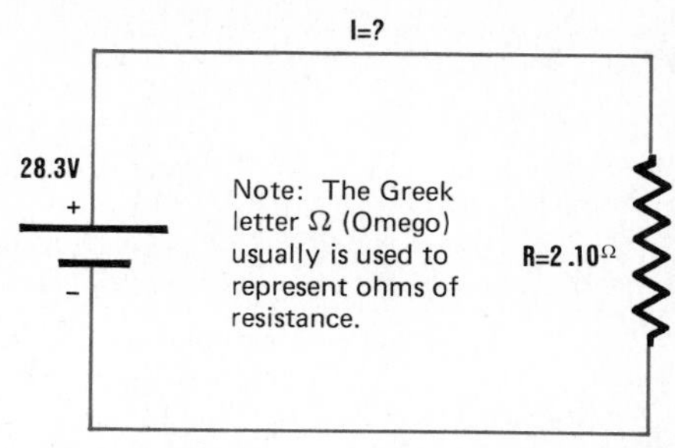

[11-46] Single series circuit using Ohm's law for solution to obtain unknowns.

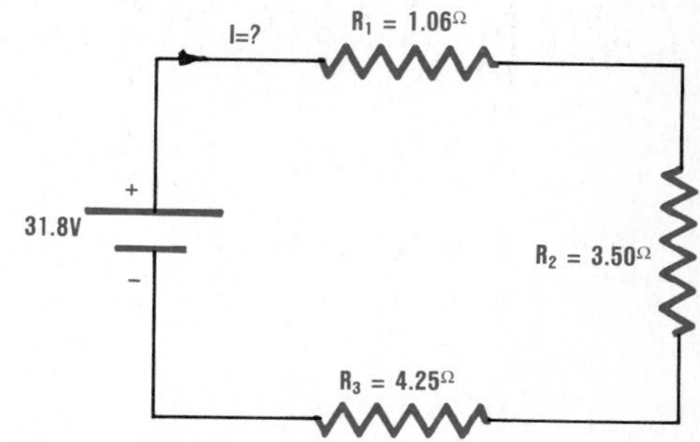

[11-47] Resistances in series.

Parallel Circuits

Figure [11-48] is a sketch of a circuit containing resistances in parallel with a voltage source. To solve for the currents in this circuit, first find the value of a single equivalent resistance that can replace the parallel set. This single equivalent can be found by the expression

$$\frac{1}{R_{\text{equiv}}} = \frac{1}{R_1} + \frac{1}{R_2}$$

$$R_{\text{equiv}} = \frac{1}{\frac{1}{R_1} + \frac{1}{R_2}}$$

$$= \frac{1}{\frac{1}{21.5} + \frac{1}{18.1}} = \frac{1}{0.0466 + 0.0553}$$

$$= \frac{1}{0.1019} = 9.83\ \Omega$$

Using this equivalent resistance in Ohm's law, we obtain

$$I_1 = \frac{V}{R_{\text{equiv}}}$$

$$= \frac{24.8}{9.83} = 2.52\text{ A}$$

Currents I_2 and I_3 can be found in several ways. For instance, electrical currents will divide in inverse ratio to the resistances. This enables a current ratio to be determined, and since the total current (2.52 A) is known, the individual currents can be found. A more

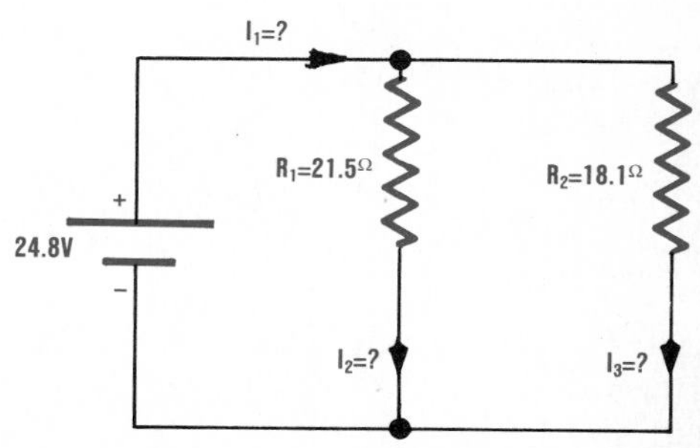

[11-48] Parallel arrangement of resistances.

universal method is to find the current in each resistance by using the *voltage drop* method. Since, in a parallel circuit, the same voltage appears across each resistance, an application of Ohm's law to each branch will permit a solution for the current.

$$I_2 = \frac{V}{R_1}$$

$$= \frac{24.8}{21.5} = 1.15 \text{ A}$$

and

$$I_3 = \frac{V}{R_2}$$

$$= \frac{24.8}{18.1} = 1.37 \text{ A}$$

As a check, $I_2 + I_3$ should add to give the total current out of the voltage source.

$$I_1 = I_2 + I_3$$

$$= 1.15 + 1.37 = 2.52 \text{ A}$$

To summarize, for series circuits, the current in all parts of the circuit is the same and the sum of the voltage drops across the resistances equals the total available voltage of the source. For parallel circuits, the voltage is the same across each parallel path, and the sum of the currents in each branch or path equals the total current supplied by the voltage source.

Series–Parallel Circuits

If a problem involving a series–parallel combination of circuit elements is given, an application of the principles shown above will provide a means of solution.

Example 11-26

Analyze the circuit shown in Figure [11-49].

Solution:

If the parallel arrangement of resistances R_2 and R_3 can be combined into a single equivalent resistance, the circuit then will be a single series circuit and a method of determining currents or voltages will be available as was used in a previous example.

The equivalent resistance of R_2 and R_3 will be

$$\frac{1}{R_{equiv}} = \frac{1}{R_2} + \frac{1}{R_3}$$

$$= \frac{1}{6.88} + \frac{1}{5.26}$$

$$= 0.146 + 0.190$$

$$= 0.336$$

$$R_{equiv} = \frac{1}{0.336} = 2.98 \ \Omega$$

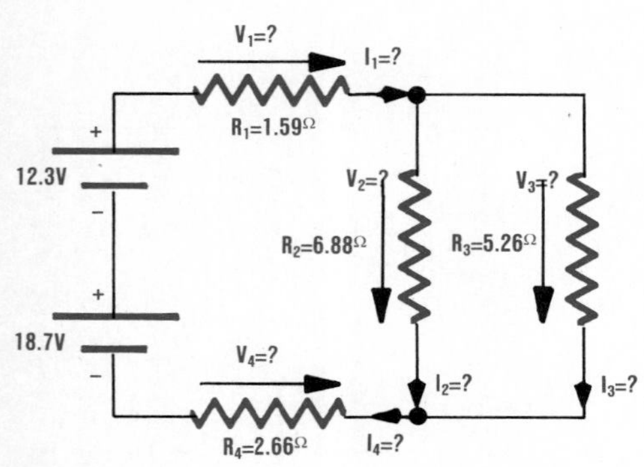

[11-49] Series–parallel arrangement of resistances.

This means that if the parallel combination were replaced by a single 2.98-Ω resistance, the current and voltage values in the remainder of the circuit would be unchanged. The circuit can then be redrawn substituting R_{equiv} for R_2 and R_3 as shown in Figure [11-50].

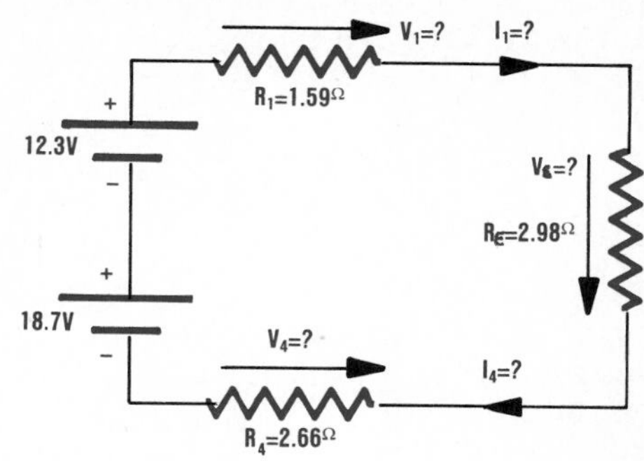

[11-50] Equivalent circuit of [11-49].

First, obtain the total voltage of the voltage sources. This is simply the sum of the individual voltages of the sources.

$$V_{total} = 12.3 + 18.7$$
$$= 31.0 \text{ V}$$

Second, find the total circuit resistance. For this circuit, it is the sum of the individual resistances in series.

$$R_{total} = 1.59 + 2.98 + 2.66$$
$$= 7.23 \ \Omega$$

Third, find the total circuit current. This is found by an application of Ohm's law using total voltage and total resistance.

$$V_1 = \frac{V_{total}}{R_{total}}$$
$$= \frac{31.0}{7.23} = 4.28 \text{ A}$$

Since, in a series circuit, the total current is the same as the current in each part, the current through each resistance also is 4.28 A. From this, we can obtain the voltage drop across each resistance by applying Ohm's law only to that part of the circuit:

$$V_1 = I_1R_1$$
$$= (4.28)(1.59) = 6.80 \text{ V}$$
$$V_E = I_1R_E$$
$$= (4.28)(2.98) = 12.78 \text{ V}$$
$$V_4 = I_1R_4$$
$$= (4.28)(2.66) = 11.42 \text{ V}$$

As a check, the sum of V_1, V_E, and V_4 should be the same as the total voltage from the voltage sources.

Fourth, referring to Figure [11-49], we can now solve for the currents I_2 and I_3. Since the voltage across the equivalent resistance was 12.78 V, this will also be the voltage across each member of the parallel set. That is,

$$V_E = V_2 = V_3 = 12.78 \text{ V}$$

The current I_2 and I_3 can be found by applying Ohm's law only to that part of the circuit.

$$I_2 = \frac{V_2}{R_2}$$
$$= \frac{12.78}{6.88} = 1.86 \text{ A}$$

and

$$I_3 = \frac{12.78}{5.62} = 2.43 \text{ A}$$

As a check, $I_2 + I_3$ should equal I_1 or I_4.

Problems

11-77. List in order of increasing unit resistance, the 10 best metallic conductors. In a word or two, give major advantages and disadvantages of using each as an electric conductor for power circuits.

11-78. A resistance of 3.65 kΩ is connected in series with 920 Ω. What is the combined resistance? If these two resistors are reconnected so that they are in parallel, what will be the equivalent resistance?

11-79. Three resistors having values of 128 Ω, 144 Ω, and 98.2 Ω, respectively, are connected in series. What will be their combined resistance? If these three resistances are reconnected so that they are in parallel with each other, what will their equivalent resistance be?

11-80. A circuit is suspected of having damaged insulation at some place in an aircraft. To check the insulation, a battery having a voltage of about 50 V is connected to the ship's metal structure and in series with the suspected circuit using a microammeter having an internal resistance of 100 Ω. If the microammeter reads 7.4 μA, what is the approximate resistance to ground of the circuit?

11-81. A battery having an internal resistance of 0.01 Ω and an open-circuit voltage of 27.6 V is connected to a starter on an aircraft. If the starter resistance while not turning is 0.10 Ω and the line resistance of the connecting wires is 0.03 Ω, what maximum current can flow through the starter? What will be the voltage across the starter at the instant of closing the starting circuit?

11-82. Power in W in a dc electric circuit is defined as the product of current in amperes and voltage in volts. If a 100-W lamp is connected to a 117-V line, what current will flow through the lamp? If a 40-W lamp is connected in parallel with the 100-W lamp, what total current will need to flow in the line supplying both lamps?

11-83. In the circuit of Figure [11-49], if the voltage of the cells is changed to an unknown amount but the current I_1 is measured to be 7.03 A, what will be the values of V_1, V_2, V_3, V_4, I_2, I_3, I_4, and total cell voltages?

11-84. A dc shunt is to be made to permit the measurement of starting currents in an automotive starter. The expected current should not exceed 200 A from a 12-V system. What should be the resistance of the shunt so that a current of 200 A through it will produce a voltage drop of 50 mV across it?

11-85. The life of an incandescent lamp varies inversely as the twelfth power of the applied voltage. If the rated life of a lamp is 800 hr at 117 V, what would be the expected life if operated continuously at 120 V? What would be the expected life if operated at 110 V?

11-86. If energy cost is 7 cents per kilowatt-hour, what will be the approximate cost of operating a 100-W lamp an average of 5 hr/day for a month?

11-87. A series circuit is made up using a 10,000-Ω resistance, a 3000-Ω resistance, and an ammeter having a resistance of 720 Ω, all connected to a battery. If the ammeter shows a current flow of 3.03 mA, what voltage is supplied by the battery? What voltage drop would exist across the 3000-Ω resistance when this current is flowing?

11-88. In the circuit of Figure [11-49] R_1 is 321 Ω, R_2 is 1080 Ω, R_3 is 844 Ω, R_4 is 112 Ω, and I_1 is 39.5 mA. What will be the amount of the applied battery voltage, voltage V_2, voltage V_3, currents I_2 and I_3, and voltage V_4?

11-89. By applications of Ohm's law, show by derivation that $P = V^2/R$ and that $P = I^2R$ can be obtained from the expression $P = VI$.

11-90. The heating element of a cookstove is rated at 300 W and 220 V. What current will flow when the element is turned on? To how many horsepower would 3000 W be equivalent (assuming dc)?

11-91. If energy costs 7 cents per kilowatt-hour, what would be the approximate cost of operating a 100-W lamp for 1 hr?

11-92. An electric iron is rated at 660 W, 110 V. What will be the approximate resistance of the heating element? Is it likely that copper wire is used for the heating element? Describe desirable properties that the heating element conductor should have (assuming dc).

Kirchhoff's Laws

Gustav Kirchhoff formulated two laws which bear his name and are very helpful in the analysis of the properties of simple circuits. The first law deals with currents, and the second law is associated with voltages in a closed loop.

Kirchhoff's first law states that the algebraic sum of the currents at any junction point (or node) must be zero. This law follows from the fact that charges cannot be stored at a junction point—whatever charges enter the point must also leave the point with their magnitudes unchanged. The first law is demonstrated for three and four currents in various directions in Figure [11-51].

Kirchhoff's second law applies to circuit elements that are part of a closed loop. It states that the algebraic sum of the potential differences across the circuit elements forming the loop must be zero. This law is based on the fact that the work involved in moving an electric charge from any given point completely around the closed loop and back to that point must be zero. Consider Figure [11-52]. Suppose that the voltage or source of electromotive force (emf) for the closed-loop circuit is 10 V. Suppose further that the emf value accounts for the small (internal) resistance at the source. It is desired to find the value of the current, I, in the closed loop and the voltages across each resistance. We will adopt three conventions: (1) the positive direction for current flow will always be clockwise in any loop regardless of the polarity at the source; (2) the arrow representing the sense of direction of the emf voltage will be drawn so that the head of the arrow always points to the positive pole; and (3) the arrow representing the sense or direction of the voltage at each resistor will be oriented so that the head of the arrow points in a direction opposite of the assumed positive current flow. Voltages will be treated as positive if the current travels in the same direction as the voltage arrow, and negative otherwise. It should be recognized that these conventions are for ease in problem solving and may not coincide with the physics of the problem. Furthermore, these conventions may differ from those suggested in other books. However, the solutions will be the same regardless of the conventions used. If negative values result for the current or voltages across the resistors when the problem is solved, then the actual direction of flow of the current and polarity of the voltages across the resistors is opposite to that which was assumed.

Applying Kirchhoff's second law to the loop of Figure [11-52], we have

$$10 - 4 \cdot I - 5 \cdot I - 6 \cdot I = 0$$

or

$$I = 0.67 \text{ A}$$

and the voltages across the resistors are

$$V_{R1} = 0.67(4) = 2.7 \text{ V}$$
$$V_{R2} = 0.67(5) = 3.3 \text{ V}$$
$$V_{R3} = 0.67(6) = 4.0 \text{ V}$$

Now consider the two-loop circuit of Figure [11-53]. Note that R_2 is shared by each loop. The net current through R_2 will be $I_1 - I_2$. Applying Kirchhoff's second law to each loop, we have

$$10 - 4I_1 - 5I_1 - 6I_1 + 5I_2 = 0$$
$$-15 - 5I_2 - 7I_2 - 8I_2 + 5I_1 = 0$$

or

$$10 - 15I_1 + 5I_2 = 0$$
$$-15 + 5I_1 - 20I_1 = 0$$

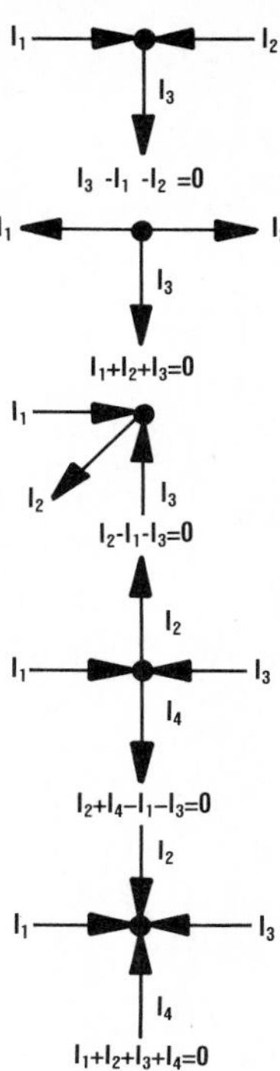

[11-51] Illustrations of Kirchhoff's first law.

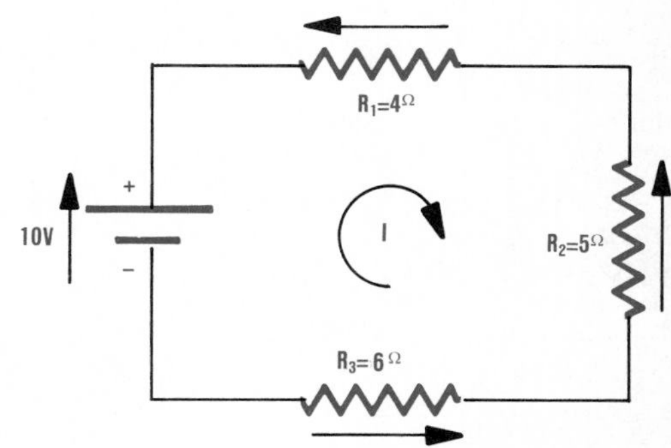

[11-52] Single-loop circuit.

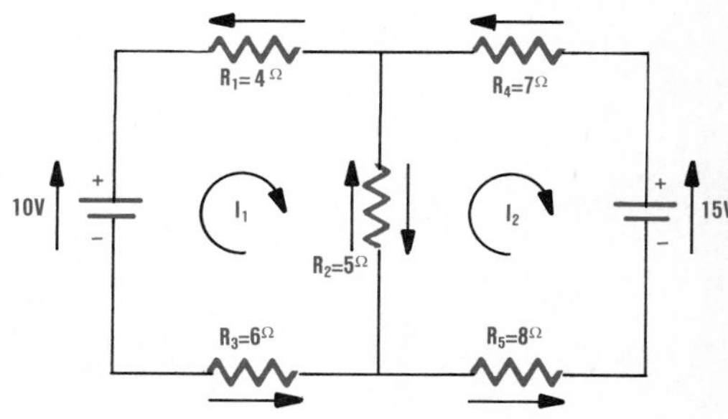

[11-53] Double-loop circuit.

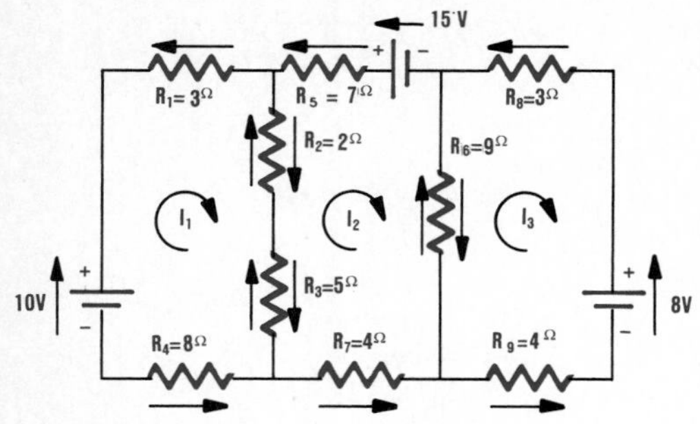

[11-54] Three-loop circuit.

Hence

$$I_1 = 0.45 \text{ A}$$
$$I_2 = -0.64 \text{ A}$$

and the voltages across the resistors are:

Resistor	***Voltage***
1	0.45(4) = 1.8
2	[0.45 − (−0.64)](5) = 5.5
3	0.45(6) = 5.5
4	0.64(7) = 4.5
5	0.64(8) = 5.1

Note that $V_1 + V_2 + V_3 = 10$ = emf for loop 1 and that $V_2 + V_4 + V_5 = 15$ = emf for loop 2.

Our third example of Kirchhoff's second law features three closed loops, each with an emf source [11-54]. The voltage equations are:

$$10 - 3I_1 - 2I_1 - 5I_1 - 8I_1 - 2I_2 - 5I_2 = 0$$
$$-15 - 9I_2 - 4I_2 - 5I_2 - 2I_2 - 7I_2 + 5I_1 + 2I_1 + 9I_3 = 0$$
$$-8 - 4I_3 - 9I_3 - 3I_3 + 9I_2 = 0$$

or

$$10 - 18I_1 + 7I_2 = 0$$
$$-15 + 7I_1 - 27I_2 - 9I_3 = 0$$
$$-8 + 9I_2 - 16I_3 = 0$$

Hence $I_1 = 0.24$ A, $I_2 = -0.81$ A and $I_3 = -0.96$ A.

Power

Electric power is determined in dc circuits[10] by the product of current and voltage. That is,

$$P = VI$$

where P is the power in watts, V is the voltage in volts, and I is the current in amperes. This expression can be applied to a part of a circuit, but then only the current and voltage in that part should be used.

Example 11-27

Refer to Figure [11-49]. Suppose that we need to determine the power dissipated in resistance R_2 and the total power required by the voltage source.

Solution:

For the power dissipated in resistance R_2, we use values only for that part.

[10]ac circuits are beyond the scope of this book.

$$P_R = V_2 I_2 \text{ (V)(A)}$$
$$= 12.78(1.86)$$
$$= 23.7 \text{ W}$$

For the battery power, we use the total voltage and current values.

$$P_B = V_B I_1$$
$$= (12.3 + 18.7)(4.28)$$
$$= 133 \text{ W}$$

By algebra it can be shown that power can also be found by these expressions:

$$P = \frac{V^2}{R} \quad \frac{(\text{volts})^2}{\text{ohms}}$$
$$= I^2 R$$

PROBLEMS

Use Kirchhoff's second law to determine the I values in each of the following problems.

11-93.

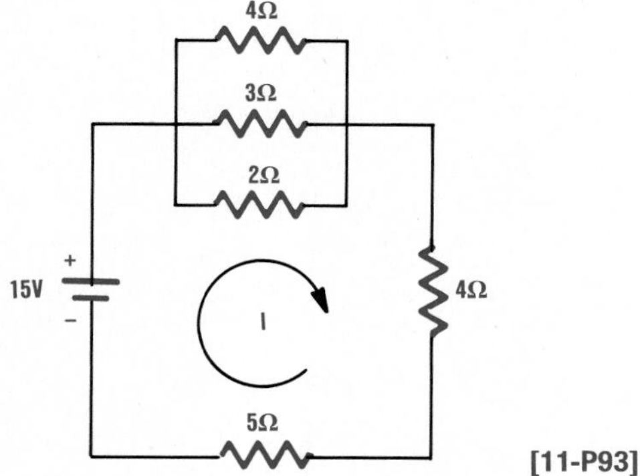

[11-P93]

11-94.

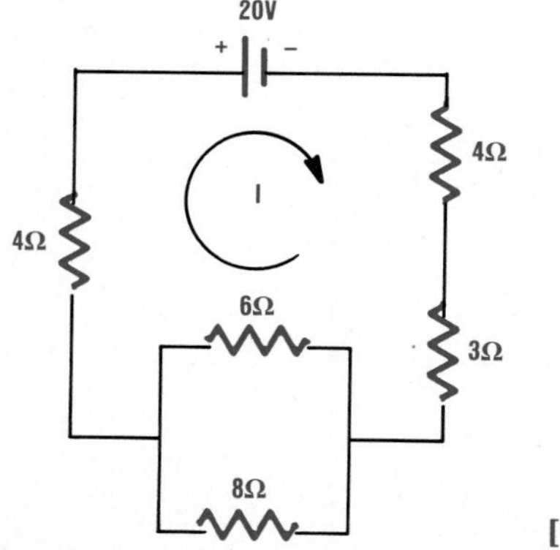

[11-P94]

11-95.

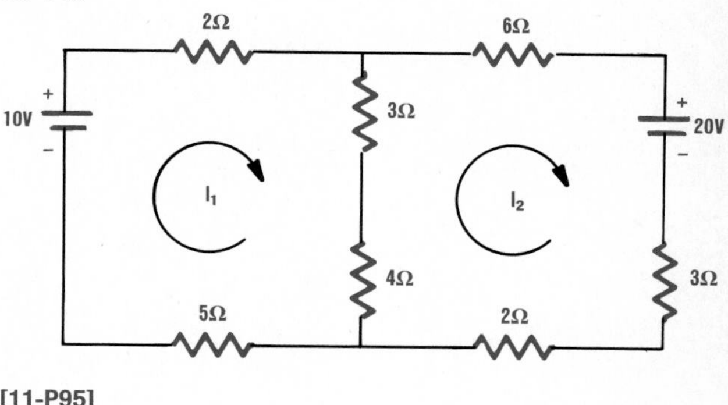

[11-P95]

11-96.

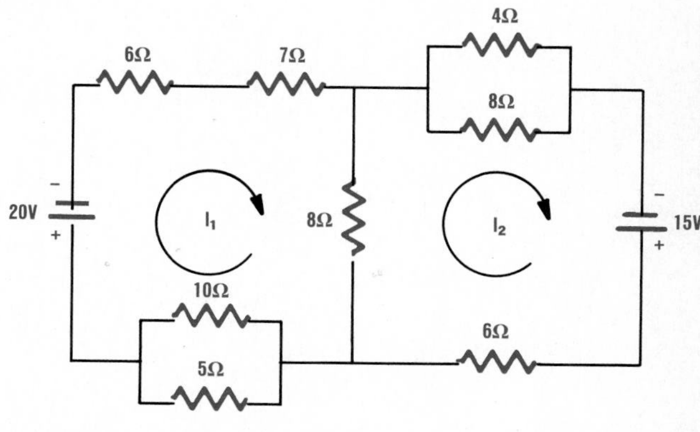

[11-P96]

11-97.

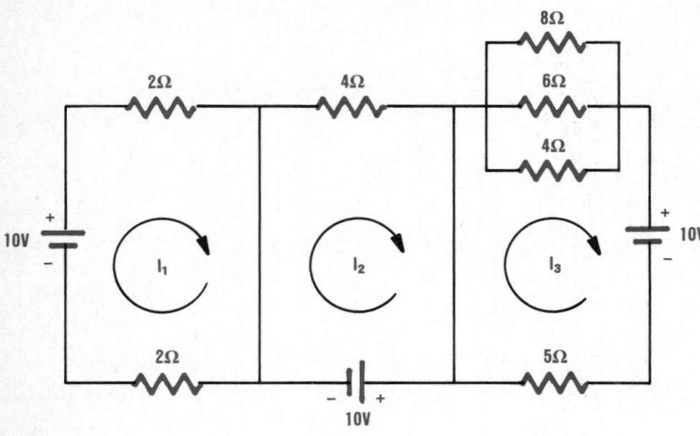

[11-P97]

11-98.

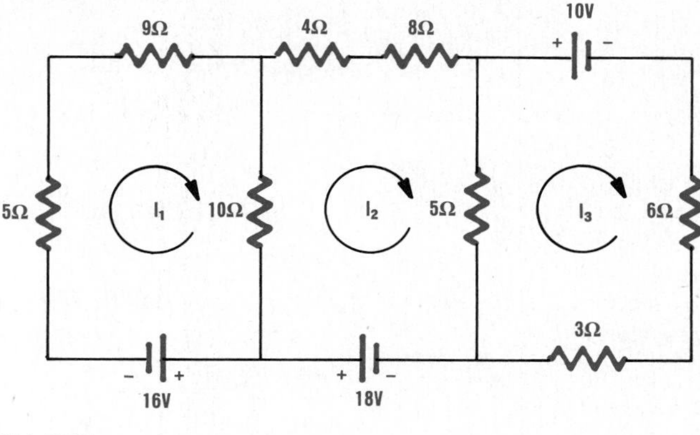

[11-P98]

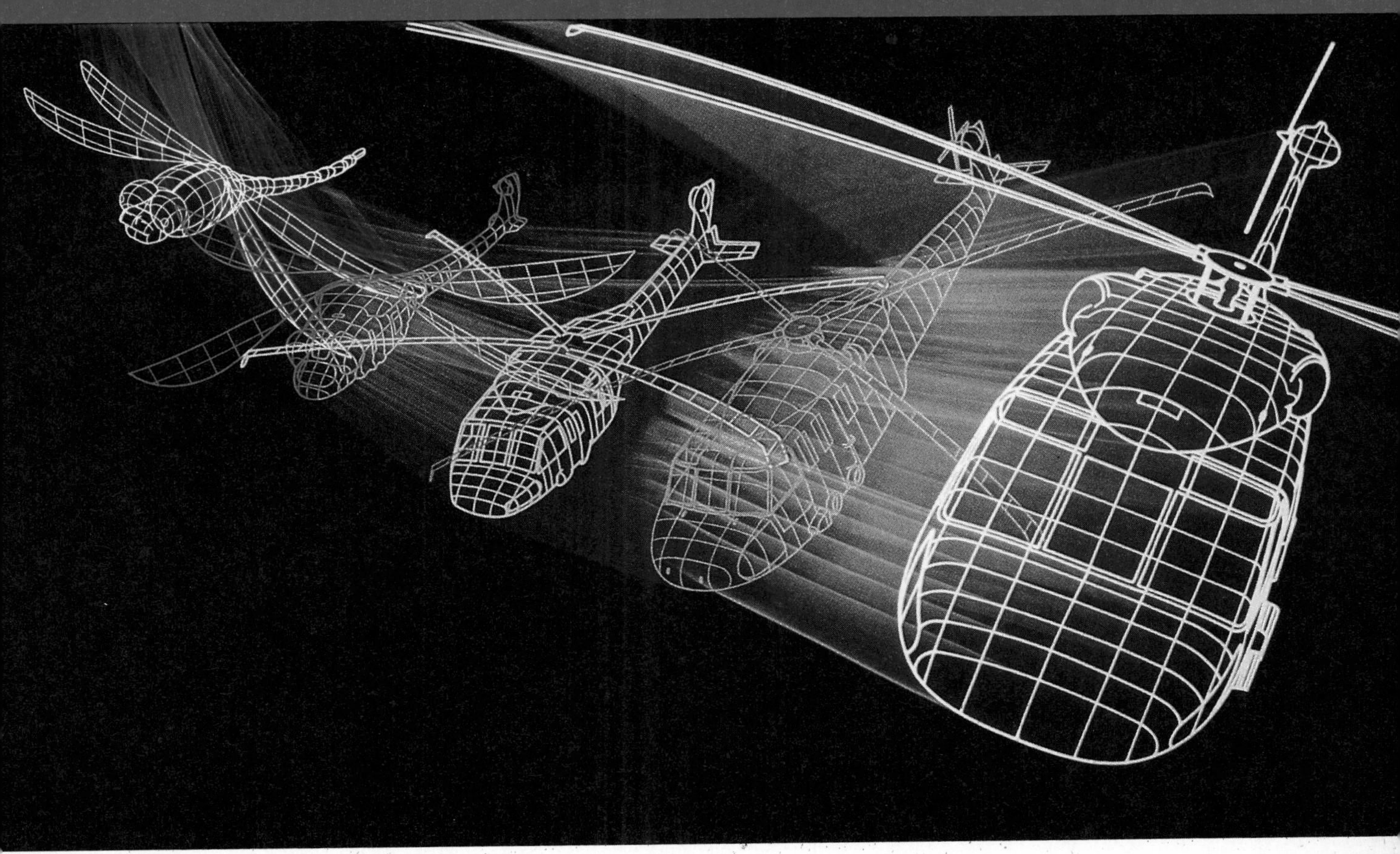

Section Three

Engineering Design

Chapter 12

The Process of Engineering Design

Through experience it has been found that the efficient and effective design of engineering elements or systems follows a logical process. This design process is actually just a more complete type of engineering problem-solving strategy than was described in Chapter 9. The solutions to engineering design problems most often take the form of the development, creation, or invention of an engineered device, structure, or system. The procedure or "plan of attack" used by the best design engineers is similar, regardless of the type of problem being solved. This chapter will develop these stages of the design process. Special attention will be given to the "blocks," or obstacles, which are frequently present and hinder problem-solving activities. These same stages of engineering design are also used in the solution of large-scale engineering problems or projects which differ from routine engineering design only in the complexity and amount of creativity and inventiveness required. Thus when the term "problem solving" is used in this chapter, it will also refer to all the efforts associated with large-scale problems, projects, or designs.

. . . the process of design, the process of inventing physical things which display new physical order, organization, form, in response to function.

Christopher Alexander, Notes on the Synthesis of Form

A scientist can discover a new star but he cannot make one. He would have to ask an engineer to do it for him.

Gordon L. Glegg

THE STAGES OF ENGINEERING DESIGN

Figure [12-1] illustrates conceptually how the stages of the design process are related. The stages are:

1. Identification of the problem.
2. Analysis.
3. Transformation.
4. Idea development.
5. Modeling.
6. Information gathering.
7. Experimentation.
8. Synthesis.
9. Evaluation and testing.
10. Presentation of the solution.

In general it has been observed that successful problem solvers, design engineers in particular, include most of these steps in their everyday efforts to attain a particular problem solution or project goal. In Figure [12-1] we can see that although the stages of the

If you make people think they're thinking, they'll love you; but if you *really* make them think, they'll hate you.

Don Marquis

[12-1] The design process.

The evolution of the concept of feedback can be traced through three separate ancestral lives: the water clock, the thermostat, and mechanisms for controlling windmills.

Otto Mayer
"The Origins of Feedback Control," Scientific American, *October 1970*

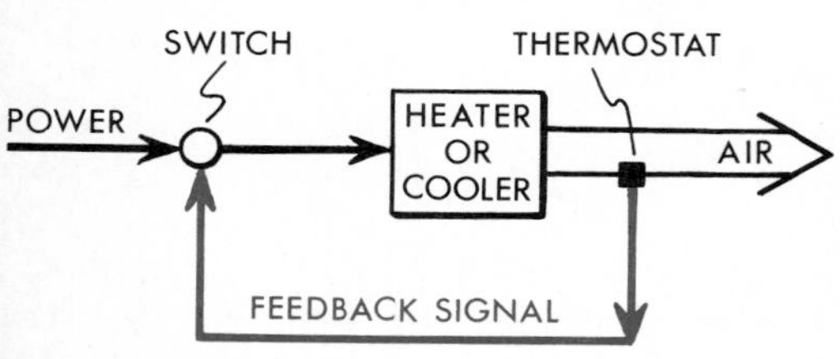

[12-2]

design process occur in sequential order, there is often a need to drop back (give a feedback signal) or "loop back" to an earlier stage (not necessarily the immediately preceding stage), or to expand or modify the extent of a stage as a result of information that has been learned. The feedbacks, backtracking, or looping to prior stages may be an infrequent occurrence when small problems are involved [12-2]. However, with large-scale problems or complex designs, there will be many returns to prior stages before the problem can be solved. Each return to a prior stage will probably result in changes or additions to the developments made in all the intervening stages between the current stage and the "return" stage. It is occasionally necessary to "jump ahead" a step or two before completing the activities at a current step. For example, the designer may find it helpful to obtain some information during the "idea development" or "modeling" steps and therefore would proceed to "information gathering" before completing either "idea development" or "modeling." The rate at which you proceed through the problem-solving cycle is a function of many factors, and these factors change with each problem. Depending on the situation, either considerable time or very little time may be spent at any stage within the cycle. Thus the design process is a dynamic and constantly changing process that provides allowances for the individuality and capability of the user. The importance of a particular stage within the process varies with the nature of the problem being solved and, in the case of engineering design, the importance of a stage also varies according to the phase of the design (see Chapter 14).

Identification of the Problem

Every problem must be understood as completely as possible before attempting to bring about a solution. The types of problems that are usually presented for solution in engineering textbooks differ markedly from those the engineer faces in real-life situations. In an actual work environment the problems are often poorly defined. In textbooks, problems are most often carefully defined. Also, in real life those who propose that there are identifiable problems that need solving are themselves frequently unsure of exactly what is wanted (the desired outcome). In fact, the precise nature of the desired outcomes may be unspecified or ambiguous. The problem solver should ask: *What is not wanted?* as well as *What is wanted?* Similarly, answers to the questions *What must be done?* and *What must not be done?* should be obtained. Problem identification also includes definitive statements about how the inputs (or the "known information") of the problem relate to the desired outcome. It may be discovered, for example, that one or more of the inputs has no effect on the desired outcome, and hence may be disregarded. The problem identification stage should also include definitive statements about the scope of the problem. That is, the problem boundaries should be defined. For example, it might be decided that a traffic control study at a large airport should be restricted to an evaluation of commercial aircraft traffic. Anticipated difficulties or constraints in solving the problem should also be identified at this stage. In large-scale problem solving or design, such constraints usually include the time, facilities, equipment, manpower, and money available for use on the project, as well as the limitations of potential sources of information. Finally, the problem solver must know how the problem solution is to be judged, since this will affect the evaluation process.

[12-3] In the design process one of the most difficult tasks is to define the problem accurately.

The engineer's first problem in any design situation is to discover what the problem really is.

What's the ideal airplane?

The ideal plane is fast as light, powerful, highly automated with gadgets that do everything but brush the automatic pilot's teeth.

Punctual and dependable, it gets in and out of airports quickly and safely.

For efficiency, it has a huge capacity. And every seat is always filled.

It's a paradise with wings, with all the comforts of home and then some.

LOUMYER

[12-4]

The mere formulation of a problem is far more often essential than its solution, which may be merely a matter of mathematical or experimental skill. To raise new questions, new possibilities, to regard old problems from a new angle requires creative imagination and marks real advances in science.

Albert Einstein, 1879–1955

One of our problems is trying to find out which way is up and which way is down.

John Young
Astronaut, Apollo Ten

Important ideas are those that lie within the allowable scope of nature's laws.

Reason can answer questions, but imagination has to ask them.

Creativity is the art of taking a fresh look at old knowledge.

All men are born with a very definite potential for creative activity.

John E. Arnold

The mind is not a vessel to be filled but a fire to be kindled.

Plutarch

It takes courage to be creative. Just as soon as you have a new idea, you are a minority of one.

E. Paul Torrance

Disciplined thinking focuses inspiration rather than constricts it.

It is better to wear out than to rust out.

Bishop Richard Cumberland

[12-5] Behold the turtle, he makes progress only when his neck is out.

James B. Conant
President, Harvard University

Analysis

The analysis stage refers to a decomposition of the problem into separate, distinct, and manageable subproblems. At this stage, consideration should be given to the order in which the problem will be solved, the functions that are involved, the techniques that will be applied, the resources that are currently available, and the restrictive time requirements. This information will also simplify the selection of the subproblems. In large-scale engineering projects, subproblem identification will greatly assist in the selection of individuals with the appropriate training and skills to manage the activities associated with each subproblem. A return (loop back) to the problem identification stage will be made for each subproblem so that clear subproblem identifications will result. At this point, timetables for achieving the desired outcomes are developed for each subproblem, and estimated completion times are noted for each of the activities to be completed. Plans are then developed for progress reporting and the execution of corrective action when difficulties occur.

Transformation

The first two stages were concerned primarily with unknown or unfamiliar conditions and circumstances. At this third stage, the problem solver moves from the unfamiliar to the familiar. Here the following questions might be asked: "Have I (or we) worked on a similar problem before?", "What does this problem have in common with other problems which I (or we) have solved or projects already completed?", "Can the problem be redefined at the problem identification stage so as to conform to already familiar patterns of problem solving?", "What information would be needed, or what could be added or deleted at the problem identification stage to make solution procedures more apparent?", and "Can the subproblem configurations be rearranged or regrouped to more closely align with familiar patterns?" The answers to these questions should lead the problem solver to:

- Identify similarities with other, more familiar problems.
- Obtain information to translate this problem to a more easily recognizable problem.
- Modify problem identifications.
- Rearrange or regroup subproblems.

Idea Development

Today the ability to think creatively is one of the most important assets that all men[1] possess. The accelerated pace of today's technology emphasizes the need for conscious and directed imagination and creative behavior in the engineer's daily routine. Creativity is a human endeavor. It presupposes an understanding of human experience and human values, and it is without doubt one of the highest forms of mental activity. In addition to requiring innovation, crea-

[1]In discussing activities of mankind—the human species—the authors use the generic word *man* to represent both male and female.

tive behavior requires a peculiar insight that is set into action by a vivid but purposeful imagination—seemingly the result of a divine inspiration that some often call a "spark of genius." Indeed, the moment of inspiration is somewhat analogous to an electrical capacitor that has "soaked up" an electrical charge and then discharges it in a single instant. To sustain creative thought over a period of time requires a large reservoir of innovations from which to feed. Creative thought may be expressed in such diverse things as a suspension bridge, a musical composition, a poem, a painting, or a new type of machine or process. Problem solving, as such, does not necessarily require creative thought, because many kinds of problems can be solved by careful, discriminating logic.

The engineer who redesigns a computer or improves an automobile engine uses established techniques and components; he or she synthesizes. Innovators are those who build something new, and who combine different ideas and facts with a purpose. We only call those individuals "creative" who originate, make, or cause to come into existence an entirely new concept or principle. (Patents are mostly the result of clever innovation rather than creative effort.) If we had to rely on creativity for patents, we would not have the over 4 million patents presently registered in the United States. All engineers must synthesize, some will innovate, but only a very few are able to be truly creative.

Years ago most American youths were accustomed to using innovative and imaginative design to solve their daily problems. Home life was largely one of rural experience. If tools or materials were not available, they quickly improvised some other scheme to accomplish the desired task. Most people literally "lived by their wits." Often it was not convenient, or even possible, to "go to town" to buy a clamp or some other standard device. Innovation was, in many cases, "the only way out." A visit to a typical midwestern farm or western ranch today, or to Peace Corps workers overseas, will show that these innovative and creative processes are still at work. However, today most American youths are city or suburb dwellers who do not have many opportunities to solve real physical problems with novel ideas.

A person is not born with either a creative or noncreative mind, although some are fortunate enough to have exceptionally alert minds that literally feed on new experiences. Intellect is essential, but it is not a golden key to success in creative thinking. Intellectual capacity certainly sets the upper limits of one's innovative and creative ability; nevertheless, motivation and environmental opportunities determine whether or not a person reaches this limit. Surprisingly, students with high IQs are not necessarily inclined to be creative. Recent studies have shown that over 70 percent of the most creative students do not rank in the upper 20 percent of their class on traditional IQ measures.

Everyone has some innovative or creative ability. For the average person, due to inactivity or conformity, this ability has probably been retarded since childhood. If we bind our hand or foot (as was practiced in some parts of the Orient) and do not use it, it soon becomes paralyzed and ineffective. But unlike the hand or foot, which cannot recover full usefulness after long inactivity, the dormant instinct to think creatively may be revived through exercise and

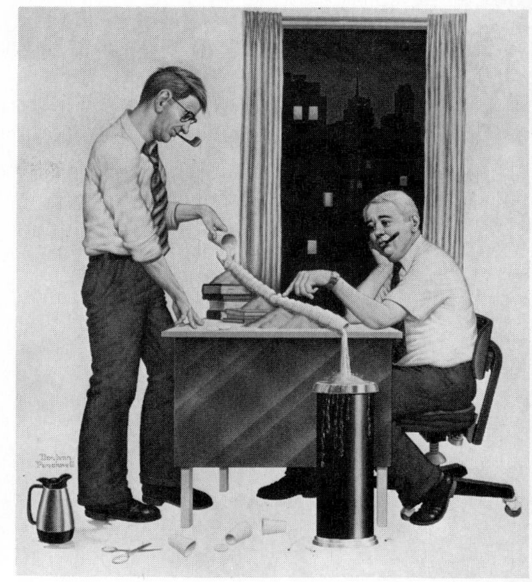

[12-6] Engineers are motivated to work at a particular task partly because of the exhilaration, thrill, special satisfaction, pride, and pleasure they get from completing a creative task. However, the path to success is not always easy, and on occasion frustration is rampant.

A man must have a certain amount of intelligent ignorance to get anywhere.

Charles F. Kettering

If you want to kill an idea, assign it to a committee for study.

. . . every idea is the product of a single brain.

Bishop Richard Cumberland

Society is never prepared to receive any invention. Every new thing is resisted, and it takes years for the inventor to get people to listen to him and years more before it can be introduced.

Thomas Alva Edison, 1847–1931

[12-7] I wonder . . .

[12-8] The mind can die from inactivity.

Creativity is man's most challenging frontier!

A child is highly creative until he starts to school.
Stanley Czurles, Director of Art Education. New York State College for Teachers

Imagination is more important than knowledge.
Albert Einstein, 1879–1955

More today than yesterday and more tomorrow than today, the survival of people and their institutions depends upon innovation.
Jack Morton, Innovation, *1969*

The age is running mad after innovation. All the business of the world is to be done in a new way. Men are to be hanged in a new way.
Samuel Johnson, 1777

An inventor is simply a fellow who doesn't take his education too seriously.
Charles F. Kettering, 1876–1958

Everybody is ignorant, only in different subjects.
Will Rogers, 1879–1935

Getting an idea should be like sitting down on a pin: it should make you jump up and do something.
E. L. Simpson

A first rate soup is more effective than a second rate painting.
Abraham Maslow
Creativity in Self-actualizing People

stimulated into activity after years of nearly suspended animation. Thus everyone can benefit from studying the creative and innovative processes and the psychological factors related to them.

Idea development is the most critical stage of design, and it is at this stage that individual creativity plays a critical role. There are those who feel that this part of the process is an art and cannot be effectively taught, or at the very least is difficult to teach. Nevertheless, a number of methods have been developed for stimulating ideas. Certain of these procedures will work satisfactorily in one situation, while at other times different methods may be needed. Let us examine some of the more successful methods of idea development. The first is *attribute listing.*

Attribute Listing. The most common method of idea generation involves listing all the attributes associated with a particular situation and then combining each attribute with other ideas for improvement. For example, all physical objects have observable attributes such as material, weight, color, shape, and size. For each attribute or feature listed we might ask "Can we modify, rearrange, eliminate, combine, magnify, minify, substitute, reverse, or adapt another attribute?" The answers to these questions may bring to mind significant improvements that could be made in the design of a system or device. The attributes of other situations not involving physical objects may also be listed. For example, suppose that the situation involves the improvement of a maintenance procedure for a manufacturing facility. The attributes of this situation would include such factors as the frequency of maintenance, the number of personnel involved, the skill levels of the personnel, the replacement policy, the provisions for spare parts, and the test and diagnostic procedures to be used when failures occur. The same questions that are asked about each attribute of a physical object would also be appropriate to be asked here. For example, *magnification* with regard to skill levels might lead us to the conclusion that more personnel should be involved. *Substitution,* as applied to replacement policy, might suggest using throwaway replacement parts rather than using repairable replacement parts, and so on.

Matrix Checklist. This method is often used to make a systematic search of possible combinations of several design parameters. In effect, the matrix checklist approach allows several system variables to be considered at the same time. In this way, the combined effects (or interaction) of the attributes of each variable may be examined. The necessary steps for implementing this type of idea search are as follows:

1. Describe the problem. This description should be broad and general, so that it will not exclude possible solutions.
2. Select the major independent-variable conditions required in combination to describe the characteristics and functions of the problem under consideration.
3. List the alternative methods that satisfy each of the independent-variable conditions selected.

4. Establish a matrix with each of the independent-variable conditions as one axis of a rectangular array. Where more than three conditions are shown, the display can be presented in parallel columns.
5. Consider the combinations of conditions that are shown in the matrix. Select the most promising for further investigation.

Let us consider a specific example to see how these steps can be applied.

1. Problem statement: A continuous source of contaminant-free water is needed.
2. List the independent-variable conditions.
   a. Energy
   b. Source
   c. Process
3. List the possible methods of satisfying each condition.
   a. Types of Energy:
      (1) Solar
      (2) Electrical
      (3) Fossil
      (4) Atomic
      (5) Mechanical
   b. Types of sources:
      (1) Underground
      (2) Atmosphere
      (3) Surface supply
   c. Types of process:
      (1) Distillation
      (2) Transport
      (3) Manufacture
4. Form a matrix.
5. Consider the resulting combinations.

A matrix for this particular set of conditions may be represented as an orderly arrangement of 45 small blocks stacked to form a rectangular parallelepiped [12-9]. Every block will be labeled with the designations selected previously. Thus block *X* in our preceding example suggests obtaining pure water by distilling a surface supply with a solar-energy power source, block *Y* means transporting water from an underground source by some mechanical means, and block *Z* recommends manufacturing water from the atmosphere using atomic power. Obviously, some of the blocks represent well-known solutions, and others suggest absurd or impractical possibilities. But some represent untried combinations that deserve investigation.

Where more than three variables are involved, computers may be used to excellent advantage. After the matrix has been programmed, the computer can print a list of all the alternative combinations. Use of the computer is especially helpful when considering a large number of parameters.

The preceding techniques of stimulating new design concepts are particularly useful for the individual engineer. But often several designers may be searching jointly for imaginative ideas about some

"I can't believe that," said Alice. "Can't you?" the Queen said in a pitying tone. "Try again; draw a long breath and shut your eyes." Alice laughed. "There's no use trying," she said, "one can't believe impossible things."

"I daresay you haven't had much practice," said the Queen. "When I was younger, I always did it for half an hour a day. Why, sometimes I've believed as many as six impossible things before breakfast."

Lewis Carroll, 1832–1898
Through the Looking Glass

We do not have to teach people to be creative; we just have to quit interfering with their being creative.

Ross L. Mooney

BEWARE! Don't become victimized by habit.

Necessity may be the mother of invention, but imagination is its father.

They can have any color they want . . . just as long as it's black.

Henry Ford, 1863–1947

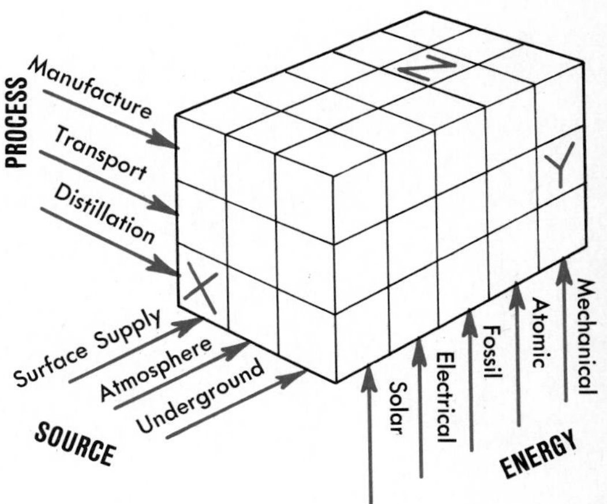

[12-9]

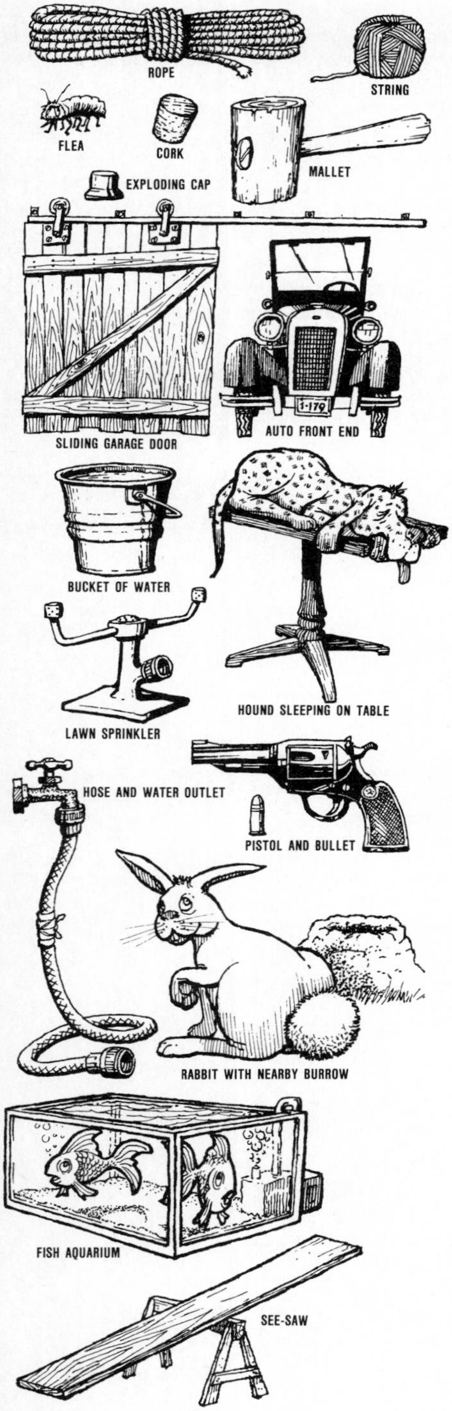

particular product. Then it is advantageous to use "brainstorming" or "synectics," the nominal group technique, or the Delphi technique.

Brainstorming. The term "brainstorming" was coined by Alex F. Osborn[2] to describe an organized group effort aimed at solving a problem. The technique involves compiling all the ideas that the group can contribute but deferring judgment concerning their worth. This is accomplished (1) by releasing the imagination of the participants from restraints such as fear, conformity, and judgment; and (2) by providing a method to improve and combine ideas the moment an idea has been expressed. Osborn points out that this collaborative group effort does not replace individual ideative effort. Group brainstorming is used solely to supplement individual idea production and works very effectively for finding a large volume of alternative solutions or novel design approaches. It has been particularly useful for stimulating imaginative ideas for new products or product development. It is not recommended where the problem will depend primarily on judgment or where the problem is vast, complex, vague, or controversial. A homogeneous "status group" of six to twelve persons seems to be best for stimulating ideas with this method. However, the U.S. armed forces have used 100 or more participants effectively. The typical brainstorming session has only two officials: a chair and a recorder. The chair's responsibility is to provide each panel member with a brief statement of the problem, preferably 24 hours prior to the meeting. He or she should make every effort to describe the problem in clear, concise terms. It should be specific, rather than general, in nature. Some examples of ideas that satisfy the problem statement may be included with the statement. Before beginning the session, the chair should review the rules of brainstorming with the panel. These principles, although few, are very important and are summarized as follows:

1. All ideas that come to mind are to be recorded. No idea should be stifled. As Osborn says, "The wilder the idea, the better; it is easier to tame down than to think up." He recommends recording ideas on a chalkboard as they are suggested. Sometimes a tape recorder can be very valuable, especially when panel members suggest several different ideas in rapid succession.
2. Suggested ideas must not be criticized or evaluated. Judgments, whether adverse or laudatory, must be withheld until after the brainstorming session, because many ideas that are normally inhibited because of fear of ridicule and criticism are then brought out into the open. In many instances, ideas that would normally have been omitted turn out to be the best ideas.
3. Combine, modify, alter, or add to ideas as they are suggested. Participants should consciously attempt to improve on other people's ideas, as well as contributing their own imaginative ideas. Modifying a previously suggested idea will often lead to other entirely new ideas.

[12-10] Imagination is the key.[3] Given these elements, can you devise a way to open a garage door automatically?

[2]Alex F. Osborn, *Applied Imagination* (New York: Scribner, 1963).

[3]Don Fabun, *You and Creativity,* (Oakland, Calif. Kaiser Aluminum & Chemical Corporation, 1968), p. 36.

4. The group should be encouraged to think up a large quantity of ideas. Research seems to indicate that when a brainstorming session produces more ideas, it will also produce higher-quality ideas.

The brainstorming chair must always be alert to keep evaluations and judgments from creeping into the meeting. The spirit of enthusiasm that will permeate the group meeting is also very important to the success of the brainstorming session. The entire period should be conducted in a free and informal manner. It is most important to maintain, throughout the period, an environment where the group members are not afraid of seeming foolish. Both the speed of producing and recording ideas, and the number of ideas produced, help create this environment. Each panel member should bring to the meeting a list of new ideas that he or she has generated from the problem statement. These ideas help to get the session started. In general, the entire brainstorming period should not last more than 30 minutes to 1 hour.

The recorder keeps a stenographic account of all ideas presented and after the session, lists them by type of solution without reference to their source. Team members may add ideas to the accumulated list for a 24-hour period. Later, the entire list of ideas should be rigorously evaluated, either by the original brainstorming group or, preferably, by a completely new team. Many of the ideas will be discarded quickly—others after some deliberation. Still others will probably show promise of success or at least suggest how the product can be improved.

Some specialists recommend that the brainstorming team include a few persons who are broadly educated and alert but who are amateurs in the particular topic to be discussed. Thus new points of view usually emerge for later consideration. Usually, executives or other people mostly concerned with evaluation and judgment do not make good panel members. As suggested previously, particular care should be taken to confine the problem statement within a narrow or limited range to ensure that all team members direct their ideas toward a common target. Brainstorming is no substitute for applying the fundamental mathematical and physical principles at the command of the engineer. It should be recognized that the objective of brainstorming is to stimulate ideas—not to effect a complete solution for a given problem.

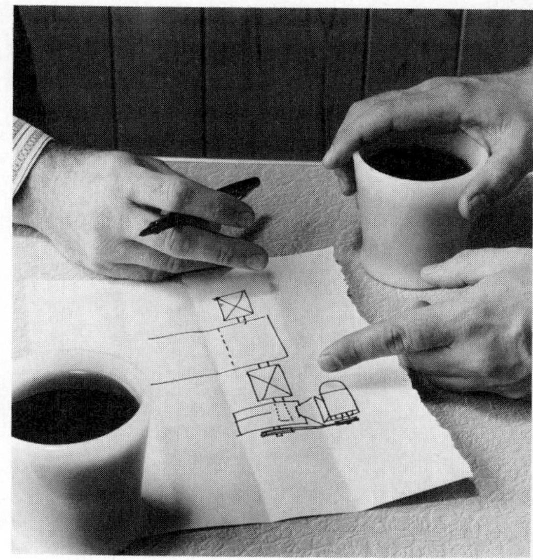

[12-11] Not uncommonly, one team member's inspired idea will set off a chain reaction of ideas from other team members.

The person who is capable of producing a large number of ideas per unit of time, other things being equal, has a greater chance of having significant ideas.
J. P. Guilford

He that answereth a matter before he heareth it, it is folly and shame unto him.
Proverbs 18:13
The Holy Bible

Depend upon it, sir, when a man knows he is to be hanged in a fortnight, it concentrates his mind wonderfully.
Samuel Johnson
September 19, 1777

Those who dream by day are cognizant of many things which escape those who dream only by night.
Edgar Allen Poe, 1809–1849

"The horror of that moment," the king went on "I shall never, never forget!" "You will, though," the queen said, "if you don't make a memorandum of it."
Lewis Carroll, 1832–1898

What good is electricity, Madam? What good is a baby?
Michael Faraday

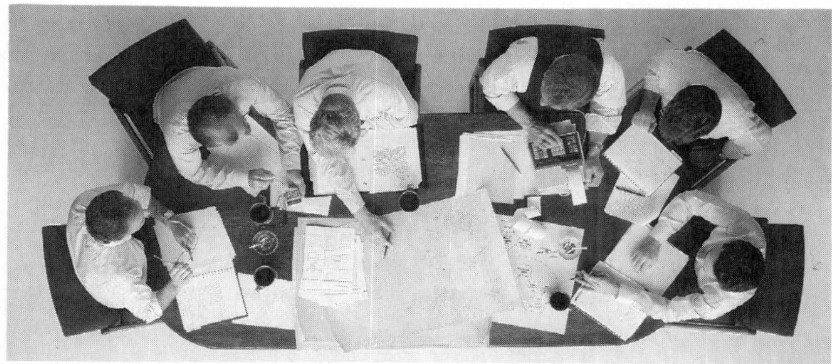

[12-12]

[12-13] Synectics can be a source of innovative solutions.

William J. J. Gordon[4] has described a somewhat similar method of group therapy for stimulating imaginative ideas, which he calls "synectics."

Synectics. This group effort is particularly useful to the engineer in eliciting a radically new idea or in improving products or developing new products. Unlike brainstorming, this technique does not aim at producing a large number of ideas. Rather, it attempts to bring about one or more solutions to a problem by drawing seemingly unrelated ideas together and forcing them to complement each other. The synectics participant tries to imagine himself or herself as the "personality" of the inanimate object: "What would be my reaction if I were that gear (or drop of paint, or tank, or electron)?" Thus familiar objects take on strange appearances and actions, and strange concepts often become more comprehensible. A key part of this technique lies in the group leader's ability to make the team members "force-fit" or combine seemingly unrelated ideas into a new and useful solution. This is a difficult and time-consuming process. Synectics emphasizes the conscious, preconscious, and subconscious psychological states that are involved in all creative acts. In beginning, the group chairman leads the members to understand the problem and explore its broad aspects. For example, if a synectics group is seeking a better roofing material for traditional structures, the leader might begin a discussion on "coverings." He or she could also explore how the colors of coverings might enhance the overall efficiency (white in summer, black in winter). This might lead to a discussion of how colors are changed in nature. The group leader could then focus the group on a more detailed discussion of how roofing materials could be made to change color automatically to correspond to different light intensities—like the biological action of a chameleon or a flounder. Similarly, the leader might approach the problem of devising a new type of can opener by first leading a group discussion of the word "opening," or he or she could begin considering a new type of lawn mower by first discussing the word "separation."

In general, synectics recommends viewing problems from various analogous situations. Paint that will not adhere to a surface might be viewed as analogous to water running off a duck's back. The earth's crust might be seen as analogous to the peel of an orange. The problem of enabling army tanks to cross a 40-ft-wide, bottomless crevasse might be made analogous to the problem that two ants have in crossing chasms wider than their individual lengths.

Synectics has been used quite successfully in problem-solving situations in such diverse fields as military defense, the theater, manufacturing, public administration, and education. Whereas most members of the brainstorming team are very knowledgeable about the problem field, synectics frequently draws the team members from diverse fields of learning, so that the group spans many areas of knowledge. Philosophers, artists, psychologists, machinists, physicists, geologists, and biologists, as well as engineers, might all serve equally well in a synectics group. Synectics assumes that someone

New things are made familiar, and familiar things are made new.

Samuel Johnson
Lives of the Poets, *1781*

It's amazing what ordinary people can do if they set out without preconceived notions.

Charles F. Kettering
Forbes' Scrapbook of Thoughts in the Business of Life

[4]William J. J. Gordon, *Synectics,* (New York: Harper & Row, 1961.)

who is imaginative but not experienced in that field may produce as many creative ideas as one who is experienced in that field. Unlike the expert, the novice can stretch his or her imagination. He or she approaches the problem with fewer pre-conceived ideas or theories and is thus freer from binding mental restrictions. (Obviously, this will not be true when the problem requires analysis or evaluation, where experience is a vital factor.) There is always present in the synectics conference an expert in the particular problem field. The expert can use his or her superior technical knowledge to give the team missing facts, or may even assume the role of "devil's advocate," pointing out the weaknesses of an idea the group is considering. All synectics sessions are tape recorded for later review and to provide a permanent record.

Many believe that brainstorming comes to grips with the problem too abruptly while synectics delays too long. However, industry is using both methods successfully today.

Although brainstorming and synectics have been successful in many problem-solving situations, each is subjected to the inherent negative syndrome of "follow the leader" that is usually present in any group activity. I. L. Janis[5] has examined the group decision-making processes. The negative effects of what he calls "groupthink" include a psychological pressure on group members to reach a consensus, blocking out other stimuli leading to other conclusions, dominance by those individuals who hold positions of rank in the group, and a lack of tolerance for opposing views. Janis suggests that the leader be absent from the group problem-solving process and that each member of the group serves as a "devil's advocate" from time to time. The nominal group technique and the Delphi technique are modifications of brainstorming that do not have the pronounced negative effects of the "groupthink" processes.

Nominal Group Technique. The nominal group technique, developed by Delbecq and Vandeven,[6] uses a person trained in the method to lead the group activities. This person does not contribute ideas to the group and is often an "outsider" who is not a member of the group. The method begins with each person silently generating ideas about the problem to be solved. These ideas are written on paper but are not shared with other group members. The person conducting the session then asks each group member sequentially to contribute a single idea for consideration by the group. These ideas are written on a large blackboard or a "flipchart" so that all members of the group can see them. As ideas are presented, there is a period of discussion concerning each one. At this time, the contributor might be asked to clarify the idea and others may comment concerning clarification. However, no evaluations are made until all ideas have been contributed. Discussions are permitted and on occasion, ideas are combined with other similar ideas. The process continues with ideas being contributed one by one until each group member has no

[5] I. L. Janis, "Groupthink," *Psychology Today,* 1971.

[6] A. L. Delbecq and A. H. Vandeven, "A Group Process Model for Problem Identification and Program Planning," *Journal of Applied Behavioral Science,* Vol. 7, No. 4 (1971).

[12-14] Engineering design is often conducted in uncharted territory.

Our doubts are traitors and make us lose the good we oft might win by fearing to attempt.
William Shakespeare, 1564–1616

Whatever one man is capable of conceiving, other men will be able to achieve.
Jules Verne

More ways of killing a cat than choking her with cream.
Charles Kingsley
Westward Ho, *1855*

Originality is just a fresh pair of eyes.
W. Wilson

It is obvious that invention or discovery, be it in mathematics or anywhere else, takes place by combining ideas.
Jacques Hadamard
An Essay on the Psychology of Invention in the Mathematical Fields

more ideas to contribute. At that time, the members are asked to silently rank (on paper) all the ideas by first choosing what the member thinks is the very best idea, and ranking it 1. Then the idea which the individual member feels is the worst idea is ranked n, or last among all n ideas. Next, the remaining best idea is ranked 2 and the remaining worst idea is ranked $n - 1$. This process continues until all ideas have been ranked. The group members then submit their rankings to be tabulated. The idea with the smallest sum of rankings is considered to be the best idea, the idea with the next smallest rank sum is judged next best, and so on. This method of idea generation has been proven to be extremely effective in a variety of problem-solving situations.

Watch your step when you immediately know the one way to do anything. Nine times out of ten, there are several better ways.

W. B. Given, Jr.

Use logic to decide between alternatives, not to initiate them.

When we mean to build, we first survey the plot, then draw the model; and when we see the figure of the house, then must we rate the cost of the erection.

William Shakespeare
King Henry IV, Part II, Act I, Sc. 3, Line 41

Seek simplicity, and distrust it.

Alfred North Whitehead

The Delphi Technique. The Delphi technique was developed at the Rand Corporation[7] as a specific means of overcoming some of the difficulties associated with brainstorming. In this technique, the participants of the group are intentionally separated. In fact, the participants may be thousands of miles apart. Each participant is given a statement of the problem to be solved and is asked to submit ideas for solution (often by return mail). Ideas are collected at a central office. Similar ideas are combined or grouped in the same category. Next, all the ideas that were sent to the central office are sent to each participant in questionnaire form. The participant is asked to rank the ideas according to importance in solving the problem. Each set of rankings is returned to the central office where the rankings are tabulated. To avoid early elimination of ideas having low rankings by most of the participants, the originators of these ideas are asked to give detailed justification of the merit of the idea in solving the problem. The results of the ranking process are then sent to all participants along with the detailed justifications for the ideas with low rank sums. In some cases, an idea previously thought to

[7] N. C. Palkey, *Delphi* (Santa Monica, Calif., Rand Corporation, 1967).

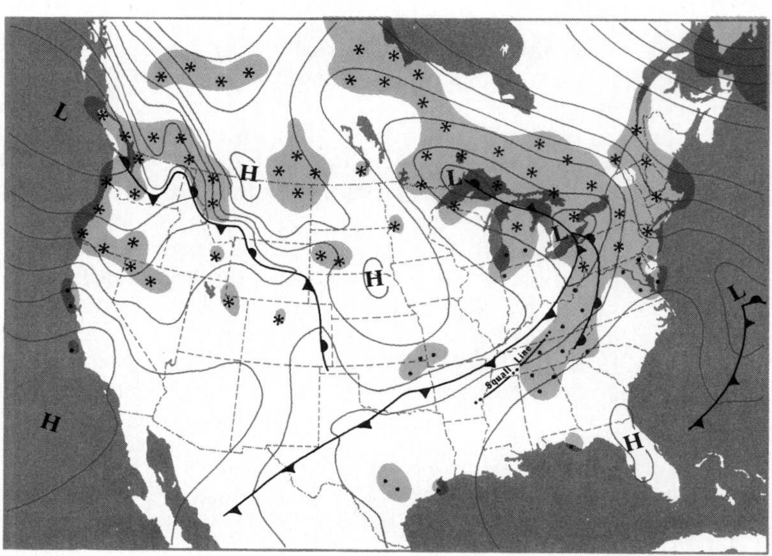

[12-15] Weather maps are models of weather conditions.

have little merit by most of the participants is reallocated a high ranking after a detailed justification is presented. The ranking, tabulation, and justification process is repeated until no changes result from the previous iteration. The Delphi technique has most often been applied to problems involving forecasting future trends and in these instances, the participants are usually individuals with considerable expertise in the problem area. However, this method has been used successfully within the confines of several companies with the participants being separated in different rooms.

Modeling

Models are representations of the problem or system that is being studied. The engineer uses modeling in several ways: to describe the system to others, to simulate the behavior of the system under a variety of operating conditions, and to predict the behavior of the system under revised operating conditions. Models are used to describe, simulate, and predict the behavior of objects, situations, environments, and events.

We are already familiar with several types of models: with maps as models for a road system or of weather conditions [12-15], with catalogs of merchandise as models of items offered for sale, with toy trains as models of full-sized railroad trains, and with diagrams of football plays [12-16] as models of how the play should be executed in a football scrimmage or game. We have a mental image (model) in our mind of the food we eat, the clothes we buy, and even of the partner we want to marry. Two characteristics, more than most others, determine an engineer's competence. The first is an ability to devise simple, meaningful models; and second is a breadth of knowledge and experience applicable to situations with which the models can be compared. The simpler the models, and the more generally applicable they are, the easier it is to predict the behavior and compute the performance of the system. Yet *models have value only to engineers who can properly analyze them*. For example, the beauty and simplicity of a model of the atom [12-17] will appeal most particularly to someone familiar with astronomy. A free-body diagram of a wheelbarrow handle [12-18] has meaning only to someone who

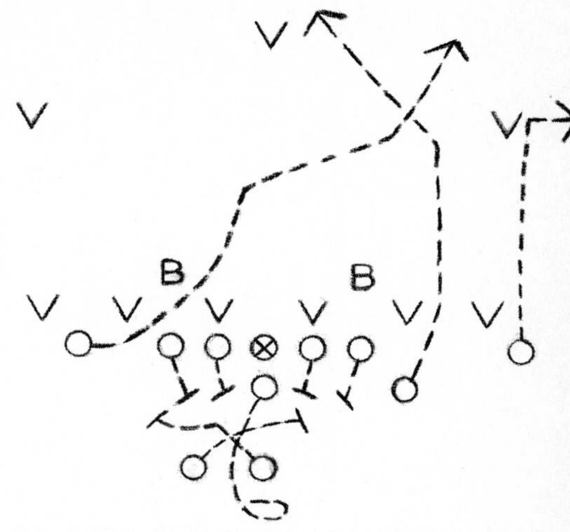

[12-16] Every football fan understands the value of diagrammed models in preparing for Saturday's "big game."

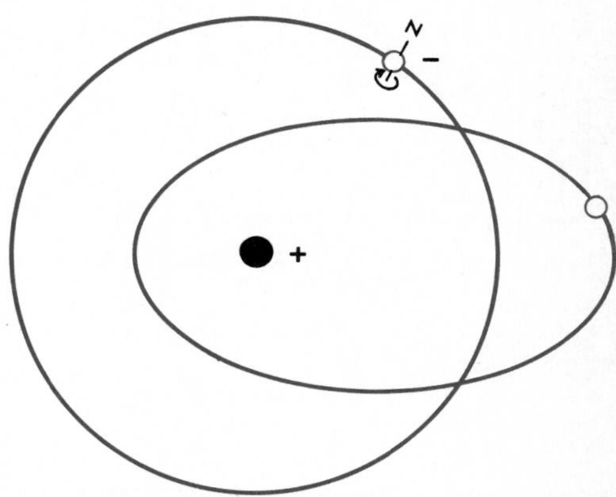

[12-17] Modified Bohr–Sommerfeldt model of the atom.

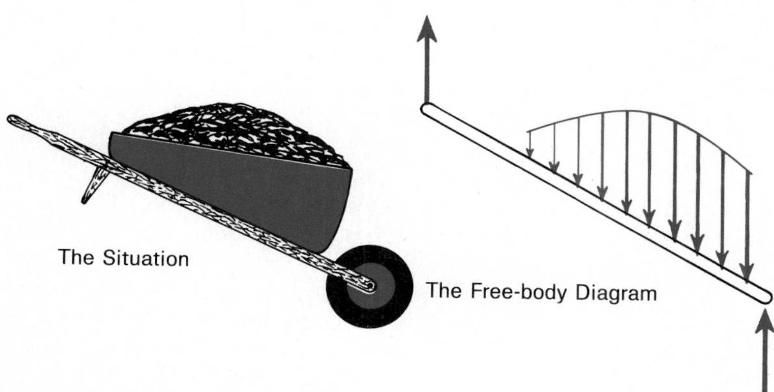

[12-18] Situations can be modeled with free-body diagrams.

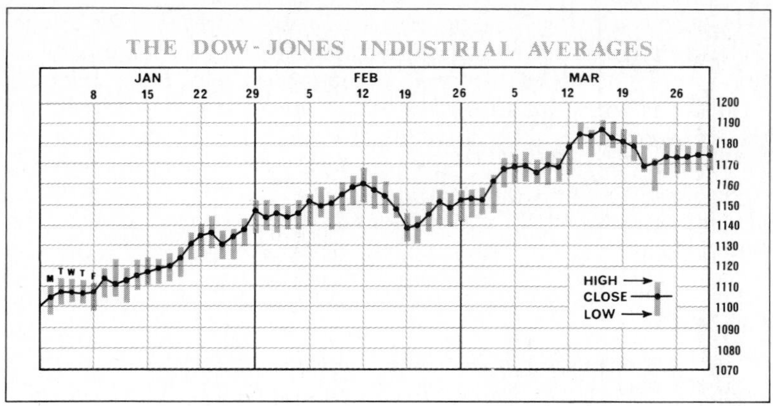

[12-19] Many people utilize financial models.

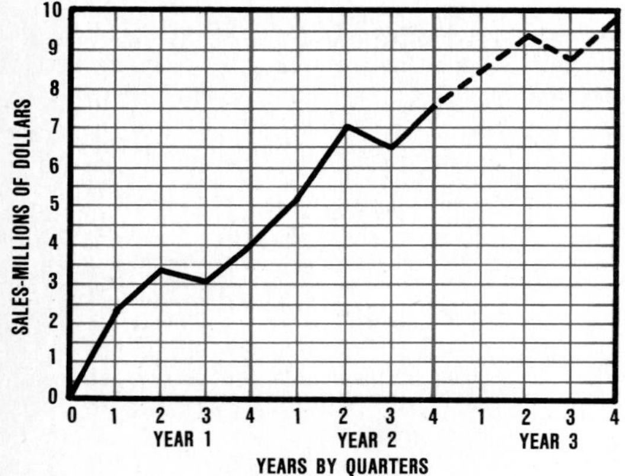

[12-20] Sales of the Apex Corporation.

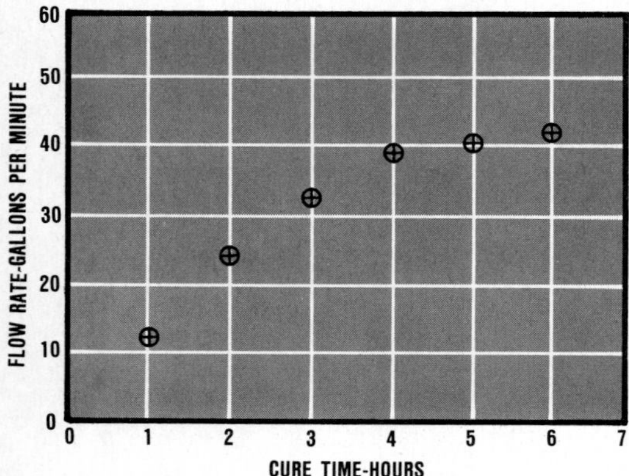

[12-21] Flow rate versus cure time.

knows how such a diagram can be used to calculate the forces exerted on the handle. Models are particularly useful to the engineer if the behavior of the system that is being studied follows known physical and/or mathematical laws and if the relationship between the inputs and the outputs of the system being modeled is known or can be approximated.

Most models may be classified into one of the following types:

Conceptual.
Graphical.
Physical.
Mathematical.
Computer.

Conceptual Models. The creation of all models begins with ideas in the mind. Indeed, many modelers can convey their ideas about the behavior of a problem or system by describing their mental images of how the system works. For example, an engineer might say "the radio communications system will be designed so that the receiver-transmitter, the antenna controller, the antenna coupler, the filter, and the antenna will all be connected in series, while a duplicate receiver-transmitter unit and antenna series system will operate in parallel." Obviously, when the problem or system being studied is complex (such as this one), conceptual models alone will not be sufficient, and other types of models must be used. The ability to make a freehand drawing (see Chapter 7) is particularly important in conceptual modeling.

Graphical Models. Foremost among graphical modeling techniques is the use of charts, functional plots, and diagrams. Charts are often used for financial analyses or to show the sales of a firm. Figure [12-19] is simply a descriptive model of the behavior of industrial stock prices over time. Figure [12-20] describes the sales of a firm over a two-year period and also predicts the sales for the next year.

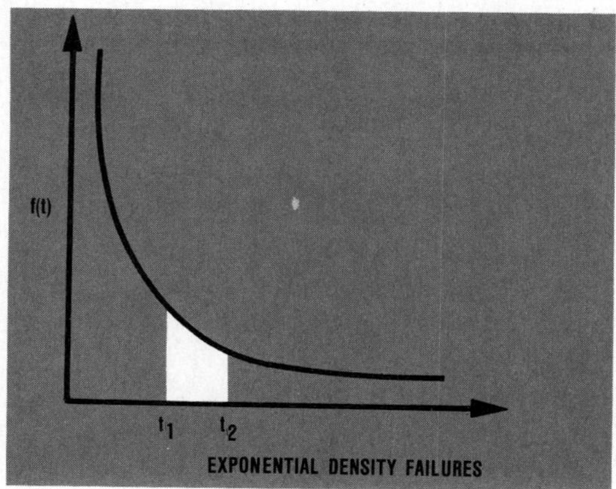

[12-22]

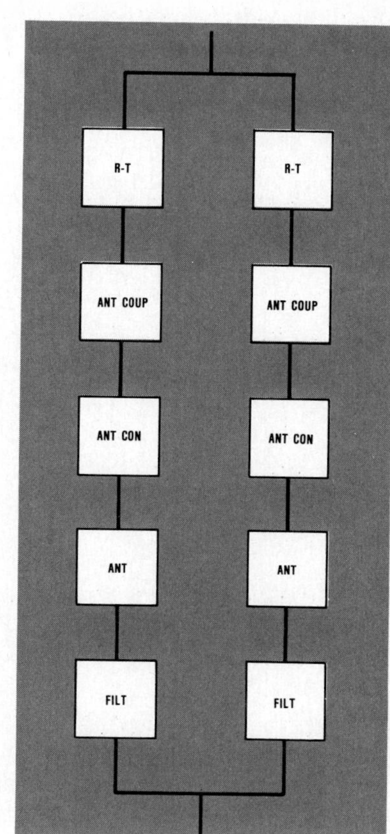

[12-23] Block diagram of a communications system.

Functional plots show relationships between variables. Sometimes the relationship is based on data gathered as a part of an experiment, while at other times, the relationship is a theoretical one that is believed to adequately describe the actual events. To obtain the plot of Figure [12-21], a chemical compound was allowed to cure (set at a constant temperature) for one, two, three, four, five, and six hours. At the end of each of these periods, a fixed volume of the compound was poured and the flow rate in gallons per minute was used as a measure of the viscosity. Figure [12-22] shows the known relationship between the time between failures, t, of an electrical device and the failure density function, $f(t)$. The failure density function describes the likelihood of failure of the device at a particular time. The probability of failure in any time interval, for example, t_1 to t_2, can be obtained by computing the area under the curve between the points t_1 and t_2. This type of model is appropriate only for those devices which in the past have empirically followed the exponential time-to-failure function.

Diagrams are also useful models for the engineer. The most popular form of diagram is the *block diagram,* where components being studied are represented as blocks. This allows relationships and interdependencies among the components to be pictured. Figure [12-23] is a diagrammatical representation in block diagram form of the radio communications system that was previously described as a conceptual model. The energy diagram is used primarily in the study of thermodynamic systems involving mass and energy flow. An example of the use of an energy diagram is given in Figure [12-24].

The electrical diagram is a specialized type of model used in the analysis of electrical problems. This form of idealized model represents the existence of particular electric circuits by using conventional symbols for brevity. These diagrams may be of the most elementary type, or they may be highly complicated and require many hours of engineering time to prepare. In any case, however, they are representations or models in symbolic language of electrical assembly. Figure [12-25] shows a schematic diagram of an FM oscillator. Notice that the diagram details only the essential parts that are neces-

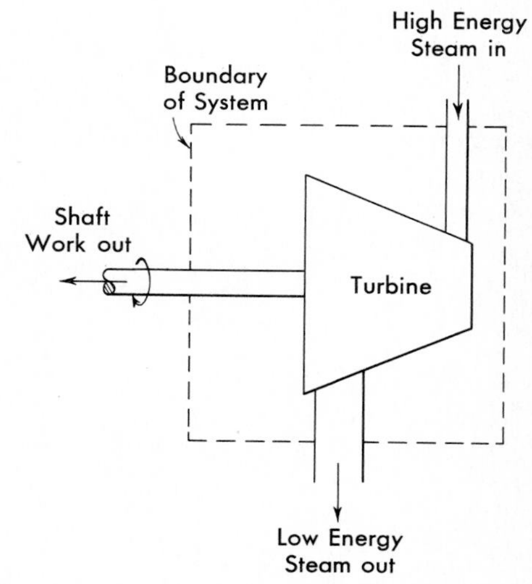

[12-24] Energy diagrams can be especially useful in understanding thermodynamic systems.

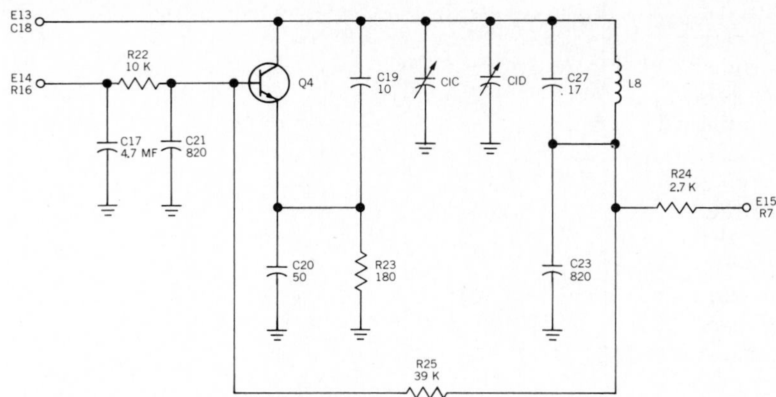

[12-25] Schematic diagram of an FM oscillator.

[12-26] The situation.

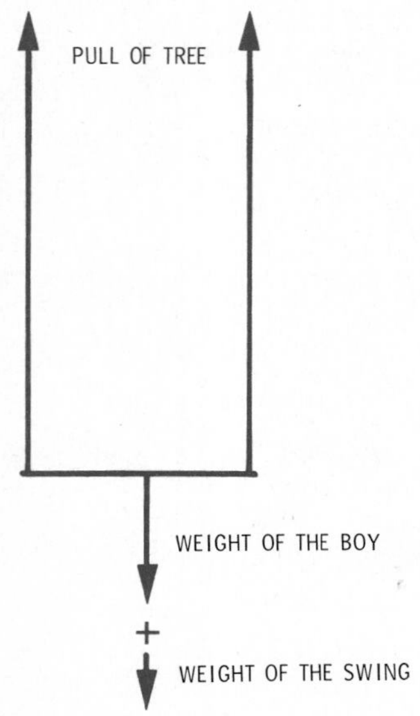

[12-27] The free-body diagram that represents the situation.

sary for electrical continuity, and thus is an idealization that has been selected for purposes of simplification.

Another popular engineering model is the free-body diagram [12-26 and 12-27]. A free-body diagram is a representation of a physical system that has been removed from all surrounding bodies or systems for purposes of examination and where the equivalent effect of the surrounding bodies is shown acting on the free body. Such a diagram may be drawn to represent a complex system or any part of it. This form of idealized model is most useful in showing the effect of forces that act on a system. The free-body diagram was discussed more fully in Chapter 11.

The computer is a valuable tool for developing graphical models. Software is available to draw charts, functional plots, and diagrams of many types. Indeed, computer graphics is so popular that a national society, the National Computer Graphics Association,[8] has been formed to accommodate those with such interests.

Physical Models. There are two general types of physical models that are employed by engineers in problem-solving efforts: scale models and analog models. Scale models are enlarged or reduced-in-size versions of their actual counterparts. They are designed to resemble quite closely the systems being modeled. Analog models are physical representations of a system that usually bear little visual resemblance to the actual system but are developed to function in much the same way as it does.

The Scale Model. Scale models are used often in various problem-solving situations, especially when the system or product is very large and complex or very small and difficult to observe. A scale model is a replica, usually three-dimensional, of the system, subsystem, or component being studied. It may be constructed to any desired scale relative to the actual design. Such projects as dam or reservoir construction, highway and freeway interchange design, factory layout, and aerodynamic investigations are particularly adaptable to study using scale models.

[8]National Computer Graphics Association, 8401 Arlington Boulevard, Suite 601, Fairfax, VA 22031.

Scale models are useful for predicting performance because component parts of the model can be moved about to represent changing conditions within the system. Of considerably more usefulness are those scale models which are instrumented and subjected to environmental and load conditions that closely resemble reality. In such cases the models are tested and experimental data are recorded by an engineer. From an analysis of these data, predictions of the behavior of the real system can be made.

By using a scale model that can be constructed in a fraction of the time, a final design can be checked for accuracy prior to actual construction. Although scale models often cost many thousands of dollars, they are a relatively minor expense when compared with the total cost of a particular project [12-30].

[12-28] Some models are successful . . . .

[12-29] . . . other models fail.

Analog Models. Analogs and similes are used to compare somesomething that is unfamiliar to something else that is very familiar. Writers and teachers have found the simile to be a very effective way to describe an idea. Engineers use analogs in much the same way that teachers use similes [12-31]. An analog, however, must provide more than a descriptive picture of what one wants to study; its action should correspond closely with the real thing. It should be mathematically similar to it, that is, the same type of mathematical expressions must describe well the action of both systems, the real and the analog.

A vibrating string is an analog of an organ pipe because the sound in an organ pipe behaves quite similarly to the waves traveling along a vibrating string [12-32]. Under certain assumptions, similar mathematical equations can describe both systems. In other words, we can compare the corresponding actions of the model of the organ pipe with a model of the vibrating string. It is the models that be-

[12-30] Design studies of large projects frequently require the use of scale models, such as this one of New York City.

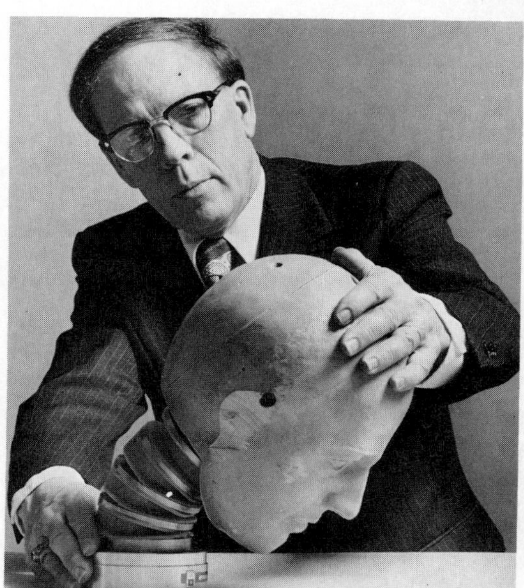

[12-31] The designs of anthropomorphic dummies used in safety restraint systems for automobile and airplane crash tests require a humanlike neck that will react in a manner analogous to the way your own neck would react.

[12-32] The sounds of an organ may be studied mathematically by considering analogously the waveform of a vibrating string or rope.

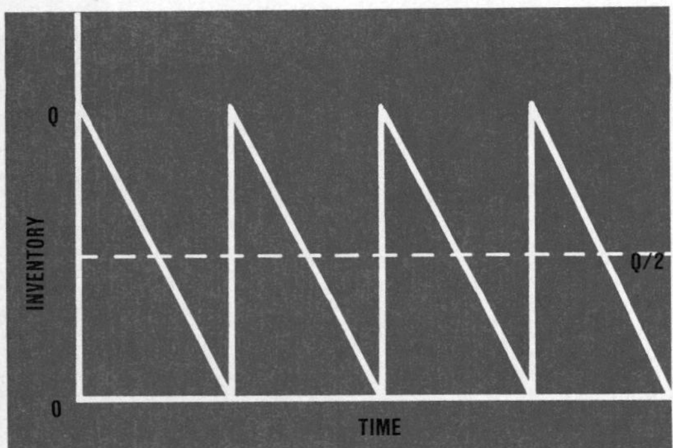

[12-33] Inventory level versus time.

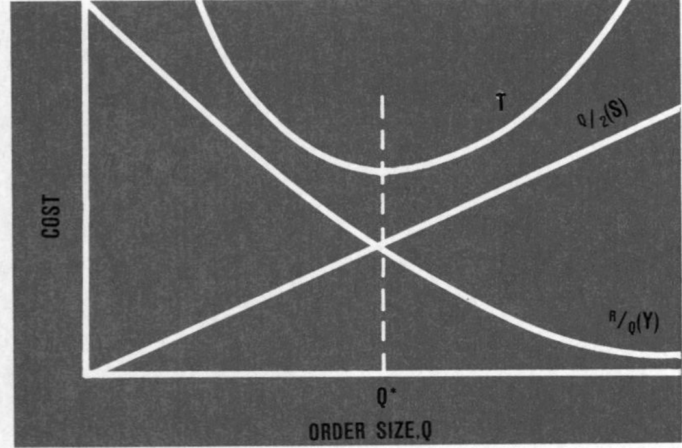

[12-34] Inventory problem costs and optimal order quantity, Q^*.

have exactly alike, not the real systems. If these models are "good" models, then under certain conditions one can perform experiments with the string and draw valid conclusions concerning how the organ pipe would behave. Since one system may be much easier to experiment with than the other, one can work with the easier system and obtain results that are applicable to both.

An example of the use of a very successful analog is the electrical network that forms an analog for complete gas pipeline systems. Using such a model, one can predict just what would happen if a large amount of gas was suddenly needed at one point along the system. Experiments with the actual pipeline would be very costly and might disrupt service. The electrical network analog provides the answers faster, cheaper, and without disturbing anyone.

Mathematical Models. Mathematical models generally take the form of equations which express the relationship between one or more independent variables (that often appear on the right side of the equation) and a dependent variable (often appearing on the left side). The variables on the right side of the equation are independent in the sense that they may assume ranges of different values chosen by the engineer. The variable on the left side of the equation is dependent on the exact values that are assumed by the independent variables. A simple mathematical model that you encountered in high school physics is Newton's second law of motion, $\mathbf{F} = m\mathbf{a}$. This mathematical model says that if a mass of 12 g is to be accelerated at 5 m/s^2, a force of 60 N will be required. (See Chapter 10 for a further discussion of this relationship.)

Another example of the use of a mathematical model is found in inventory control. Industrial engineers frequently wish to determine the lot sizes of raw material, Q, to be ordered such that the total cost of ordering and storing the raw materials until use will be minimized. Suppose that R units of raw material will be required during the next year and that the material will be ordered in lots of size Q. That is, each time that an order is placed with the vendor of the raw material, Q units will be the size of the order. Thus the number of orders per year will be R/Q. Suppose that the cost of placing an order (labor, paperwork, and telephone charges), including the transportation charges, is \$$Y$ per order. Then the total annual cost of placing and receiving each order is R/Q multiplied by Y.

The raw material must be stored in inventory before it is used. Suppose that the material is used at a constant rate and that orders are placed just in time to be received when the inventory level reaches zero. This situation is represented in Figure [12-33]. From Figure [12-33] we see that the average number of units in inventory during the year is $Q/2$. If it costs \$$S$ per unit to store an item in inventory, the average annual cost of storage is $Q/2$ multiplied by S. Therefore, the mathematical model for the total annual cost, T, of ordering and storing the raw material is given by

$$T = (R/Q)Y + (Q/2)\,S \tag{12-1}$$

The graph of T is given in Figure [12-34]. Notice that T is the sum of the $(R/Q)Y$ and $(Q/2)S$ curves. Q^* is the solution to the problem. It is the value of the order size, Q, which results in a minimum total cost, T. This mathematical model may be plotted graphically to obtain the solution. Another way of solving this mathematical model is to use calculus, which is beyond the scope of this text. The solution is

$$Q^* = \sqrt{\frac{2RY}{S}} \qquad (12\text{-}2)$$

Computer Models. A computer model can be defined as any combination of mathematical and logical relationships that have been written in computer code to represent a problem to be solved or a system to be studied. Often problems that can be solved analytically (i.e., usually mathematical techniques) will be coded for computer solution if the analytical solution is expected to be too time consuming. For example, in Figure [12-36] consider the problem of finding the path from node (circle) 1 to node 7, which results in the minimal cost of travel. The numbers with arrows indicate the dollar costs of travel between nodes in the direction of the arrow. There are several analytical procedures that can be used to solve this problem. All are time consuming. One way of using the computer to solve the problem and minimize the calculation time is to input all the data concerning costs between pairs of nodes and to input which nodes are connected to each other. The computer can then be programmed to find the least-cost path by constructing all possible paths from node 1 to node 7, totaling the cost associated with each path, and then selecting the path with the least cost. Any computer model that solves problems by trying all possible solutions, as in this example, is said to be using the "brute force" technique. Incidentally, if you wish to solve this problem by brute force or any other method, the answer is given at the bottom of this page.[9] Not all problems are solvable by the brute-force method. If it were possible to program a computer to play a game of chess using the brute-force method, the computer would be programmed for each move, to select the best move based on the analysis of all possible moves, followed by all possible moves by the opponent, followed by all possible moves by the computer, followed by all possible moves by the opponent, and so on. In this way the computer would be "thinking" dozens of moves ahead. (To-

[9]Best path 1-2-4-3-6-5-7; cost = \$26.

[12-35] Computer-generated solid model of a crankshaft.

Who so boasteth himself of a false gift is *an analog to* clouds and wind without rain.

Proverbs 25:14
The Holy Bible

The battle of Waterloo was won on the playing fields of Eton.

Arthur Wellesley, Duke of Wellington
From Fraser, Words on Wellington, *1889*

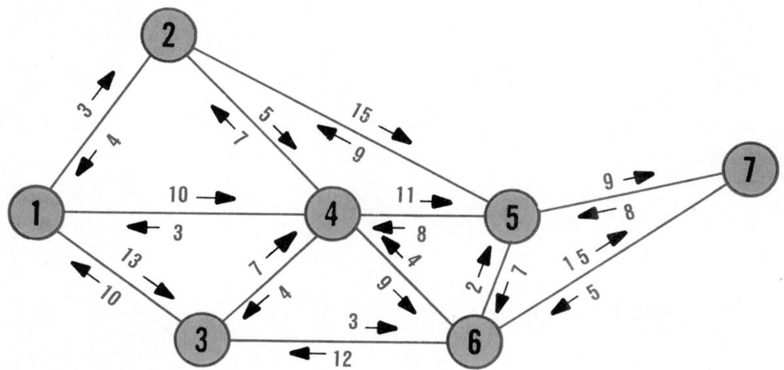

[12-36]

[12-37] Chess-playing robots rely on heuristics.

day's chess masters are limited to thinking no more than three or four moves ahead of the current move.) With the computer each move ahead is analyzed all the way to the completion of the game. There are millions of combinations of moves that must be analyzed just to determine a single move on the next play. The world's fastest computer would take thousands of years to do this. Hence chess-playing computer programs do not use brute force; instead, they use the method of "heuristics." A *heuristic* is a rule of thumb. Heuristics do not guarantee an optimal solution, but if they are applied correctly, they will generally produce results that are quite satisfactory. Chess-playing heuristics would involve, for example, only sacrificing a pawn if a major piece of the opponent could be put in jeopardy within four moves. Another heuristic might involve always "castleing" whenever the king's knight and bishop have been lost [12-37].

The most popular computer modeling technique is *computer simulation*. In computer simulation, computer code is developed to represent the behavior of a system. The model is tested to ensure that it functions in the same manner as the system being simulated. The experiments or simulations are conducted using the model in order to predict the behavior of the real system under different operating conditions or test various conditions of system behavior [12-38].

As an example of a computer simulation procedure, consider the conveyor belt system shown in Figure [12-39]. A large paper mill had six papermaking machines which formed the paper into large rolls. The rolls were pushed by hand to the conveyor belt, where they were moved to one of two elevators (E), shown at the bottom of the drawing. The company planned to add a seventh paper machine and wanted to know if the existing conveyor system could handle the increased number of rolls after adding machine 7. If not, a new parallel conveyor system would have to be installed. Before the simulation program could be devised, a vast amount of data was needed. For example, each machine generated different numbers of rolls of different sizes at different times during the day. The probability of machine breakdowns and stoppages on the belt also had to be determined. In operation, the rolls had to be weighed individually (at W) and strapped with a steel band (at S). The belt had to be stopped for each of these operations.

After the data were obtained, a simulation model was developed and tested with the existing six machines. The validity of the simula-

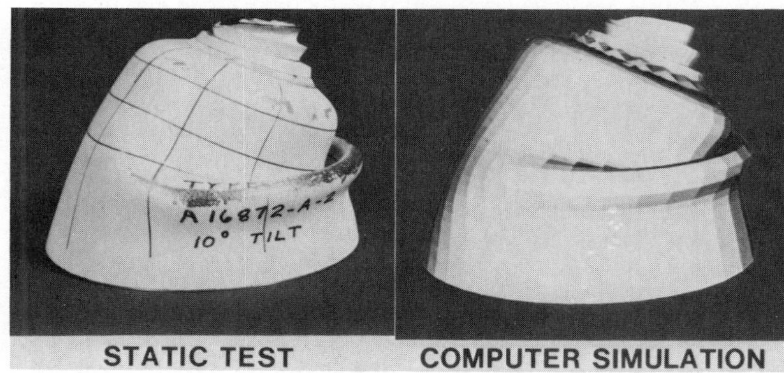

[12-38] Computer simulation and actual test result made on a nose-section part of an aircraft.

tion program was established when it was able to predict a production of rolls of paper at the same rate as the actual machines produced them. It not only had to "move" these rolls on the belt at the same speed that the actual rolls were moved, but it also had "to cause" breakdowns of the paper machines and delays on the belt (due to weighing or strapping) in the same manner that they actually occurred. In this way 45 days of production were simulated in approximately three hours of computer time. The total number of rolls of each size processed through the conveyor system were compared with typical production (rolls processed through the system) for a 45-day period. The final values were indeed similar. Finally, the new machine, number 7, was added to the model. The simulation was then performed using a two-week production period. Results indicated that although there was considerable blockage in front of machines 1 and 2 (since they are at the end of the flow on the belt and rolls could not be loaded at various times because the conveyor space in front of these machines was occupied), the existing conveyor system could accommodate the additional rolls from machine 7.

This result was important information for the managers of the paper mill. They were able to act on the information provided by the simulation and install the new paper machine without adding an additional conveyor belt. This revised conveyor belt system was installed with the new machine at a cost of $800,000. If the simulation had been in error, and the second belt had been installed after machine 7 was operational, the cost would have almost doubled due to work stoppages and movement of equipment. Fortunately, the computer model provided the correct answer and the existing belt was able to accommodate the production of all seven machines.

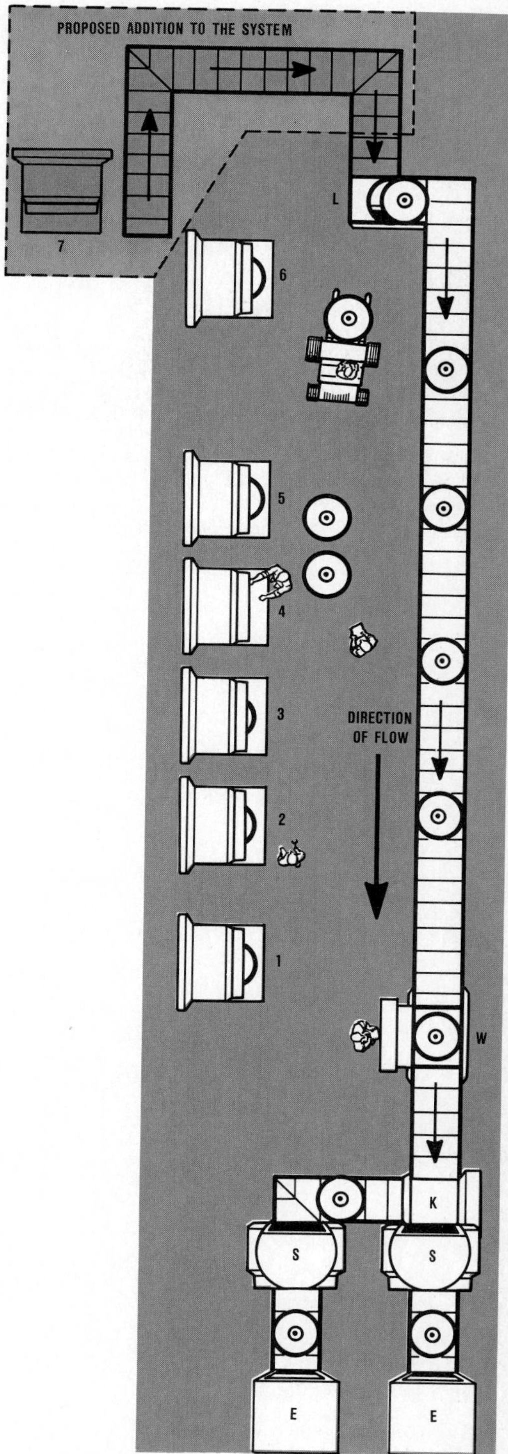

[12-39] Conveyor system in paper mill.

We have discussed modeling in some detail, as it may be unfamiliar to the beginning engineering student. This rather extensive treatment of the topic does not imply that modeling is more important than any other stage of the design process. The next stage, information gathering, is very dependent on the five preceding stages.

Information Gathering

Novice problem solvers often begin the problem-solving process at this stage. There is a tendency to want to gather data and identify information to be used in the solution process much too early to be of benefit. Often a lot of information is obtained at a great expense of time, resources, and money only to discover that much of the information that has been gathered is of little value. The gathering of information should always be preceded by problem identification, analysis, and transformation. In some cases, information is needed in the modeling stage. Where this is the case, information gathering would precede modeling.

There are two aspects of information gathering. One is concerned with obtaining information that relates specifically to matching the solution procedures identified in the idea development stage with the models that were developed in the previous stage. Here a search is made for detailed information regarding exactly how the techniques and solution procedures selected earlier will be applied to

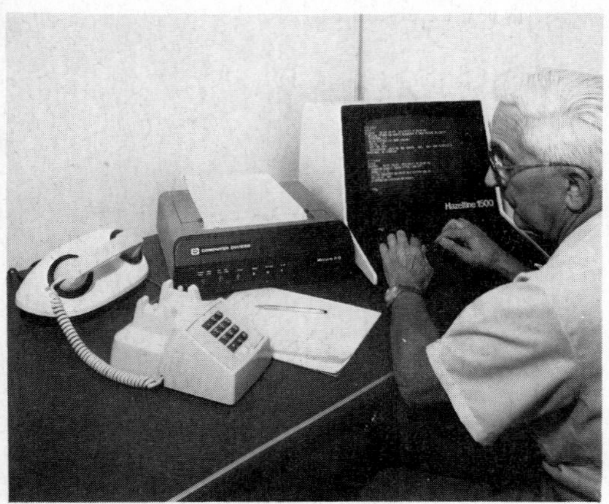

[12-40] Computerized retrieval systems utilizing millions of pages of data are a significant improvement for the engineer over more traditional library systems.

each of the subproblems. Sources of this type of information include technical reports published as a result of government and university research and development, trade journals, patents, handbooks, catalogs, and engineering indexes. Before writing any computer programs, the problem solver should investigate the availability of existing software relating to current needs. There are voluminous numbers of computer programs available in the public domain at little or no cost or for a nominal fee. Others can be purchased at higher costs. Membership in a user's group for a specific type of microprocessor may access help in the identification of software appropriate for a particular application. Many professional journals have sections of each issue devoted to descriptions (and sometimes listings) of computer programs for application in the particular field of the journal. Whenever possible, the problem solver should make use of the experience of others.

The other aspect of information gathering relates to identifying sources of data and/or collecting data for use in the experimentation stage (the next stage) of the problem-solving process. Data sources and the nature of the data are unique for each problem, and specific suggestions cannot be offered on how all data should be obtained. However, in general, the data gathered should be highly representative of the process or problem being studied. Data should be collected such that they can be analyzed easily. If the data are to be used in making inferences about a process, they should be both random and independent. Data selected from a process should be a sample of all the possible data generated by that process. The sample is considered to be random if each data element in the process from which the sample is drawn has the same likelihood of being a member of the sample. The data of the sample are considered to be independent if the selection of a single data element does not influence nor is influenced by the selection of any other data element.

Fool me once, shame on you; fool me twice, shame on me.

Chinese proverb

The information-gathering stage helps us determine how one or more of the solution procedures is to be applied to each subproblem. It also helps us determine the identification of sources of data to be applied with each procedure.

Experimentation

Within each subproblem, experiments should be conducted to determine how well the procedures operate with simulated data. When necessary, modifications are made to various design procedures and sometimes loops are made back to prior stages to generate alternative procedures, more trial data, and so on. Mathematical models may need to be adjusted in light of their predictions; scale models may be changed; or, if computer simulation models are used, comparisons of their outputs with the sample data might reveal that modifications to the code are needed.

The experimentation stage should develop reliable solution procedures for each subproblem. It remains to be determined how effective each subproblem solution procedure will integrate with other subproblem procedures as overall solution to the problem is approached. This question is addressed in the next stage, synthesis.

Synthesis

The word synthesis means "putting together." For example, a team is synthesized from a group of players, and a song is synthesized from a set of musical notes. In this stage, the solution to the whole problem will be obtained by the process of combining the results of each of the subproblem solution procedures. The synthesis process may be likened to solving a jigsaw puzzle [12-41]. Like the puzzle solver, the problem solver may discover that there are missing pieces, and there are also pieces that will not fit.

In problem solving the missing pieces are usually areas of the original problem which were not identified during the analysis stage. Therefore, a return to this stage is required. A piece that will not fit is usually a subproblem that was poorly identified. It can also be a subproblem with a poor solution procedure or it could be the result of an inappropriate model or insufficient data. Once again, a loop back to a prior stage or stages is required.

One can usually expect difficulties at the synthesis stage due to the fact that most subproblems are not independent of each other. That is, what happens with one subproblem affects or is affected by what happens with one or more of the other subproblems. Another way of saying this is to state that there is usually some interaction or dependency between the subproblem inputs and their solutions. The experienced problem solver is aware of many of these dependencies at the time of analysis, and devises solution procedures and models to account for the dependency. But even in such cases, it is only when synthesis is performed that additional, unexpected dependencies become apparent. This is where the pay-off of computerization and CAD/CAM brings big dividends! Interdependencies can be discovered early in the process.

A preliminary indication of how well the synthesis is working is obtained by a return to the experimentation stage using the combined trial data with the synthesized solution procedure. But it is only by testing the designer's solution in the actual problem environment that the effectiveness of the engineer's design can be determined. This evaluation and testing is the next stage of the problem-solving process.

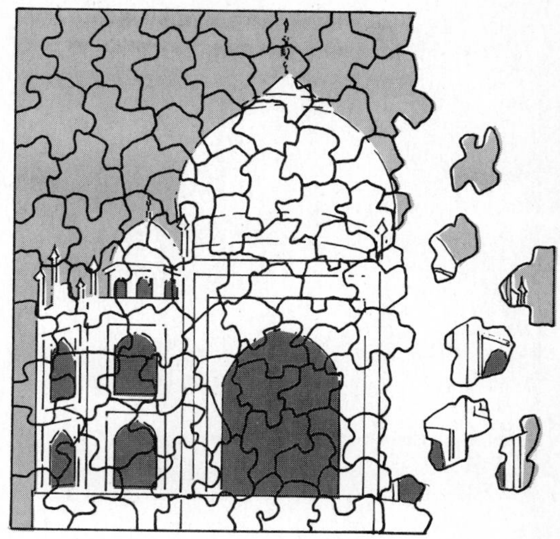

[12-41] The synthesis process may be likened to solving a jigsaw puzzle.

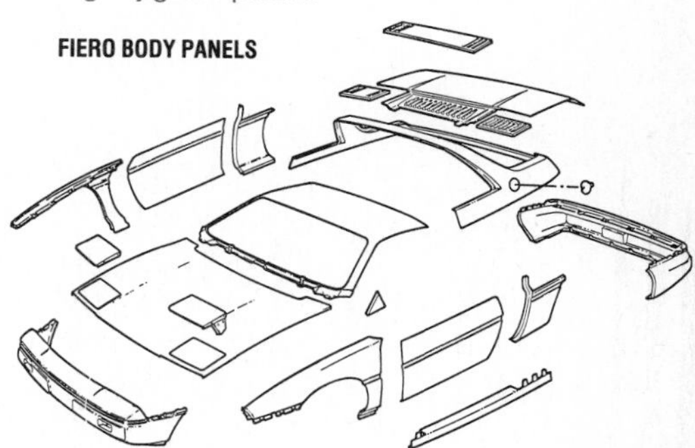

[12-42] The assembly of . . .

[12-43] . . . automobile body panels . . .

[12-44] . . . is a process of synthesis.

Evaluation and Testing

The process of making judgments about the validity of a problem solution or an engineering design is called *evaluation*. The evaluation will result in conclusions about the merits of a problem solution or engineering design. The conclusions may range from simple statements such as "It worked" or "It didn't work" or "It partially worked" . . . to detailed and very technical engineering reports.

Testing refers to the sequence of activities that the engineer uses to secure evidence necessary to make evaluations. Thus testing is a means of proving or disproving the solution or design. Testing provides results that can be compared using specified criteria. Evaluation uses the results of testing to draw conclusions about the worthiness of the problem solution or engineering design.

[12-45] The synthesis of quality tested components does not always produce a good design.

[12-46] Some tests need to be continued until the part actually fails.

No idea is so outlandish that it should not be considered with a searching but at the same time with a steady eye.

Winston Churchill, 1874–1965

An idea, in the highest sense of that word, cannot be conveyed but by a symbol.

Samuel Taylor Coleridge, 1772–1834
Biographia Literaria

In the evaluation and testing stage, the engineer's synthesized solution is operated in an actual working environment. The problem solver has been aware throughout each stage of the process of the nature and conditions of each test of the problem solution or design and of the criteria that will be used in the evaluation. Testing usually requires measurements and the collection of data. For this reason it is important that the problem solver be familiar with the measurement process. Knowledge of measurements is not restricted to equipment and techniques, but should also involve the study of accuracy and precision of measurements as well as measurement errors.

With regard to engineering design, the evaluation process consists of one or more tests of performance, economy, reliability, safety, environmental effects, and other important criteria that are identified in the specifications. For example, the specifications might stipulate that "the motor speed must be at least 1500 revolutions per minute, materials used in the design must not exceed $9000, no failures are permitted in 48 hours of testing, the device must function at −20°C, the design must adhere to safety standards A-78 and A-111," and so on. Specifications are generated throughout the design process. A further discussion of design evaluation criteria is given in Chapter 14.

Presentation of Solution

A final stage remains in the problem-solving process. The proposed solution or completed design must be "sold" to the client or decision makers of the organization. The manner in which the presentation is structured and organized can be critical to approval or disapproval. It is not unusual to hear reports of excellent solutions to engineering problems being turned down as a result of a lack of preparation, or a

[12-47] There are times when the precision of a design is the difference between life and death.

poor presentation, or both. Proper preparation must be matched with skilled presentations to assure a good chance of acceptance.

Preparation involves anticipation of all events that will take place during both the written and oral phases of presentation: anticipation of which models will be most useful, anticipation of what questions will be asked by the decision makers, as well as which alternative solutions should be prepared if unforeseen difficulties are encountered with the original solution. An alert problem solver does all the "homework" necessary to ensure that a good problem solution is not defeated because of lack of attention to appropriate details in the presentation of results. A written report is usually given to the decision makers before an oral presentation is scheduled. This gives an opportunity for the decision makers to review all the details of the solution procedure and to formulate questions regarding matters where more explanation is needed or questions about difficulties that exist in the solution. Sometimes the written report is returned to the problem solver for revision before the oral presentation is made. In any event, the success of the problem or project solution is very much dependent on the communications skills of the problem solver. Engineers who desire to improve their communications skills must first be good readers. Typically, those who are most effective in communicating their thoughts to others are also avid readers. Engineers who communicate well are not only well versed in their technical area of interest by reading technical journals and reports, but are also conversant in topics involving local, regional, national, and international affairs. Furthermore, many engineers have a broad range of interests so as to allow time to speak with others intelligently on topics involving sports, literature, history, music, and the arts. If you believe that you need more breadth in nontechnical areas, improve your reading habits. Begin by reading a newspaper every day. Devote attention to the headline, editorial, and business pages. Magazines such as *Time, Newsweek,* and *U.S. News and World Report* contain excellent stories about current events. If you have an interest in music, art, history, or literature, cultivate that interest with reading in these areas. This will make you a more complete person and others will recognize this extra dimension in your personality.

Written Reports. There may have been a time when engineers were not required to write reports, that is, when most of them worked with small groups of people and they could let their ideas be known by word of mouth or by circulating an occasional sketch or drawing. These times are gone for all but a very few engineers. Most engineers today work in large organizations. They cannot be "heard" unless their ideas are written down in proposals and their findings are recorded in reports. This does not mean that the spoken word and the drawing or sketch have lost their importance, but rather that they must be supplemented by the written word. Therefore, it is important for you to know how best to communicate your ideas to a reader: how to put your best foot forward with a client whom you may never see or with the company vice-president to whom you will report.

As an engineer, your writing is often directed toward other engineers; however, occasionally you may be called upon to write for an

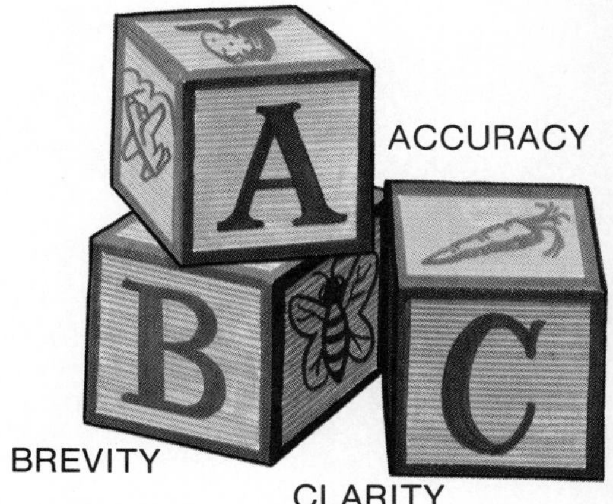

[12-48] Accuracy, brevity, and clarity are fundamental to good communication.

Short words are best and the old words, when short, are best of all.
Sir Winston Spencer Churchill, 1874–1965

There are three things to aim at in public speaking: first to get into your subject, then to get your subject into yourself and lastly to get your subject into your hearers.
Gregg

[12-49] If you have a message to tell, be enthusiastic in its presentation.

[12-50] Remember, you should not expect that all listeners will be interested in hearing your observations.

audience unfamiliar with technical terms. Then you must be able to express your thoughts in terminology that can be understood by an intelligent layperson. During the years to come, when engineers must solve the problems of society, such as the urban crisis, air and water pollution, energy, food production, and so on, the need for cooperation between technical and nontechnical persons becomes increasingly important. The engineer must learn to communicate with people of all types of background and must be able to state views clearly and concisely.

Oral Communication. Although written reports are important, you must not neglect your oral presentations. In preparing a speech or oral presentation, first make an outline of the principal ideas that you wish to project. Place them in a logical sequence and prepare your illustrations and similes, but do not attempt to write every word of your speech. Few things are more likely to put an audience to sleep than a speaker who reads a speech. If you tend to be nervous, memorize the first sentence or two, which will get you started, and then use notes only as reminders for the sequence of your talk and to make sure that you have said everything that you wanted to say. Visual aids, such as slides and overhead transparencies, in addition to illustrating a point, also serve as prompts to the speaker so that little memorization is required. Since an audience can best follow simple ideas, it is rarely advisable to present mathematical developments in a speech unless it is to an audience of mathematicians. Nor is it often useful or desirable to delve into the circuitous routes that were used during the development of the idea or the research that is being presented. The audience is interested in the results and in the usefulness of the results for their own purposes. All of us are interested primarily in our own life and work, and the better a speaker can convince us that his or her findings are useful to us, the more successful we believe the speaker to be. Therefore, in preparing a speech, first, find out to whom you will be speaking, and then ask yourself what it is that you can give to the audience that is useful to them. What will they remember after you have stopped speaking?

Your speeches will be successful if you fulfill these four goals of effective public speaking:

1. Command attention. (This can be done if you achieve the other three goals.)
2. Be confident. (Make a few strong points.)
3. Know the material. (Repeat, restate.)
4. Be enthusiastic. (An audience will forgive a speaker many things but will not forgive being bored.)

The successful speech, like a successful athletic contest, requires practice and rehearsal. In practicing, use a "sparring partner"—a person not afraid to criticize or interrupt and ask questions when something is not clear. Go over a speech with your "sparring partner" again and again until you are sure that you could present it even if you lost all your notes.

OVERCOMING BLOCKS TO CREATIVE THOUGHT

No chapter on problem solving and design methodology would be complete without mention of obstacles that tend to occur during the idea generation stage of problem solving and suggestions concerning how these obstacles may be overcome. In an excellent thought-provoking book, Adams[10] has outlined a set of mental blocks that inhibit the creative process. These blocks may be classified as:

Perceptual blocks.
Cultural blocks.
Environmental blocks.
Emotional blocks.
Intellectual and expressive blocks.

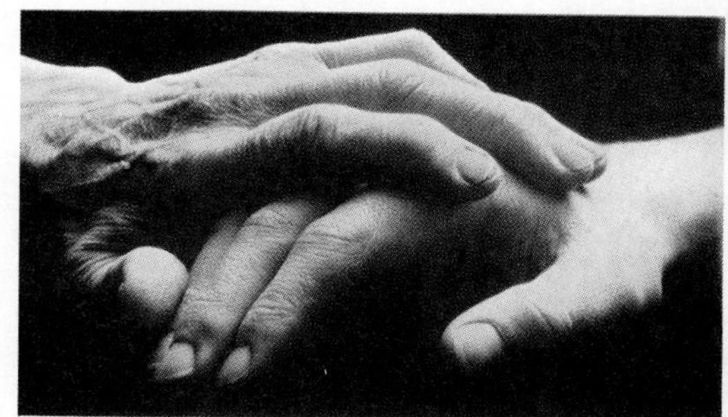

[12-51] Communication is complex. An idea may be transmitted and acknowledged between earth and moon in a matter of seconds . . .

[12-52] . . . while a similar transmission and acknowledgment between individuals standing within arm's length may take years.

Examples of *perceptual blocks* include difficulty in isolating the problem, a tendency to delimit the problem area too closely, an inability to see the problem from various viewpoints, seeing what one wants to see—stereotyping, saturation with the problem, and a failure to use all sensory inputs. Consider the story of the man whose automobile developed a flat tire outside the grounds of a state mental institution [12-53]. He pulled the car to a stop just off the roadway and adjacent to a steep embankment. He removed the lug nuts from the wheel and placed them inside the removed hub cap. Just after he replaced the flat tire with a spare, he accidentally kicked the hub cap and sent it and the contents far down the embankment, which was covered with heavy brush. The four lug nuts were not recoverable. He was sitting on the ground by his car, lamenting his woes, when he was approached by an inmate of the institution, who was obviously a trustee who had freedom to walk about the grounds and the periphery of the institution. The inmate inquired about his problems and the man related his story. The inmate replied, "Why don't you remove a single lug nut from each of the other three wheels, and place these three lug nuts on the fourth wheel, resulting in three lug nuts on each wheel, which would be sufficient for you to travel to the next town where you could purchase four new lug nuts?" The traveler was astounded. "Why didn't I think of that?", he exclaimed, "How can a man like you solve a problem that baffled me?" "Look mister," the inmate replied, "I may be crazy, but I'm not stupid."

The traveler had a perceptual block. In his mind, the source of lug nuts did not include those already being used on the other three wheels. Problem solvers can be trained to overcome perceptual blocks by not limiting the boundaries of the problem too tightly, by stepping back and seeing the problem from a number of different viewpoints, and by considering rather obscure solutions as well as conventional ones. These ideas can be summarized by asking the problem solver to think in various shades of gray, rather than trying to see everything as either black or white. A recent newspaper article

[12-53] . . . now it seems to me that you have a simple problem . . .

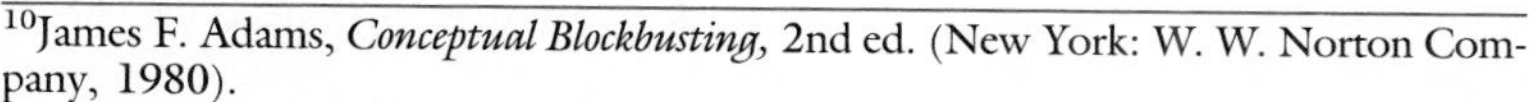

[10]James F. Adams, *Conceptual Blockbusting,* 2nd ed. (New York: W. W. Norton Company, 1980).

[12-54] . . . all undressed, and nowhere to go!

cited an example that seems somewhat ridiculous, but it illustrates the point. Some engineers were asked to design an automated chicken processing plant. One engineer in the group (obviously having read about overcoming perceptual blocks) extended his thoughts beyond the factory itself to consider the raw material, the chicken. One of the more difficult tasks in automating the process was the task of defeathering the chickens. This engineer was reported to have employed several veterinarians and geneticists to develop a featherless chicken so that the complete feather removing operation could be eliminated [12-54]. It has been reported that they are still working on the problem; rumor has it that some progress has been made, although the birds are said to taste a little like armadillos. The point is that uninhibited thinking is frequently needed in creative activities.

Types of *cultural blocks* include having the feeling that institutions and qualitative judgments are bad, preferring tradition over change, believing that all problems can be solved by scientific thinking and money, failing to express a sense of humor, and allowing oneself to be governed by habits. Many discoveries in the world of science, rather than being the result of scientific thinking, were serendipitous events (a serendipitous event is an unexpected positive result). Radium was left in a drawer with a photographic plate and other items. Later, someone discovered that there was an image on the plate: hence discovery of the x-ray process. Engineers at National Cash Register Company were examining coating processes for papers and ribbons when it was discovered that fragrances or scents could be trapped beneath the plastic compound. Hence the discovery of encapsulization, the process behind today's popular "scratch and sniff" advertisements and greeting cards. Recently, 5-fluorouracil (5-FU) was being injected beneath patients' skin for treatment of Parkinson's disease. It was discovered that patients who also suffered from a type of skin cancer had the cancer lesions disappear. Now, 5-FU in ointment form is a rather common treatment for basal cell skin cancer. Each of these discoveries was "accidental." They were made because the investigators were willing to accept new avenues of thought. In our Western culture, a heavy emphasis is placed on "left-brain" activities. The right-side functions of the body are governed by the left side of the brain, which controls verbal, mathematical, and logical tasks as well as reasoning. The right side of the brain controls spatial and geometric perceptions, synthesis of ideas, sensitivities, and feelings. Refer to Chapter 7 for a more complete discussion of these principles. Designers are encouraged to remove cultural blocks by exercising more of their "right brain": that is, to practice visualization, the process of creating mental images of shapes and forms, and rotating and translating these images in many ways. The visualization process has been shown to be very effective in many problem-solving efforts.

Examples of *environmental blocks* are being bothered by distractions, experiencing a lack of trust among colleagues, and experiencing a lack of cooperation and support from colleagues. However, we can be trained to work among distractions. Indeed, many complaints were registered in a city by people living near the railroad tracks when the trains no longer passed by their houses. They had adapted to the noisy distractions and had made them a part of their daily routine until a point was reached where they preferred the distractions to silence. In like manner we can adapt to distractions surrounding our work. Overcoming a lack of trust and support among our colleagues is not an easy task. A starting point would be to develop a sense of openness when dealing with colleagues. Their distrust and lack of cooperation may be due to fear that your activities will affect their secure position in the company or fear that by helping you, their own chances for success will be diminished. Conveying the feeling of openness will sometimes help to overcome these group difficulties.

Emotional blocks include the fear of risk, the fear of failing, an inability to tolerate disorder, inability to incubate ideas, not being able to perform well when there is no challenge, excessive zeal, and making judgments too early in the problem-solving process. Charles Lindbergh had little fear of failure when he became the first pilot to make a trans-Atlantic flight in 1927. It took more than 20 years to develop television sets for the home after the first pictures were transmitted in the 1920s. In this case, there was obviously no rush to generate ideas, and no early judgments were made in the problem-solving process.

Those who have *intellectual* and *expressive blocks* may try to solve the problem by the incorrect "language" (e.g., mathematical rather than visual), they may have inadequate language skills to express and record ideas, they may make inflexible or inadequate use of problem-solving strategies, or they may suffer from a lack of information or incorrect information.

With regard to solving problems by incorrect methods, we are often tempted to use a certain technique whether or not it is best for the situation. Consider today's widespread use of personal computers. There are many practical uses for these devices in dozens of everyday activities. Some business managers and hobbyists are so enamored with these devices that they put them to use on problems that could be solved much more easily by other methods.

Many of the blocks discussed above are not easily overcome, but merely recognizing that they exist will aid the thought processes of the problem solver as the creative component is exercised.

PROBLEMS

12-1. After reviewing the ecological needs of your hometown, state three problems that should be solved.

12-2. Give three examples of *feedback* that existed prior to 1800.

12-3. Talk to an engineer who is working in design or development in industry. Describe two situations in work where he or she has not been able to rely on *theoretical textbook solutions* to solve problems. Why was he or she forced to resort to other means to solve the problems?

12-4. Describe an incident where a person or group abandoned a course of action because it was found that they were spending time working on the wrong problem.

12-5. List the properties of a kitchen electric mixer.

12-6. List the properties of the automobile that you would like to own.

12-7. Make a matrix analysis of the possible solutions to the problem of removing dirt from clothes.

12-8. For 10 minutes solo brainstorm the problem of disposal of home wastepaper. List your ideas for solution.

12-9. List five types of models that are routinely used by the average American citizen.

12-10. Diagram the model of the football play that made the longest yardage gain for your team this year.

12-11. Draw an energy system of an ordinary gas-fired hot-water heater.

12-12. Draw an energy system representing a simple refrigeration cycle.

12-13. Draw an energy system representing a "perpetual motion" machine.

12-14. Draw an electrical circuit diagram containing two single-pole, double-throw switches in such a manner that a single light bulb may be turned on or off at either switch location.

12-15. Arrange three single-pole, single-throw switches in an electrical circuit containing three light bulbs in such a manner that one switch will turn on one of the bulbs, another switch will turn on two of the bulbs, and the third switch will turn on all three bulbs.

12-16. Describe three situations where a scale model would be the most appropriate kind of idealized model to use.

12-17. How can engineering help solve some of the major world problems?

12-18. Discuss some of the inventions that contributed to the success of man's first lunar exploration.

12-19. Write a paragraph entitled "Fiction Today, Engineering Tomorrow."

12-20. Propose a method and describe the general features of a value system whereby we could replace the use of money.

12-21. List five problems that might now confront the city officials of your hometown. Propose at least three solutions for each of these problems.

12-22. Cut out five humorous cartoons from magazines. Recaption each cartoon such that the story told is completely changed. Attach a typed copy of your own caption underneath the original caption for each cartoon.

12-23. Propose a title and theme for five new TV programs.

12-24. The following series of five words are related such that each word has a meaningful association with the word adjacent to it. Supply the missing words. Example:

girl	blond	hair	oil	rich
a. astronaut	______	______	______	engineer
b. rain	______	______	______	auto
c. college	______	______	______	book
d. football	______	______	______	radio
e. food	______	______	______	energy

12-25. Suggest several "highly desirable" alterations that would encourage personal travel by rail.

12-26. What are five ways in which you might accumulate a crowd of 100 people at the corner of Main Street and Central at 6 A.M. on Saturday?

12-27. "As inevitable as night after day"—using the word "inevitable," contrive six similar figures of speech. "As inevitable as . . ."

12-28. Name five waste products, and suggest ways in which these products may be reclaimed for useful purposes.

12-29. Recall the last time that you lost your temper. Describe those things accomplished and those things lost by this display of emotion. Develop a strategy to regain that which was lost.

12-30. You have just been named president of the college or university that you now attend. List your first 10 official actions.

12-31. Describe the best original idea that you have ever had. Why has it (not) been adopted?

12-32. Discuss an idea that has been accepted within the past 10 years but which originally was ridiculed.

12-33. Describe some design that you believe defies improvement.

12-34. Describe how one of the following might be used to start a fire: (a) scout knife; (b) baseball; (c) pocket watch; (d) turnip; (e) light bulb.

12-35. At night you can hear a mouse gnawing wood inside your bedroom wall. Noise does not seem to encourage him to leave. Describe how you will get rid of him.

12-36. Write a jingle using each of these words: cow, scholar, lass, nimble.

12-37. You are interviewing young engineering graduates to work on a project under your direction. What three questions would you ask each one to evaluate his or her creative ability?

12-38. Describe the most annoying habit of your girlfriend (boyfriend). Suggest three ways in which you might tactfully get this person to alter that habit for the better.

12-39. Suggest five designs that are direct results of ideas that have been stimulated by each of the five senses.

12-40. (**Note to the instructor:** The following problems are related. They are intentionally vague and ill-defined like most real-life problems. Their purpose is to stimulate creativity and imaginative solutions, to permit students to find out for themselves, make assumptions, test them, compare ideas, build models, and prepare reports—written or oral—to convince a nontechnical audience. Give only as much aid or additional information as you believe to be absolutely essential. Additional problems for this setting may suggest themselves.)

You are a Peace Corps volunteer (or a small team) about to be sent to a village of about 500 people in a primitive, underdeveloped country. The village lies 3000 feet below a steep escarpment in a valley through which a raging river flows. The river is about 80 feet wide, 4 to 8 feet deep, and too fast to wade or swim across. On your side of the river there is the village of mud huts in a clearing of the hardwood forest. The trees are no more than 40 feet tall. At the foot of the escarpment there is broken rock. Across the river there is another village, which cannot be reached except by a very long path and a difficult river crossing upstream. There are other villages on top of the escarpment. The people are small, few over 5 feet 6 inches tall. They live mostly by hunting, gathering, and fishing, although they could trade to their benefit with the people across the river and on the escarpment if communication were easier.

Before you leave for your assignment, you should try to

find solutions to one or more of the following problems:

a. How to improve communication, trade, and social contact between the two villages on each side of the river.
b. How to transport goods easily up and down the escarpment. There is a path up the escarpment, but it is steep, dangerous, and almost useless as a trade route.
c. Suggest a better way of hunting than with the bow and arrows now used. A crossbow has been suggested to be more powerful, easier to aim, and more accurate. Evaluate these claims and provide design criteria.
d. Provide for lighting of the huts. The villagers now use wicks dipped in open bowls of tallow. Can you improve their lamps so that they burn brighter, smoke less, and do not get blown out in the wind?

For each of these problems, select criteria for evaluating ideas. Choose several different solutions, check them against the criteria, and pick the best one; develop this idea by analysis and testing until you know how it will work. All the while, keep track of and test your assumptions whenever possible. Finally, prepare a way to convince the villagers of the value of your idea.

Chapter 13

Engineering Design Philosophy

Development of a Project

1. Enthusiasm.
2. Disillusionment.
3. Panic.
4. Search for the guilty party.
5. Punishment of the innocent party.
6. Fame and honor for the non-participants.

OMNI, July 1980

Philosophy may seem like a forbidding word to you. It should not. Literally, the word *philosophy* means love of wisdom. However, today the word has two meanings. It may also be thought of as a personal attitude toward life. Your attitude toward life, in no small way, governs the pattern of your everyday behavior. In a broader sense, it is common to speak of a philosophy of history, which might serve as a guidepost to indicate how history might be studied to glean relevant information to help us interpret today's events. Similarly, a philosophy of science outlines both what science can or cannot do to benefit humankind. Many companies have a philosophy of conducting business which indicates how the company will develop and maintain a relationship with its customers. Although such philosophies do not generally provide details concerning what should be done in specific cases, they do provide guidance relating to what is important in conducting affairs in specific disciplines. An engineering design philosophy, therefore, should indicate what considerations are important in the design process as well as what principles, concepts, and general methods are appropriate as the engineer attempts to satisfy varying design criteria.

This chapter presents a philosophy of design and identifies the critical aspects of design that must be considered if a design project is to be successful. The philosophical aspects of design may be thought of as requirements emanating from four points of view. Each point of view in effect represents a separate and distinct zone of influence. These zones are end user, control, system, and effectiveness. They are illustrated in Figure [13-1]. The design must achieve a balance as a result of the influences in each zone. It must continue to maintain its equilibrium in order to be a successful design. When the engineer makes decisions about how to satisfy a particular design requirement, each decision is very dependent on how the other design requirements are to be satisfied. All the requirements may be thought of as interrelated subproblems such as those discussed in the "analysis" stage of the design process (Chapter 12). Let us consider each of these zones of influence. The first zone includes those requirements most closely associated with the end user.

END USER REQUIREMENTS

The term "end user" has been in used in Figure [13-1] to convey the idea that the designer must be responsive to the concerns of the

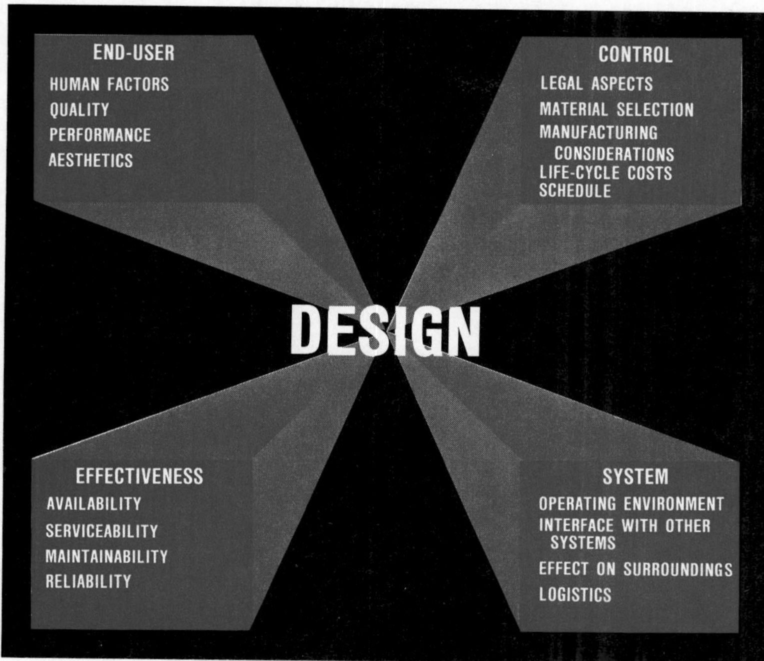

[13-1]

ultimate user of the device or system, the customer. Often the demands of the end user include a desire for attractiveness—a concern for properties such as style, form, and proportion. Thus engineering designs should have *aesthetic* appeal. Another obvious user requirement is *performance*. The device or system must do what it is supposed to do and do it well. *Quality* may be thought of as "fitness for use." The end user demands products that are defect-free, *human factors* should mean provisions for safety, comfort, and ease of use.

These end user requirements for design will now be presented in some detail.

Aesthetics

Aesthetics have often been referred to as beauty and the values that we attach to beauty. It would be useful to people of all walks of life, particularly to the engineer as well as the artist, if one could discover a mathematical formula that describes beauty and its applications to design. This might be possible if beauty were an identifiable entity, external to the observer. Then its fundamental ingredients could be isolated, analyzed, and assigned values. The resulting values could then be applied to new designs and all observers should generally agree that the new design was indeed beautiful. If such a simplification is possible, it has yet to be accomplished. However, there are some clues that tell us whether something is aesthetically acceptable or not. People continually disagree among themselves concerning the question, "What is beautiful?" If they happen to agree that a particular object is beautiful, their reactions will likely be of different intensities. Thus, for the engineer, problems in aesthetics and beauty

The Aesthetic Moment

"The aesthetic moment" is "that flitting instant, so brief as to be almost timeless, when the spectator is at one with the work of art. He ceases to be his ordinary self, and the picture or building, statue, landscape or aesthetic actuality is no longer outside himself. The two become one entity."

Bernard Berenson
The Pocket Dictionary of Art Terms (edited by Mervyn Levy)

Certain musical intervals have been found to be particularly harmonious (pleasing) to the ear: the octave, the major third, and the major sixth.

An aesthetic feeling is the emotion generated by the interaction between an object of beauty and an observer.

H. E. Huntley
The Divine Proportion, *1970*

are quite unlike problems in mathematics that can have a single, irrefutable answer.

Beauty has two recognizable attributes—the emotional and the intellectual. The first of these is related to the personality of the observer, to the period in which he or she lives, and to the surrounding environment. Emotion is judged mainly through intuitive processes. The primary reason that a meaningful formula for beauty has not been accepted is that this subjective component, emotion, is unique to each of us. It is this undefinable, intuitive aspect of our personality that provides us with the means to gain an emotional appreciation of an object or of a particular design.

The second quality of beauty, intellect, is less subjective, since it involves one's mental ability to understand clearly and completely the function of a design. Specifically, it measures the suitability of an object for its intended use, taking into account particular specifications and conditions that have been imposed by materials and other identifiable constraints. This aspect of beauty provides one with the ability to rationally appreciate an object.

[13-2] We are living in a world of beauty but how few of us open our eyes to see it?

Lorado Taft

The qualities of beauty cannot be measured independently because they are intrinsic components of the concept of form. Throughout the years, however, it has become evident that certain factors, particularly those concerning the intellectual qualities of beauty, can be identified and their effects consciously applied to the design of new objects. The location of mass can be determined, pleasing relationships of color have been established, and the proportions of forms have been documented. If these aesthetic qualities and many others are considered and incorporated into the final plan during the preliminary stages of the design, a far better product will result, because a unified and more aesthetically satisfying design can be achieved.

[13-3]

The elements that contribute to an aesthetically pleasing design will be introduced here so that you may see that although no mathematical equation exists for determining aesthetic value, the elements of aesthetics are identifiable. Furthermore, definitive statements may be made to guide the designer toward the proper application of these elements so that the best aesthetic properties may be realized.

Function. The concept of *function* is probably the most important and the most easily understood principle of aesthetics. A famous architect, Louis Sullivan, is credited with the coining of the phrase "form follows function" [13-4 and 13-5]. It means, in effect, that a product or design should possess a form descriptive of the actual function that the product is intended to perform. This concept is very important, because many products today are quite complex and their function is difficult for the user to appreciate intellectually. It is important for the buyer of a product to recognize, by means of an application of aesthetic principles, what the product is intended to do and how it is to be used. The intellectual capability of the engineer must be forced to deal simultaneously with the purpose for which an object is being made, with the aesthetic considerations, the capacities of the material, the tools, and the design process—in short, all the related aspects that must be controlled to assure completion of the design.

[13-4]

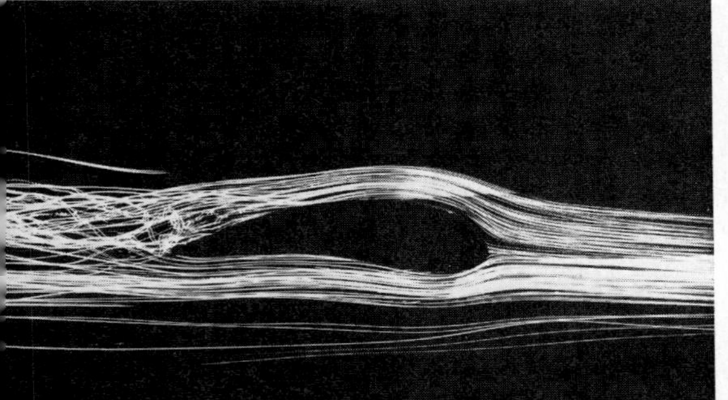

Form. Design is the process of inventing a real, operational system that displays a new physical order, organization, or form in response to function. Form involves the overall visual appearance of an object. It consists of many elements, such as line, mass, space, balance, proportion, contrast, and color. These elements are organized to give a design its unique form. The arrangement of these elements is either disorganized and unattractive [13-6] or orderly and attractive [13-7]. A good form is one that successfully imparts to the user an understanding of what the designer intended the object to do; a good form clearly identifies itself and its function.

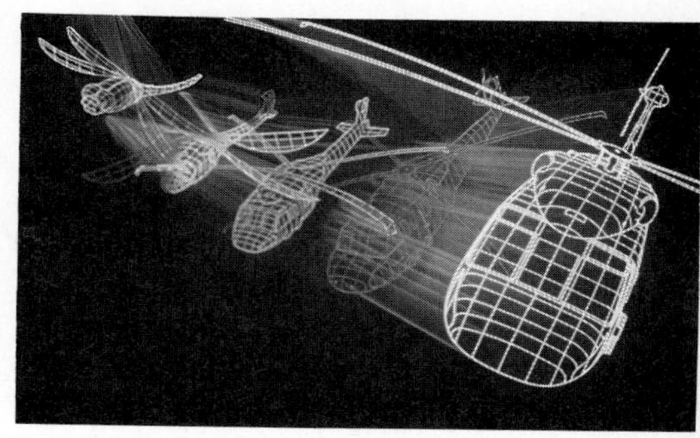

[13-5]

Line. The basic element used in the creation of a form is the line. This is particularly true of two-dimensional representations of form, which is the shorthand of design. Like beauty, the line has aspects that appeal to both intellect and emotion. On the intellectual scale, a line can be straight or curved; it can have length, thickness (sometimes referred to as weight), and it can have orientation. A gentle curve, strong, undecided, delicate, lively, weak, bold, or harsh lines are some of the many expressions that may be used to describe the emotive characteristics of lines [13-8]. The engineer must recognize that a line is more than a simple connection between points in space. A line also defines the separation between two dissimilar objects or conditions. Using this conceptual definition, two colors, different textures, dissimilar materials, variations in space, or any other elements that have a visual contrast can also be used to define a line [13-9 and 13-10]. The designer who can effectively use the emotional as well as the intellectual qualities of a line will be able to create a design with strong aesthetic appeal. Chapter 7 describes the functions of a line in visualization.

[13-6]

Mass. Mass is that aspect of aesthetics that imparts a sense of weight or heaviness to an object—a sense of solidity. When looking at an object for the first time, many of us may have said "That just

[1]C. W. Valentine, *The Experimental Psychology of Beauty* (London: Methuen, 1962).

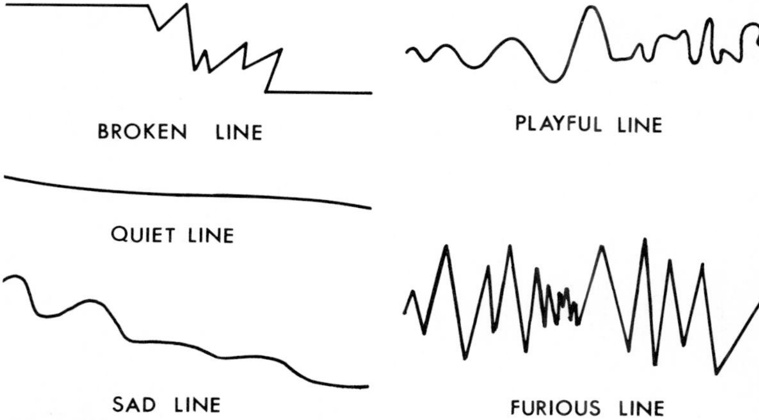

[13-8] Line personalities.[1]

[13-7]

[13-9] Farm contours define lines of visual separation.

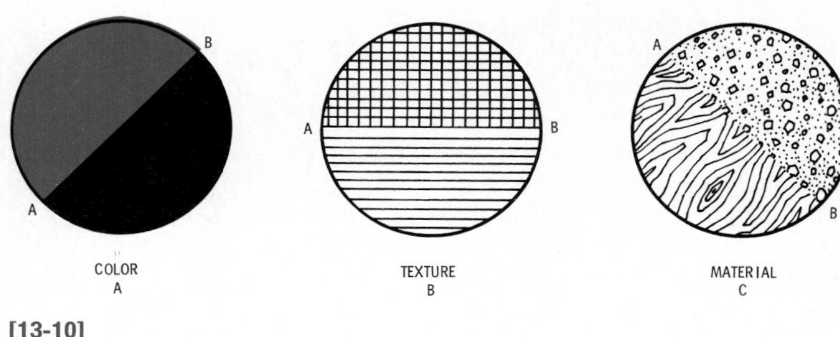

[13-10]

doesn't look strong enough" or "I wonder if that will last?" These feelings are often related to one's emotional assessment of the product based on traditional expectations of the mass of the materials or the elements used in this design. For example, what might appear to be flimsy or fragile if made of one material may appear to be adequately strong if made of another material.

Space. To the designer, space refers not only to the infinite reaches beyond the atmosphere of the earth, but also to the real environment that exists within the visual field. It is the total environment in which you perceive an object to exist. The psychologists have defined space as having a figure–ground relationship. The figure is the object, three-dimensional or two-dimensional, that attracts and holds one's attention. It is viewed relative to the environment that surrounds its, which is referred to as the ground. For the engineer, the figure is often a solid mass occupying a positive space. The remaining, encompassing space is the ground. It may be referred to as a negative space. Space when considered in this manner becomes manageable, and is as responsive to the concepts of aesthetics as any other element. The visually contrasting edge between positive and negative space defines a line. Negative areas, such as holes, apertures, and voids, are as much a part of design as the positive masses of material.

The effective apportionment of space is a difficult and sometimes confusing matter because the elements of a figure–ground relationship, particularly in two dimensions, can sometimes interchange their relative positions [13-12]. Of course, with a three-dimensional object it is impossible for such an exchange of space to occur physically, but if the engineer does not consider the object in relation to its surrounding environment, a visual translation could occur. For example, camouflage is the intentional use of visual translation. Imagine the problems that might develop if an operator of a high-speed router were not visually certain of the figure–ground relationship of the cutting blade and the working surface.

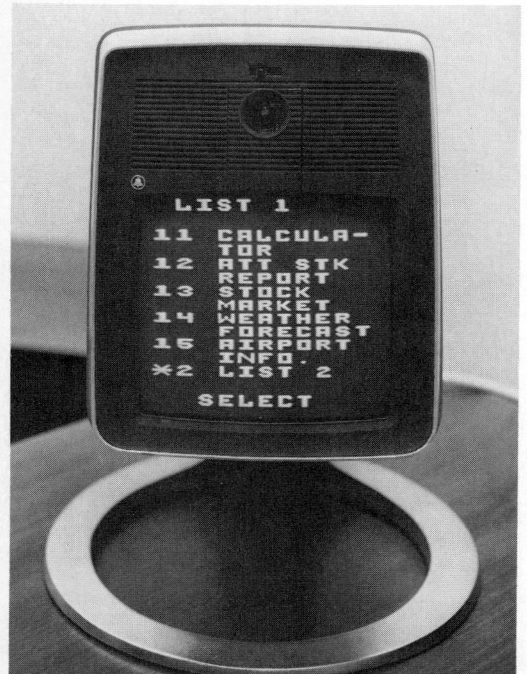

[13-11] The Picturephone—an example of the use of positive and negative space.

Balance. Balance is an important consideration in design. There are two types of balance, symmetrical and asymmetrical. Symmetrical balance is the more easily understood, and it can be measured to some extent. Symmetry is the balance of mirror figures about a point, line or plane [13-13]. It is a simple matter for one to measure the length, width, and other pertinent dimensions on one side

[13-12] Is it a chalice . . . or . . . is it two profiles?

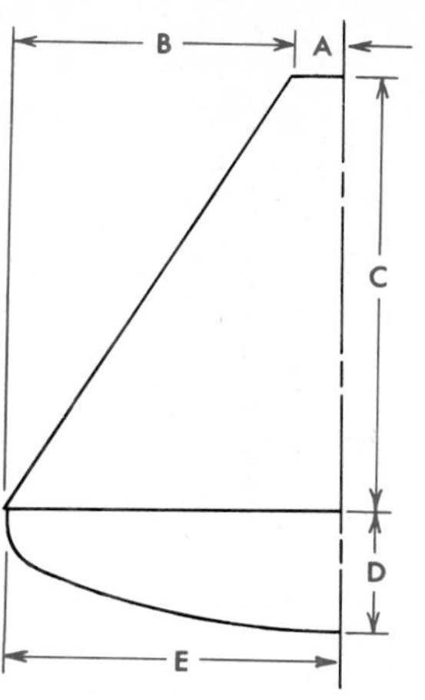

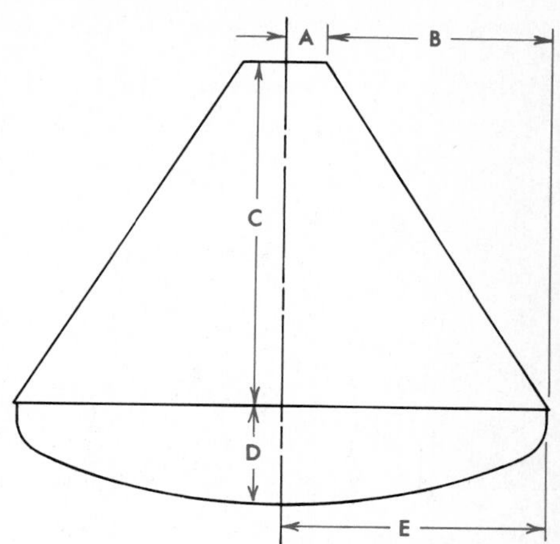

[13-13] Symmetrical balance.

of the center line of a design and to repeat the measurements on the other side. By using this technique, one is assured of achieving a symmetrically balanced design of the object [13-14]. The usual emotional response to symmetrical balance is "stability" or "satisfaction with the status quo."

Asymmetrical balance, sometimes referred to as occult balance, is much more subtle, but it usually stimulates a dynamic and emotional response. Acceptable asymmetric balance depends largely on the emotional qualities of aesthetics and the personalities of the designer and the viewer. Occult balance is somewhat like the concept of torque or moment effect in physics. The difference is that varying centers of interest are involved instead of varying forces (weights) and lengths of moment arms [13-15]. A form may be in equilibrium, but the elements that will contribute to the feeling of "equilibrium" are not always easy to identify or understand. The examples using the relative position and differences in color are indicative of elements that one might encounter when developing an asymmetrical design.

It is important for the engineer to be able to achieve asymmetrical balance in a design, because the development of a new product or

[13-14] The Apollo space capsule—a symmetrically balanced design.

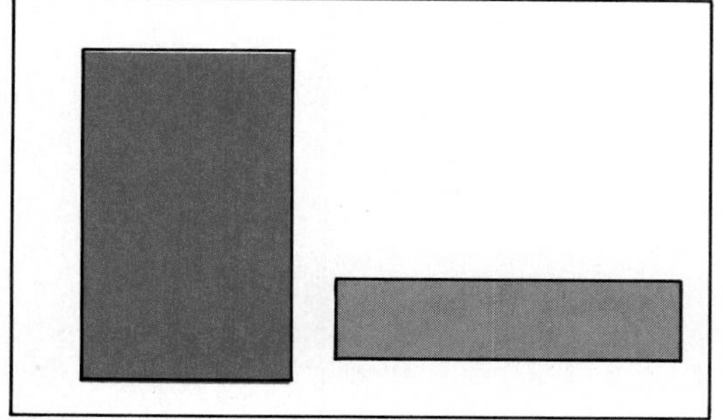

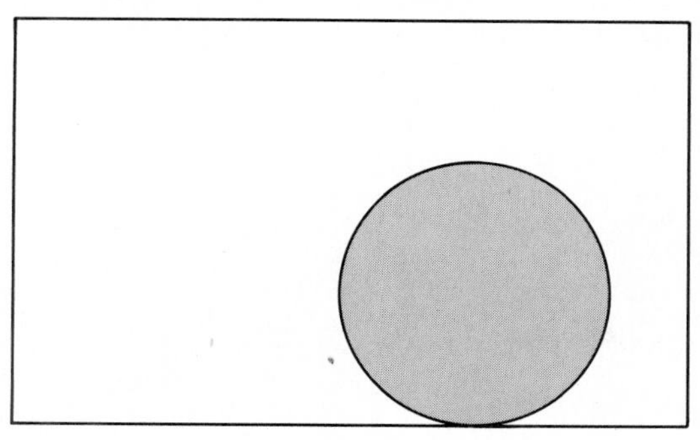

[13-15] Asymmetrical balance.

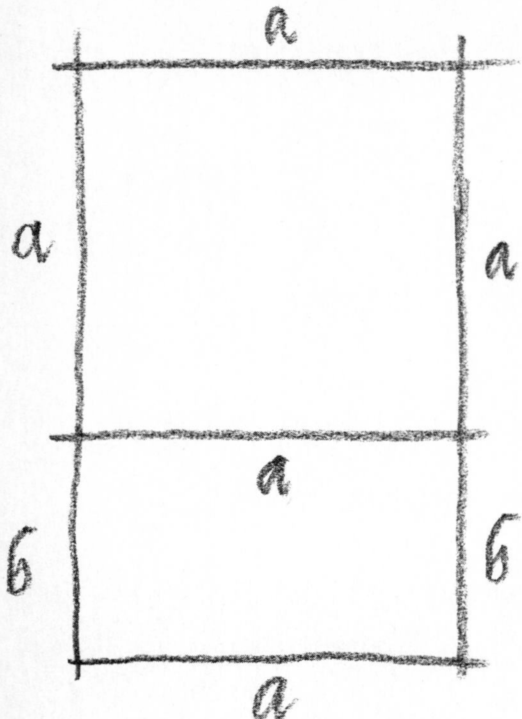

[13-16] The divine proportion.

Proportion is not only found in numbers and measurement but also in sounds, weights, times, positions and in whatsoever power there may be.
Leonardo Da Vinci

technique is not always compatible with the concept of symmetrical balance. The sewing machine, the machine tool lathe, and the sporting rifle are examples of products that have effectively created a sense of equilibrium and of form through asymmetrical balance.

Proportion. Proportion relates the overall dimensions to the dimensions of smaller elements or parts within the form. In the simplest sense, it is the ratios of the parts to the whole. Proportion is one of the most effective elements for achieving aesthetical balance and the concept of unity. The simplest proportions to use and control are regular geometric figures, such as circles, equilateral triangles, and squares. However, these shapes are monotonous if they are used exclusively. Variety helps.

Complete systems of aesthetics have been developed around the mathematical relationships that exist between the length and width dimensions of rectangles. One of the most popular and effective concepts using rectangles is *modular construction*. It is used in architecture, particularly mass housing, in electronics, such as computer circuitry, and in machine design involving component subassemblies.

A system of proportions is a common means of developing an aesthetically pleasing design, as well as a functionally efficient one. Since the days of Pythagoras and Plato, and perhaps even before, human beings have tried to establish a universally acceptable proportioning system. One of the popular systems to produce an orderly relationship between parts is believed to have been used in building the Greek Parthenon. This system has been called the *divine proportion, golden ratio,* or *golden rectangle*. This ratio is believed by some to be the most pleasing visual arrangement for a rectangle, and several systems for proportioning form have evolved using it as the base. It is a rectangle whose proportions are such that when a square is removed from either end, the remaining rectangle will be proportional to the original rectangle [13-16].

To discover the secret of the golden ratio ($x : 1$), and hence be able to construct golden rectangles [13-17], we can mathematically equate the ratio of width to length in the original rectangle to the ratio of width to length in the remaining rectangle. The result would be

$$\frac{1}{x} = \frac{x-1}{1} \quad \text{or} \quad x^2 - x - 1 = 0$$

The solution to this equation is obtained by applying the quadratic formula and selecting the positive root.

$$x = \frac{-(-1) + \sqrt{1 - 4(1)(-1)}}{2}$$

$$= \frac{1 + \sqrt{5}}{2}$$

$$= 1.6180339 \quad \text{or} \quad 1.618$$

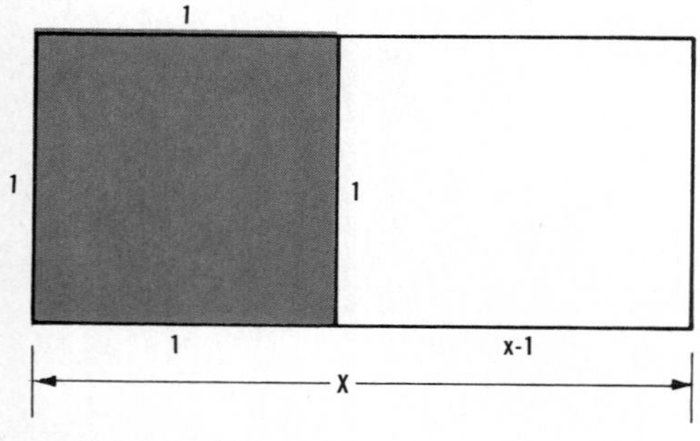

[13-17] Golden rectangle.

In the literature the Greek letter phi(ϕ) is often used to denote the value of the golden ratio. There is a mathematical relationship between ϕ and a sequence of numbers attributed to the famous

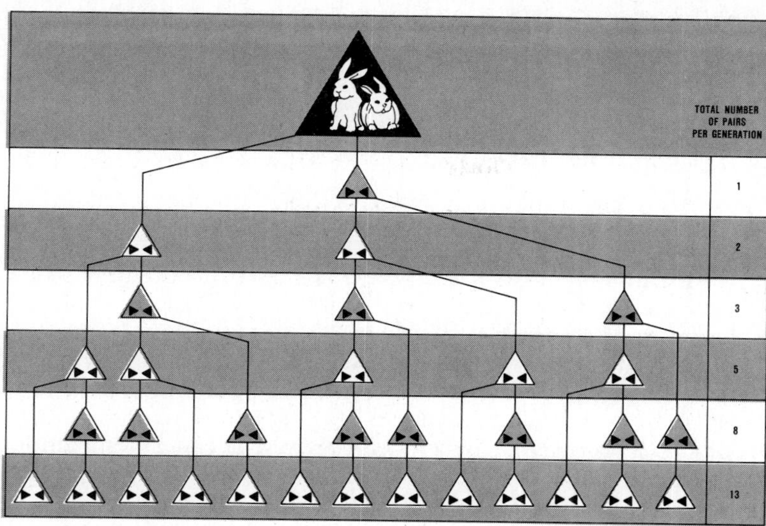

[13-18] Rabbits and generations.

mathematician Leonardo Fibonacci (also called Leonardo of Pisa), who was born in 1175. Fibonacci originated and solved a mathematical puzzle which resulted in the number sequence bearing his name. The puzzle was phrased something like this:

"Suppose that a single pair of rabbits is bred and the pair produces a pair of young after one month and a second pair the following month and then stops reproducing. Suppose that each pair of offspring in turn does the same. How many *new* pairs of rabbits are there at the end of each month?" Figure [13-18] illustrates the solution. You will see that each number in the right column is the sum of the preceding two numbers. If we define $F_0 = 1$ and $F_1 = 1$, we may write the expression

$$F_{n+1} = F_n + F_{n-1}$$

where
$n = 1, 2, 3, \ldots$. Writing the Fibonacci sequence beginning at F_0, we have

1, 1, 2, 3, 5, 8, 13, 21, 34, 55, 89, 144, 233, 377, 610, 987, 1597, 2584, 4181, . . .

It can be seen that F_n/F_{n-1} approaches $\phi = 1.618$ as n gets very large. There are many other interesting properties of Fibonacci numbers which appear in mathematics and in nature. H. E. Huntley[2] has written an excellent treatise on the subject.

It was discovered long ago, perhaps by Pythagoras's Egyptian predecessors, that the pleasing ratio of the golden ratio plays a very important role in the morphology of the natural world, both organic and inorganic. Forms of crystals, shells, plants, and even parts of the human body seem to follow this proportion closely.

Other systems of proportioning have also found enthusiastic followers. A number of these systems are based on the premise that, in

[2]H. E. Huntley, *The Divine Proportion* (New York: Dover, 1970).

. . . if we are conscious of a law of proportion, and then slightly deviate from it, to avoid its precision, we shall produce a more beautiful effect.

Sir Herbert Read
Art and Industry, *1953*

If . . . we propose to create beauty in our own handiwork, it is a fair conclusion that we should not forget those same divergencies from mathematical exactitude which the Greeks were so careful to recognize in the lines and columns of the Parthenon; and we should realize that one great factor in the beauty of art and of Nature consists in those same subtle variations which have molded vital forms since life began.

Theodore Andrea Cook
The Curves of Life, *1914*

[13-19] Honeybees build hexagonal cells with trihedral bases. This design is not only the strongest possible structure for a mass of adjacent cells, but it is also the one that requires the least amount of labor and wax.

There is a very significant characteristic of the application of the spiral to organic forms; that application invariably results in the discovery that nothing which is alive is ever simply mathematical. In other words, there is in every organic object a factor which baffles mathematics—a factor which we can only describe as Life. The nautilus is perhaps the natural object which most closely approximates to a logarithmic spiral; but it is only an approximation; the nautilus is alive and, therefore, it cannot be exactly expressed by any simple mathematical conception; we may in the future be able to define a given nautilus in the terms of its differences from a given logarithmic spiral; and it is these differences which are one characteristic of life.

Theodore Andrea Cook
The Curves of Life, *1914*

[13-20] This radiograph of the *chambered nautilus* (a sea shell) shows clearly the logarithmic spiral pattern of growth of the successive chambers. As the shell grows, the chamber sizes increase, but their shape remains unchanged.

nature, the growth of things into particular shapes results because of forces that act in accordance with well-defined laws of mathematics and physics. The honeybee's comb is an example of this logic [13-19].

Other plants and animal forms also assume mathematically definable configurations, for example, the lens of a water beetle's eye, snowflakes, and seashells. The latter "children of nature" follow curvilinear rather than rectangular growth patterns, with the logarithmic spiral being the most common [13-20].

Another system of proportioning is based on the pear shape. This shape occurs in nature when there is a balance between the forces of gravity and surface tension.

The engineer may be tempted to copy a system such as one of these to satisfy the aesthetic requirements of a particular design and by so doing believe that he or she has made an object that will be judged to be beautiful. This is a false premise and such temptations should be shunned. If a proportioning system is needed in a design, the engineer is advised to establish a system based on the concepts of function, form, and unity as it applies to the object being developed, rather than to adopt arbitrarily some "so-called" established system. Remember, the honeybee does not use a honey container designed for use by a chicken [13-21].

Contrast. The fact that you are able to read this textbook is due in part to the aesthetic element of contrast. The contrast between the black letters and the white page improves the legibility of the text. Contrast is the same quality as difference: difference in color, line, mass, or some other element.

Contrast, as a design element, is useful for drawing attention to or emphasizing a component or combination of components within a design. Conversely, reducing visual contrast may diminish an unsightly or distracting aspect of a design [13-22]. Many machines have specific areas for the control mechanisms, such as start-stop buttons, levers, and wheels. By applying the principle of contrast, the engineer draws attention to the location of the controls and thereby increases safety and efficiency.

Reducing the contrast is particularly desirable when conditions exist that are beyond the control of the engineer. If it is too expensive to enclose exposed pipes and ductwork in factory offices, they can be painted the same color as adjacent surfaces.

Color. Color is probably the element that is most commonly associated with aesthetics, and its use readily demonstrates the two qualities of aesthetic appreciation, intellect and emotion. Intellectually, colors are described as those sensations that are produced in the brain as a result of light waves impinging on the retina of the eye. The visible spectrum consists of electromagnetic radiation with wavelengths between 380 millimicrometers (violet) and 760 millimicrometers (red). Within this rather limited range, it is estimated that the human eye can distinguish 10 million different colors. We are, therefore, quite sensitive to color.

Colors have an emotional quality, a psychological effect. Properly applied, these effects can greatly enhance the appearance as well as the functional suitability of an object. Artists have been aware of many of these effects and have used them successfully for many centuries. Since it is difficult to measure or explain some of these effects technically, some examples of the more commonly known color effects will be discussed.

Some hues have predictable emotional effects. Yellow is considered friendly, happy, warm, cheerful, sunny, and is associated with the spring season. Orange tends to be gay, warm, vivacious, and outgoing. Red is exciting, arresting, and lively. Orange and red are normally related to the autumn season. Green is relaxed, pastoral, earthy, reposed, and reflects the spring and summer seasons. Serene, cool, and melancholy are effects stimulated with blue hues. Blue hues are often associated with the sea and the winter season. Violet is royal, splendid, elegant.

The appearance of the design form may be enhanced by a proper application of color. In general, one should avoid excess in the number of colors applied to a design. Too many colors on a product can create a garish and offensive impression. However, colors can accentuate the form of a product. Misapplied color can do much to destroy an otherwise successful design [13-23].

Unity. Unity is another word for order or harmony. In aesthetics, unity is a harmonious combination of parts. The use of similar proportions and shapes and identical decorative motifs are examples of ways to provide unity in design. Using elements that are similar through all the products of a line is a typical means of expressing unity and of developing product identification for a manufacturer. Unity in a product means that it appears to be complete. Nothing is missing nor are there any superfluous elements. Further-

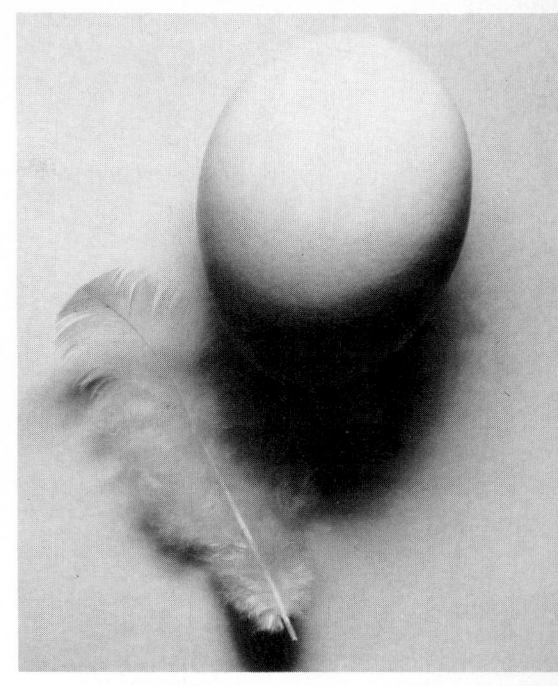

[13-21] Based on function and biological "manufacturing techniques", the egg is Nature's perfect package. Its basic form has remained virtually unchanged for over 300 million years.

The Bower birds of Australia and New Guinea build bowers for courtship with brightly coloured fruits or flowers which are not eaten but left for display and replaced when they wither. . . . They stick to a particular colour scheme. Thus, a bird using blue flowers will throw away a yellow flower inserted by the experimenter, while a bird using yellow flowers will not tolerate a blue one.

W. H. Thorpe
Science, Man and Morals

[13-22] Creatures in nature rely on visual contrast for self preservation.

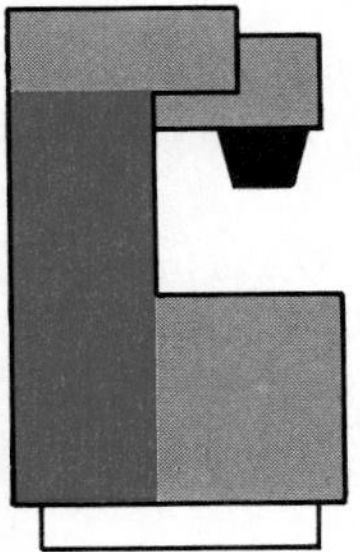

(a) Color tends to destroy the concept of the form.

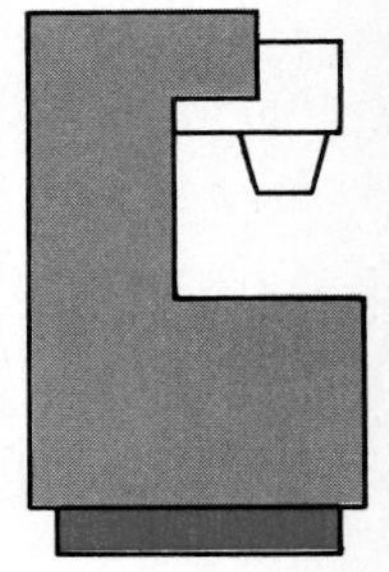

(b) Color enhances the concept of the form.

[13-23]

[13-24] Too much of one thing can lead to monotony.

more, there is a proportionate relationship between all the parts contributing to an orderly whole. A designer achieves unity by regarding an object as a total composition and not as an assembly of individually designed parts. Many engineers have designed efficient and functionally attractive subassemblies that, when joined to make the whole, failed to satisfy the concept of unity. A need for variety in design is obvious, but at the same time, the effective application of variety is an extremely difficult task. Too much variety results in the destruction of unity, because the relationship of the visual elements is lost. The designer must be able to recognize and create this delicate balance of unity with variety within the framework of an object's form.

Styling. Styling, decoration, ornamentation: these are different names for one of the most discussed and least understood aspects of aesthetics. Styling, the application of ornamentation, is often in direct conflict with the concepts of function, form, and unity [13-25]. All too frequently styling is applied to an object after the function and form of a design have been fully developed. A good design does not need styling. If the function and form of a product have been developed improperly, it is impossible to achieve unity by applying decoration or ornamentation—in other words, by styling.

Styling is closely associated with marketing products. It is a common device used to promote the sale of goods, but it should not be confused with the aesthetic quality that results from proper function, form, and unity. Styling applied to products reflects the buying public's immediate interest and concept of "what is beautiful." The basic engineering design of the automobile has not changed for several years, but styling revisions in the outward appearance have changed annually. These "new styles" are created each year to generate consumer interest.

Styling has much in common with commercial art. Many of the techniques are the same, as well as the motives and purpose. Styling

Beauty is unity in variety.

J. Bronowski
Science and Human Values, *1964*

I shall define beauty to be harmony of all parts in whatsoever subject it appears, fitted together with such proportions and connections that nothing could be added, diminished or altered but for the worst

Alberti

. . . The first quality that we demand in our sensations will be order, without which our sensations will be troubled and perplexed, and the other quality will be variety without which they will not be fully stimulated.

Roger Fry
Essay in Aesthetics

Differences in education are mainly responsible for differences in taste.

H. E. Huntley
The Divine Proportion

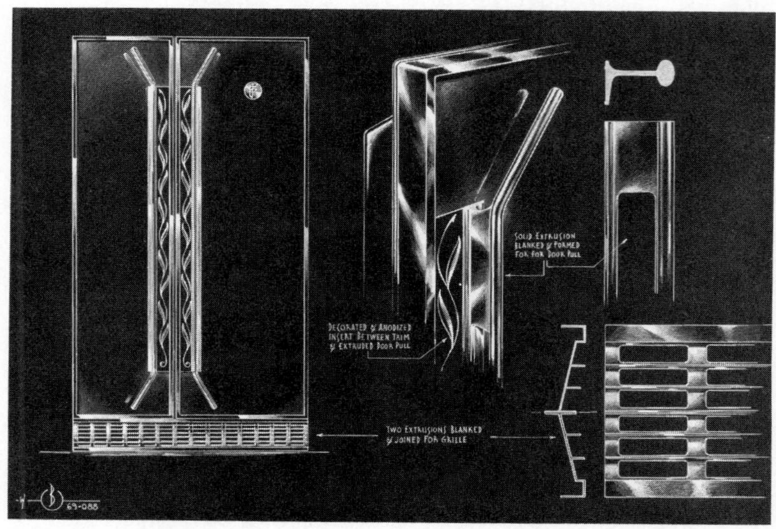

[13-25] As with illicit love, *styling* is a pleasure of the moment and tarnishes with time.

is concerned primarily with outward appearance. It is the development of a form with little or no functional consideration. It is applied decoration that relies primarily on the emotional qualities of aesthetics. It might be considered the reverse of the dictum "form follows function" because after a "style" has been created that is judged to be acceptable, the functional elements may be modified or fitted to it. Engineers should avoid styling their designs, and they must be able to combat the influences of styling in the development of new or existing products.

[13-26] All new products, such as this improved golf club, should be tested in use and their performance measured.

Performance

Engineering designs are developed to meet one or more purposes or specifications. These specifications should be clearly defined during the problem identification stage of the design process. Many of them are based on requirements of the end user of the design. Performance of the designed product or system is always a paramount consideration in achieving the design objectives. The extent to which the design specifications have been met is determined by performance tests. This requires that a portion of the designer's efforts be directed toward ensuring that the device or system can pass the tests. A performance test simply shows whether a design does what it is supposed to do. In addition to certifying the capability of the product to perform its functions, it also measures the skill of the engineer and the validity of the assumptions that were made in the design analysis.

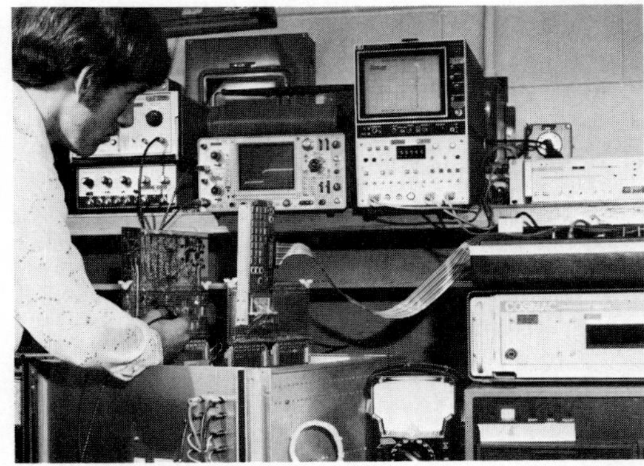

[13-27] Testing microprocessor circuitry for performance.

For a structure, satisfactory performance may mean being able to meet a certain load-carrying capacity while maintaining a specified maximum deflection. For an engine, satisfactory performance might be achieved if the efficiency (output-to-input ratio) can be kept at or above some predetermined value. Often, the design engineer must conduct experiments to determine which values of certain design variables will result in optimal or acceptable performance levels. Usually, there are limits associated with each measured design variable which cannot be exceeded if the performance criteria are to be met. These limits are referred to as the *design specifications*.

Performance testing generally does not wait until the design is completed. It follows step by step with the design. For example, the heat shield of the space capsule is tested in the supersonic wind tunnel to see if it can withstand the aerodynamic heating for which it was designed; the parachute is tested to see if it supports the capsule at just the right speed; structural members are tested for strength and stiffness; and instruments are checked to show if they indicate what they are supposed to measure.

Performance tests may require special testing apparatus, such as supersonic wind tunnels and space-simulation chambers. They always need careful planning and instrumentation to assure that the tests measure what is really needed—a proof of the validity of the design. Figure [13-28] illustrates the design specifications that have been determined as a result of an experiment which measured performance over a range of values of a design variable. L and U refer to the lower and upper specification limits. The particular design variable must be held within these limits to achieve performance in the acceptable range. These specification limits will be transmitted to the

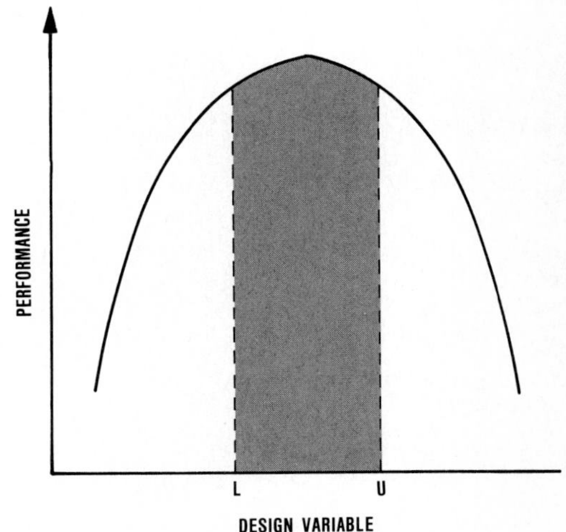

[13-28] Specifications for values of the design variable are often set by experiments measuring performance.

[13-29] You oaf! You misread the scale. I wanted a toy! What could we ever do with a horse this big?

[13-30] Quality is strongly related to durability and error-free, failure-free operation.

[13-31] At the end of the manufacturing operation the parts are returned to the palette station for quality control.

manufacturing department. They provide information about the manufacturing processes that must be used to keep the design variable within these limits.

Quality

Quality is another requirement of the end user. It can best be defined as "fitness for use." Although we usually believe that quality requires maximum durability and infrequent failures or breakdowns, a design can be considered to be one of high quality if it satisfies all the performance criteria that have been specified.Conversely, if the design does not do what it is supposed to do, it can hardly be considered to be a quality product. Thus, if you purchase an automobile primarily for fuel economy, you will not believe that you have purchased a product of high quality if the fuel consumption is considerably lower than the promised specified value at the time of purchase even though the automobile may demonstrate a low frequency of repair. A popular expression among design engineers is: "Quality cannot be inspected into a product; it must be built into the product." This means that regardless of the attention devoted to inspecting and testing products before they are sold so that the "defectives," may be separated from the "non-defectives," inspection and testing cannot change the quality of the product. These measures can only prevent the defectives from reaching the end user. Quality must be achieved during the manufacturing process. The role of the designer in this process is to set proper specification limits for the design variables. Specification limits are based not only on performance criteria, but also on a knowledge of the manufacturing methods that will be used, including how the product is to be inspected and tested; therefore, quality is achieved through the joint efforts of design, manufacturing, and quality control engineers, each of whom must know something about the job of the others.

Quality is strongly related to durability and error-free, failure-free operation. Today's consumers are more quality conscious than ever before. We have entered the age of consumerism, where users demand and expect high quality with every purchase. With many products, such demands and expectations are made possible by keen competition in the marketplace. Much of this competition today comes from foreign competitors. For example, many observers believe that Japan's recent success in manufacturing electronics, automobiles, and shipbuilding has been made possible by strict adherence to the principles of quality in manufacturing. Many of the quality control techniques that are used in Japan were developed in the United States, but it has taken the success of the Japanese to reawaken a concern for product quality among American designers and manufacturers. Concerted efforts directed toward achieving quality in manufactured products are required if the United States is to be reestablished as the world leader in the design and manufacture of high-quality products.

Quality may also be thought of as a grade or level of a product's performance. For example, we often see different versions of the same design advertised as "good, "better," and "best." Such classifications are made in regard to the perceived level of product quality. Some integrated-circuit manufacturers classify microchips in three quality levels, with the highest level being reserved for use in military and aerospace applications, the next highest quality level being reserved for specialized industrial buyers, and the lowest acceptable level of solid-state products being used in such items as electronic toys and games. In this case, each chip of a particular quality level will work acceptably in the environment for which it was intended. For example, the chip used for toys and games will not be tested to the temperature extremes that might be associated with an aerospace application of the same chip. Because of this, the toy chip need not possess all the characteristics of the chip that will be used in aerospace applications.

Quality will be achieved by attention to the effectiveness requirements of design—reliability, availability, serviceability, and maintainability. These requirements are discussed later in this chapter.

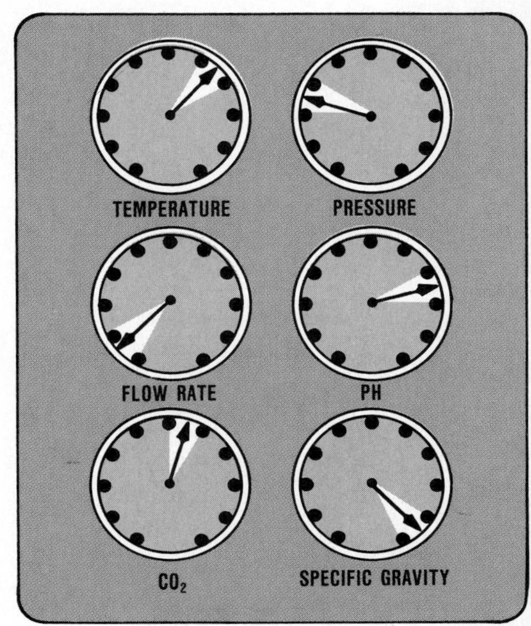

[13-32] Original control panel design.

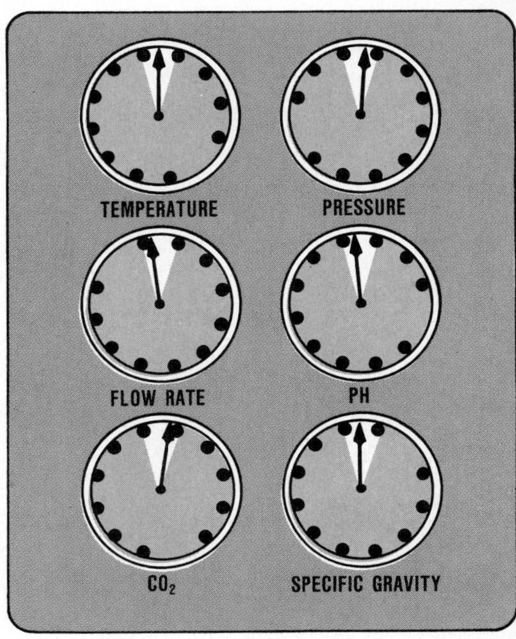

[13-33] Improved control panel design.

Human Factors

There are many designs in use today that have disregarded the needs and comforts of the person who uses them. It is important that all engineers have a fundamental understanding of human factors so that they can participate effectively as team members in the design of highly complicated human–machine systems. The engineer should design instruments that are easy and safe to operate and that give clear signals and make possible rapid and accurate operator decisions. Dials and gauges should be easy to read at a glance. Figure [13-32] shows a set of dials such as might be used to monitor several critical system variables in a control room—variables such as temperature, pressure, flow rate, and so on. The curved region on each dial indicates the range of acceptable values. If the pointer, shown with an arrow, is in the acceptable range, the control room operator has no need to take corrective action. If the pointer points outside the allowable region, the operator must initiate some corrective control procedures. Often, these actions must be initiated as soon as possible to prevent damage to equipment, reinstate a nonconforming product, or prevent injury to personnel. In Figure [13-32], the radial location of the range of acceptable values, which is called the *null point,* varies from dial to dial. Because of this, departures from the null are difficult to detect at a glance. A preferred redesign of the control panel dials is shown in Figure [13-33]. Here the null points are located in the same visual position, providing a more rapid detection of out-of-control conditions. In addition, when an out-of-control condition is reached, each dial might be accompanied with a flashing red light and perhaps an alarm system to warn the operator.

Designing for safety is mandatory—not only from a viewpoint of concern for human welfare, but also to lessen the possibility of costly liability suits. As an example, some machines which perform cutting and stamping operations are designed to require that both the operator's hands be on control levers, even though an alternative design could require the use of only one hand. With the two-hand design it is impossible for an operator's hand or arm to be in the cutting or stamping area while the machine is operating.

"I am only an average man, but I work harder at it than the average man." *Theodore Roosevelt, 1858–1919*

Man is not the largest of the animals nor are his vision, his hearing or his sense of smell the most acute; he is not the strongest of animals, the most fleet of foot nor even the longest-lived; but he is superlative in the quality of his central nervous system and he possesses a hand which does its bidding.
J. C. Boileau Grant
Method of Anatomy, *1952*

The desire for safety has been responsible for establishing many of the design criteria used in industry. The transportation industry today is working to make severe impacts survivable and to make severe impacts tolerable. A good motto for the engineer to adopt is "delethalize design!" Designing for emergency situations is an important part of safety. There are three important criteria to remember in designing emergency-warning equipment.

1. Get the user's attention. A number of ideas have been tried with varying degress of success. Red lights, blinking lights, horns, bells, or even a soft human voice can be effective at certain times and under certain circumstances.
2. Indicate the course of action that should be taken and the degree of emergency that exists. Also, indicate the time allowed for action.
3. Use a separate, independent, and testable power source for emergency systems.

It is important for engineers to follow design rules and principles that have proven useful in controlling injuries and illnesses related to product use. The American Society of Mechanical Engineers provides the following advice to designers.[3]

> It must be kept in mind that there is a basic difference between an accident and an injury. An accident is a general term that refers to an unexpected event—it may have fortunate or unfortunate consequences for the person involved. Finding a $20 bill while walking in the street and slipping a few minutes later on a patch of ice are both accidents. But only the second event will cause you an injury.

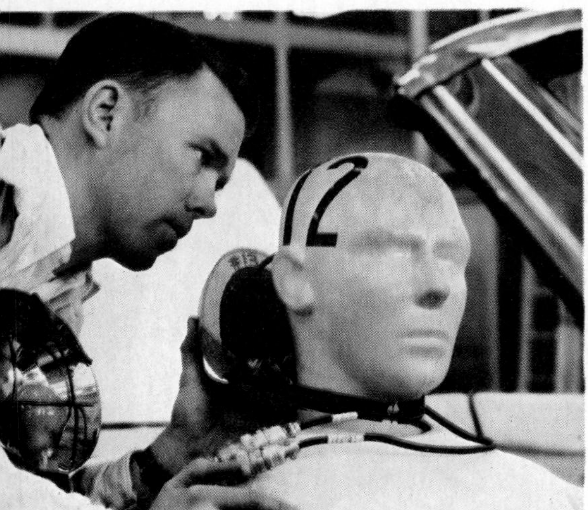

[13-34] Man is the only animal that eats when he is not hungry, drinks when he is not thirsty, and makes love at all seasons.

[3] *An Instructional Aid for Occupational Safety and Health in Mechanical Engineering,* (New York: American Society of Mechanical Engineers, 1984), p. 1.

There are six basic guidelines that a designer can apply to maximize the safety level of products or manufacturing processes. The National Safety Council has published this list in descending order of effectiveness. One should rely on the highest concept attainable, but if this is not possible, the very next one shown should be used. In brief these are:

1. Eliminate the hazard from the product or process by altering its design, material, usage, or maintenance method.
2. Control the hazard by capturing, enclosing, or guarding it at its source.
3. Train personnel to be aware of the hazard and to follow safe procedures to avoid it.
4. Provide adequate warnings and instructions in appropriate forms and locations.
5. Anticipate common areas and methods of abuse and take steps to eliminate or minimize the consequences associated with such actions.
6. Provide personal protective equipment to shield personnel against the hazard.

These six rules are interrelated and more than one can be used in a specific situation. For instance, training and warnings can often supplement machine guarding and the use of personal protective equipment.

[13-35] There are environments that are unforgiving. Safety is of more than minor concern to these astronauts.

Table 13-1 illustrates an even more general approach to the matter of hazard reduction.

Comfort, accessibility, and ease of use are other important human considerations in design [13-36]. Human factors engineers at AT&T Manufacturing have recently redesigned a pair of pliers used in a crimping operation so that less stress is placed on the operator's forearm muscles. Component parts that may need to be serviced or replaced must be placed in positions where service and repair personnel can reach them with minimum difficulty. Ease of use is further facilitated with operational instructions being written in simple terms and easy-to-understand language, numerous diagrams, and "fool-proofed." Instructions are fool-proofed through an iterative process where versions of the instuctions are presented to naive users (those not familiar with the product of design) who are asked to assemble or use the device. The users are asked to comment about difficulties they encounter. These comments are then used to rewrite new versions of the instructions. The process continues until no errors are present and ease of use has been demonstrated.

The physical and biological influences and limitations of the human–machine system are the direct result of the design constraints of the physical environment. Human size, strength, and biological tolerances are particularly important considerations (see Table 13-2). However, these influences are measured in different types of units. Size is measured in units of length, strength in terms

> The right formulation of the relationship between man and machines is not one of competition, but of cooperation.
>
> *Ward Edwards*
> Innovation, *1969*

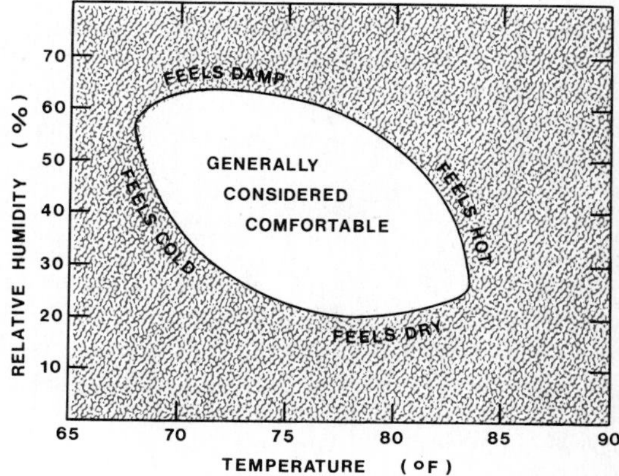

[13-36] Comfort zone.[4]

[4]W. E. Woodson and D. W. Conover, *Human Engineering Guide for Equipment Designers* (Berkeley: University of California Press, 1964), pp. 2–227.

The most general survey shows us that the two foes of human happiness are pain and boredom.
Arthur Schopenhauer
Essays. Personality; or, What a Man Is, *1853*

Time, with all its celerity, moves slowly to him whose whole employment is to watch its flight.
Samuel Johnson
The Idler, *No. 21*

Isolation is the sum total of wretchedness to man.
Thomas Carlyle
Past and Present, *1795–188l*

TABLE 13-1 TECHNIQUES AVAILABLE FOR HAZARD REDUCTION

Rule of Thumb	*Example*
1. Prevent the creation of the hazard in the first place	Prevent production of dangerous materials such as nuclear waste
2. Reduce the amount of hazard created	Reduce the lead content of paint
3. Prevent the release of of hazard that already exists	Pasteurizing milk
4. Modify the rate of spatial distribution of the hazard released at its soucre	Quick-acting shutoff valves
5. Separate in time or space the hazard and that which is to be protected	Storage of flammable materials in an isolated location
6. Separate the hazard and that which is to be protected by imposing a material barrier	Containment structures for nuclear reactors
7. Modify certain relevant basic qualities of the hazard	Using breakaway roadside poles
8. Make what is to be protected more resistant to damage from the hazard	Making structures more fire- and earthquake-resistant
9. Begin to counter the damage already done by the environmental hazard	Rescuing the shipwrecked
10. Stabilize, repair, and rehabilitate the object of the damage	Rebuilding after fires and earthquakes

Source: W. Haddon, "The Basic Strategies for Reducing Damage from Hazards of All Kinds," *Hazard Prevention,* September–October 1980, pp. 8–12.

[13-37] All automobiles are not comfortable to drive.

of the force that the body can apply, and so on. Much of our surrounding environment can be measured in units of length—for example, desk and counter heights, doorway and passageway widths, and arm and hand working spaces. It is apparent that one should not design small, closely spaced footpedals for use by populations with wide feet, nor small drawer pulls to be used by people wearing thick gloves (such as bakers), nor require women to perform a task that is difficult and dangerous to them (such as handling a 10-lb casting at arm's length). Many design problems are dependent on the dimensions of the human body, such as the reach and strength of its limbs, its endurance, and the range of these factors relative to sex and age. Therefore, designers must have at their disposal a complete "data bank" when a design is to be used by human beings. The study of the shapes, weights, and measurements of people is called

TABLE 13-2 MAN VERSUS MACHINE

Functions that men perform better than machines	*Functions that machines perform better than men*
Sensory functions: Human capacities often surpass those of instruments, especially where the minimum absolute energy for sensory detection is concerned, within the visual and auditory range.	*Speed and power:* Machines can be devised to make movements smoother, faster, and with greater power than men.
Perceptual ability: A man is very good at sizing up complex situations quickly, especially if data are presented with adequate pictorial or familiarly patterned displays.	*Computation:* When the rules of operation—the postulates—are built in, machines are more efficient computers than men.
Alertness: With adequate provisions for activity and interest the human is alert for changes, impending changes, and can often prevent impending undesirable consequences.	*Short-term storage:* Machines can be built which can store quantities of information for short periods of time and erase these memories to make place for a new operation.
Flexibility: Where flexibility is required to provide insurance against complete breakdowns in emergencies, human beings can play an important role.	*Complexity:* A complex machine is capable of carrying on more different activities simultaneously than man.
Judgment and reasoning: Where it is impossible to reduce all operations to logical preset procedures, men are needed to make judgments. This is particularly true when it is required to assemble a set of facts from different sources in arriving at problem solutions	*Long-term memory:* Machines (computers) are more efficient than men in tasks that require long-term memory needed in handling unique problems.

Source: Human Factors Bulletin 55-4H (New York: Flight Safety Foundation)

anthropometry. One of the most complete and accurate anthropometric studies done in this country is that by Hertzberger, et al.[5] Selected parts of this work are included in Appendix IV.

Everyone complains of his memory, but no one complains of his judgment.

Duc de la Rochefoucauld
Reflections, *1678*

CONTROL REQUIREMENTS

Brain. An apparatus with which we think that we think.

Ambrose Bierce
The Devil's Dictionary

The next set of design requirements are those that impose degrees of control on the design. These requirements place boundaries or limits on what the designer can accomplish compared to what could be done with limitless time and resources and no binding legal restrictions. The first of the control requirements to be considered are the legal aspects.

Legal Aspects

Foremost among the legal requirements of design is the problem of dealing with patents. Patents are granted by the government for a limited period of time (usually 17 years). All but the patent holder are enjoined from making, selling, or using a patented device without permission. Such permission can be granted in the form of a license or permit. The license or permit is valid for a specified period of time and may or may not be renewable. Patents impact the de-

[5]H. T. E. Hertzberg, G. S. Daniels, and E. Churchill, *Anthropometry of Flying Personnel—1950,* WADC Technical Report 52-321, USAF, Wright Air Development Center, Wright-Patterson AFB, Ohio, September 1954.

signer in two ways: the designer must make sure that the design or invention is not in violation of existing patents, and steps should be taken to obtain a patent for the device. A patent search will precede the filing of an application for a new patent and the process can be time consuming and costly. It can best be accomplished with the aid of a reputable patent attorney.

Many of today's designs involve both computer hardware and software. Computer software can be protected by copyright. A copyright is another form of agreement with the government whereby others are not permitted to use in any form the written (or in the case of software, coded) material of the copyright holder. Recent legislation extends some copyrights for as many as 50 years beyond the life of the original copyright holder.

There are other legal considerations that have an impact on design. When structures are involved, local building codes also require compliance. Designs that use hazardous materials must comply with state and federal regulations and laws. For example, asbestos materials recently have been more carefully regulated. There are many places where they may not be used at all. Whenever the use of asbestos is permitted, the design must consider safeguards required by law. Designers of children's toys must also pay particular attention to laws regarding shapes and sizes.

Materials Selection

Materials cannot be selected without knowledge of how the product is to be manufactured. Other considerations in determining which materials might be selected for a design include:

1. Physical properties such as crystal structure, density, melting point, and viscosity.

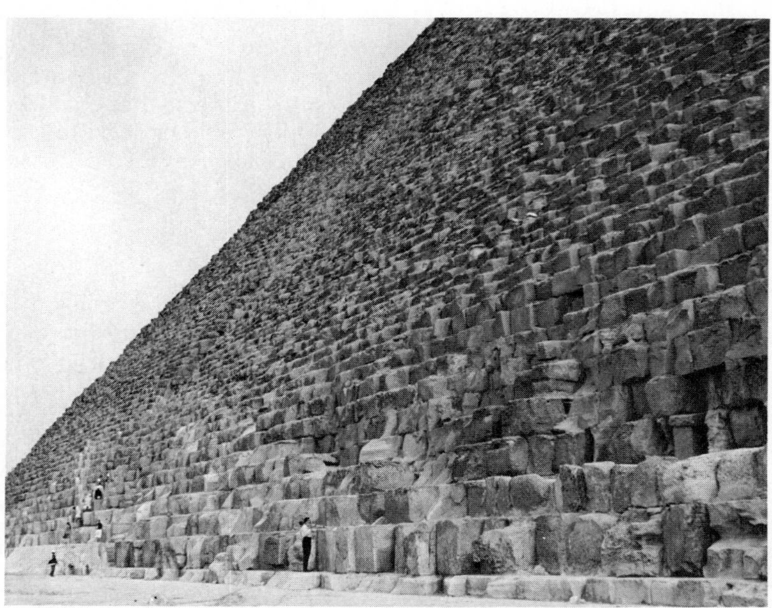

[13-38] "Whatever an engineer builds . . . he uses materials of construction." The Great Pyramid of Egypt is an example of the durability of properly chosen building materials.

[13-39]

2. Electrical properties such as conductivity and resistance.
3. Mechanical properties such as hardness, yield strength, and modulus of elasticity.
4. Thermal properties, including conductivity and specific heat.
5. Chemical properties such as oxidation, corrosion, and stability.
6. Wear properties such as abrasion and friction coefficient.
7. Fatigue properties, including buckling, creep, and thermal fatigue.
8. Fabrication properties such as hardenability, heat treatability, machinability, and weldability.

Sources of information about material properties include handbooks, technical journals, trade literature, and company reports. With knowledge of the manufacturing processes, the decision to select one material over another is usually made on the basis of performance, cost, failure analysis, and value analysis. The performance specifications indicate the functional requirements of the product and are based on the need the product must satisfy as well as the consequences of failure. The cost of a material (dollars per pound) is sometimes a valid criterion for comparison. However, some materials (like plastics) are usually compared on a cost per unit volume basis. Laboratory tests and published reports from suppliers may be used to select materials on the basis of their abilities to withstand various modes of failure. Value analysis or value engineering is usually a team activity intended to identify and eliminate unnecessary costs without sacrificing the performance, quality, or reliability of the design. A value analysis team is typically composed of representatives from areas such as manufacturing, design, sales, quality engi-

[13-40] Good engineering design can do much to overcome inherent limitations of engineering materials. Who would have believed, for example, that a 12,000-lb truck could be supported by a 32-ft.-long paper bridge? The bridge above was built to span a gorge in Nevada to demonstrate the strength properties of paper. It was designed so that it could support a 35-ton load, or that its length could be expanded to 80 ft.

neering, and research and development. The term *value* is intended to include not only the monetary cost of a component but also refers to its ability to function satisfactorily. The value analysis team, when evaluating materials and component parts, asks such questions as:

- What does it do?
- Can we do without it?
- Is there another part or material that could do what this part does, and do it better?
- If this part is a special part, can we modify the design and replace it with a standard part?
- Is there a less expensive way to make this part?
- Can we buy this part less expensively than we can make it?
- How long will this part last?
- Is the part worth its cost?

Suppose that a design is expected to be in use for, at most, five years. One possible outcome of a value analysis of the component parts of a design might be to substitute shorter-lived parts of similar performance, quality, and reliability characteristics in lieu of other parts that could be used in the design which are expected to last much longer than five years.

The selection of materials is also related to the requirement of manufacturing considerations, which is discussed in the next section.

Manufacturing Considerations

The manner in which the product or design will be produced is crucial to the design engineer. As mentioned earlier in this chapter, design specifications are established not only with regard to obtaining optimal performance, but also considering the extent to which the manufacturing process is capable of meeting these specifications. The selection of materials for use in the design is related to the nature of the production process. There is a two-way relationship between material selection and production processes. The choice of material will dictate the type of machines, equipment, and processes used in manufacturing, and conversely, manufacturing facilities already in place will guide the selection of materials.

The design engineer works closely with production engineers to determine design specifications and material capabilities relating to manufacturing. For example, the designer must understand how material properties of metals and metal alloys are affected by casting, forming, cutting, welding, joining, and finishing. When plastics are involved, injection molding, thermoforming, and blow-molding techniques and their effects on the choice of materials must be studied. Similarly, special processes in glass manufacturing and woodworking must be known to the designer who uses these materials. Packaging is a part of the manufacturing process, and sometimes the package becomes a part of the design. Packaging of the final product must be designed to protect the product from damage during shipping. The materials and techniques of packaging (as well as a con-

[13-41] The designer must account for increased automation in manufacturing.

sciousness of the cost of packaging) must also be familiar to the design engineer.

As a designer you must also understand the capabilities of the manufacturing process. You must realize that there is variability in every manufacturing process and this variability must be considered when design specifications are established. Variability implies that even the most sophisticated computer-controlled manufacturing processes cannot make every part to exact dimensions time after time. Tools and equipment exhibit gradual wear which will change the dimensions of some of the design variables. The material itself may change from piece to piece, and this variability must also be considered. When human operators are involved, differences in skill levels will result in measured values of a design dimension which will also exhibit variability. When the designer sets the upper and lower specification limits for a variable based on performance (see Figure [13-28], the difference between these limits is called the *tolerance* and is a measure of how much the dimension of the variable may change and still provide adequate performance. The tolerance should be compared with the process variability (due to tool and equipment wear, change in materials, operator skills, etc.) to see if the specifications are realistic. If the tolerance is greater than the process variability, there should be few problems in manufacturing the part in question to specifications. However, if the tolerance is less than or equal to the process variability, parts not meeting specifications are likely to be made unless either the tolerance is increased or the process variability is reduced. Usually, attempts are made to reduce the process variability by better control of materials, frequent checks for tool and equipment wear, better selection and training of operators, and so on, rather than to change the design specifications.

Life-Cycle Cost

Before the advent of life-cycle cost programs, control of the cost of a design was directed toward the cost of materials used in the design and the labor, energy, and overhead costs required to produce the design. That is, cost containment was strictly a "front-end" process which considered only the costs incurred in the development and manufacture of a design. In the late 1960s, the U.S. government began to require that its contractors (suppliers) consider the "downstream" costs as well. The downstream costs refer to costs incurred while the design is in the hands of the user. These costs include operations costs, maintenance costs, training costs, tools and test equipment costs, energy costs during use, costs of spare parts, inventory, and material support, and retirement and disposal costs. In short, these costs encompass everything that may be spent on the design from its inception until it is not longer in use. Obviously, these costs are incurred by the user and are not a part of the design and manufacturing costs which are incurred by the seller. One might ask, "Why should the seller be concerned with money spent on a design after it is in the hands of the user?" There are many valid answers to this question other than "because a government directive requires such considerations in design." Designing to life-cycle cost

[13-42] Life-cycle cost analysis is an important aspect of every aircraft design.

includes modularization of design components so that when failures occur in use, a replacement module may be quickly exchanged with the module containing the failed part or device so that the design is once again operational and the module with a failure may be repaired without affecting the "up time" or availability of the system. The design to life-cycle cost philosophy also requires a design which can easily be accessed for use with test and diagnostic equipment for easy repair. It is clear that a design with features like those just mentioned will have great appeal to the potential buyers of the design. For this reason designs considering life-cycle costs tend to have a competitive edge over those not providing features that reduce the ownership costs.

During the Vietnam war, helicopter rotor blades had to be replaced at frequent intervals due to damage in combat. At the time, each helicopter had a left rotor and a right rotor. There were many instances when a right rotor was damaged and when a search for a spare was made only left rotors were available in the spares inventory—and vice versa. Engines were difficult to remove, special tools were required for field servicing, skilled personnel were required for maintenance and repair, and maintenance and repair instructions were very difficult to comprehend. These problems were eliminated in the design of the Army's Blackhawk helicopter, which featured identical and thus interchangeable rotor blades, an engine that could be removed by two people using a simple hoist and serviced and repaired using common tools and equipment, and detailed, easy-to-read instructions which featured numerous informative diagrams. The design to life-cycle cost program has worked well in government use, and it is now a part of the design philosophy of many companies in the private sector as well. This is particularly true of products that use solid-state electronics components.

The following life-cycle costs must be considered in the design phase:

- Acquisition Costs, including research, development, and design costs.
- Manufacturing and construction costs.
- Distribution costs, including packaging, transportation, and storage costs.
- Operations costs.
- Maintenance costs.
- Tools and test equipment costs.
- Logistics support costs.
- Training costs.
- Inventory costs of spares and other materials.
- Retirement and disposal costs.

For each alternative design being considered, the design engineer will estimate these costs on a year-by-year basis from the time that design planning begins until the estimated time that the design will no longer be in use. Then alternative designs may be compared on the basis of net present worth (see Chapter 8) to determine the design with the smallest life-cycle cost. When the present worth method of engineering economy is applied to year-to-year costs during the life cycle, the process is often refereed to as "discounted cash flow."

Schedule

Nearly every design has a due date. Often, the time allowed for completion of the design is obtained by "working backward" from the scheduled completion date of the entire project, which culminates with delivery of the finished product to the end user. The final completion date may be documented by contract with the user, or it may be a function of promises made by salespeople. Too often, delivery dates are determined by what the competition promises or is expected to do. In any event, the designer must schedule activities associated with all phases of the design in such a way as to ensure timely completion. The phases of engineering design are detailed in Chapter 14, and at that time design planning techniques will be discussed with reference to establishing and meeting schedules.

SYSTEM REQUIREMENTS

The term "system" is used here to include the complete set of conditions having an influence on the design and influenced by the design after it is installed and made operational. Outcomes concerning the requirements described below must be anticipated and accounted for by the design engineer.

Operating Environment

The operating environment is the physical environment within which the design will operate. Factors such as temperature, pressure, humidity, vibration, shock, wind, salt spray, dust, dirt, rain, mildew, and radiation will often have a negative effect on both performance and reliability unless the design has been developed with a thorough understanding of these conditions. The designer must first know what levels or ranges of each of the environmental factors may be expected. These levels are often dependent on where the design will be installed—arctic, tropics, deserts, rain forests, or temperate climate; mountains or level terrain; indoor or outdoor; above ground or underground; in aircraft or in space; and so on. Once the locations are known, the environment at the locations must be determined. The likelihood of shocks or disturbances to the system as a result of other systems in the immediate locale should be estimated. If the installation locations and operating environments are many and varied, the designer must decide if it is feasible to develop a single design to accommodate conditions at any of the locations and their corresponding environment or if a different design for each location–environment combination (or for certain groups of location–environments) would be preferred.

The design engineer usually has access to a large body of information about environmental factors and their effect on performance

[13-43] Climate test chambers are used to stimulate extreme operating environments.

and reliability. For example, paints and other surface finishes have been exposed to sun, wind, and rain for long periods of time, thereby furnishing knowledge concerning how such finishes behave under both normal and unusual weather conditions. Similarly, an extensive body of knowledge exists on how chemicals deteriorate or "corrode" construction materials. These and other environmental factors are under continuing study. Therefore, unless the engineer is faced with an unusual environment, he or she can often (but not always) find pertinent information in the literature to predict how a particular environment is going to affect the design.

However, there are many instances where additional tests are needed. Tests are particularly important where two or more environmental effects work together, such as moisture and heat, or chemicals and vibrations. The result of such effects may not be predictable from either of the individual effects acting by themselves. Thus we know that a vibrating environment in a saltspray atmosphere can cause corrosion fatigue at a rate far higher than that which might have been predicted from either the vibration or the saltwater corrosion taken independently.

With the advent of space travel, one of the most intensively studied types of environment is *space*. When far from the earth's atmosphere, a body in space will be in a nearly complete vacuum. However, it will be exposed to a variety of types of radiation and to meteoric dust from which the earth's atmosphere normally provides protection. The radiation effects may be severe enough to attack electronic circuits seriously and cause deterioration of transistors and other electronic devices. The meteoric dust, although generally quite fine, travels with speeds of 10,000 to 70,000 mi/hr and has sufficient energy to penetrate some of the strongest materials. Within the last few years some ways have been found to simulate in the laboratory both the high radiation and the presence of meteoric dust, and (on earth) to subject space equipment to these kinds of attacks.

Interface with Other Systems

We have mentioned above that the designer must know about other systems that are to be located in the same locale as the system being designed so as to determine environmental effects caused by these systems. The other systems may vibrate extensively, generate heat, smoke or vapors, discharge water or chemicals, and so on. Beyond the question of harmful effects of other systems, the designer is often faced with the task of designing a system which is intended to interface or work with other systems to achieve project goals. It is this specific problem that is addressed in this section.

Let's say that a hypothetical design engineer is developing system *A*. *A* must interface with systems *B* and *C*. Perhaps the output from *B* will be an input to *A*, and the output from *A* will be an input to *C* or *B*, and the output of *C* is an input to *A* [13-44]. There are obviously many other combinations of input–output relationships among the three systems, and there are possible additional cases, such as where all three systems must furnish inputs to a fourth system, or when the three systems receive input from other systems and must deliver a specified output, and so on. However, let us consider

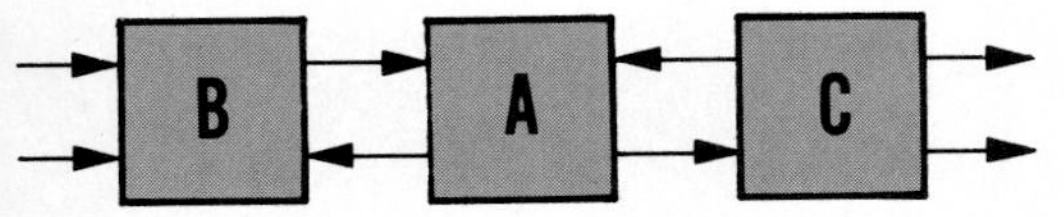

[13-44] Interacting systems *A*, *B*, and *C*.

this simple situation for illustrative purposes. The designer must know the number and form of the outputs from *B*. It is possible that a transformation of energy form must be made to receive the outputs from *B* in order to put them in a form that can be used by system *A*. For example, a transducer may be needed to convert light energy from *B* to electrical signals to be used by *A*. Similarly, the designer must know the number and form of the inputs of system *C* so that the design of *A* will generate outputs that can be received by *C*. In addition to energy, there are many other input–output relationships that must be considered in the design of interacting systems. Among these are designing to accommodate failure of one or more of the interacting systems, determinating rates of material flow, and if computers are used with the system, the rate of information flow will be an important design parameter. Systems that are interfaced by computer control are becoming more commonplace in today's technology. When the systems are used for manufacturing, the term *computer-integrated manufacturing* (CIM) is used. It is not always the case where each design that is to be provided as part of an integrated system originates within the same organization. Often, several companies must work together on integrated system designs although they may be in separate locations, including the case where organizations contributing designs are in different parts of the world.

Companies may also compete with each other to supply designs which are part of integrated systems. The term *compatability* is a prominent part of current advertising literature. We read announcements such as: "This software is IBM-compatible," meaning that the software package will run on IBM computers as well as other computers which themselves are also IBM-compatible. Telephone communications systems are designed for compatibility. Designing for compatibility means specifying commonly used input and output devices and, whenever computers are involved, choosing popular operating systems and language versions.

Effect on Surroundings

Not all of the outputs of an engineering design study are desirable. Those outputs that constitute a disruption to ecological systems must be controlled. An ecological system is one in which biological entities such as human beings, animals, insects, plants, and other living things interact with their surroundings. In nonhuman ecological systems, a balance has always existed—that which is produced is consumed without significant changes to the system. For centuries, human ecological systems functioned in much the same way, with only minor changes to the surroundings being a result of human activities. The industrial revolution of the nineteenth century and subsequent technological developments has significantly changed both human and nonhuman ecological systems. Air and water, once pure and clean, are now often contaminated with by-products of human progress in technology. In some countries, solid waste is left in huge piles throughout the countryside. Strange noises fill the air and interfere not only with conversations, but with people's emotional stability as well. More recently, human-made radiation has been found to damaging cell tissues of unsuspecting workers. For decades, the ecological disruptions remained unchecked. It was not until the 1960s and 1970s that federal legislation demanded clean air, clean water, and noise control. As a result of the Clean Air Act (1963), Air Quality Act (1967), Water Quality Act (1965), Clean Water Act (1972), Solid Waste Disposal Act (1965), Noise Control Act (1972), and subsequent amendments, attempts are being made to protect ecological systems. Agencies such as the Environmental Protection Agency (EPA) and Occupational Safety and Health Administration (OSHA) have been created to regulate and monitor activities of those who impose a threat to the natural environment. The Atomic Energy Commission (AEC) and the U.S. Public Health Service have established standards for radiation protection.

What can the design engineer do to help? The designer must not only create innovative and sensible technology but also learn to recognize the most frequent causes of pollution and understand how to remove them or make them harmless.

The best way to remove fine particles or droplets from the air is to avoid putting them in the air in the first place. At one time, lead particles were present in nearly all automobile exhaust. Lead-free gasoline is now being used in most automobiles. Some electric power companies have found that it was more economical to change from coal to oil or natural gas than to remove fly ash from their stacks. Now, if the combustion process cannot be altered to make it smoke- or dust-free, most of the particles must be trapped before they enter the atmosphere. Devices for trapping and removing air particulates range from simple bag filters (similar to the air bag used on most vacuum cleaners) to electrostatic precipitators in which particles are charged electrostatically and then attracted to oppositely charged plates, where they settle out.

Solutions to water quality problems follow the same guidelines as those for air quality. Avoid discharging chemicals into water systems [13-45]. Some chemical by-products of industrial processes may be neutralized by combining them with other chemicals. At one time, chemical and radioactive waste was dumped and buried in the

[13-45] Avoid discharging chemicals into water systems.

ground without thought of seepage into underground water systems. Such disposal methods are now prohibited.

Today, solid waste may be burned under controlled conditions or buried in a sanitary landfill, where alternate layers of waste and dirt are placed on top of one another. Landfill sites are becoming scarce and thus the choice is narrowing to incineration. Some wastes, such as aluminum and certain other metals, glass, and paper, may be recycled if the cost is not prohibitive. Nuclear waste continues to create a most serious problem. It has been stored underground in lead-shielded containers. The nuclear waste dumps in use today will remain toxic at least until A.D. 12,000.

Noise is best cured at its source. Examples are the design of better mufflers for cars and motorcycles, the use of rubber-tired wheels for subways (underground railways) in place of steel wheels, and the aerodynamic design of air ducts to minimize sharp corners around which the air can whistle. If the source itself cannot be removed, the next step is to enclose it in a sound-absorbing box or enclosure, so that little of the noise can escape [13-46], or to deflect the noise in a direction (such as up) where it will cause no harm. This technique is used in turbine and compressor installations, in test cells for aircraft engines, and along the road through towns with new rapid transit systems.[6] Least effective, although sometimes the only alternative, is to surround the noise recipients (the public) with sound-absorbing walls or other sound-protection devices. This is done with the control rooms in noisy machine shops, with earmuffs for airplane mechanics, and with soundproofing in houses, apartments, and office building in our noisy cities. Acoustically satisfactory building construction is now possible, although it is practiced all too rarely. Occasionally, a harmless but objectionable noise that cannot be silenced can be "hidden" by another, pleasing sound, like that of a splashing water fountain. However, this practice is not recommended unless all else fails.

[13-46] Anechoic (antiecho) chambers are used to test the sound-reflective characteristics of objects and materials.

Logistics

Logistics are activities which are used to support the design after it is operational. The process of logistics is concerned with managing the resources associated with the design. A simplified definition of the purpose of logistics is "Logistics should include everything that is necessary to get the right thing to the right person at the right place at the right time in the most cost-effective manner possible."

The "right thing" requires spares provisioning for the design and includes cataloging of spares and all documentation of parts. Other things required for an operational design are: test equipment, maintenance equipment, and tools that are used to monitor operations; testing and diagnosing system abnormalities; and performing routine maintenance and repair of the system when necessary. Additionally, logistics must include the supply of all consumable materials used with the design.

The "right person" is the maintenance engineer, serviceperson, or inventory manager. The necessary skill requirements of these indi-

[6]V. Salmon, "Noise in Mass Transit Systems," *Stanford Research Institute Journal,* No. 16 (September 1967), pp. 2–7.

viduals must be identified and special training in working with the design must be provided. Manuals must be written to identify and document all maintenance requirements.

Dependable transportation is a key to delivery at the "right place." The logistics engineer must determine which methods are most effective and result in minimal costs. The location of warehouses and service centers is also an important consideration. Once the locations where the design is to be installed are known, the locations of supply warehouses and repair centers are established. The task of determining which service centers and warehouses will service or supply which installations is not an easy one. The cost of moving parts and materials must be balanced with the need for prompt delivery of consumable materials and replacement parts and rapid delivery of repaired parts.

Delivery at the "right time" means getting the material, equipment, and parts to the terminal point when they are required. Early deliveries are not always desirable. Materials and parts that arrive before they are needed must be stored so as to be protected from the elements, guarded against theft, and so on. Storage can be very expensive. Many manufacturers are employing the JIT, or "just in time," principle, which involves coordination among suppliers and ensures that parts and materials arrive at the address where they are needed and at the *exact* time that they are needed—no sooner, no later. When JIT systems are working properly, there is no cataloging of inventories and hence fewer problems in retrieving materials from storage.

The design engineer is obviously not responsible for all of the logistics activities. However, it is critical that logistics be planned during the design phase. The prime equipment required for the design must be integrated with the test and support equipment. It must be designed for testability, maintainability, and repairability. Often, training and service manuals are written under the designer's supervision. It is impossible to separate logistics and its implications for a successful design from the next set of design requirements, which relate to effectiveness.

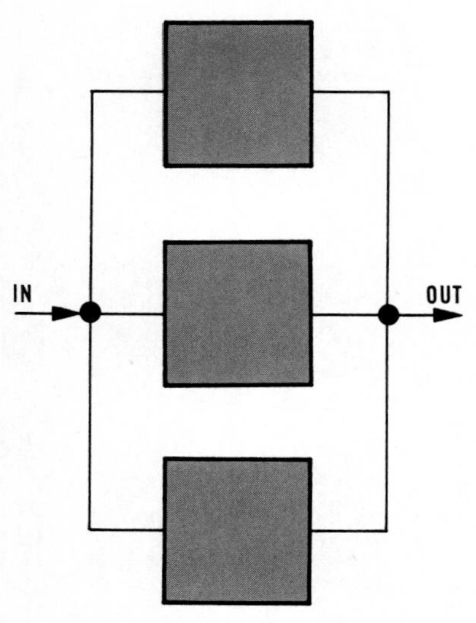

[13-47] Three identical components operating in parallel.

EFFECTIVENESS REQUIREMENTS

A system that fails infrequently can be tested and repaired when it does fail. It can be maintained in an efficient manner. This can best be achieved as a result of the design process and not after the design has been completed. The effectiveness parameters of reliability, maintainability, serviceability and availability must be an integral part of the design process.

Reliability

Reliability is the probability that a device or system will perform its function adequately for a specified period of time under the operating conditions encountered. Thus, reliability is a number between zero and 1 which reflects the likelihood that the design will survive for a given operating period. A reliability of 1.00 is essentially a theoretical goal since there are many factors in any operating situation which limit a design's performance. Nevertheless, with proper attention and planning, many designs have achieved reliabilities very close to 1.00.

One method of achieving high reliabilities with design efforts is through the use of redundancy. Redundancy refers to the employment of duplicate components or of additional components that perform the same function as the original component in either an active or stand-by mode. Active or parallel redundancy means that the redundant components are energized or are placed in an operating mode at the same time as the original components. Figure [13-47] shows three components in parallel. If only one part is needed for the system to function, two failures may occur and the system will still be operational. In this arrangement, all the components are operating whenever the system operates. Redundant computers have been used in many of the NASA space missions. In a standby redundant system [13-48], only one unit is operating at any given time. When the operating component fails, a switch (*S*) is activated which puts the next standby component in an operational mode. Some aircraft are equipped with hydraulic brakes as the principal system, with electric brakes in standby or backup position in case the hydraulic system fails, and with mechanical brakes as still another backup for the electrical system. The emergency brake on your automobile may be thought of as a standby device for the operating system.

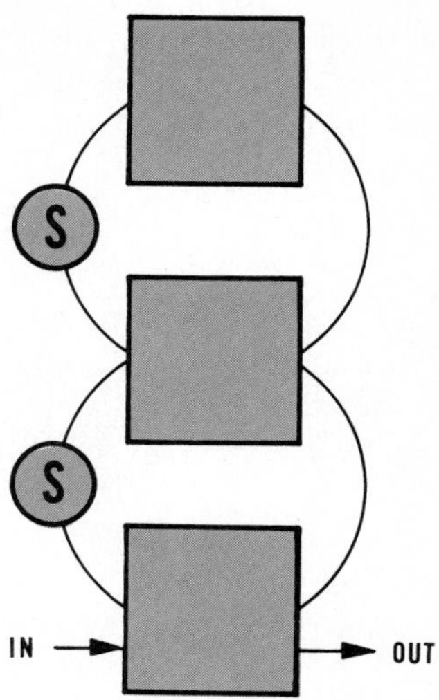

[13-48] Single component operating with two identical components in standby mode.

Other ways to reach high reliability levels in design are through the use of simplicity and standardization. Simplicity is achieved by using the fewest possible number of parts in the design. In general, the simpler the design, the higher the reliability. Standardization involves the use of well-known components which have been debugged and have proven their use in other applications. As mentioned earlier, the human element must also be considered. The designer should take precautions to ensure that it is virtually impossible to assemble or use the design in a manner that may cause subsequent failure.

Another step that can be taken by the designer to assure high reliability is the practice of derating. In derating, the designer selects components that have been rated to operate at much higher levels of temperature, voltage, or other operating or environmental stresses than the stresses which are expected to be encountered in actual use. The practice of derating is very common with electronic components. These components may be studied to determine their failure rate (failures per unit time) as a function of stress level. The results of one such study might be a series of failure-rate curves, as shown in Figure [13-49]. The failure rate per hour is plotted on a logarithmic scale against operating temperature in °C for a range of rated voltages. The graph indicates that much lower failure rates can be obtained by selecting devices that will operate at lower voltages than their rated values.

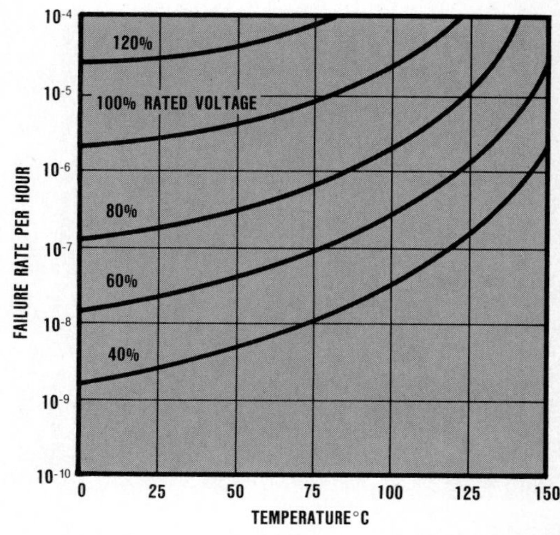

[13-49] Failure-rate curves for derating analysis.

Many electronic components, and some nonelectric components as well, exhibit a failure rate over time much like the curve of Figure [13-50]. The shape of this curve gives it the popular name, the *bathtub curve of reliability*. For the first portion of operating life, the failure rate starts at a high value and then decreases steadily until a

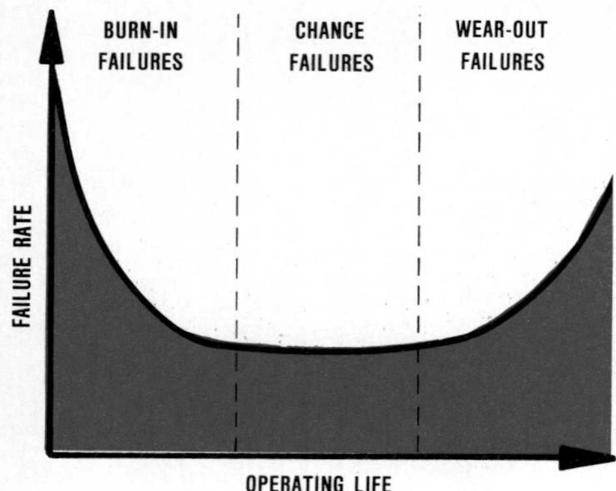

[13-50] Bathtub curve of reliability.

"leveling off" occurs. This period is called the *burn-in* period, for reasons soon to be discussed. The reason for the decreasing failure rate in this period is that substandard components due to defects in materials, poor workmanship or bad assemblies, and so on, will fail in this interval until only reliable components remain. Then follows a period where the failure rate is relatively constant. This period is called the period of *useful life* or the period of *chance failures*. Components fail in this interval due to unforeseen causes, such as sudden stresses which may occur randomly. At the end of this second period, components begin to fail due to physical deterioration because of their prolonged use. This period is called the period of *wear-out failures*. Reliability engineers use their knowledge of the bathtub curve to keep components in operation only during the period of useful life (chance failures). Components are "burned in" or energized at high stress levels (usually, a range of temperature is used in the case of electronic components, so that the substandard components will be identified through failures). Components surviving the burn-in period are those which are installed in the design, so that all the components being used are known to be capable of surviving high stress levels. These components will continue to be used in the design until they reach the start of the wear-out period, at which time they will be replaced with new components—again only those which have already been burned in. Thus reliability engineers on the design team may improve reliability of a device by requiring that all components go through a period of burn-in before they are installed in the device.

Material selection, mentioned earlier as a design control requirement, is also crucial to attaining reliability goals. Materials must be selected for their ability to withstand the mechanisms of failure as well as for performance reasons.

Components or devices are often subjected to one or more reliability tests before they can be approved for use in the final design. These tests are necessary to ensure that reliability goals can be met. One such test requires that components demonstrate that they can exceed a minimum mean time to failure (the reliability goal). Components are placed on test stands and activated or energized until a predetermined number fail or until a prescribed time is reached. With knowledge of the statistical time-to-failure distribution, reliability engineers can predict, based on either the elapsed time (if a fixed number of failures are to be observed) or a count of the number of failures (if a fixed time limit is used), whether or not the devices attained the minimum acceptable mean time to failure. Another type of test is the life test, in which a sample of identical devices is tested until all fail. The time to failure is recorded for each device and is used with information about the statistical distribution of failures to make inferences about the operating life of all the devices of that kind. In some cases where insufficient time is available for a life test, an accelerated test may be used. In an accelerated test, the devices are subjected to abnormally severe environmental conditions or to operating conditions above normal levels, or to both. With knowledge of the effects of such accelerated conditions, inferences can be made about when the device would have failed under normal environmental and/or operating conditions.

[13-51] Reliability test on the floor section of a C-17 transport plane.

High values of reliability can sometimes be achieved only by increasing life-cycle cost or decreasing performance. Sacrifices in reliability, cost, and/or performance may have to be made to achieve overall design goals. These sacrifices are called "trade-offs" and are discussed in Chapter 14.

Maintainability

Maintainability is the probability that a maintenance task on a device or system can be completed in a specified interval of time. Maintenance tasks are either preventive or corrective. Preventive maintenance is performed to keep a system or equipment in satisfactory operating condition through systematic maintenance actions such as checkouts, cleaning, lubrication, or removal and replacement (even though the device has not failed). Some preventive maintenance is done at fixed intervals, such as every three months, while other preventive maintenance may be done when certain milestones are reached, such as when the machine has logged 1000 hours or when an automobile has been driven 10,000 miles since the last maintenance action.

Corrective maintenance is performed on a nonscheduled basis to restore a failed equipment or system to operating status. A good preventive maintenance program will avoid the need for corrective maintenance for most systems.

Like reliability, maintainability must be planned for during design. Maintainability design is the collection of inherent features in the design which support the maintenance function. Maintainability design provides functions that ensure diagnosis of a failure, replacement of a failed part, and verification that a correct repair has been made. Replaceability of a failed part is an important part of maintainability design [13-52]. The part must be accessible, and tools required for replacement must be carefully considered. The design must also include procedures to verify that a correct repair or fix has been made, including checks on surrounding or dependent equipment. For example, it is possible that the repair of one part may induce a failure on another part. Proper design procedures can minimize the likelihood of this happening. Design decisions also include such considerations as whether to use modular or nonmodular construction. Modularization requires more design effort, but it reduces the time required for diagnosis and replacement when a fault is detected in the field. The exact location of the fault need not be known; the only requirement will be to find the module in which the fault occurs. This module may then be exchanged with another module of the same kind to restore the system to operational status. The module containing the fault is then repaired at the site, returned to the factory, or sent to a service center for correction. Another decision that must be made during design is to determine whether to design parts or modules to be repaired or to design them to be discarded and replaced with new parts or modules. Also, test equipment should be selected during the design phase. The designer must choose whether to build in test components with the design or to provide fittings on the design so that it can easily be tested using external test equipment. Built-in testing equipment usually reduces the time required for diagnosis, but most often this is more costly

[13-52] Replaceability of a failed part, such as was the case with the Westar VI communications satellite in 1984, is not always an easy task.

than designing for the use of external test equipment. Modular design and the use of proper tools for repair were mentioned earlier, and hence the decisions to be made regarding maintainability in design are strongly influenced by the "downstream" costs associated with maintenance.

Maintenance strategies should be developed during design. Maintenance strategies are closely related to logistics. Maintenance strategy considerations include labor skill-level determinations, training of repair personnel, location of repair centers, travel itineraries, and the determination of the number and the location of spares. The maintenance strategy established must be based on a thorough product life-cycle cost analysis and must conform to product reliability and availability requirements.

Serviceability

Serviceability is defined as the ease with which a device or system can be repaired. Unlike reliability, maintainability, and availability, which are probabilities, serviceability has not been quantified. However, it does have an influence on repair time—the better the serviceability, the less repair time needed. Serviceability can be thought of as a combination of maintainability design features and maintenance strategies. Serviceability is described in terms of modular design, built-in test equipment or connectors, accessibility, response to requests for spares, training of repairpersons, and so on. It is a general term denoting how well the design team has provided for ease of repair. Those who service their own automobile are aware that some manufacturers have not given enough attention to the serviceability aspect.

Availability

Availability is the steady-state or long-term probability that a device or system is operating. It is usually thought of as the percent of "up time." Availability may be measured by tracking the operation of a device or system over a long period. For example, consider a copying machine used in an office. Suppose that over 125 eight-hour working days (1000 hours), the machine was operating or awaiting operation for 831 hours, was serviced by repairpeople for 35 hours, was cleaned and refilled with fluid or powder for 4 hours, and was in a failed condition awaiting the arrival of a repairperson for 130 hours. Based on this study, the availability of the copying machine would be estimated at 83.1 percent. Since this was the percentage of the time that the machine was operational or available for use, whether it was actually used or not.

[13-53] Some designs, such as this communications satellite shown being launched, are expected to operate for many years without failure.

It appears that availability is strongly related to reliability and maintainability, and this is certainly the case. The more reliable the design, the less it breaks down, and hence the more "up time" that is accumulated. If the maintainability design and maintenance strategy are efficient, the device or system will be restored to operating status in a short period of time and availability will be increased.

Availability goals are set during the design phase and must be predicted on the basis of reliability requirements, maintainability goals, and maintenance strategies.

PROBLEMS

13-1. Name five products or engineering designs that are aesthetically appealing to you on an intellectual basis; name five that appeal to you on an emotional basis.

13-2. From a recent engineering design magazine or trade journal, select three examples of what you consider as good functional design. Be prepared to discuss the reasons for your choice. Do the three examples satisfy all the requirements of aesthetics?

13-3. Construct a $n \times n$ matrix ($n \geq 3$) row by row whose entries are any set of consecutive Fibonacci numbers, excluding F_0. Calculate the determinant of this matrix. Try another matrix constructed of consecutive Fionacci numbers and compute the determinant. What is your conclusion? Verify your conclusion mathematically.

13-4. Consider the difference of the squares of any two Fibonacci numbers which have one intervening Fibonacci number (i.e., their subscripts differ by two). Calculate this difference for several sets of Fibonacci numbers whose subscripts differ by two. What is your conclusion?

13-5. Find an issue of a design engineering or other engineering journal which describes a performance test. Write a report on the test.

13-6. Quality has been defined as fitness for use. It has also been defined as conformance to specifications. Which definition do you consider to be more appropriate? Why?

13-7. Describe the role of the design engineer, the manufacturing engineer, and the quality engineer with respect to the problems of determining specifications of a component part.

13-8. Name five products familiar to you whose design does not adequately consider human factors. Indicate why this is so.

13-9. List ways in which you might improve the safety of a three-wheeled, handlebar-steered all-terrain vehicle. Accompany your suggestions with sketches.

13-10. Some companies and many governmental agencies are required to purchase equipment on the basis of lowest bid. Illustrate by example, referencing as many of the design requirements mentioned in this chapter as possible, why this is not a good practice. Typical examples might include a police automobile purchased by a city, or a road grader, scraper, or roller purchased by a county department of highways.

13-11. Suppose that you are requested to design an electronic home security system. Rank order the following factors in order of importance to you. Explain why you ranked each as you did:

Aesthetics.

Serviceability.

Availability.

Human factors.

Logistics.

13-12. Repeat the process of Problem 13-11 if you are to design a paper-cup dispenser for a home bathroom.

13-13. Repeat the process of Problem 13-11 if you are to design a programmable electric lawn mower which will cut grass in lawns of any geometrical shape and is neither ridden nor pushed, but operates automatically.

13-14. Visit a shop that repairs at least three different types of video cassette recorders (VCRs) of the VHS type. On the basis of interviews with employees, write a report about three different manufacturers, commenting on maintainability, serviceability, and logistics (with specific reference to the availability of spare parts).

13-15. Contrast a gasoline-powered and a battery-operated automobile with respect to the following requirements:

Performance.

Life-cycle costs.

Effect on surroundings.

Reliability.

13-16. Estimate the availability of a copying machine in a department of your college or in any office by interviewing a person familiar with the operating history of the device (such as a secretary). Discuss reasons for poor availability (if it exists) and make recommendations for improvement.

13-17. The end of the life cycle of a personal computer is more likely to occur because of obsolescence than because of wear-out. Name five other products which are more likely to be replaced by improved devices before they wear out.

13-18. Schedule is an important design requirement. Given a tight schedule to complete a design, which of the design requirements mentioned in this chapter are most likely to be ignored or given little attention because of an impending deadline? Why is this so?

13-19. Contrast an aluminum automobile engine with one made of cast iron with respect to:

Availability.

Length of the life cycle.

Operating environment.

Interface with other systems.

Performance.

Reliability.

Human factors.

13-20. With respect to the design requirements of this chapter, contrast the design of a portable gasoline-powered electric generator that will be used in Fairbanks, Alaska, with one to be used in Phoenix, Arizona.

Chapter 14

Engineering Design Phases

. . . the process of design, the process of inventing physical things which display new physical order, organization, form, in response to function.

Christopher Alexander
Notes on the Synthesis of Form

A scientist can discover a new star but he cannot make one. He would have to ask an engineer to do it for him.

Gordon L. Glegg

Much of the human history has been influenced by developments in engineering, science, and technology. When progress in these fields was impeded, the culture of the era tended to stagnate and decline; the converse was also true. Although many definitions have been given of "engineering," it is generally agreed that the basic purpose of the engineering profession is to develop technical devices, services, and systems for the use and benefit of people. The engineer's design is, in a sense, a bridge across the unknown between the resources available and the needs of humankind [14-1].

Regardless of field of specialization or complexity of a problem, the method by which the engineer does his or her work is known as the engineering design process. This process is a creative and iterative approach to problem solving as detailed in Chapter 12. It is creative because it brings into being new ideas and combinations of ideas that did not exist before. It is iterative because it brings into play the cyclic process of problem solving, applied over and over again as the scope of a problem becomes more completely defined and better understood.

Thus a design engineer must be a creative person, one who will try one idea after another without becoming discouraged. In general, the designer learns more from failures than from successes, and the final design will usually include compromises and departures from the "ideal" that is desired.

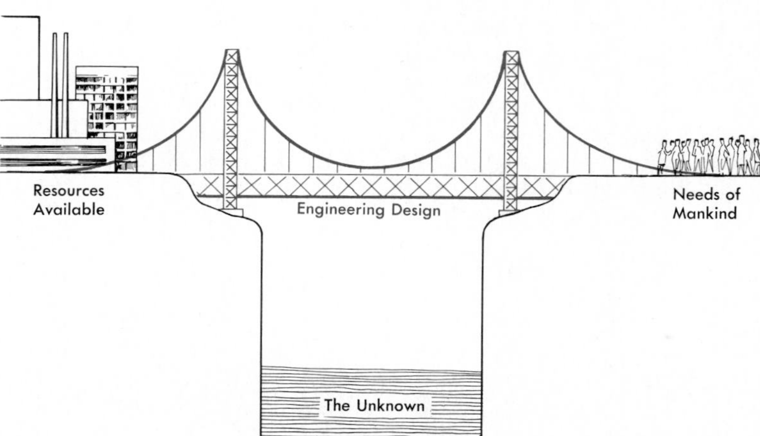

[14-1] The engineer's design is a bridge across the unknown, using available resources to provide an improved condition of life.

A final engineering design is usually the product of the inspired and organized efforts of more than one person. The personalities of good designers vary, but certain characteristics are strikingly similar. Among these will be the following:

1. Technical competence.
2. Understanding of nature.
3. Empathy for the requirements of others.
4. Active curiosity.
5. Ability to observe with discernment.
6. Initiative.
7. Motivation to design for the pleasure of accomplishment.
8. Confidence.
9. Integrity.
10. Willingness to take a calculated risk and to assume responsibility.
11. Capacity to synthesize.
12. Persistence and sense of purpose.

Certain design precepts and methods can be learned by study, but the ability to design cannot be gained solely by reading or studying. The engineer must also grapple with real problems and apply pertinent knowledge and abilities to finding solutions. Just as an athlete needs rigorous practice, so an engineer needs practice on design problems to gain proficiency in the art. Such experience must necessarily be gained over a period of years, but now is a good time to begin acquiring some of the requisite fundamentals.

PHASES OF ENGINEERING DESIGN

Most engineering designs go through three distinct phases:

1. The feasibility study.
2. The preliminary design.
3. The detailed design.

Chronologically, these phases are executed in the sequence indicated in Figure [14-2]. The amount of time spent on any phase is a function of the complexity of the problem and the restrictions placed upon the engineer—time, money, or performance specifications.

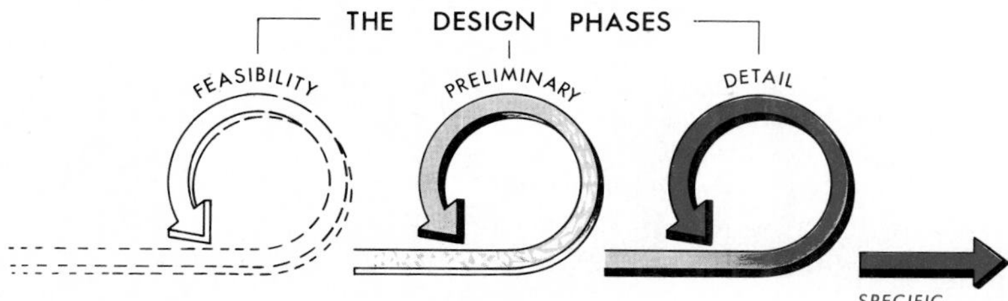

[14-2]

The Feasibility Study

The feasibility study is concerned with the following:

1. Definition of the elements of the problem.
2. Identification of the factors that limit the scope of the design.
3. Evaluation of the difficulties that can be anticipated as probable in the design process.
4. A realistic appraisal of the return (profit) on investment.
5. Consideration of the consequences of the design.

The objectives of the feasibility study are to discover possible solutions and to determine which of these appear to have promise and which are not feasible, and why.

The ideas and possibilities that are generated in early discussions should be checked for the following:

1. Acceptability in meeting the specifications.
2. Compatibility with known principles of science and engineering.
3. Compatibility with the environment.
4. Compatibility of the properties of the design with other parts of the system.
5. Comparison of the design with other known solutions to the problem.

Each alternative is examined to determine whether or not it can be physically achieved, whether its potential usefulness is commensurate with the costs of making it available, and whether the return on the investment warrants its implementation. The feasibility study is in effect a "pilot" effort whose primary purpose is to seek information pertinent to all possible solutions to the problem. After the information has been collected and evaluated, and after the undesirable design possibilities have been discarded, the engineer still may have several alternatives to consider—all of which may be acceptable.

During the generation of ideas, the engineer has intentionally avoided making any final selection in order to have an open mind for all possibilities and to give free rein to the thought processes. Now the number of ideas must be reduced to a few—those most likely to be successful, those that will compete for the final solution. The number of ideas retained will depend on the complexity of ideas and the amount of time and labor that he or she can afford to spend during the preliminary design phase. In most design situations the number of ideas remaining at the end of the feasibility study will vary from two to six.

At this point no objective evaluations are available; the discarding of ideas must depend to a large extent upon experience and judgment. There are few substitutes for experience, but there are ways in which judgment can be improved. For example, decision processes based on the theory of probability can be employed effectively. Analog and digital simulations are particularly useful to the engineer in this early comparison of alternatives.

Which of you, intending to build a tower, sitteth not down first, and counteth the cost, whether he have sufficient to finish it?

Luke 14:28
The Holy Bible

There can be no economy where there is no efficiency.

Benjamin Disraeli 1804–1881

Capitalism and communism stand at opposite poles. Their essential difference is this: The communist, seeing the rich man and his fine home, says: "No man should have so much." The capitalist, seeing the same thing, says: "All men should have as much."

Phelps Adams

In some instances, it will be more convenient for the engineer to compare the expected performance of the component parts of one design with the counterpart performances of another design. When this is done, it is very important to consider if the component parts create the optimum effect in the overall design. Frequently, it is true that a simple combination of seemingly ideal parts will not produce an optimum condition. It is not too difficult to list the advantages and disadvantages of each alternative, but the proper evaluation of such lists may require the wisdom of Solomon. Economic feasibility is also a requirement of all successful designs.

[14-3] In a market enterprise society, customers young and old daily exercise judgments affecting the success or failure of the product.

Economics. Economics is that social science that is concerned primarily with the description and analysis of the problems of production, distribution, and use of goods and services. In the United States today, products are not sold because they have been made; they are made because they have been sold. And they are sold because there is a demand for them. Demand means that at a specific level of product price, a certain number of units can be marketed. The number that can be marketed will determine (in general) the production facilities and processes that are needed. Any economic problem always has two aspects: production and distribution. Every society has had to fashion some kind of system to produce the goods and services that its members need or want. Also, every society has had to fashion some type of system to distribute the goods and services that are produced. At different times, and in different places, the ways in which a particular society has solved its economic problem have varied sharply.

Of all human powers operating on the affairs of mankind, none is greater than that of competition.

Senator Henry Clay
Address before the U.S. Senate, February 2, 1832

Money never starts an idea; it is the idea that starts the money.

W. J. Cameron

Historically, there have been three solutions or methods of controlling the market:

1. Tradition.
2. Command.
3. Market enterprise (sometimes called free enterprise).

Often these solutions do not have sharp boundaries. Each type exists somewhere in the world today, and aspects of each of them can be observed in every society.

[14-4]

The Life Cycle of New Products. The life span of a new product depends on the type of product and may vary from a few years to many decades. In congressional hearings on the drug industry a few years ago, one of the major pharmaceutical manufacturers testified that 95 percent of all the products in its current catalog were less than five years old. Because of such competitive conditions within the pharmaceutical industry, there is tremendous pressure to develop and market new products at a very high rate. However, this also means a very rapid rate of obsolescence of the products that are currently in production. At the other end of the spectrum we have what are referred to as producers' goods, such as water turbines for power generation or the heavy steel rolls for billet mills in the steel

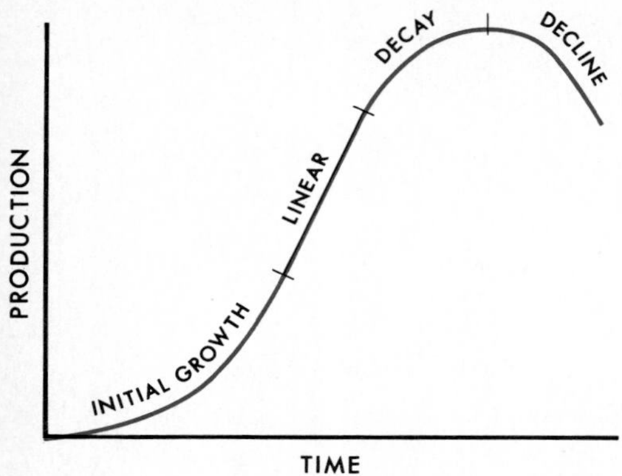

[14-5] Typical growth history of successful products.

industry. Tools such as these are more likely to be objects of continual improvement rather than of overnight obsolescence.

The life of most successful products follows a trend similar to that shown in Figure [14-5]. This type of growth curve is seen often. It describes the typical growth of populations—of people, plants, and animals. Frequently, it also describes new-product life cycles, such as those of the steamboat, the steam locomotive, the automobile, and television. Such a curve has four basic sections. In the early stages it rises at a slow but increasing rate. Then there follows a period during which production is proportional to the passage of time. We usually call this section the linear portion of the curve. Then there follows a section late in time, when production is still increasing, but at a decreasing rate. Finally, as with the buggy and then with the steam locomotive, production declines.

What can the designer do to prevent this decay? As time passes and production increases, continuous improvement should be made in the design. New materials should be tested, new technical improvements sought, and perhaps new packaging and marketing strategies investigated. By such modifications the product's usefulness may be extended to new operating environments or perhaps totally new uses can be found for it in its present environment. The addition of color has greatly extended the linear portion of the TV life cycle. Combination of TV with computers and with telephones may extend that life even further.

After the designer has completed the product prototype, and after a manufacturable version of it has been developed, it is time to turn to improvement, simplification, and cost reduction. In these ways the onset of incipient decline can be markedly delayed. This is one reason that design is such an exciting activity—it is always immersed in action.

I will build a motor car for the great multitude . . . so low in price that no man . . . will be unable to own one—and enjoy with his family the blessing of pleasure in God's great open spaces.

Henry Ford, 1863–1947

When dealing with people, remember you are not dealing with creatures of logic, but with creatures of emotion, creatures bristling with prejudice and motivated by pride and vanity.

Dale Carnegie

When shallow critics denounce the profit motive inherent in our system of private enterprise, they ignore the fact that it is an economic support of every human right we possess and without it, all rights would soon disappear.

Dwight D. Eisenhower

Many times an economical design with a predictable life is superior to an expensive design with an indefinite life. . . .

Gordon L. Glegg
The Design of Design, *1969*

Industry prospers when it offers people articles which they want more than they want anything they now have. The fact is that people never buy what they need. They buy what they want.

Charles F. Kettering, 1876–1958

The Preliminary Design Phase

With alternatives narrowed to a few, the engineer must select the design to be developed in detail. The choice is easy if only one of the proposed designs fulfills all requirements. More often, several of the concepts appear to meet the specifications equally well. The choice then must be made on such factors as economics, novelty, reliability, and the number and severity of unsolved problems.

Since it is difficult to make such comparisons subjectively without introducing personal bias, it is useful to prepare an evaluation table. All the important design criteria are listed, and each is assigned an importance factor. There will always be both positive and negative criteria. Then each design is rated as to how well it meets each criterion. This rating should be done by somebody who is not aware of the value assigned to each importance factor, so that no undue influence is present.

Let us apply this procedure to our city transportation problem, particularly to the selection of the propulsion system. Let us assume that the ordinary automobile engine has already been discarded because it is unable to meet air pollution requirements, and that the choice has narrowed to one of three types of engines: the gas turbine, the electric motor, and the steam engine. We will then enter these as designs 1, 2, and 3 in a table and assign values to the various positive and negative design criteria (Table 14-1). For example, the gas turbine and electric motor rate low on "novelty" for they are well developed, but an automobile steam engine could rate high if it used modern thermodynamic principles. On "practicability" the electric motor rates higher than the others, for it requires the least service and provides the easiest and safest way to power a small vehicle. This table is completed to the best ability of the engineer for each of the criteria.

Then the engineer "blanks out" the ratings and assigns "importance" factors to each of the criteria (Table 14-2). For example, it may be appropriate to rate practicality much higher than novelty.

Finally, the ratings and importance factors are multiplied and added, yielding a final rating for the three systems (Table 14-3), which, in this case, favors the electric motor drive. Although others may come up with different ratings, the method minimizes personal bias.

After selecting the best alternative to pursue, the engineer should make every effort to refine the chosen concept into its most elementary form. Simplicity in design has long been recognized as a hallmark of quality. Simple solutions are the most difficult to

TABLE 14-1 EVALUATION OF PROPULSION SYSTEMS
[Importance (I) Varies from 1 (Small Importance) to 5 (Extreme Importance); Rating (R) Values are 3 (High), 2 (Medium), 1 (Low), and 0 (None)].

Importance Design criteria	Importance	Design 1: gas turbine		Design 2: electric		Design 3: steam	
	I	R	$R \times I$	R	$R \times I$	R	$R \times I$
Positive							
a. Novelty		0		1		3	
b. Practicability		1		3		2	
c. Reliability		2		3		1	
d. Life expectancy		2		2		2	
e. Probability of meeting specifications		2		3		2	
f. Adaptability to company expertise (research, sales, etc.)[a]		1		1		1	
g. Suitability to human use[a]							
h. Other[a]							
Total positive score							
Negative							
a. Number and severity of unresolved problems		1		2		3	
b. Production cost		3		1		2	
c. Maintenance cost		1		1		2	
d. Time to perfect		1		1		3	
Environmental effects[a]		1		0		1	
Other[a]							
Total negative score							
Net score							

[a]May not always be pertinent.

TABLE 14-2 EVALUATION OF PROPULSION SYSTEMS
[Importance (I) Varies from 1 (Small Importance) to 5 (Extreme Importance)].

Design criteria	Importance I
Positive	
a. Novelty	2
b. Practicability	5
c. Reliability	5
d. Life expectancy	3
e. Probability of meeting specifications	4
f. Adaptability to company expertise (research, sales, etc.)[a]	3
g. Suitability to human use[a]	N.A.
h. Other[a]	
Total positive score	
Negative	
a. Number and severity of unresolved problems	3
b. Production cost	4
c. Maintenance cost	4
d. Time to perfect	4
Environmental effects[a]	4
Other[a]	
Total negative score	
Net score	

[a]May not always be pertinent.

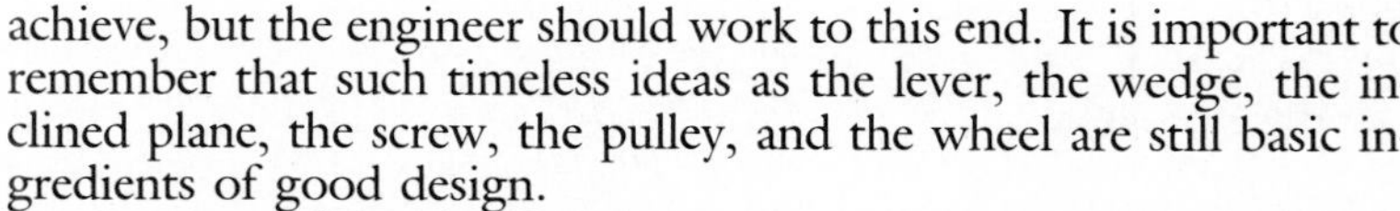

achieve, but the engineer should work to this end. It is important to remember that such timeless ideas as the lever, the wedge, the inclined plane, the screw, the pulley, and the wheel are still basic ingredients of good design.

In terms of the electric drive vehicle, this means that initially the designer will strive for a single motor, directly driving the rear wheels, and a battery that can be recharged in each parking area. It may later be discovered that a smaller motor at each wheel is preferable, that a geared-down, high-speed motor is more efficient than a direct-drive motor, or that an on-board electric generator is preferable to a rechargeable battery. The design engineer will start with the simplest ideas [14-6].

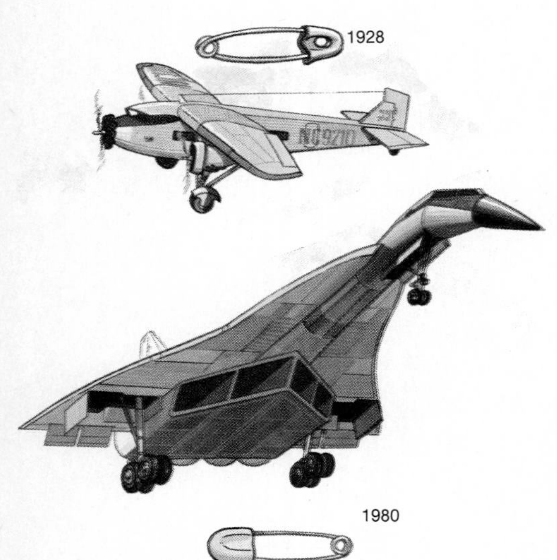

[14-6] Simple designs often have a longer life than complex designs.

Once the design concept has been selected, the engineer must consider all the component parts—their sizes, relationships, and materials. In selecting materials, strengths, dimensions, and the loads to which they will be exposed must be considered. In this sense, the engineer is analogous to the painter who has just chosen the subject and now must select colors, shapes, and brush strokes and put them together in a pleasing and harmonious arrangement. The engineer, having selected a design concept that fulfills the desired functions, must organize his or her components to produce a device that is not only pleasing to the eye but is economical to build and operate.

Engineers must make sure that their designs do not interfere with or disturb the environments, that it agrees with man and nature. We are especially reminded of these responsibilities when we encounter foul air, polluted streams, and eroded watersheds. Environmental effects are increasingly important criteria in the design of engineering structures, as evidenced by the concern about such projects as the transAlaska pipeline, the supersonic jet transport, and facilities for the disposal or reclamation of industrial and human waste. As the earth's natural resources are depleted, the engineer will be under increasing pressure to provide technical assurances that no harm is done to the environment.

The process of anticipating hazards, determining the degree of acceptable risk and then selecting the alternative with the best benefit/risk ratio is called *risk analysis*. Risk analysis must be a part of every engineering design. The risk analysis process can be grouped into two major areas: *risk assessment* includes determining the numeric probability of an adverse effect, and *risk management* involves the study of the consequences of risk and the implementation of actions to reduce risk. Details of risk analysis may be found in Chapter 6.

TABLE 14-3 EVALUATION OF PROPULSION SYSTEMS

[Importance (I) Varies from 1 (Small Importance) to 5 (Extreme Importance); Rating (R) Values are 3 (High), 2 (Medium), 1 (Low), and 0 (None)].

Importance Design criteria	Importance I	Design (1) gas turbine R	R × I	Design (2) electric R	R × I	Design (3) steam R	R × I
Positive							
a. Novelty	2	0	0	1	2	3	6
b. Practicability	5	1	5	3	15	2	10
c. Reliability	5	2	10	3	15	1	5
d. Life expectancy	3	2	6	2	6	2	6
e. Probability of meeting specifications	4	2	8	3	12	2	8
f Availability to company expertise (research, sales, etc.)[a]	3	1	3	1	3	1	3
g. Suitability to human use[a]	N.A.						
h. Other[a]							
Total positive score		—	32	—	53	—	38
Negative							
a. Number and severity of unresolved problems	3	1	3	2	6	3	9
b. Production cost	4	3	12	1	4	2	8
c. Maintenance cost	4	1	4	1	4	2	8
d. Time to perfect	4	1	4	1	4	3	12
Environmental effects[a]	4	1	4	0	0	1	4
Other[a]							
Total negative score		—	27	—	18	—	41
Net score		—	5	—	35	—	−3

[a]May not always be pertinent.

The designer must consider such factors as heat, noise, light, vibration, acceleration, air supply, and humidity, and their effects upon the physical and mental well-being of the user. For example, although it would be desirable to accelerate to top speed as quickly as possible, there are human comfort limits on acceleration that should not be exceeded. Controls must respond rapidly, have the right "feel," and not tire the driver. The suspension system must be "soft" for a comfortable ride, but stiff enough for good performance on curves. Automatic heating and air conditioning will probably be required in most parts of the country.

By now, the picture of the vehicle has become clearer, and the chief engineer can delegate the preliminary design of components to various engineers or designers in the organization. Someone will be working on the drive train, another on the wheels and suspension, a third on the battery. Then there are the speed control systems, the interior layout, and perhaps three or four other components, such as access protection, recharging (if electric), and systems for redistributing the cars that must be developed.

Trade-Off Analysis. Chapter 13 has emphasized that as the design evolves, a number of critical decisions must be made concerning matters such as determining the means to attain high performance, control life-cycle costs, meet the design schedule, and achieve reliability, maintainability, and availability goals. Unfortunately, improvement in the level of one of these requirements often results in a decrease in the quality levels of one or more of the other requirements. For example, the design that yields the best operating performance measures may result in a failure rate that cannot be tolerated, or the design that incorporates sophisticated built-in testing and diagnostic equipment for maintainability improvement may extend the first costs beyond a maximum specified value.

To resolve problems such as these, compromises must be made. The nature of these compromises cannot be known until the interaction effects of the variables associated with the requirements in question are modeled. *Interaction* refers to the effect that changes in one of the variables has on another. The modeling process and the resulting choices of values or ranges of values for the design requirements variables is referred to as *trade-off analysis*. Trade-off studies are an important part of the preliminary design phase.

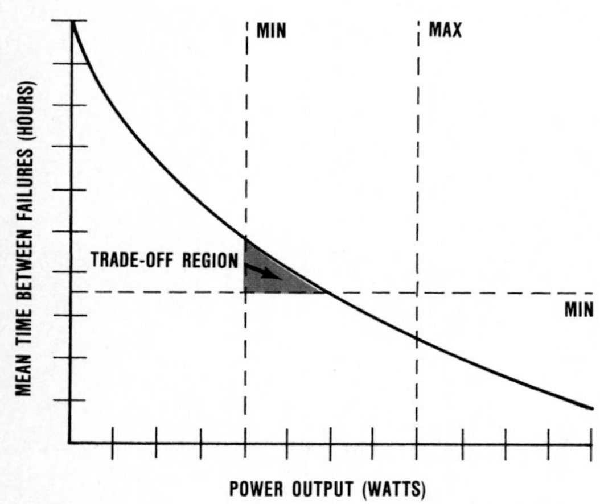

[14-7] Results of trade-off study: Reliability vs. Power Output.

The models used to investigate the relationships between design requirement variables may assume any of the forms discussed in Chapter 12. Quite often, a mathematical model is chosen to represent the relationship between pairs of variables. Figure [14-7] depicts reliability, measured in mean time between failures (MTBF), as a function of the power output. The design requires a minimum mean time to failure while the power output must assume a value between two specified limits in order for the design to perform satisfactorily. The shaded area, called the trade-off region, represents acceptable values of both reliability and the performance variable, power. The design engineer will select a power output value that gives the best performance but does not result in an estimated MTBF below its minimum value. The relationship between life-cycle cost and system availability for a particular design is shown in Figure

[14-8]. It is required that the system be available at least 90 per cent of the time. However, as the availability increases, so does the life-cycle cost. The design features associated with availability must be decided on so as to keep both the availability above 0.90 and the life-cycle cost below its maximum budget. Figure [14-9] shows that a value for design weight must be selected in the shaded region so that minimal power output is achieved without exceeding the maximum allowable weight.

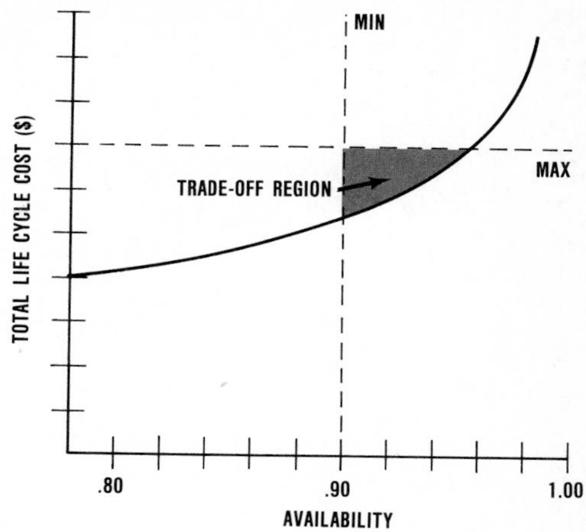

[14-8] Results of trade-off study: Life Cycle Cost vs. Availability.

The Detailed Design Phase

Detailed design begins after determination of the overall functions and dimensions of the major members, the forces and allowable deflections of load-carrying members, the speed and power requirements of rotating parts, the pressures and flow rates of moving fluids, the aesthetic proportions, and the needs of the operation—in short, after the principal requirements are determined. The models that were devised during the preliminary selection process should be refined and studied under a considerably wider range of parameters than was possible originally. The designer is interested not only in normal operation, but also in what happens during startup and shutdown, during malfunctions, and in emergencies. The range of the loads that act on a design and how these loads are transmitted through its parts as stresses and strains must be evaluated. The effects of temperature, wind, and weather and of vibrations and chemical attack should be considered. In short, the range of operating conditions for each component of the design and for the entire device must be determined.

Design engineers must have an understanding of the mechanisms of engineering: the levers, linkages, and screw threads that transfer and transform linear and rotating motion; the shafts, gears, belts, and chain drives that transmit power; and electrical power generating systems and their electronic control circuits.

With today's wide range of available materials, shapes, and manufacturing techniques, with the growing array of prefabricated devices and parts, the choices for the design engineer are vast indeed. How should you start? What guidelines are available if you want to produce the best possible design? It is usually wise to begin investigating that part or component which is thought to be most critical in the overall design—perhaps the one that must withstand the greatest variation of loads or other environmental influences, the one that is likely to be most expensive to make, or the most critical in operation. You may find that operating conditions limit your choices to a few possibilities.

At this stage, as the designer you will encounter many conflicting requirements. One consideration tells you that you need more power, another that the motor must be smaller and lighter. Springs should be stiff to minimize road clearance; they should be soft to give a comfortable ride. Windows should be large for good visibility, but small for safety and high body strength. The way to resolve this type of conflict is called *optimization*. It is accomplished by assigning values to all requirements and selecting that design which maximizes (optimizes) the total value.

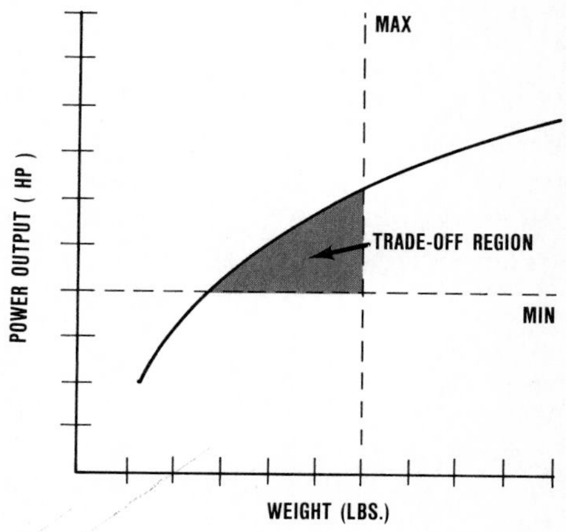

[14-9] Results of trade-off study: Power Output vs. Weight.

Materials and stock subassemblies are commercially available in a specific range of sizes. Sheet steel is commonly available in certain thicknesses (gauges), electric motors in certain horsepower ratings, and pipe in a limited range of diameters and wall thicknesses. Generally, the engineer should specify commonly available items; only rarely will the design justify the cost of a "special mill run" with off-standard dimensions or specifications. When available sizes are substantially different from the desired optimum size, the engineer may have to revise his or her optimization procedure.

A civilization is both developed and limited by the materials at its disposal.
Sir George Paget Thomson

To illustrate, let us look at the design of a meteorological rocket. At an earlier point in the design process the fuel for this rocket will have been chosen. Let us assume that it is a solid fuel, a material that looks and feels like rubber, burns without air, and when ignited produces high-temperature, high-pressure gases which are expelled through the nozzle to propel the rocket. The rocket consists principally of the payload (the meteorological instruments that are to be carried aloft), the nose cone which houses the instruments, the fuel, the fuel casing, and the nozzle. If we can estimate the weight of the rocket and how high it is to ascend, we can calculate the requirements.

The most critical design part is the fuel casing, that is, the cylindrical shell which must contain the rocket fuel while it burns. It must be strong enough to withstand the pressure and temperature of the burning fuel, and strong enough to transmit the thrust from the nozzle to the nose cone without buckling and without vibrating. The shell must also be light. If the casing weighs more than had been estimated originally, more fuel will be needed to propel the rocket. More fuel will produce higher pressures and higher temperatures inside the casing. This, in turn, will require a stronger casing and even more weight. This additional weight requires still more fuel, and the spiral continues [14-10].

Let us assume that we decided to use a high-strength, high-temperature-resistant steel for our casing. Our calculations indicate

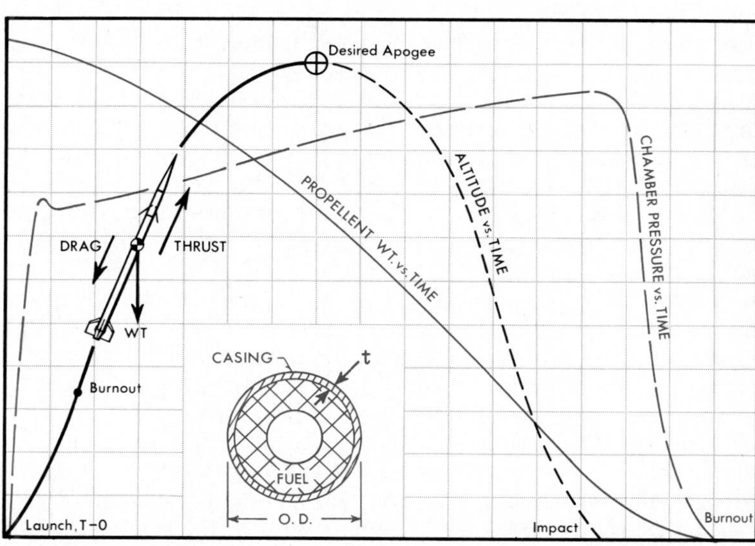

[14-10] Multiple variables must be considered in trade-off evaluations.

its wall thickness to be not less than 0.28 in. Our steel catalog tells us this steel is generally available in sheet form only in thicknesses of $\frac{1}{4}$ and $\frac{3}{16}$ in. If we use the thicker sheet, the casing weight will increase by 2.7 per cent; then we must recalculate the amount of fuel required, the pressures and stresses in the casing, and consequent changes in the dimensions of the rocket. Will the $\frac{1}{4}$-in. material withstand the resultant higher stresses? Can we improve its strength by heat treating? If we choose the thinner material, must we provide the casing with extra stiffeners (rings that will reduce the stresses in the casing shell)? In either case, the original design must be altered until the stresses, weights, pressures, and dimensions are satisfactory. Similar design procedures will be followed in designing the nose cone, the nozzle, and the launching gear for the rocket.

It is important to understand that this example is typical of the design process. Design is not a simple straightforward process but a procedure of trial and error and compromise until a well-matched combination of components has been found. The more the engineer knows about materials and about ways of reducing or redistributing stresses (in short, the more alternatives that are available), the better the structural design is likely to be.

Consider, as another example, that as the engineer you have been asked to design the gear-shift lever for a racing automobile. The gearbox has already been designed, so you know how far the shifting fork (the end that actually moves the gears in the gearbox) must travel in all directions. You also know how much force will be required at the fork under normal and abnormal driving conditions. You will need to refer to anthropometric[1] data to learn how much force the healthy driver can provide forward, backward, and sideways, and what her reach can be without distracting her eye from the road. With all this information you can choose the location of the ball joint, the fulcrum of the gear-shift lever, and the length of each arm of the lever. You may decide to use a straight stick or you may find that a bent lever is more convenient for the driver [14-11]. Before you finalize this decision you may build a mock-up and make experiments to determine the most convenient location. Next, you must select the material and the cross-sectional shape and area of the lever. Since it is likely to be loaded evenly in all directions, you may find that a circular or a cruciform cross section is most suitable. You must decide between a lever of constant thickness and a lighter, tapered stick (with the greater strength where it is needed—near the joint), which is more costly to manufacture.

Next, you will consider the design of the ball joint, which transmits the motion smoothly to the gearbox and provides vibration isolation so that the hand of the driver does not shake. It is difficult to find just the right amount of isolation which will retain for the driver the "feel" that is so essential during a race. Finally, you will need a complete understanding of lubricated ball joints and proficiency in testing a series of possible designs.

The final component in this design is the handle itself, which should be attractive to look at and comfortable to grip. Here again

[14-11] You may find that a bent lever is more convenient for the driver.

[1]Anthropometry is the study of human body measurements, especially on a comparative basis. See Appendix IV.

anthropometric data can tell you much, yet you will be well advised to make several mock-ups and to have them tested for "feel" by experienced drivers.

During the design process, you will have made a series of sketches (somewhat like those in Chapter 7) to illustrate to yourself the relative position of the parts that you are designing. Now you or your drafter will use these sketches to make a finished drawing. This will consist of a separate detail drawing for each individually machined item, showing all dimensions, the material from which it is to be made, the type of work to be performed, and the finish to be provided. There will also be subassembly and assembly drawings showing how these parts are to be put together.

The detail design phase will include the completion of an operating physical model or prototype (a model having the correct layout and physical appearance but constructed by custom techniques), which may have been started in an earlier design phase. The first prototype usually will be incomplete and modifications and alterations will be necessary. This is to be expected. Problems previously unanticipated may be identified, undesirable characteristics may be eliminated, and performance under design conditions may be observed for the first time. This part of the design process is always a time of excitement for everyone, especially the engineer.

The final phase of design involves the checking of every detail, every component, and every subsystem. All must be compatible. Much testing may be necessary to prove theoretical calculations or to discover unsuspected consequences. Assumptions made in the earlier design phases should be reexamined and viewed with suspicion. Are they still valid? Would other assumptions now be more realistic? If so, what changes would be called for in the design?

As one moves through the design phases—from feasibility study to detail design—the tasks to be accomplished become less and less abstract and consequently more closely defined as to their expected function [14-12]. In the earlier phases, the engineer worked with the design of systems, subsystems, and components. In the detail design phase he or she will also work with the design of the parts and elementary pieces that will be assembled to form the components.

In the previous phase of engineering design, a large majority of the people involved were engineers. In the detail phase this is not necessarily the case. Many people—metallurgists, chemists, tool designers, detailers, drafters, technicians, checkers, estimators, and manufacturing and shop personnel—will work together under the direction of engineers. These technically trained support people probably will outnumber the engineers. The engineer who works in this phase of design must be a good personnel manager in addition to being expert in the area of technical responsibility. Resulting successes may be measured largely by his or her ability to bring forth the best efforts of many people.

The engineer should strive to produce a design which is the "obvious" answer to everyone who sees it, once it is complete. Such designs, simple and pleasing in appearance, are in a sense as beautiful as any painting, piece of sculpture, or poem, and they are frequently considerably more useful to our well-being.

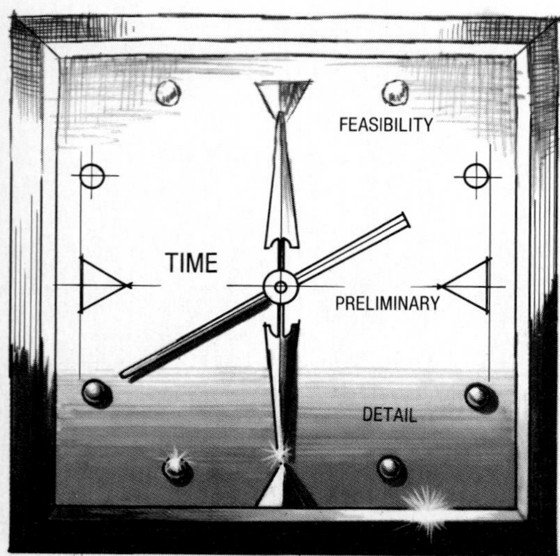

[14-12] Changes in level of abstraction varies with time.

DESIGN REVIEWS

Formal design reviews are a part of each phase of engineering design. These reviews are made for each level of the design as well—component, subsystem, and system. They are usually required by corporate management and are planned and scheduled like many other design activities. Such design reviews are conducted by a team of specialists who are not directly involved in the design project. These specialists may be thought of as support personnel for the design. Design engineers are present at the review meetings and interact extensively with the support personnel. Members of the design review team are usually representatives from the manufacturing, quality, safety, financial, and marketing areas of the organization. When the occasion dictates, a representative from the company's legal department may also be a part of the design review team. A record is made of the events of each meeting and assignments and follow-ups are made which require action by team members. The minutes provide an official record of the design decisions that were made and the reasons for making them.

The purpose of a design review is to evaluate the proposed design with respect to its ability to perform satisfactorily, be manufactured in an efficient manner and at a reasonable cost, achieve quality and reliability goals, and provide for customer safety, comfort, and fitness for use. It offers the advantages of expert opinions and outside viewpoints. It provides a check or an audit on the current design activities. The criteria used to make evaluations during the design review are often determined by the requirements of the customer, company objectives, equipment requirements, budget requirements, and experience with similar designs.

Design reviews provide an opportunity to identify and correct problems before they can seriously affect the successful completion of the design. They allow for an exchange of information between the design group and the support personnel. Each group obtains a better understanding of the other. For example, design engineers are kept aware of new manufacturing equipment and recent improvements in manufacturing techniques. Design reviews help to prevent the tendency to rush a design into manufacturing before it is ready. The design review helps to reduce the risk of failing to complete the design on schedule.

THE PLANNING OF ENGINEERING PROJECTS

In every walk of life, we notice and appreciate evidence of well-planned activities. You may have noticed that good planning involves more than "the assignment of tasks to be performed," although this frequently is the only aspect of planning that is given any attention. Planning in the broad sense must include the enumeration of all the activities and events associated with a project and a recognition and evaluation of their interrelationships and interdependencies. The assignment of tasks to be performed and other aspects of scheduling should follow.

Since "time is money," planning is a very important part of the implementation of any engineering design. Good planning is often the difference between success and failure, and the young engineering student would do well, therefore, to learn some of the fundamental aspects of planning as applied to the implementation of engineering projects.

In 1957, the U.S. Navy was attempting to complete the Polaris Missile System in record time. The estimated time for completion seemed unreasonably long. Through the efforts of an operations research team, a new method of planning and coordinating the many complex parts of the project was finally developed. The overall saving in time for the project amounted to more than 18 months. Since that time, a large percentage of engineering projects, particularlythose which are complex and time consuming, have used this same planning technique to excellent advantage. It is called PERT (Program Evaluation and Review Technique).

Though this be madness, yet there is method in it.
Shakespeare, 1564–1616

PERT enables the engineer in charge to view the total project as well as to recognize the interrelationships of the component parts of the design. Its utility is not limited to the beginning of the project; rather, it continues to provide an accurate measure of progress throughout the work period. Pertinent features of PERT are combined in the following discussion.

How Does PERT Work?

Basically PERT consists of events (or jobs) and activities arranged into a time-oriented network to show the interrelationships and interdependencies that exist. One of the primary objectives of such a network is to identify where bottlenecks may occur that would slow down the process. Once such bottlenecks have been identified, extra resources such as time and effort can be applied at the appropriate places to make certain that the entire process will not be slowed. The network is also used to portray the events as they occur in the process of accomplishing missions or objectives, together with the activities that necessarily occur to interconnect the events. These relationships will be discussed more fully below.

The Network. A PERT network is one type of pictorial representation of a project. This network establishes the "precedence relationships" that exist within a project. That is, it identifies those activities that must be completed before other activities are started. It also specifies the time that it takes to complete these activities. This is accomplished by using events (points in time) to separate the project activities. In other words, project events are connected by activities to form a project network. Progress from one event to another is made by completing the activity that connects them. Let us examine each component of the network in more detail.

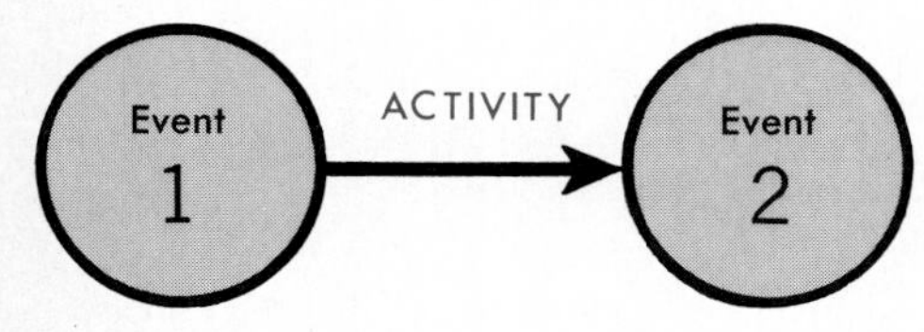

[14-13]

Events. An event is the start or completion of a mental or physical task. It does not involve the actual performance of the task. Thus events are points in time which require that action be taken or that decisions be made. Various symbols are used in industry to designate events, such as circles, squares, ellipses, or rectangles. In this book circles, called *nodes,* will be used [14-13].

Events are joined together to form a project network. It is important that the events be arranged within the network in logical or time sequence from left to right. If this is done, the completion of each event will occupy a discrete and identifiable point in time. An event cannot consume time and it cannot be considered to be completed until all activities leading to it have been completed. After all events have been identified and arranged within the network, they are assigned identification numbers. Since events and activities may be altered during the course of the project, the logical order of the events will not necessarily follow in exact numerical sequence, 1, 2, 3, 4, 5, and so on. The event numbers, therefore, serve only for identification purposes. The final or terminal node in the network is usually called the *sink,* while the beginning or initial node is called the *source*. Networks may have varying numbers of sources and sinks.

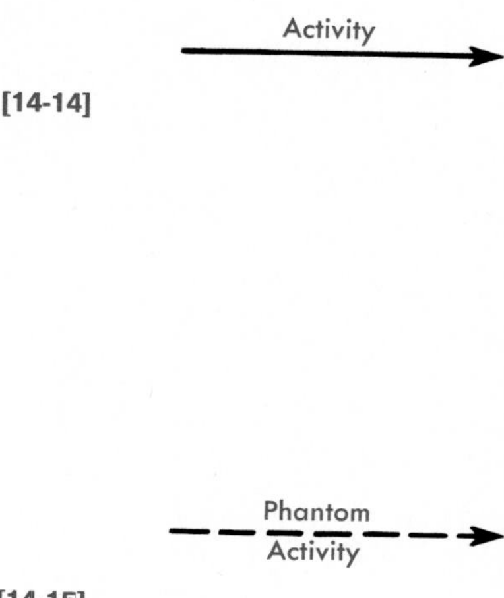

[14-14]

[14-15]

Activities. An activity is the actual performance of a task and, as such, it consumes an increment of time. Activities separate events. An activity cannot begin until all preceding activities have been completed. An arrow is used to represent the time span of an activity, with time flowing from the tail to the point of the arrow [14-14]. In a PERT network an activity may indicate the use of time, labor, materials, facilities, space, or other resources. A phantom activity may also represent waiting time or "interdependencies." A phantom activity, represented by a dashed arrow [14-15], may be inserted into the network for clarity of the logic, although it represents no real physical activity. Waiting time would also be noted in this manner. Remember that:

Events "happen or occur."

Activities are "started or completed."

The case of Mr. Jones getting ready for work each morning can be examined as an example.

Event	***Activity***
1. The alarm rings	
2. Jones awakens	*A.* Jones stirs restlessly. *B.* Jones nudges his wife. *C.* Jones lies in bed wishing that he didn't have to go to work.
3. Wife awakens	*D.* Wife lies in bed wishing that it were Saturday.
4. Jones's wife gets up and begins breakfast.	*E.* Wife cooks breakfast.
Meanwhile.	
5. Jones begins morning toilet.	*F.* Jones shaves, bathes, and dresses.
6. The Joneses begin to eat breakfast.	*G.* The Joneses eat part of their breakfast.
7. Jones realizes his bus is about to pass the bus stop.	*H.* Jones jumps up, grabs his briefcase, and runs for the bus.
8. Jones boards bus.	*I.* Wife goes back to bed.
9. Wife falls asleep.	

[14-16]

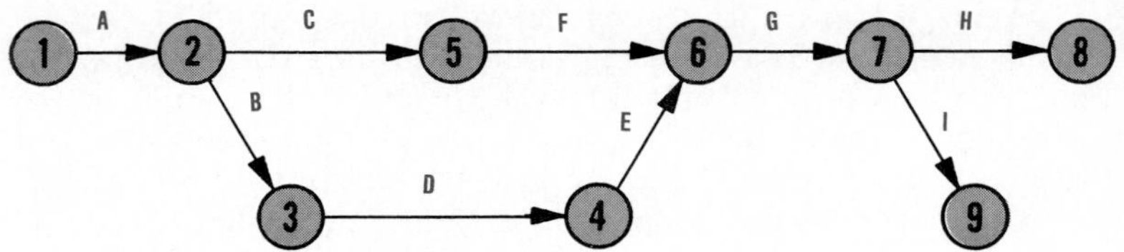

His PERT network can now be drawn as shown in Figure [14-16]. This is a very elementary example, but it does point up the constituent parts of a PERT network. Note that Jones and his wife must wait until he is dressed (F) and the breakfast cooked (E) before they can eat.

In a PERT network each activity should be assigned a specified time for expected accomplishment. The time units chosen should be consistent throughout the network, but the size of the time unit (years, workweeks, days, hours, etc.) should be selected by the engineer in charge of the project. The time value chosen for each activity should represent the weighted average of the various times that the activity would take if it were repeated many times.

The project manager asks the engineer or individual in charge of each activity to supply three time estimates for that activity—an optimistic estimate, t_o, a pessimistic estimate, t_p, and an estimate of the most likely time, t_m, that will be required to complete the activity. In estimating, t_o, consideration is given to factors that would allow the activity to be finished with virtually no problems—no work stoppages, material will arrive on time, no substandard materials will be delivered that must be replaced, perfect weather conditions will prevail, and so on. To develop a t_p value, those factors which will hinder a timely completion are assumed to occur—a labor strike, substandard materials to be replaced, delays in shipments, unavailability of manufacturing equipment and key personnel, and so on. The most likely time, t_m, is simply the "best guess," with all factors considered, of how long it will take to finish the activity. Experience has shown that when these estimates were compared with the actual time required, the best estimate or expected time, t_e, for the completion of the activity is given by

$$t_e = \frac{t_o + 4t_m + t_p}{6} \tag{14-1}$$

There is also a theoretical basis for the determination of the expected time. It is the mode (or most frequently appearing value) of a beta probability distribution.[2] In the examples that appear on the next few pages, the reader should recognize that times given for each activity have already been computed from equation (14-1). The optimistic and pessimistic times are also used to estimate the variability of each activity time. These variability estimates, in turn, are used with the expected time to complete the entire project and the project due date to obtain the probability that the project will be completed on time. These calculations are not discussed in this text.

[2]A beta probability distribution is a mathematical function whose shape has been shown to describe or "fit" the actual times requested to complete the thousands of tasks that were studied.

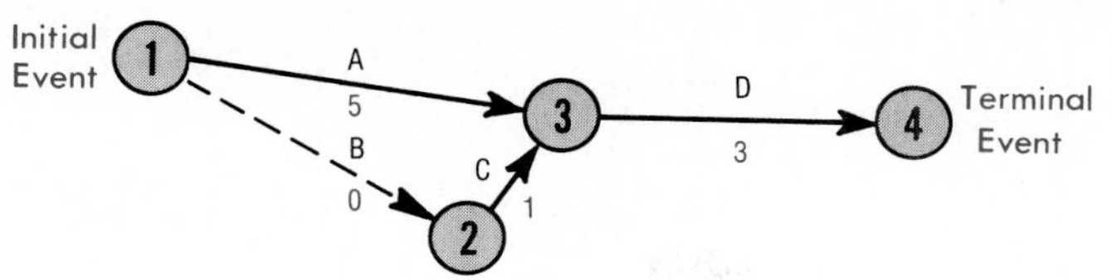

[14-17]

By using the network of events and activities and by taking into account the times consumed by the various activities, a critical path can be established for the project. It is this path that controls the successful completion of the project, and it is important that the engineer be able to isolate it for study. Let us consider the PERT network in Figure [14-17], where the activity times are represented by arabic numbers and are indicated in days. Activities represent the expenditure of time and effort. For example, activity A (from event 1 to event 3) requires five days and is likely devoted to planning the project, while activity C requires one day and may represent the procurement of basic supplies. Activity B is a phantom activity inserted to comply with the rule that all events must be preceeded by the initial event. Event D requires three days and may represent the expected time to do the actual work using the supplies. Event 1 is the beginning of the project and event 4 is the end of the project. The first step in locating the critical path is to determine the "earliest" event time (T_E), the "latest" event times (T_L), and the "slack" time ($T_L - T_E$).

Earliest Event Times

The earliest expected time of an event refers to the time, T_E, when an event can be expected to be completed. T_E for an event is calculated by summing all the activity duration times from the beginning event to the event in question *if the most time-consuming route is chosen.* To avoid confusion, the T_E times of events are usually placed near the network as arabic numbers within rectangular blocks. For reference purposes the beginning of the project is usually considered to be "time zero." In Figure [14-18], T_E for event 3 would be $\boxed{0} + 5 = \boxed{5}$ and T_E for event 4 would be $\boxed{0} + 5 + 3 = \boxed{8}$. However, there are two possible routes to event 4 ($A + D$, or $C + D$). The *maximum* duration of these event times should be selected as the T_E for event 4. Summing the times, we find

By path $A + D$: $\boxed{0} + 5 + 3 = \boxed{8}$. Select as T_E for event 4.

By path $C + D$: $\boxed{0} + 1 + 3 = \boxed{4}$.

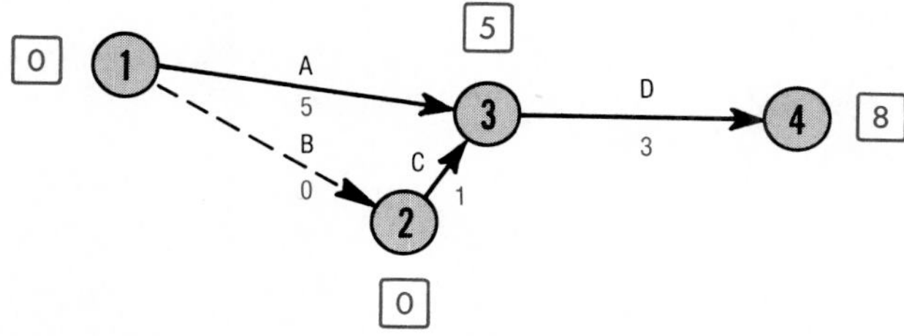

[14-18]

[14-19]

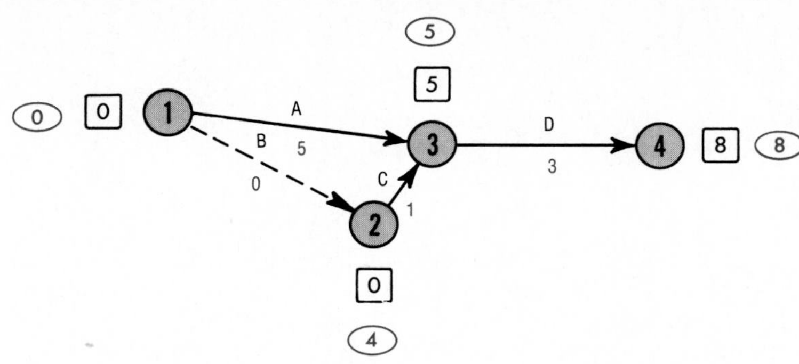

Latest Event Times

Method is like packing things in a box; a good packer will get in half as much again as a bad one.

Cecil

In the latest expected time of an event, T_L, refers to the latest time at which an event may occur, assuming that the entire project is kept on schedule. T_L for an event is determined by beginning at the terminal event and working backward through the various event circuits, subtracting the activity duration *assuming the most time-consuming route is chosen*. The resulting values of T_L are recorded as arabic numbers in small ellipses located near the T_E times. Thus, in [14-19] T_L for event 3 would be (8) − 3 = (5); for event 2, (8) − 3 − 1 = (4); and for event 1, (8) − 3 − 5 = (0).

Remember that T_L is determined to be the *minimum* of the differences between the succeeding event T_L and the intervening activity times. Also, in calculating T_L values, one must always proceed backward through the network—from the point of the arrows to the tail of the arrows.

Slack times

The *slack* time for each event is the difference between the latest event time and the earliest possible time ($T_L - T_E$). Intuitively, one may verify that it is the "extra time that an event can slip" and not affect the scheduled completion time of the project. For example, in Figure [14-19] the slack time for event 2 is (4) − [0] = 4. For this reason activity C may be started as much as 4 days late and still not cause any overall delay in the minimum project time of 8 days.

The Critical Path

The *critical* path through a PERT network is a path that is drawn from the initial event of the network to the terminal event by connecting the events of zero slack. The *critical path* would be shown connecting events 1-3-4 [14-20]. Slack times for each event are indicated as small Arabic numbers that are located in triangles adjacent to the events.

Remember that the *critical path* is the path that controls the successful completion of the project. It is also the path that requires the most time to get from the initial event to the terminal event. Any event on the critical path that is delayed will cause the final event to

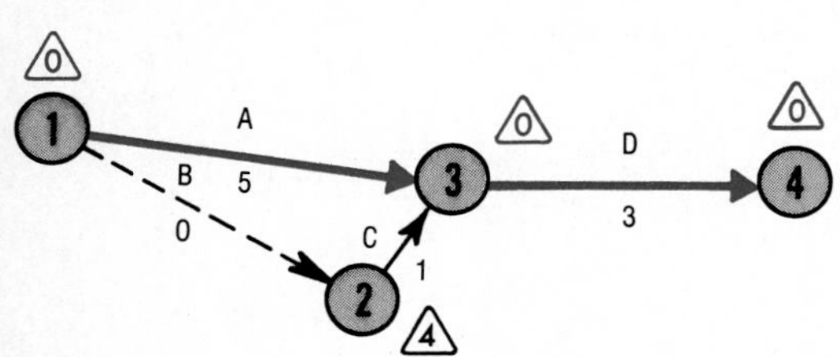

[14-20]

be delayed by the same amount. Conversely, putting an extra effort on noncritical activities will not speed up the project.

Although calculations in this chapter have been done manually, it is conventional practice to program complex networks for solution by digital computer. In this way thousands of activities and events may be considered, and one or more critical paths can be located for further study. A PERT computer program is presented in Chapter 19. Finally, the PERT network should be updated periodically as the work on the project progresses [14-21].

The following example will show how a typical PERT diagram is analyzed. It should be noted here, however, that in real-life situations the most difficult task is to identify the precedence relationships that exist and to draw a realistic network of the events and activities. After this is accomplished, following through with a solution techniques becomes a relatively routine task.

[14-21] A proper evaluation of PERT will help the engineer to schedule all subcontracts in proper sequence and especially not to allow one work assignment to be pushed ahead of others prematurely or to lag behind unnecessarily.

Example 14-1

In the PERT network diagram of Figure [14-22], assume that all activity times are given in months and that they exist as indicated on the proper activity branch. Find the earliest times, T_E, the latest times, T_L, and the slack times for each event. Identify the critical path through the network.

Solution:

See Figure [14-23].

It is usually advisable to construct a summary table of the calculations.

Event ○	**Path**	T_E □	**Path**	T_L ⬭	**Slack, $T_L - T_E$** △	**On critical path**
1	—	0	7-6-2-1	0	0	√
2	1-2	2	7-6-2	2	0	√
3	1-3	3	7-6-5-3	5	2	
4	1-4	1	7-6-5-4	4	3	
5	1-2-5	6	7-6-5	7	1	
6	1-2-6	10	7-6	10	0	√
7	1-2-6-7	11	—	11	0	√

The critical path then is 1-2-6-7 [14-24]. This means that as the project is now organized it will take 11 months to complete.

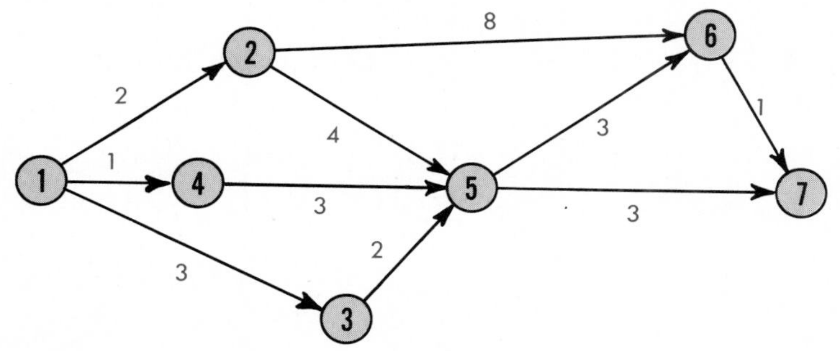

[14-22]

[14-23]

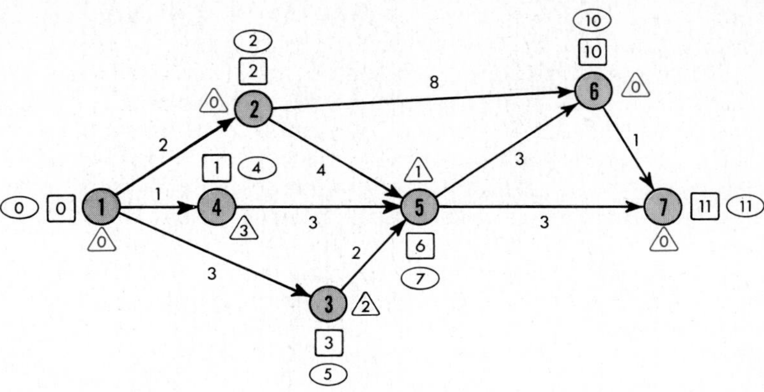

[14-24]

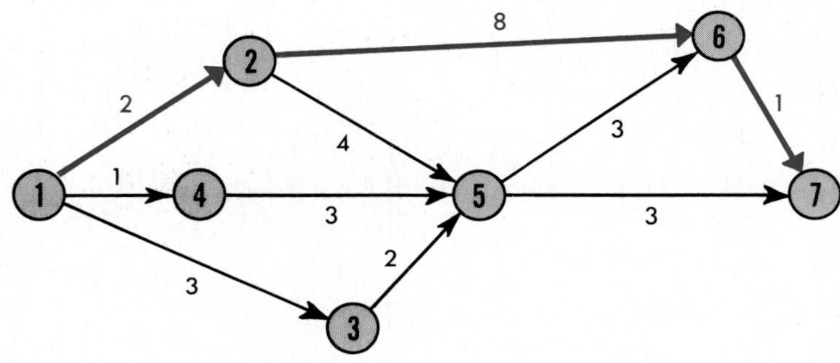

PROBLEMS

14-1. Consider the network in Figure [14-P1]. Find T_E, T_L, slack times, and the critical path through the network.

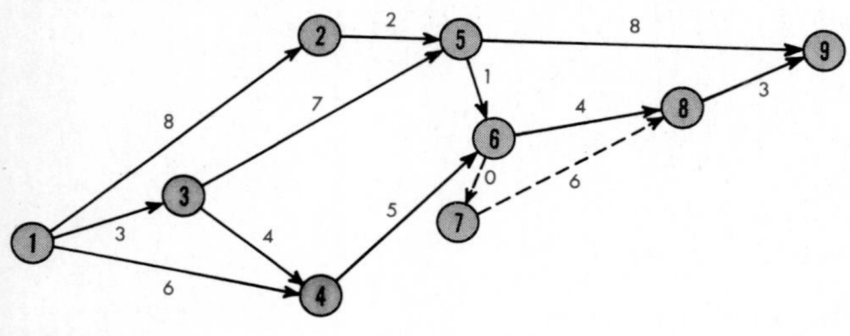

[14-P1]

14-2. In Figure [14-P1] what effect on project length would the following changes have?

a. Decrease activity 1-2 to six days.
b. Decrease activity 5-9 to one day.
c. Decrease activity 3-4 to two days.

14-3. Explain why "phantom activities" are necessary, and give an example of one.

14-4. Given the following tabular information, determine the PERT network and its critical path.

Activity	*Preceding activity*	*Time*
A	None	5
B	None	3
C	A	1
D	B	4
E	B	3
F	E	7

14-5. For some general process with which you are familiar, construct a PERT network. Be sure to label all events and activities.

14-6. Find the critical path in Figure [14-P6] and explain its significance here.

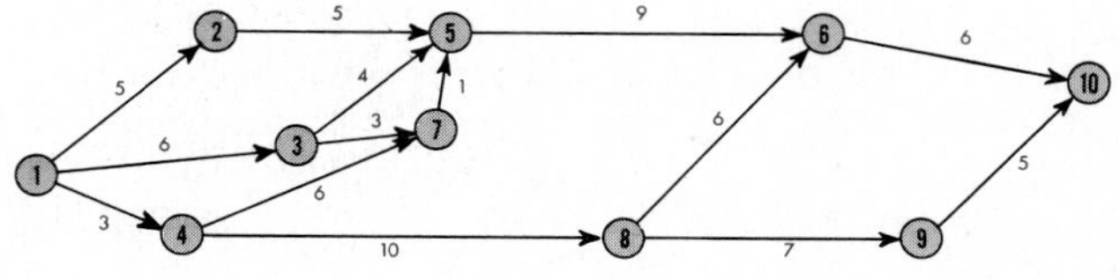

[14-P6]

14-7. a. Does a decrease in an activity time on the critical path always decrease the project time correspondingly? Why or why not? *Hint:* See Problem 14-6.

b. Does an increase in an activity time on the critical path always increase the project time correspondingly? Why or why not? *Hint:* See Problem 14-4.

14-8. Given the PERT network in Figure [14-P8], when is the earliest possible project completion time?

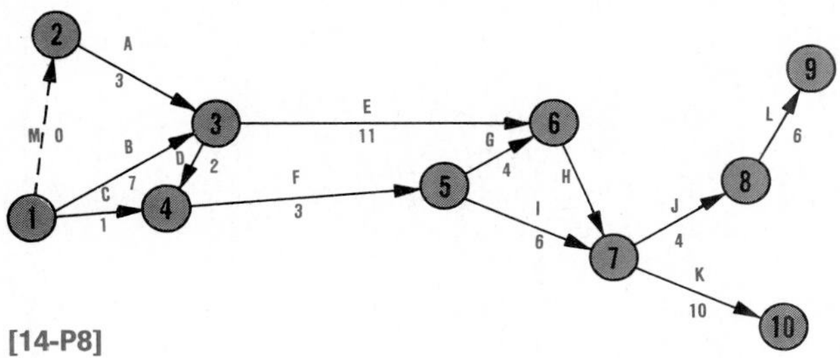

[14-P8]

14-9. If you have extra resources to allocate to one activity in Figure [14-P8], where would you put those resources and why? How might this affect the expected duration time of the project?

14-10. One thing that a PERT analysis does not consider is the allocation of limited resources (see Problem 14-9). How can this inability affect the usefulness of a PERT analysis?

14-11. Estimate the number of drugstores in the United States. Give reasons for your estimate.

14-12. Estimate the number of liters of water of the Mississippi River that pass New Orleans every day. Show your analysis.

14-13. In 100 words or less, describe how a household water softener works.

14-14. By the use of simple sketches and a brief accompanying explanation, describe the mechanical operation of a household toilet.

14-15. By the use of a diagrammatic sketch show how plumbing in a home might be installed so that hot water is always instantly available when the hot-water tap is opened.

14-16. Analyze and discuss the economic problems involved in replacing ground-level railroad tracks with a suspended monorail system for a congested urban area.

14-17. Discuss the feasibility of railroads offering a service whereby your automobile would be carried on a railroad car on the same train on which you are traveling so that you might have your car available for use upon arrival at your destination.

14-18. Discuss the desirability of assigning an identifying number to each person as soon as they are born. The number could, for example, be tatooed at some place on the body to serve as a social security number, military number, credit card number, and so on.

14-19. Using local gas utility rates, electric utility rates, coal costs, fuel oil costs, and wood costs, what would be the comparative cost of heating a five-room house in your home community for a winter season?

14-20. Discuss the advantages and disadvantages of having a TV channel show nothing but market quotations, except for brief commercials, during the time the New York stock market and the Chicago commodity market are open.

14-21. You are called to Alaska to consider the problem of public buildings that are sinking in permafrost due to warm weather. What might you do to solve this problem?

14-22. You are located on an ice cap. Ice and snow are everywhere but no water. Fuel and equipment are available. How can you prepare a well from which water can be pumped?

14-23. Assemble the following items: an ink bottle, a marble, a yardstick, an engineer's scale of triangular cross section, five wooden matches, a pocket knife, a candle, a pencil, and a key. Now, using as few of the objects as possible, balance the yardstick across the top "knife-edge" of the engineer's scale in such manner that soon after being released, and without being touched again, it unbalances itself.

14-24. Explain the operation of the rewind mechanism for the hand cord of a home gasoline lawnmower.

14-25. Devise a new method of feeding passengers on airplanes.

14-26. List the consequences of everyone being able to read everyone else's mind.

14-27. At current market values, determine the number of years that would be necessary to regain the loss of money (lost salary plus college expense) if one stayed in college one additional year to obtain a master's degree in engineering. What would be the number of years necessary to regain the loss by staying three years beyond the bachelor's degree to obtain a doctorate in engineering?

14-28. Estimate the number of policemen in (*a*) New York City, and (*b*) the United States.

14-29. Estimate the number of churches of all faiths in the United States.

14-30. Explain how the following work:

a. An automobile differential.
b. A toggle switch.
c. An automatic cutoff on gasoline pumps.
d. A sewing machine needle when sewing cloth.
e. A refrigeration cycle which does not depend upon electricity.

14-31. Using six equal-length sticks construct four equilateral triangles.

14-32. Estimate the number of aspirin tablets now available in the United States.

14-33. A cube whose surface area is 6 mi^2 is filled with water. How long will it take to empty this tank using a 1000 gal/min pump?

14-34. From memory sketch (*a*) a bicycle, (*b*) a reel-type lawn mower, (*c*) a coffee pot, (*d*) a saltwater fishing reel, and (*e*) a rifle.

14-35. Make something useful from the following items: a piece of corrugated cardboard 12 in. × 24 in., 6 ft of string, 3 pieces of chalk, 10 rubber bands, a small piece of gummed tape, 3 tongue depressors, 5 paper clips, and 7 toothpicks.

14-36. Propose some way to eliminate the need for bifocal glasses.

14-37. Design a device that can measure to a high degree of accuracy the wall thickness of a long tube whose ends are not accessible.

14-38. Design a man's compact travel kit that can be carried in the inside coat pocket.

14-39. Design a home-type sugar dispenser for a locality where the average rainfall is 100 in./yr.

14-40. Design a new type of men's apparel to be worn around the neck in lieu of a necktie.

14-41. Design a new type of clothespin.

14-42. Design a new fastener for shirts or blouses.

14-43. Design a personal monogram.

14-44. Design a device to aid federal or civil officers in the prevention or suppression of crime.

14-45. Design a highway system and appropriate vehicles for a country where gasoline is not obtainable and where motive power must be supplied external to the vehicle.

14-46. Design an electrical system for a home that does not receive its energy from a power company or a storage battery.

14-47. Design a device for weighing quantities of food for astronauts who are enroute to the moon.

14-48. Design a machine or process to remove Irish potato peelings.

14-49. Design a "black-eyed pea" sheller.

14-50. Design a corn shucker.

14-51. Design a trap to snare mosquitoes alive.

14-52. Design the "ideal" bathroom, including new toilet fixtures.

14-53. Design a toothpaste dispenser.

14-54. Design a woozle.

14-55. Design a device that would enable paralyzed people to read in bed.

14-56. Design a jiglike device that an amateur "do-it-yourself" home workman could use to lay up an acceptably straight brick wall.

14-57. Design a device to retail for less than $20 to warn "tail-gaters" that they are too close to your automobile.

14-58. Devise a system of warning lights connected to your automobile that will warn drivers in cars following you of the changes in the speed of your car.

14-59. You live in a remote community near the Canadian border, and you have a shallow well near your home from which you can get a copious supply of water. Although the water is unfit for drinking or irrigation, its temperature is a constant 64°F. Design a system to use this water to help heat your home.

14-60. Design and build a prototype model of a small spot welder suitable for use by hobby craftsmen. Prepare working sketches and make an economic study of the advisability of producing these units in volume production.

14-61. Design some device that will awaken a deaf person.

14-62. Design a coin-operated hair-cutting machine.

14-63. Design a two-passenger battery-powered Urbanmobile for use around the neighborhood, for local shopping center visits, to commute to the railway station, and so on. The rechargeable battery should last for 60 mi on each charge. Provide a complete report on the design, including a market survey and economic study.

14-64. Design some means of visually determining the rate of gasoline consumption (mi/gal) at any time while the vehicle is in operation.

14-65. Design a device to continuously monitor and/or regulate automobile tire pressures.

14-66. Design a novel method of catching and executing mice that will not infringe the patent of any other known system now on the market.

14-67. Design a new toy for children ages 6 to 10.

14-68. Design a device to replace the conventional oarlocks used on all rowboats.

14-69. Devise an improved method of garbage disposal for a "new" city that is to be constructed in its entirety next year.

14-70. Design and build a simple device to measure the specific heat of liquids. Use components costing less than $3.

14-71. Design for teenagers an educational hobby kit that might foster an interest in engineering.

14-72. Design a portable traffic signal that can be quickly put into operation for emergency use.

14-73. Design an egg breaker for kitchen use.

14-74. Design an automatic dog-food dispenser.

14-75. Design a device to mix body soap in shower water automatically as needed.

14-76. Design an improved keyholder.

14-77. Design a self-measuring and self-mixing epoxy glue container.

14-78. Design an improved means of cleaning automobile windshields.

14-79. Design a noise suppressor for a motorcycle.

14-80. Design a collapsible bicycle.

14-81. Design a tire-chain changer.

14-82. Design a set of improved highway markers.

14-83. Design an automatic oil-level indicator for automobiles.

14-84. Design an underwater means of communication for skin divers.

14-85. Design a means of locating lost golf balls.

14-86. Design a musician's page turner.

14-87. Design an improved violin tuning device.

14-88. Design an attachment to allow a motorcycle to be used on water.

14-89. Design a bedroll heater for use in camping.

14-90. Design an improvement in backpacking equipment.

14-91. Design an improved writing instrument.

14-92. Design a means of disposing of solid household waste.

14-93. Design a type of building block that can be erected without mortar.

14-94. Design a means for self-cleaning of sinks and toilet bowls.

14-95. Design some means to replace door knobs or door latches.

14-96. Design a simple animal-powered irrigation pump for use in developing nations.

14-97. Design a therapeutic exerciser for use in strengthening weak or undeveloped muscles.

14-98. Design a Morse-code translator that will allow a deaf person to read code received from radio receivers.

14-99. Design an empty-seat locator for use in theaters.

14-100. Design a writing device for use by armless people.

14-101. Design and build an indicator to tell when a steak is cooked as desired.

14-102. Design a device that would effectively eliminate wall outlets and cords for electrical household appliances.

14-103. Design the mechanism by which the rotary motion of a 1-in.-diameter shaft can be transferred around a 90° corner and imparted to a $\frac{1}{2}$-in.-diameter shaft.

14-104. Design a mechanism by which the vibratory translation of a steel rod can be transferred around a 90° corner and imparted to another steel rod.

14-105. Design a device or system to prevent snow accumulation on the roof of a mountain cabin. Electricity is available, and the owner is absent during the winter.

14-106. Using the parts out of an old spring-wound clock, design and fabricate some useful device.

14-107. Out of popsicle sticks build a pinned-joint structure that will support a load of 50 lb.

14-108. Design a new device to replace the standard wall light switch.

14-109. Design and build a record changer that will flip records as well as change them.

14-110. Design a wheelchair that can lift itself from street level to a level 1 ft higher.

14-111. Design a can opener that can be used to make a continuous cut in the top of a tin can whose top is of irregular shape.

14-112. Design and build for camping purposes a solar still that can produce 1 gallon of pure water per day.

14-113. Design, build, and demonstrate a device that will measure and indicate 15 sec of time as accurately as possible. The device must not use commercially available timing mechanisms.

14-114. Few new musical instruments have been invented within the last 100 years. With the availability of modern materials and processes, many novel and innovative designs are now within the realm of possibility. To be marketable over an extended period of time such an instrument should utilize the conventional diatonic scale of eight tones to the octave. It could, therefore, be utilized by symphonies, in ensembles, or as a solo instrument using existing musical compositions. You are the chief engineer for a company whose present objective is to create and market such a new instrument. Design and build a prototype of a new instrument that would be salable. Prepare working drawings of your model together with cost estimates for volume production of the instrument.

14-115. Design some means of communicating with a deaf person who is elsewhere (such as by radio).

14-116. For a bicycle, design an automatic transmission that will change gears according to the force applied.

14-117. Design a "decommercializer" that will automatically cut out all TV commercial sounds for 60 sec.

14-118. Design a solar-powered refrigerator.

14-119. Design a small portable means for converting seawater to drinking water.

14-120. Design a fishing lure capable of staying at any preset depth.

14-121. Design an educational toy that may be used to aid small children in learning to read.

14-122. Design some device to help a handicapped person.

14-123. Design a heating and cooling blanket.

14-124. Design a portable solar cooker.

14-125. Design a carbon monoxide detector for automobiles.

14-126. Design a more effective method for prevention and/or removal of snow and ice from military aircraft.

14-127. Design a "practical" vehicle whose operation is based on the "ground-effect" phenomenon.

14-128. Design a neuter (neither male nor female) connector for quick connect and disconnect that can be used on the end of flexible hose to transport liquids.

14-129. Design an electric space heater rated from 10,000 to 50,000 Btu/hr for military use in temporary huts and enclosures.

14-130. There is need for a system whereby one device emplaced in a hazardous area (minefield or other denial area) would interact with another device issued to each soldier, warn him of danger, and send guidance instructions for him to avoid or pass through the area of safety. Design such a system.

14-131. Develop some method to rate and/or identify the presence of rust spots when coatings fail to protect metal adequately. Present visual methods are unreliable and variable in results.

14-132. Develop a system whereby diseases of significance could be diagnosed rapidly and accurately.

14-133. Design a strong, flexible, lumpless, V-belt connector.

14-134. Design an inexpensive system for keeping birds out of ripening fruit trees.

14-135. Design a replacement for the paper stapler which will not puncture the paper.

14-136. Design and construct a vehicle that will carry a payload across the classroom floor as far as possible.

Specifications

1. The vehicle is to be powered by a conventional spring-activated household mousetrap about 4.75 cm × 10 cm.
2. The maximum dimensions of the vehicle are 20 cm long, 15 cm wide, and 15 cm high.
3. Payload must be in one piece and removable for weighing. The maximum dimensions for the payload are 5 cm × 5 cm × 5 cm.
4. The amount of the payload to be carried, materials of construction, and vehicle design are your responsibility.

Testing

A maximum of 1 min will be allowed to position the vehicle and prepare it for the run. The vehicle must cover at least a distance of 5 m or it will be disqualified.

Evaluation

A total of five quantities are included in the evaluation. They are:

P = payload weight (newtons)
L = total distance traversed by the vehicle from the starting line (meters)
T = total elapsed time to traverse 5 m (seconds)
W = vehicle weight (newtons)
C = cost of the vehicle at $0.10 per newton

The overall value of the vehicle will be determined by the following formula:

$$V = \frac{P \times L^2}{T \times W \times C}$$

To complete the individual design project, after testing prepare a report containing the calculation of the value of your vehicle, a copy of the preliminary sketch, and a brief analysis of your reasoning for the particular design you used in the construction. Discuss the advantages and limitations of your design. How might the design be improved?

14-137. Design a new type of fishing lure.

14-138. Design and construct a powered, self-controlled surface vehicle that will negotiate a "figure 8" course on a smooth, horizontal surface whose dimensions are 1 m × 2 m.

Specifications

Construct the vehicle from the following materials:

1. Balsa wood and/or cardboard ≤5.0 mm thick, not to exceed 1500 cm^2.
2. Cotton thread (no nylon!) ≤20 gauge, not to exceed 30 cm in length.
3. Balsa wood cement; not more than one small tube.
4. Maximum of four standard-size paper clips.
5. Maximum of four circular rubber bands. The original width of each band must not exceed 4 mm, and the unstretched length of the elongated oval must not exceed 10 cm.

Cost Schedule

Balsa wood or cardboard	at $1/cm^3
Cotton thread	at $3/cm
Rubber bands	at $50 each
Standard office paper clips	at $20 each

Quantity	Cost
Total cost	

Evaluation

Three quantities are to be evaluated in determining the value (V) of the vehicle. They are:

W = weight of vehicle (newtons)
C = cost of vehicle (dollars)
f = fraction of "figure 8" course successfully negotiated

$$V = \frac{f \times 10^6}{W + C}$$

To complete the individual design project, after testing prepare a report containing the calculation of the value of your vehicle, a copy of the preliminary sketch, and a brief analysis of your reasoning for the particular design you used in the construction. Discuss the advantages and limitations of your design. How might the design be improved?

14-139. Design a bridge to span 450 mm between supports, with a 30-mm-wide roadway and having a vertical road clearance of at least 20 mm. (Suspension bridges must have an actual roadway and vertical "tower" supports.) Construct the bridge to support a load that will be applied to the center of its roadway, midway between the two supports.

Construction Materials

Bridge construction materials are limited to the following:

Ordinary soda straws	at $0.01/mm
Plain cardboard (maximum thickness 2 mm)	at $0.01/m^2
Corrugated cardboard (maximum thickness 6 mm)	at $0.02/m^2
Cotton string (no nylon)	at $0.01/mm
Ordinary round, wooden toothpicks	at $0.50 each
Standard office paper clips	at $1.00 each

Quantity	Cost
Total cost	

Note: Adhesive paste, glue, epoxy, casein, etc. (no tape) can be used to join the materials at no charge. However, this material will serve only the function of fasteners to connect structural members.

Evaluation

Three quantities are to be evaluated in determining the value (V) of the bridge. They are:

W = weight of the bridge (newtons)

L = load applied at center of bridge that will deflect the center of the bridge 40 mm—or to failure, if that occurs first (newtons)

C = cost of construction materials (dollars)

$$V = \frac{L \times 10^2}{W + C}$$

To complete the individual design project, after testing prepare a report containing the calculation of the value of your bridge, a copy of the preliminary sketch, and a brief analysis of your reasoning for the particular design you used in the construction. Discuss the advantages and limitations of your design. How might the design be improved?

14-140. Design a nozzle or device that will utilize a round rubber balloon to drive a boat, transporting a mass of your choice, selected from those made available by the instructor. The channel of water used to float the boat is 5 in. wide, 2 in. deep, and 120 in. long.

Boat Construction

The boat may be made of any material and is to be constructed as shown in Figure [14-P140]. The balloon shall be a new, ordinary rubber balloon. The boat may not be altered in any way (including driving nails or wires into the surface). However, tape, rubber bands, and so on, may be used to attach a balloon supporting structure. The propulsion device must be an original design. It will be attached to the boat via the 1 in. × 2 in. × $\frac{1}{2}$ in. slot on the stern.

Testing

The balloon must be inflated to 7 in. in diameter at the start of the test (use a go–no go wire gauge for measurement). Allow a maximum of 2 min for a test run, including preparation for test and actual test. The boat may be placed anywhere in the test trough at start. Use a pencil as a gate to block the boat from moving prematurely. Air is to be leaving balloon before the gate is lifted.

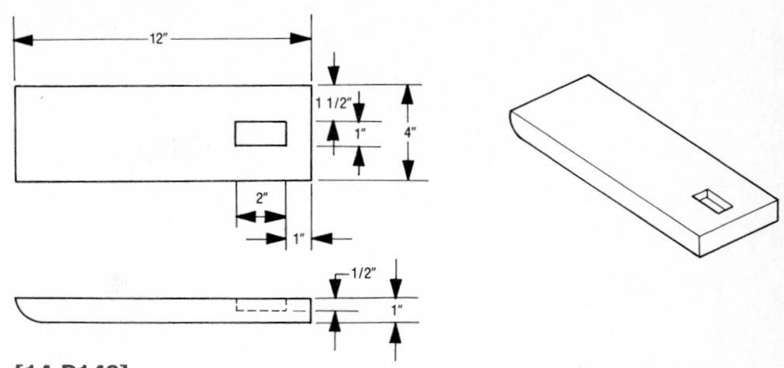

[14-P140]

Evaluation

Three quantities are to be evaluated in determining the design performance (P). They are:

M = (number of steel washers carried) × $20/980 g_m$

L = distance (cm) front of boat travels from time "zero" to time T

T = time (sec) from point where gate is lifted to point when all the air is exhausted from the balloons, or when the boat strikes the end of the trough—whichever occurs first.

$$P = \frac{ML}{T}$$

To complete the individual design project, after testing prepare a report containing the calculation of the value of your boat, a copy of the preliminary sketch, and a brief analysis of your reasoning for the particular design you used in the construction. Discuss the advantages and limitations of your design. How might the design be improved?

14-141. Design a structure that will fit around an object having a parabolic profile. The specific dimensional constraints are as follows:

- A The structure must be designed so that a parabolic profile, 4 in. × 6 in., shown at A in Figure [14-P141], may pass completely through the space between the structure and the loading cable plane.[3]
- B, C, D Overall dimensions of the structure cannot exceed 10 in. × 10 in. × 4 in.
- E Maximum thickness of the structure cannot exceed 3 in.
- F Extension of the base of the structure beyond the loading cable cannot exceed 3 in.
- G Overhand of the top of the structure beyond the loading cable cannot exceed 1 in.
- H, I Dimensions H and I are left to the discretion of the designer, provided that they do not make the overall height of the structure exceed 10 in.

[3] Adapted from a problem used at Carnegie-Mellon University.

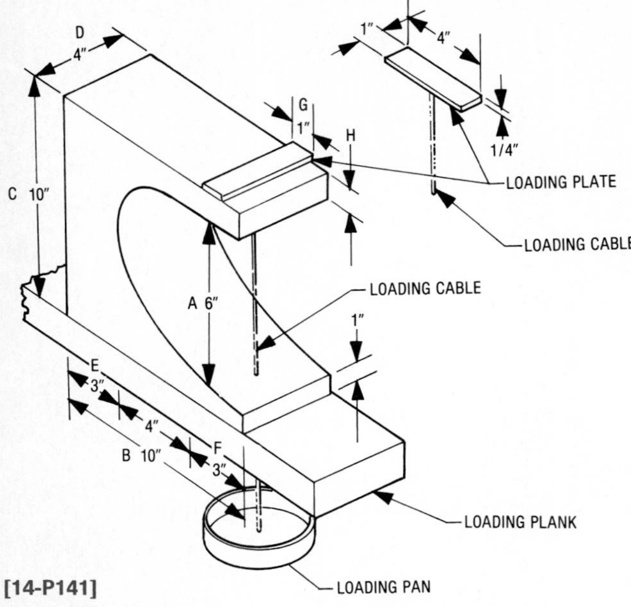

[14-P141]

Materials

Only balsa wood and glue may be used as materials. Maximum cross section for sticks is $\frac{3}{4}$ in. × 1 in. Maximum cross section for sheets is $\frac{3}{8}$ in × 4 in. The use of glue will be restricted to areas actually joining two pieces of wood. The wood, other than at joints, may not be coated or impregnated with lacquer, shellac, paper, or any other material. The use of glued laminations of sheets of balsa wood is prohibited. There may be no concealed elements in the structure.

Testing

The load will be applied in a vertical direction by a cable attached to the center of a 4 in. × 1 in. × $\frac{1}{4}$ in. steel loading plate, which will rest on top of the structure. The loading cable will pass through an opening provided in the base of the structure and through an opening in the loading plank. The base of the structure must be self-supporting and rest on the loading plank. The top of the structure must be designed to accommodate the loading plate with a cable attached so that it may be easily inserted or placed on the structure without the use of any clamps, screws, glue, and so on. Failure under load will be considered to be the point where the structure collapses, can no longer carry additional load, or the specified parabolic profile will no longer clear the structure and the plane of the loading cable. The load carried by the structure will include the weight of the loading plate and cable.

Evaluation

The quality of performance (P) of the structure will be determined by two factors, as follows:

L = load necessary for failure (lb)
W = weight of structure (lb)

$$P = \frac{L}{W}$$

To complete the individual design project, after testing prepare a report containing the calculations of the value of your structure, a copy of the preliminary sketch, and a brief analysis of your reasoning for the particular design you used in the construction. Discuss the advantages and limitations of your design. How might the design be improved?

14-142. Design, construct, and test a wind-powered unit that will lift a stationary $\frac{1}{3}$-kg weight a distance of 1 meter [14-P142]. The wind source is a medium-sized electric fan, not to exceed $\frac{1}{8}$ hp. Your design should have a base capable of being held on the table with a laboratory C-clamp. The fan will be positioned 1 meter from the edge of the test table, as shown. Your design must receive its energy from the air movement of the fan.

Construction Materials

Any materials may be selected for your design, including the type of string that will be used as the lift cable. However, you cannot use any industrial-fabricated components for your design.

Evaluation

The design performance (P) depends on three factors, as follows:

h = height of weight lifted (meters)
t = time of the lift (seconds)
w = weight of the design, including string (newtons)

$$P = \frac{h}{tw}$$

To complete the individual design project, after testing prepare a report containing the calculation of the value of your unit, a copy of the preliminary sketch, and a brief analysis of your reasoning for the particular design you used in the construction. Discuss the advantages and limitations of your design. How might the design be improved?

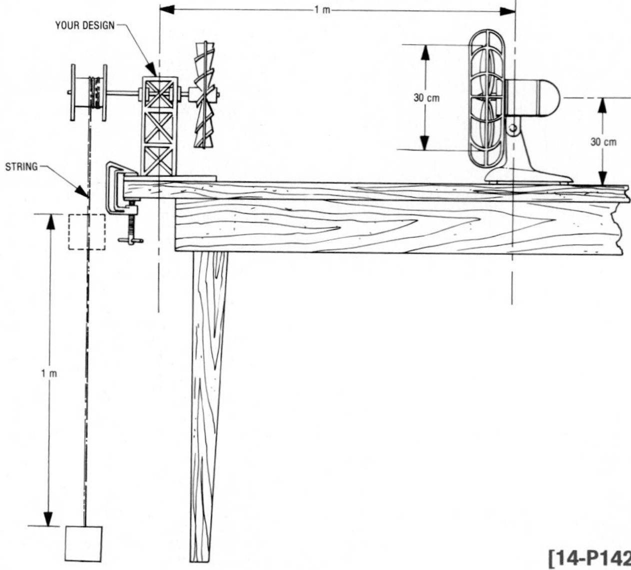

[14-P142]

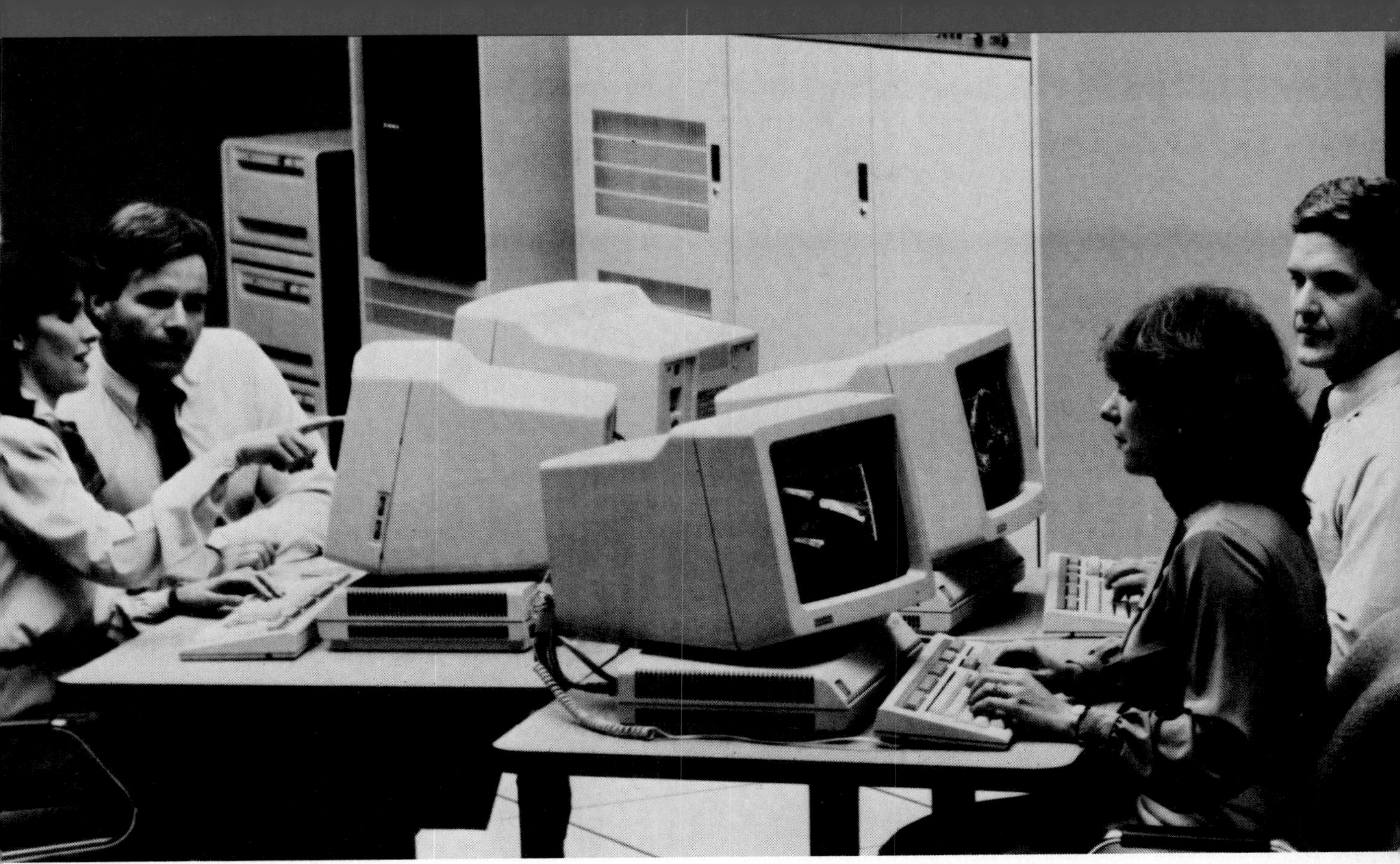

Section Four

Computers in Engineering

Chapter 15

Digital Computers

THE HISTORY OF COMPUTING

Numbering Systems

Numbering systems have been in evidence since man began to inhabit the earth. One of the simplest representations of numbers was the use of tally sticks [15-1]. These pieces of wood were scored with notches to record a certain count, and it was usual to split the notched piece into two parts, with each person involved in a transaction keeping one part.

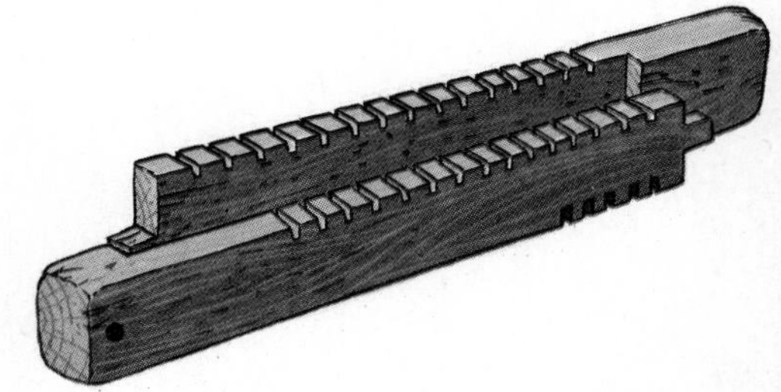

[15-1] Tally sticks were frequently notched to record a count and then split in halves of which each party kept one.

Translations of Egyptian hieroglyphs have revealed a numbering system that may have been in use at the time of the design of the great pyramids. This system used seven symbols to represent powers of 10 and these were combined right to left in a desired arrangement [15-2 and 15-3]. In this system the number $153{,}864 = +1 \times 10^5 + 5 \times 10^4 + 3 \times 10^3 + 8 \times 10^2 + 6 \times 10^1 + 4 \times 10^0$.

[15-2] Egyptian hieroglyphic symbols.

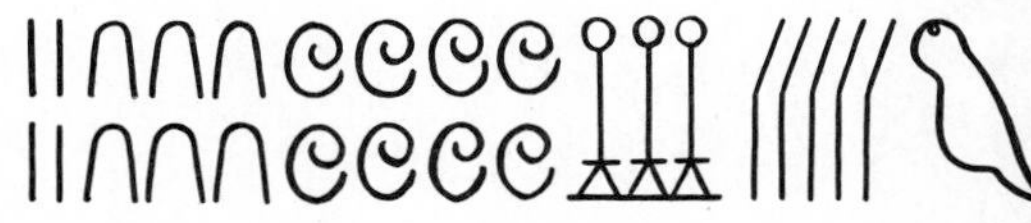

[15-3] The number represented is 153,864.

Roman Numerals. By the thirteenth century, the Roman numeral system was in use throughout western Europe and was reasonably well suited for the mathematical operations of addition and subtraction. This system was based on the use of the letters I, V, X, L, C, D, and M to represent symbolically the numbers 1, 5, 10, 50, 100, 500, and 1000,[1] respectively. As in the Egyptian system, the value was computed as the sum of the values in a string of symbols. But unlike the Egyptian system, the symbols were written with the large symbols at the left and smaller at the right. For example, VI would mean 5 + 1, or 6. Later, modifications to the system permitted a lower-value symbol to appear at the left of a higher-valued symbol. This indicated that the quantity represented was to be subtracted from the overall value. Thus IV would mean $5 - 1 = 4$. Some examples of the use of this system[2] are:

MCMLIX is 1959.

MCDXCII is 1492.

MCMLXXXVII is 1987.

The concepts of zero and negative numbers were not incorporated into the Roman system. In fact, they were not understood. Also, fractions, although used since ancient Egyptian and Babylonian times, were not conveniently handled in the Roman system. And, of course, the operations of multiplication and division were

[1]Originally, the Romans wrote ⊂|⊃ to represent 1000.

[2]W. Gellert et al., eds., *The VNR Concise Encyclopedia of Mathematics* (New York: Van Nostrand Reinhold, 1975), p. 19.

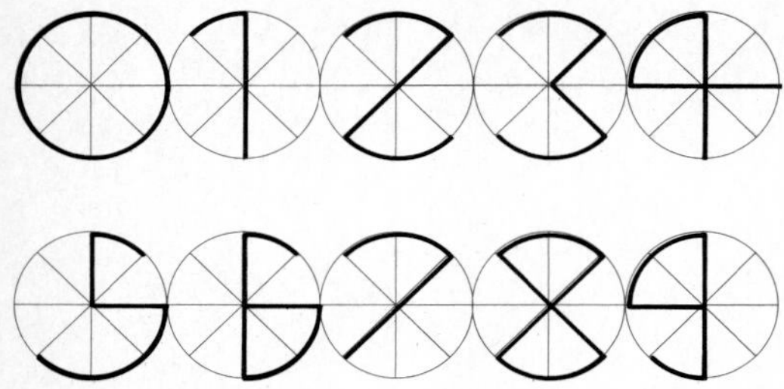

[15-4] The Arabic numbering system.

not easily done. Most of western Europe had no idea that the routine counting and measuring schemes were in any need of revision.

Stimulated by the Renaissance, the rapid development of science and engineering led quite naturally to arithmetic operations being required on a much larger scale. Such calculations involved multiplication, division, exponentiation, root extraction, and uses of trigonometric functions.

Mathematical developments during this period provided a foundation for many complicated types of computation. For example, sines and cosines could be computed to any desired degree of accuracy using the Maclaurin or Taylor series of mathematics, but the task of doing so was tedious, time consuming, and error prone. Mistakes in calculation were easy to make, difficult to correct, and sometimes remained undiscovered for long periods of time.

Arabic Numerals. The Arabic system of numbers had been introduced in Italy in the thirteenth century and was slowly making its way across Europe. It had been in use in the Arab world since about the ninth century, apparently having originated in India.[3] This system introduced the ideas of zero, negative numbers, and decimal fractions. But more important still, it was based entirely on the idea of place notation—the idea that the position of a digit in a sequence of digits, as well as its value, could be used in forming a number. Thus, with the Arabic system [15-4], numbers of extremely large or small size could be represented compactly and to any desired degree of significance. Further, with the Arabic system, addition and subtraction by pencil and paper became much more convenient than it had ever been.

Computational Techniques

The abacus [15-5] has been in use for at least 5000 years. Prior to the introduction of the Arabic number system it was widely used in computing sums and differences, so day-to-day business activities did not suffer from lack of simple computation capability. When the Arabic system became widespread the abacus began to disappear from much of the world scene. The last outpost for the use of this computational aid is east Asia, where it is still used daily by shopkeepers and merchants.

Even though the Arabic system allowed a more compact notation and speeded up hand calculations of products and quotients, better computational methods were still needed. These needs led to a steady development in computational techniques and schemes. Since addition and subtraction could be performed quite easily, for many years it was desirable to transform the operations of multiplication and division into addition and subtraction by the use of logarithms

[15-5] The abacus was one of man's first calculating machines.

[3]There is more than one theory about the origin of our own numbers, commonly referred to as Arabic or Hindu-Arabic. One of the more interesting has been related by Jan J. Tuma, Professor of Engineering, Arizona State University. According to this version, an Arabic prince is said to have consulted an Indian seer to obtain a more functional number system. The Wise Man reasoned that perfection is a circle and for this reason all parts of a perfect numbering system would be made of parts of a circle. He offered the system shown in Figure [15-4].

(developed by John Napier about 1600). The computation and publication of extensive mathematical tables became very important, and special techniques were developed for interpolating within the tables to get required values.

The invention of logarithms also led to computational devices known as *slide rules*. Slide rules were *analog* computational devices using the fundamental idea of logarithms. (Analog devices can be distinguished from *digital* devices since they use physical quantities such as length, angle, or voltage as *analogies* for arithmetic operands. *Digital* devices, in effect, manipulate numeric representations or *digits* of the operands.) In the case of slide rules, all operands were represented either as length on appropriately calibrated sticks [15-6] or as angles on circular disks.

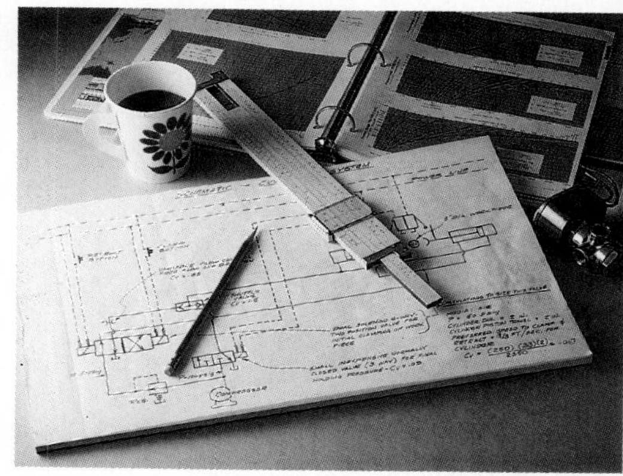

[15-6] The slide rule, now essentially replaced by the hand-held calculator, was the primary computational tool of the engineer for decades.

Slide rules became the essential tool of computation for engineers and scientists and for a great number of years were used extensively when approximately three significant digits[4] of computational accuracy were required in multiplication, division, and raising of powers. However, the user was required to record intermediate results when any additions and subtractions were needed. Engineers or scientists requiring more accuracy than that available from slide rule calculations had to fall back on the use of logarithms or of mechanical calculators to get the required accuracy.

Digital Computers

We turn briefly now to the history of digital computers, for we will devote the remainder of this chapter, as well as the next four, to the operation and capabilities of these machines. Although it is difficult to pinpoint exactly when such machines began to develop, it seems certain that it started sometime in the seventeenth century with the work of Blaise Pascal (1623–1662) and Gottfried Wilhelm Leibniz (1646–1716). Although both of these men are, perhaps, better known for other contributions to science and mathematics,[5] both worked to develop computing machines. Their purpose was to reduce the drudgery of lengthy calculations that were becoming more and more a part of their society.

Pascal pursued the elusive dream in order to help his father function in a job as a government tax assessor. Leibniz, apparently an eternal optimist, devoted attention to calculators because he felt, "It is unworthy of excellent men to lose hours like slaves in the labor of calculation which could safely be relegated to anyone else if machines were used."

Perhaps even more than the work of Pascal and Leibniz, the efforts of Charles Babbage (1792–1871) were exceptionally significant in the birth of computers. Babbage set out to build what he called the *difference engine*. He had hoped that it would relieve people of the repetitive and time-consuming calculations of such things as constructing tables of logarithms.

[15-7] The first machine to be controlled by punched cards was Falcon's loom (1728).

[4]Significant digits and significant figures are discussed in Chapter 9.

[5]It was Leibniz who perfected the binary numbering system consisting only of 0's and 1's. It was so inefficient for manual use that it lay idle for nearly three centuries before being revived for use by computers. We talk more about binary numbers and their use later in this chapter.

[15-8] Charles Babbage's difference machine was a forerunner of modern computing systems.

[15-9] Ada Lovelace, the daughter of poet Lord Byron, developed some of the first software for computing systems.

In 1820, he constructed a pilot model [15-8] which, two years later, helped him to secure funding for a prototype from the British government—the first known government funding of computer science. But the larger machine jammed easily, became inoperable regularly, and was largely unsuccessful.

The Analytical Engine. Undaunted by the fact that his mechanical difference engine was not successful, Babbage kept working. In fact, he spent the rest of his life trying to develop the *analytical engine,* which would have been the first true computer, although a mechanical one. It could carry out calculations, store the results, and print the output. Even more, its instructions could be changed. That is, it could be programmed. It is noteworthy that the device "read" its inputs by interpreting holes that had been punched into cards.

Babbage's difference engine had attracted the support of a young lady by the name of Augusta Ada Byron Lovelace [15-9], a mathematician and the daughter of the poet Lord Byron. Her interest, after first seeing this machine, was aptly described by a friend:[6]

> While the rest of the party gazed at this beautiful instrument with the same sort of expression and feeling that some savages are said to have shown on first seeing a looking glass or hearing a gun, Miss Byron, young as she was, understood its workings and saw the great beauty of the invention.

Babbage and Lovelace joined forces in developing Babbage's newly conceived analytical engine. Babbage became the hardware developer while Lovelace played the role of the software developer. In spite of her vices of opium and gambling, Lovelace was at least 100 years ahead of her time in understanding how computers might function. Among the concepts she envisioned, but which were not to be realized in her lifetime, were the concept of the subroutine or a set of reusable program steps (discussed in Chapter 17), looping or executing a set of instructions over and over (discussed in Chapter 16), and branching (also discussed in Chapter 16). She even speculated about whether some form of intelligence could ever be cultured in computing machines.

In trying to keep their project alive, Babbage and Lovelace apparently considered sales of the machine for such uses as playing tic-tac-toe and chess, and betting the ponies. But alas, the machine was not to be. The technology of the times could not produce the more than 50,000 precision parts that were needed to make the analytical engine a success.

The hardware developer lived to the age of 79, but the software specialist died when she was only 36. Ada Lovelace was honored posthumously by the Pentagon, which named a new "superlanguage" after her in 1970. *Ada* is now a registered trademark of the U.S. government.

Many people followed in the footsteps of Babbage, several having been enticed by Lovelace's writings entitled, "Observations on Mr. Babbage's Analytical Engine." As technology advanced and more and more minds were put to work, the analytical engine be-

[6]See: The Lunch Group and Guests, *Digital Deli* (New York: Workman, 1984).

came a reality. In fact, mechanical calculators became so reliable and serviceable that they became a necessity for anyone doing calculations. Such machines were in widespread use into the early 1970s.

The work of numerous other people contributed to the development of the modern digital computer, but we have space to mention only some of it. Much of this work was funded by governmental agencies, particularly the armed services during the time of war. The impetus for this support was the need to better determine the flight of artillery shells and to break secret communication codes of the enemy.

The Hollerith Counting Machine. One significant, entirely nonmilitary, milestone in computer development was that of the Hollerith counting machine [15-10]. In the early 1890s, Herman Hollerith won a design competition for a counting machine to be used by the U.S. Census Bureau. His counter used punched cards that were a little larger than a dollar bill. Various data from the census were encoded on the cards in the form of punched holes. The holes in the cards allowed electrical circuits to be completed as the cards were sent through Hollerith's machine. The electrical pulses that were generated triggered counters, thus indexing the counters to the next number.

With Hollerith's machine, the 1890 census information on 62,622,250 people was compiled in a mere one month. This was an astounding increase in speed over the 7 years that it had taken to compile the 1880 census. In 1896, Hollerith left the Census Bureau to form the Tabulating Machine Company, which later became the Computing Tabulating and Recording Company. In 1913, Thomas

[15-10] Herman Hollerith designed a counting machine for the U.S. Census Bureau in the early 1890s.

[15-11] The Mark I, developed at Harvard University in 1944, was the first fully automatic electromechanical computer.

J. Watson took over the helm of Hollerith's company and, in 1924, transformed it into the International Business Machines Corporation, or IBM.

The Mark I. The first fully automatic electromechanical computer was the Mark I, developed in 1944 at Harvard University under the direction of Howard Aiken. The project had the support of both the U.S. Navy and the emerging IBM Corporation. The purpose of the Mark I was to calculate the trajectories of artillery shells for the armed services.

The Mark I consisted of over 1 million components, 500 miles of electric wire, and 3000 noisy electromechanical switches or relays [15-11]. It was 51 ft long and 8 ft wide and, as Grace Murray Hopper, a newly graduated mathematician assigned to the project by the Navy, pointed out, "You could walk around inside her."

The computer was programmed by changing the coded holes punched into a paper tape read by the machine. It could perform one multiplication per second or add three eight-digit numbers per second.

The ENIAC. The task of computing artillery firing tables was so important to the armed services during World War II, that the U.S. government pursued several projects concurrently with the Mark I development. At the University of Pennsylvania, John W. Mauchly, J. Presper Eckert, Jr., and J. G. Brainerd designed the Electronic Numerical Integrator and Computer (ENIAC), which used the new technology of *vacuum tubes* instead of electromechanical relays.

Although this computer was a giant in physical size [1-23], consisting of forty 2-ft by 4-ft panels, it had minuscule capability compared to even the smallest of home computers today in spite of its 18,000 vacuum tubes. In fact, it would have been hard pressed to outperform today's hand-held calculators. It could perform 300 multiplications per second or 5000 additions per second, considerably faster than the Mark I. The ENIAC could store only 700 bits (we discuss bits later in this chapter) in programmable memory[7] and 20,000 bits in read-only memory.[8] Its speed, considered lightning-like at its time, was a factor of 20 to 40 times slower than the typical home computers that are available today.

The ENIAC took more than three years to build and the war was over by the time it solved its first problem. Shortly after the ENIAC was completed, Mauchly and Eckert formed their own company, the Electric Control Corporation, the name of which they later changed to the Eckert-Mauchly Computer Corporation.

In 1949, Grace Murray Hopper joined with Eckert-Mauchly Computer Corporation, which was soon to start work on the UNIVAC computer. It was at this time that Hopper wrote the first *compiler,* which dramatically changed the way computers were programmed (we discuss compilers in Chapter 16).

In 1955, the Eckert-Mauchly Computer Corporation merged with the Sperry Corporation to form the Sperry-Rand Corporation and the *UNIVAC* computer became the first commercially available computer.

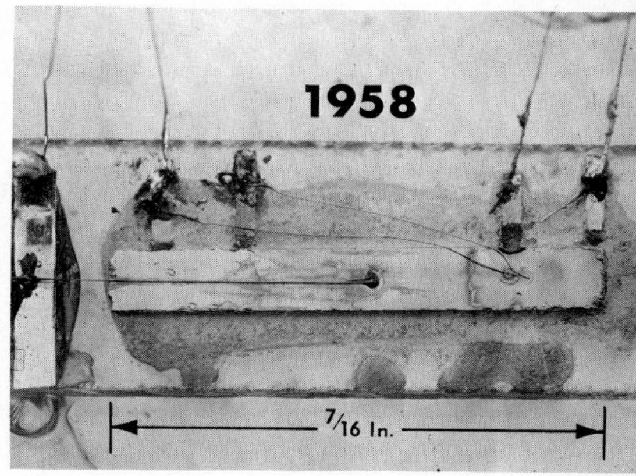

[15-12] The first transistor to be used as a design component was a crude device by today's manufacturing standard, but it worked . . . and it opened a whole new world of solid-state electronics.

The Appearance of the Transistor. The age of miniaturization began with the invention of the *transistor* by William Shockley, John Bardeen, and Walter Brattain at Bell Laboratories [1-24].

Transistors, made of materials called semiconductors, act as small, quiet, low-power-consuming switches. As their prices dropped and availability increased, they replaced the vacuum tubes and electromechanical relays in computers, making possible smaller machines that required much less electrical power to operate [15-12].

Even though the price of transistors dropped, the problem of inserting them into circuit boards in a production environment was difficult to solve. Also, their size, although less than that of vacuum tubes, remained the chief characteristic limiting the physical shrinking of the computer.

The Integrated Circuit. Significant advances in miniaturization were made with the development of the *integrated circuit* in 1963 [15-13]. This technology allowed a large number of transistors and interconnecting circuits to be placed on a small semiconductor "chip." This eliminated the need for hand insertion of individual transistors on the circuit boards and provided for shortened current paths between the transistor switches. These improvements paved the way for smaller and faster computers.

LSI and VLSI. By the mid-1970s, technology had advanced to the point that engineers were able to place thousands of circuits on a single semiconductor chip. Although the numbers are not absolute, any chip with more than about 1000 transistors on it is called a *large-scale integration,* or LSI, chip. It was an LSI chip of about 6000 circuits that spawned the hand-held calculator, for example. The digital watch was another product that made use of LSI chips.

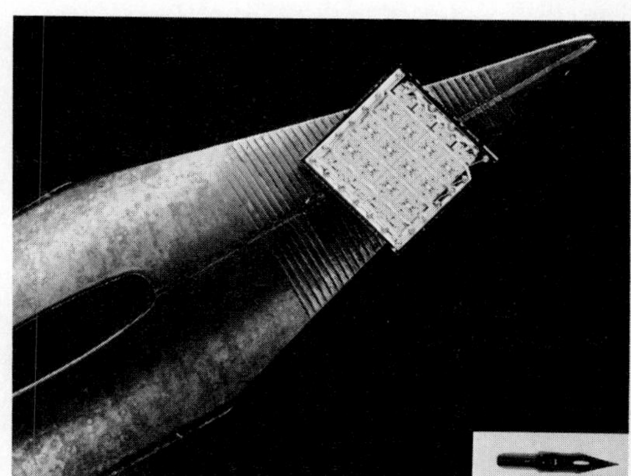

[15-13] In 1969 placing 64 complete electronic memory circuits on a chip of silicon was considered to be a major accomplishment. Such a chip is shown for size comparison on the nib of a pen.

[7]We will call this type of memory RAM, for random access memory.

[8]We will call this type of memory ROM, for read-only memory.

[15-14] Wrestling with a computer his own size, this ant attacks an NSC-3200 32-bit microprocessor with computing functions requiring an entire roomful of equipment in the 1960s.

When the number of transistors on each chip is between 30,000 and 1,000,000, the technology is generally referred to as *very large scale integration,* or VLSI. Once integration efforts reached that level, the computer on a chip, or the *microprocessor,* was inevitable [15-14]. By 1980, as many as 300,000 transistors on chips less than 1 mm^2 were available. *Ultra large scale integration,* or *ULSI,* is a term used to identify those semiconductor chips with more than 1 million transistors. These should be available in mass production by the year 2000.

What is the lesson to be learned from this history? Certainly, one lesson must be that it is naive to project a limit to this technology. All previous projections have underestimated our advances. For example, it is said[9] that Aiken, the designer of the Mark I computer discussed above, once stated that just six of his computers would be enough to fulfill the world's computational needs.

However, there is a pattern that emerges. It is the historical pattern of computers being developed, usually at a significant expense, but then quickly being surpassed and replaced by much more powerful computers—computers followed very quickly by more capable computers. The pace of this development quickens with time [15-15].

People who disdain computers, probably wish that Aiken's prediction had come true. However, computers are here to stay—they are a fact of life. What is necessary now is a concerted effort to make them simpler to operate so that we can all participate in, and benefit from, their use. We need to obscure the advances in technology from end users so that the users can continue productive work while technology is permitted to advance at its ever-increasing pace.

COMMON ELEMENTS OF ALL COMPUTING MACHINERY: HARDWARE, SOFTWARE, AND FIRMWARE

I never could make out what those damned dots (decimals) meant.

Lord Randolph Spencer Churchill
Quoted in Lord Randolph Churchill II, *by Winston Churchill.*

Can you imagine a computing system that would accept your information, perform the required calculations, but never provide you the answer? Or, how many computers might be sold that could perform intricate and difficult calculations but not allow you to input unique information for your own problem? These questions and others lead us to the conclusion that there must be several important components for any useful computing system, whether it be the human brain, a mechanical invention, or an electronic device. These are:

- A device for inputting information (numbers, text, sight, taste, hearing, etc.).
- A device for storing information (including the input data, intermediate numbers generated during calculations, and the final results).
- A device to process or execute all the necessary calculations.

[9] J. Presper Eckert, "Who Needs Personal Computers?" in *Digital Deli,* by The Lunch Group and Guests (New York: Workman, 1984). Eckert does say that Aiken later retracted the statement.

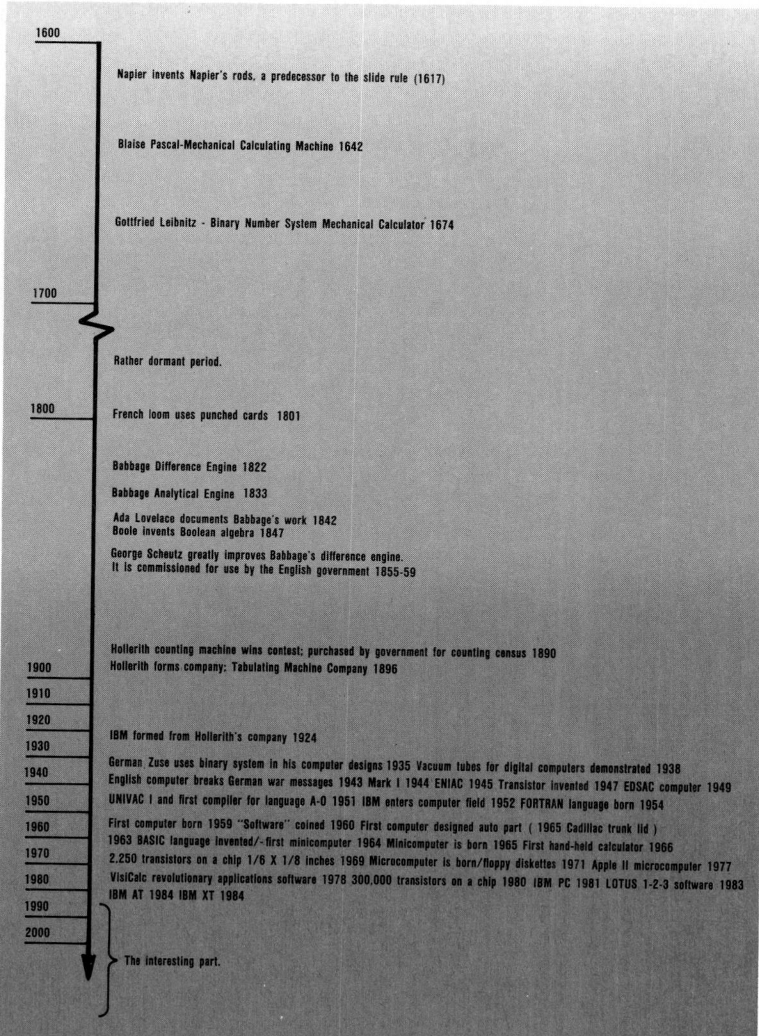

1600
Napier invents Napier's rods, a predecessor to the slide rule (1617)
Blaise Pascal-Mechanical Calculating Machine 1642
Gottfried Leibnitz - Binary Number System Mechanical Calculator 1674
1700
Rather dormant period.
1800
French loom uses punched cards 1801
Babbage Difference Engine 1822
Babbage Analytical Engine 1833
Ada Lovelace documents Babbage's work 1842
Boole invents Boolean algebra 1847
George Scheutz greatly improves Babbage's difference engine.
It is commissioned for use by the English government 1855-59
Hollerith counting machine wins contest; purchased by government for counting census 1890
1900
Hollerith forms company: Tabulating Machine Company 1896
1910
1920
IBM formed from Hollerith's company 1924
1930
German Zuse uses binary system in his computer designs 1935 Vacuum tubes for digital computers demonstrated 1938
1940
English computer breaks German war messages 1943 Mark I 1944 ENIAC 1945 Transistor invented 1947 EDSAC computer 1949
1950
UNIVAC I and first compiler for language A-0 1951 IBM enters computer field 1952 FORTRAN language born 1954
1960
First computer born 1959 "Software" coined 1960 First computer designed auto part (1965 Cadillac trunk lid)
1963 BASIC language invented/-first minicomputer 1964 Minicomputer is born 1965 First hand-held calculator 1966
1970
2,250 transistors on a chip 1/6 X 1/8 inches 1969 Microcomputer is born/floppy diskettes 1971 Apple II microcomputer 1977
1980
VisiCalc revolutionary applications software 1978 300,000 transistors on a chip 1980 IBM PC 1981 LOTUS 1-2-3 software 1983
IBM AT 1984 IBM XT 1984
1990
2000
The interesting part.

[15-15] A time-line of computer development.

- A device to communicate the results to the user.
- A set of instructions that provides the directions to the devices for performing all the tasks.

Without any one of the items listed above, a computing system is severely limited in its usefulness.

The *physical* devices that are used to perform the foregoing tasks comprise the *hardware* of the system [15-16]. As computer systems have advanced through the years, the term *computer hardware* has come to refer to a diverse and ever-increasing number of devices. Devices for entering information into the computer range from crude mechanical card readers interpreting holes punched in cards, to sophisticated devices using laser technology and holography

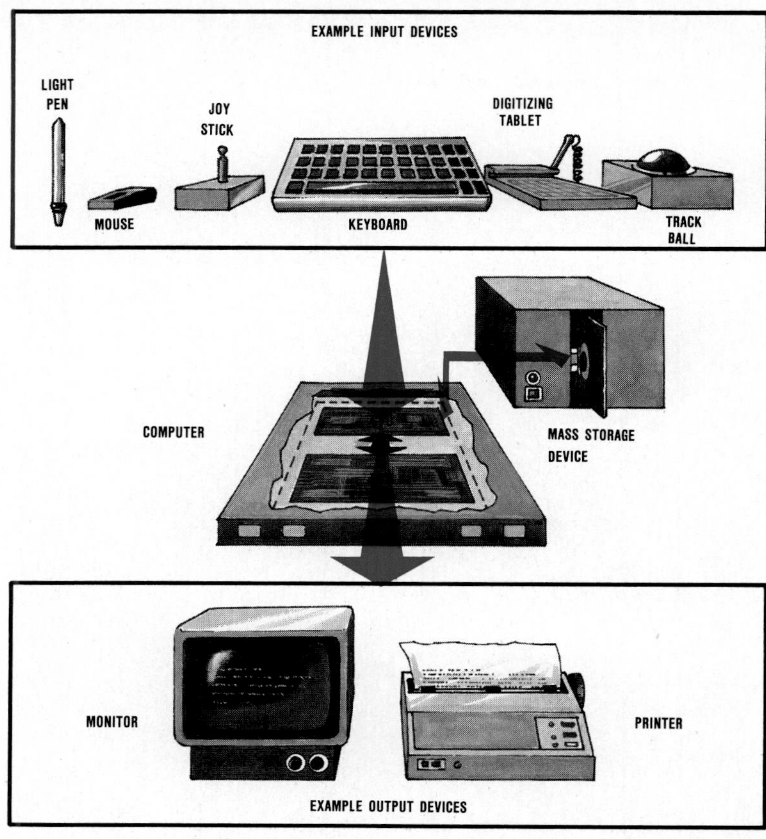

[15-16] Computer hardware.

"See if you can read it, Watson," said he with a smile.

It contained no word, but this little line of dancing men.

"If you use the code which I explained," said Holmes, "you will find that it simply means 'Come here at once.' I was convinced that it was an invitation which he could not refuse . . .

"And so, my dear Watson, we have ended by turning the dancing men to good when they have so often been the agents of evil . . . "

Sherlock Holmes
The Adventures of the Dancing Men

which "check you out" at the supermarket. Later in this chapter we discuss many of the more common devices that make up most of today's computing systems.

The one remaining element of the computing system, however, is the set of instructions that directs all of this hardware to solve some user specified problem. This *instruction set* is a most important component of the computer system, for without it, the computer would simply be a collection of hardware unable to solve any kind of problem. This set of instructions is generally of two types: *software,* instructions that are loaded into the machine by the operator, and *firmware,* instructions that are an inextricable part of certain elements of the hardware. Both software and firmware enable the user to direct the computer to solve a wide range of problems simply by specifying the appropriate instructions from the computer's instruction set in the proper order. The process of establishing the software instruction set is known as "programming," and all of Chapters 16, 17, and 18 are devoted to it.

In the beginning section of this chapter we discussed several types of computing machinery that have been developed and used. We will now narrow our discussion to electronic digital computing systems. The word digital represents the fact that these devices use *binary digits*—digits that can represent only one of two values. Binary digits, often referred to simply as *bits* (from *B*inary dig*ITS*), can be considered to be the "atoms" of the information world. They represent the most elemental particle or building block of information processing. All information in digital computers is made up of varying length *strings* of such bits.

The physical representation of binary digits in computers or in their peripheral equipment takes many forms: a hole may or may not be punched in a particular location on a paper tape or card,[10] a particular place on a magnetic record can be magnetized in one of the two polarities—north or south, a transistor can be either turned on (conducting at full current) or turned off (conducting no current), or a switch can be set in either the *off* or *on* position.

Regardless of the physical representation of the bits, the mathematical representations of the two binary states are constructed only with the two numeric digits 0 (zero) and 1 (one) [15-17]. Internally, digital computers then perform all of their tasks by reading, interpreting, and outputting multidigit sequences or strings of these binary zeros and ones.

Digital computers typically handle these strings of bits or binary digits in groups called *bytes*. The 8-bit string 01101000 represents one possible 8-bit byte. *Bytes are almost universally defined to consist of 8 bits.* The term *word* is often used interchangeably with the term byte, although it is usually used to represent a collection of several bytes that have some significance. In the section on central processors below, we will discuss the use of other length strings of bits, usually always multiples of 8 such as 16, 24, and 32. These binary strings of bits, or sequences of one or more bytes, are often referred to as *words*.

The counterpart to the digital computer is the *analog computer*. Analog computers do not rely on digits, binary or otherwise, to physically represent information within the computer or its attendant hardware. Such physical representation is accomplished through the use of continuously variable signals—signals that are not limited to only one or two discrete values. Examples are the rotation or partial rotation of shafts in mechanical devices, variable resistors (as opposed to the transistor in digital computers), and voltage levels in electrical devices. For most applications, electrical analog computers, once a strong competitor of digital computers, are now overshadowed by digital devices.

In the next section we discuss in more detail the major parts of the hardware that make up the modern digital computer. This is important since the hardware places overall limitations on what can be accomplished on a particular system. We believe that you need to be familiar with its function in order to better understand what is happening when you try to communicate with any device. You will also be introduced to some of the important computer jargon so that you may talk intelligently with other users.

The "black box" approach presented here ignores the intricate inner workings of the computer, such as the internal logic of the

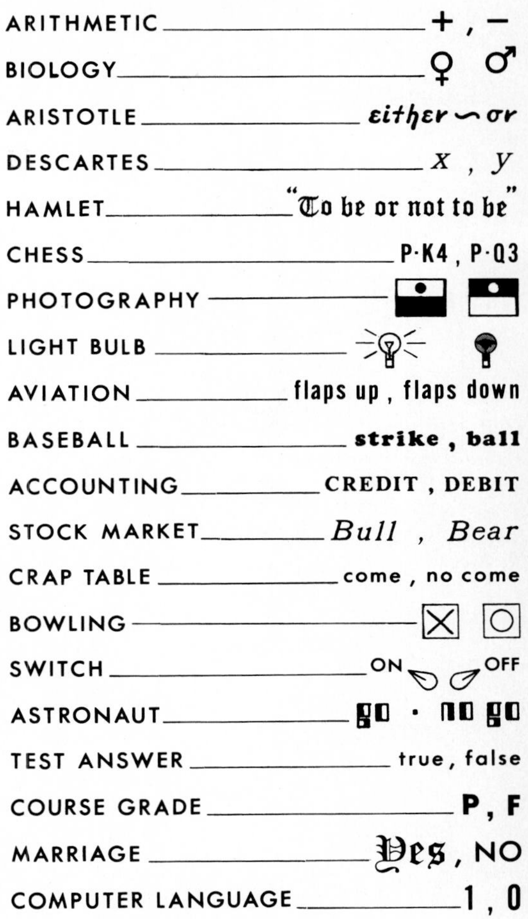

ARITHMETIC	+, −
BIOLOGY	♀ ♂
ARISTOTLE	either ~ or
DESCARTES	x , y
HAMLET	"To be or not to be"
CHESS	P-K4 , P-Q3
PHOTOGRAPHY	
LIGHT BULB	
AVIATION	flaps up , flaps down
BASEBALL	strike , ball
ACCOUNTING	CREDIT , DEBIT
STOCK MARKET	Bull , Bear
CRAP TABLE	come , no come
BOWLING	☒ ○
SWITCH	ON OFF
ASTRONAUT	go · no go
TEST ANSWER	true, false
COURSE GRADE	P, F
MARRIAGE	Yes, NO
COMPUTER LANGUAGE	1 , 0

[15-17] Many things in life are binary.

[10] Punched paper tapes and cards, once an inseparable part of digital computing, have both almost totally succumbed to advancing storage technologies. They will soon take their place, along with the historical computer devices discussed in the first section of this chapter, in science and technology museums.

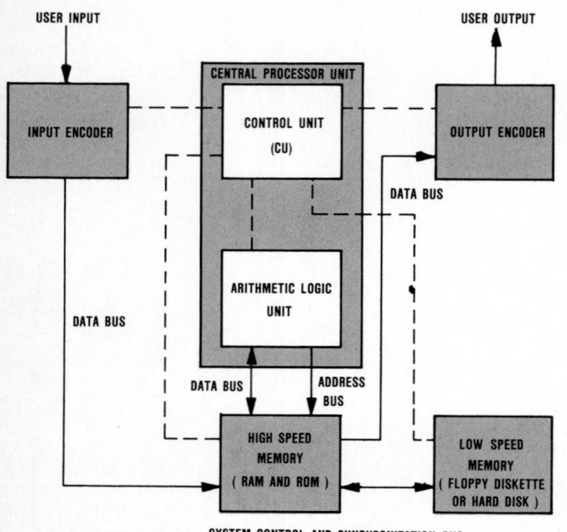

[15-18] A schematic diagram showing the operational arrangement of most digital computers.

processor. Our goal is not to turn you into a computer scientist but to transform you into an intelligent computer user. If we stimulate you to further study, we will have fulfilled our intent.

GENERAL DIGITAL COMPUTER HARDWARE

The Central Processor

Figure [15-18] is a schematic of the operational arrangement of most digital computers. The heart and soul of the system is the *central processor unit* or *CPU*. It is in this device that the most giant of strides has been made in reducing the size, cost, and electrical power consumption and in increasing the speed of computing.

The CPU is the home of the *arithmetic and logic unit,* or *ALU,* which performs basic arithmetic and logical operations (addition, subtraction, comparison of numbers, etc.). It is where the "number crunching" is done. However, it must be told by some other element in the system where to get the numbers or information it is to use, exactly what it is to do with that information, and where it should put the results.

The ALU performs its operation on strings of binary digits temporarily stored in a very limited number of "registers"[11] and places the results in similar "registers." In modern machines, the actual tasks of the CPU are carried out in the thousands of transistor switches and *gates*[12] in the VLSI circuits mentioned at the beginning of this chapter. The creation and design of these complicated circuits are the objects of "logic simulation" studies in the fields of electrical and computer engineering. Ironically, such studies are possible only with the aid of the modern computer. "We build on what has gone before."

When a user "runs a program," the *control unit (CU)* in the CPU is the administrator whose task it is to coordinate all the operations that must go on throughout the entire system. This includes coordinating the incoming and outgoing information or data, as well as providing all the necessary instructions to the ALU. The CU obtains both the data to be used and the instructions to be executed from the high-speed memory. It also places results back into the high-speed memory.

High-speed memory is divided into equal parts with each part having an *address* that is specifiable as a string of binary digits. Stored in these addressable parts of memory are binary strings of numbers that represent either instructions to the machine or data (this includes data to be used in some calculation or data that resulted from some operation). The CU addresses each location in memory over the address bus and retrieves instructions and data over the data bus. Of course, it is important that the CU get a valid, executable instruction when it expects one and a valid data string when it expects one.

[11]A register is nothing more than a temporary storage location that the ALU can use during arithmetic operations, or locations that the CPU uses to temporarily store bytes of data when it needs to "fetch" from or store to the high-speed memory.

[12]A gate is an electronic circuit that provides an output based on the input of one or more signals.

A most important part of the task of the CU is keeping straight what information in high-speed memory is data and what locations contain instructions to be executed. Since the computer can only store and process groups of bytes consisting of 0's and 1's, the CPU's instruction set must consist of combinations of these values, just as the data must consist of similar combinations.

You can imagine the consequences if the control unit were to lose track of what addresses in high-speed memory contain data and what addresses contain instructions for execution. If the CU begins to fetch instructions from locations containing data, or if it has problems interpreting what is supposed to be instructions, the computer usually quits functioning, and we experience what is often called a "system crash." The only way of recovering is to restart the machine from scratch.

As soon as the computer is turned on or restarted, the CU in the CPU will attempt to fetch an instruction from a fixed address in high-speed memory and begin execution. This location in memory is usually a location that has a sequence of instructions that is permanently stored there for specifically starting up the computer system. This area of memory is usually called "read-only memory" (ROM) since it can only be read and not erased or written over with new information. Once these instructions begin executing, they direct the CPU to load further instructions into high-speed memory and eventually perform the bulk of the task of starting up the computer. This process is often called "booting" the computer since it "pulls itself up by its bootstraps" by loading the necessary instructions to get itself started. At the end of the booting process, there are enough instructions loaded into memory and initially executed to allow the machine to load, on command, a user's program.

It should be pointed out that all instructions retrieved from memory for execution must consist only of items that the CPU was designed to interpret. Each CPU has its own way of interpreting the sequences of 0's and 1's that represent an executable instruction; thus each CPU has its own unique instruction set. Software written for one CPU will generally not execute on another CPU. We say more about this in the next chapter.

CPUs are classified both by the number of bits that can be used in addressing memory over the address bus and by the number of bits they can shuttle across the data bus. Both of these buses can be thought of as multiwire cables or multilane highways, with each wire or lane carrying a bit [the bits in the form of voltage pulses (bit = 1) or no voltage pulses (bit = 0), are transmitted in parallel over this cable or highway, one bit per wire or lane] [15-19].

Perhaps we should explore briefly how the computer makes sense out of the strings of binary numbers with which it works. Let us assume that we are dealing with a machine that works with single bytes or 8-bit strings (its word length, or length of binary string that it will hold in its registers, is 8 bits). In the decimal (base 10) system there are 256_{10} numbers encompassed by the smallest 8-bit binary number (00000000) and the largest (11111111) (see Table 15-1 for the decimal (base 10), hexidecimal (base 16), and octal (base 8) equivalents of the binary numbers 00000000 through 01111111).

When the bytes are to be interpreted as instructions, there could

TABLE 15-1 AMERICAN STANDARD CODE FOR INFORMATION INTERCHANGE (ASCII) CODES AND THEIR INTERPRETATIONS

ASCII Code				Character or control	Mnemonic name	Better description	ASCII Code				Character	ASCII Code				Character
Dec	Hex	Oct	Binary				Dec	Hex	Oct	Binary		Dec	Hex	Oct	Binary	
000	00	00	00000000	^@	NUL	Null	043	2B	53	00101011	+	086	56	126	01010110	V
001	01	01	00000001	^A	SOH		044	2C	54	00101100	(Comma) ,	087	57	127	01010111	W
002	01	01	00000010	^B	STX		045	2D	55	00101101	(Hyphen) -	088	58	130	01011000	X
003	03	03	00000011	^C	ETX		046	2E	56	00101110	(Period) .	089	59	131	01011001	Y
004	04	04	00000100	^D	EOT		047	2F	57	00101111	/	090	5A	132	01011010	Z
005	05	05	00000101	^E	ENQ		048	30	60	00110000	0	091	5B	133	01011011	[
006	06	06	00000110	^F	ACK		049	31	61	00110001	1	092	5C	134	01011100	\
007	07	07	00000111	^G	BEL	Bell	050	32	62	00110010	2	093	5D	135	01011101	]
008	08	10	00001000	^H	BS	Backspace	051	33	63	00110011	3	094	5E	136	01011110	^
009	09	11	00001001	^I	HT	Horizontal tab	052	34	64	00110100	4	095	5F	137	01011111	(UnderScore) _
010	0A	12	00001010	^J	LF	Line feed	053	35	65	00110101	5	096	60	140	01100000	(bkwd ') `
011	0B	13	00001011	^K	VT	Vertical tab	054	36	66	00110110	6	097	61	141	01100001	a
012	0C	14	00001100	^L	FF	Formfeed	055	37	67	00110111	7	098	62	142	01100010	b
013	0D	15	00001101	^M	CR	Carriage Return	056	38	70	00111000	8	099	63	143	01100011	c
014	0E	16	00001110	^N	SO	Shift out	057	39	71	00111001	9	100	64	144	01100100	d
015	0F	17	00001111	^O	SI	Shift in	058	3A	72	00111010	(Colon) :	101	65	145	01100101	e
016	10	20	00010000	^P	DLE		059	3B	73	00111011	(Semicolon) ;	102	66	146	01100110	f
017	11	21	00010001	^Q	DC1		060	3C	74	00111100	<	103	67	147	01100111	g
018	12	22	00010010	^R	DC2		061	3D	75	00111101	=	104	68	150	01101000	h
019	13	23	00010011	^S	DC3		062	3E	76	00111110	>	105	69	151	01101001	i
020	14	24	00010100	^T	DC4		063	3F	77	00111111	?	106	6A	152	01101010	j
021	15	25	00010101	^U	NAK		064	40	100	01000000	@	107	6B	153	01101011	k
022	16	26	00010110	^V	SYN		065	41	101	01000001	A	108	6C	154	01101100	l
023	17	27	00010111	^W	ETB		066	42	102	01000010	B	109	6D	155	01101101	m
024	18	30	00011000	^X	CAN	Cancel	067	43	103	01000011	C	110	6E	156	01101110	n
025	19	31	00011001	^Y	EM		068	44	104	01000100	D	111	6F	157	01101111	o
026	1A	32	00011010	^Z	SUB	Cancel	069	45	105	01000101	E	112	70	160	01110000	p
027	1B	33	00011011	^[	ESC	Escape	070	46	106	01000110	F	113	71	161	01110001	q
028	1C	34	00011100	^\	FS		071	47	107	01000111	G	114	72	162	01110010	r
029	1D	35	00011101	^]	GS		072	48	110	01001000	H	115	73	163	01110011	s
030	1E	36	00011110	^~	RS		073	49	111	01001001	I	116	74	164	01110100	t
031	1F	37	00011111	^	US		074	4A	112	01001010	J	117	75	165	01110101	u
032	20	40	00100000	-		Space	075	4B	113	01001011	K	118	76	166	01110110	v
033	21	41	00100001	!			076	4C	114	01001100	L	119	77	167	01110111	w
034	22	42	00100010	"		Quotes	077	4D	115	01001101	M	120	78	170	01111000	x
035	23	43	00100011	#			078	4E	116	01001110	N	121	79	171	01111001	y
036	24	44	00100100	$			079	4F	117	01001111	O	122	7A	172	01111010	z
037	25	45	00100101	%			080	50	120	01010000	P	123	7B	173	01111011	{
038	26	46	00100110	&			081	51	121	01010001	Q	124	7C	174	01111100	¦
039	27	47	00100111	'		Apostrophe	082	52	122	01010010	R	125	7D	175	01111101	}
040	28	50	00101000	(			083	53	123	01010011	S	126	7E	176	01111110	(tilda) ~
041	29	51	00101001	)			084	54	124	01010100	T	127	7F	177	01111111	
042	2A	52	00101010	?			085	55	125	01010101	U					

In the "ASCII Code" columns, "Dec" denotes decimal number; "Hex" denotes hexidecimal number; "Oct" denotes octal number; "Binary" denotes binary number.
In "Character or Ctrl" column, ^ denotes "Control" (e.g., ^G is Control-G).
"Mnemonic Name" column gives ASCII standard abbreviations for ASCII control codes 0 through 31_{10}.
"Character" columns give the control code or common keyboard character the ASCII code represents.

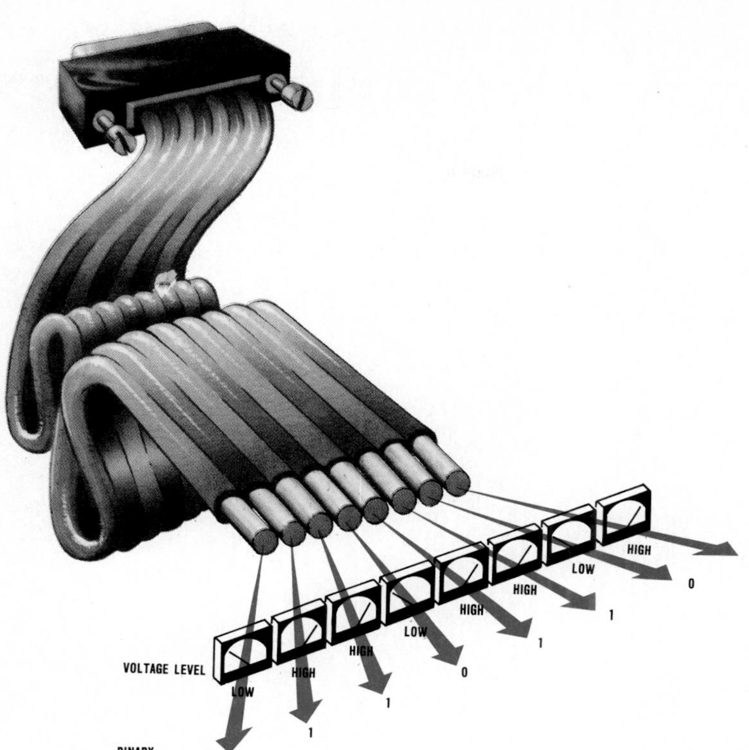

[15-19] Eight parallel wires, each transmitting a bit, can be used to send 1 byte (8 bits) of information.

be 256_{10} different instructions possible in the instruction set for an 8-bit word.[13] There is no standard for what instruction is associated with each binary number; as we have just pointed out, each CPU has its own built-in, unique relationship between binary numbers and instructions.

The representation of data by the eight binary digits available in a one-byte string is considerably more standard, although not universal. Here the problem is one of representing all the useful characters (numbers, lower- and uppercase letters of the alphabet, and other assorted symbols) necessary for carrying out useful tasks. The most widely used standard is that adopted as the American Standard Code for Information Interchange (ASCII). However, the IBM Corporation has adopted its own standard for its mini and mainframe computers called the EBCDIC character set. Table 15-2 compares these two systems for the uppercase A–Z. Table 15-1 lists the ASCII "codes" for the binary numbers from 00000000 through 01111111 (a total of 128 different ASCII code numbers and associated characters). Table 15-1 lists the ASCII codes in decimal (base 10), binary (base 2), hexidecimal (base 16), and octal (base 8), together with the symbols they represent.

Of course, more numbers and larger instructions can be represented by larger strings of binary numbers. For example, with a 16-

[13]The subscripts here denote the number base being used. The $_{10}$ is for base 10 or decimal, $_{16}$ is for base 16 or hexidecimal, and $_{2}$ is the base 2 or binary numbers.

TABLE 15-2 EXTENDED BINARY-CODED-DECIMAL INTERCHANGE CODE (EBCDIC) COMPARED TO ASCII (SEE TABLE 15-1) FOR THE CAPITAL LETTERS

Character	EBCDIC			ASCII		
	Dec	Hex	Binary	Dec	Hex	Binary
A	193	C1	11000001	065	41	01000001
B	194	C2	11000010	066	42	01000010
C	195	C3	11000011	067	43	01000011
D	196	C4	11000100	068	44	01000100
E	197	C5	11000101	069	45	01000101
F	198	C6	11000110	070	46	01000110
G	199	C7	11000111	071	47	01000111
H	200	C8	11001000	072	48	01001000
I	201	C9	11001001	073	49	01001001
J	209	D1	11010001	074	4A	01001010
K	210	D2	11010010	075	4B	01001011
L	211	D3	11010011	076	4C	01001100
M	212	D4	11010100	077	4D	01001101
N	213	D5	11010101	078	4E	01001110
O	214	D6	11010111	079	4F	01001111
P	215	D7	11011000	080	50	01010000
Q	216	D8	11011001	081	51	01010001
R	217	D9	11011010	082	52	01010010
S	226	E2	11100010	083	53	01010011
T	227	E3	11100011	084	54	01010100
U	228	E4	11100100	085	55	01010101
V	229	E5	11100101	086	56	01010110
W	230	E6	11100110	087	57	01010111
X	231	E7	11100111	088	58	01011000
Y	232	E8	11101000	089	59	01011001
Z	233	E9	11101001	090	5A	01011010

"Dec" denotes decimal number; "Hex" denotes hexidecimal number; "Oct" denotes octal number; "Binary" denotes binary number.

bit word (smallest number 0000000000000000_2 and largest number 1111111111111111_2) there are $65{,}536_{10}$ different numbers possible.

Many low-cost, high-quality "home computers" make use of 8-bit microprocessors (8-bit words) that can both address memory and transfer data in 8-bit, or one-byte, strings. The emergence of 16-bit microprocessors stimulated the use of microcomputers in the business world because they would address more memory (they can hold much larger numbers in their address registers than 8-bit CPUs) and they could accommodate a much larger instruction set. Thus larger and more complex software could be accommodated. The 16-bit processor was also responsible for awakening the engineering community to the power of the small computer in the first significant way. Intel Corporation's 8088 (used in the IBM PC and the host of PC compatibles) and the Motorola Incorporated's 6800 processors have been the most popular for this type of machine. However, these particular 16-bit processors communicate over an 8-bit data bus. Thus they must move two bytes (8 bits each) to make use of the 16-bit capability in the processor.

The Intel Corporation's 80286 microprocessor, which is now popular in many machines, has 24 address channels (three-byte words), giving it much more memory addressing capability than

16-bit CPUs and a much larger potential instruction set. Again, larger, more complex programs can be accommodated. However, the 80286 moves data over the data bus in 16-bit, or two-byte, strings of binary numbers.

CPUs having 32-bit addressing capabilities are also common. Some (like the Motorola Incorporated's 68000 microprocessor) use a 16-bit data bus and some (like the Motorola 68010) use a full 32-bit data bus. The Digital Equipment Company uses its own 32-bit processor in its popular VAX line of medium-size computers. CPUs of even larger addressing capability are used in the super minicomputers and in mainframe computers.

Storage

High-Speed Memory. High-speed memory nearly always contains a section of volatile memory commonly referred to by any of the following names: *random access memory (RAM), read/write memory, main memory,* or *core*. It is from this memory that the CPU gets its instructions for performing its work. That is, the user's program must be loaded into this memory before it can be executed.

The word "volatile" is used to emphasize that information stored in RAM is lost when the computer is turned off or suffers a power failure. This type of memory requires electrical power in order to function.

Random access memory (RAM) implies that any part of memory can be accessed as easily and as quickly as any other, as long as addresses are specified. Only the address of the memory location is required to read or write to that location. This leads to very fast access times and increased computation speeds. The name "read/write" highlights the fact that you can both read from and write to this type of memory. The last commonly used name, "core," stems from the era, not too long ago, when high-speed memory was made of ferritic (iron-based) "cores," doughnut-shaped objects that could be magnetized by an electrical current in a wire running through the hole. Today, semiconductor material is used in lieu of ferritic material.

Research in storage techniques is continually increasing storage density. The 256-kilobyte memory chip, capable of storing 256 kilobytes of information, is now replacing the 64-kilobyte chip. Soon 512- and 1024-kilobyte chips will be commercially available.

High-speed memory almost always includes some *read-only memory (ROM)*. ROM, like RAM, can be randomly addressed and accessed by the CPU, but information cannot be written to it. ROM does not normally lose its contents when power is shut off. What is stored in ROM can be reaccessed when power is restored.

ROM contains software that is encoded into a storage element which is a part of the hardware. This encoding is either done when the ROM is manufactured or added by special machinery in the field. ROM usually has stored, as a minimum, the basic instruction sets needed by the CPU for performing very basic operations.

Some brands of hardware have elaborate instruction sets in ROM for executing predetermined program steps. It was pointed out earlier that software embedded in the hardware is called firware.

For programs that are executed often, placing them in firmware offers fast-loading software without the need for external storage.

Low-Speed Memory. We are using the term "low-speed memory" in this text for a wide range of storage devices that are used to save or hold files, data, and programs. A synonymous term often used is *mass storage*. Included in the definition are all storage forms for which the individual storage elements are not directly addressable by the CPU. This means that when material is retrieved from or placed into the low-speed storage, the interaction with the CPU is much slower than when material is temporarily stored in RAM. Mass storage is also distinct from semiconductor RAM because it does not lose its contents when electrical power is removed. It is nonvolatile. Although equipment does fail occasionally, the contents of the low-speed storage can normally be recovered when power is restored.

Punched cards and punched paper tape served for decades as a means of mass storage. Magnetic storage media have improved so much in recent years that they are the predominant type of low-speed memory.

There are essentially two different types of common magnetic mass storage devices. These are:

1. Devices that can be removed from the computer.
2. Devices that are essentially a fixed part of the hardware.

Floppy Disks. The most common type of removable storage is the *diskette* (sometimes referred to as *floppy disk* or soft disk). These consist of a thin, flexible, circular disk which is coated with a magnetic material and encased in a permanent paper or fiber jacket. When mounted in the "diskette drive" on the computer and rotated within the paper jacket, the diskette's magnetic states are sampled or changed by the drive's read/write head. The head glides along the diskette surface accessing the diskette through a slot in the paper jacket [15-20].

Floppies are convenient to use because of their size, their storage density, and their low cost. However, because of their relatively low spin speeds when mounted and the wide dimensional tolerances they must have, their access time (the time required to read or write to the disk) is relatively slow.

Floppies store their information within "sectors" on concentric "tracks" as demonstrated in Figure [15-20]. Unfortunately, the exact format of stored information has been standardized only on 8-in. (20-cm)-diameter floppies, recording on only a single side, under the CP/M operating system.[14] For example, $5\frac{1}{4}$-in. (approximately 13 cm)-diameter floppies, another common size, are used at various storage densities from 120 kilobytes per diskette (approximately $120{,}000 \times 8$ bits) to 1.2 megabytes of information per diskette. A formatted 360-kilobyte diskette would hold, for example, the characters (letters, numbers, punctuation, etc.) from about 100 pages of typed, single-spaced, $8\frac{1}{2}$- by 11-in. paper.

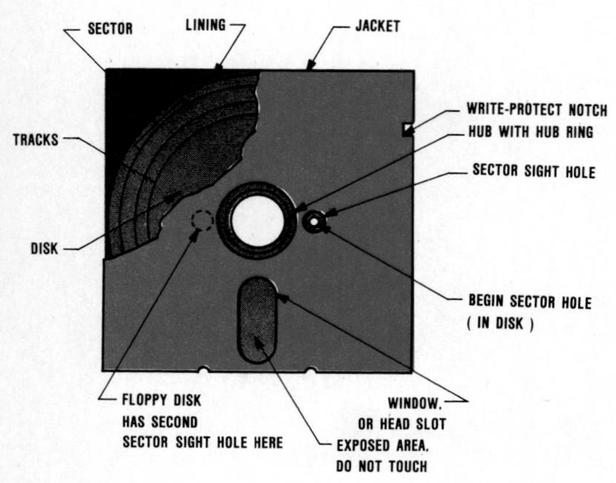

[15-20] A floppy disk.

[14]Operating systems are discussed briefly in Chapter 16.

New diskettes must be "formatted" for the particular brand of hardware on which they will be used. This "formatting" process, which sets the number of sectors and tracks, is carried out by executing a software program. For all but the 8-in. (20-cm) CP/M-based floppies, this formatting is unique to each type of hardware, making it difficult if not impossible to move a formatted floppy from one brand of hardware to another as a means of transferring data.

Part of one of the tracks on a diskette always contains a directory where information is stored on the file name, size, date of modification, and exact storage location on the diskette of each file. The system updates this information automatically when diskette activities (reads and writes) are necessary. An operating system command (operating systems are discussed in the next chapter) is usually provided to enable a user to query this directory for a list of the information in the directory.

Floppies generally store their information in a manner that allows *random access* to the stored material. That is, when a certain file is required by the user, the system permits that file to be accessed without reading through the other files stored on the diskette.

Disk Packs. Removable disk "packs" are often used on larger computers. These usually consist of several magnetically coated disks, similar to phonograph records, that are attached to a cylindrical hub. The hub is attached to a "drive" and spun about its axis at a high rate of speed, typically 40 revolutions per second. Read/write heads, passing near the surface of the disks, perform read and write operations with appropriate parts of the magnetic surfaces. It is common for any position of a disk to come under a magnetic read/write head every 25 msec, producing data transfer rates of 500,000 bytes per second.

The disk packs can be demounted, so that others may be mounted and used. Figure [15-21] shows a typical magnetic disk drive unit with a disk pack mounted. Disk packs, like floppy disks, are random-access-type storage.

Magnetic Tape. Magnetic tape, either mounted on reels or in cartridges, is another form of removable magnetic storage. On large computers, this is the form of demountable storage used most often. Magnetic tape libraries of large computer centers often have thousands of tapes on file.

Magnetic tapes have their widths divided into what are called "tracks," most often nine in number, with each of the parallel tracks running the length of the tape. At any one location along the tape, the bits of information can be encoded across the tape with one bit per track. All but one of the bits are used in making up a byte while the remaining one is used for error checking. A tape reading head, monitoring the tracks, simultaneously senses all the bits across the tape.

The layout of the data on a magnetic tape is said to be *sequential* rather than random access. That is, one byte follows another along the tape. If the computer is required to search a magnetic tape for certain information, it must start at the beginning of the tape and progress through the data until the desired information is reached.

[15-21] Magnetic disk drive with disk pack.

[15-22] For clarity, the cover has been removed from the head/disk assembly of this hard disk.

Thus sequential searches will be slower than finding data in the individually addressed locations in RAM.

The number of frames per inch (fpi)—often called bytes per inch, or simply *bpi*—is called the *tape density*. Typical tape densities are 800, 1600, and 6250 bpi. When the tape density is 6250 bpi and the tape is running at a speed of 75 in., the data transfer rate (read or write) is nearly 500,000 bytes per second. At this rate it would take about 0.2 sec to transfer the information contained in this chapter.

A simple calculation would show that a 2400-ft tape, using a density of 6250 bpi, could hold 180,000,000 bytes, or about half of the information in a 30-volume encyclopedia. However, because of blank gaps that must be left periodically to accommodate acceleration and deceleration of the tape, much of the tape is unused. True storage in practice is well below the maximum possible of 180,000,000 bytes.

Hard Disks. Nonremovable disks (sometimes referred to as hard disks) offer improved performance over floppies. Since they are generally made for permanent installation, they have closer tolerances and can spin at higher speeds [15-22]. Thus reads and writes with the disk can be accomplished at much higher speeds. Hard disks operate in much the same fashion as the removable disk packs discussed above.

On small computers, it is not uncommon to find hard disks as small as approximately 13 cm × 15 cm × 4 cm in size (5 in. × 7 in. × 1.5 in.) and capable of storing 20 megabytes (approximately 20,000,000 × 8 bits) of information (about two volumes of a 30-volume set of encyclopedias). Of course, much larger units are common on larger computers.

Like the removable storage floppies, hard disks have directories stored on the disk, so that current information (file names, exact location on the disk, size, and date of last changes) on the files stored can be maintained.

Input Encoders

We have already discussed the computer world as being a binary one—either on or off, 0 or 1. Our physical world, however, is essentially an analog one. That is, the engineer talks about such things as the stress level in a bridge column, the voltage level in a circuit, or the temperature level in some object. The observable or measurable quantities with which engineers deal are nearly all continuously variable.[15] They are not restricted to discrete values. Therefore, if these measurable quantities are to be represented in computations on a digital computer, some means must exist for converting the inputs of physical measurables into digital signals. This is the purpose of the input encoders. They take nonbinary data and encode it into binary representation.

[15]We point out when you study modern or atomic physics you will find that, on a microscopic scale, this is not quite true. But on the scale that most engineering work is done, the permissible discrete steps are so close together as to appear continuous.

Keyboards. *Keyboards* are the most often used data encoders, their source of input data being the human brain. They represent digital input devices more than the other input encoders discussed below, since keys are either pushed, or they are not. There is no "in between" [15-23].

When any one of the keys on the keyboard is activated, signals representing the binary string of the appropriate keyboard character are generated within the computer. When the characters are to be shown on the display screen,[16] for example, the binary string generated (e.g., the ASCII codes shown in Table 15-1) on keyboard activation is sent to a character ROM that determines which little dots on the screen will be illuminated.

[15-23] Keyboards are used to input data and text.

The *cursor control keys* that allow the user to move the active or input area around the screen are also found on most computer keyboards. The *cursor* is that character, cross-hair, or symbol on the screen that shows you where the input or active area is located.

Many keyboards have *function keys,* sometimes referred to as softkeys, that can often be programmed to do special things. These might be used to carry out several commands at once, save to disk data, execute programs that are located in RAM, or activate certain features of the software being run.

Some keyboards also have a *number pad* that facilitates the entry of numbers into the computer. The number pad is a collection of keys with numerical labels arranged in a similar manner to the keys on a common calculator. These are separate from the ordinary number keys that are located across the top of the keyboard.

The arrangement of the keys on the keyboard was chosen many years ago when typewriters were first invented and introduced. This arrangement is not the most efficient layout of the keys based on the frequency of use of the letter in the English alphabet. Instead, it was chosen purposely to slow down the typist so that various mechanical parts in the typewriter would not interfere with one another. The conventional keyboard is called the *Qwerty* keyboard.

As improved typewriters were invented and, certainly, as the electronic typewriter emerged, there was less and less need to retain this awkward arrangement. Studies by Dvorik have led to the definition of the *Dvorik* keyboard layout, which would greatly increase typing speeds. It places the most often used keys in the most convenient locations. So far, however, tradition has prevailed, and there has been no significant shift to this new, efficient layout.

User-controlled input encoders other than the keyboard include the mouse, the track ball, the joystick, the touch screen, the digitizing tablet, and the light pen. All these devices are *pointing devices* for cursor control, not text entry devices.

The Mouse. The *mouse* is a small hand-held pointing device that is moved around on a flat surface [15-16]. There is an approximate one-to-one relationship between cursor movement and mouse

[16]Keyboard entry generally does not go immediately to the screen, but goes instead to RAM. The user and/or the software often can determine whether the keyboard input is also displayed on the screen. When it is sent to the screen, keyboard entry is generally said to be "echoed" to the screen.

movement; *displacing* the mouse *displaces* the cursor. The direction of the displacement of the mouse determines the direction of displacement of the cursor. Mechanical mice usually roll on balls which activate electrical sensors that in turn send information to the computer through a connecting cord. Optical mice usually sense light reflected from a mirrored pad on which the mouse is operated. The light originates in a source attached to the mouse.

Trackballs. *Trackballs* are very closely related to mice. In fact, they are essentially inverted mechanical mice. However, instead of moving an object over a surface, the operator uses his or her fingers to rotate a ball within a socket. *Rotation* of the ball produces *displacement* of the cursor on the screen. The direction of motion of the cursor is determined by the direction of rotation of the ball. Trackballs do not require a clear area next to the computer as does the mouse. Clear areas are sometimes hard to find on a busy desk.

Thumbwheels. *Thumbwheels* are used in pairs: the rotation of one wheel determines either the vertical or the horizontal cursor displacement. The rotation of the other wheel determines the cursor displacement in the direction mutually perpendicular to the displacement caused by the first wheel. The wheels are merely variable resistors across which either the voltage or current change according to the rotation of the wheel.

Joysticks. The *joystick* is quickly recognized by the video game enthusiast. One end of the joystick is attached to a stationary socket, while the other end can be manipulated by the operator [15-16]. The *speed* of motion of the cursor on the screen is proportional to the *displacement* of the free end of the stick from the static position. Thus, it is not a displacement–displacement device like the mouse or the trackball. The direction of the cursor on the screen is related to the direction of displacement of the free end of the joystick.

Touch Screens. *Touch screens* have sensors, usually optical or acoustical, at the periphery of the screen that are sensitive to the placement of an object (your finger, perhaps) near the screen. The *displacement* of the objects near the screen determines the *displacement* of the cursor on the screen. The cursor follows the object. Due to the sensing technique and the finite size of pointing objects, the accuracy to which the cursor can be positioned is less than that obtained with most other cursor positioning devices.

Digitizing Tablets. *Digitizing tablets* are somewhat like mice in that there is a "pointer" that determines the position of the cursor. However, the "pointer," in the case of the digitizing tablet, is operated over a sensitive surface, not over an ordinary desk top or optically reflecting pad [15-16]. *Displacement* of the pointer over the sensitive surface produces *displacement* of the cursor on the screen.

Light Pens. The *light pen* is typically a small pointer that is approximately the size of a pencil. It is connected to the computer through an electrical wire that provides the power for a light and/or electrical source located in the end of the pointer [15-16]. The pointer source is moved on the exterior of the screen until it is located at the spot where the cursor is desired. When the source is activated the cursor movement is executed. The *position* of the pen is related to the *position* of the cursor.

Analog-to-Digital Converters. As pointed out previously, the physical world is almost totally an analog world. There are many instances where it is desirable to have data from analog instrumentation entered into a digital computer. Laboratory or pilot-plant monitoring, data acquisition, and data reduction are examples of situations for which direct computer entry is desirable. Conversion of such data can be carried out in devices known as *analog-to-digital (A-to-D) converters*. These devices can accept continuously varying signals and convert them to digital representations for analysis in digital computers.

Exotic Encoders. We list here several technologies that are possible now, but because of cost, are not currently in wide use. As these develop and costs decrease, they have the potential of wide acceptance. These include voice recognition and encoding devices (eagerly awaited for entering text and numerical data into the computer without typing) and image encoding devices (video cameras, optical character readers, laser scanners, etc.). Many supermarkets are already making use of the latter technologies for speeding the data entry at the checkout line.

Output Decoders

Few of us can rapidly peruse binary data (strings of binary digits) and comprehend the meaning. Our senses all give us the ability to monitor and detect analog signals: shades of gray, not just black and white; awareness of amazingly broad levels of sound, not just quiet and loud; and an astonishing sense of touch. We therefore learn the most about a set of information if we can experience it in these analog ways.

We need to decode the binary output of computers to other forms of output that we can readily grasp and understand. Fortunately, there are several ways that this can be accomplished with existing equipment. We will discuss a few of these devices in the sections that follow. In our descriptions, we are including what are termed the *device drivers*. These are interfaces that help decipher the binary codes generated by the software so that they can be displayed on the final output device. These device drivers may include both hardware and additional software.

CRTs. *Cathode ray tube (CRT)* is a generic name often applied to the TV-like display screens associated with computers [15-16]. *Monitor* and *video display tube (VDT)* are other terms often used to identify the CRT. Screen displays are temporary ones, since information displayed on the screen will be lost if the screen is overwritten or the power is turned off. We are including the interface in the description of the monitor.

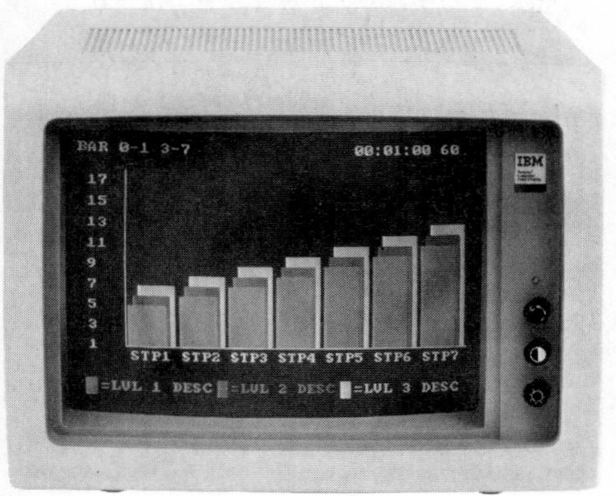

[15-24] A screen resolution of 320 × 200 on this eleven-inch diagonal screen makes lettering and diagonal and curved lines look rather ragged.

[15-25] A screen resolution of 640 × 400 on this eleven-inch diagonal screen makes all lettering and lines sharp and distinct.

Computers, being digital devices, do not display continuous patterns on CRT screens. The screen is always broken into pieces or *pixels,* each of which can be controlled—turned on or off, made to blink on and off, or instructed to display any of a very limited set of colors. The number of pixels that can be addressed on the screen largely determines the quality of the display. The term *resolution* is used to describe the potential quality: 320 × 200 pixels (i.e., 320 horizontally by 200 vertically) on a 12-in. (diagonal) screen usually give a very grainy appearance, while 640 × 480 pixels make displays appear to be almost continuous. Figures [15-24 and 15-25] illustrate two different screen resolutions.

There are two important types of CRTs in use, these being distinguishable by their abilities to display information. One is the *text* monitor, on which only textual characters (letters, numbers, punctuation) can be displayed. For example, when the CPU desires to print on the screen an answer of, for example, 50, it sends one byte, an ASCII 50_{10}, 32_{16}, or $00\ 110\ 010_2$, to the interface (see Table 15-1). The interface then takes this code into a table, usually located in a ROM on the interface, to find which of the screen dots must be illuminated in order to display the character "50." Notice that the CPU has only to deal with one byte in order to display a text character that encompasses many pixels.

The other type of display screen is that of the *graphics* monitor. It is not limited to the ROM-defined ASCII characters; it can display almost any type of design as long as it is told which pixels to illuminate. These necessary instructions must come through the CPU from the software instructions in RAM. On a monochrome graphics screen (e.g., a black-and-white display), the individual bits in a binary string may be used to control the illumination—1 bit per pixel with the digit 1 being on and the digit 0 being off. This necessarily involves more information being handled by the CPU (run times increase) and more space in RAM being required for the instructions (more memory is required).

For *color graphics* displays, additional information, over and above that required by monochrome displays, must be supplied. Not only must pixels know whether or not they are to be illuminated, they must also know what color they are to display. If 256 colors are to be available, for example another byte (8-bit string) per pixel must be supplied for color coding. For high-resolution screens, for example 1024 × 1024 pixels, that would entail keeping track of over 1 million (1024^2) bytes of just color coding information (one byte per pixel). In this case, performance will degrade and memory requirements will increase.

Printers. *Printers* are hard-copy devices that produce permanent records of the computer output. *Hard-copy terminals* are special types of printers that can be used for both entering information and recording output. They are particularly good for beginning programmers because they provide the user with a record of both the input and the output.

Fixed-font printers have character shapes defined by the printer hardware. They often bear names such as *impact printer, daisy-wheel printer,* and *letter-quality printer.* When the computer sends to the printer the proper binary code (e.g., an ASCII code from Table 15-1) for a particular character, the printer prints the predefined shape it has available. These printers cannot be used to print true graphics. They can only be used for *character graphics;* that is, they can only approximate the desired output using the predefined characters available to them.

In *dot-matrix printers,* printing is done by placing small dots on the paper in appropriate order or shapes such that the shapes resemble normal characters. In character or text printing mode, the printer receives a binary code (e.g., an ASCII code) from the computer, uses this to retrieve the proper character dot pattern from a stored table of dot patterns, and prints the pattern.

Most dot-matrix printers are also capable of printing in the *graphics mode,* whereby any arbitrary dot pattern can be printed. However, the computer must instruct the printer as to how the dots must be arranged, much like the computer must tell a graphics monitor which pixels to illuminate. Much more information has to be passed to the printer, resulting in degraded performance.

Laser printers use the binary output from the computer to control a small laser beam that is swept over the paper being processed. The beam sensitizes the paper, causing it to attract a fine powder, or toner, in the sensitized regions. With further processing, the toner essentially becomes permanently bonded to the paper, resulting in the printed page. The laser printer offers enhanced speed, versatility, and improved quality over that obtainable with either fixed-font or dot-matrix impact printers.

Plotters. *Plotters* [15-26], like printers, produce hard-copy records. However, unlike printers, they do not operate on binary codes that are translated into characters. Instead, the binary codes relayed to a pen plotter tell it where to take the pen, when to put the pen down on the writing surface, where to take the pen while it is down, and when to pick the pen up from the writing surface. The instructions that do this come from the software instructions in the computer's RAM.

Fast plotters do not actually make use of pens and mechanical linkages. In these, the incoming binary information defines locations to be sensitized on the final medium (usually paper) or on a drum over which the medium passes. These sensitized areas then pick up toner to give substance to the lines and areas that make up the plot.

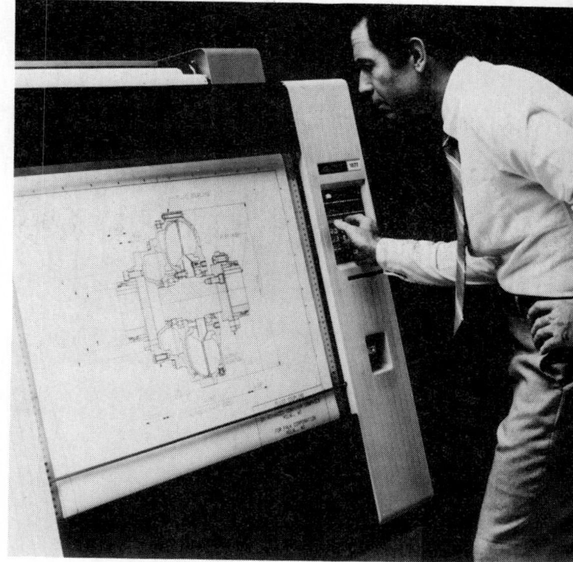

[15-26] Plotters produce hard-copy records.

Digital-to-Analog Converters. Digital computers are often used to control some physical process, such as changing the air/fuel ratio in a carburetor, controlling the speed of a motor, or controlling the temperature in an oven. These are uses where an analog (continuously variable) output from the computer is desired rather than a series of binary digits. Here, a *digital-to-analog (D-to-A) converter* can be used to make the conversion from the digital signals of the digital computer to the required analog signal. Such devices are commercially available.

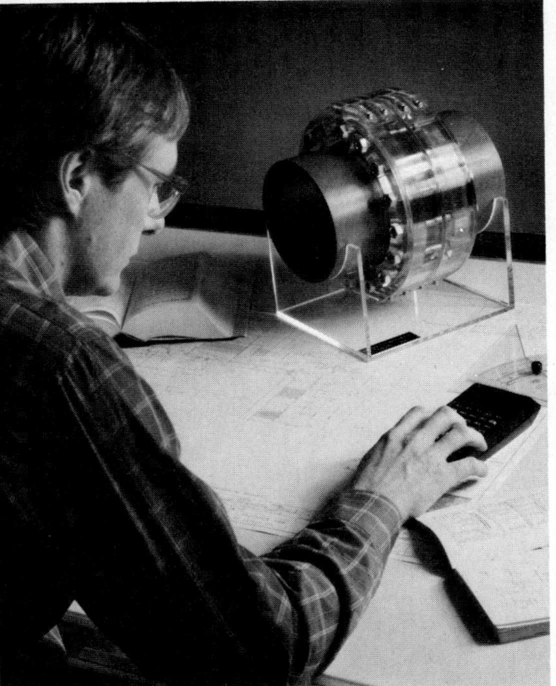

[15-27] Hand-held calculators continue to serve engineers as the prime calculating device for many computations.

It would be better if students had brains in their heads before they put them in their pockets.
Newsweek *(in reference to calculators)*

[15-28] This hand-held programmable calculator can be programmed in BASIC. The peripheral devices include a printer and cassette tape drive.

COMPUTERS: HAND-HELD CALCULATORS TO MAINFRAME COMPUTERS

Digital computers span a large range of sizes, shapes, and types. The smaller sizes include the digital hand-held calculators, some small enough to fit into your wallet or purse. At the large end of the scale, there are huge super computers that require large rooms, large amounts of electrical power, and water cooling.

Traditionally, digital computers have been grouped into the following categories: calculators, microcomputers, minicomputers, and mainframe computers. These groupings are more a result of historical development stages than of functionality. Today, only the calculator is distinct enough from the others to merit a class by itself. Considering the pace at which the technology is developing, even this distinction may soon disappear. However, we will talk briefly about each type.

Hand-Held Calculators

The "keyboard" encoder of the four-function pocket calculator typically consists of 10 number keys, the function keys of +, /, ×, ÷, =, and perhaps some storage and retrieval keys. The output decoder usually consists of a register with either light-emitting diode (LED) display or liquid-crystal display (LCD). If there is storage capability, then there is RAM. A CPU is always present to carry out the calculations. Thus, there exist all the hardware elements of the digital computer. However, the calculator is generally based on the decimal or base 10 system, rather than the binary or base 2 system.

The calculator also differs from other digital computers (micro, mini, etc.) in the origin of the instruction set that tells the CPU what to do. With the calculator, the instruction set originates in the operator's brain and is conveyed through the finger actuating the keypad [15-27]. In other computers, the instruction set is stored in RAM. In the calculator, the CPU spends most of its time during a calculation waiting on the operator to input the individual instructions that make up any computation. In other computers, the CPU proceeds quickly through all the instructions stored in any one program; most of the time is spent waiting on the next program to be loaded into RAM and executed.

Of course, as every student knows, one other large difference between calculators and the rest of the digital computer family is price. As we will see, price is often used to distinguish between other members of the family. Of course, calculators typically do not have all of the peripherals (printers, plotters, digitizers, etc.) available for the larger computers, but this is one reason they are so portable.

Programmable calculators [15-28] begin to blur the one main source of demarcation, that of the source of the instructions. In the programmable calculator, the operator keys into the device's memory all of the steps necessary to perform a given calculation and then initiates the execution of those instructions. For these devices, only typical RAM size—and price—separate them from other members of the computer family.

Calculators became popular when internal memory was still very expensive. As a result, most programmable calculators store only a small number of keystrokes—a few hundred at most—because they do not have very large memories. However, with rapidly decreasing costs for memory, calculator technology and computer technology will probably soon blend together. Students may soon have available a medium-power computer in a package the size of a notebook.

[15-29] This microcomputer features a 32-bit CPU and monochrome graphics screen.

Microcomputers

Microcomputers [15-29] are the newest members of the digital computer family. They were born in the infancy of small CPUs and have matured from a hobbyist's toy to a respected tool in business and industry. They are most often thought of as small machines that conduct single tasks for one user at a time. Computational speed has typically been slower than minicomputers and, certainly, mainframes. However, recent advances in memory and CPU capabilities are changing this condition.

Microcomputers that can execute more than one task at a time (called *multitasking*) for more than one user (called *multiuser* systems) are appearing. Also, computational speed is approaching that of many minicomputers. Only two things distinguish them from other members of the computer family: physical size (they are somewhat smaller than minicomputers and considerably smaller than mainframes) and price. Microcomputers are available in the range of a few hundred dollars to about $10,000.

Minicomputers

Minicomputers [15-30], products of the advances made in CPU design, chronologically represent the first departure from large mainframe technology. These machines were less expensive, smaller, and easier to maintain than previous mainframes. The minicomputer allowed the placing of computing power in the physical locations where it was needed. In large companies, it was welcomed by some users as a means of breaking management's hold on centralized computer centers. For smaller companies, it offered an economical alternative to the rental of computer time from other organizations.

Today, what the minicomputer sometimes lacks in computational speed for a particular computation compared to a mainframe, it more than makes up in its ease of use and smaller queues.[17] Minicomputers typically cost in the range of tens to hundreds of thousands of dollars, considerably more than minicomputers, and substantially less than mainframes.

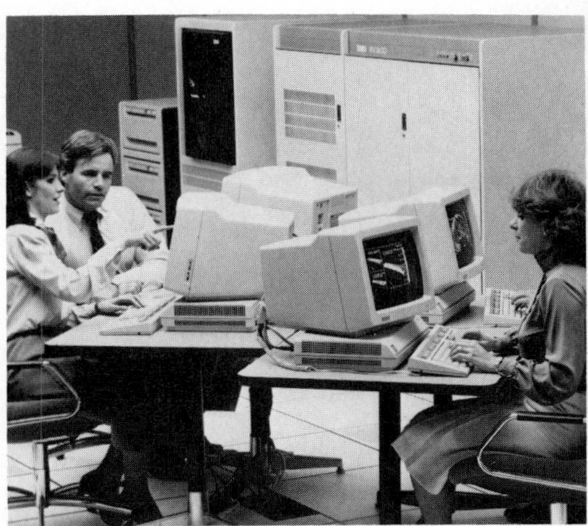

[15-30] This 32-bit minicomputer supports a multiuser environment.

Mainframe Computers

Mainframes [15-31] are typically characterized by several megabytes of main memory, the capability of addressing four or more bytes,

[17] *Queue* is a British term synonymous with "waiting line" and is a part of computer jargon. There is the "input queue" for the jobs waiting their turn to get into the computer, and the "output queue" of jobs awaiting printing or plotting.

[15-31] This large mainframe has a rather benign appearance that masks its speed and power.

hundreds to thousands of kilobytes of fixed disk storage, and multiple magnetic tape read/write units. They allow *time sharing;* that is, they service many users at any one time. Mainframes are usually supported and maintained by a specialized staff employed strictly for that purpose. They are also operated in secured areas. The users are typically not co-located but access the machine through remote terminals.

Large mainframe computers are *not* a dying breed, although some of their work load has been taken over by the micro- and minicomputers. Ironically, the micro- and minicomputer revolution could, in fact, soon begin to increase the load on large machines. As more and more engineers become computer literate through the ease of use and availability of the small machines, they will continue to seek more and more computing power. All these new users may soon discover, perhaps rediscover, the power and speed of the larger machines.

There are many uses that will probably not be assumed by the micros and minis. These include the execution of large design and analysis programs that consume huge amounts of core and large computation times and the maintenance of large corporate data bases. The small computers may be used to access pieces of these larger tasks, but it seems unlikely that they will replace the mainframes.

As stated earlier, price is one of the distinguishable differences between the micro, the mini, and the mainframe. Mainframes are at the high end of the scale; they usually cost in the millions of dollars.

COMPUTERS AND THE ENGINEER

Man is still the best computer that we can put aboard a spacecraft—and the only one that can be mass-produced with unskilled labor.

Wernher von Braun

The progress being made in the field of computer hardware is phenomenal. From 1981 to 1984 alone, the performance of CPUs increased by a factor of 2.5, the cost of RAM decreased by more than a factor of 10, and the storage density on floppy diskettes improved by a factor of nearly 10. The price-to-performance ratio for CPUs is projected to improve continually at 15 to 20 percent per year.

These rapid advancements are driving a change that is occurring in the engineering profession. The engineer is becoming more and more inextricably entwined with the computer. Without an understanding of the machine, its role, and its use, today's engineer is obsolete.

But the world will never live by computers alone. They are just tools. The engineer who learns to use these tools and not to abuse their use will be well served by them. Computers will never "engineer" new designs. What they will do is assist in the engineering process. They will calculate, compare, organize, and reduce error. However, without wise and seasoned operators, the computer will just as easily miscalculate, miscompare, and disorganize. If aspiring engineers take the time to learn the art and science of engineering, and at the same time learn a little about the computer, they will be able to continue to function in this new age.

PROBLEMS

15-1. Convert the following Roman numerals to Arabic decimal numbers:
a. CCCLI.
b. XLIV.
c. MMMDCV.
d. CMXCIX.

15-2. In library books that deal with the history of computing, determine what Helmut Schreyer's contribution to computing was.

15-3. In library books on computing or numbering systems, identify the person who perfected the binary numbering system.

15-4. In library books on computers or numbering systems, learn how to convert from binary numbers to decimal numbers.

15-5. Convert the following binary numbers to decimal equivalents.
a. 00000001
b. 00000010
c. 00000100
d. 00001000
e. 11111111
f. 00101010
g. 01100111

15-6. What is the largest make (manufacturer and model number) of computer at your institution?
a. What CPU does it use?
b. How many bytes of RAM are installed?
c. How many bits does it hold in its internal address registers? (This is referred to as its word length.)
d. How many bits does it shuttle simultaneously over the data bus?
e. How many bytes of disk storage are available "on-line?"
f. How many tape drives does it have? What is the preferred tape density?
g. Is it operated in a time-share mode?
h. Would you call it a micro, mini, or mainframe?

15-7. If you use floppy diskettes on your computer or if floppy diskettes can be used on the computers at your institution:
a. What (physical) size are they?
b. How many bytes can be stored on these diskettes?
c. Assuming one character per byte, estimate the number of pages of single-spaced text that could be stored on one diskette.

15-8. Does your computer facility have any hard-copy terminals available for your use? What are their disadvantages?

15-9. Does your computer facility have any CRT terminals available for your use?
a. Are they capable of graphics display?
b. Do they display textual characters only?
c. How many columns of text characters can be displayed on the screen?
d. How many rows of text can be displayed on the screen?

15-10. Does your computer facility have both hard-copy terminals and CRTs available for your use? If so, which do you prefer, and why?

15-11. What input encoders are available at your computer facility?

15-12. What output encoders are available at your computer facility?

15-13. Do you own a hand-held calculator? Is it programmable? If so, how many program steps will it store in its memory?

15-14. Seek out information on the Dvorik keyboard and compare it with the Qwerty keyboard. The most often used letter in the English alphabet is the "e." Compare its placement on the two keyboards. What are the advantages and disadvantages of each?

15-15. Does your institution offer any courses in:
a. Logic simulation?
b. Microprocessor design?
c. Microprocessor applications?

15-16. What CPU is used in:
a. The IBM PC microcomputer?
b. The IBM AT microcomputer?
c. The Apple Macintosh microcomputer?
d. The Digital Equipment Company's Rainbow microcomputer?
e. The Zenith Z150 or Z100 microcomputer?

Chapter 16

Communicating with the Computer

SOFTWARE, ALGORITHMS, AND COMPUTER PROGRAMS

As pointed out in Chapter 15, *software* is the set of instructions that tells the computer where to get the information that it needs to process, exactly what to do with that information, and where to put the intermediate and final results. It must exist in order for the computer to do its job.

The word *algorithm*[1] is sometimes used in referring to computational processes. In its most general use, an algorithm is a step-by-step procedure that leads to a solution of an entire class of problems in a *finite* number of steps. The mathematical formulation or mathematical model of a subject, whether in explicit or implicit form, then, is not an algorithm. But the step-by-step procedure for solving it is. Thus software, because it is a collection of step-by-step instructions, is essentially a very detailed computer-based algorithm.

Whereas the term *algorithm* can be used to describe any degree of detail of the solution steps, the term *computer program* is often used to represent the collection of the actual *lines of code* or sequential lines of computer steps that are written by a programmer in order to implement the solution on the computer. Therefore, "programs" and "software" are more nearly synonymous than are "algorithms" and "software."

This chapter is intended to demonstrate the rudiments of algorithm construction and includes the preparation for computer analyses. We will then discuss some general, existing software that is useful to engineers. Such software is often referred to as *applications* software. In Chapter 17 we cover some techniques for writing code for specialized situations where existing software is inadequate or nonexistent.

---

[1]Algorithm is a relatively new word with a long genealogy. It originated in the ninth century with a Middle Eastern mathematician named Mohammed ben Musa, but called *al-Khwarizmi* (the man from Khwarizm—a province of the old Islamic World, now a part of Uzbek and Turkmen USSR). His book *al-Jabr*—from which the word algebra evolved—was instrumental in the introduction of Arabic numerals and algebraic ideas into Europe. His name led to the word *algorism*, meaning the art of calculating with nine figures, zeros, fractions, irrational numbers, and so on. The word *algorithm* was originally just an alternative spelling of *algorism* but is now strongly entrenched in computer literature with the meanings given above.

SOLVING PROBLEMS WITH THE COMPUTER: LAYING OUT A PLAN

Define the Problem to Be Solved

Without a problem, there is no need for computation! But if a problem exists, who is to define its extent and its method of solution? Certainly not the computer, for it is a machine that is only able to carry out very specific instructions. It is the engineer, trained and skilled at problem definition as well as problem solution, who must make these decisions.

Quite often, a perceived problem changes as the problem is being defined. That is, what is first thought to be *the problem* may turn out to be less significant than some other feature originally neglected or thought to be unimportant. Without the problem definition step, it is quite possible to deduce a brilliant solution, but perhaps a solution to the wrong problem. This is as true for problems that require the use of the computer as for problems that do not. As your career in engineering develops, you will find that there exists a huge amount of software which can be best described as "answers awaiting questions."

What we hope you will learn from this chapter and the next is the idea that it is possible to write software that answers existing questions. But for you to do so, you must first have your questions well defined.

Establish the Inputs and Outputs

As the problem definition begins to take shape, it should become clear that while there are certain "knowns" or *inputs*, there are also certain "unknowns" or *outputs*. It should be obvious that in communicating with (or programming) the computer, you must know what these inputs and outputs are going to be. If you have studied Chapter 11, you have already had some experience at determining inputs and outputs. Chapter 19 will give you more practice at these skills. We only point out here that the user has to determine these quantities, not the computer.

Establish a Plan or Framework

Once the problem definition is well in hand and as the inputs and outputs are being finalized, it is time to begin the task of formulating a solution. We wish to concentrate in some detail here on the computer implementation plan that is likely to be a part of any problem solution.

Creativity in Programming. It is generally true in technical problem solving that if there is one way to solve a particular problem, there are also several more. So it is with computer-oriented problem solving. You will find that there is enough flexibility in the structuring of computer programs to give rise to a nearly infinite number of ways that a program can be constructed and still arrive at

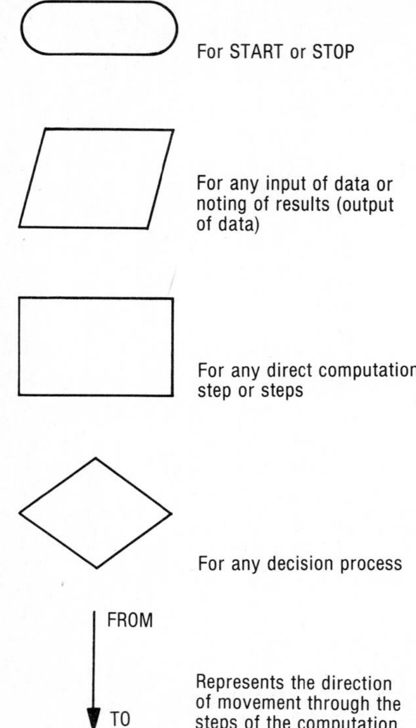
For START or STOP
For any input of data or noting of results (output of data)
For any direct computation step or steps
For any decision process
FROM
TO
Represents the direction of movement through the steps of the computation.

[16-1] Flowchart symbols.

the correct numerical answer. However, all of the possible solutions are not equally fast in computing or equally efficient in the use of computer memory or as intellectually satisfying. Nor are they equally easy to use.

The first duty of the programmer is to develop a plan or framework for the computer solution. This task is very much a right-brain-oriented one;[2] it draws on the writer's creativity. It is very helpful to develop the ability to visualize the overall framework to aid in grasping the direction, the intent, and the limitations of the plan. It must be decided how the program is to "flow" and how the interactions with the user (i.e., inputs and outputs) are to be handled.

Most people find that as ideas develop in their minds, they must do something to capture and document them. Often, sketching or diagramming is an effective way of doing this. In the next section, we address *flowcharting*, the practice of sketching or diagramming the logic structure of programs. However, since there is some controversy in the programming community over the usefulness of flowcharts, we also present a section on a more recent technique for capturing and documenting program structure called *pseudocode*.

Flowcharts as Aids to Visualization. An effective way of capturing a computer programming idea is to use a *flowchart*. This visual aid allows the originator of an idea to better understand and communicate the overall concepts and "flow" within the program. Flowcharts can be constructed to give the broadest overview of the process or to illuminate the finest detail of calculation. The choice depends on the needs of the prospective users.

Flowcharts will be introduced with the four symbols or blocks shown in Figure [16-1]. Lines are used to connect the blocks to show the steps of the algorithm. Arrows should be used at the forward ends of the lines to show the direction of "flow." Sometimes arrows are omitted when using the convention that flows are always downward or to the right unless otherwise indicated.

Simple Flowcharts. A simple flowchart, such as that shown in Figure [16-2], is appropriate for very simple algorithms that involve computational steps that are executed one after the other, from start to finish. Figure [16-2] documents the program "flow" for computing the minimum distance of separation along the surface of a sphere of two points on the sphere (often called the great circle distance). The formula is

$$d = \arccos[\sin \phi_1 \sin \phi_2 + \cos \phi_1 \cos \phi_2 \cos(\lambda_1 - \lambda_2)]c \qquad (16\text{-}1)$$

where ϕ_1, λ_1 = latitude and longitude, respectively, of the first place on the sphere

ϕ_2, λ_2 = latitude and longitude, respectively, of the second place on the sphere

c = coefficient to convert the angular distance given by the arccosine function to a curvilinear distance

[2]Right-brain and left-brain functions were discussed in Chapter 7.

$$c = 60 \text{ nautical miles/deg}$$
$$= 69.05 \text{ statute miles/deg}$$
$$= 111.1 \text{ km/deg}$$

Figure [16-2] does not give the actual program code or steps that would comprise the actual computer algorithm, but it does provide the necessary detail for any user to understand the process being used.

You may recognize that a simple flowchart for the straightforward problem in Figure [16-2] really serves the same purpose as would a detailed list of the steps that you might follow in arriving at an answer. However, problems being solved with the computer seldom have such a simple step-by-step structure as that of the great-circle-distance calculation. The use of flowcharts becomes even more important when decision branching points are needed and looping can occur.

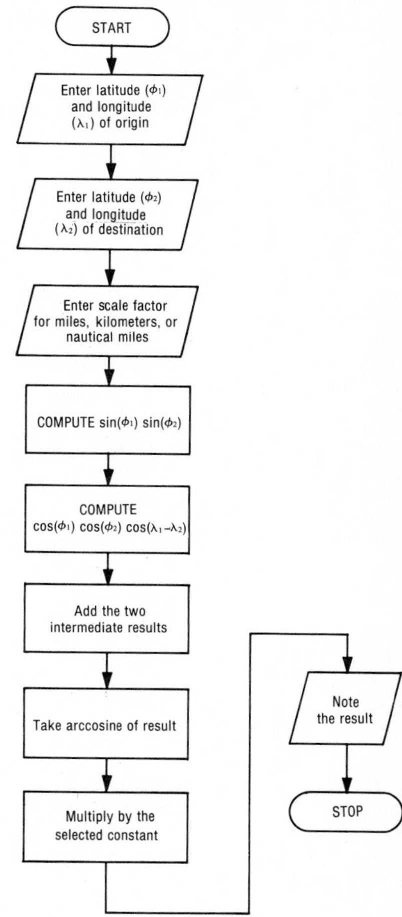

[16-2] Single flowchart: great circle distances.

Branching Flowcharts. Figure [16-3] demonstrates a flowchart for an algorithm where decision branching points are encountered. The algorithm leads to the solution of a second-order algebraic equation

$$ax^2 + bx + c = 0 \quad (16\text{-}2)$$

by the binomial theorem,

$$x = \frac{-b \pm (b^2 - 4ac)^{1/2}}{2a} \quad (16\text{-}3)$$

The first block after START in Figure [16-3] calls for the three coefficients, a, b, and c, to be entered for later use. Since the balance of the solution depends on the value of the discriminant ($b^2 - 4ac$), the discriminant must be calculated and branching done accordingly. If $b^2 - 4ac > 0$, there are two real roots and the right branch of the flowchart must be followed. If $b^2 - 4ac = 0$, there is only one real root and the central branch is taken. Finally, if $b^2 - 4ac < 0$, the roots are imaginary and the left branch is followed.

If you are solving the problem on a hand-held nonprogrammable calculator, you must perform the branching manually. When you see the value of the discriminant in the calculator's register, you can make the decision on how to proceed with the remainder of the calculations.

If you are working with a programmable hand-held calculator, or with a more sophisticated computer, you will want to include the branching in the program steps. We will demonstrate how this is done later in this chapter. Notice that computations with branching are much more easily visualized by flowcharts than would be the case by looking at a numbered list of steps.

Looping Flowcharts. In the example of Figure [16-3], branching took place, but none of the branches caused a return to an earlier part of the process. Branches that cause a return to an earlier part of the process create *loops*.

[16-3] Branching flowchart: the quadratic equation.

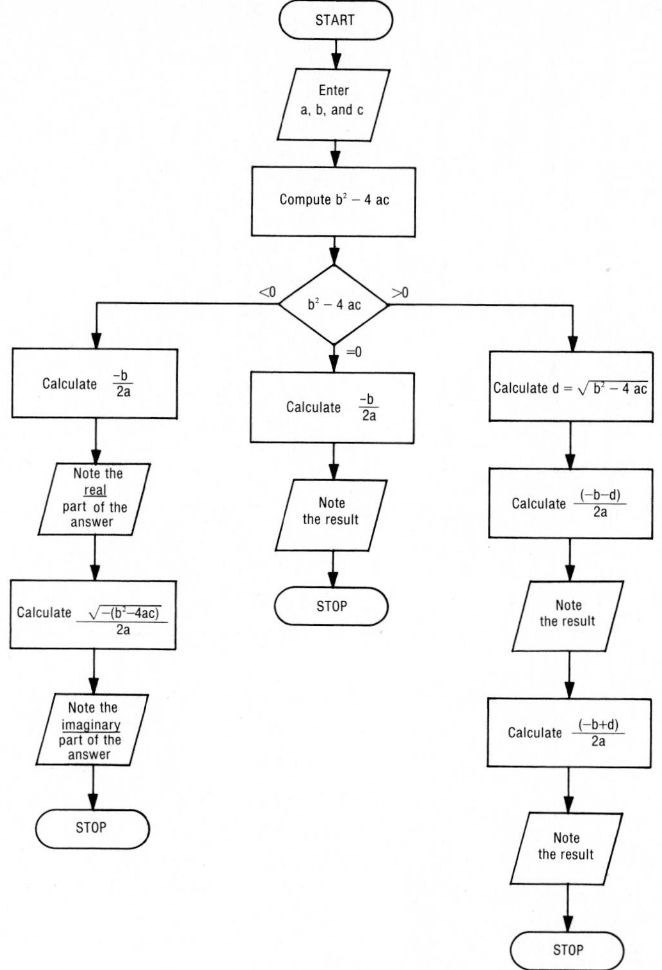

Helpful Hints!

Flowcharts are easy to understand and less likely to contain errors if the following conventional practices are observed:

- The START block should be located at the top of the flowchart, either centered or located to the left side of the page.
- An arrow should be placed at the end of every line or where two lines join. Arrowheads should not be used at the beginning of lines.
- No effort should be made to place STOP statements at the bottom of the diagram. These should appear on the flowchart where they logically occur.
- The information written in the blocks should be brief but descriptive of the process that is to take place. Combinations of English and algebra can be used.
- The lines leading from decision blocks must clearly be representative of the condition that must be met for that line or branch to be followed. It must be definitive in terms of the decision statement within the block.
- A single line should enter each block (except for the START block, which can not be entered) from the top. If more than one line enters a block, their paths should be joined outside the block (see Figure [16-5]).
- Departing lines (except for those from decision blocks) should always be from the bottom. There are no exits from a STOP block.
- Lay out flowcharts so as to minimize the crossing of lines.

The ability to create loops adds a tremendous amount of power to problem solving. However, its use can be a two-edged sword. If you are not careful, you can cause *infinite looping* and have no way to stop the process. In this case, your computer would just keep grinding away at the calculations, loop after loop—forever, unless some operator or automatic system action is taken to stop the process. Avoidance of infinite looping is another reason why you should become proficient in flowcharting and should check your logic all the way through to each solution end point. In this way, programming will become a straightforward implementation of your flowchart logic.

Since loops in programs could lead to never-ending computations, an additional decision variable and branch are often used to ensure that the looping will terminate.

Figure [16-4] gives a broad overview flowchart of the process of solving implicit equations. However, it is so broad that it is useful only as a framework for designing algorithms for treating any single-variable implicit equation. A more detailed flowchart would be needed to treat a particular problem.

Note that in the flowchart shown in Figure [16-4] there are two decision points. The first decision block is a test of whether or not the result achieved to that point is close enough to be acceptable, while the second decision block terminates the process without a solution. The latter course of action is necessary because some implicit equations do not have solutions and others do not respond well to poor estimates. For this reason, it is necessary to have an alternative exit from the process. You may already have deduced from this example that branching with looping algorithms is much more easily visualized by flowcharting than by looking at a listing of numbered statements.

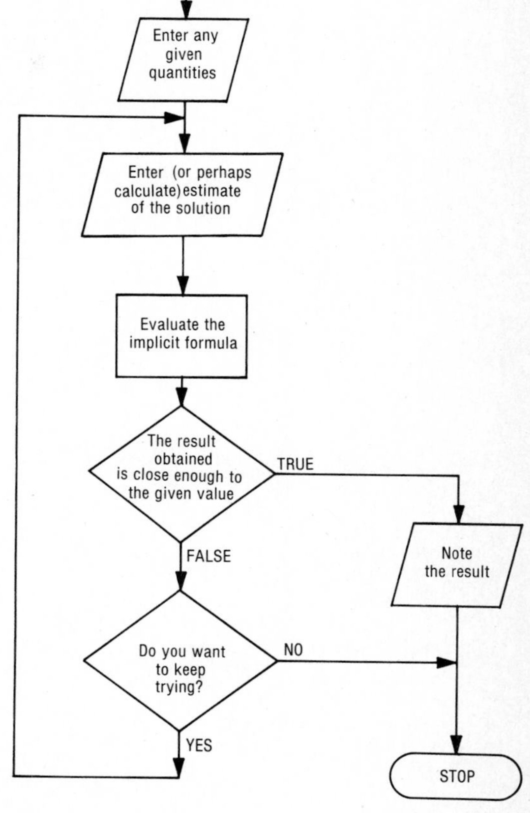

[16-4] Looping flowchart: an overview of the general implicit problem.

A Branching and Looping Example. Consider the following formula for the area (A) of a segment of a circle (of radius r) as a function of the angle (α, in radians) which it subtends at the center of the circle [16-5].

If we are interested in the fraction (f) that the segment represents of the total area of the circle, we can divide A by the total area (πr^2) to produce

$$f = \frac{r^2(\alpha - \sin \alpha)/2}{\pi r^2} = \frac{\alpha - \sin\alpha}{2\pi} \qquad (16\text{-}4)$$

For any value of the central angle(α), it is a simple matter to compute f. However, given f, it is not a simple matter to find α, since the equation cannot be solved explicitly for α. Here, an iterative approach will be required. The iterative process is one that, given a first estimate of the solution, uses it to get a better estimate. Successive iterations are made through this process (looping) until the estimates *converge*; that is, they change less and less each time through the iteration. When the change in successive estimates becomes small enough to be useful (a branching is convenient here), the iteration is stopped and the results are displayed.

Starting the estimate: We could make some "arbitrary" guess at a starting point, but there is a more logical procedure that could be used. In calculus, you will soon become acquainted with a mathematical relationship called the MacLaurin series, to represent the value of the sin α—if you do not know it already. It is

$$\sin \alpha = \alpha - \frac{\alpha^3}{3\,!} + \frac{\alpha^5}{5\,!} - \frac{\alpha^7}{7\,!} + \cdots \qquad (16\text{-}5)$$

Knowing this, we can use the first two terms as an approximation for the value of sin α. This will give us a good starting guess, even though we know it will not give us the exact answer. Now we have reduced our formula to the following:

$$f = \frac{\alpha - \sin \alpha}{2\pi} \simeq \frac{\alpha - (\alpha - \alpha^3/3\,!)}{2\pi} = \frac{\alpha^3}{2\pi 3\,!} = \frac{\alpha^3}{12\pi} \qquad (16\text{-}6)$$

From this we can deduce that $\alpha \simeq (12\ \pi f)^{1/3}$. So for any value of f in which we may be interested, we can produce a useful first estimate of α.

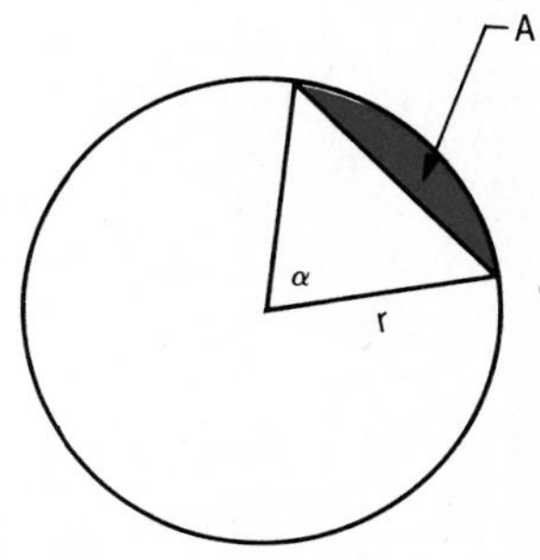

[16-5] The shaded area is the segment of a circle.

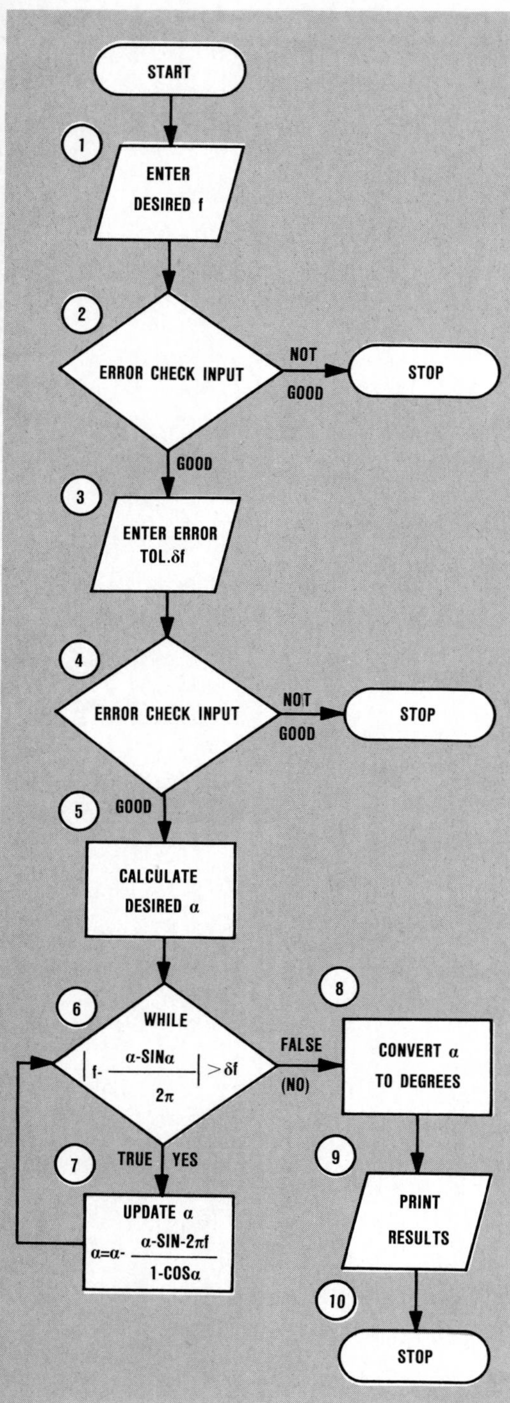

[16-6] A possible flowchart to calculate the angle α in [16-5], given *f*, the ratio of the area of the segment to the total area of the circle.

Refining the estimate: We can now use a technique known as the Newton–Raphson or Newton's[3] method for improving our estimate of α. This technique says that if x_{old} is an existing approximation to the root of the equation

$$g(x) = 0 \tag{16-7}$$

a new and better approximation for x can be obtained from the relationship[4]

$$x_{\text{new}} = x_{\text{old}} - \frac{g(x_{\text{old}})}{g'(x_{\text{old}})} \tag{16-8}$$

This technique is often used in numerical work.

In our problem, we are looking for the root of the equation (here f is considered to be a known constant)

$$g(\alpha) = \frac{\alpha - \sin\alpha}{2\pi} - f = 0 \tag{16-9}$$

from which we get

$$g'(\alpha) = \frac{1 - \cos\alpha}{2\pi} \tag{16-10}$$

Thus, a new approximation for α is obtained from an old one by the relationship

$$\alpha_{\text{new}} = \alpha_{\text{old}} - \frac{\alpha_{\text{old}} - \sin\alpha_{\text{old}} - 2\pi f}{1 - \cos\alpha_{\text{old}}} \tag{16-11}$$

Figure [16-6] demonstrates a flowchart of one possible algorithm for solving this problem. The user must supply the value of f for which Equation (16-4) is to be solved; this is done in step 1 (shown in Figure [16-6]). Also in step 1, the user must specify the relative error that will be permitted. That is, in this iterative problem, the program must know when the user is willing to quit improving on his or her successive approximations to the answer. There are many ways that this can be done. In block 6 we have chosen to require that α_{new} produce an estimate of f that is within Δf of the actual f specified in block 2.

Note the looping from block 7 back to the beginning of block 5 when the f based on α_{new} is not sufficiently close to the desired f in the branching block 7. Each cycle in the loop computes a better value of α based simply on its most recent calculation of that quantity since α is updated in block 8 before the loop is executed each time. Without this updating, convergence would never occur.

We have not protected against any divergence of the solution. That is, if for some reason the Δf values became larger and larger instead of smaller and smaller, we could be in an *infinite loop* from which there would be no end. Good programming practices protect

[3]This method was used by Isaac Newton to solve a cubic polynomial in his 1686 book *Principia*. You will probably discuss the technique in a calculus course or in a course on numerical analysis.

[4]Here $g'(x)$ is the derivative of $g(x)$ with respect to x. If you have not yet had a course in calculus, we ask that you accept some of this discussion on faith.

against these unforeseen things. Protection here might consist of another branching from block 7 if Δf starts to increase (Δf would have to be stored each time through the loop in order to make comparisons).

Now that the overall algorithm for this problem is determined, the next step would be to program the algorithm to carry through the process on a computer. In Chapter 17 we show how this is done.

General Observations on Flowcharts. We can now make a few general observations regarding flowcharting. The advantage of using a flowchart to define an algorithm is that it forces unambiguous thinking. There must be no uncertainty about which path is to be taken. For this reason there can be only one starting point on a flowchart (a START block), and that block can have only one exit (line) departing from it. The only block that permits more than one departing flow line is the diamond-shaped decision block. These blocks must specify the specific conditions under which each exit branch will be selected for continuation of the computation. Finally, a STOP block can have no departing (exit) lines. Note that more than one stopping block is permitted and that flow path lines can join.

Pseudocode. As mentioned above, programmers are not in universal agreement on the value of flowcharts in displaying the logic structure of programs. In recent years the concept of *pseudocode* has received recognition as a replacement for flowcharts. Pseudocode is a generic form of programming in which the program structure is laid out in a step-by-step fashion in words rather than the attendant, strict syntax required by all programming languages. Figure [16-7] gives a very general idea of pseudocode. More specific examples are given later in this chapter. Notice in Figure [16-7] that there are no sketches—no boxes with linking lines, and so on—as there would be in flowcharts. You may note that pseudocode is very readable. It can also be extremely precise and unambiguous.

Some people who write computer programs can go directly from conception in the mind to pseudocode. These people seem to be able to function quite well without the need for flowcharts or sketches. However, there are other programmers who find the flowchart invaluable in establishing the structure of programs, especially for programs that are quite large or complex.

Pseudocode seems to have developed in parallel with the evolving practice of writing structured programs—programs that flow from the top down in a natural way.[5] A reason for its success may be that modern programming practices generate code that carries out calculations or performs the duties of the program in a way that seems natural to most programmers. Programmers can therefore let pseudocode "flow" in the same natural manner that they would use to solve the problem by hand. Once the pseudocode or generic program is written, the task is more or less a mechanical one that involves converting the pseudocode into actual computer language.

Since flowcharts and pseudocode are both currently used by active programmers, we present a summary of both techniques. We

[5]We shall say more about "structured programming" in Chapter 17.

```
READ f, the fraction desired
If f < 0.0 or f > 1. THEN
  PRINT error message
  STOP
END IF
READ Δf, the permissible error
IF Δf < 0. THEN
  PRINT error message
  STOP
ENDIF
α SET TO (12πf)^(1/3)
WHILE | f− [(α−sin α)/2π]| > Δf
  Compute updated α
END WHILE
  Convert α to degrees
PRINT α
STOP
END
```

[16-7] Pseudocode: segment of circle—solving equation (16-4).

have already discussed flowcharts. We will demonstrate more on pseudocode in the programming sections of Chapter 17. It will be up to you to determine which method best suits your abilities and needs.

Selecting the Appropriate Resources. After you have taken your program plan through its initial stages, you must devote some attention to the resources that you will bring to bear on the problem. On what machine will you implement your plan? Will you use an existing program, or must you write new programs to solve the problem?

Small problems with few iterations and little branching might be best solved on a programmable hand-held calculator. Since these calculators vary considerably from manufacturer to manufacturer, and since the manufacturer's instruction books provided with each machine generally cover the programming steps, we will not detail their use here.

Generally, you must assess the type of computers that are available to you (micros, minis, or mainframes) and learn what software exists for these machines. There are thousands of applications software packages in existence, so do not write your own code if someone has already done it for you. Prepackaged programs range all the way from modules that can be incorporated into or used by programs you write yourself to packages that will do everything that you may need to do. We discuss some of the very general programs in the next section. Through experience and exposure as your career advances, you will learn about other, more specific programs.

You may not always find existing routines that do what you need done. You may decide that you can improve on existing packages or you may find that your problem is unique enough to require original programming. In these cases, you must invest some time and effort into writing the programs you need. In Chapter 17 we discuss programming languages so that you can be as original as you please.

In any event, you need to apply "appropriate technology" to the solution of your problem. For example, do not use the mainframe computer just to add two numbers. Similarly, do not use your hand-held calculator to solve 20 equations with 20 unknowns. Be conscientious of the resources at your disposal and use them judiciously.

SOLVING PROBLEMS WITH THE COMPUTER: FOLLOWING THE PLAN

Once a solution plan is laid out, it must be implemented. The implementation stage, unlike the layout stage, is quite regimented. This stage involves actually instructing the computer either through programs that you write yourself or through application programs.

Since instructions to the computer must conform *exactly* to the style and syntax that the CPU is expecting, the implementation stage is very analytical and symbolical. It is essentially left-brain dominated (we discussed left-brain abilities in Chapter 7). Thus there is a distinct difference from the right-brain-dominated plan formulation stage, which calls for spatial relationships, intuitive insight, and holistic views.

Essentially, computer programming consists of (1) learning the very strict structure or syntax of the programming language that has been chosen to be used for the problem solution, and (2) learning how to order the possible statements that make up the language so that an overall plan is accomplished in an efficient manner.

In conversational languages, there are rules of grammar and spelling that must be followed in order for communication to take place between a speaker or writer and his or her audience. But these two necessities are just sets of analytical rules. What is usually important—where the real creativity generally surfaces—is the thought that originated the communication. In programming, the plan is analogous to the thought; the syntax and structure represent the spelling and grammar.

Working through an Operating System

In Chapter 15 we discussed starting or "booting" the computer system. In that operation, certain instructions are loaded into a reserved portion of RAM so that other instructions (e.g., those contained in a user-supplied program) can be loaded and executed. The initial set of boot instructions form the working part of what is called the *operating system*.

Included in the operating system are procedures for loading a program from low-speed memory into RAM and executing it, copying a program from one form, or part, of low-speed storage to another form, or part, of low-speed storage (e.g., from floppy to floppy) and presenting a list of the files in a certain part of low-speed memory, to name only a few. In multitasking systems, the operating system keeps track of the multiple tasks that it is doing. In multiuser or time-sharing systems it must coordinate the jobs of the many users for which programs are being executed.

Thus the operating system is the software that supervises and orchestrates the overall activity of the computer. In a well-designed operating system, users are only aware of services that are being provided (storage of programs and data, execution of programs, translation of programs, record keeping, etc.). Users are not exposed to all the details of system supervision and integration.

To accomplish a task on the computer, the user must communicate with it through operating system *commands*, commonly called *control statements* or *job control statements*. These commands do not have to be unique to a certain CPU (e.g., the command to copy files could be the same from CPU to CPU), but the actual machine instructions that are carried out when the command is issued are unique to each CPU.

There are many operating systems in use today and, unfortunately, each system seems to have its own set of commands or control statements. Examples are CP/M, MS DOS, PC DOS, UNIX, RT-11, RSTS, VM, VMS, and CMS. On mainframe and minicomputers, the operating systems are usually unique to each manufacturer. In the early stages of the microcomputer revolution, each brand of hardware also seemed to have its own operating system.

However, 8-bit micros soon stabilized on CP/M as an operating system while on 16-bit machines, DOS became the de facto standard. Stabilization and standardization of an operating system makes programs much easier to move from system to system and lessens the burden on users. Stabilization leads to greater availability of software, while standardization promotes an increase in the number of people attempting to use computers.

The UNIX operating system has been implemented on nearly all sizes of computers. For this reason, it may eventually become the de facto standard for all computers. The major advantage of such an adoption is that users would then have at their disposal a set of operating commands that would be the same whether they were on a microcomputer, a minicomputer, or a mainframe.

Applications Packages

We consider applications programs to be software for which the detailed program steps are already written or coded. In using an applications program, we would not have to resort to programming languages (to be discussed later in this chapter) in order to write a program that would solve a particular problem. The use of such programs generally saves considerable time and may even do a better job than software that you would write yourself in any reasonable time.

Seemingly, there is an almost infinite number of existing software packages, with more appearing everyday. For example, in 1984, there were 12,000 software companies writing programs for microcomputers alone. At that rate, if what you need is not yet written, it probably will be some day—if you can wait.

Spreadsheets. One of the most versatile of all applications programs is the *electronic spreadsheet* or just *spreadsheet*. The story of its invention is the epitome of all the "university-students-become-famous" stories that abound. Spreadsheets were born not in an engineering college or in a computer science department, but in a business school. As is so often the case, necessity was the mother of invention.

Daniel S. Bricklin and Robert M. Frankston, as accounting students, became convinced that there was a better way to handle accounting data than to enter them by hand on the standard ruled sheets divided into rows and columns. Tedious entry, complicated and repetitive calculations, and the ever-present possibility of propagation of errors, all necessitated hours—even days—of effort to complete the sheets. Then, if a professor (or executive) wanted to see what the profits would be if, for example, labor costs went up x percent and product price went up y percent, the calculations would have to be completely redone. Burn the midnight oil!

The emerging microcomputer appeared to Bricklin and Frankston as being ideally suited to simplifying this task. If fed the necessary numbers and told the necessary calculations to perform, the computer could easily produce the final results. Of course, this concept in itself was not revolutionary, for that is what computers had been doing for 30 years. What was revolutionary was the way that Bricklin and Frankston went about solving the problem.

First, they divided the CRT screen into rows and columns just like the accountant's spreadsheet. Next, they permitted the user to enter a number or a label into any one of the boxes or "cells" formed by the intersection of a given row and a given column. The *piece de resistance* was that they also permitted formulas to be entered into cells—formulas that derived their inputs from other cells on the "sheet" and displayed the results on the screen in the cells where the formulas were stored.[6] Thus, the screen or sheet looked just like the accountant's paper sheets, displaying all of the numbers and labels that were entered and the results of any calculations that were required with the entered numbers. For example, sales could be totaled, expenses could be summed, and profits could be calculated.

What was so revolutionary about the electronic spreadsheet was its interactive nature made possible by the computer. Quick answers could be obtained for "what if?" types of questions. If company management wanted to know if profits could support a 10 percent raise in salaries, the accountant only had to reenter a few numbers and all the formulas were recalculated almost instantaneously.

Bricklin and Frankston were so successful in implementing their ideas on a 32-kilobyte RAM Apple microcomputer (which, at the time, was a new product on the market) that they are now widely credited with being largely responsible for the microcomputer infusion into the business world. Their newly developed software was marketed under the name *VisiCalc.*[7]

Besides VisiCalc, today's most common spreadsheets bear names such as Multiplan,[8] Lotus 1-2-3,[9] and SuperCalc,[10] to mention only a few.

Spreadsheets all have commands, such as copy, move, insert row or column, delete row or column, and edit, which are used to construct and alter entries on the sheet. There are also formatting commands which allow the user to clarify and improve the appearance of the sheet. Input and output (I/O) support is always provided to allow storing, retrieving, and printing sheets.

The versatility and convenience of the spreadsheet idea is now penetrating the engineering world. The functions available for use in formulas include most of those needed for engineering calculations (e.g., sine, cosine, maximum, minimum, average, etc.). Searches for tabular data, capabilities for iterative problem solving, statistics, and table generation are some of the more useful features.

Spreadsheets now commonly support graphics. That is, the numbers displayed on the electronic sheet can also be easily displayed in graphical form for better visualization of the data.

---

[6]Only a number or a label or a formula could be entered in a cell, not a combination of the three.

[7]Trademark of VisiCorp.

[8]Trademark of Microsoft Corporation.

[9]Trademark of Lotus Development Corporation.

[10]Trademark of Sorcim/IUS Micro Software.

Some spreadsheets have also been integrated with other tasks often requested by the user, such as data bases, word processing, and capabilities for communications with other computers. These enhanced capabilities have come about because of better CPUs and larger RAMs in microcomputers.

A spreadsheet can be considered to be both an applications package and a programming language. Although most of the code exists in the spreadsheet software, users have to supply or "program" their own formulas or equations. More detailed information on the use of these tools in engineering is given in Chapter 18.

Specialized Applications. The subject of specialized application software is too numerous and too broad to cover in any degree of detail in this text. It extends from the spreadsheets discussed in the preceding section, to equation solvers, project schedulers, specialized design programs, and large-scale generalized design programs. The latter class includes computer-aided design/computer-aided manufacturing (CAD/CAM) packages. We discuss only two specific categories in this section.

In the first category, we include programs that provide the framework for a solution and require the user to supply some piece of the framework. The microcomputer revolution has produced some major contributions in this category, starting with the spreadsheet discussed in the preceding section, but it is probable that the best is yet to come. Three examples, TK!Solver,[11] FORMULA/ONE,[12] and muMath/muSimp,[13] are probably early indicators of the future trend.

The first two of these, TK!Solver and FORMULA/ONE, solve simultaneous linear and nonlinear algebraic equations. These application packages, are significant because they free the user from the tasks of algorithm construction and program development. For example, if a mathematical model is developed in the engineering design process, then, assuming that the equations meet the restrictions of the programs, the designer can go directly to the solution. The designer simply types in the equations and commands the programs to solve them. The equations do not have to be entered with the unknowns on the left side of the equations, as is required by nearly all other programming forms. The programs take care of all those details.

TK!Solver and FORMULA/ONE are convenient for doing parameter analyses because they allow the use of tables of inputs and then form tables of the corresponding outputs. Graphing of the inputs and outputs can also be accomplished.

muMath/muSimp is a rational mathematics package that can do algebraic manipulation, matrix algebra, trigonometry, and differential and integral calculus.

The second category of applications package of which students need to be aware covers the mathematics and statistical *libraries* that are generally available on any large computer installation. The user

---

[11]Trademark of Software Arts, Incorporated.

[12]Trademark of Alloy Computer Products, Inc.

[13]Trademark of The Soft Warehouse

provides the framework for the solution in the form of a program and the libraries are used within that framework. These packages consist of many subprograms or subroutines that can be called by high-level languages in order to perform calculations that are common to many different problems (high-level languages are discussed in the next section; subprograms and subroutines are discussed in Chapter 17).

For example, if users wanted to solve a set of simultaneous equations, they might calculate the coefficients in the equations and then use an equation solver from a mathematics and statistical library to solve the equations. Similarly, users could perform integration, curve fitting, matrix inversion, statistical analysis, and other tasks by first writing programs that define, collect, or calculate preliminary data and then turning the bulk of the work over to a library routine.

Doing It from Scratch

When existing applications programs will not fill your need, there is no alternative but to write your own program. Although we call this section "Doing It from Scratch," we do not necessarily mean that you must start with the binary instructions that can be understood by your CPU. But you must start programming in some language that the machine can understand or decipher. We now take a brief look at some programming languages.

Machine Language. The only instructions that a computer can accept *directly* are those written in *machine language*. Machine language uses strings of binary numbers to represent the individual instructions and addresses of operands. These are the instructions that must reside in RAM and be processed by the CPU in order to carry out any type of operations on the computer.

As pointed out in Chapter 15, each model or make of central processor unit has its own machine language—its own set of instructions that must match its capabilities. Knowledge of one machine language is generally not directly transferable to another machine.

Instructions keyed into hand-held calculators are essentially machine language instructions. These calculations usually assign an integer to each key that identifies the row and column position of the key on the keyboard. As the calculator is programmed, the integers representing the keys—not the names of the keys—are stored in the program memory. You can see these stored values by recalling the program memory one step at a time.

Although machine language programming might be suitable for small CPU machines, it is unthinkable for programs on larger machines because the numeric code associated with each instruction must be memorized or found in a manual, and the addresses of all operands must be assigned and controlled by the programmer. This is a mind-numbing, often error-prone process.

Specialists in machine language programming must spend considerable time in training. Unfortunately, the skills acquired in this training are not highly portable. When a new CPU comes along, a new set of rules must be learned.

Fortunately, only the people who need to write machine-oriented compilers or assemblers—discussed below—are those who need to work in this language. However, these people make life possible for the rest of us.

Compilers. Soon after computers first came into use, it became obvious that a more human-oriented form of programming than machine language had to be devised. What was needed was a way of allowing users to program in forms (characters, symbols, and syntax) that were more meaningful to them, and then providing them with a way to translate these into instructions the machine could use. Since the computer was exceptionally able in carrying out laborious and tedious tasks, it was an ideal candidate for accomplishing the intricate job of translating human-oriented forms of programming into the machine language forms it needed. That is, special programs were devised that instructed the computer to prepare a machine language program for itself.

Programs that translate other programs into machine language programs are called *compilers, translators,* or *processors*. Compilers are supplied by computer manufacturers and by companies that specialize in designing such software. Any minicomputer or mainframe computer installed to serve a number of users will have a number of compilers in internal memory (usually ROM) or stored on peripheral equipment. These can be called into use—executed, in computer terminology—by appropriate job control statements.

Languages that need to be translated into machine language are generally classified into two categories. *Interpretive* languages are those that get translated, line by line, as they are being executed. No complete, compiled version is saved; when the program is re-executed, each individual program step must be recompiled again. *Compiled* languages are those in which all the program steps are compiled together when the programming is finished and prior to execution. At run time, the compiled version—often called a *load module*—is executed.

The programming language BASIC,[14] discussed below under "High-Level Languages," is the best known example of an interpretive language. Interpretive languages are convenient to use because they are usually coupled with a built-in editor that is activated automatically when a compilation error is encountered. They also go into execution more quickly than compiled languages, because the sometimes lengthy process of compiling a complete program for a compiled language is not required. However, interpretive languages execute more slowly than compiled languages, since before each step is carried out, it must first be compiled. Since scientific and engineering programs often rely heavily on looping—doing the same calculations over and over until convergence is obtained—slow execution can be a severe detriment.

Assembly Language. The first significant step above machine language programming is to teach the machine to recognize program steps written in symbolic rather than a binary machine language code. This is done by designing a special compiler called an

[14]There are "compiled" versions of BASIC commercially available.

assembly language program, symbolic program, or simply an *assembler*. Assemblers are machine oriented in the sense that, in general, each assembler statement is compiled into one machine language statement. Because of the machine orientation, knowledge of the symbolic statements used for one computer is only of partial help in writing symbolic programs for another model.

Assemblers are written with mnemonic names in place of the binary codes representing operators or operand addresses. The *mnemonic* names—that is, names that bring to mind the operand—are assigned by the designers of the compiler. Accordingly, A may represent addition, S subtraction, STO store, and so on. This symbolism is much easier to remember than strings of binary numbers that might actually represent these same operations in machine language.

Symbolic names are also used for memory locations. It is the task of the assembly language programmer to select names that are both acceptable to the compiler and are also mnemonic in that they resemble the English name for the quantity.

This book will make no attempt to describe the details of any assembler. Students should realize, however, that assemblers are an important link in the "education" of a computer. One of the first things a computer manufacturer must do when a new model is introduced is to ensure that an assembler is made available. The assembler must, of course, be written in machine language, because a computer just coming off the assembly line understands only machine language. The assembler then becomes the basic tool for *systems programmers* in writing higher-level compilers, editors, and operating systems.

Fortunately, most of us do not have to program in either machine language or assembly language. There are easier ways of using the machine that suffice. However, with these easier ways also come penalties. When we take the easier route, we have less freedom in deciding how we will do things, in how we will take advantage of the various parts of the computer, in how efficiently we use the available memory, and in how fast we can conduct our computations. For flexibility and speed, it is hard to beat a well-written assembly language code.

High-Level Languages. Two major developments were needed for electronic digital computation to attract the broad base of users it has today. First, it was necessary to design programming languages and their associated compilers so that the programs could be written in terms of the problem to be treated rather than in terms of an elementary instruction set hardwired into the computer. Second, it was necessary to standardize the languages to the greatest extent possible so that users would not need to learn a new language each time they had occasion to use a different computer. That is, *machine-independent* or *portable, problem-oriented* languages were needed.

Grace Murray Hopper legitimized work in the first of these two areas in 1952, with her invention of the first compiler for a language known as A-2. Work since that time has resulted in the creation of a number of high-level languages that have received wide acceptance. Such languages not only make programming easier but permit the programmer to adapt programs to most any computer simply by learning a few job control statements and ensuring that there is available a compiler for each new machine.

The following are a few of the high-level languages you are likely to encounter.

FORTRAN. FORTRAN, an abbreviation for FORmula TRANslation, was the first widely accepted high-level language. It was designed so that its statements would visually resemble conventional algebraic notation, as much as the computer symbols of that time permitted. For instance, the multiplication of two variables, A and B, to yield a third variable, C, is represented by the following line of code:

```
C = A * B
```

Similarly, addition is represented in FORTRAN by

```
C = A + B
```

The FORTRAN statement

```
C = A * (B + D)
```

is an instruction to give C the value of A multiplied by the sum of B and D. These simple instructions are much easier to follow and understand than would be equivalent abstract and tedious machine language code. A FORTRAN compiler does the work of translating these instructions into machine language. Since each FORTRAN instruction produces about 10 machine language instructions when "compiled," writing in FORTRAN represents a considerable savings in time and complexity.

FORTRAN originated at the IBM Corporation in the mid-1950s and has been periodically upgraded to new standards. FORTRAN II was the first version widely used. FORTRAN IV, a more standard version of FORTRAN, published by the American National Standards Institute (ANSI) in the late 1960s, was the basis of most FORTRAN textbooks for the next decade. Much existing code is written in FORTRAN IV; indeed, it is still in use in many places.

FORTRAN 77, the newest ANSI standard version, came into being in 1978. This latest version contains the same basic structure as FORTRAN IV, and most programs written in FORTRAN IV will compile properly with a FORTRAN 77 compiler. However, FORTRAN 77 has some additional features that allow it to become a language much more capable of supporting structured programming principles. Structured programming is discussed in more detail in Chapter 17.

FORTRAN is not necessarily restricted to scientific applications just because of its algebraic orientation. It is sufficiently powerful to handle many business problems as well and is often used in their solution.

BASIC. BASIC, an acronym for Beginner's All-purpose Symbolic Instruction Code, was designed primarily as a teaching language using printing or display (CRT) terminals. It was invented by Dartmouth professors John Kemeny and Thomas Kurtz in 1963. Because of its availability, ease of use, and small RAM requirements,

BASIC became *the* language for microcomputers as these electronic marvels have gone from hobbyist's toys to truly sophisticated computers. Now, many beginning programming courses, particularly in high schools, are taught in BASIC. This language has many FORTRAN-like features and, in particular, retains the algebraic-like appearance.

Although BASIC has evolved well beyond the beginner's level in capability, most versions of BASIC are not as convenient as FORTRAN for solving complex analytical problems. For most versions, the major limitations are its inability to make use of generalized subprograms or subroutines, and the slow execution speeds. However, there are several versions of BASIC now available which overcome these problems. In the future, BASIC may indeed be a complete language for technical use.

Others. There are a large number of other high-level languages in use today. These include Pascal, C, COBOL, ALGOL, PL/1, LISP, and Ada, to name only a few. These represent major survivors and maturing newcomers in the programming language evolution. Hundreds of other languages have been developed, but most have failed to survive the competitive environment. High-level languages, once well established, take on a life of their own—much like natural languages and systems of dimensions (dimensions were discussed in Chapter 10). They are not easily displaced even when they retain defects or are subject to theoretical criticism.

FORTRAN remains the major language for scientific and engineering computation. It has been more than adequate for most tasks, although there are doubts that it will suffice in the long term as we grow into a computer graphics world. The technical community will probably continue to court FORTRAN, since it needs to support its huge inventory of FORTRAN programs. It would be quite expensive to rewrite the hundreds of thousands of existing programs in some other language. Languages designed to make FORTRAN obsolete do appear from time to time, but widespread acceptance and controlled evolution have kept FORTRAN abreast of the latest programming concepts and computer technologies.

It is for the reasons noted above that we will discuss some of the intricacies of FORTRAN in Chapter 17. However, in a side-by-side presentation with FORTRAN, we will also discuss BASIC, so that those students with only one of these languages available to them can enjoy the fun of commanding the computer. The first language is always the most difficult to learn, so whether you study the FORTRAN or the BASIC instruction below, you should subsequently find that learning the other is much easier.

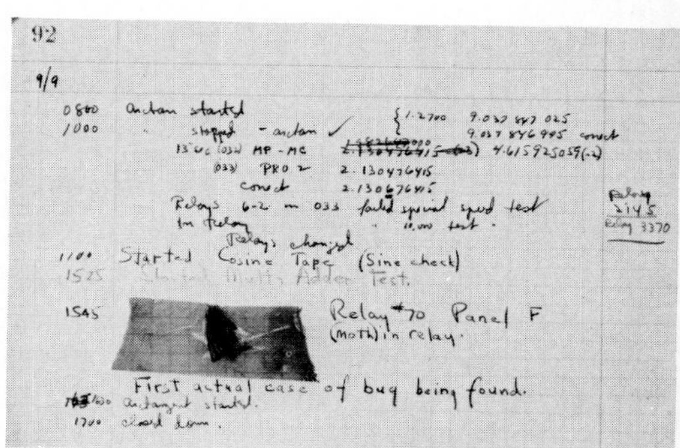

[16-8] The term "bug," meaning a flaw in a program, originated in the early days of computing when the machinery included many electromechanical relays. After the Mark II computer quit one day (September 9, 1947) while Grace Murray Hopper was working on it, she found the cause to be an insect (a bug) which had been caught in one of the relays. She taped the bug into her logbook with the notation, "First actual case of bug being found." This is a photo of that page, which is now preserved in the Naval Museum at Dahlgren, Virginia.

Checking, Debugging, and Misuse

Of course, the most important aspect of any solution technique is that it provide correct answers. However, long and complicated computer solutions often become impossible to check absolutely. All but the simplest of codes have "bugs" [16-8] that can produce erroneous, if not unexpected, results. Therefore, all computer programs must be checked and double checked.

The most often used form of checking is to have the program perform calculations for a situation (e.g., a set of data) in which you know the solution. In writing a program it is usually advisable to initially include a large number of steps that supply intermediate answers in any computerized solution. When the calculation procedure is confirmed by checking the intermediate results, these extra steps can be removed.

Programs are always subject to misuse. In any analyses, there are always assumptions that must be made in the modeling. When computer programs are used in the solution, they, too, include these assumptions. It is easy to forget about all of these original assumptions once the programming is finished and to use the program for problems that violate the assumptions that were made. This is especially true in applications programs—programs where the user does not write code but only inputs the data for a specific problem and then collects the output.

We presented a discussion of errors and mistakes in Chapter 9. Many of the points discussed there also apply to computer calculations.

PROBLEMS

16-1. The allowable load on steel columns is generally computed from a formula selected by evaluation of a slenderness ratio, ℓ/k, where ℓ is the length of the column and k is the least radius of gyration in the same units. (The radius of gyration is a property of the cross section of the column and varies for different shapes. For example, the radius of gyration of a solid cylindrical column is one-fourth of the diameter.) The American Bridge Company at one time used the following formulas with the slenderness ratio acting as a decision variable:

Slenderness ratio	*Allowable load (psi)*
0–60	13,000
60–120	19,000 − 100 (ℓ/k)
120–200	13,000 − 50 (ℓ/k)

Draw both a general flowchart and a detailed flowchart for computing the allowable load, given ℓ and k. Follow through your methodology and determine the maximum allowable loads for each of the following columns.

Length	*Radius of gyration*
(a) 24 ft	6.3 in.
(b) 16 in.	$3\frac{1}{2}$ in.
(c) 600 in.	9.4 in.
(d) 14.6 m	8.16 cm

16-2. Assume that you take your little black book of dates and enter the names and phone numbers into a computer file. You now want to be able to enter either a name and retrieve the phone number, or enter a phone number and retrieve a name. There are only 100 names, but they consist of a last name, comma, space, first name, as in *James, Tony*. Draw a general flowchart that could serve as a starting point for coding this problem.

16-3. What is the name of the operating system on the computers that you use? Is it capable of handling multiple users?

16-4. For the operating system on your computer, what is the job control statement that would allow you to:

a. Obtain a listing of the files that you have stored in low-speed memory.

b. Copy a file from one part of low-speed memory to another part of low-speed memory.

c. Format a floppy diskette.

d. Execute a load module by the name of DRAW.

16-5. Does your institution offer a course in machine language programming? If so, to what computer is it specific?

16-6. What does the word *mnemonic* mean?

16-7. Does your institution offer courses in assembly language programming? If so, to what computer is it specific?

16-8. What compilers are available on the computers at your institution?

16-9. What applications programs are available on the computers at your institution?

16-10. What interpretive languages are available on the computers at your institution?

16-11. In which high-level languages does your institution offer programming courses?

16-12. Using library resources, write a one-page paper on the super language Ada. Why was it invented? For whom is it intended?

Chapter 17

Programming in FORTRAN and BASIC

There will be many times in your engineering career when you will need to construct your own custom-made applications codes. In the following pages we cover the rudiments of programming with FORTRAN and BASIC, two high-level programming languages, in order to equip you for situations such as these.

There are various features of these two languages that are product dependent. That is, depending on what FORTRAN compiler or what BASIC interpreter you will be using (in other words, who wrote the software that compiles or interprets your programs), you may find small departures from the details described here. When in doubt, always check the manual that describes the product you are using.

We will not cover any of the details of how you generate your program code (generally called the source code) or what you must do to get your code to execute on the computer you will be using. Your instructor or computer center will have to supply those details.

PROGRAMMING HABITS

It is extremely important to develop good programming habits. In fact, your programming habits are far more important than all the intricacies and fancy features of the language that you first learn. For this reason, and because of space limitations in this text, we will cover only enough FORTRAN and BASIC to get you started. Once you learn the fundamentals, you can extend your programming skills without formal instruction.

There are two desirable habits that you should acquire. The first habit is that of documenting your code. This includes adding a sufficient number of comments in the actual code itself, using comment statements which are covered below, as well as keeping documentation that is separate from the program. In the latter case, you are urged to keep up-to-date flowcharts and/or pseudocode, a written list of the inputs and outputs, a written cursory discussion of the type of calculation procedures you have used, and a list of the equations that you are solving.

The second habit that you should develop is to keep your program as simple as possible. Experience tells us that programmers never document their work well enough. (However, we have not given up on trying to teach students to document.) There never seems to be enough time to document programs while they are being written and debugged. Flowcharts, for example, are seldom updated with changes that are invariably made in programs. Incorrect documentation is little better than no documentation.

In cases where the documentation is poor, incorrect, or nonexistent, the only recourse in debugging or deciphering the code is to literally "read" it. Simple programs are the easiest to follow, with or without documentation. The word "simple" is used here to mean easy to follow—flowing almost naturally from one part to another—but not necessarily being unsophisticated. To keep a program simple, you should let it flow in a *top-down* fashion. Commonly referred to as *structured programming,* this top-down flow means that program *execution* should progress from the beginning of the program to the end without jumping around within the program. Good code is also written in this way so that it can be checked out and debugged in a similar fashion.

Structured programming is now a well-established concept that is worthwhile to adhere to as closely as a language will allow. Much of the code written prior to this concept is referred to as *spaghetti code* because, during execution, it jumps around throughout the program and does not simply flow from the beginning of the program to the end. It is as if there were loose ends everywhere, making the program extremely difficult to follow or debug.

Under the top-down strategy, the main program contains the overall logic and should be written first. It might flow, in very general pseudocode, in the following way:

```
input the data,
do the following:
  calculate,
end the do,
check the results,
if such-and-such
  do this,
if not
  do that,
compute other things,
output the results.
```

If the code becomes lengthy, the programmer should not hesitate to use subprograms (discussed below) for things such as the generic program parts, *this* and *that* in the pseudocode above.

Neither FORTRAN nor BASIC demand that programs be structured in top-down fashion. Therefore, the idea is more one of philosophy than of required execution. We urge you to adopt this philosophy. Your programming tasks will greatly benefit from it. The steps of programming can quickly get complicated and confusing. If you do not make a special effort to keep the steps simple, you are likely to find yourself confused and unable to complete your task.

CONSTANTS, VARIABLES, AND OPERATIONS

Before we proceed into too much programming detail, let us consider some general aspects of storing information and carrying out arithmetic operations that are the same for both FORTRAN and BASIC.

Constants

There are five *types* of fixed quantities or *constants* that can be used in FORTRAN and BASIC: *REAL, INTEGER,* (*CHARACTER* in FORTRAN or *STRING* in BASIC), *DOUBLE PRECISION,* and *COMPLEX* (FORTRAN only). Because of space limitations, we will not discuss the last two in this list.

In the equation

$$X = 3.14159/Z + 21 \qquad (17\text{-}1)$$

the constants are 21 and 3.14159. These are numbers that would not change if a computer program that contained the equation were executed. It may not make any difference to you whether[1] (21) or (21.) is the value used in the calculation of an X given a value of Z. However, to the computer, *there is a critical difference* between (21) and (21.), and it is important that you understand this difference.

Whole numbers or *integers* can easily be represented as binary numbers. For example, 21_{10} is the same as 10101_2. Therefore, the binary representation of the number 21_{10} could be stored in the computer, since the computer can only process binary information. There are, of course, some limitations as to the size of numbers that can be stored in this binary manner. If the memory registers in the computer's CPU have an 8-bit (one-byte) limitation, the largest integer that could possibly be stored is $11111111_2 = 255_{10}$. However, one of the bits is used to denote the sign of the integer (plus or negative), which leaves only 7 bits for integer representation. The largest signed integer would then be 1111111_2, which is 127_{10}. In an n-bit (n/8-byte) machine, the limit is $(2^{n-1} - 1)$. In a mainframe computer having a 32-bit word length (four bytes), the largest integer that can be accommodated is $2{,}147{,}483{,}647_{10}$. Although this may at first appear to be a very large number, it is a severe limitation for many engineering calculations. On some compilers there is a way to increase this limit, but even this would not solve the problem of representing numbers that have a fractional part.

The number 3.14159 in Equation (17-1), of course, is not an integer and cannot be represented by a binary number in the manner described in the preceding paragraph. Numbers like 3.14159 are termed *floating-point* or *REAL* numbers and are stored differently than INTEGERS.

One simple method of representing a floating-point number (say, 12.34) is first to express it as a number less than 1, called the mantissa (here .1234) multiplied by 10^x, where x is some signed INTEGER exponent (here, +2). The collection of numbers constituting the mantissa (without the decimal point, here 1234) can then be represented as an INTEGER or binary number. The resulting floating-point representation will then consist of a series of bytes that represent the integer equivalent of the mantissa and a series of bytes (usually one) which represent the exponent. Table 17-1 shows

---

[1]We are using parentheses here only to make a clear distinction between when the decimal point is included with the number and when it is not.

TABLE 17-1 DECIMAL PRECISION AND RANGE OF SELECTED BINARY REGISTERS

	REAL Numbers *(single precision)*		*(double precision)*	
Register length. Used as follows:	32 bits	36 bits	64 bits	72 bits
Sign of the value	1	1	1	1
Fractional part of the value	24[a]	27	56[a]	60
Sign (or bias) of the exponent	1	1	1	1
Value of the exponent	7	7	7	10
Equivalent significant decimal digits[b]	7.2	8.1	16.8	18.0
Maximum magnitude for decimal exponent[c]	38	38	38	308

[a]This includes one so-called hidden digit.
[b]Equivalent significant decimal digits = $\log_{10}(2) \times$ (number of bits in the fractional part).
[c]Maximum decimal magnitude = $\log_{10}(2) \times 2^n$, where n is the number of bits assigned to the exponent.

how the two types of real numbers are represented. *Single-precision* numbers, in a 36-bit word computer, usually consist of four bytes:[2] three bytes for the fractional part and one for the exponent. However, for this word size, *double-precision* numbers allocate seven bytes (see footnote 2) for the fraction and usually one byte for the exponent. By allocating more bytes to the mantissa, more accuracy is obtained in the floating-point representation of each number and in any resulting calculations using double-precision numbers.

It is important that you understand that the computer *must* know if the data it is processing are INTEGER or REAL. To distinguish between INTEGER and REAL constants, *a programmer* simply includes the decimal point for REAL constants or omits it for INTEGER constants. Thus (3.14159) is REAL, (21) is INTEGER, and (21.) is REAL.[3] This means that when their binary representations are compared, (21) is never equal to (21.). Most compilers allow *mixed-mode arithmetic* (i.e., arithmetic operations that involve both REALs and INTEGERs), but it should only be used with extreme caution.[4]

[2]This number is not universally valid. Check the number for your installation.

[3]In many BASICs, if you entered a floating-point number that did not have a decimal part, such as (21.) into a program, the decimal point would show as a ! when listed back later, for example as (21!). The ! character is a variable name suffix that denotes single-precision real numbers. It is just telling you that the number entered into the code is a single-precision floating-point number. This does not happen with floating-point numbers that have a decimal part, such as (3.14159).

[4]Some languages are much more lenient in their treatment of mixed-mode expressions. Most BASIC interpreters, for example, will allow integer values to be mixed in with floating-point numbers in expressions by simply translating all of the integers to their floating-point equivalents prior to evaluating the expressions.

The third type of constant is the CHARACTER. It is used for representing strings of characters (combinations of letters of the alphabet, numbers, and certain symbols) that the programmer does not intend to use as numeric strings. Such strings are referred to as *alphanumeric strings* and are useful for entering and printing labels, titles, and so on. Typically, these strings are stored in the computer as a collection of bytes, each byte representing a code for one of the characters in the string (see, e.g., the ASCII character codes in Table 15-1). The *programmer* tells the computer that a particular alphanumeric string is to be a CHARACTER constant by surrounding the desired string with single quotes (e.g., 'TITLE') in FORTRAN or double quotes (e.g., "TITLE") in BASIC.

Variables

The parameters X and Z in Equation (17-1) represent *variables* that can be assigned values during execution. That is, their values can change at various times while computations are being made. Each variable is coded into a program using a unique name chosen, within broad limits, by the programmer. We will discuss these limits below. Recall that in order for the computer to perform any calculations using the ALU (arithmetic and logic unit) it must first fetch the appropriate data from locations in memory. It must also be able to interpret the binary information that has been fetched. From our previous discussions of INTEGER, REAL, and CHARACTER constants it should be obvious that this interpretation must involve knowing whether to use the binary information fetched from storage directly as a number (an INTEGER), to use some of the bits as a "mantissa" and others as an exponent (a REAL number), or to use the bits as a representation of a code for a certain character (a CHARACTER). Therefore, as in the case of constants, the computer must know whether variables in the program are to have INTEGER, REAL, or CHARACTER values (the same assortment available for constants).

This necessary *declaration* of variable types can be done in either of two ways in FORTRAN and BASIC. The first, and preferable way, is to use *declaration statements* (we will discuss declaration statements under nonexecutable statements below). The second declaration method is to make use of appropriate variable names. Both FORTRAN and BASIC permit the latter declaration technique, but most other languages (such as Pascal and C) do not. For this reason, using appropriate names without declaring types for all variables is poor programming practice.

In FORTRAN, declarations via declaration statements are easily made, but in BASIC there are some limitations on declarations by this technique. As a result of the latter shortcoming, we almost always have to resort to declaration through variable names in BASIC. But as long as we follow the appropriate declaration rules of each language, the FORTRAN compiler or the BASIC interpreter takes care of all the necessary interpretations that must be made on the binary data stored in memory.

Operations

The following arithmtic operations can be performed in both FORTRAN and BASIC:

- *Exponentiation* (signified by ** in FORTRAN or ^ in BASIC):

  X^2 is written as X**2 (FORTRAN) or X ^ 2 (BASIC)

- *Multiplication* (signified by * in both FORTRAN and BASIC):

  $2 \cdot Y$ is written as 2*Y

- *Division* (signified by / in both FORTRAN and BASIC):

  $X \div 3$, or $\frac{X}{3}$ is written as X/3

- *Addition* (signified by + in both FORTRAN and BASIC):

  $3 + X$ is written as 3 + X

- *Subtraction* (signified by − in both FORTRAN and BASIC):

  $3 - X$ is written as 3 − X

In statements where more than one arithmetic operation is performed, there is an established set of *operator precedence* rules that are invoked automatically. These rules specify that:

1. Expressions within parentheses are first evaluated using order rules 2 to 5 below.
2. All exponentiations are performed first.
3. All multiplications and divisions are performed next.
4. All additions and subtractions are performed next.
5. Within any of rules 1 to 4, the machine processes the instructions in left-to-right order. For example:

   2*Y**3/X is computed as $(2Y^3)/X$

   2*Y/X*Z is computed as $(2Y/X) \cdot Z$ or $2YZ/X$

   2*X+Y*Z is computed as $(2X)+(YZ)$

Exponentiation deserves some extra discussion. In this operation; it *is* permissible to

- Raise an INTEGER to an INTEGER power, provided that the power is greater than 0.
- Raise a REAL to an INTEGER power and raise a REAL to a REAL power, although it is less efficient and more time consuming than raising a REAL to an INTEGER power.

But it *is not* permissible to

- Raise in INTEGER to a REAL power, due to the way in which integers are stored.

ARRAYS

We have discussed representing variables involved in computations by using a unique name for each variable we want to represent. To represent a wide variety of variables, we could then invent a large number of names, one for each value. In many cases, this would be acceptable, particularly if each variable was not related to any of the other variables in some fashion.

Imagine the situation, however, if a table of information consisting of 100 rows and 100 columns of data needed to be stored in the computer—the user would have to come up with 10,000 unique variable names! For such situations, it is desirable to have the ability to reference the entire table under one name. Each *element*, or member, of the table could then be referenced by specifying the table name and the row and column number of the desired entry as follows:

name(row,column)

where *name* is a unique name for the table that meets all the rules for variable names.

The expression

X = TOLRNC (3,41) (17-2)

when executed, would take a copy of the value stored in the third row and the forty-first column of table TOLRNC and assign it to the variable X (we will say more about the assign statements later).

If the table consisted of only one column, we could access each element simply by using

name(row)

In addition, if the table was more than one page long (e.g., a table with three independent variables such as the tables found in Appendix III, we could express a typical entry using

name(row,column,page)

As you can see, this is much easier than devising large numbers of unique variable names!

What we have just described is the implementation of *arrays* or *subscripted variables* in both FORTRAN and BASIC. Tables that consist of only one column are called *one-dimensional arrays*. Similarly, tables with rows and columns are called *two-dimensional arrays* and tables with rows, columns, and pages are called *three-dimensional arrays*. The "row, column, page" values are defined as *subscripts* or *indexes* of the array and are restricted to INTEGER constants, INTEGER variables, or INTEGER-valued expressions (arithmetic expressions permitted).[5]

Engineering computer programs tend to be very *recursive* in nature. That is, they tend to do things over and over again in carrying out their computations. The looping and branching flowchart of Figure [16-7] shows this behavior quite well. In recursive calculations, then, it is usually convenient to use arrays for variables that are computed or involved in such calculations, since only the subscripts would have to change, and not the variable name, during each iteration.

The computer, during compilation, can count the number of nonarray or *un*subscripted variables in a program in order to know

---

[5]Many FORTRAN compilers default to the requirement that the smallest number possible for an array subscript is 1, but some FORTRAN compilers allow users to override the default and set their own lower limit. In most versions of BASIC, the default lower limit is 0, but users can override it and set it to 1.

how many memory locations will be required. However, the computer does not like to count the array or subscripted variable elements. But it still needs to know how many memory locations to set aside for storing these. Therefore, the programmer must tell the computer the maximum number of elements it is to expect for each array.[6] Since this is done differently in FORTRAN than in BASIC, we will discuss the mechanics of this later.

Of course, the programmer must also declare the type of the subscripted arrays (INTEGER, REAL, or CHARACTER) so that the binary string of numbers stored for each element gets interpreted correctly when computations are made.

STATEMENTS

FORTRAN and BASIC programs, indeed programs written in any language, consist of *statements* or *lines of code*. These are the smallest units that carry a complete thought or instruction to the computer. They are analogous to sentences in conversational or natural languages and to cells in spreadsheets (spreadsheets are discussed in Chapter 18). Statements, generally, can be classified into the following five types:

1. Nonexecutable or specification statements.
2. Input/output statements (I/O).
3. Assign or arithmetic statements.
4. Program flow control statements.
5. Comments.

The statements that will probably be most easily recognized are the assign or arithmetic statements, for they resemble algebraic equations. These contain the symbols (** or ^, *, /, +, −) that denote arithmetic operators as described previously. However, we shall discuss all five types of statements in more detail.

An Example Program

Before delving further into the details of programming, it is instructive to examine some actual code. As an example, let us use one of the problems introduced in Chapter 16. The problem involves determining the size of the central angle whose segment forms a given fraction of the area of the circle; in particular we wish to find α given f from Equation (16-4). The flowchart for a possible algorithm is given in Figure [16-7]. This same problem will be solved with a spreadsheet in Chapter 18 (Example 17-9).

A possible coding of this problem is shown in Topic 17-1. We have coded this problem in a very simple way, to avoid confusing the new programmer. Of course, the problem is not complicated; thus we would not expect the code to be complex. However, the present

[6]For example, in Equation (17-2), the computer would have to have at least 3×41 spaces reserved for the array TOLRNC.

programs are not very robust,[7] in that they do not check on the validity of the inputs and do not recover from operator error. The programs are also not very "user friendly" in their present form.[8]

Two things should be apparent in studying the programs in Topic 17-1. One is that there is a great deal of similarity between the FORTRAN code and the BASIC code. The other is that it is not too difficult to "read" the code from top to bottom.

Each line in the programs is a statement. The FORTRAN program starts out by telling the machine that the variables to be used will be of type REAL. (There is not a convenient way to do this in BASIC, so we will allow the names to determine the type.)

Both FORTRAN and BASIC then assign a value of 3.14159 to the variable PI in what looks like an algebraic relationship. PI = 3.14159 (these are assign statements; that is, they assign a variable a value). Since PI never again appears to the left of an equal sign (i.e., it never gets *re*assigned), it becomes, in essence, a variable of constant magnitude.

PI, not the constant 3.14159, is entered into equations in the programs in order to make the equations look more familiar, or more mnemonic, to the reader. You will find 2.*PI appearing in the equation for ALPHA just below line 10 in the FORTRAN program or in line 170 in the BASIC program (Topic 17-1). That product could have been entered simply as 6.28318, but the significance of the true origin of the term (i.e., $2 \cdot \pi$) would probably be lost to any prospective user who must "read" the code.

After establishing the value of PI, the programs next accept input data. Notice that, in FORTRAN, there is no prompting of the user for the data; the user must know what to enter and when. The input data include the value of the fraction for which the user desires to know the subtended angle and the convergence tolerance. The latter quantity tells the program how close it must come to the desired fraction in order to satisfy the desires of the user.

In FORTRAN, the data are collected by **READ** statements. In the BASIC program, the data are requested and collected by the **INPUT** statements.

Next, a first guess is made for ALPHA prior to seeking a more accurate value through iteration. The iteration itself is conducted within a **WHILE** loop, a structure or arrangement of statements that we will discuss shortly. Unfortunately, FORTRAN 77 does not have a **WHILE** statement. Since nearly all other modern high-level languages have such a construct, we believe it is important to pretend that FORTRAN does also. That is what we have done here.

Finally, when the iteration process produces a value of the fraction that is acceptable, the results are listed by the output decoders (CRTs or printers), so the user can see them. The outputs are pro-

[7]Robustness is an important property of good programs. It is synonymous with "idiot proof," but it is perhaps better described as "the ability to take a licking and keep on ticking." If a robust code receives bad data on input, it is able to recover and ask the user to try again. The code we present in Topic 17-1 is not robust, in that if it receives a fraction that is larger that 1.0 or smaller that 0, it either "bombs" or gives incorrect results.

[8]"User friendliness" is the trait of working in harmony with the user, anticipating his or her needs, echoing the input data, making clear what is needed next, and clearly labeling the output.

duced by the **PRINT** statements, although **WRITE** statements, in FORTRAN, or **LPRINT** statements, in BASIC, could have been used.

Before we go into more detail concerning the intricacies of FORTRAN and BASIC, we would like to point out that the programs presented above flow from top to bottom in a natural way, making them relatively easy to "read." Even as a new programmer, perhaps you can grasp the essence of the code. The programs are not filled with statements that require the execution order of the program to jump to random places within the code. Of course, in this case, the problem being solved is rather simple, making it particularly easy to write simple code.

Suppose, for the remainder of this selection, that the problem were of such a complexity that it would be difficult to keep the

Topic 17-1 FORTRAN

Example Program

```
C THIS PROGRAM SOLVES F=(ALPHA-SIN(ALPHA))/(2*PI) FOR ALPHA, GIVEN AN F
C THE NEWTON-RAPHSON ITERATIVE TECHNIQUE IS USED WITH THE CONVERGENCE CRITERIA
C  THAT  ABS(F-(ALPHA-SIN(ALPHA))/2*PI) .LE. DELF
C                               DLE 12-2-84
C
      REAL F, FC, DELF, ALPHA, PI, ALPHAD
      PI = 3.14159
C
C   ENTER THE FRACTION DESIRED
      READ (5,*) F
C
C   ENTER THE ERROR TOLERANCE
      READ (5,*) DELF
C
C   CALCULATE THE FIRST GUESS FOR  ALPHA
      ALPHA=(12.*PI*F)**(1./3.)
C
C **** WHILE TO ITERATE FOR ALPHA
10    CONTINUE
      IF(ABS(F-(ALPHA-SIN(ALPHA))/(2.*PI)).GT.DELF) THEN
          ALPHA=ALPHA-(ALPHA-SIN(ALPHA)-2.*PI*F)/(1.-COS(ALPHA))
          GO TO 10
      END IF
C **** ENDWHILE
C
C WRAP-UP THE CALCULATIONS AND OUTPUT RESULTS
      ALPHAD=ALPHA*180./PI
      FC=(ALPHA-SIN(ALPHA))/(2.*PI)
      PRINT *, F, FC, ALPHA, ALPHAD
      END
```

various major parts of the code simple (the parts such as the initialization, the collection of inputs, the iteration, and the listing of outputs). In this case, if we were to code the problem entirely in one main program, it could become quite complex and difficult to decipher. For this reason, we should consider using *subprograms*. Let's use this technique and recode the same problem of finding the central angle containing a given fraction of the area of a circle. It might take the form shown in Topic 17-2.

Notice in Topic 17-2, that the main program serves only to initialize the problem and to control the order in which the subprograms are called and used. Each subprogram (the subprograms being used here are referred to as *subroutines*) carries out one of the important steps of the overall program and returns to the line in the code that follows the call (as will be shown in Topic 17-7, the call, in

Topic 17-1 BASIC

Example Program

```
10 PI =3.14159
20 INPUT " INPUT THE FRACTION DESIRED  ",F
80       INPUT " ENTER THE CONVERGENCE TOLERANCE DSIRED  ", DELF
120 REM
130 REM CALCULATE THE FIRST GUESS FOR ALPHA
140 ALPHA=(12*PI*F)^(1/3)
160 WHILE ABS(F-(ALPHA-SIN(ALPHA))/(2*PI)) > DELF
170    ALPHA=ALPHA-(ALPHA-SIN(ALPHA)-2*PI*F)/(1-COS(ALPHA))
180 WEND
190 REM
200 REM WRAP-UP THE CALCULATIONS AND OUTPUT RESULTS
220          ALPHAD=ALPHA*180/PI
230          FC=(ALPHA-SIN(ALPHA))/(2*PI)
310          PRINT F, FC, ALPHA, ALPHAD
350 END
```

Topic 17-2 FORTRAN

Example Program

```
C THIS PROGRAM SOLVES F=(ALPHA-SIN(ALPHA))/(2*PI) FOR ALPHA, GIVEN AN F
C THE NEWTON-RAPHSON ITERATIVE TECHNIQUE IS USED WITH THE CONVERGENCE CRITERIA
C  THAT  ABS(F-(ALPHA-SIN(ALPHA))/2*PI) .LE. DELF
C                                 DLE 12-2-84
C
      REAL F,DELF,ALPHA,ALPHAD,FC,PI
      PI = 3.14159
      CALL READIN(F,DELF)
C
C CALCULATE THE FIRST GUESS FOR  ALPHA
      ALPHA=(12.*PI*F)**(1./3.)
      CALL ITERATE(ALPHA,F,DELF)
C
C WRAP-UP THE CALCULATIONS AND OUTPUT RESULTS
      ALPHAD=ALPHA*180./PI
      FC=(ALPHA-SIN(ALPHA))/(2.*PI)
      CALL OUTPUT(ALPHA,F,ALPHAD,FC)
      END

      SUBROUTINE READIN(G,DEL)
      REAL G, DEL
      READ (5,*) G
      READ (5,*) DEL
      RETURN
      END

      SUBROUTINE ITERATE(ALP,FR,DE)
      REAL F, ALPHA, DELF, PI
C **** WHILE TO ITERATE FOR ALP
10    CONTINUE
        IF(ABS(FR-(ALP-SIN(ALP))/(2.*PI)).GT.DE) THEN
           ALP=ALP-(ALP-SIN(ALP)-2.*PI*FR)/(1.-COS(ALP))
           GO TO 10
        ENDIF
C **** ENDWHILE
      RETURN
      END

      SUBROUTINE OUTPUT(ALPHA,F,ALPHAD, FC)
      REAL F, FC, ALPHA, ALPHAD
      PRINT *, F, FC, ALPHA, ALPHAD
      RETURN
      END
```

Topic 17-2 BASIC

Example Program

```
10 PI =3.14159
20 REM COLLECT THE INPUT INFORMATION
30 GOSUB 520
40 REM PERFORM THE ITERATION
50 GOSUB 720
60 REM WRAP-UP THE CALCULATIONS AND OUTPUT RESULTS
70 GOSUB 820
80 END
520 INPUT " INPUT THE FRACTION DESIRED  ",F
530 PRINT " THE ENTERED NUMBER WAS  ",F
540 IF(F<0) THEN PRINT "THE FRACTION MUST NOT BE NEGATIVE": END
545 REM ELSE
550     IF(F>1) THEN PRINT "THE FRACTION MUST NOT BE GREATER THAN 1.": END
560 REM ENDIF
565 REM ENDIF
580 INPUT " ENTER THE CONVERGENCE TOLERANCE DESIRED  ", DELF
590 PRINT " THE ENTERED NUMBER WAS  ",DELF
600 IF(DELF<0) THEN PRINT " THE CONVERGENCE TOLERANCE MAY NOT BE NEGATIVE": END
610 REM ENDIF
620 RETURN
720 REM CALCULATE THE FIRST GUESS FOR ALPHA
730 REM
740 ALPHA=(12*PI*F)^(1/3)
750 PRINT "START ALPHA", ALPHA
760 WHILE ABS(F-(ALPHA-SIN(ALPHA))/(2*PI)) > DELF
770 ALPHA=ALPHA-(ALPHA-SIN(ALPHA)-2*PI*F)/(1-COS(ALPHA))
780 WEND
790 RETURN
800 REM
810 REM CALCULATE ALPHA IN DEGREES
820 ALPHAD=ALPHA*180/PI
830 FC=(ALPHA-SIN(ALPHA))/(2*PI)
870 PRINT
880 PRINT " OUTPUT, WHERE: F IS DESIRED, FC IS CALCULATED FRACTION"
890 PRINT "        ALPHA IS IN RAD., ALPHA(DEG) IS IN DEG."
900 PRINT
905 PRINT "  F               FC             ALPHA        ALPHA(DEG)"
910 PRINT F, FC, ALPHA, ALPHAD
960 RETURN
```

Topic 17-3 FORTRAN

Statement Numbers and Statement Form

STATEMENT NUMBERS

Statement numbers in FORTRAN, if present, must be placed at the beginning of the statement of which they are a part. They may be located anywhere in the first five columns or character locations. This placement, of course, limits their values to numbers from 1 to 99999. You may *not* use any number more than once (sorry, no duplicate statement numbers).

Statement numbers on sequential statements in the program do not need to be in numerical order. That is, statement 30 could appear later in the program than statement 500.

Acceptable examples are as follows:

```
         1         2         3
123456789012345678901234567890..........
10    CONTINUE
 278  CURRENT=VOLT/RESIST
    3 FORMAT(5F10.3)
         GOTO 10
      KP=PCO2/(PCO*PO2**0.5)
```

Notice that the last two statements above do not have statement numbers. In FORTRAN, statement numbers are required in only certain instances, not on all statements. For example:

FORMAT statements need to be numbered so that they can be referenced by input/output statements (see Topic 17-7).

GOTO statements (Topic 17-11) divert execution to a specified statement number.

DO statements (Topic 17-10) must specify what statement numbers end the DO loops.

LENGTH OF STATEMENTS

FORTRAN permits actual statement lengths that are longer than you should ever use. In some FORTRANs this can be as many as 1320 characters, although this may vary from compiler to compiler. This is probably much longer than you would care to debug. It is better from a management standpoint to break long calculations into several shorter ones.

Statements must be located in columns 7 through 72. If you need more space than can be accommodated on one line, you may continue your entry on the next line (again somewhere in columns 7 and 72) if you put a nonblank, *nonzero* character in column 6 of the continuation line. The character that you use in column 6 does not enter your calculation, but serves only to tell the compiler that the present line is a continuation of the previous one.

(Continues on page 508)

Topic 17-3 BASIC

Statement Numbers and Statement Form

STATEMENT NUMBERS

Statement numbers in BASIC must be placed at the beginning of the statement of which they are a part. They must start in the first column or character location. Each statement *must* have a number and the numbers *must* increase monotonically from the first statement to the last. You may *not* use any number more than once (sorry, no duplicate statement numbers). Numbers must generally be in the range 0 to 65529. Acceptable examples:

```
          1         2         3
123456789012345678901234567890..........
10 CONT
278 CURRENT=VOLT/RESIST
301 GOTO 10
```

Unacceptable examples:

```
Unacceptable examples are:
          1         2         3
123456789012345678901234567890..........
      KP=PC02/(PC0*P02**0.5)  no statement number
90 X=A*B                         numbers not monotonically
10 INPUT X                         increasing
```

There is usually a RENUMber command to renumber the statements automatically when you request it.

LENGTH OF STATEMENTS

BASIC statement lengths are generally limited to 255 characters, although this will depend on the BASIC you are using. Check in the manual that describes the BASIC you have. The length will probably be longer than you should ever use and much longer than you would care to debug. It is better from a management standpoint to break long calculations into several shorter ones.

Statements start one space after the statement numbers and proceed from there. They may use any of the remaining columns on the screen or printer.

Examples are:

```
          1         2         3
123456789012345678901234567890..........
10 BERM = 3*LENGTH/HEIGHT
20         BERM=3*LENGTH/HEIGHT
30 INTRST=(BAL-PROFIT(1)+LOSS(3)*RATE*EXP(TIME/CONST)+ EVERY*THING*LEFT
```

FORTRAN, is issued by a statement that says **CALL** *subrtname*; in BASIC, the call is invoked by the statement *Line_no1* **GOSUB** *Line_no2*). The logic of the program, again flowing from the top down, is easy to follow even though each of the subroutines could get quite complex. In trying to find an error that exists in entering data, for example, a user would know that the error most often would be located in the subroutine that handles the input data.

Topic 17-3 FORTRAN

Statement Numbers and Statement Form (Continued)

Example:

```
         1         2         3
123456789012345678901234567890..........
10    BERM = 3*LENGTH/HEIGHT
 10         BERM=3*LENGTH/HEIGHT
   10 INTRST=(BAL-PROFIT(1)+LOSS(3)*RATE*EXP(TIME/CONST)+
     @ EVERY*THING*LEFT
```

Characters typed in columns 73 through 80 are ignored by FORTRAN compilers. Of course, the three statements that have been numbered 10, in the examples above, could not all appear in the same program.

Topic 17-4 FORTRAN

Comments

In FORTRAN, any statement that has a "C" (capital C) in the first column is assumed to be a comment and is ignored when the program is compiled. Examples of proper and improper comment statements are as follows:

Proper:

```
         1         2         3
123456789012345678901234567890..........
C ---- THIS IS A SAMPLE OF A COMMENT STATEMENT
C   IF YOU ARE HERE IT IS BECAUSE THERE ARE SIXTEEN ROOTS
*   YOU ARE HERE
```

Improper:

```
         1         2         3
123456789012345678901234567890..........
  ---- THIS IS NOT A COMMENT STATEMENT-YOU WILL GET AN ERROR WHEN COMPILING
5     HELLO='HELLO'
c     LOWER CASE C'S DO NOT SIGNIFY COMMENTS
```

Statement Line Numbers (Topic 17-3)

One very significant and visual difference between the FORTRAN and the BASIC programs in Topics 17-1 and 17-2 is the use of statement numbers. They appear only on *certain* statements in the FORTRAN program and on *all* statements in the BASIC version. The rules are stated and examples given for each language in Topic 17-3.

Comment Statements (Topic 17-4)

From the FORTRAN and BASIC programs listed in Topics 17-1 and 17-2, it is apparent that there are some lines of code that serve no other purpose than to provide clarification to the reader. However, this is an extremely important function, especially to those persons who try to use the code without the benefit of any supporting documentation. In FORTRAN, these statements contain a C in the first column, while in BASIC they contain the letters REM after the line number. The rules of comment statements are given in Topic 17-4.

The comment statements are ignored during compilation or interpretation and hence do not affect program speed or, in the case of compiled languages, final machine language program size.

Topic 17-4 — BASIC

Comments

In BASIC, any statement that has a "REM," an abbreviation for remark, following one column after the statement number is a comment. You can also add a comment to every line by placing a ' (single quote) after the last useful entry on a statement line and following this with the comment you wish to make. Examples of proper and improper comment statements are as follows:

Proper:

```
         1         2         3
123456789012345678901234567890..........
10 REM ---- THIS IS A SAMPLE OF A COMMENT STATEMENT
20 REM IF YOU ARE HERE IT IS BECAUSE THERE ARE SIXTEEN ROOTS
30 Y=5*EW  ' THIS IS ALSO A VALID COMMENT
40 ' THIS IS ALSO A VALID COMMENT
```

Improper:

```
         1         2         3
123456789012345678901234567890..........
10---- THIS IS NOT A COMMENT STATEMENT-YOU WILL GET AN ERROR
30     HELLO='HELLO'
50     THERE IS NO REM ON THIS LINE FOLLOWING THE NUMBER
```

Variable Names (Topic 17-5)

Both FORTRAN and BASIC have restrictions on variable names that must be observed. In Topic 17-1, these variable names are F, FC, DELF, ALPHA, PI, and ALPHAD. The rules for naming the variables are listed in Topic 17-5. Some of the rules can be overridden. You will notice in Topic 17-5 that variable names can generally be quite long (check with the instruction manual for the compiler you are using, however).

It is good practice to keep your variable names both short and mnemonic. The former will cut down on the number of typing mistakes when composing your code, while the latter will allow you to relate your code better to the problem you are solving.

Nonexecutable or Specification Statements (Topic 17-6)

Nonexecutable statements or specification statements are those statements which set up specifications that are necessary for the execution of the program. In Topic 17-1, the FORTRAN makes use of the

Topic 17-5 FORTRAN
Variable Names

Variable names in FORTRAN generally must begin with a letter (A–Z) and may include up to 40 alphanumeric characters, depending on your compiler. However, name length is compiler dependent, so check the instruction book for the compiler you are using if you really want to use seven or more characters. Most compilers accept at least six characters in a name. Some are limited to six. *For maximum portability of your code to other machines, you should limit yourself to a maximum of six.*

Unless specifically declared in nonexecutable statements, REAL and INTEGER variables (as opposed to CHARACTER variables) will have their type specified by the first letter of their name. The defaults are:

Names beginning with:	Default to:
A through H	REAL
I through N	INTEGER
O through Z	REAL

Lowercase letters are generally not recognized unless they appear in the program as character strings enclosed in single quotes ‘like this’. However, some compilers do have an option that can be invoked when compiling that will convert lowercase letters to uppercase (except for those letters that appear within single quotes).

nonexecutable **REAL** statement. The simple BASIC code does not contain a nonexecutable statement.

Such things as declaring data and variable types (**INTEGER, REAL, CHARACTER,** etc.) and reserving memory space for arrays (**DIMENSION**) that will be used in the code, among others, are included in this category of nonexecutable statements. We will cover only declaration statements and **DIMENSION** statements here; other nonexecutable statements are beyond the scope of this text.

Since specifications have to be established before computation begins, these nonexecutable statements must be collected at the beginning of the program. The order of the statements is not important, however. The statements do not do any of the computation work and are processed only once each time the program is executed. Several nonexecutable statements and some examples are covered in Topic 17-6. Note that in FORTRAN, typing and **DIMENSION**ing can be accomplished in the type declaration (**REAL, INTEGER**) statement. Also, if you will be using **CHARACTER** variables, you must declare, in a **CHARACTER** statement, the number of individual characters that will be strung together

Topic 17-5 — BASIC

Variable Names

Variable names in BASIC must begin with a letter (A–Z) and generally may include up to 40 alphanumeric characters. However, name length is language dependent, so check the instruction book for the BASIC you are using if you really want to use more than seven characters.

Lowercase letters are usually recognized by most BASIC interpreters, but check yours to make sure.

Type definitions for all variables that start with specified first letters of the alphabet can be declared by the DEFSTR, DEFSNG, and DEFINT statements discussed in Topic 17-6. Type declarations (INTEGER, REAL, CHARACTER) for individual variable names can also be established by the use of type definition suffixes. The latter technique always take preference over the former when both are used. For the latter technique, the suffixes are:

Suffix	Type
%	Two-byte integer
!	Single-precision REAL
#	Double-precision REAL
$	Character or string

Variables whose names have none of these suffixes default to single-precision REAL. Examples:

TAG% is a two-byte integer.

TAG! is a single-precision REAL variable.

TAG# is a double-precision REAL variable.

TAG$ is a character (alphanumeric) string.

TAG is a single-precision REAL variable.

Topic 17-6 FORTRAN

Nonexecutable Statements

The following are some of the allowable nonexecutable statements permitted by FORTRAN.

```
         1         2         3
123456789012345678901234567890..........
```

Statement	Use
`REAL` *list*	To declare variables in *list* to be REAL. Subscripted variables DIMENSIONed within the *list* do not need to be, and cannot be, listed in a separate DIMENSION statement.
`INTEGER` *list*	To declare variables in *list* to be INTEGER. Subscripted variables DIMENSIONed within the *list* do not need to be, and cannot be, listed in a separate DIMENSION statement.
`CHARACTER*`*n* *list*	To declare variables in *list* to be strings of characters *n* bytes long. If *n is omitted, one-byte characters are assumed (there is one character per byte, e.g., see Table 15-1).
`DIMENSION` *list*	To reserve storage space for the *array* variables in *list* that are not DIMENSIONed in a type statement (see examples below).
`IMPLICIT` *type* (*list*)	Declares all variables whose first letter is in *list* to be of type *type*, where *type* is any of CHARACTER, REAL, or INTEGER.

Examples:

```
         1         2         3
123456789012345678901234567890..........
```

Statement	Use
`IMPLICIT INTEGER(A-G)`	Declares all variables whose first letter is any of A, B, C, D, E, F, G to be of INTEGER type.
`REAL STRAIN(2,6),STRAIN`	This does two things: It first declares STRESS and STRAIN to be REAL variables. Second, it declares STRAIN to be a subscripted 2 x 6 array (12 elements in the array). STRAIN should not, and cannot, appear in a separate DIMENSION statement when it is dimensioned in another type statement, as it is here.
`DIMENSION MOLE1(5),MASS1(5)`	Saves storage space for subscripted array elements MOLE1(1), MOLE1(2), MOLE1(3), MOLE1(4), MOLE1(5), MASS1(1), MASS1(2), MASS1(3), MASS1(4), and MASS1(5). MOLE1 and MASS1 are one-dimensional arrays, each having five elements. MOLE1 and MASS1 cannot be dimensioned in any other type statement.
`CHARACTER*3 A,B,C`	Declares A, B, and C to be character strings of length three bytes (three individual characters long)

Other nonexecutable statements that are beyond the scope of this book:

COMMON, COMPLEX, DATA, DOUBLE PRECISION, EQUIVALENCE, EXTERNAL, INTRINSIC, LOGICAL, PARAMETER, SAVE

Topic 17-6 BASIC

Nonexecutable Statements

The following are some of the allowable nonexecutable statements permitted by BASIC.

Statement	Use
`123456789012345678901234567890..........`	
no DEFSTR *list*	To declare variables whose names begin with the letter or letters in the *list* to be character or string.
no DEFSNG *list*	To declare variables whose names begin with the letter or letters in the *list* to be single precision REAL.
no DEFINT *list*	To declare variables whose names begin with the letter or letters in the *list* to be INTEGER.
no DIM *list*	To declare variables in *list* to be DIMensioned variables of dimensions given (see examples below).

Examples:

`123456789012345678901234567890..........`	
`10 DEFSNG S,U-W`	This declares all variables whose names begin with the letters S, U, V, and W to be single-precision variables unless such a name has one of the declaration suffixes given in Topic 17-5.
`20 DIM MOLE1(5),MASS1(5)`	DIMensions MOLE1 and MASS1 to be subscripted one-dimensional arrays, each having five elements.

Other nonexecutable statements that are beyond the scope of this book:

COMMON.

to define the variable (each byte represents 1 character, i.e., a,b,...,z,A,...,Y,Z,0,1,...,9).

Executable Statements

Executable statements are those that actually participate in some way in performing the computations, in controlling the flow the program, or in the inputting or outputting of information. These statements may be processed thousands of times in each use of the program or they may not be executed at all for a particular application. We devote special attention to most of the possible executable statements in the remaining part of this chapter.

Input/Output Statements (Topic 17-7). We mentioned in conjunction with the programs of Topic 17-1 that there are program statements that read data into the computer and there are statements that print data. The **READ, WRITE,** and **PRINT** statements in FORTRAN, and the **INPUT, READ, PRINT** and **LPRINT** statements in BASIC are examples of I/O (an abbreviation for input/output) statements. There are other I/O statements available in both languages, but the ones listed above will serve you well in initiating your programming career.

Of course, not all of the data used in a computer program is sent in and out by I/O statements. Only data that will change from use to use should be entered and retrieved in this fashion. Data that do not change every time a program is executed should be made a permanent part of the program by writing it into the code.

Topic 17-7 discusses the intricacies of the I/O statements for both languages and gives some examples. After you have mastered this material you may find the programming guides printed on the front and back text end papers to be useful for refreshing your memory in the future.

Assign or Arithmetic Statements. Assign statements are the ones that set a variable equal (=) to something else. Take, for example, the assign statement given by (either FORTRAN or BASIC, with VOLT assumed to be **REAL**)

```
210 VOLT = 23.2                                        (17-3)
```

It is important to recognize that this form of statement is slightly different from the conventional concept of equality used in mathematics, although it looks very much the same. You must remember that the computer thinks in terms of memory storage locations and not variables. That is, since the variable VOLT is being used in the program of which Equation (17-3) is a part, the computer assigns a storage location to it. The computer also assigns a storage location to the number 23.2 and loads that location with the binary (but floating-point) representation of the number.

The assign statement in Equation (17-3) tells the computer to make a *copy* of whatever it finds in the storage location assigned to the right-hand side and place the copy in the storage location assigned to the left-hand side. Thus immediately after statement 210 is executed, the value stored in the memory location assigned to VOLT will be 23.2 (in floating-point binary form, of course), regardless of what was stored for VOLT before executing the step. At first glance, that may sound like mathematical equality, but it is not necessarily the case.

We illustrate the difference between the assign statement and a mathematical equality with the following line of code (either FORTRAN or BASIC, with ICOUNT assumed to be **INTEGER**):

```
180 ICOUNT = ICOUNT + 1
```
(17-4)

This statement would have been rejected by your very first algebra teacher if you had submitted it on an exam, because it is certainly not a mathematical equality. How could anything be equal to itself plus 1? But that is not how the computer interprets this statement. This assign statement instructs the computer to make a copy of the value currently stored in the memory location for ICOUNT and a copy of the value stored in the location assigned to the integer 1 (which will be 1), add the two values together, and then store the sum in the memory location reserved for ICOUNT. If ICOUNT has a value of 36 before statement 180 is executed, it will have a value of 37 after execution.

In a similar manner, the instruction

```
301 TEMP = OLDTEMP
```
(17-5)

instructs the computer to make a copy of the value stored in the memory location reserved for OLDTEMP and to place the copy in the location reserved for TEMP. The value stored for OLDTEMP is not changed, but the value stored for TEMP is changed if TEMP and OLDTEMP were not equal prior to the execution of the statement.

Assign statements can be considerably more complex since the right-hand side can take on any algebraic form that is permitted by the variables you are using and the operations (exponentiation, multiplication, division, and so on, mentioned previously). For example, the statement (in FORTRAN or BASIC)

```
92 DURESS = (1. + FORCE/AREA)*EXP(E*Z/L)
```
(17-6)

takes copies of the values stored in the locations for FORCE, AREA, E, Z, and L, along with the intrinsic function **EXP** (discussed below), and carries out the calculation dictated by the operator precedence rules. The value obtained is then stored in the memory location assigned to the variable DURESS. The operator precedence rules yield a calculation that is equivalent to that obtained in the algebraic equation

$$\text{DURESS} = \left\{1. + \frac{\text{FORCE}}{\text{AREA}}\right\} \text{EXP}\left\{\frac{\text{E} \cdot \text{Z}}{\text{L}}\right\} \tag{17-7}$$

CHARACTER (in FORTRAN) or **STRING** (in BASIC) variables can be set in a manner similar to non-**CHARACTER**s. For example, the statement

```
49 NAME = 'DON'        in FORTRAN
```
(17-8a)

for

```
49 NAME% = "DON"       in BASIC
```
(17-8b)

(Text continues on page 525)

Topic 17-7 FORTRAN

Input/Output

Input statements tell the computer when to read, where to find, and how to interpret information that is to be supplied to the program as it is being executed or run. Output statements tell the computer when and where to write and how to format information that the programmer would like to have archived or sent to the world outside the program.

INPUT

In FORTRAN, input statements consist of only one form, called the **READ** statement. Its syntax is

```
         1         2         3
123456789012345678901234567890123456.......
stno1 READ(unit#,fmt#,END=stno2,ERR=stno3) var1,var2,var3....
```

where *stno1* is a valid statement number (e.g., if some part of the program must branch to this **READ** statement, the statement must possess a statement label); *unit#*, called the "unit number or device number," contains the information on where the computer is to find the data to be read; *fmt#* is the statement number of the line of code, called the **FORMAT** statement, that will tell the processor how to interpret the data; *stno2* and *stno3* are valid statement numbers of statements that can be found elsewhere in the program (**,END** = *stno2* and/or **,ERR** = *stno3* may be omitted from this statement as desired); and *var1,var2,var3* is a list of variables that are to be assigned values when this read statement is executed.

Unit numbers: Most computer installations have a number of peripheral devices which can be used as data sources. Among these are magnetic tapes and disks, and keyboards. An integer constant or simple-integer variable is used as the *unit#* to identify to the program the peripheral device which will hold the data that are to be read. Check your local site information for the unit numbers that apply to your installation. An asterisk (*) used in place of a number or variable for the *unit#* designates that the default unit number (usually the terminal on which you are working, i.e. the keyboard) will be used. Check your local site information for the default units at your installation.

FORMAT number: The computer needs information to establish the spaces or columns in which data items are to be found and how the data are to be converted (e.g., **REAL, INTEGER,** or **CHARACTER**) for storage in the memory registers. This information must be quite detailed and it is usually convenient to place it elsewhere in another statement in the program. The statement that contains this information is called a **FORMAT** statement; we will discuss **FORMAT** statements in more detail below. The *fmt#* is the statement number of this **FORMAT** statement. The substitution of an asterisk (*) in place of a number for the *fmt#* denotes a list directed or free format, which is discussed below. The use of the free format reduces the amount of detail that must be supplied by the programmer but at the same time, it reduces the flexibility in organizing output information.

(Continues on page 518)

Topic 17-7 BASIC

Input/Output

Input statements tell the computer when to read, where to find, and how to interpret information that is to be supplied to the program as it is being executed or run. Output statements tell the computer when and where to write and how to format information that the programmer would like to have archived or sent to the world outside the program.

INPUT

In BASIC, input statements consist of at least three types: the **INPUT**, **READ**, and **INKEY** statements. The syntax of each is

```
         1         2         3         4
1234567890123456789012345678901234567890.......
no INPUT[;][alphanumeric;]var1,var1,var3.....
no READ var1,var2,var2.......
no char_strg=INKEY$
```

where the brackets [] enclose optional items (the [] are never included), the *nos* are valid statement numbers, *alphanumeric* is an alphanumeric string that is to be printed out as a prompt before data are read, *var1,var2,var3* is a list of variables that are to be assigned values when this read statement is executed, and *char__strg* is a valid character or string variable name.

The **INPUT** statement retrieves data from the keyboard during execution. When the **INPUT** statement is executed, a question mark is printed to the terminal and the program awaits the typing of data followed by a carriage return (or enter). The *alphanumeric* list (as defined in BASIC by enclosing in double quotes, " "), if present, is a prompt that is printed out on the user's terminal ahead of the question mark. If a semicolon is present after the ending double quote for the alphanumeric list, the question mark is suppressed. The optional semicolon that can follow immediately after the word **INPUT** causes the cursor to remain on the present line; that is, no carriage return is produced after the data have been entered. The entered data, which must be separated from one another by commas, are assigned sequentially to the *vari*s. The number and type (integer, real, or character) must correspond to the number and *vari*s included in the *var1,var2,var3,. .* list; otherwise, an error message results. Character strings that are being inputted do not have to be enclosed in double quotation marks unless the strings include commas or leading or trailing blanks.
Examples:

```
         1         2         3
123456789012345678901234567890123456.......
30 INPUT "TYPE IN A VALUE FOR X";X
35 INPUT "ENTER Y,Z";Y,Z
590 INPUT A,B,C,D,E,F%,G$
```

30 INPUT "TYPE IN A VALUE FOR X";X	*prints TYPE IN A VALUE FOR X and awaits entry of a value of X*
35 INPUT "ENTER Y,Z";Y,Z	*prints ENTER Y,Z and awaits entry of Y and Z*
590 INPUT A,B,C,D,E,F%,G$	*prints a ? and awaits entry of these seven variables*

(Continues on page 519)

Topic 17-7 FORTRAN

Input/Output (Continued)

END=specifier: If present, the **END**=*stno2* clause transfers execution to *stno2* if an End-of-File (EOF) is encountered when attempting to read the input data.

ERR=specifier: If present, the **ERR**=*stno3* clause transfers execution to *stno3* if an error of some sort is encountered when attempting to read the input data.

Examples:

```
          1         2         3
123456789012345678901234567890123456.......
   23 READ(*,249) A,B,X(4)
      READ(5,*) RUB,DUB,(FLUB(I),I=1,5)
```

OUTPUT

In FORTRAN, output statements consist of two common types. The syntax of each is

```
          1         2         3
123456789012345678901234567890123456.......
stno1 PRINT fmt#,var1,var2,var3.....
stno1 WRITE(unit#,fmt#)var1,var2,var3....
```

The former output statement is used to output to the default unit number, which is usually the screen when working on a VDT, or the line printer when printed information is diverted to the line printer. The latter output statement is used when the programmer desires to output the required information to something other than the default unit number. In fact, the use of an asterisk for the *unit#* in the **WRITE** statement designates that output should occur to the default unit number and makes the **WRITE** under this condition equivalent to the **PRINT** statement. In both forms, the other features *(stno1,fmt#,var1,var2,var3* have the same meaning as they do in the **READ** statement. An asterisk in place of the *fmt#* again denotes list directed or free format style.

Examples:

```
          1         2         3
123456789012345678901234567890123456.......
231   PRINT *,A,B,(X(I),1=1.6)
 232  WRITE(6,42) RUB,DUB,FLUB
  239 PRINT 83, TIME,TEMP,HUMID
```

THE LIST DIRECTED OR FREE FORMAT SPECIFIER

A convenient substitute for a number for the *fmt#* that generally appears in a **READ, PRINT,** or **WRITE** statement is the asterisk (*). This denotes the list directed or free format which frees the user from the need to specify completely all the details of how the data are to be arranged and declared. On output, the data are spaced apart in a man-

(Continues on page 520)

Topic 17-7 BASIC

Input/Output (Continued)

The **READ** statement retrieves data from the **DATA** statement, which must be another statement in the program. The syntax of the data statement is

```
         1         2         3         4
1234567890123456789012345678901234567890.......
no DATA cons1,cons2,cons3.....
```

where *cons1,cons2,cons3..* is a list of comma-separated constants (numeric or string) that are assigned, on a one-for-one basis, to the *var1,var1,var3..* list in the **READ** statement. However, the constants must agree with the variable type to which they are being assigned; otherwise, an error diagnostic will result.

DATA statements may be placed at any location in the program and may contain as many constants as will fit on a line. **READ** statements that are executed in the program treat the data in **DATA** statements as sequential lists; later **READ** statements begin selecting data where the previous **READ** statement ended. If all the data on one **DATA** statement are read, the reading continues at the next **DATA** statement encountered. If there are fewer data available than there are requests to **READ**, an error diagnostic results.

A **RESTORE** statement, the syntax of which is

```
         1         2         3         4
1234567890123456789012345678901234567890.......
no RESTORE st
```

allows the rereading of the data in **DATA** statements starting again at statement number *st*.

Examples:

```
         1         2         3
1234567890123456789012345678901.......
10 READ A,B,C%,D$,E$             reads two REALS, one INTEGER, two Strings
20 DATA 19.,1874.,10,HELP        the source of data for all but one variable in READ above
30 DATA OK                       source of fifth variable in READ above
40 RESTORE 30                    the next READ will start with OK in statement 30
50 READ WRNG$,F,G,H%,I%          WRNG$ is read from statement 30
60 DATA 7111.,17.,17,21          F,G,H%,I% are taken from here
```

INKEY$** is essentially a function that, when used in the syntax shown at the beginning of this topic, continuously samples the keyboard input. If any key is actuated, **INKEY$ returns the single byte character represented by that key. If no key has been activated, **INKEY$** returns a null string (a string of zero bytes). **INKEY$** is very useful for entering single characters into a program without the need for terminating the input with a carriage return, as is required in the **INPUT** and **READ** statements. To accomplish this, a string variable must be assigned the value returned by **INKEY$** as shown near the beginning of this topic, and an **IF** statement must be invoked as shown in the examples below.

(Continues on page 521)

Topic 17-7 FORTRAN

Input/Output (Continued)

ner determined by the compiler in use. On input, care does not have to be given to what columns the data must occupy as is necessary in formatted input/output. Rather, the various numbers and alphanumeric strings representing the data need simply to be separated by commas or spaces.

For example,

```
           1         2         3
123456789012345678901234567890123456.......
      REAL A,B,C
      INTEGER I
      READ(*,*) A,B,C,I
```

would read any of the following data from the default unit number (in the fourth line, only the first four numbers would be read):

```
           1         2         3
123456789012345678901234567890123456.......
1.2,2.4,6.1,5
1.2 3.4 6.1 5
1,2,3,5
1,2,3,5,10,30
```

but would abort on

```
           1         2         3
123456789012345678901234567890123456.......
1,2,3,5.2
```

since 5.2 is not a valid **INTEGER** number.

THE FORMAT STATEMENT

The purpose of the **FORMAT** statement is to describe the details of the form of the input or output data (e.g., how the data are to be placed across the page or screen) and how these data are to be converted for storage, if inputs, or display, if outputs (e.g., **REAL, INTEGER,** or **CHARACTER** variables). The syntax of the **FORMAT** statement is

```
           1         2         3
123456789012345678901234567890123456.......
stno  FORMAT(fld1,fld2,fld3,...)
```

Here *stno* is the mandatory statement number (by which this **FORMAT** is referenced in the **READ, PRINT,** or **WRITE** statements; that is, the *fmt#* reference in the discussions of the **READ, PRINT,** or **WRITE** statements above) and *fld1, fld2, fld3,. .* is a list of fields (number of columns or spaces and, for **REAL** numbers, the number of digits in the decimal part of a number) reserved for each element in the list of varia-

(Continues on page 522)

Topic 17-7 BASIC

Input/Output (Continued)

This string variable may be converted to a numerical value with the **VAL** function (see Topic 17-14).

Examples:

```
         1         2         3         4
1234567890123456789012345678901234567890.......
50 PRINT "DO YOU WANT TO DO MORE (Y/N)"      see description of PRINT statement below
60 A$=INKEY$                                 string A$ set equal to function INKEY$
70 IF A$="" GOTO 60                          checks to see if a key has been pushed
80 IF A$="n" OR A$="N" THEN END              ENDs for certain keys—N or n
85 IF A$<>"y" OR A$<>"Y" GOTO 50             recycles if unexpected key
                                             (<>) means "not equal to"
90 PRINT "SELECT A NUMBER FROM  1  TO  9"    continues if Y or y was chosen
100 A$=INKEY$
110 IF A$="" GOTO 90
120 NUM=VAL(A$)                              converts string to numeric
125 IF NUM < 1 OR NUM > 9 GOTO 90            checks for allowable result
130 PROD=NUM*PRNCPL
```

OUTPUT

There are six different ways of outputting information from a BASIC program which we will discuss. These are the **PRINT, ?, WRITE, PRINT USING, LPRINT,** and **LPRINT USING** statements. The syntax of each is

```
         1         2         3         4
1234567890123456789012345678901234567890.......
no PRINT val1,val2,val3...[;]
no ? val1,val2,val3...[;]
no WRITE val1,val2,val3...[;]
no PRINT USING v$; val1,val2,val3...[;]
no LPRINT val1,val2,val3...[;]
no LPRINT USING v$; val1,val2,val3...[;]
```

where *v$* is a special alphanumeric string discussed further below, and the *val1,val2,val3,..* list is the same as that discussed under *Input* above, except that here, any of the *val*is could be an alphanumeric or string constant (defined in double quotes " ").

Printing to the Terminal: **PRINT, ?, WRITE,** and **PRINT USING** all send output the user's terminal. **PRINT**, and **?** are identical in their function. They and the **WRITE** statement are similar in that each allows very little freedom in how the data are formatted. The **PRINT USING** statement gives the user the ability to specify the format of the output.

In the **PRINT** and **?** statements, a comma used to separate the *var*is gives the field of each outputted variable 14 spaces and places each number at the beginning of it's field (a blank space precedes positive numbers; a minus precedes negative numbers). If the number to be

(Continues on page 523)

Topic 17-7 FORTRAN

Input/Output (Continued)

bles being referenced in **READ, PRINT,** or **WRITE** statements. These fields contain "edit descriptors" which determine type, size, and resolution of numbers to be transferred. Allowable edit descriptors, or *fld*'s are:

fAs Allocates f consecutive fields of s spaces or columns each for **CHARACTER** variables. (The f specifier need not be included when its value is 1.) On output in **PRINT** or **WRITE** statements, the **CHARACTER** variables are left justified within the fields. Inputting of **CHARACTER** variables is best done in list directed input (free formatting) by enclosing the variables in single quotes ('). Examples are 5A2, A6, 2A8.

fEs.d Allocates f consecutive fields of s spaces or columns each for **REAL** numbers. (The f specifier need not be included when its value is 1.) Each **REAL** number is to be displayed in scientific notation as a fractional number between 0.1 and 1.0 multiplied times 10 raised to some power. On output in **PRINT** or **WRITE** statements, there will be an **E** in place of the number 10, three digits (generally) allocated to the exponent on the 10 and its sign, and d digits to the right of the decimal point in the fractional part. The desired output will be right justified in the field. The number s includes the spaces for the **E**, the exponent and its sign, the decimal point, and if the number is negative, a space for the leading sign. Thus s should always exceed d by at least 6. If it does not, the field will be filled with asterisks instead of the desired number. On input, this type of *fld* or edit descriptor functions in an identical way to the fFs.d *fld* described below. Examples are E12.3, 3E9.2, 5E10.3.

fFs.d Allocates f consecutive fields of s spaces or columns each for **REAL** numbers. (The f specifier need not be included when its value is 1.) Each **REAL** number is to have d digits to the right of the decimal point. On output in the **PRINT** or **WRITE** statements, each number will be right justified in its respective field. The value of s must include a space for the decimal point and if the number to be processed is negative, a space for the leading sign. Thus s should exceed d by at least 1 and usually 2 as a minimum. If it does not, the field will be filled with asterisks instead of the desired number. On input in **READ** statements, a **REAL** number without a decimal point is interpreted as having the d rightmost digits in the field as the fractional part. If a decimal point is present, it takes precedent in determining the fractional part regardless of where the number is within the field. Allowable numbers may also be followed by an E and a signed integer constant to denote powers of 10 which are part of the number. Examples are F10.5, 3F9.2, 5F12.6.

fIs Allocates f consecutive fields of s spaces or columns each for **INTEGER** numbers. The middle character in

(Continues on page 524)

Topic 17-7 BASIC

Input/Output (Continued)

printed requires fewer than seven digits, it is displayed in floating-point form. Otherwise, it is displayed in scientific form. All spaces in the field not used for display of the number are blank filled. A semicolon used to separate the *vari*s causes each *vari* to be printed within one space of the last *vari*; that is, most of the blank spaces will be eliminated from the print line. The optional semicolon [;] used at the end of the list prevents a carriage return from being initiated at the end of the print line. It is used when the programmer wants output from the next **PRINT** statement to be printed as an extension of the previous output.

The **WRITE** statement is identical to the **PRINT** statement with a comma-delimited *var1,var2,var3...* list except that commas are inserted between printed fields.

The **PRINT USING** statement gives maximum control over the format of the output. The format specifiers are included in the alphanumeric string *v$*, which must be enclosed in double quotes (""). The following special characters may be used to format the fields:

#	The number sign (#) is used as a designee of digit places in numeric output. These are used in a *format string* in conjunction with a decimal point (.) to designate the manner in which numbers are to be displayed. Examples are #.#, ##.#, ##.##, ####.###.
+	Used at the beginning (or end) of a format string, a plus sign (+) forces the sign of the number to be displayed before (or after) the number. Examples are +##.#, +###.###, ###.###+.
−	Used at the end of a format string, a minus (−) forces negative numbers to be displayed with a trailing minus sign. Examples are ##.#−, ###.###−.
^^^^	Used at the end of a format string, four carets (^) force numbers to be displayed in scientific notation. One digit position is used to the left of the decimal sign to print a space (for positive numbers) or a minus (for negative numbers). Examples are ##.## ^ ^ ^ ^, ###.#### ^ ^ ^ ^.
_	The underscore (_) in a format string causes the character that follows the underscore to be displayed on the terminal. Examples are ##.##_%, _$###.##.
!	When used with a character string field, the exclamation sign (!) permits only the first character in the requested string to be displayed.
n spaces\\	Two backslashes separating *n* spaces can be used with character string fields to display only the first 2+*n* characters in the requested string. Examples are \ \, \ \, \\.
&	The ampersand (&) is used with character string fields to display the whole string requested.

The final optional semicolon [;], when included, suppresses a carriage return that is normally inserted. The next printing statement will

(Continues on page 525)

Topic 17-7 FORTRAN

Input/Output (Continued)

this edit descriptor is an upper case "i". (The f specifier need not be included when its value is 1.) On output in **PRINT** or **WRITE** statements, the **INTEGER**s are right justified in their fields. If the number to be processed exceeds the space allotted through s, the field is filled with asterisks instead of the desired number. On input for the **READ** statement, each **INTEGER** should be right justified within its field. Examples are I4, 5I4, 10I2.

sX — Allocates a field of s spaces or columns that are considered to contain all blanks. Examples are 10X, 5X, 2X.

'alphanumeric string' — When used as one of the *vars* in a **PRINT** or **WRITE** statement (see above) or as one of the *flds* in a **FORMAT** statement outputs the text and/or numbers (i.e., the *alphanumeric string*) between the single quotes (''). Examples are 'PRINT THIS', 'or THIS', or 'Even THIS'.

/ — Signifies the end of the present line and initiates printing on a new line.

' ' — (a blank as an alphanumeric character) Printed as the first character in a line, initiates a single space before printing (this is almost always the desired action).

'0' — (a zero) Printed as the first character in a line, initiates a double space before printing.

'1' — (a one) Printed as the first character in a line, initiates a page eject on a printer so that all subsequent printing will be done from the top of a new page.

'+' — (a plus sign) Printed as the first character in a line, suppresses any spacing before printing, thus causing overprinting of the previous line.

Example **FORMAT**s:

```
         1         2         3
123456789012345678901234567890123456.......
      READ(*,*,END=700,ERR=800) START,STOP
    4 READ(5,19) START2,STOP2
   19 FORMAT(4X,2F10.2)
      ANSWER=22.222222
      ITER=31
      PRINT 30, ANSWER,ITER
30    FORMAT('1','THIS IS THE ANSWER ',F10.4,' IN ',I5,' ITERATIONS')
35    WRITE(*,40)
   40 FORMAT(/,' GOOD job')
      PRINT 45, ANSWER,ITER
45    FORMAT(//,5X,' THIS IS REALLY THE ANSWER',F6.2,' IN',I3,' ITERATIONS')
```

(Continues on page 526)

stores the ASCII string DON in the storage locations assigned to the variable called NAME[9] or NAME%.

Before using variables on the right-hand side of an assign statement you must make sure that they are given initial values. This practice is a part of good programming principles. A variable will have received a value in its store if

- It was listed as one of the arguments in an input (**INPUT** or **READ**) statement that was executed,
- It has appeared on the left side of an assign statement with previously defined variables or constants on the right side, or
- (In FORTRAN) it appeared in a **DATA** statement (this nonexecutable or specification statement is not covered in this book).

Topic 17-7 — BASIC

Input/Output (Continued)

start where the present one left off instead of at the start of the next line.

Examples:

```
           1         2         3         4
1234567890123456789012345678901234567890......
1  W=146794.4436
2  X=1.344
3  Y=2.34
4  Z=99.347
5  B$="GOODBYE"
10 PRINT "HELLO"                          prints HELLO
15 ' "HELLO"                              same as statement 10
20 PRINT "HERE IS X ";X;" AND Y ";Y       prints HERE IS X 1.344 and Y 2.34
30 ' "HERE IS X ";X;" AND Y ";Y           same as statement 20
40 WRITE X,Y,Z                            prints 1.34,2.34,99.35 (note rounding)
50 PRINT USING "###.## ";X,Y,Z            prints 1.34 2.34 99.35
55 PRINT USING "!";B$                     prints G
60 PRINT USING "\  \";B$                  prints GOOD
70 PRINT USING "_$###.##";Z               prints $ 99.35
80 PRINT USING "##.#^^^^";W               prints 1.47E+05 (note rounding and space for sign)
```

Printing to the Line Printer: The **LPRINT** and the **LPRINT USING** statements mimic the **PRINT** and **PRINT USING** statements except that the output is sent to the line printer (if present) instead of the terminal. See the descriptions of the **LPRINT** and **LPRINT USING** statements above under *Printing to the Terminal* for more details.

---

[9]In FORTRAN, the variable NAME has to be declared of type CHARACTER of length at least three bytes. That is, at the beginning of the program, among the nonexecutable statements, would have to be the statement (see Topic 17-6)

```
CHARACTER*3 NAME
```

Following one of these procedures for each of the variables in your program assures that the variables will have the values that you expect and want them to have. Obviously, this is extremely important to define variable values before you use them to determine the values of other variables in the programs.

Most computer systems preset all memory locations to zero (0) before loading the next program for execution. Thus, if you executed statement 180 in Equation (17-4) without setting the value of ICOUNT to the starting value you desired, you will find ICOUNT set to 1 after execution, which may or may not be what you intended.

There are a few computer systems that do not initialize all storage locations to zero before loading the next program. These systems allow the memory locations to retain the contents that were present when the previous program finished. Thus, if you do not purposely give values to variables used in your program, you may find that data from the preceding job are being used in your program. In these cases, the best than can happen is that your program will error off ("bomb"). You do not even want to think about the worst that can happen.

Topic 17-7 FORTRAN

Input/Output (Continued)

would read the two lines

```
         1         2         3
123456789012345678901234567890123456.......
56.75,45.96
56.75     45.96
```

or equivalently the two lines

```
         1         2         3
123456789012345678901234567890123456.......
56.75 45.96
      5675      4596
```

and produce, at the start of a new page (caused by the '1' in **FORMAT** 30), on the default unit number

```
         1         2         3         4         5
12345678901234567890123456789012345678901234567890.......
 THIS IS THE ANSWER    22.2222 IN    31 ITERATIONS

 GOOD job


      THIS IS REALLY THE ANSWER 22.22 IN 31 ITERATIONS
```

Program Flow Control Statements

Up to this point the program details discussed are necessary but not entirely sufficient to provide the powerful programming tools needed in science and engineering. With only the topics discussed so far, we cannot compare values, force branching, or make programming decisions. These capabilities are the function of the *flow control statements*. We will cover the major ones—the ones that have counterparts in other modern languages that you might encounter later in your career.

The flow control statements include the **IF-THEN-ELSE** and its near relatives the **IF-THEN** and the **IF**, the **WHILE**, the **DO** (FORTRAN) or **FOR** (BASIC), the **GOTO**, the **CONTINUE** (FORTRAN), and the **END**.

The example codes that were shown in Topics 17-1 and 17-2 use some of these program control statements. For each of the constructs, the flowcharts and pseudocode are similar for both FORTRAN and BASIC; therefore, we will present these before discussing the details pertinent to each language.

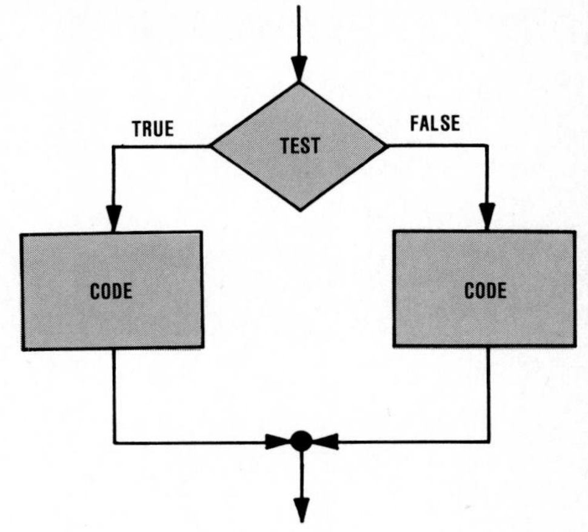

[17-1] The flowchart for the IF-THEN-ELSE structure

```
IF test is true
  do these things
ELSE
  do these things
END IF
```

[17-2] Pseudocode for IF-THEN-ELSE.

IF-THEN-ELSE/IF-THEN (Topic 17-8). Suppose that within a program you have written, you calculate the chemical composition for a mixture of chemicals that has reached equilibrium. Suppose further that you wanted to perform one of two possible sets of calculations depending on the value of the composition of one of the chemical species. This is an ideal place to use the **IF-THEN-ELSE** construction. The flowchart for this useful programming structure is shown in [17-1]. In pseudocode language, this construct appears as in Figure [17-2].

The condition that is specified after the **IF** (see Topic 17-8) is called the *test*. Many times the control of the program execution requires a sequence of statements to be executed when the test condition is **TRUE** or satisfied and an alternative set of statements to be executed when the test condition is **FALSE** or not satisfied. On completion of the activities along either branch, program execution is to continue from a common point. As you can see from Figure [17-2], the **IF-THEN-ELSE** construct is ideal for this situation.

Topic 17-8 gives details of the **IF-THEN-ELSE** structure for FORTRAN and BASIC. Often, there is a need for the **IF-THEN** structure without the **ELSE**. It is useful when a section of code is to be executed under certain conditions but there is no alternative. There is an **IF-THEN** structure to handle these situations. Since it is closely related to the **IF-THEN-ELSE** construct, we also detail it in Topic 17-8.

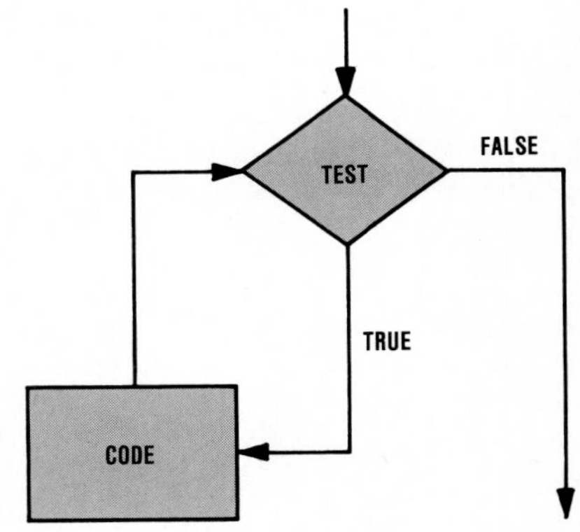

[17-3] The flowchart for the WHILE structure.

WHILE (Topic 17-9). To further accommodate the recursive nature of many engineering problems, it is convenient to have available a construct that performs a *test*, much like the **IF-THEN-ELSE** construct does, but then initiates a looping exercise which continues as long as the *test* is satisfied. As you can see from the flowchart [17-3] and pseudocode [17-4], this is exactly what the **WHILE** statement does.

```
WHILE test is satisfied
do these things
END WHILE
```

[17-4] Pseudocode for WHILE.

Topic 17-8 FORTRAN

IF-THEN-ELSE Structure

The logical **IF-THEN** construct in FORTRAN is

```
         1         2         3
1234567890123456789012345678901234567890..........
      IF (test) THEN
                  lines of code to be executed if test is true
                  .........
                  ........

      ELSE
                  lines of code to be executed if test is false
                  .........
                  ........

      END IF
```

If the expression *test* is true or satisfied, the lines of code between the THEN and the ELSE are executed. If the expression *test* is false or not satisfied, the lines of code between the ELSE and the END IF are executed.

The logical **IF-THEN** construct is identical to the **IF-THEN-ELSE** except that the **ELSE** and the lines of code following it do not appear. The **END IF** always appears.

IF-THEN-ELSEs can be nested; that is, **IF-THEN-ELSE**s can be placed within **IF-THEN-ELSE**s. It is good practice, when possible, to nest the **IF-THEN-ELSE**s not as **IF-THEN-(IF-THEN-ELSE)-ELSE**s, but as **IF-THEN-ELSE-(IF-THEN-ELSE)**s.

THE *TEST*

The *test* is an expression involving variables, constants, or functions that can be evaluated. The comparisons that can be made in the *test* are

Comparison	Syntax
Equal	.EQ.
Less than	.LT.
Greater than	.GT.
Greater than or equal	.GE.
Less than or equal	.LE.
Not equal	.NE.

along with the logical operators

And	.AND.
Or	.OR.

In *tests*, you should always avoid equality checks (=) of **REAL** numbers against any value. Because of the way **REAL** numbers are stored, such direct equalities will rarely be true. Instead, use **ABS**(*real variable-*

(Continues on page 530)

Topic 17-8

IF-THEN-ELSE Structure

The **IF-THEN-ELSE** construct in BASIC is one of two types:

```
         1         2         3
1234567890123456789012345678890..........
stno IF test THEN clause1 ELSE clause2

stno IF test GOTO stno1 ELSE clause1
```

In these BASIC statements *stno* and *stno1* are statement numbers, *test* is a condition upon which action is based, and *clause1* and *clause2* are BASIC statements or groups of statements separated by commas.

If the expression *test* is true or satisfied, *clause1* is executed. If the expression *test* is false or not satisfied, *clause2* is executed.

The logical **IF-THEN** construct is identical to the **IF-THEN-ELSE** except that the **ELSE** and *clause2* do not appear.

IF-THEN-ELSEs can be nested; that is, **IF-THEN-ELSE**s can be placed within **IF-THEN-ELSE**s. It is good practice, when possible, to nest the **IF-THEN-ELSE**s not as **IF-THEN-(IF-THEN-ELSE)-ELSE**s, but as **IF-THEN-ELSE-(IF-THEN-ELSE)**s.

The *test*: *test* is an expression involving variables, constants, or functions that can be evaluated. The comparisons that can be made in the *test* are

Comparison	Syntax
Equal	=
Less than	<
Greater than	>
Greater than or equal	>= (or =>)
Less than or equal	<= (or =<)
Not equal	<> (or ><)

along with the logical operators

And	AND
Or	OR

In tests, you should always avoid direct equality checks (=) of REAL numbers against any value. Because of the way REAL numbers are stored, such direct equalities will rarely be true. Instead, use ABS *(real variable - comparison value)* < *small number.* See BASIC Topic 17-14 (Intrinsic Functions) for a description of the ABS function.

The logical operators AND and OR allow two conditions to be made a part of the *test*. For example, assuming *cond1* and *cond2* are two separate conditionals, a *test* of (*cond1*) **AND** (*cond2*) is true, or satisfied, if *both cond1* and *cond2* are true. A *test* of (*cond1*) **OR** (*cond2*) is true, or satisfied, if *either cond1* or *cond2* is true. The parentheses that we have included in these *tests* are not necessary, but we recommend them for clarity where the logic becomes complicated.

(Continues on page 531)

BASIC has a legitimate **WHILE** statement but, unfortunately, FORTRAN 77 does not. We believe that this is such an important programming tool that we should devise or concoct an illegitimate **WHILE** for use in FORTRAN. In this book, for FORTRAN, we will follow the suggestion of John L. Lowther in Daniel McCracken's book[10] for simulating the **WHILE**. This construct is shown in Topic 17-9.

DO (FORTRAN)/*FOR* (BASIC) (Topic 17-10). The **DO/FOR** statement is an extremely useful programming structure in

[10]D. D. McCracken, *Computing for Engineers and Scientists with FORTRAN 77* (New York: Wiley, 1984).

Topic 17-8 FORTRAN

IF-THEN-ELSE Structure (Continued)

comparison value) .LT. *small number.* See FORTRAN Topic 17-4 (Intrinsic Functions) for a description of the **ABS** function.

The logical operators **.AND.** and **.OR.** allow two conditions to be made a part of the *test*. For example, assuming *cond1* and *cond2* are two separate conditionals, a *test* of (*cond1*) **.AND.** (*cond2*) is true, or satisfied, if *both cond1* and *cond2* are true. A *test* of (*cond1*) **.OR.** (*cond2*) is true, or satisfied, if *either cond1* or *cond2* is true. The parentheses and spaces that we have included in these *tests* are not necessary, but we recommend them for clarity where the logic becomes complicated.

Examples:

```
         1         2         3
123456789012345678901234567890..........
      IF(A.GT.B) THEN
          A=B
      ELSE
          A=-B
      END IF

      IF (A.LT.C.AND.B.LT.C) THEN
           CALL LOWER(A,B,C)
      ELSE
         IF(A.GT.C.AND.B.GT.C) THEN
            CALL UPPER(A,B,C)
            ELSE
               CALL MIDDLE(A,B,C)
         END IF
      END IF

      NOGO=0
      IF (ABS(VOLTS-35.5).LT.0.00001) THEN
          NOGO=1
      END IF
```

FORTRAN/BASIC. In one line of code it provides for the care and maintenance of a running variable that can control the execution of groups of statements. These groups of statements will be executed over and over again until the running variable reaches its user-specified limit.

DO/FORs are particularly useful in processing subscripted variables, since a variable named in the **DO/FOR** loop could have only its subscript changed each time through the loop. Its subscript could very easily be related to the value of the running variable.

DO/FORs are closely related to **WHILEs** except that **WHILEs** do not maintain their own running variable that can be automatically used to start and stop the looping inherent in both constructs, or that can be used in expressions and subscripts within the loops. **DO/FORs** are discussed specifically in Topic 17-10.

GOTO (Topic 17-11). It is sometimes desirable to direct program execution to another part of the code or to force program execution to jump so that a specific part can be executed. We often wish to do this if certain conditions or tests are met. For example, suppose that you were calculating the value of a variable Z in a **DO/FOR** loop and that you wanted to exit the loop if Z acquired a certain value. An **IF** statement might be used to monitor the value of Z each time through the loop. If Z ever becomes equal to the desired value, a **GOTO** statement would be used to direct execution to another piece of code outside the loop. We will cover the details of the **GOTO** construct for both FORTRAN and BASIC in Topic 17-11.

We believe that the use of **GOTOs** has to be carefully monitored because **GOTOs** are one of the primary tools that make spaghetti code easy to write. Ideally, structured programming would not contain any **GOTOs**, but in FORTRAN and BASIC, particularly FORTRAN, this construct is sometimes necessary and useful.

CONTINUE/CONT (Topic 17-12). The **CONTINUE** statement in FORTRAN and the **CONT** statement in BASIC serve separate and distinct purposes in the two languages. We describe their use in Topic 17-12.

(Text continues on page 538)

Topic 17-8 — BASIC

IF-THEN-ELSE Structure (Continued)

Examples:

```
          1         2         3
1234567890123456789012345678901234567890..........
25 IF (A > B) THEN A = B ELSE A = -B
34 IF Z% = 4 SEARCH = 2*INTERVAL
40 IF (A < C) AND (B < C) THEN GOSUB 1100 ELSE IF (A > C) OR (B > C) THEN GOSUB 2000 ELSE GOSUB 3000
61 IF (ABS(VOLTS-35.5) < .00001) THEN NOGO%=1 ELSE NOGO%=0
```

Topic 17-9 FORTRAN

WHILE

FORTRAN 77 does not contain a **WHILE** construct. However, it can be simulated in FORTRAN with the following type of structure:

```
         1         2         3
1234567890123456789012345678901234567890..........
C - - WHILE test
stno  IF(test) THEN
        lines of code to be executed if test is true
        ...........
        GO TO stno
        END IF
C - - END WHILE
```

where *stno* is some appropriate statement number and *test* is some expression that can be evaluated (see *test* under FORTRAN Topic 17-8). If the expression *test* is true or satisfied, the lines of code between the **THEN** and the **END IF** are executed. There may be occasions when you will want to insert an **ELSE** clause in this construct. This is permitted.

Examples:

```
         1         2         3
1234567890123456789012345678901234567890..........
      .............
      T=0
C - WHILE T=0
30    IF(T.EQ.0) THEN
        PRINT *,' My Name'
        GO TO 30
      END IF
        ..........
```

the above will keep printing out, on the default unit,

My Name

My Name

My Name

....

until $T \neq O$ or until the operator terminates the program.

```
         1         2         3
1234567890123456789012345678901234567890..........
        ....
      T=1
C - WHILE T.EQ.1
100   IF(T.EQ.1) THEN
        READ(5,200) FORCE(I)
        IF(ABS(FORCE(I)).LT.1.E-10) THEN
          T=0
        END IF
        GOTO 100
      END IF
C - END WHILE
        ....
```

This keeps reading values of **FORCE** from unit 5 until it encounters a zero. This could also be done with a **DO**; see FORTRAN Topic 17-10.

Topic 17-9 BASIC

WHILE

The **WHILE** construct in BASIC has the following structure:

```
         1         2         3
1234567890123456789012345678901234567890..........
no2 WHILE test
 .....lines of code to be executed if test is true
no1 WEND
```

where *no1* and *no2* are some appropriate BASIC statement numbers and *test* is an expression that can be evaluated (see *test* under BASIC Topic 17-8). If the expression *test* is true or satisfied, the *lines of code* are executed.

Examples:

```
         1         2         3
1234567890123456789012345678901234567890..........
.............
50  T=0
60 WHILE T=0
70   PRINT *,' My Name'
80 WEND
.............
```

the above will keep printing out, on the default unit,

My Name
My Name
My Name
....

until T ≠ O or until the operator terminates the program.

```
         1         2         3
1234567890123456789012345678901234567890..........
 ....
20 T=1
25 I=1
30 WHILE T = 1
40   INPUT FORCE(I)
50   IF ABS(FORCE(I)) < 1.E-10) THEN T=0
55   IF ABS(FORCE(I)) > 1.E-10) I=I+1
60 WEND
 ....
```

This keeps reading values of **FORCE** from unit 5 until it encounters a zero. This could also be done with a **FOR**; see BASIC Topic 17-10.

Topic 17-10 FORTRAN

DO

The form of the **DO** construct is

```
         1         2         3
1234567890123456789012345678901234567890..........
      DO stno var = no1,no2,no3
              lines of code to be executed
              .......
              .......
stno  CONTINUE
```

where *stno* is an appropriate statement number, *var* is an appropriate variable name, *no1* is the starting value to use for the "running" variable *var*, *no2* is the ending value to use for *var*, and *no3* is the amount by which *var* is to be incremented each time the *lines of code to be executed* are executed. If the *,no3* is omitted, *var* is indexed by 1 each time through the loop. This construct need not end with a **CONTINUE** statement, although it must end on a numbered *executable* statement. However, we strongly recommend that the **CONTINUE** statement (see Topic 17-12) always be used as the loop terminating statement.

Examples:

```
         1         2         3
1234567890123456789012345678901234567890..........
      DO 500 MIKE=1,7,2
        DECIBEL(MIKE)=LOG(NOISE)
 500  CONTINUE
```

The above **DO** loop computes **DECIBEL(1), DECIBEL(3), DECIBEL(5),** and **DECIBEL(7)** sequentially as it progresses through the loop four times.

```
         1         2         3
1234567890123456789012345678901234567890..........
      SUM=0.
      DO 36 I=1,4
        Q(I)=A(I)*V(I)
        SUM=SUM+Q(I)
 36   CONTINUE
```

The loop above sequentially computes **Q(1), Q(2), Q(3),** and **Q(4)** and sums these values as the loop is executed four separate times.

Topic 17-10 BASIC

FOR

The form of the **FOR** construct is

```
          1         2         3
1234567890123456789012345678901234567890..........
no FOR var = no1 TO no2 STEP no3
...     lines of code to be executed
.......
no4 NEXT
```

where *no* and *no1* are appropriate statement numbers, *var* is an appropriate variable name, *no1* is the starting value to use for the "running" variable *var*, *no2* is the ending value to use for *var*, and *no3* is the amount that *var* is to be incremented by each time the *lines of code to be executed* are executed. If the *,no3* is omitted, *var* is indexed by 1 each time through the loop.

Examples:

```
          1         2         3
1234567890123456789012345678901234567890..........
.....
50 SUM=0.
60  FOR I=1 TO 4
70    Q(I)=A(I)*V(I)
80    SUM=SUM+Q(I)
90  NEXT
```

The **FOR** loop above sequentially computes **Q(1)**, **Q(2)**, **Q(3)**, and **Q(4)** and tallies their **SUM** as they are computed.

```
100 FOR MIKE=1 TO 7 STEP 2
120        DECIBEL(MIKE)=LOG(NOISE)
140 NEXT
```

The **FOR** loop above computes **DECIBEL(1), DECIBEL(3), DECIBEL(5),** and **DECIBEL(7)**, sequentially as the loop is executed five times.

Topic 17-11 FORTRAN

GOTO

The **GOTO** statement in FORTRAN has the form

```
         1         2         3
1234567890123456789012345678901234567890..........
      GOTO stno
```

where *stno* is a valid statement number of an executable statement **FORMAT** statements (listed under Topic 17-7) and the other nonexecutable statements (listed under Topic 17-6) are not valid statements to cite in a **GOTO**.

Example:

```
         1         2         3
1234567890123456789012345678901234567890..........
         .....
C WHILE (SIMULATED)
32    IF(X.GT.0) THEN
         .....
         GOTO 32
      END IF
```

Topic 17-12 FORTRAN

CONTINUE

The syntax of the **CONTINUE** statement is

```
         1         2         3
1234567890123456789012345678901234567890..........
stno  CONTINUE
```

where *stno* is a valid statement number. One use of this statement is as the final statement of a **DO** loop (see Topic 17-10). However, there are other uses (see Topic 17-13).

Topic 17-11 BASIC

GOTO

The **GOTO** statement in BASIC has the form

```
         1         2         3
1234567890123456789012345678901234567890..........
stno GOTO no1
```

where *stno* and *no1* are valid BASIC statement numbers. If the statement number *no1* happens to be an executable statement, execution continues there. If *no1* is a statement number for an nonexecutable statement, execution continues at the next executable statement after the nonexecutable statement.

Example:

```
         1         2         3
1234567890123456789012345678901234567890..........
.....
248 GOTO 530
.....
530 PRINT X+5
.....
680 GOTO 530
.....
800 END
```

Topic 17-12 BASIC

CONT

In some interpretive BASICs, the **CONT** statement is used in the "direct mode" to resume execution after a Ctrl-Break has been issued. It is most often used in program debugging. It is not useful as a normal statement within a BASIC program; it is typed from the keyboard. Check the version of BASIC you are using to locate more information on this statement.

END (Topic 17-13). All FORTRAN programs must end with an **END** statement in order to compile and execute without error. BASIC programs should end with an **END** statement in order to properly close all files that may be in use by the program. The syntax of this statement is given in Topic 17-13.

SUBPROGRAMS

Subprograms are useful for executing calculations and tasks that are easily separated from the remainder of the program. They are also useful for handling calcualtions that are executed more than once within an algorithm. We demonstrated the use of the subroutine, one type of subprogram, in Topic 17-2.

As the end paper programming guides on FORTRAN and BASIC that accompany this book show, there are two types of subprograms: subroutines and functions. In FORTRAN, subroutines and functions are quite similar in their construction. However, there are some important differences that need to be clarified.

Functions

As shown on the end paper prgramming guides, there are three types of functions: *intrinsic* functions, *statement* functions, and *external* functions. The latter two could be termed *user-supplied* functions.

Topic 17-13 FORTRAN

END

The syntax of the **END** statement is

```
         1         2         3
1234567890123456789012345678901234567890..........
      END
```

This statement must appear as the last statement in any program. It must not have a statement number. If you desire to send program execution to this statement, precede it with a numbered **CONTINUE** statement (see Topic 17-12).

Example:

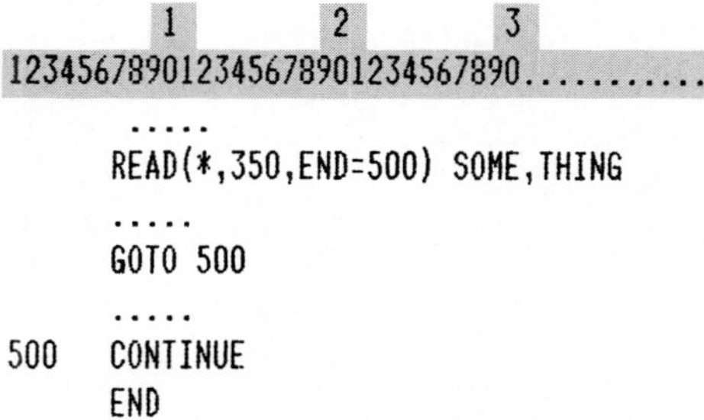

```
         1         2         3
1234567890123456789012345678901234567890..........
       .....
      READ(*,350,END=500) SOME,THING
      .....
      GOTO 500
      .....
500   CONTINUE
      END
```

Intrinisic Functions (Topic 17-14). The intrinsic functions are those that are included in the library of functions that are a part of any good language system. Common ones include those shown in Topic 17-14. The *function names* shown in Topic 17-14 are usually regarded as reserved by the language system (FORTRAN or BASIC) and should not be used by programmers for any purpose other than representing the intended functions. Notice that all functions require arguments, or values, that must be supplied before the function can be evaluated. Notice also that functions always return a value if used correctly. For example, consider the following lines of code involving the sine of an angle of 30° (FORTRAN or BASIC):

```
...............
...............
40 X = 30.*3.14159/180.
50 Y = SIN(X)
...............
```

Nearly all compilers or interpreters recognize SIN(X) as a special name for which they are to supply the algorithm for the calculation of the sine of the argument. The *argument* of the SIN here is X; it must have a value before SIN(X) can be evaluated. In this example, we have assigned X the value of 30° (it is converted to radians since nearly all trigonometric relationships evaluated by the computer assume that angles are always expressed in radians). After the com-

Topic 17-13 — BASIC

END

The syntax of the **END** statement in BASIC is

```
         1         2         3
123456789012345678901234567890..........
stno END
```

where *stno* is a valid, sequential statement number in BASIC. This statement should be the last executable statement in any program.

Example:

```
         1         2         3
123456789012345678901234567890..........
.....
180 END
```

Topic 17-14 FORTRAN

Intrinsic Functions

(Because of the many dialects of FORTRAN, this list is not complete.)

FUNCTION	RESULT
ABS(*rn/v*)	absolute value of *rn/v* (returns a REAL; see IABS below)
ACOS(*rn/v*)	arccos (in radians) of *rn/v* (returns a REAL)
AINT(*rn/v*)	truncates to a REAL the decimal part of REAL *rn/v*
ALOG(*rn/v*)	natural logarithm [(ln) or ($\log_e$)] of *rn/v* (returns a REAL)
ALOG10(*rn/v*)	$\log_{10}$(*rn/v*) (returns a REAL)
AMAX0(*in/v list*)	maximum value in the list (returns a REAL; see MAX0 below)
AMAX1(*rn/v list*)	maximum value in the list (returns a REAL; see MAX1 below)
AMIN0(*in/v list*)	minimum value in the list (returns a REAL; see MIN0 below)
AMIN1(*rn/v list*)	minimum value in the list (returns a REAL; see MIN1 below)
AMOD(rn/v_1,rn/v_2)	REAL remainder from division of rn/v_1 by rn/v_2 (see MOD below)
ANINT(*rn/v*)	same as AINT except rounds the number
ASIN(*rn/v*)	arcsin (in radians) of *rn/v* (returns a REAL)
ATAN(*n/v*)	arctangent (in radians) of *rn/v* (returns a REAL)
ATAN2(rn/v_1,rn/v_2)	arctangent (in radians) of quotient [rn/v_1]/[rn/v_2] *(returns a REAL)*
CHAR(in/v)	the CHARACTER whose ASCII code is *in/v*
COS(*rn/v*)	cosine of *rn/v* where *rn/v* is in radians (returns a REAL)
COSH(*rn/v*)	hyperbolic cosine of *rn/v* (returns a REAL)
EXP(*rn/v*)	$e^{(r\,n/v)}$ (returns a REAL)
FLOAT(*in/v*)	converts *in/v* to REAL variable
IABS(*in/v*)	absolute value of *in/v* (returns an INTEGER; see ABS above)
ICHAR(c)	ASCII code for (1 byte) CHARACTER c
INT(*rn/v*)	converts (truncates) *rn/v* to INTEGER
MAX0(*in/v list*)	maximum value in the list (returns an INTEGER; see AMAX0 above)
MAX1(*rn/v list*)	maximum value in the list (returns an INTEGER; see AMAX1 above)
MIN0(*in/v list*)	miminum value in the list (returns an INTEGER; see AMIN0 above)
MIN1(*rn/v list*)	minimum value in the list (returns an INTEGER; see AMIN1 above)
MOD(in/v_1,in/v_2)	INTEGER remainder from division of in/v_1 by in/v_2 (see AMOD above)
NINT(*rn/v*)	rounds *rn/v* to nearest INTEGER
REAL(*in/v*)	REAL number representation of *in/v*
SIN(*rn/v*)	sine of *rn/v* where *rn/v* is in radians (returns a REAL)
SINH(*rn/v*)	hyperbolic sine of *rn/v* (returns a REAL)
SQRT(*rn/v*)	square root of *rn/v* (returns a REAL)
TAN(*rn/v*)	tangent of *rn/v* where *rn/v* is in radians (returns a REAL)
TANH(*rn/v*)	hyperbolic tangent of *rn/v* (returns a REAL)

where:

- *c* is a one byte CHARACTER.
- *cs* is a CHARACTER string.
- *in/v* is an INTEGER number or variable.
- *list* is a list of numbers or variables, with each item separated by commas.
- *rn/v* is a REAL number or variable.

Examples:

```
          1         2         3
123456789012345678901234567890..........
      ADJ=HYPOT*COS(THETA)
      TAN=ATAN(OPPOS/ADJ)
      PERIOD=2*PI*SQRT(ALENGTH/GRAVITY)
      THE_MAX=AMAX1(A(1),A(2),A(3),A(4))
```

Topic 17-14 BASIC

Intrinsic Functions

(Intrinsic functions in BASIC can be rather numerous, especially if the dialect in use addresses graphics. Listed here is only a limited set.)

FUNCTION	RESULT
ABS(*rn/v*)	absolute value of *rn/v*
ATN(*n/v*)	arctangent (in radians) of *rn/v*
CHR$(*in/v*)	the CHARACTER whose ASCII code is *in/v*
CINT(*rn/v*)	rounds *rn/v* to nearest INTEGER
COS(*rn/v*)	cosine of *rn/v*, where *rn/v* is in radians
EXP(*rn/v*)	$e^{(r\ n/v)}$
FIX(*rn/v*)	converts (truncates) *rn/v* to INTEGER
INSTR(n,cs_1,cs_2)	INTEGER value for first occurrence of substring cs_2 within cs_1 starting at position n
INT(*rn/v*)	same as FIX
LEN(*cs*)	length of *cs*
LOG(*rn/v*)	natural logarithm [(ln) or ($\log_e$)] of *rn/v*
RND(*in*)	returns a random number; *in* may be omitted
SGN(*rn/v*)	returns the sign of *rn/v*
SIN(*rn/v*)	sine of *rn/v*, where *rn/v* is in radians
SQR(*rn/v*)	square root of *rn/v*
STR$(*irn/v*)	string value of *irn/v*
STRING$($c_1,c_2$)	a string consisting of ASCII CHARACTER c_2 repeated c_1 times
TAN(*rn/v*)	tangent of *rn/v*, where *rn/v* is in radians

where:

c	is a one byte CHARACTER.
cs	is a CHARACTER string.
in	is an INTEGER number.
in/v	is an INTEGER number or variable.
irn/v	is an INTEGER or REAL number or variable.
rn/v	is a REAL number or variable.

Examples:

```
         1         2         3
1234567890123456789012345678890..........
50 ADJ=HYPOT*COS(THETA)
75 TAN=ATN(OPPOS/ADJ)
90 PERIOD=2*PI*SQR(L/G)
```

puter carries out the computations prescribed by the SIN algorithm, the value of Y will be set to 0.5. Thus, when a system-supplied intrinsic function is encountered by a compiler, or an interpreter, an algorithm in the library is used to calculate the value of the function. Users need not concern themselves with the details, as long as the number and size of the arguments are correct.

Now consider a pair of program lines similar to the two used in the preceding section (FORTRAN or BASIC):

```
................
................
20 X = 30.*3.14159/180.
25 Y = MYCALC(X)
................
```

MYCALC is not an intrinsic function of most compilers and interpreters, so it is not recognized as a reserved name. Therefore, the language system cannot automatically supply an algorithm for computing its value. MYCALC does, however, appear to the compiler or interpreter as either a subscripted variable or a function. If you intend MYCALC to be a subscripted variable, you simply need to include it in a DIMENSION statement.

Topic 17-15 Statements Functions — FORTRAN

The form of the statement function in FORTRAN is

```
         1         2         3
1234567890123456789012345678901234567890..........
      name(dummy1,dummy2,....)=expression
```

where *name* is the symbolic name of the function as it will be used in the program, *dummy1, dummy2,* and so on, the dummy variables that will be replaced with the variable names, in a one-to-one correspondence, when the function is used in the program, and *expression* is an algebraic/trigonometric collection of constants, intrinsic functions, and the variables *dummy1, dummy2,* and so on. The function *name* must be declared as to type (**INTEGER, REAL,** or **CHARACTER**) or it must conform to the default rules for variable types (see Topic 17-5). The statement function must appear among the nonexecutable statements at the beginning of the program.

Examples:

```
         1         2         3
1234567890123456789012345678901234567890..........
```

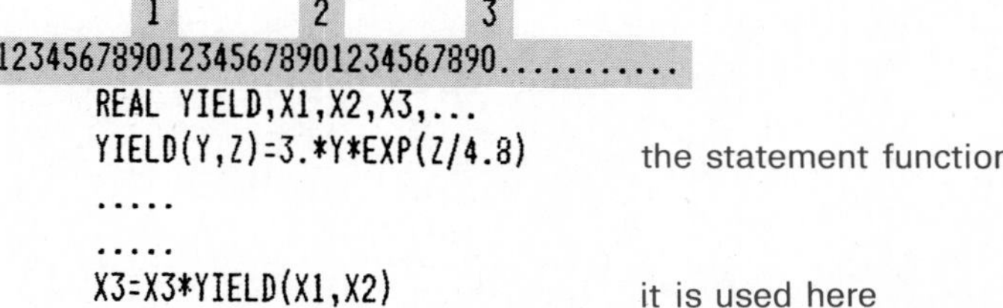

```
      REAL YIELD,X1,X2,X3,...
      YIELD(Y,Z)=3.*Y*EXP(Z/4.8)        the statement function
      .....
      .....
      X3=X3*YIELD(X1,X2)                it is used here
```

If you intend MYCALC to be a function, however, you must not include it in a DIMENSION statement but, instead, supply the algorithm for calculating its value. Thus MYCALC would be a *user-supplied* function; these are described below.

Statement Functions (Topic 17-15). For functions that can be expressed in only one line of code, both FORTRAN and BASIC provide *statement functions* that can be included in the nonexecutable statements of a program, or subprogram, in which the function is used.

As an example, suppose that MYCALC(X) was computed from X as $(X^2 + 31 \cdot X)$; then we could place a statement function in among the nonexecutable statements that would provide the computer the instructions for calculating its value. Then, when MYCALC(X) was needed in some computation, it could be used as if it were known, as in the assign statement 25 below (FORTRAN or BASIC):

```
...............
...............
24 X = 2.37
25 Z = MYCALC(X) + 24.
...............
```

Topic 17-15 BASIC

Statement Functions

The form of the statement function in BASIC is

```
         1         2         3
123456789012345678901234567890..........
no DEF name(dummy1,dummy2,....)=expression
```

where *name* is the symbolic name of the function as it will be used in the program, *dummy1, dummy2,* and so on, are dummy variables that will be replaced with the variable names, in a one-to-one correspondence, when the function is used in the program, and *expression* is an algebraic/trigonometric collection of constants, intrinsic functions, and the variables *dummy1, dummy2,* and so on. The function *name* must be consistent with the type of data that the function is to produce. The statement function must appear among the nonexecutable statements at the beginning of the program.

Examples:

```
         1         2         3
123456789012345678901234567890..........
10 DEF FNYIELD(Y,Z)=3.*Y*EXP(Z/4.8)    the statement function
.....
50 X3=X3*FNYIELD(X1,X2)             it is used here
.....
```

Topic 17-16 FORTRAN

External Functions

External functions are seperate program entities from the main program. The first noncomment line in an external function must be of the form

```
        1         2         3
1234567890123456789012345678901234567890..........
      type FUNCTION name(dummy1,dummy2,dummy3,...)
```

where *type* is one of **REAL, INTEGER,** or **CHARACTER**, the *name* is the symbolic name by which the function will be used in the program which needs it, and *dummy1, dummy2, dummy3,* and so on, are dummy variables that will be replaced, in a one-to-one correspondence, with the variables that are specified by the programmer when the function is used in the program which needs it. The function *name* must be declared as to type (**INTEGER, REAL,** or **CHARACTER**) or it must conform to the default rules for variable types (see Topic 17-5).

The function subprogram must define the function before returning. There must be a **RETURN** or **END** statement in the function subprogram. The syntax of the **RETURN** is

```
        1         2         3
1234567890123456789012345678901234567890..........
      RETURN
```

Example:

```
        1         2         3
1234567890123456789012345678901234567890..........
      PROGRAM MAIN
      REAL ALLOY,ANNEAL,PRCT1,PRCT2,TEMP    note ANNEAL
      ...                                     is declared
      ALLOY=ANNEAL(TEMP,PRCT1,PRCT2)        the function
      ...                                     appears here
      END

      REAL FUNCTION ANNEAL(T,P1,P2)   start of function
      REAL TEMP,P1,P2,Q,R             subprogram
      Q=21.7
      R-19.0
      ...
      ANNEAL=P1/P2*EXP(Q/RT)   function is defined before returning
      RETURN                   if omitted, the END statement
      END                        serves the same purpose
```

Functions return a value and, like any other variable, must be declared (**REAL, INTEGER,** and **CHARACTER**).[11] Also, functions are not limited to a single argument, but may have several arguments.

External Functions (Topic 17-16): FORTRAN Only. If the function you want to compute *cannot* be expressed in one line of code, FORTRAN (but *not most* BASICs) allows you to use an *external function* subprogram. As shown in the end paper programming guide programming summary, function subprograms are separate entities much like subroutines.

Subroutines

As shown on the end paper programming guide, subroutines are another type of subprogram. Although there are three types, we shall discuss in detail only *user-supplied* FORTRAN subroutines.

User-Supplied Subroutines (Topic 17-17). Suppose that you had a 40-line computational procedure that had to be executed at 10 different places within a program. One way of coding this would be to simply insert 400 (i.e., 40 × 100) lines of code into the main program. But this is not a totally satisfactory solution since it adds length and may add memory storage requirements to the code. Also, if the computational procedure needs changing, all 10 occurrences would have to modified. Subroutines provide an alternative.

Lines of code that fit nicely into a separate package may be bundled into subroutines that can be given a name in FORTRAN, and a unique line number in BASIC. Augusta Ada Byron Lovelace first established the concept of the subroutine in 1842, when she streamlined the writing of her code for Charles Babbage's analytical engine with the notation, "Here follows a repetition of operations 13 to 23." The modern use of subroutines was demonstrated in Topic 17-2, for both FORTRAN and BASIC.

[11]That is, the computer treats the function just like any other variable and reserves a storage location, in its name, for saving its value. The computer must know whether to store the information as an INTEGER, a REAL number, or a CHARACTER.

Topic 17-16 — BASIC

External Functions

External functions are not permitted in most BASICs.

Topic 17-17 FORTRAN

User-Supplied Subroutines

The user-supplied subroutine is a separate program entity much like the external function. The first non comment line of the subroutine must be of the form

```
         1         2         3
1234567890123456789012345678901234567890..........
      SUBROUTINE name(dummy1,dummy2,...)
```

where *name* is name you want associated with this subroutine and *dummy1, dummy2, dummy3,* and so on, are dummy variables which will be replaced, on a one-for one bases, with the variable listed with the subroutine name when it is called in the calling program. The subroutine name, unlike the external function, does not need to be declared as to type, because it does not return a value as does the function. There must be a RETURN or an END statement in the subroutine.

The subroutine is invoked in a program by a subroutine CALL whose syntax is

```
         1         2         3
1234567890123456789012345678901234567890..........
      .....
      CALL name(var1,var2,var3,...)
      .....
```

where *var1, var2, var3,* and so on, (the inputs and outputs to the subroutine) are constants or variables that will be substituted for the dummy variables in the SUBROUTINE statement when it is executed. The *number* and *type,* but not necessarily the names, of argument variables in the call (i.e., the *var1, var2, etc.*) must match those of the argument names in the SUBROUTINE statement (i.e., the *dummy1, dummy2, etc.*)

Examples:

```
         1         2         3
1234567890123456789012345678901234567890..........
      REAL INTI_VEL,TIME,G,DISTANCE
      ....
      ....
      INIT_VEL=4.
      TIME=37.
      G=9.8
      CALL PROJECT(INIT_VEL,TIME,G,DISTANCE)
      PRINT *,DISTANCE
      ...
      END
```

(Continues on page 548)

Topic 17-17 BASIC
User-Supplied Subroutines

In most BASICs, a user-supplied subroutine enters the program just like most other groups of statements. The subroutine does not stand separate from the main program. It is identified by a beginning line number and not a name.

The subroutine is invoked in a program by a subroutine call whose syntax is

```
         1         2         3
1234567890123456789012345678890..........
stno GOSUB no1
```

where *stno* is an appropriate statement number and *no1* is the line number of the first line of the subroutine.

Example:

```
         1         2         3
1234567890123456789012345678890..........
...
....
50 INIT_VEL=4.
60 TIME=37.
70 G=9.8
80 GOSUB 2000
90 PRINT *,DISTANCE
.....
150 END
.....
2000 REM SUBROUTINE FOR COMPUTING DISTANCE
2010 DISTANCE=-G*TIME**2/2+INIT_VEL*TIME
2020 RETURN
```

The program above uses the subroutine to calculate a value for the variable DISTANCE, knowing the variables INIT_VEL, TIME, and G. After the subroutine is called, the value of DISTANCE is available and can be printed out. That is, three of the arguments (INIT_VEL, TIME, and G) are inputs to the subroutine and one argument (DISTANCE) is an output.

Note that dummy variables cannot be used in the subroutine as is done in FORTRAN. The same variable names have to be used in both the subroutine and the part of the program that calls it for execution to proceed properly.

SOME EXAMPLE PROBLEMS

We will now demonstrate some of the principles of programming and work some example problems. First, let us redo the problem that was programmed in Topic 17-1. We mentioned earlier when introducing that topic that the code displayed there was not very robust or user friendly.

Example 17-1:

Solve the segment of the circle problem in a more robust and and friendly fashion than was done in Topic 17-1. This problem is expressed in Equation (16-4), where we desire to solve for α, given an f.

Solution:

The code for this problem is given below. Compared to the code in Topic 17-1, the present version prompts the user for the appropriate inputs, checks to see that the inputs are within reasonable bounds, and gives back a more informative output (however, the output could be improved considerably—we will leave that to you).

FORTRAN version:

```
         1         2         3         4         5         6         7         8
12345678901234567890123456789012345678901234567890123456789012345678901234567890
C THIS PROGRAM SOLVES F=(ALPHA-SIN(ALPHA))/(2*PI) FOR ALPHA, GIVEN AN F
C THE NEWTON-RAPHSON ITERATIVE TECHNIQUE IS USED WITH THE CONVERGENCE CRITERIA
C  THAT  ABS(F-(ALPHA-SIN(ALPHA))/2*PI) .LE. DELF
C                            DLE 12-2-84
C
       IMPLICIT CHARACTER*16(A-Z)
       REAL F, FC, DELF, ALPHA, PI, ALPHAD
       PI = 3.14159
C     ENTER THE FRACTION DESIRED
C
       PRINT *, ' ENTER THE FRACTION DESIRED
       READ (5,*) F
       PRINT *, 'THE ENTERED VALUE WAS   ',F
C
       IF(F.LT.0.0) THEN
         PRINT *, ' THE FRACTION MUST NOT BE NEGATIVE'
         STOP
       ELSE
         IF(F.GT.1.0) THEN
            PRINT *,' THE FRACTION MUST NOT BE GREATER THAN 1.0'
            STOP
         ELSE
```

Topic 17-17 FORTRAN

User-Supplied Subroutines (Continued)

```
SUBROUTINE PROJECT(V,T,G,D)
REAL D,G,T,B
....
D=-G*T**2/2+V*T
RETURN
END
```

The program above uses the subroutine PROJECT to calculate a value for the variable DISTANCE, knowing the variables INIT_VEL, TIME, and G. After the subroutine is called, the value of DISTANCE is available and can be printed. That is, three of the arguments (INIT_VEL, TIME, and G) are inputs to the subroutine and one argument (DISTANCE) is an output.

```
CC ENTER THE ERROR TOLERANCE
C
            PRINT *, ' ENTER THE CONVERGENCE TOLERANCE DESIRED
            READ (5,*) DELF
            PRINT *, ' THE ENTERED VALUE WAS  ',DELF
            IF (DELF.LT.0.) THEN
               PRINT *,' THE CONVERGENCE TOLERANCE MAY NOT BE NEGATIVE'
            ELSE
C CALCULATE THE FIRST GUESS FOR  ALPHA
               ALPHA=(12.*PI*F)**(1./3.)
C **** WHILE TO ITERATE FOR ALPHA
10                CONTINUE
                IF(ABS(F-(ALPHA-SIN(ALPHA))/(2.*PI)).GT.DELF) THEN
                ALPHA=ALPHA-(ALPHA-SIN(ALPHA)-2.*PI*F)/(1.-COS(ALPHA))
                  GO TO 10
               ENDIF
C **** ENDWHILE
C
C WRAP-UP THE CALCULATIONS AND OUTPUT RESULTS
               ALPHAD=ALPHA*180./PI
               FC=(ALPHA-SIN(ALPHA))/(2.*PI)
               PRINT *
               PRINT *
               WRITE(6,15)
15          FORMAT(' OUTPUT IN THE ORDER OF: F, FC, ALPHA, ALPHA(DEG)')
               PRINT *
               PRINT *,' WHERE: F IS DESIRED, FC IS CALCULATED FRACTION'
               PRINT *,'         ALPHA IS IN RAD., ALPHA(DEG) IS IN DEG.'
               PRINT *
               PRINT *, F, FC, ALPHA, ALPHAD
             ENDIF
          ENDIF
        ENDIF
       PRINT *,' RESTART PROGRAM FOR ANOTHER F.'
       END
```

We should note several things about this FORTRAN program. First, we begin by declaring all variables to be of type CHARACTER (the IMPLICIT CHARACTER statement at the beginning of the code). Then we declare the type for the REAL and INTEGER variables that we will use in the program. This is a useful practice. We have seen many student-written codes (and we have this problem ourselves from time to time) wherein undeclared variables are allowed to default to the type determined by the first letter of the variable name (see Topic 17-5), but where the use of the variable was incompatible with the default type. By declaring all variables to be of type CHARACTER, unless specified otherwise, we will get an error message when we try to use an undeclared variable for a number in an expression. In this way we will not get any unexpected surprises that might arise when we use an INTEGER where we should have used a REAL number. The method we have chosen here forces us to declare all of the numerical variables before we use them, a practice that is demanded by other languages.

Second, we have used a bare minimum of statement numbers, which is often a good sign that we are conforming to good structured programming principles. The program is structured top down, first prompting the user for the appropriate input, then checking the value of the coefficient on f (variable F in the program) for being less than or greater that zero, and then checking the convergence tolerance (DELF) for being less that zero.

One bad feature is that only one calculation is performed each time the program is executed. Example 17-2 describes a way of remedying this flaw.

BASIC version:

```
         1         2         3         4         5         6         7         8
12345678901234567890123456789012345678901234567890123456789012345678901234567890
10 PI =3.14159
20 INPUT " INPUT THE FRACTION DESIRED  ",F
30 PRINT " THE ENTERED NUMBER WAS  ",F
40 IF(F<0) THEN PRINT "THE FRACTION MUST NOT BE NEGATIVE": END
50 REM ELSE
60     IF(F>1!) THEN PRINT "THE FRACTION MUST NOT BE GREATER THAN 1.": END
70 REM ELSE
```

```
80        INPUT " ENTER THE CONVERGENCE TOLERANCE DESIRED  ", DELF
90        PRINT " THE ENTERED NUMBER WAS  ",DELF
100       IF(DELF<0!) THEN PRINT " THE CONVERGENCE TOLERANCE MAY NOT BE NEGATIVE"
: END
110 REM  ELSE
120 REM CALCULATE THE FIRST GUESS FOR ALPHA
130 REM
140            ALPHA=(12!*PI*F)^(1/3)
150            PRINT "START ALPHA", ALPHA
160            WHILE ABS(F-(ALPHA-SIN(ALPHA))/(2!*PI)) > DELF
170              ALPHA=ALPHA-(ALPHA-SIN(ALPHA)-2!*PI*F)/(1-COS(ALPHA))
180            WEND
190 REM
200 REM WRAP-UP THE CALCULATIONS AND OUTPUT RESULTS
210 REM
220            ALPHAD=ALPHA*180!/PI
230            FC=(ALPHA-SIN(ALPHA))/(2!*PI)
280            PRINT " WHERE: F IS DESIRED, FC IS CALCULATED FRACTION"
290            PRINT "        ALPHA IS IN RAD., ALPHA(DEG) IS IN DEG."
300            PRINT
310            PRINT F, FC, ALPHA, ALPHAD
320 REM    ENDIF
330 REM   ENDIF
340 REM ENDIF
350 END
```

Declaring variables is considerably different in BASIC than in FORTRAN. We have chosen not to make a sweeping declaration such as we did in the FORTRAN code above; therefore, the suffix attached to the variables will determine the type. Since we have no integer or string variables in the problem, there are no variable names that end in \$ or %. The names we have chosen will all default to single precision, REAL type. In BASIC, there is no easy way to declare *individual* variables with a nonexecutable statement such as the REAL, INTEGER, or CHARACTER statements in FORTRAN.

Next, let us examine the solution of a quadratic equation using the binomial theorem. We constructed the flowchart for this problem in Chapter 16 ([16-3]); we will also solve this problem with a spreadsheet in Example 18-7.

Example 17-2

Write a program to read the coefficients *a*, *b*, and *c*, of a quadratic equation of the form $ax^2 + bx + c = 0$, compute the solution(s), and return to the read again for collecting data on the next equation to be solved.

Solution:

If $a = 0$, the solution is a single root of $-c/b$. However, if $a \neq 0$, the solution depends on the value of the discriminant, $b^2 - 4ac$. Let us write the pseudocode for this problem in preparation for coding it.

```
t = 0
While t = 0
Read a, b, c
Calculate discriminant = b^2 - 4ac
If a = 0 then
  One root, -c/b
Else
  If disc < 0
    Real part -b/2a
    Imaginary part (-disc)^(1/2)/2a
  Else
```

```
        If disc > 0
          1st root, (−b + disc^(1/2))/(2a)
          2nd root, (−b − disc^(1/2))/(2a)
        Else
          One root, −b/(2a)
        Endif
      Endif
    Endif
    End While
    End
```

Pseudocode, when correct, makes actual coding very easy. For clarity, we have nested the IF-THEN-ELSEs with each new set being indented from the last set. We chose the IF-THEN-ELSE construct here, because it very adequately solves the problem and is easy to "read."

FORTRAN version: Again, FORTRAN 77 does not have a WHILE statement, so we will have to synthesize one (Topic 17-9). The FORTRAN code for this is

```
         1         2         3         4         5         6         7         8
12345678901234567890123456789012345678901234567890123456789012345678901234567890
      IMPLICIT CHARACTER*16(A-Z)
      REAL DISC,A,B,C,R1,R2
      INTEGER T
      T=0
C--WHILE T>0
10    IF(T.EQ.0) THEN
        READ(*,*,END=20,ERR=20) A,B,C
C ECHO THE INPUT
        PRINT *,A,B,C
       IF(ABS(A).LT.1.E-10.AND.ABS(B).LT.1.E-10.AND.ABS(C).LT.1.E-10)
     @    STOP 'ALL COEFFICIENTS ZERO - NO EQUATION'
        DISC=B**2-4.*A*C
        IF(ABS(A).LT.1.E-10) THEN
C - - - -  EQ IS LINEAR, WITH ONE ROOT
          R1=-C/B
          PRINT *,' ONE ROOT = ',R1
        ELSE
C - - - -  EQ IS QUADRATIC
          IF(DISC.LT.0) THEN
C - - - - -  IMAGINARY ROOTS
            R1=-B/2./A
            PRINT *,' IMAGINARY ROOT, REAL PART = ',R1
            R2=SQRT(-DISC)/2./A
            PRINT *,'                     IMAGINARY PART = ',R2
          ELSE
            IF(DISC.GT.0) THEN
C - - - - - - -REAL ROOTS - THERE ARE 2 OF THEM
              R1=(-B+SQRT(DISC))/2./A
              PRINT *,'  FIRST REAL ROOT = ',R1
              R2=(-B-SQRT(DISC))/2./A
              PRINT *,' SECOND REAL ROOT = ',R2
            ELSE
C - - - - - - -ONE REAL ROOT (DISC=0)
              R1=-B/2./A
              PRINT *,' 1 REAL ROOT',R1
            END IF
          END IF
        END IF
      GO TO 10
      END IF
C--END WHILE
20    CONTINUE
      STOP
        END
```

As in Example 9-1, we have again used a bare minimum of statement numbers, which is often a good sign that we are conforming to good structured programming principles. The program is structured top down, first checking the value of the coefficient on x^2 for being nonzero. Then the program calculates the discriminant and exhaustively checks all possibilities for this important group of numbers. Notice that the pseudocode and the FORTRAN program look very much alike. Once the pseudocode is written, actually coding becomes more-or-less mechanical.

Notice also that checks on the coefficient "a," a REAL number, being equal to zero are handled carefully. For example, since "a" is being stored as a REAL or floating-point number, it will probably never be exactly equal to what you might expect, although it should be very close. Therefore, you need to give a little tolerance when checking its value against some absolute number. That is what we are doing in lines 10 and 13 (counting from the top). *This is always true of REAL numbers. Never make an absolute comparison with REAL numbers, for absolute comparisons are seldom fulfulled. This is not true of INTEGER and CHARACTER variables for they are stored differently in the computer than are REALs.*

As long as you keep inputting values for a, b, and c, the program continues to calculate and display the roots and to ask for more inputs. Data entry is ended, the program terminates, and you are returned to the operating system when 0's are entered for a, b, and c.

BASIC version: Let us now look at how this program might be coded in BASIC. BASIC presents a small problem because the IF-THEN-ELSE construct is limited to one statement (see Topic 17-8). We can code from the pseudocode presented earlier, but we can not do a lot of calculation in each section of the IF-THEN-ELSE statements. We have chosen to keep the IF-THEN-ELSE constructs but to use subroutines (GOSUB statements, Topic 17-17) to actually carry out the calculations and to print the roots.

There are, of course, many ways to code the problem. We have again chosen not to make a sweeping declaration such as we did in the FORTRAN code above. The BASIC code we have chosen is

```
         1         2         3         4         5         6         7         8
12345678901234567890123456789012345678901234567890123456789012345678901234567890
10 T%=0
20 WHILE T%=0
30   INPUT;"INPUT A,B,C  ",A,B,C
31   PRINT
32   PRINT
33 REM TERMINATE IF A=B=C=0
35   IF ABS(A)<1E-10 THEN IF ABS(B)<1E-10 THEN IF ABS(C)<1E-10 THEN END
40   DISC=B^2-4!*A*C
45   PRINT "DISCRIMINANT IS  " DISC
46   PRINT
50   IF ABS(A)<1E-10 THEN GOSUB 500
60   IF DISC<-1E-10 THEN GOSUB 600 ELSE IF DISC>1E-10 THEN GOSUB 700 ELSE GOSUB
800
65   PRINT
66   PRINT
70 WEND
500 PRINT "ONLY ONE ROOT, EQUATION IS LINEAR - NOT QUADRATIC"
505 PRINT "ONLY ROOT IS " (-C)/B
510 RETURN
599 REM --------------------------------
600 REM - IMAGINARY ROOTS
605 PRINT "IMAGINARY ROOTS:   REAL PART IS " (-B)/A/2!
610 PRINT "                 IMAGINARY PART IS " SQR(-DISC)/2!/A
620 RETURN
699 REM --------------------------------
700 REM - TWO REAL ROOTS
710 PRINT "REAL ROOTS:        ONE ROOT IS " (-B+SQR(DISC))/2!/A
720 PRINT "                 OTHER ROOT IS " (-B-SQR(DISC))/2!/A
730 RETURN
799 REM --------------------------------
800 REM ONLY ONE REAL ROOT, DISCRIMINANT IS ZERO
810 PRINT "ONLY ONE (REAL) ROOT " (-B)/2!/A
820 RETURN
830 END
```

Just as in the FORTRAN code above, the program keeps calculating as long as you continue to enter data on a, b, and c. If you enter 0's for these three coefficients, the program terminates and returns you to the operating system.

We now pose a problem that will allow us to demonstrate both the use of subscripted variables and the use of the DO (or FOR) loop.

Example 17-3:

Suppose you were required to write a program that would accept as many as 20 numbers entered from the keyboard, and then provide back to the user a list of the numbers together with the average of all numbers and the difference between each number and the average. You may assume that none of the numbers will be zero, so that you may use this number (zero) for terminating the input process and initiating the calculation of the average. The program should continually ask for new sets of numbers to average, until such time that a zero is entered as the first number in the sequence. This should signify to the program that the equations are to be terminated.

Possible pseudocode for this problem might take the following form:

```
While T = 0
  Do as many as 20 times
    Read in a number
    If first number = 0
      Then terminate
    Else
      If an entry is zero
        Exit the Do
      Else
        Add this number to the sum of the previous numbers
        Keep a count of the number of numbers entered
      End If
    End If
  End the do
  Average = Sum of numbers/number of numbers
  Do for each entered nonzero number
    Print number, number minus average, average
  End the do
End the while
```

FORTRAN version: Coded in FORTRAN, the problem might look as follows:

```
         1         2         3         4         5         6         7         8
12345678901234567890123456789012345678901234567890123456789012345678901234567890
      IMPLICIT CHARACTER*10 (A-Z)
      REAL NUM(20),SUM,AVG
      INTEGER I,T
      T=0
C--WHILE T=0
10    IF(T.EQ.0) THEN
       PRINT *,' '
       PRINT *,' FIND AVERAGE OF AS MANY AS 20 NUMBERS'
       PRINT *,'   Enter a zero to signify the end of entries'
       PRINT *,'   If first number is zero, the program terminates'
       PRINT *,' '
       SUM=0.
C - - -READ IN THE NUMBERS
       DO 20 I=1,20
         PRINT *,' NUMBER ',I,'    '
         READ(*,*) NUM(I)
```

```
C - - - - -TERMINATE IF FIRST NO. IS ZERO
           IF(I.EQ.1.AND.ABS(NUM(I)).LT.1.E-10) THEN
             STOP 'TERMINATING'
           ELSE
C - - - - - -DON'T READ MORE IF NO. IS ZERO
             IF(ABS(NUM(I)).LT.1.E-10) THEN
                        GOTO 30
             ELSE
C - - - - - - -SUM 'EM UP AND KEEP GOING
               SUM=SUM+NUM(I)
             END IF
           END IF
20      CONTINUE
        I=I+1
30      AVG=SUM/(I-1)
        PRINT *,' HERE ARE NUMBERS, DEVIATIONS, AND AVERAGE
        DO 40 J=1,I-1
        PRINT *, NUM(J),(NUM(J)-AVG),AVG
40      CONTINUE
        GOTO 10
        END IF
C--END WHILE
        END
```

This problem offers a prime opportunity to make use of a subscripted array, since many numbers are to inputted and each one gets the same type of treatment. We need to save each entered number in order to compute its deviation from the average. If we did not use a subscripted variable, we would have to introduce as many as 20 different variable names in order to code this problem.

We have to be careful of the values of the running variable in the first DO loop, especially if we wish to use it as a counter for tabulating the number of numbers entered. If a zero input is encountered on input, the counter (here I in the first DO loop) has already indexed one too many times. If a zero is not encountered and a full 20 numbers get entered, the counter just matches the correct number of inputs (here 20). If the latter happens, we add 1 to the value of the counter, before we divide the sum of the numbers by (I-1).

BASIC version: Coded in BASIC, this problem might take a form such as that shown below.

```
         1         2         3         4         5         6         7         8
12345678901234567890123456789012345678901234567890123456789012345678901234567890
10 DIM NUM(20)
20 PRINT
30 PRINT
40 PRINT " INPUT UP TO 20 NUMBERS, ONE AT A TIME"
50 PRINT "    Enter a zero to signify the last number has been entered"
60 PRINT "    If the first number entered is a zero, the program terminates"
62 PRINT
64 SUM=0
70 FOR I%=1 TO 20
80    PRINT " NUMBER " I%;
90    INPUT NUM(I%)
91 REM TERMINATE IF FIRST NO. IS ZERO
95    IF I%=1 THEN IF NUM(I%)=0 THEN END
97 REM STOP READING NO. IF YOU FIND A ZERO
100   IF NUM(I%)=0 GOTO 130
103 REM SUM 'EM UP AS YOU GO
105   SUM=SUM+NUM(I%)
110 NEXT
120 I%=I%+1
130 AVG=SUM/(I%-1)
140 PRINT "HERE ARE NUMBERS, DEVIATIONS, AND AVERAGE"
142 FOR J=1 TO I-1
146 PRINT NUM(J),(NUM(J)-AVG),AVG
148 NEXT
150 END
```

This BASIC program does what the preceding FORTRAN program did in almost exactly the same way.

Our purpose here has been to give you a rudimentary understanding of programming, touching on both FORTRAN and BASIC. We do not have enough space to make master programmers out of you. However, you should now have enough working knowledge to allow you to pick up more details on your own, with perhaps the help of only a programming manual.

PROBLEMS[12]

17-1. Find the character set of the language you will be using. Find the entire character set for the computer you will be using.

17-2. Find the maximum number of bytes that may be stored in a "word" of your computer.

17-3. Use the program of Topic 17-1 as a beginning exercise in the use of the computer. Prepare the program and data using a terminal and a related editor. Find out what job control cards or lines are needed to make up the complete job. Execute the job using the data exactly as given.

17-4. Upgrade the program of Problem 17-3 using the program of Example 17-1. What improvements do you notice?

17-5. Find the maximum number of binary digits (n) that can be used to represent an INTEGER for your computer. Compute the largest integer that this permits.

17-6. Find out the number of binary digits used to represent the fractional part of a REAL number for your computer. Using the formula given in the second footnote of Table 17-1, compute the maximum amount of significance (in terms of decimal digits) that your computer can give to a REAL number.

17-7. Each of the following is (A) a valid FORTRAN:BASIC INTEGER constant, (B) a valid FORTRAN:BASIC INTEGER constant, (C) a valid FORTRAN:BASIC INTEGER variable name, (D) a valid FORTRAN:BASIC REAL variable name, or (E) none of the above. Identify each. If you find that it is (E), explain what seems to be wrong with it.

a. 0.0:0.0
b. WRONG:WRONG!
c. MIKE:MIKE%
d. PIKE:PIKE!
e. AB17:AB17
f. A*B:A*B
g. POMEGRANATES:POMEGRANATES
h. 5000.762:5000.762
i. 5000.762E-8:5000.762E-8
j. 5000.762E+85:5000.762E+85
k. ALPHA0:ALPHA0!
l. ALPHAO:ALPHAO!
m. −17,456.2:−17,456.2
n. -4E0:-4E0
o. -4E-0:-4E-0
p. -4.E+0:-4.E+0
q. TEST:TEST!
r. 1492AD:1492AD!
s. THE WHO:THE WHO
t. ARRAY 1:ARRAY 1
u. MATRIX 2:MATRIX 2%
v. 5 280:5 280!
w. 5,280

17-8. Each of the following is (A) a valid INTEGER FORTRAN:BASIC expression, (B) a valid REAL FORTRAN:BASIC expression, (C) an FORTRAN:BASIC expression with an improper mixing of INTEGER and REAL modes or types—but otherwise valid, or (D) not a valid FORTRAN:BASIC expression. Identify each. If it is (D), explain what the defect seems to be.

a. 95:95
b. AB17:AB17!
c. AB*17:AB!*17
d. AB*17.0:AB!*17!
e. A**2 + B*B:A^2 + B*B
f. JOE + MOE:JOE% + MOE%
g. J0E*M0E:J0E*M0E
h. (A + B)(C + 17.3):(A + B)(C + 17.3)
i. (((A)))/2.0:(((A)))/2.0
j. (95):(95)
k. SIN(4):SIN(4)
l. SQRT(946.0):SQR(946.0)
m. ATAN2(Y,X):ATN(Y)
n. A*B**3:A*B^3
o. A/B/C:A/B/C
p. A/(B/C):A/(B/C)
q. ((A/B)/C):(A/B)/C)
r. A1*X**2 + A2*X + A3:A1*X^2 + A2*X + A3
s. (A*X**2 + B)/(C*Y + D):(A*X^2 + B)/(C*Y + D)
t. (I + J)(IJ − K):(I% + J%)(I%J% − K%)
u. 2 + (I − 1)*Q:2 + (I% − 1)*Q!
v. SIN(4.0)COS(4.0):SIN(4!)COS(4!)
w. ATAN(Y/X):ATN(Y/X)
x. ALOG(I + J):ALOG(FLOAT(I% + J%))

17-9. Write valid FORTRAN or BASIC expressions for the following algebraic expressions.

a. $a + b + c + d$
b. $a + b - c + d$
c. $(a + b)/(c + d)$
d. $\dfrac{ax^2 + \mathrm{bx} + \mathrm{c}}{dy + f + \cos z}$
e. $\sqrt{s(s - a)(s - b)(s - c)}$
f. $\cos 75°$
g. $\tan^{-1}(y/x)$
h. e^2
i. $\sec(z + 10^6)$
j. $\dfrac{a/c}{x + \dfrac{\cos z}{a + b/c}}$
k. $i + j - (k - m)$
l. $v_1 + i_2r_2 + \dfrac{rv}{r_3}$
m. $\dfrac{i}{1 - \dfrac{1}{(1 + i)^n}}$
n. $1/[1 - (1 + i)^{-n}]$

17-10. Express in algebraic notation the formulas implied by the following FORTRAN: BASIC expressions.

a. X + Y/W + Q:X + Y/W + Q
b. Z1*Z2/Z3*Z4:Z1*Z2/Z3*Z4
c. EXP(−2.0*PI*AL/ALMAX):EXP(−2.0*PI*AL/ALMAX)
d. 3.0*Y**2/Z**3:3.0*Y^2/Z^3
e. A*SIN(X)*COS(X):A*SIN(X)*COS(X)
f. INDEX − INDEX/2*2:INDEX − INDEX/2*2
g. A**1.06*B**2.73*C*.25

[12]FORTRAN:BASIC means if you are studying FORTRAN use what is on the left of the ":", if you are studying BASIC use the expressions on the right of the ":".

17-11. Write FORTRAN:BASIC assignment statements for each of the following:

a. $f = \frac{c}{2}\sqrt{\left(\frac{p}{l}\right)^2 + \left(\frac{q}{w}\right)^2 + \left(\frac{r}{h}\right)^2}$

[This formula is used in acoustics for finding the normal modes of resonance (f) in a rectangular room ($l \times w \times h$). c is the velocity of sound. p, q, and r are nonnegative integers representing different modes of resonances.]

b. $p = P + \frac{1}{2}\rho V^2\left[1 - \left(\frac{v}{V}\right)^2\right]$

(This formula is used in aerodynamics.)

c. $t_1 = \frac{R_t - R_0}{R_{100} - R_0}(100) + \delta\left(\frac{t_2}{100} - 1\right)\frac{t_2}{100}$

(This formula is used in thermometry.)

17-12. Given the following formula from spherical trigonometry:

$$\cos c = \cos a \cos b + \sin a \sin b \cos \gamma$$

where the lowercase letters represent the sides of the triangle and the Greek letter represents an angle of the spherical triangle. Assume that angles are given in degrees and are to be presented in degrees.

a. Write FORTRAN:BASIC assignment statements for c.
b. Write FORTRAN:BASIC assignment statements for γ.

Remember: You must do the mathematical manipulation before you write the assignment statement.

17-13. Prepare output statements for printing the variables listed below (include the appropriate FORMAT statements for FORTRAN).

a. $x = 17.0$, $y = 87.6$, $z = 97.23$, $k = 6$
b. $q_1 = -47.973$, $q_2 = -13.76 \times 10^{-3}$, $q_3 = 0.03 \times 10^6$
c. $i = 4{,}976$, $j = -003$, $k = 4{,}724$
d. $k = 6$, $z = 97.23$, $y = 87.6$, $x = 17.0$
e. $x_0 = 4.73\text{E}2$, $y_0 = 9.417\text{E}1$, $z_0 = 0.3176\text{E}3$

17-14. Prepare input statements for reading the variables listed in Problem 17-13 (include the FORMAT statement for FORTRAN if you use a FORMATted READ).

17-15. Assume that you have been instructed to prepare a data form that resembles a handwritten one that your boss has been using. To avoid confusing him, it should look like that shown below. Prepare the output statements necessary (include the FORMAT statement if you code it in FORTRAN).

```
a.     97Ø  -14  19.25  174
b.    174ØØØØ  19.25ØØØ   97.ØØØØ  -14
c.   .174E+Ø2 .1925E+Ø2  .97E+Ø2 -  14
d.  741925-1497
e.     17.4ØØ -34   -14     ASDFG   97 1925
```

[17-P15]

17-16. Write straight-through programs for the formulas given in Problem 17-11. These should include inputting the data and outputting the results. Have the program prompt the user for the correct data and identify the results when they are printed out.

17-17. Execute these programs from Problem 17-15 using some sample data. Verify that your programs give the correct results (you might use your hand-held calculator in the verification).

17-18. Write expressions for the *test* in the IF-THEN-ELSE construct (Topic 17-8) that will check for the following.

a. $i = 0$
b. i greater than or equal to 1
c. i greater than 0 (If i is INTEGER, does this differ from part b?)
d. $z < 17.3$
e. $(a + b)/2 \geq \sqrt{ab}$
f. $z_{min} \leq z \leq z_{max}$
g. x and y are within 0.005 of each other
h. k_6 is not more than seven counts larger than k
i. k_6 and k are not more than seven counts apart
j. The relative difference between e^z and cos $(3a + b)$ does not exceed 1 per cent of e^z.
k. $0 \leq a \leq a_{max}$ and $0 \leq b \leq b_{max}$
l. $x > x'$ or $y > y'$
m. a within 0.05 of a_{max} but $a \neq a_{max}$
n. q does not have a fractional part

17-19. Modify the segment of a circle code (Example 17-1) so that it does not stop after each solution, but rather, asks for a new input when finished with the previous calculations. Have the program stop execution if the user enters a fraction, f, of 0.

17-20. Given the following lines of code:

```
24 I = 1
25 .......
......... intervening statements not including any
......... branching or looping
.........
37 I = I + 1
39 (see below)
40 .......
```

If statement 39 is for FORTRAN (with two following statements):

```
IF (I.LE.17) THEN
GOTO 25
ENDIF
```

or for BASIC (the nonexecutable statement DEFINT I is assumed):

```
IF I < 17 THEN GOTO 25
```

how many times will statement 25 be executed before statement 40 is reached?

17-21. Given the following lines of code:

```
30 I = ISTART
35 .......
40 I = I + ISTEP
45 (see below)
50 .......
```

where statement 45 is given by for FORTRAN (along with the next two statements):

```
45 IF (I.LE.ISTOP) THEN
   GOTO 25
   ENDIF
```

for BASIC (where the declarations statement DEFINT I is assumed):

```
45 IF I < ISTOP THEN GOTO 25
```

Give a step-by-step method for calculating the number of times statement 25 will be reached during an execution of these statements. Illustrate your method with:

a. ISTART = 0	b. ISTART = 1	c. ISTART = 15
ISTOP = 14	ISTOP = 0	ISTOP = 17
ISTEP = 5	ISTEP = 4	ISTEP = −1

17-22. Obtain all real roots of the following equations on the digital computer using Newton's method. Use a convergence criteria of 0.0001. (The derivatives are shown in [ ].) The program should prompt the user for a value of the first guess.

a. $f(x) = 7.3x^3 - 20x^2 + 14x - 3 = 0$
   $[f'(x) = 21.9x^2 - 40x + 14]$
b. $f(x) = 6x^4 - 2x - 3.6 = 0$
   $[f'(x) = 24x^3 - 2]$
c. $f(x) = 4 + 6.3x^{-1} + 4.32x^{-2} - 7.91 \times 10^2x^{-3} = 0$
   $[f'(x) = -6.3x^{-2} - 8.64x^{-3} + 23.73 \times 10^2x^{-4}]$

17-23. Write a DIMENSION or DIM statement that provides storage locations in RAM for a vector named VECTOR with 75 locations, and for a 25-row, six-column array named ARRAY. How many memory locations will be reserved by this DIMENSION or DIM statement?

17-24. Suppose that you were asked to expand the program in Example 17-3 to accommodate 1000 numbers. What changes would you make in the code to allow this?

17-25. Determine the validity of the following DO:FOR statements. You may ignore the lack of statement numbers on the BASIC statements. If something is wrong, explain what it is.

a. DO 918 I = 1,19:FOR I = 918, 1, 19
b. DO427 JK = 7, 14, 3:FOR JK = 7 TO 14 STEP 3
c. DO 76, INDEX = ISTART, ISTEP,3:FOR INDEX = ISTART, ISTEP,3
d. DO 987683 17 = 4, 31, 6:FOR 987683 = 17 TO 4 STEP 31, 6
e. DO 6 I = 1.3:FOR I = 1.3
f. DO 1 K = KSTART, 16, KSTOP:FOR K = KSTART TO 16, STEP KSTOP
g. DO 2 J = 0.6:FOR J = 0 to 6 DOWN TO 2

17-26. Modify the programs in Examples 17-1 to 17-3 so that they produce appropriate labels for each of the output values.

17-27. Modify the program of Example 17-1 to incorporate two statement functions. One of these is to be used for the computation of initial guesses for α, and the other for the updated α.

17-28. Divide the program of Example 17-2 into a main program and a SUBROUTINE subprogram.

17-29. Write a program that will integrate an arbitrary (but well-behaved) function between a and b, using the approximate formula

$$\int_a^b f(x)\,dx \approx [f(a) + \sum_{i=1}^{n-1} f(a + ih) + f(b)]\,\frac{h}{2}$$

$$\text{where } h = \frac{b-a}{n}$$

The function arguments are to represent a, b, and n. The function to be integrated is to be a statement function. Execute it under a variety of conditions for which you can compute the correct result. This should build your confidence in the ability of your subprogram to integrate functions for which you cannot compute exact results. Some test integrals might be

$$\int_0^{\pi} \sin x\,dx = 2 \qquad \int_{0.1}^{1} \frac{dx}{x} = \ln 1 - \ln 0.1$$

Chapter 18

Programming on Spreadsheets

The origin of spreadsheets was discussed in Chapter 16, where it was projected that they have just begun to have an impact on the engineering community. As pointed out there, spreadsheets are applications programs that can be programmed to solve specific problems. The use of the term "programmed" in this sense means that the user must adhere to the commands and rules of syntax that the spreadsheet program establishes in order to solve any problem. This sounds very much like the process of following the necessary rules and syntax set forth by BASIC or FORTRAN when writing computer programs. The primary difference, however, is the framework within which users can solve problems and view results when using a spreadsheet. Although not as flexible as BASIC or FORTRAN, the spreadsheet can often provide quicker, more thorough, and much easier solutions to interpret (e.g., it can graphically display results) than specialized FORTRAN or BASIC computer programs.

The small trigonometric Table 18-1 was constructed on a spreadsheet in about 15 minutes. What is even more impressive is that within another 5 minutes, Figure [18-1] was plotted to graphi-

TABLE 18-1

Angle (deg)	*Sin*	*Cos*	*Sin*Cos*
0	.000	1.000	.000
5	.087	.996	.087
10	.174	.985	.171
15	.259	.966	.250
20	.342	.940	.321
25	.423	.906	.383
30	.500	.866	.433
35	.574	.819	.470
40	.643	.766	.492
45	.707	.707	.500
50	.766	.643	.492
55	.819	.574	.470
60	.866	.500	.433
65	.906	.423	.383
70	.940	.342	.321
75	.966	.259	.250
80	.985	.174	.171
85	.996	.087	.087
90	1.000	.000	.000

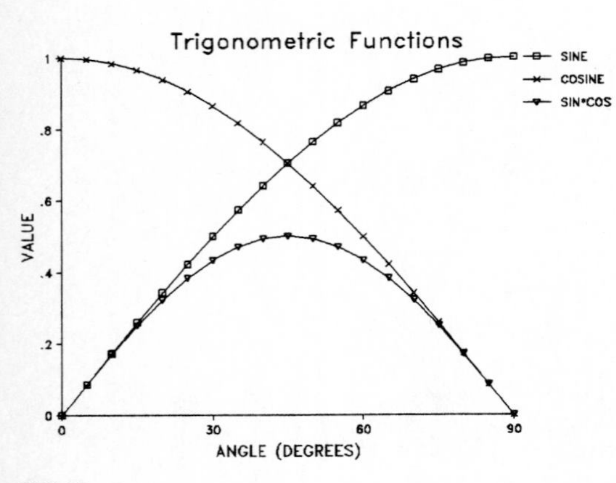

[18-1]

cally show the behavior of the three functions tabulated in Table 18-1 (the functions are the sine, the cosine, and the product of sine and cosine). We include this table and figure not just to show the versatility of the spreadsheet, but to whet your appetite for its use. Indeed, the spreadsheet is much more versatile than can be displayed here. More examples are developed later. But first we need to explain the concept of the spreadsheet and how it functions.

We will give a generic discussion of spreadsheet programming but we will supplement it with details on both *SuperCalc3* (Release 2, or later)[1] and *Lotus 1-2-3*.[2] In addition to the normal capabilities found in almost every spreadsheet, both of these have the ability to accommodate iterative solutions and both include the added enhancement of graphics. Iterative capability is important in many engineering problems of the implicit type—problems that will not yield to direct solution techniques. Also, graphing capability can add a whole new dimension to the understanding of data and results. The special examples, where the details of SuperCalc3 and Lotus 1-2-3 are discussed, will be called *Topics*.

THE CONCEPT OF THE CELL

Spreadsheets are workspaces that consist of a matrix of many rows and columns. They are much like ruled paper accounting forms, although the actual rulings are not visible on the computer screen.

The intersection of each row and each column is the location of a *cell*, the smallest unit that contains information in a spreadsheet. This information can consist of text, numeric data, or formulas. As might be expected, there must be a way to specify the type of information that is to go into a cell and this is covered in detail in the following pages. In most spreadsheets the columns are distinguished by alphabetic letters, starting in the leftmost column with A and progressing with A, . . . ,Z, AA, . . . , AZ, BA, . . . , BZ, and so on. The rows are generally designated by sequential numbers, with row 1 being at the top of the sheet. Cells are then *addressed* by their [column letter] [row number] as in A1 or D73.[3]

A cell address can be thought of as a variable name representing a location containing a value much like a variable name in BASIC or FORTRAN represents an address in memory (or RAM) where the current value of the variable is stored. Whenever a variable name is used in a BASIC or FORTRAN formula or expression, the value of the variable is fetched from the address in memory that has been assigned to that variable name by the compiler or interpreter. So cell addresses, such as A1 or D73, can be used in expressions to represent

SuperCalc®3
RELEASE 2

[1]Trademark of Sorcim/IUS Micro Software.

[2]Trademark of Lotus Development Corporation.

[3]The major exception to this is Multiplan (trademark of Microsoft), which uses numbers to designate both rows and columns. Absolute cell addresses are specified with an R before the row number and a C before the column number, as in R3C7 for the cell at the intersection of the third row and the seventh column. Relative cell addresses are specified with relative locations placed in square brackets, such as in R[-1]C[+2] denoting the cell located one row up (-) and two columns to the right (+) of the cell that contains this relative address.

the value that each cell contains, much like variable names are used in FORTRAN or BASIC! Couple this capability with that of being able to store expressions or formulas, not just values, in cells and we have touched on the real power of the spreadsheet. This should become clearer after we discuss formulas below.

Cell addresses are also required in spreadsheet commands to carry out those commands on a particular cell or group of cells. Spreadsheet commands are actions that let you take control of the spreadsheet; they are discussed in a separate section below.

Range of Cells

There is often a need to describe a number of contiguous cells in a shorthand notation. In practice, the most useful collection of contiguous cells is one in which the cells form a rectangle, the rectangle generally being called a *range*. A range of cells is generally denoted by specifying the address of the upper left-hand corner and the address of the lower right-hand corner of the range, separated by the *range delimiter* or *operator*. For example, the range A1.C5[4] refers to cells A1, A2, A3, A4, A5, B1, B2, B3, B4, B5, C1, C2, C3, C4, C5. Both A1 and A1.A1 refer to the single cell A1.

Example 18-1

State the range of cells that are visible in the worksheet screens (Screens 2) of Topic 18-1.

Solution:

The range is A1.H12. On most screens, newly loaded blank worksheets will actually display about 20 working rows and eight columns. We have shown only 12 rows in Topic 18-1, for brevity.

Size Limitations and the Working Area

Current spreadsheet manufacturers commonly advertise that they can address a rather large number of cells (see Topic 18-1). However, the actual number of cells that can be used is determined by the size of RAM and the complexity of a particular sheet and is usually much smaller than the potential maximum.

Most spreadsheet problems require fewer cells than the maximum permitted by reasonable size RAMs (e.g., 256K). *However, for efficient use of memory it is important when you use spreadsheets that you confine your work to the top left corner of the sheet, working in the smallest rectangle possible*. We will refer to this rectangle as the *working area* of the sheet.

The Screen as a Window

It is not unusual for the working area of a spreadsheet to contain more cells than can be displayed at one time on the limited size

---

[4]Some spreadsheets, including Lotus 1-2-3, use a period,".", as the ***range delimiter*** or ***operator*** and some, including Multiplan by Microsoft, use a colon, ":". Others, including SuperCalc3, can use either. We will use the period as the range delimiter in the range specification.

screen of most CRTs. Most programs use the screen as a *window* through which you can see only part of the worksheet. The screen windows in Topic 18-1 are viewing columns A through H and rows 1 through 12. Figure [18-2] graphically illustrates the potential spreadsheet, the working area, and the screen window.

The screen window can be moved around on the sheet to view different areas. We will describe how this is done in the next section. It is also possible to split the CRT screen in order to view two[5] different parts of a sheet at the same time by opening a second window simultaneously with the first one.

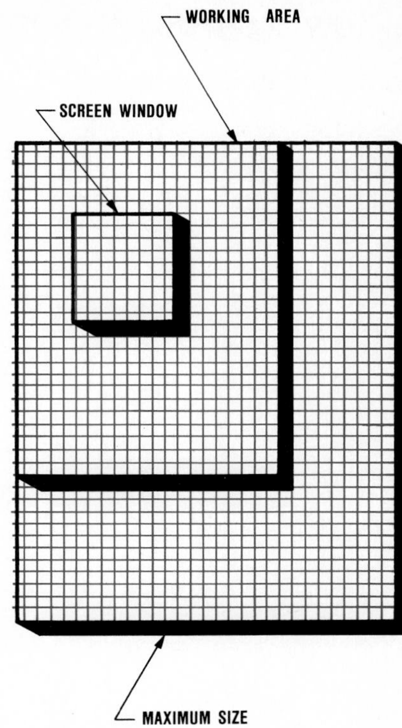

[18-2] Relationship between maximum size sheet, the working area, and the screen window. The working area should be in the upper left corner of sheet. The screen window can be taken anywhere on the sheet to view the cells.

THE CURRENT CELL AND ENTERING DATA

Through the window afforded by the screen, the user can enter any one of the following three types of data into the current cell:

1. An alphanumeric string.
2. A number.
3. A formula.

The current or active cell is that cell in which the cursor is currently located, the cursor being the large highlighted rectangle that always appears in one of the cells located in the screen window. The cursor can be moved to any desired cell in order to make that cell active. This is accomplished by using the cursor control keys (either the keys on the keyboard with the arrows on them, or the S-E-D-X cursor diamond control keys actuated by holding down the Control key and touching S for left, E for up, D for right, or X for down). In Topic 18-1 the cell B2 is the active cell because it contains the cursor (the highlighted box in the active worksheet). That same cell also contains the text or label (an alphanumeric string) "Cell B2".

If you desired to make the cell M20 the active one, you would use the cursor control keys to move the cursor to the right and down on the sheet. When the edge of the window is reached, the sheet will begin to scroll by the window, taking the window where ever you desire. There is another way to reach a cell quickly and that is with the aid of the GoTo command or the GoTo key. It is discussed briefly with the other commands below.

Most spreadsheets reserve several lines in the screen display for aiding users in programming and using the sheet efficiently. We have identified these areas of the screen in Topic 18-1, by surrounding them in dotted rectangles (the dotted rectangles are our addition and do not appear in actual use). The first of these is the *current cell status line,* which always gives the address of the current cell (in Topic 18-1 note that it reads "B2" since the cursor is in cell B2) and its contents (note the "Cell B2" in the status lines of Topic 18-1). It may also contain some information that is peculiar to the particular spreadsheet you are using. We shall discuss the other two special lines on the screen later in their specific context.

[5]Some spreadsheets, such as Multiplan by Microsoft, allow more than two windows to be open.

Topic 18-1 Lotus 1-2-3

Opening Screens and Ultimate Size

Shown below are the opening screen and a sample spreadsheet screen for Lotus 1-2-3. In Lotus 1-2-3, the opening screen represents the Lotus Access System and contains a menu of items that can be selected. The menu items are listed in the horizontal line beginning with 1-2-3 and ending with Exit. When the term 1-2-3 is highlighted in the menu (1-2-3 is the default item), and <Enter> will begin the 1-2-3 spreadsheet program.

1-2-3's spreadsheets can be as large as 256 columns by 8192 rows. Potentially, 1-2-3 can have up to 2,097,152 cells.

Screen 1

```
1-2-3  PrintGraph  Translate  Install  View  Exit
Enter 1-2-3 -- Lotus Worksheet/Graphics/Database program

                    1-2-3 Access System
               Lotus Development Corporation
                      Copyright 1985
                    All Rights Reserved
                         Release 2

The Access System lets you choose 1-2-3, PrintGraph, the Translate utility,
the Install program, and A View of 1-2-3 from the menu at the top of this
screen.  If you're using a diskette system, the Access System may prompt
you to change disks.  Follow the instructions below to start a program.

o  Use [RIGHT] or [LEFT] to move the menu pointer (the highlight bar at
   the top of the screen) to the program you want to use.

o  Press [RETURN] to start the program.

You can also start a program by typing the first letter of the menu
choice.  Press [HELP] for more information.
```

Screen 2

```
B2: ^Cell B2              <-- Current Cell Status Line
                                  <-- Command Status Line
                                  <-- Prompt Line
        A        B        C        D        E        F        G        H
 1
 2           Cell B2
 3
 4
 5
 6
 7
 8
 9
10
11
12
```

Topic 18-1 SuperCalc3

Opening Screens and Ultimate Size

Shown below are the opening screen and a sample spreadsheet screen for SuperCalc3. An <Enter> takes you from the opening screen to the spreadsheet.

SuperCalc3 spreadsheets can occupy up to 127 columns and 9999 rows, providing, potentially, up to (127 × 9,999 = 1,269,873) cells.

Screen 1

```
                              SuperCalc3(tm)
                              Version  2.1
                               I B M   P C
                             (WITH 8087 NDP)
                           S/N-000000, IBM DOS

                             Copyright 1985

                 COMPUTER ASSOCIATES INTERNATIONAL, INC.


Enter "?" for HELP or <RETURN> to start.
F1 = Help; F2 = Cancel; F9 = Plot; F10 = View
```

Screen 2

```
   |  A  ||  B  ||  C  ||  D  ||  E  ||  F  ||  G  ||  H  |
 1
 2          Cell B2
 3
 4
 5
 6
 7
 8
 9
10 This gives the current cell address
11 |      This denotes current cell Text is Centered
12 |      |       This says current cell holds text which is Cell B2
   ↓      ↓        ↓
  B2      TC    Text="Cell B2                        ←Current Cell Status
Width:  9  Memory: 304  Last Col/Row:B2    ? for HELP  ←Prompt Line
  1>                                                   ←Command Status Line
F1 = Help; F2 = Cancel; F9 = Plot; F10 = View
```

```
  A1              Text:"123-45-6789
Width: 11  Memory:220 Last Col/Row:B2
  1>
```

[18-3] Notice the entry on the current cell status line has the " symbol preceeding it. This is the text definition or prefix character in SuperCalc3. This symbol was entered before the SS # for cell A1. It was not included when the same SS # was entered in cell A2, resulting in the SS # being interpreted as a formula and the result being displayed in the cell. If the cursor were located in cell A2, the current cell status line would show 123-45-6789, without any " preceding it.

Entering Text or Labels

In most cases, you will enter text or labels into the current cell simply by typing the desired text and pressing the Enter key.[6] As you are typing, the text is displayed on the current cell status line. You may use the backspace key to correct typing mistakes.

Text entered into the cell in this manner will generally be left justified as a default. That is, the text will begin at the left edge of the active cell and proceed to the right.

If the text overfills the cell (i.e., if the text string has more characters than there is space in the active cell), some spreadsheets will display the overflowing text in the next cell(s) to the right if that (those) cell(s) is (are) empty.[7] An example of this is shown in Figure [18-3], where the entry in cell B1, which is "←SS # Entered as Label," flows over into cells C1 and D1. If the adjacent cells to the right are not blank, only the part of the text that will fit into the active cell will be displayed.

In some spreadsheets, the text or label string will always be truncated to fit the length of the cell. Of course, there is usually more than one way to solve a problem. In a later section we discuss the command that allows you to change the width of individual columns.

Text can also be displayed centered in the active cell or right justified. But the method by which this is accomplished depends on the actual spreadsheet being used. See Topic 18-2 for the details in SuperCalc3 or Lotus 1-2-3.

In cases where the alphanumeric string being entered appears to the program as a formula or a number (the latter is a very specific formula), the string is usually not entered as a text, but as a formula or number. For example, if you desired to enter a social security number 123-45-6789 as a label, but simply typed 123-45-6789

[6] The Enter key will vary depending on the computer you are using. On IBM PCs and compatibles, this is the key marked ↵.

[7] On some spreadsheets, such as SuperCalc3, the adjacent cells to the right must be truly empty (we discuss the blank or erase commands later), which includes not being formatted.

Topic 18-2 — *Lotus 1-2-3*

Current Cell Status Line Editing

In 1-2-3, the cursor control keys behave somewhat like the enter key, entering the text into the active cell, but then moving the cursor on to the next cell.

Beginning the label or text string with a ' will cause the text to be left justified, ^ (shift 6) centers the string, and " right justifies the string. To change the alignment of previously entered text in 1-2-3, you may either retype and reenter the text, or you may edit (we discuss editing the cell contents later) the ', ^, or " that is already there (these symbols are not visible when text is displayed in a cell). To enter strings that otherwise would appear as formulas or numbers, simply precede the string with any of (', ^, ").

⟨Enter⟩, the number -6711 would appear in the active cell for many spreadsheets [18-3]. That is because these programs would believe you were typing in a formula that consisted of 123 minus 45 minus 6789, the result of which is -6711. To enter special cases such as these as text or labels, precede the text with the appropriate text definition or prefix character (see Topic 18-2).

Example 18-2

Using an actual spreadsheet, enter the following labels into various cells in the window of the screen:

January	Energy	J7
122nd Airbourne	COUNT	@modular

Solution:

Take the cursor to the cell where you want the label entered, type in the label, and execute the entry with the <Enter> key. The label should appear in the cursor cell on the sheet. Special problems you may encounter here are:

SuperCalc3: You should encounter no problems here except with J7, which is a valid cell address and makes the program think that a formula is intended. Since cell J7 is probably empty, a 0 (zero) appears in the cell where you entered J7. If you enter "J7, you should be able to enter the label J7. There is a command toggle / G " which when on will make SuperCalc3 think that every entry, unless preceded by a ", is a formula.

Lotus 1-2-3: The labels that cause problems here are J7, 122nd Airbourne and @modular. Lotus 1-2-3 thinks all of these are formulas, because they start with a cell address, a number, and a function symbol (the @; see Topic 18-3), respectively. Preceding each of these three with one of the text prefix characters will cure the problem (see Topic 18-2).

Entering Numbers

On nearly all spreadsheets, numbers are entered simply by typing the number into the current cell status line and entering. Usually, no

Topic 18-2 — SuperCalc3

Current Cell Status Line Editing

In SuperCalc3, you may use the cursor control keys to pass back over typed text in order to reach an earlier mistake, and the Del and Ins keys, as well as the backspace key, for editing.

Text formatting is done with the /F command (discussed later). To enter strings that otherwise would appear as formulas or numbers, just precede the string with a double quote (") for SuperCalc3.

If errors are made in formula entry and the formula gets entered as a label or text, you must either repeat the entry or edit the cell entry to both correct the mistake and to remove the " that precedes the entry. Edit is discussed under Command Mode.

number definition character need precede the number itself. Anytime the first character typed on the current cell status line is a number (0 through 9), most programs assume that what will follow is either more numbers or a formula. The numbers can be integers with no decimal point or decimal part, fixed point (i.e., have a decimal point and a decimal part), or be entered in scientific notation using an E to denote the start of an exponent of 10 (e.g., 7.5E13 for 7.5×10^{13}). In sophisticated spreadsheets, these numbers can have as many as 15 significant figures and be as large as 10^{63} or as small as 10^{-64}, although this will vary from spreadsheet to spreadsheet.

Numbers are generally right justified in the current cell, although some programs allow you to reformat to left justified. The display format of a number [i.e., whether it is preceded by a dollar sign ($), is displayed in exponential notation, or has a fixed number

Topic 18-3

Intrinsic Functions

	SuperCalc3[a]	***Lotus 1-2-3***[a]
Trigonometric functions		
cosine (*c/v* in radians)	COS*(c/v)*	@COS*(c/v)*
sin (*c/v* in radians)	SIN*(c/v)*	@SIN*(c/v)*
tangent (*c/v* in radians)	TAN*(c/v)*	@TAN*(c/v)*
Inverse Trigonometric functions		
arc cosine (returns radians)	ACOS*(c/v)*	@ACOS*(c/v)*
arc sine (returns radians)	ASIN*(c/v)*	@ASIN*(c/v)*
arc tangent (returns radians)	ATAN*(c/v)*	@ATAN*(c/v)*
4-quadrant arc tangent (returns radians)		@ATAN2*(c/v)*
Exponentiation and Logarithms		
exponentiation, $e^{(c/v)}$	EXP*(c/v)*	@EXP*(c/v)*
$\log_e$	LN*(c/v)*	@LN*(c/v)*
$\log_{10}$	LOG*(c/v)*	@LOG*(c/v)*
Constants		
pi (3.14159+)	PI	@PI
Arithmetic functions		
absolute value	ABS*(c/v)*	@ABS*(c/v)*
count non-blank cells	COUNT*(range)*	@COUNT*(range)*
maximum value in a range	MAX*(range)*	@MAX*(range)*
minimum value in a range	MIN*(range)*	@MIN*(range)*
division remainder	MOD($c/v_1, c/v_2$)	@MOD($c/v_1, c/v_2$)
square root	SQRT*(c/v)*	@SQRT*(c/v)*
sum of values in range	SUM*(range)*	@SUM*(range)*
Statistical functions		
average of values in range	AV*(range)*	@AVG*(range)*
standard deviation of values in range	-	@STD*(range)*
variance of values in range	-	@VAR*(range)*
Financial functions		
internal rate of return	IRR*([guess,]range)*	@IRR*(guess,range)*
future value	FV*(pmt,int,per)*	@FV*(pmt,int,per)*
net present value	NPV*(disc,range)*	@NPV*(c/v,range)*
payment	PMT*(prin,int,per)*	@PMT*(prin,int,per)*
present value	PV*(pmt,int,per)*	@PV*(pmt,int,per)*

of digits after the decimal point] is controlled by the cell format command (discussed further below).

There is a distinct difference between a series of digits entered as a number and a series of digits entered as text (i.e., preceded by the appropriate text definition or prefix character). Numbers entered into a cell can be used by formulas that are entered into any other cell and their true value will be used. That is, if a formula in some cell takes a 2 that has been entered in another cell as a number 2, and adds 1 to it, the result will be 3. On the other hand, all text or label strings, including strings of digits entered as text, default to zero when inadvertently used by formulas that reside in other cells. Thus, if the formula mentioned above takes a 2 that has been entered in another cell as text or a label 2, and adds 1 to it, the result will be 1 [18-4].

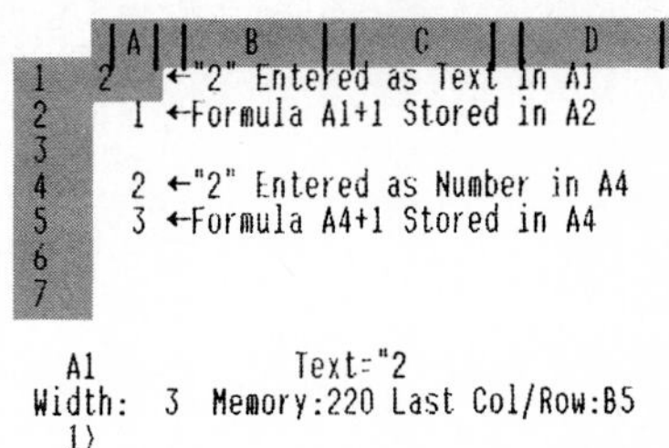

[18-4] Notice that on the current cell status line, the 2 in cell A1 was entered as text or label. (How can you tell?) Notice that it does not work in the formula of cell A2. Textual entries default to 0 when used in formulas on most spreadsheets (SuperCalc3).

SuperCalc3 and Lotus 1-2-3

	SuperCalc3[a]	***Lotus 1-2-3***[a]
Logical functions		
if-then-else	IF($cond,c/v_1,c/v_2$)[b]	@IF($cond,c/v_1,c/v_2$)[c]
and	AND($cond_a,cond_b$)[d]	#AND#
or	OR($cond_a,cond_b$)[e]	#OR#
not	NOT(*cond*)[f]	#NOT#
true (returns 1)	TRUE	@TRUE
false (returns 0)	FALSE	@FALSE
not available	NA	@NA
error	ERROR	@ERR
Miscellaneous functions		
integer part (truncates)	INT(*c/v*)	@INT(*c/v*)
random number generator	RAN	@RAND
rounded value	ROUND(*c/v,places*)	@ROUND(*c/v,places*)
table lookup	LOOKUP(*c/v,range*)	@HLOOKUP(*c/v,range,row#*)
		@VLOOKUP(*c/v,range,column#*)
Arithmetic and relational operators		
add	+	+
subtract	−	−
multiply	*	*
divide	/	/
raise to power	^ (or) **	^
percent	%	
equal to	=	=
not equal to	<>	<>
less than	<	<
less than or equal to	<=	<=
greater than	>	>
greater than or equal to	>=	>=

[a]*c* - reference to a single Cell; *c/v* - a Cell reference or a Value; *cond* - some algebraic Condition (e.g., A5 > 5.); *disc* - Discount rate; *guess* - a Guess; *int* - Interest; *per* - Period or term; *places* - number of Places after the decimal; *pmt* - Payment; *prin* - Principal; *range* - a Range of cells (e.g., B3.C6); *v* - a Value entered here.
[b]The function IF takes the value of c/v_1 if *cond* is satisfied or the value of c/v_2 if *cond* is not satisfied.
[c]The function @IF takes the value of c/v_1 if *cond* is satisfied or the value of c/v_2 if *cond* is not satisfied.
[d]The function OR takes the value of 1 if *either* $cond_a$ or $cond_b$ are satisfied. Otherwise, the value of OR is 0 (zero).
[e]The function AND takes the value of 1 if *both* $cond^a$ and $cond^b$ are satisfied. Otherwise, the value of AND is 0 (zero).
[f]The function NOT takes the value of 1 if *cond* is not satisfied. Otherwise, the value is 0 (zero).

Example 18-3

Enter the following numbers into cells A1, B2, C3, respectively, of a spreadsheet:

2	-4.7	9.666666E17

Solution:

Take the cursor to cell A1, type 2 into the current cell status line, and push the <Enter> key.

Take the cursor to cell B2, type -4.7 into the current cell status line, and push the <Enter> Depending on the formatting of the cell and your equipment, this negative number may appear in (), or it may display in a different color than positive numbers.

Take the cursor to cell C3, type 9.666666E17 into the current cell status line, and push the <Enter> key. Since the default cell width is typically nine characters on most spreadsheets, this number probably got shortened in the display (we cover the column-width command later, which will allow you to change the width of a column). Was it rounded or truncated? Although the number may have been shortened in the display, the full number is retained in memory and used in any calculations.

Entering Formulas

The ability to enter formulas into the cells of a spreadsheet is probably the most powerful attribute of spreadsheet programming. A formula is typed into the current cell status line and then entered into the current cell by actuating the <Enter> key. Once a formula is entered in a cell, only the number that results from the evaluation of the formula is displayed on the sheet. In addition, when that cell is referenced in other formulas (by using the cell address, e.g., A1 of D78), the result of the formula is used rather than the formula itself.

Unfortunately, this trait of displaying only the results means that by looking at the sheet, you cannot tell whether the numbers displayed are results that have been calculated from formulas or are simply numbers that have been entered.

The current cell status line plays an important role here, for if a formula has been entered in the current cell, the formula will be displayed on the status line. It is often helpful to remember that the current cell status line always shows what is actually stored in the current cell, not just what is displayed in that cell on the sheet.

Operations Possible. Nearly all common spreadsheets allow for the following arithmetic operations to be performed within any formula:

- Exponentiation (signified by either ^ or **).
- Multiplication (* and division (/).
- Addition (+) and subtraction (−).

Most spreadsheets (including both SuperCalc3 and Lotus 1-2-3) follow the order preference rules adopted by most programming languages, which specify:

- Exponentiation is done first.
- Multiplication and division are next.
- Addition and subtraction are last.
- Within any of the groups above, apparent conflicts are resolved in left-to-right order (e.g., c/d*e multiplies the quotient of c/d by e).

However, there are a few spreadsheets that use as a preference rule that *all expressions* are evaluated from left to right, so be careful. In any case, parentheses can be used to override the rules above, because expressions within parentheses are always evaluated first. Therefore, the liberal use of parentheses is always wise.

Constants and Variables. Both constants and variables can be entered in formulas. Similar to BASIC and FORTRAN expressions, constants are values that do not change from one evaluation of the expression to the next. In Equation (18-1), the constant values are 3 and 4.6; these values will not change with each evaluation of the expression. The size and number of significant figures permitted are the same as discussed in the preceding section on numbers.

Variables, on the other hand, are quantities that you expect to change during the use of the spreadsheet. They are represented by a cell address, as described earlier. Some spreadsheets also enable the user to give a specific name to a cell and these user-defined names can also be used in formulas in lieu of the actual cell address. For example, if you wanted to perform calculations with the formula

$$y = 3x^2 + 4.6 \qquad (18\text{-}1)$$

you might enter this formula into cell A2. But before doing this, you must decide where the formula will find the value of x to use in the calculation. You cannot enter the value of x into cell A2 without destroying the formula you wish to put there. Assuming that we decide to use 5 as the desired value of x, we might choose to enter this number into cell A1 on the sheet. The formula we enter in cell A2 is then

3*A1^2 + 4.6 (18-2)

If we have not made an error in entry and if we have already entered the 5 in cell A1, the result of 79.6 will appear in cell A2 almost immediately [18-5].

Once the formula was entered in cell A2, the spreadsheet "recalculated" the current sheet. In this case, the sheet consisted of the cell A1, which contained the number 5, and the cell A2, which contained the formula given in Equation (9-1). By specifying the cell address A1 in the formula contained in A2, we have effectively instructed the spreadsheet to "fetch" the contents of A1 each time the spreadsheet is recalculated and Equation (9-1) is evaluated. The actual format of the numbers (i.e., how many decimal places, etc.) will depend on how the cells are formatted (formatting is discussed later).

On the sheet, only two numbers will be showing, a 5 in cell A1 and 79.6 in cell A2. With the cursor in cell A1, the entry on the current cell status line will read 5. With the cursor in cell A2, the

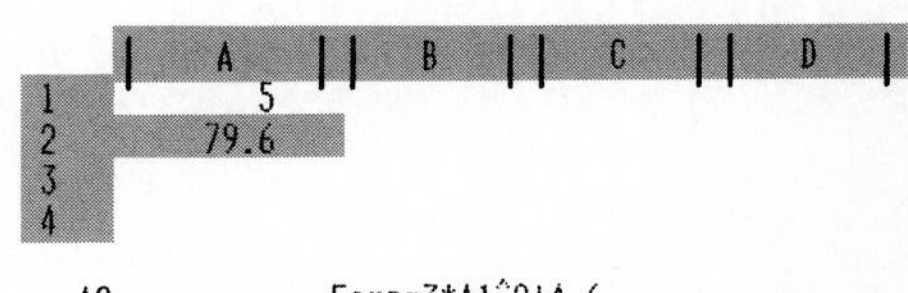

[18-5] Notice on the current cell status line that a formula has been entered in cell A2, but the result is displayed on the spreadsheet (SuperCalc3).

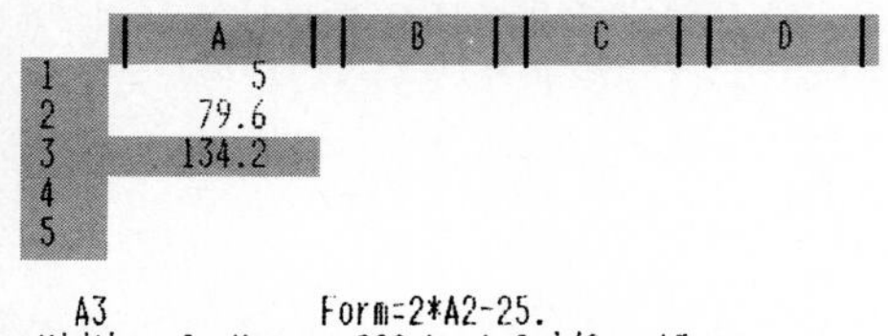

[18-6] The formula in cell A3 references the result of the formula that resides in cell A2. See Figure [18-5] to verify that a formula which references the value in cell A1 is stored in cell A2 (SuperCalc3).

entry on the current cell status line will be the formula, 3*A1^2 + 4.6. Thus, by looking at the current cell status line, you can distinguish between what is actually stored in each cell and what is displayed in the cell location on the sheet. *Remember, a formula can be stored but only the result is displayed on the sheet.*

Using Results from Other Formulas. Formulas, of course, can depend on the results of other formulas. Spreadsheets accommodate this simply by allowing the results of other formulas to be used by referencing the cell address of the results. For example, assume that we were interested in calculating

$$\alpha = 2y - 25 \tag{18-3}$$

where y is the result of Equation (18-1) for which we have already prepared a spreadsheet method of solution [cell A2 contains y, and A1 contains x, the variable that appears in Equation (18-1)]. In cell A3 of our existing sheet, we would enter the formula

$$2{*}A2 - 25 \tag{18-4}$$

As soon as it is entered, we should see 134.2 appear in cell A3 on the sheet (see Figure [18-6]). Again, the formula shows on the current cell status line, as long as the cursor is located in cell A3.

Example 18-4

Express the equation

$$\left(1 + \frac{3 + z^2}{y^5}\right) \frac{z}{x}$$

in spreadsheet formula form, where x will be placed in cell A1, y in B1, z in C1. Enter the formula into D1.

Solution:

Move the cursor to cell D1 and type into the current cell status line the formula

(1 + (3 + C1^2)/B1^5)*C1/A1

followed by an <Enter>. If you have not made an entry mistake and if nonzero values have been placed in cells A1 and B1, you should see the results in cell D1. If A1 and/or B1 are empty or 0 (zero), you will get an error appearing on the sheet in cell D1, telling you that something is wrong. (You are dividing by zero.) Just type in some nonzero values for A1 and B1, and the error message should go away. If you made an entry error, you will have to retype the equation without error or wait until we discuss the edit command to fix it.

Relative versus Absolute Cell Addresses. Depending on the spreadsheet you are using, cell addresses embedded in formulas [e.g., in Equation (18-4)] may be "relative" addresses or they may be "absolute" addresses. Relative addresses used in formulas are interpreted strictly in terms of the relative location of the cited address with respect to the cell holding the citation.

In the example of the preceding section, cell A3 cited cell A2, which is straightforward. Should either SuperCalc3 or Lotus 1-2-3 be the spreadsheet in use, the citation in A3 is really interpreted as meaning the following.

> Bring to cell A3 the result (not the formula, but the result) found in the cell located zero columns over and one row up (the relative distance between A3 and A2), multiply it by 2, subtract the number 25 from that product, and place the result in cell A3.

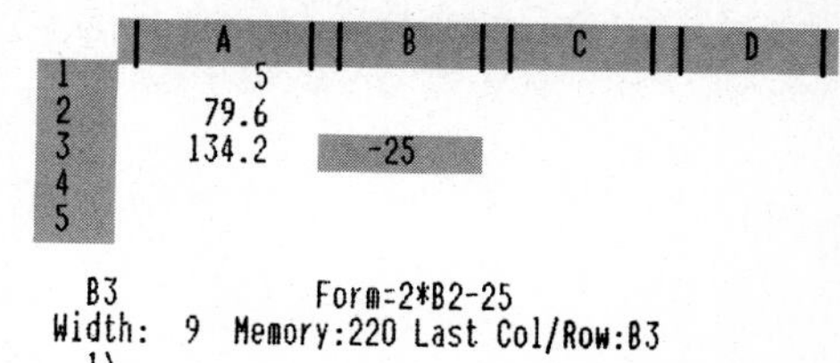

[18-7] The formula in cell B3 was replicated or copied from cell A3 using relative cell addresses. Cell A3 has the same formula as it did in Figure [18-5] (SuperCalc3).

The tricky part comes if we want to replicate or copy the formula into other cells on the sheet. Once you begin to understand spreadsheets, you will probably find replicating or copying quite useful, but it has to be done with some awareness if the newly copied formula are to give correct results.

If we were to replicate or copy[8] the contents of cell A3 to cell B3, we might be surprised to see that the number 134.20 did not appear in B3, but that (−25.) did appear (assuming that cell B2 is blank). If we check what is actually stored in cell B3 by taking the cursor to that cell and observing what appears on the current cell status line, we find that the formula has changed slightly from what it was in cell A1 before replicating or copying [18-7]. It now reads

$$2{*}B2 - 25. \qquad (18\text{-}5)$$

not

$$2{*}A2 - 25. \qquad (18\text{-}6)$$

as we might have expected. What has happened is that the *relative* address to cell A2 stored in A3 was copied as a *relative* address to cell B2 (the cell located zero columns over and one row up) when placed in cell B3.

In practice, it is relative addresses that are most useful due to the way replicating or copying are often used in setting up a worksheet. Relative addresses can cause problems at times, especially when you want a number of copied formulas to all reference a single cell for some common data input. For this you need to use an absolute cell reference—or you need to be very careful.

Most spreadsheets provide a special notation for entering absolute cell addresses into formulas. For example, Lotus 1-2-3 allows you to embed a "$" sign ahead of the column letter and/or the row number in an address to signify that the column and/or row reference is absolute and should always be treated as such when copying the reference to another cell. A2 is an absolute address to cell A2; $A2 is an absolute reference to column A and a relative address to row 2 (the latter depends on the relative row displacement from the citing cell). A$2 is a relative reference to column A and an absolute reference to row 2.

SuperCalc3 provides for absolute cell addressing in a somewhat indirect but quite adequate manner. Cell references in any formula are entered using the typical (column, row) specification. Then if you want to copy a formula to any other cell, there is an option in the replicate command that allows you to specify whether the references in the formula being replicated are to be treated as relative or absolute.

When replicating or copying formula on any spreadsheet, you should always ask yourself: *"Are the relative cell addresses implicit in*

[8]Replication or copying are spreadsheet commands and are discussed in more detail later.

the formula being copied appropriate to the new location to which it is being copied, or are absolute cell references required?"

Relative versus absolute cell addressing is a difficult concept to understand. As a result, it is one of the concepts that most often leads to spreadsheet errors. To master the concept, you should practice with some simple problems.

Calculation and Recalculation of Formulas. It is typical for most spreadsheets to *calculate* or *recalculate* all formulas in all cells every time the entry in any one cell is changed, regardless of whether or not all the formulas depend on the changed cell. If there are a large number of formulas on a spreadsheet, the recalculations could take an annoying amount of time and slow the data entry. However, this *automatic recalculation* mode usually can be overridden in order to improve data-entry speed. That is, a *manual recalculation* mode can be invoked, whereby numbers can be entered without recalculating new values for the formulas. However, the user must remember to force a recalculation of the sheet at the completion of the data entry task; otherwise, the results of any formula will not reflect the new data.

In the example displayed in Figure [18-5], assuming that the 5 for x is already entered in A1, the result of the formula for y was calculated as soon as it was entered. That is why the value of 79.6 appeared in cell A2 so quickly. If you were now to take the cursor back to cell A1 and enter a 4 in place of the 5 (just enter a 4 on the current cell status line and push the <Enter> key—the 5 will be overwritten), a new value of 52.6 appears in cell A2 almost immediately if the automatic recalculation mode is in effect. What a delight!

When a sheet is *calculated* or *recalculated,* the calculations generally start from the upper left corner (cell A1) and proceed downward and to the right. That is, the formula in cell A1 will generally be calculated before the formula in cell Z50. The more advanced spreadsheets allow several recalculation modes to be instituted. Most allow at least a choice of either recalculation by rows or recalculation by columns.

Of course, the calculation process, when initiated, may encounter a cell formula that references values from cells that are below or to the right of the current cell. Row-wise or column-wise recalculation procedures may not be able to handle these situations.

Some spreadsheets[9] provide a *natural* recalculation pattern which proceeds through the computations keeping all the dependent cells uncalculated until the cells on which they depend have been calculated. This allows users to be more flexible in their placement of formulas throughout the sheet. It particularly gives the technical user more freedom.

Circular References. With the natural mode of recalculation, the capability for any cell to reference any other cell, regardless of its position on the sheet, is assured. However, the special case of two cells referencing each other requires special attention. For example, if

[9]This includes Lotus 1-2-3, SuperCalc3, and Multiplan.

cell A1 contains a formula that references cell B1, and cell B1 contains a formula that references cell A1, then it would appear that the calculations may be "going in circles."

Such situations are referred to as *circular references* and may be intentional or accidental. Most sheets give a warning to the user when a circular reference exists so that if the circular reference is unintentional, it can be corrected. If unintentional circular references exist, they usually lead to significant errors if uncorrected.

Problems of an *iterative* nature make use of circular references, as will be demonstrated in the sample problems at the end of this chapter. Therefore, circular references can be intentional. In iterative problems, equations or formulas must be solved over and over again, with the solution being updated during each iteration until the final result *converges* or stops changing.

Intentional circular references are resolved by forcing the sheet to be recalculated a sufficient number of times so that the formula results in the circular cells converge or stabilize. Some spreadsheets allow the user either to set the number of sheet recalculations to execute in an attempt to establish *convergence,* or to set a *convergence tolerance* to strive for in resolving the circular references. This should become clearer in Example 18-9.

The Functions Available. One of the features that make formula construction so convenient is the availability of a large number of established *functions* that can be used. For example, suppose you want to enter into cell A3 a formula that represents the equation

$$z = \sin\theta + \exp x \tag{18-7}$$

with ($\theta = \pi/4 =$) 3.14159/4 radians being located in cell A1[10] and ($x =$) 0.1 located in cell A2. With the cursor in cell A3, you would type into the current cell status line the formula[11] (see [18-8])

$$\text{SIN(A1)} + \text{EXP(A2)} \tag{18-8}$$

and follow it with an ⟨ Enter ⟩ . The result of the calculation, 1.812277, will be displayed almost immediately in cell A3 of the sheet, but again, the format of the answer will depend on how the cell has been formatted.

Thus the mathematical formula for the trigonometric sine function and the exponential function do not have to be entered into cells that require their computation. These mathematical formulas are available, under specific[12] and reserved names, as a part of the

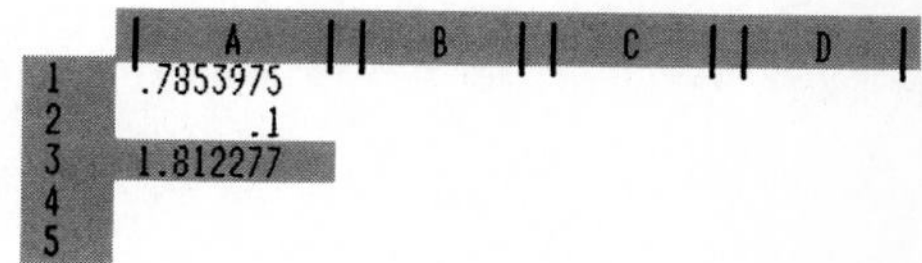

[18-8] From the current cell status line you can see that we have entered into cell A3 a formula that requires sines and exponentials. The result magically appears in the cell on the spreadsheet. The spreadsheet recognizes the names SIN and EXP, among others, and supplies the means to calculate them (SuperCalc3).

[10]It is important to remember that in almost all computer work, the trigonometric functions require their arguments to be in *radians,* not degrees. In Equation (18-8) the argument of the sine could have been (A1∗3.14159/180.) if the θ had been entered in cell A1 in degrees.

[11]Some spreadsheets require the "@" sign to be the first character in any function name.

[12]The specific names used by most spreadsheets are very close to those used by most programming languages, such as FORTRAN and BASIC (described in Chapter 17). Some spreadsheets do require that the symbol "@" precede the more common function name.

spreadsheet package. Common functions on most spreadsheets include

Trigonometric functions:
- Cosine of a number.
- Sine of a number.
- Tangent of a number.

Inverse trigonometric functions:
- Arccosine of a number.
- Arcsine of a number.
- Two-quadrant arctangent.
- Four-quadrant arctangent of a number.

Exponential and logarithms:
- Exponential of a number.
- Log_e of a number.
- Log_{10} of a number.

Constants: Pi.

Arithmetic functions:
- Absolute value of a number.
- Division remainder.
- Maximum of values in a range.
- Minimum of values in a range.
- Number of nonblank values in a range.
- Square root of a number.
- Sum of values in a range.

Logical functions:
- If-then-else with operators $=, \leq, \geq, <, >, \neq$, NOT, AND, OR.

Statistical functions:
- Average of nonblank values in a range.
- Standard deviation of nonblank values in a range.
- Variance of nonblank values in a range.

Financial functions:
- Net present value.
- Internal rate of return.
- Present worth.

Miscellaneous functions:
- Integer part of a decimal number.
- Rounded-value part of a decimal number.
- Random number generator.
- Table lookup.

Specific but limited examples for both SuperCalc3 and Lotus 1-2-3 are given in Topic 18-3. Help screens in both spreadsheets enumerate all of the available functions.

Example 18-5

Investigate the ability of your spreadsheet to do natural-order recalculation by placing the sine of x in cell A1 and the value of x in cell B2. x is to be entered in degrees.

Solution:

Move the cursor to cell B2 and enter, for example, 45. Then move the cursor to cell A1 and enter SIN(B2*PI/180), remembering that you need the argument in radians, not degrees (in some spreadsheets, you will need to use @SIN instead of SIN; see Topic 18-3). Did an answer of 0.707 appear in cell A1? The recalculation command that allows you to choose the recalculation mode is discussed in the next section.

TAKING CHARGE OF THE SHEET COMMAND MODE OPERATION

So far we have discussed how to move around on spreadsheets, how to enter labels and data, and how to enter equations or formulas. These constitute what may be called the *data entry* mode of operation. However, nearly all spreadsheets have a *command mode* of operation which is used to make alterations to the sheet itself or to the entries on the sheet. The command mode puts you in charge of the sheet.

The most common way of entering the command mode is to actuate the "/" or slash key.[13] This key activates a command mode status line which generally shows a menu of items that can be carried out (see Topic 18-1 for the location of the command status line). In some spreadsheets, this menu is inhospitable to new users in that it consists of only single-letter representations of the commands that can be chosen. Figure [18-9] lists the initial command menu from SuperCalc3. You select the command you want by typing the single-letter representation. To ease this unfriendly environment of SuperCalc3, the SuperCalc3 authors provide easy access to a help screen (help screens are discussed later), which gives a more user-friendly description of the commands available. That screen is given in Figure [18-10].

In some spreadsheets, the menus consist of whole-word representations of the commands available. Lotus 1-2-3, for example, gives a command status line or menu line and a prompt line. It appears in Figure [18-11]. The first line shows the menu items that are immediately accessible. The second line displays the submenu items that can be chosen if the highlighted item in the first line menu were to be chosen. Different items in the first menu can be chosen by using the Cursor Control keys to space over to the desired item and then actuating the ⟨ Enter ⟩ key (the cursor has been moved to the item Copy in Figure [18-11]). Once you become familiar with the menus and submenus, all you have to do is actuate the key corresponding to the first letter in the menu item you wish to select and the program responds. This single-keystroke technique has become a commonly used one.

[13]The major exception to this is Multiplan by Microsoft, which uses the escape key (generally labeled *Esc*) to access the command mode menu. In Multiplan this menu is always visible at the bottom of the screen, on what is the equivalent of a command mode status line.

```
Enter A,B,C,D,E,F,G,I,L,M,O,P,Q,R,S,T,U,V,W,X,Z,/,?
```

[18-9] The initial command menu for SuperCalc3.

[18-10] The help screen for the main command menu of SuperCalc3 (see Figure [18-9] for the main menu).

```
   Slash Commands                              SuperCalc3 AnswerScreen
/Arrange       Sorts cells in ascending or descending order.
/Blank         Removes or empties contents of cells or graphs.
/Copy          Duplicates graphs or contents and display format of cells.
/Delete        Erases entire rows or columns.
/Edit          Allows editing of cell contents.
/Format        Sets display format at Entry, Row, Column, or Global levels.
/Global        Changes global display or calculation options.
/Insert        Adds empty rows or columns.
/Load          Reads spreadsheet or portion from disk into the workspace.
/Move          Inserts existing rows or columns at new positions.
/Output        Sends display or cell contents to printer, screen or disk.
/Protect       Prevents future alteration of cells.
/Quit          Ends the SuperCalc3 program.
/Replicate     Reproduces contents of partial rows or columns.
/Save          Stores the current spreadsheet on disk.
/Title         Locks upper rows or left-hand columns from scrolling.
/Unprotect     Allows alteration of protected cells.
/View          Displays data as Pie, Bar, Line, Area, X-Y or Hi-Lo graph.
/Window        Splits the screen display.
/XeXecute      Accepts commands and data from an .XQT file.
/Zap           Erases spreadsheet and format settings from workspace.
//             Additional commands  //D accesses Data Management options.
 Press any key to continue
```

In keystroke sequences that end with a user input (such as a cell reference or range reference), an ⟨Enter⟩ is usually required to terminate the entry and execute the command.

We will demonstrate the use of many of the commands in the example problems later in this section.

As important as the command keystrokes is the *clear* or *cancel key*. This key either cancels the entire command sequence that you have started (e.g., SuperCalc3) or backs you up one menu with each press (as in Lotus 1-2-3 and SuperCalc3).

> In SuperCalc3, there two cancel keys. The F2 function key or the Ctrl Z (hold down the key labeled Ctrl, push the Z, and then release both keys) cancels the menu and all the submenus currently accessed. The Backspace key backs you up either one item or one keystroke in the current menu sequence.
>
> In Lotus 1-2-3, the Escape key (labeled Esc) is the cancel key on IBM PC and compatibles. It cancels the current menu and returns you to the previous menu or submenu that you were in.

In the following sections we give you a brief introduction to commands found in most spreadsheet programs. The list is far from exhaustive; we cover only those that you will need to get started. Others you will have to pick up from built-in help files (discussed later) or from more extensive sources such as the program manuals.

Note: Most of the commands discussed below carry out actions in ranges or groups of cells. To make your job easier, many spreadsheets are written to anticipate where you want the action initiated. You can help make this noble feature work for you by taking the

[18-11] Initial command menu for Lotus 1-2-3.

```
Worksheet  Range  Copy  Move  File  Print  Graph  Data  Quit
Copy a cell or range of cells
```

cursor to the cell that will be first involved in the action, prior to invoking the desired command. We reiterate this in a few of the command descriptions below.

Global Settings (Topic 18-4)

On most spreadsheets there is a series of commands that allow you to set features (formats, column widths, protection, recalculation mode, etc.) for the entire sheet. There is always a default set of features active when you first load a new, blank sheet, but the global commands allow you to override these defaults. The number and type of features that can be set in this manner varies widely, however.

We have given some of the global settings special attention below. These include erasing the entire sheet, setting all column widths, formatting all cells, and setting the recalculation mode.

With the global settings as defaults, features for individual cells, columns, rows, or ranges can be changed with the commands that follow.

Arranging or Sorting the Entries (Topic 18-5)

There is often a need to sort the rows of a spreadsheet based on the entries in a certain column, or to sort the columns based on the entries in a certain row. This is typically done with the *arrange* or *sort commands*. Sorting can usually be done in ascending or descending order, depending on your preference. Alphabetical ordering of textual entries is usually truly alphabetical, but on some spreadsheets it is actually done according to the ASCII codes for the characters. In the latter case, all the uppercase letters are placed ahead of lowercase letters, permitting, for example, "Zebra" to be listed ahead of "deAnza."

It is always good practice to save or file a copy of your worksheet (see "Saving and Reloading Worksheets" below) before arranging or sorting, because it may be difficult to get back to where you started once you have invoked the arrange or sort command. Also, formulas that rely in a complicated way on other cells for inputs may give strange results after arranging or sorting.

Blanking or Erasing Cell Entries (Topic 18-6)

When a spreadsheet is first loaded into a computer, none of the cell locations have anything stored in them. Thus, there is no memory (RAM) required in order to store the contents of the empty cells.

As soon as something (e.g., text, numbers, or formulas) is placed in any of the cells, memory (RAM) is then required to retain it. If it is later desired to remove the contents of the cell, the cell can be returned to its original truly empty state with the *blank* or *erase command*. Once blanked or erased, the cell will no longer require memory storage locations for its contents.

You may think that overwriting the contents of a cell with spaces would return the cell to its truly empty state. However, spaces are one-byte binary characters (ASCII 32_{10}; see Table 15-1) which require memory just like any other character. Therefore, you should

Topic 18-4 — Lotus 1-2-3

Global Settings Commands

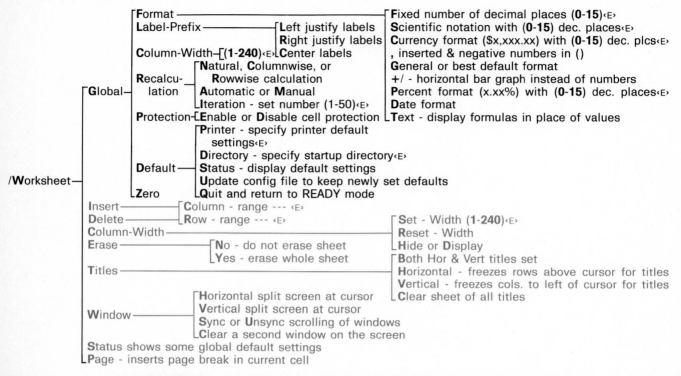

- /**W**orksheet
  - **G**lobal
    - **F**ormat
      - **F**ixed number of decimal places (**0-15**)‹E›
      - **S**cientific notation with (**0-15**) dec. places‹E›
      - **C**urrency format ($x,xxx.xx) with (**0-15**) dec. plcs‹E›
      - **,** inserted & negative numbers in ()
      - **G**eneral or best default format
      - **+**/ - horizontal bar graph instead of numbers
      - **P**ercent format (x.xx%) with (**0-15**) dec. places‹E›
      - **D**ate format
      - **T**ext - display formulas in place of values
    - **L**abel-Prefix
      - **L**eft justify labels
      - **R**ight justify labels
      - **C**enter labels
    - **C**olumn-Width
      - (**1-240**)‹E›
    - **R**ecalcu-lation
      - **N**atural, **C**olumnwise, or **R**owwise calculation
      - **A**utomatic or **M**anual
      - **I**teration - set number (1-50)‹E›
    - **P**rotection
      - **E**nable or **D**isable cell protection
    - **D**efault
      - **P**rinter - specify printer default settings‹E›
      - **D**irectory - specify startup directory‹E›
      - **S**tatus - display default settings
      - **U**pdate config file to keep newly set defaults
      - **Q**uit and return to READY mode
    - **Z**ero
  - **I**nsert
    - **C**olumn - range --- ‹E›
    - **R**ow - range --- ‹E›
  - **D**elete
    - **C**olumn - range --- ‹E›
    - **R**ow - range --- ‹E›
  - **C**olumn-Width
    - **S**et - Width (**1-240**)‹E›
    - **R**eset - Width
    - **H**ide or **D**isplay
  - **E**rase
    - **N**o - do not erase sheet
    - **Y**es - erase whole sheet
  - **T**itles
    - **B**oth Hor & Vert titles set
    - **H**orizontal - freezes rows above cursor for titles
    - **V**ertical - freezes cols. to left of cursor for titles
    - **C**lear sheet of all titles
  - **W**indow
    - **H**orizontal split screen at cursor
    - **V**ertical split screen at cursor
    - **S**ync or **U**nsync scrolling of windows
    - **C**lear a second window on the screen
  - **S**tatus shows some global default settings
  - **P**age - inserts page break in current cell

Topic 18-5 — Lotus 1-2-3

Arrange or Sort Data Command Keystrokes

- /**D**ata
  - **F**ill, with sequential nos., a range ---‹E›
    - Start #, Step #, Stop # ‹E›
  - **T**able generator - construct a
    - **1** parameter table - range --- ‹E› Input cell range --- ‹E›
    - **2** parameter table - range --- ‹E› Input cell 1 range --- ‹E› Input cell 2 range --- ‹E›
    - **R**eset table ranges and disable Table (F8) key
  - **S**ort
    - **D**ata-Range --- ‹E›
    - **P**rimary key - Col. in Data-Range to sort on ---
    - **S**econdary key - Col. in Data-Range to break ties in primary sort --- ‹E›
    - **R**eset to cancel sort range and keys
    - **G**o ahead and sort
    - **Q**uit and return to READY mode
  - **Q**uery
    - **I**nput the Data Record Range --- ‹E›
    - **C**riterion range --- ‹E›
    - **O**utput range for extracted records --- ‹E›
    - **F**ind and highlight each record matching Criteria
    - **E**xtract all matches to Output range
    - **U**nique same as Extract except no duplicates
    - **D**elete records that match criteria
      - **C**ancel
      - **D**elete
    - **R**eset/cancel input, Criterion, Output ranges
    - **Q**uit and return to READY mode
  - **D**istribution (do distributions of data) — where data values range --- ‹E› — where bins are located range --- ‹E›
  - **M**atrix **I**nvert, **M**ultiply
  - **R**egression - **X**-Range, **V**-Range, **O**utput Range, **I**ntercept **R**eset, **G**o, **Q**uit
  - **P**ause - Split data from ASCII file into separate columns of labels and numbers

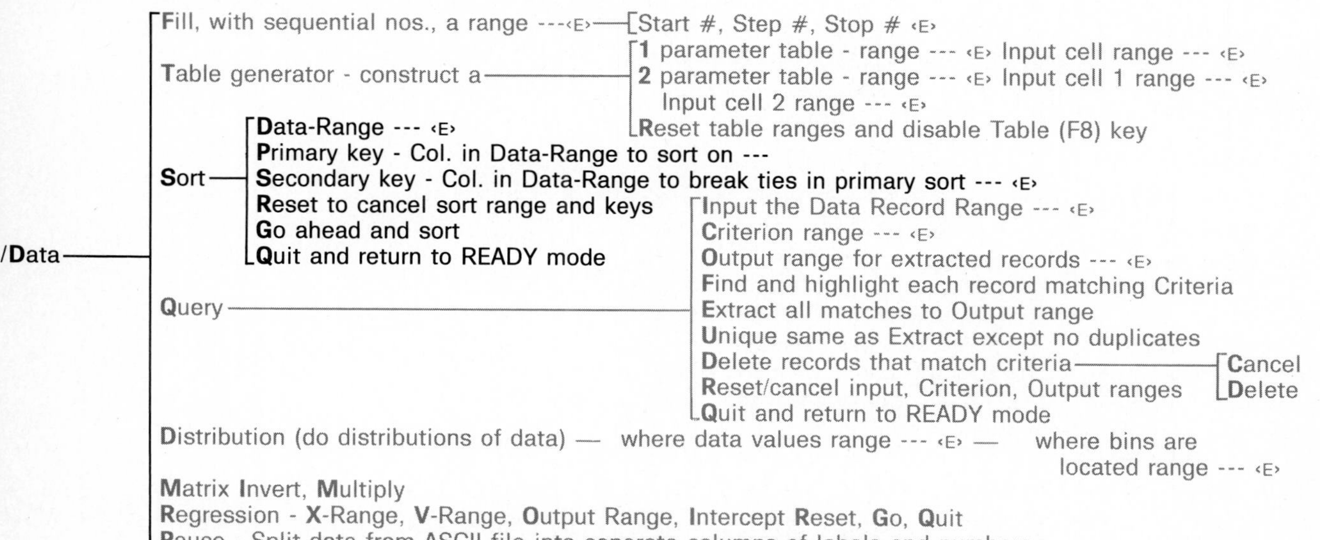

Topic 18-4 SuperCalc3

Global Settings Commands

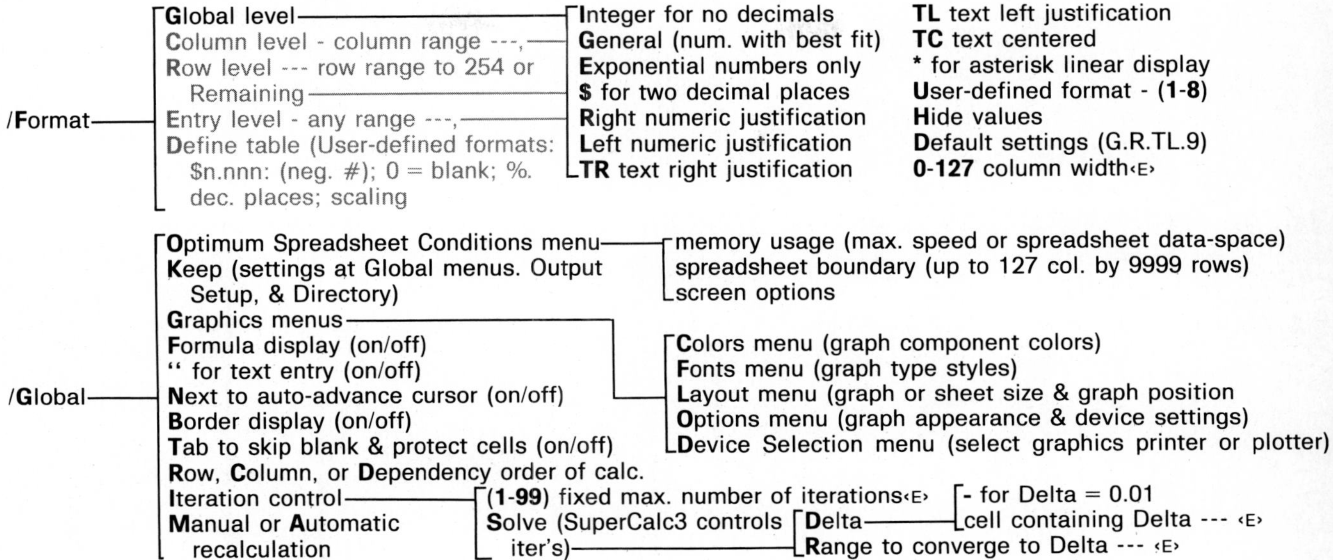

Topic 18-5 SuperCalc3

Arrange or Sort Data Command Keystrokes

/**A**rrange
- **R**ow
  - row number ---
    - ‹E› for entire row; ascending sort; no adjust
  - ‹E› current row
    - , col. range ---,
- **C**ol
  - col. number
    - , row range --- ,
  - ‹E› current col.
    - ‹E› for entire column; ascending sort; no adjust

Range options: **A**scend / **D**escend → **Y**es adj. / **N**o adj. → ‹E› primary / , secondary → row number, / col. letter, → **A**scend / **D**escend always blank or erase any cells you no longer need. Be careful, for once cells are blanked or erased, their previous contents are irretrievable.

Changing the Column Widths (Topic 18-7)

All spreadsheets display a default column width when the program is first loaded into memory. Most often this width is nine characters.

Topic 18-6 — Lotus 1-2-3

Blanking or Erasing (Entire Sheet) Command Keystrokes

- /**W**orksheet
  - **G**lobal
    - **F**ormat
      - **F**ixed number of decimal places (**0-15**)‹E›
      - **S**cientific notation with (**0-15**) dec. places‹E›
      - **C**urrency format ($x,xxx.xx) with (**0-15**) dec. plcs‹E›
      - **,** inserted & negative numbers in ()
      - **G**eneral or best default format
      - **+**/ - horizontal bar graph instead of numbers
      - **P**ercent format (x.xx%) with (**0-15**) dec. places‹E›
      - **D**ate format
      - **T**ext - display formulas in place of values
      - **H**idden
    - **L**abel-Prefix
      - **L**eft justify labels
      - **R**ight justify labels
      - **C**enter labels
    - **C**olumn-Width — (**1-240**)‹E›
    - **R**ecalcu- lation
      - **N**atural, **C**olumnwise, or **R**owwise calculation
      - **A**utomatic or **M**anual
      - **I**teration - set number (1-50)‹E›
    - **P**rotection — **E**nable or **D**isable cell protection
    - **D**efault
      - **P**rinter - specify printer default settings‹E›
      - **D**irectory - specify startup directory‹E›
      - **S**tatus - display default settings
      - **U**pdate config file to keep newly set defaults
      - **Q**uit and return to READY mode
    - **Z**ero
  - **I**nsert / **D**elete
    - **C**olumn - range --- ‹E›
    - **R**ow - range --- ‹E›
  - **C**olumn-Width
    - **S**et - Width (**1-240**)‹E›
    - **R**eset - Width
    - **H**ide or **D**isplay
  - **Erase**
    - **No** - do not erase sheet
    - **Yes** - erase whole sheet
  - **T**itles
    - **B**oth Hor & Vert titles set
    - **H**orizontal - freezes rows above cursor for titles
    - **V**ertical - freezes cols. to left of cursor for titles
    - **C**lear sheet of all titles
  - **W**indow
    - **H**orizontal split screen at cursor
    - **V**ertical split screen at cursor
    - **S**ync or **U**nsync scrolling of windows
    - **C**lear a second window on the screen
  - **S**tatus shows some global default settings
  - **P**age - inserts page break in current cell

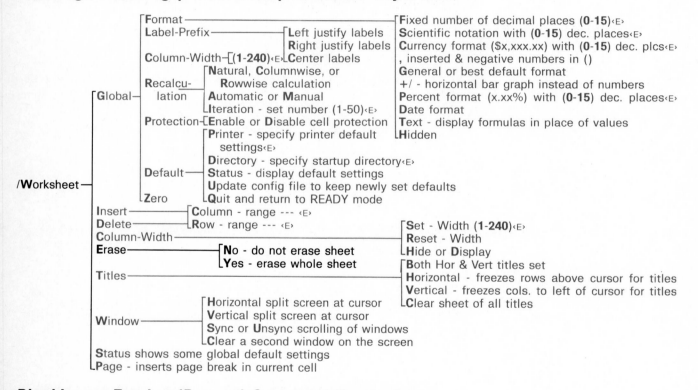

Blanking or Erasing (Ranges) Command Keystrokes

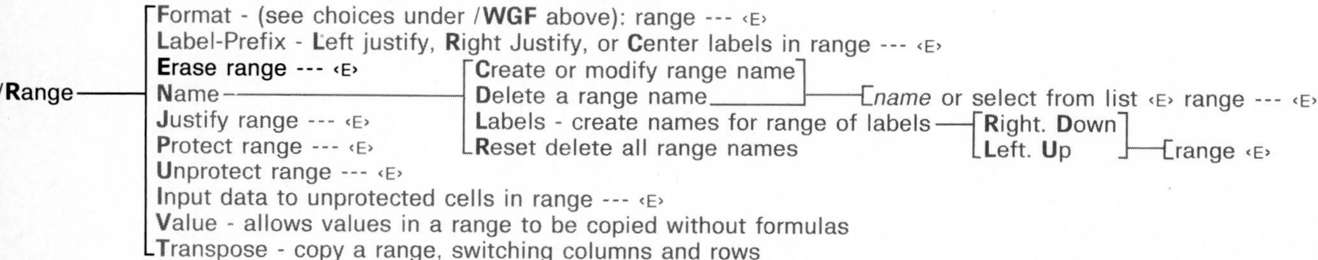

However, nearly all spreadsheets allow you to set the width of the columns to the value you desire. Usually, you can do this on a global basis (i.e., change all the columns on the sheet), or you can do this on a column-by-column basis. You cannot change the width of individual cells in a column without changing the width of all the cells in that column.

We narrowed column A in Figure (18-3) to three spaces in order to avoid a lot of blank spaces between the left-justified entry in cell

Topic 18-6 SuperCalc3

Blanking or Erasing (Entire Sheet) Command Keystrokes

/**Z**ap
- **Y**es to delete current spreadsheet; retains settings of Global menus, Output Setup, & Directory
- **N**o to cancel this command
- **C**ontents, same as Yes, but also retains User-defined format table settings

Blanking or Erasing (Ranges) Command Keystrokes

/**B**lank
- range --- <E>
- <E> for current cell
- graph range

Topic 18-7 Lotus 1-2-3

Column Width (All Columns) Command Keystrokes

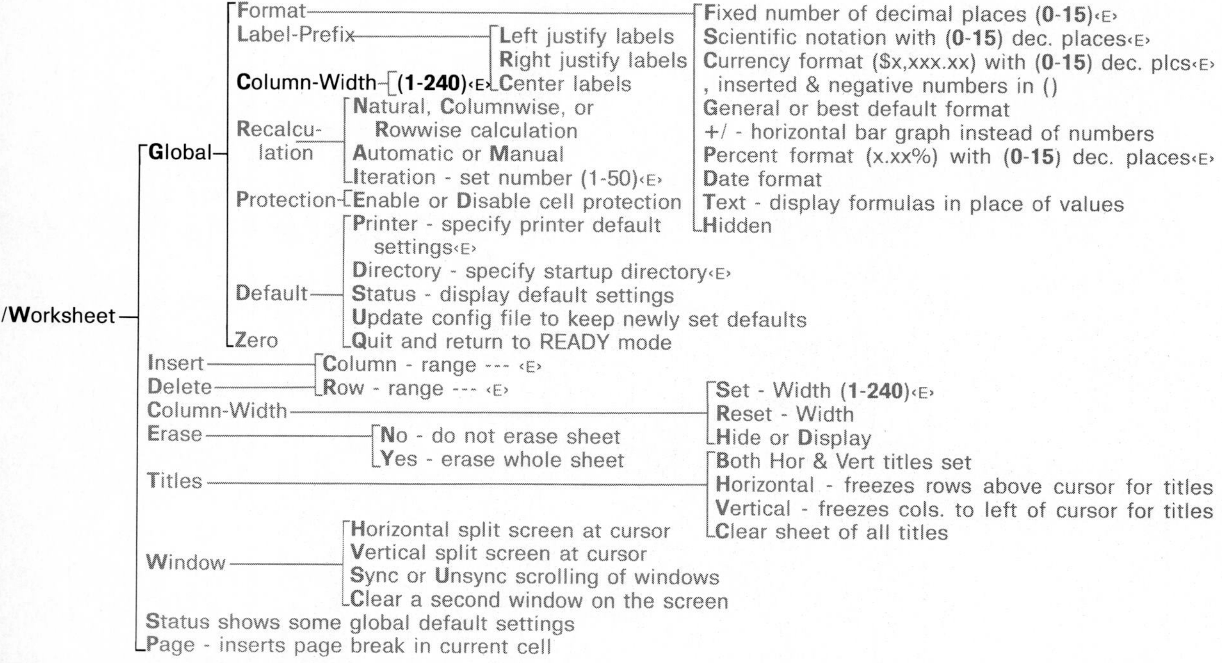

Column Width (Individual Columns) Command Keystrokes

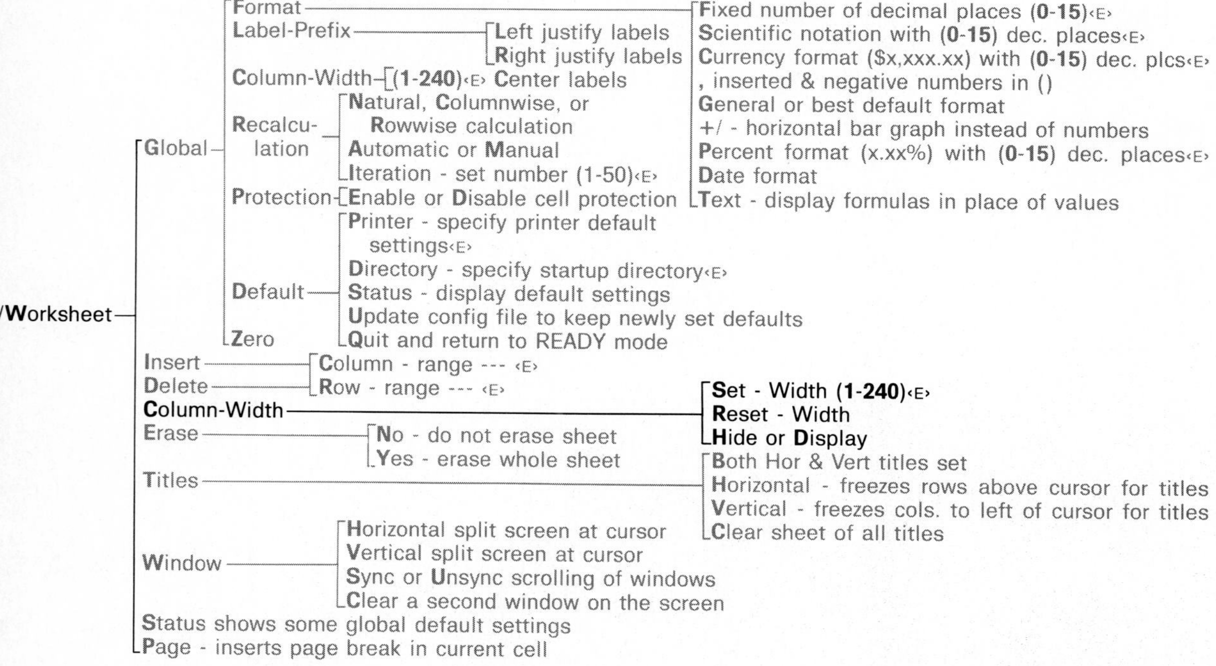

Topic 18-7 SuperCalc3

Column Width (All Columns) Command Keystrokes

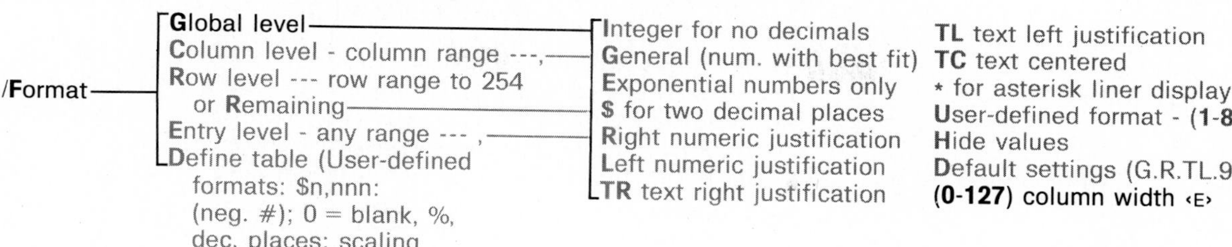

Column Width (Individual Columns) Command Keystrokes

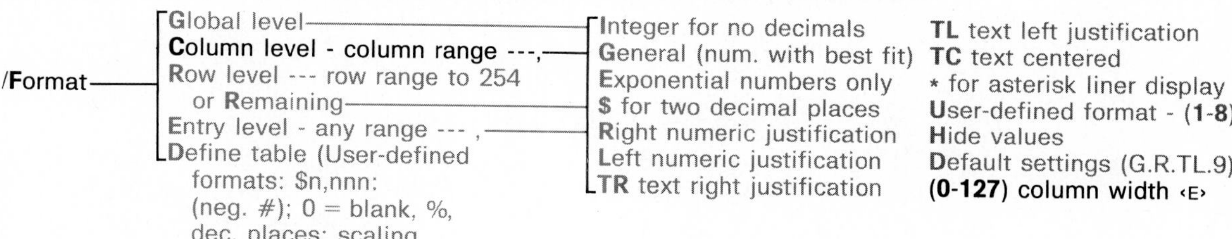

A1 and the right-justified entry in cell A4. Do you remember why A1 is left justified and A4 is right justified?[14]

Editing the Cell Contents (Topic 18-8)

There are two ways to change what is stored in a particular cell. One is to reenter the contents by typing the new value into the current cell status line and actuating the <Enter> key. This works fine for short entries but becomes error prone for long entries.

When the entries become long, it is desirable to be able to edit the contents with out retyping the whole entry. This can be done with the *edit command*.

The edit command generally brings the cell contents (label, number, or formula) into the command status line so that edit keys can be used to alter them. These edit keys typically include the Cursor Control keys, the Backspace key, the Delete key, and the Insert key.

Formatting Cells (Topic 18-9)

We have mentioned several times previously that worksheet cells can be formatted to alter the manner in which data and labels can be displayed on the sheet. Nearly all spreadsheets include a *format command* that provides the means for doing this. Formatting can generally be done on a global basis (i.e., to all the cells in the working area[15]), to a group or range of cells, or to any individual cell. Common choices include:

1. Text or labels:
   a. Left justified (usually the default).
   b. Right justified.
   c. Centered.

[14]The contents of cell A1 was entered as text which is left justified by default. The contents of cell A4 was entered as a number; numbers are right justified by default. The numbers displayed as results from formula (cells A2 and A5) are also right justified by default.

[15]On nearly all spreadsheets it is usually unwise to format all the cells that can be accommodated by the memory, for formatting generally takes up memory space and reduces the maximum number of cells available. It is also unwise to format all the cells that are in the working area, even though the working area is often substantially smaller than the maximum possible sheet. It is good practice to format no more cells than is necessary for clearness and appearance.

Topic 18-8 — *Lotus 1-2-3*

Cell Editing Command Keystrokes

F2 function key on IBM PC's and compatibles invokes editing of the current cell.
The Cursor Control keys take you any place in the editing line.
The Backspace key deletes the character to the left of the cursor.
The Delete key deletes the character over the cursor.
Typing any character inserts that character where the cursor is located.

2. Numbers and formula results:
   a. Right justified (usually the default).
   b. Left justified.
   c. Fixed point rounded to a specified number of places after the decimal point.
   d. Integer (truncated decimal parts).
   e. Floating point (e.g., 3.1E06).
   f. With or without "$" sign.
   g. With or without embedded commas (e.g., 1,000 or 1000).
   h. With negative numbers enclosed in parentheses.
   i. String of *'s, +'s, or −'s replacing the number.

GoTo Command or Key (Topic 18-10)

You may occasionally have a need to jump quickly to a given cell, especially when the given cell is not currently in the window of the screen. This can be done with either the *GoTo* command or the *GoTo key*. Most spreadsheets have one or the other available to aid in your efficient use of large sheets.

Inserting or Deleting Rows and Columns (Topic 18-11)

For problems that are solved repeatedly, it is typical for worksheets to grow from a modest beginning to a very elaborate solution sheet. Quite often it is necessary to add a row or a column in order to better maintain an orderly work space.

Generally an *insert command* is used to add rows or columns within the working area in order to make more cells available. The *delete command* is used to remove rows or columns from the sheet. Most spreadsheets are programmed to anticipate that you will want the row or column added or deleted where the cursor is located. Thus things are made much easier if you place the cursor in the row or column where you want the new one placed before issuing the insert or delete command.

On most spreadsheets, inserting and deleting rows or columns will automatically alter cell addresses used in formulas so that they reflect the changes in the location of the references caused by the new rows or columns.

Topic 18-8 — SuperCalc3

Cell Editing Command Keystrokes

/ **Edit** ——— any cell --- ‹E›
‹E› for current cell

The Cursor Control keys take you any place in the editing line.
The Backspace key deletes the character to the left of the cursor.
The Delete key deletes the character over the cursor.
The Insert key acts as a toggle. You must first toggle the Insert key on before you can insert new characters. The new characters are inserted at the cursor location. When the Insert key is toggled off you do not insert, but you *overtype* the characters where the cursor is located.

Topic 18-9 Lotus 1-2-3

Formatting (All Cells) Command Keystrokes

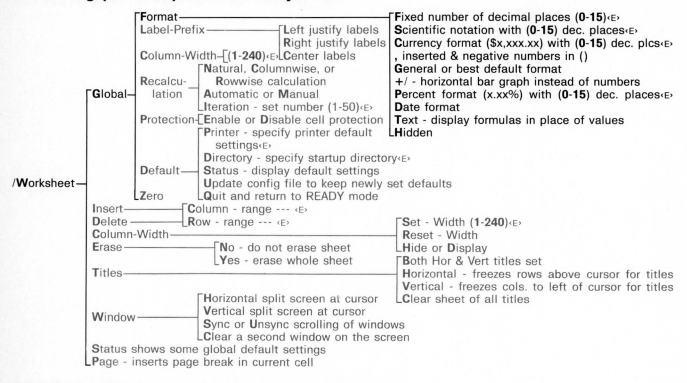

- /Worksheet
  - Global
    - Format
      - Fixed number of decimal places (0-15)‹E›
      - Scientific notation with (0-15) dec. places‹E›
      - Currency format ($x,xxx.xx) with (0-15) dec. plcs‹E›
      - , inserted & negative numbers in ()
      - General or best default format
      - +/ - horizontal bar graph instead of numbers
      - Percent format (x.xx%) with (0-15) dec. places‹E›
      - Date format
      - Text - display formulas in place of values
      - Hidden
    - Label-Prefix
      - Left justify labels
      - Right justify labels
      - Center labels
    - Column-Width
      - (1-240)‹E›
    - Recalcu- lation
      - Natural, Columnwise, or Rowwise calculation
      - Automatic or Manual
      - Iteration - set number (1-50)‹E›
    - Protection
      - Enable or Disable cell protection
    - Default
      - Printer - specify printer default settings‹E›
      - Directory - specify startup directory‹E›
      - Status - display default settings
      - Update config file to keep newly set defaults
      - Quit and return to READY mode
    - Zero
  - Insert
    - Column - range --- ‹E›
    - Row - range --- ‹E›
  - Delete
    - Column - range --- ‹E›
    - Row - range --- ‹E›
  - Column-Width
    - Set - Width (1-240)‹E›
    - Reset - Width
    - Hide or Display
  - Erase
    - No - do not erase sheet
    - Yes - erase whole sheet
  - Titles
    - Both Hor & Vert titles set
    - Horizontal - freezes rows above cursor for titles
    - Vertical - freezes cols. to left of cursor for titles
    - Clear sheet of all titles
  - Window
    - Horizontal split screen at cursor
    - Vertical split screen at cursor
    - Sync or Unsync scrolling of windows
    - Clear a second window on the screen
  - Status shows some global default settings
  - Page - inserts page break in current cell

Formatting (Range of Cells) Command Keystrokes

- /Range
  - Format - [see choices under WGF above]; range --- ‹E›
  - Label-Prefix - Left justify, Right Justify, or Center labels in range --- ‹E›
  - Erase range --- ‹E›
  - Name
    - Create or modify range name
      - *name* or select from list ‹E› range --- ‹E›
    - Delete a range name
      - *name* or select from list ‹E› range --- ‹E›
    - Labels - create names for range of labels
      - Right. Down. Left. Up
        - range ‹E›
    - Reset delete all range names
  - Justify range --- ‹E›
  - Protect range --- ‹E›
  - Unprotect range --- ‹E›
  - Input data to unprotected cells in range --- ‹E›
  - Value - allows values in a range to be copied without formulas
  - Transpose - copy a range, switching columns and rows

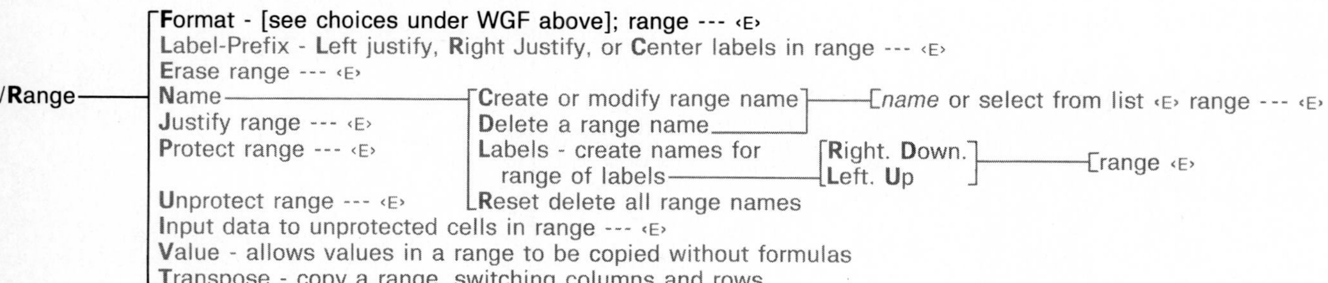

Topic 18-10 Lotus 1-2-3

GoTo Command Keystrokes

On IBM PC and compatibles
F5 function key [you supply the cell address to go to]

Topic 18-9 SuperCalc3

Formatting (All Cells) Command Keystrokes

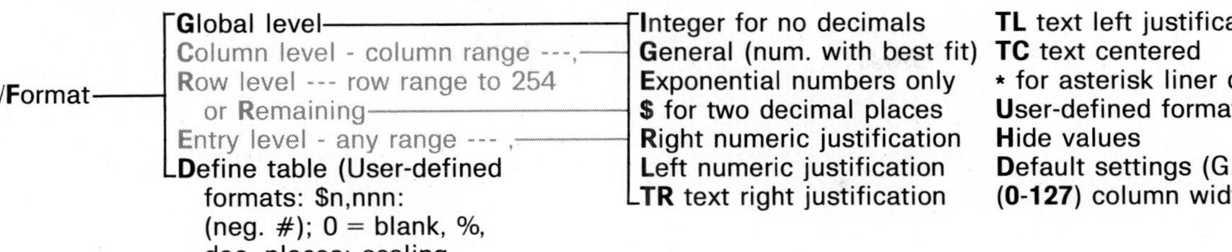

Formatting (Range of Cells) Command Keystrokes

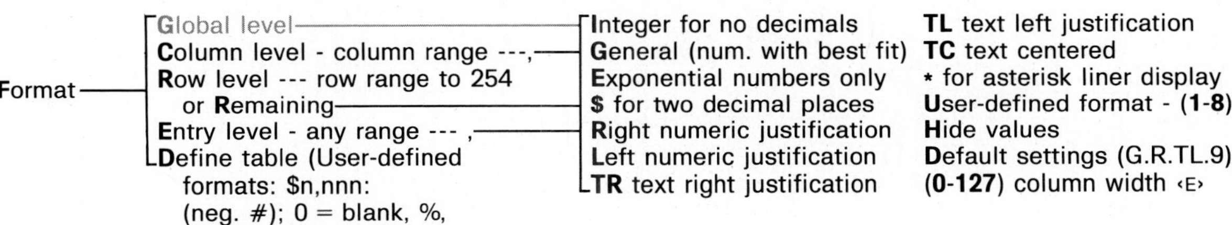

Topic 18-10 SuperCalc3

GoTo Command Keystrokes

= [you supply the cell address to go to]

Topic 18-11 Lotus 1-2-3

Insert Row or Column Command Keystrokes

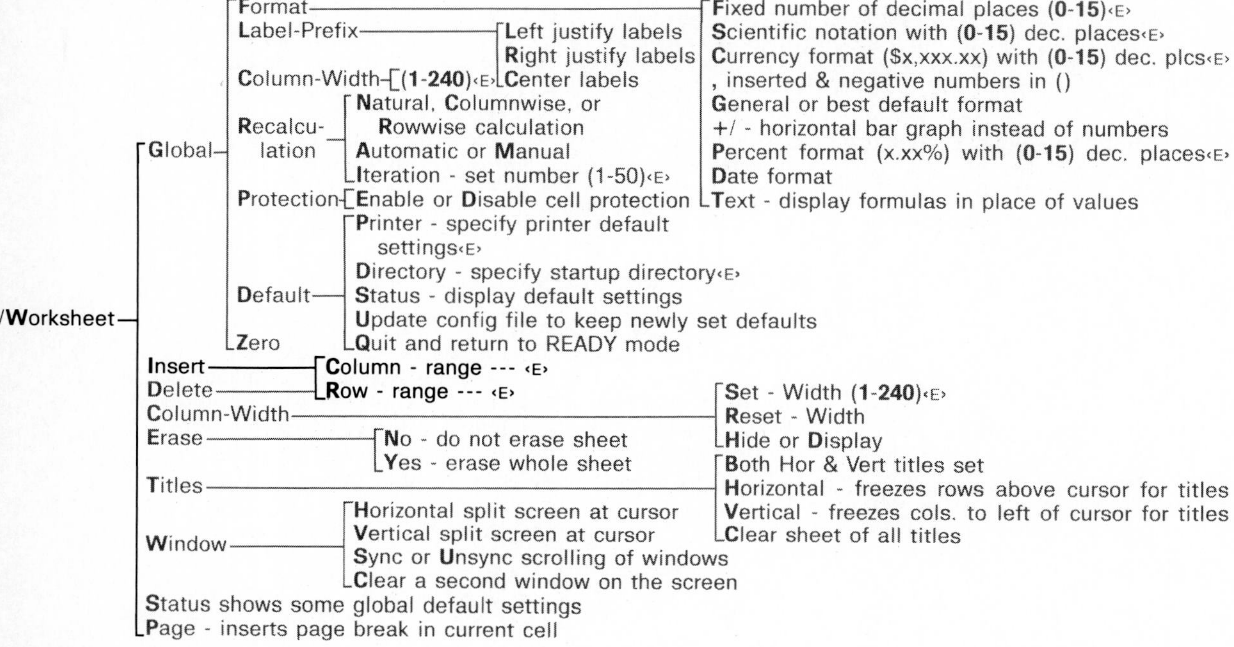

Delete Row or Column Command Keystrokes

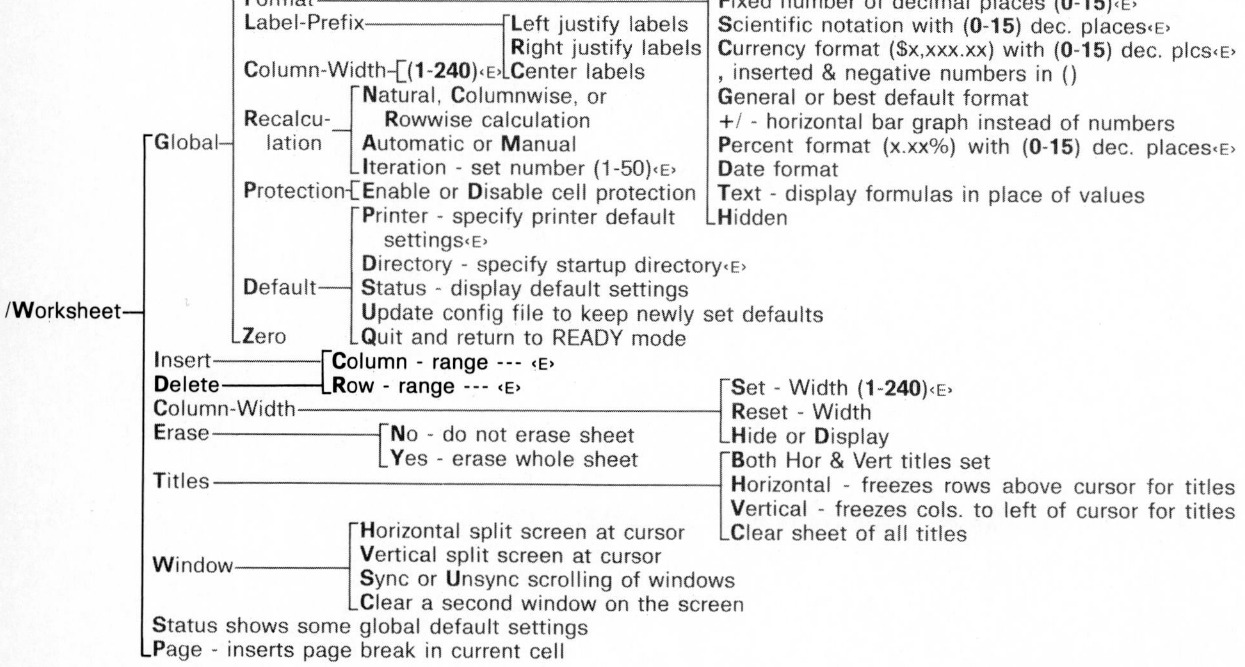

Topic 18-11 SuperCalc3

Insert Row or Column Command Keystrokes

/Insert — **R**ow — row range (to insert one or more empty rows) --- <E>
/Insert — **C**olumn — column range (to insert one or more empty columns) --- <E>

Delete Row or Column Command Keystrokes

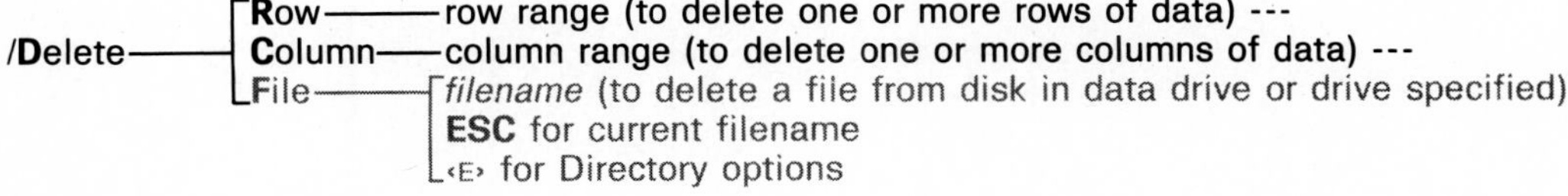

/**D**elete — **R**ow — row range (to delete one or more rows of data) ---
/**D**elete — **C**olumn — column range (to delete one or more columns of data) ---
/**D**elete — **F**ile — *filename* (to delete a file from disk in data drive or drive specified)
— **ESC** for current filename
— <E> for Directory options

Moving Cells (Topic 18-12)

For the same reasons that you may need to add or delete rows or columns to your sheet, you may also want to move existing cell contents to different locations. This is accomplished with the *move command.* Some spreadsheets can move only rows or columns, so check the one you are using.

This command, like the insert and delete command, will generally alter cell addresses used in formulas so that they reflect the changes in the location of these references caused by the move. The command also anticipates what range of cells you will want to move. It helps to have the cursor in the upper left-hand corner of the range to be moved when the command is invoked.

Printing Worksheets (Topic 18-13)

All useful spreadsheets provide a means of making hard-copy output from the sheet. This is typically done with a *print command.* The more sophisticated sheets allow you to print headers and footers, choose column and row borders, turn on or off the column and row identifiers, and send control sequences to the printer for special effects.

Topic 18-12 *Lotus 1-2-3*

Moving Cells Command Keystrokes

/**M**ove——[FROM range --- ‹E› TO range --- ‹E›

Topic 18-13 *Lotus 1-2-3*

Printing Command Keystrokes

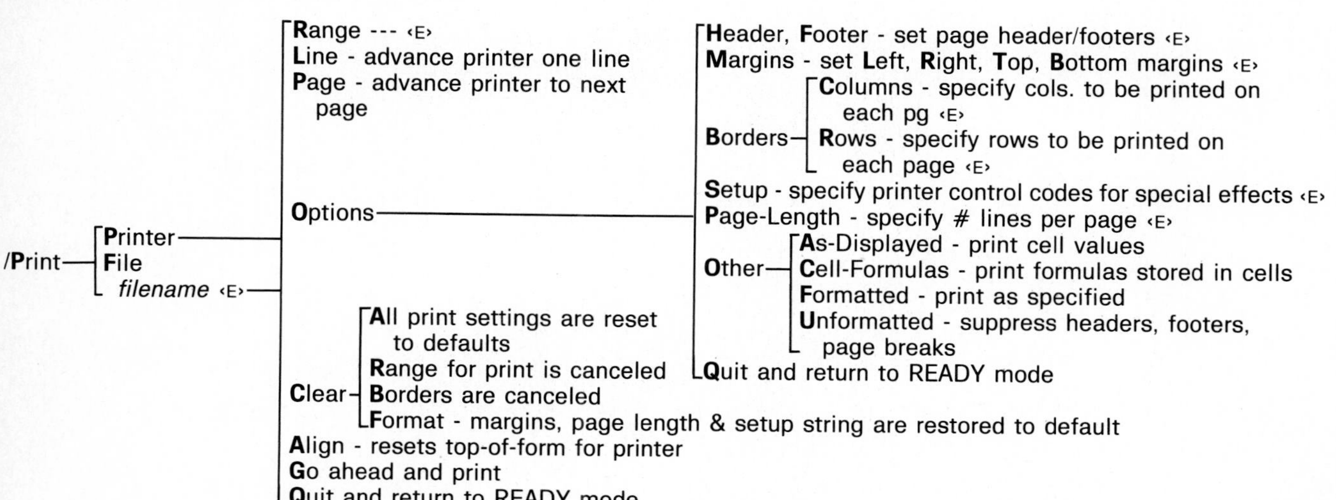

Often, you may also elect to print to a file instead of directly to the printer. The file can be printed later if desired.

Quitting the Session (Topic 18-14)

Essentially all spreadsheets have a *quit* command that allows you to exit from the spreadsheet and return to the operating system of the computer on which you are operating. These commands nearly always quiz you about your true intentions of quitting as a reminder to you to save your spreadsheet if you have not already done so. The reminders are also welcome when you have accidentally invoked the quit command, for they usually offer you a chance to resume your spreadsheet work instead of quitting.

Recalculating the Sheet (Topic 18-15)

As mentioned previously, spreadsheets can calculate all the formulas on the sheet in several ways. You can generally decide when to recalculate, although the default is generally to recalculate automatically any time that one of the entries has changed. You may want to change this default in order to speed up data and formula entry on

Topic 18-12 — *SuperCalc3*

Moving Cells Command Keystrokes

Move —— **R**ow from row range ---, to row number --- (will be top row if move is up: bot. row if move is down)
Column - from col. range ---, to col. letter --- (will be left col. if move is left: right col. if move is right)

Topic 18-13 — *SuperCalc3*

Printing Command Keystrokes

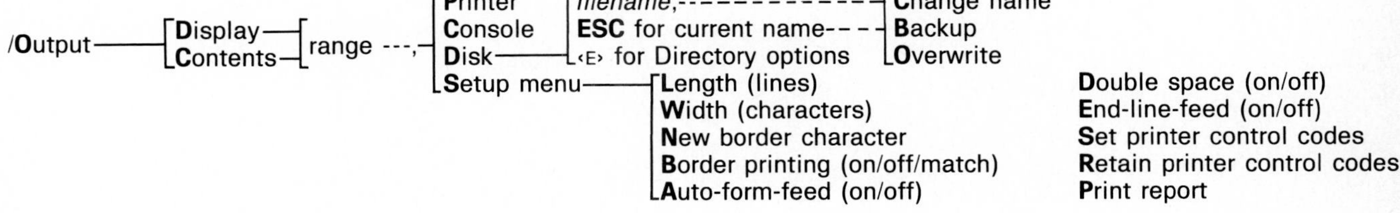

larger sheets. It can be changed from "automatic" to "manual" or back, with the *recalculation command*.

In the manual mode of recalculation, there is a designated *recalculation key* which, when depressed, will force an immediate recalculation of the sheet. It is very easy to forget to force recalculation when you are using the manual mode. Since your formulas never get updated until you do so, you need to stay alert to this potential problem.

On those spreadsheets that allow the "natural" recalculation mode, this is generally the default scheme. It can generally be changed to row-wise or column-wise recalculation order, although there is little reason ever to do this. With the natural recalculation

Topic 18-14 — Lotus 1-2-3

Quit Command Keystrokes

/**Q**uit — **N**o - do not end session
/**Q**uit — **Y**es - end session

(Be sure you have saved your work)

Topic 18-15 — Lotus 1-2-3

Recalculation Mode Command Keystrokes

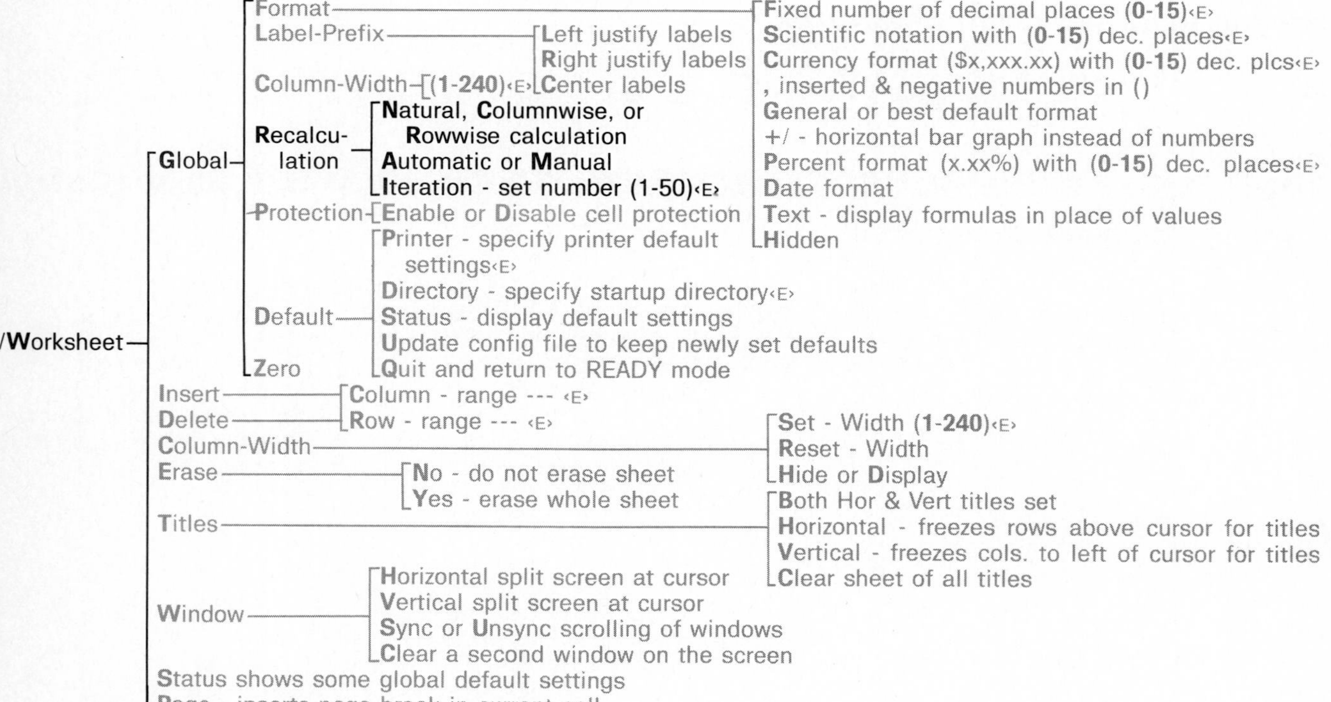

F9 FUNCTION KEY IS THE FORCE RECALCULATION KEY (IBM PC AND COMPATIBLES)

order, it is often desirable in technical problems that require iteration, to change the number of calculation iterations to be performed each time the sheet is calculated. This number can be changed (it is usually set to 1 as a default) with the recalculation command.

Replicating or Copying Cell Contents (Topic 18-16)

Of great benefit for making lists of numbers and formulas are the *replicate and copy commands*. With these commands you can place in other cells something that you have entered in a single cell. This saves a tremendous amount of time over that required to enter similar contents in many cells individually.

Topic 18-14 — SuperCalc3

Quit Command Keystrokes

```
       ┌Yes to exit from SuperCalc3 (does not save current work)
/Quit──┤No to cancel this command                                ┌program filename ‹E›
       └To quit & re-load SC# or load another program────────────┤
                                                                 └‹E› for Directory options
```

Topic 18-15 — SuperCalc3

Recalculation Mode Command Keystrokes

```
         ┌Optimum Spreadsheet Conditions menu────┬memory usage (max. speed or spreadsheet data-space)
         │Keep (settings at Global menus.        │spreadsheet boundary (up to 127 col. by 9999 rows)
         │  Output Setup, & Directory            └screen options
         │Graphics menus ─────────┐
         │Formula display (on/off)│                ┌Colors menu (graph component colors)
         │" for text entry (on/off)                │Fonts menu (graph type styles)
/Global──┤Next to auto-advance cursor (on/off)└────┤Layout menu (graph or sheet size & graph position)
         │Border display (on/off)                  │Options menu (graph appearance & device settings)
         │Tab to skip blank & protect cells (on/off)└Device Selection menu (select graphics printer or plotter)
         │Row, Column, or Dependency order of calc.
         │Iteration control──────────┬(1-99) fixed max. number of
         │                           │  iterations ‹E›                         ┌– for Delta = 0.01
         │Manual or Automatic        └Solve (SuperCalc3 controls iter's)──┬Delta──┴cell containing Delta --- ‹E›
         └  recalculation                                                └Range to converge to Delta --- ‹E›
```

! KEY IS THE FORCE RECALCULATION KEY (IBM PC AND COMPATIBLES)

These commands often are programmed to anticipate what cell or cells are to be replicated or copied. To make best use of this anticipation, you should generally place the cursor in the cell you wish to copy before initiating the command.

As we have mentioned previously, you must know whether the cell addresses that exist in the formula to be copied should be copied as a relative cell address or as an absolute cell address. Unless you take specific action, they generally will be copied as relative addresses by default. Since most copying is done in situations where relative cell addresses are desired, this works ideally. For example, assume that we use cell A1 as an input cell for a quantity that was referenced in a formula in cell B1. Now try to visualize replicating or copying the formula in B1, down column B into cells B2 through B10, in order to see the output from 10 different inputs we will enter into cells A1 through A10. If the formula A1^2 resided in the original cell B1, then after replicating, the formula in cell B2 will automatically be A2^2, that in B3 will be A3^2, and so on. This is, of course, exactly what we needed, relative cell addresses and not absolute addresses.

Saving or Reloading the Sheet (Topic 18-17)

Spreadsheets would not be so widely used if they could not be saved for use at a later time. If all data and formulas had to be reentered each time you wanted to rework the data, spreadsheets would never take the place of programming languages. However, spreadsheet programs generally have *file saving* and *file retrieving commands* for storing and reloading worksheets.

As with most file storage work, you must give a worksheet a name when you store it. This name must be consistent with the naming rules of the operating system you are using.

The more sophisticated spreadsheets allow you to store your entire worksheet sheet or only some range or part of the sheet. The default storage format is generally binary and unique to the product you are using. If the binary format is used, it is improbable that you will be able to read or load these files into any other brand of spreadsheet or other computer program. However, most programs allow you to choose other storage formats in case you need to access the data from other programs.

Topic 18-16 — Lotus 1-2-3

Replicating or Copying Command Keystrokes

/**C**opy——[FROM range --- ‹E› TO range --- ‹E›

When you retrieve or load previously saved files into the program, you must, of course, know the name under which the file was stored. If it was stored in a format other than binary, you may be required to take special action, depending on the product you are using. We will say more later about loading ASCII files into spreadsheets.

Windows: Opening and Closing (Topic 18-18)

The *window command* allows the user to open a screen window to view two portions of the worksheet at the same time on the screen. This is convenient for large sheets where the input cells are more than a screen away from the output or results cells. With a window open, both the input and output cells can be viewed on the same screen, for example. The window command can also be used to close the window. Additional options under the window command often allow you to link together the scrolling of the windows.

A designated *window change key* allows you to change the location of the cursor from one window to the other.

FINDING HELP ON-SCREEN

When actuated, the *help key* generally brings you a screen full of useful information. It is generally one of the function keys, often the F1 key or the "?" key. The help level is usually between what the terse menu items bring to mind and what you will find in the instruction manual for the program.

Actuated when using one of the commands, it brings information concerning that command. Actuated when in the data entry mode, the help key generally brings you information about entering labels, data, and formulas.

GRAPHING

Graphing of data from a spreadsheet is, perhaps, second in importance only to the ability to enter formulas in the cells. Graphing allows you to put your eyes to work in interpreting your data. You quickly gain new insights and understanding that are often not possible by just looking at the maze of numbers on the sheet.

Topic 18-16 — SuperCalc3

Replicating or Copying Command Keystrokes

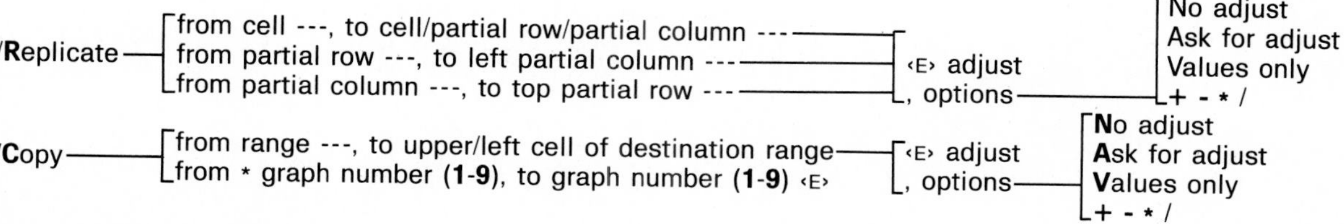

Only a few spreadsheets include graphing capabilities as an integral part of the spreadsheet package. Although there are stand-alone graphing packages that can do plotting of spreadsheet files that have been stored, the integral packages make it possible to take quick looks at the numbers displayed on the sheet and to permit quick changes to be made in cells. We will concentrate here on the integral packages. Therefore, this section is less generic than the previous topics. We will give a very concise overview of how the graphing is carried out and then continue the more detailed Topics sections covering graphing in SuperCalc3 and Lotus 1-2-3.

The *graphing commands* are typically a part of the Command Mode of operation and can be initiated from the menu accessed with the "/" key. In SuperCalc3, the proper menu item is **V** for View; in Lotus 1-2-3 it is **G**raph.

Topic 18-17 Lotus 1-2-3

Saving and Reloading File Command Keystrokes

File Saving

/**F**ile —
- **R**etrieve a worksheet: *filename* or select from list ‹E› — **C**ancel this command / **R**eplace preexisting file of this name
- **Save a worksheet: *filename* or select from list ‹E›** — **Cancel this command / Replace preexisting file of this name**
- **C**ombine — **C**opy cells from preexisting file / **A**dds values from preexisting file / **S**ubtracts values in preexisting file — **E**ntire File / **N**amed Range - range --- ‹E› — filename or choose from list ‹E›
- **X**tract — **F**ormulas / **V**alues — from present worksheet to *filename* or select from list: range ‹E› — **C**ancel this command / **R**eplace preexisting file of this name
- **E**rase — **W**orksheet file from disk / **P**rint file from disk / **G**raph file from disk — file to erase or select from list ‹E› — **N**o - cancel command / **Y**es - erase file
- **L**ist - **W**orksheet, **P**rint, **G**raph, **O**thers files on default drive
- **I**mport comma separated .PRN files as — **T**ext - Enter each line as a single, separate label / **N**umbers - Enter each line as numbers and quoted labels — *filename* ‹E›
- **D**irectory or disk desired as a default ‹E›

File Retrieving

/**F**ile —
- **Retrieve a worksheet: *filename* or select from list ‹E›** — **C**ancel this command / **R**eplace preexisting file of this name
- **S**ave a worksheet: *filename* or select from list ‹E› — **C**ancel this command / **R**eplace preexisting file of this name
- **C**ombine — **C**opy cells from preexisting file / **A**dds values from preexisting file / **S**ubtracts values in preexisting file — **E**ntire File / **N**amed Range - range --- ‹E› — filename or choose from list ‹E›
- **X**tract — **F**ormulas / **V**alues — from present worksheet to *filename* or select from list: range ‹E› — **C**ancel this command / **R**eplace preexisting file of this name
- **E**rase — **W**orksheet file from disk / **P**rint file from disk / **G**raph file from disk — file to erase or select from list ‹E› — **N**o - cancel command / **Y**es - erase file
- **L**ist - **W**orksheet, **P**rint, **G**raph, **O**ther files on default drive
- **I**mport comma separated .PRN files as — **T**ext - Enter each line as a single, separate label / **N**umbers - Enter each line as numbers and quoted labels — *filename* ‹E›
- **D**irectory or disk desired as a default ‹E›

Choosing the Type of Graph

The type of graph you wish to plot is typically specified by selection of the proper menu choice under the graph command (the / **V** in SuperCalc3 or / **G**raph in Lotus 1-2-3). Most spreadsheets that support graphics include the following types of graphs:

- *Pie charts:* Each data point of a set is displayed as a slice of pie whose size is proportional to the ratio of that point to the sum of all the data.

- *X-Y plots:* This is the most common engineering data plot. Each data pair is plotted as a point on the graph. One limitation is that multiple data sets must generally all share the same set of independent variables.

Topic 18-17 SuperCalc3

Saving and Reloading File Command Keystrokes

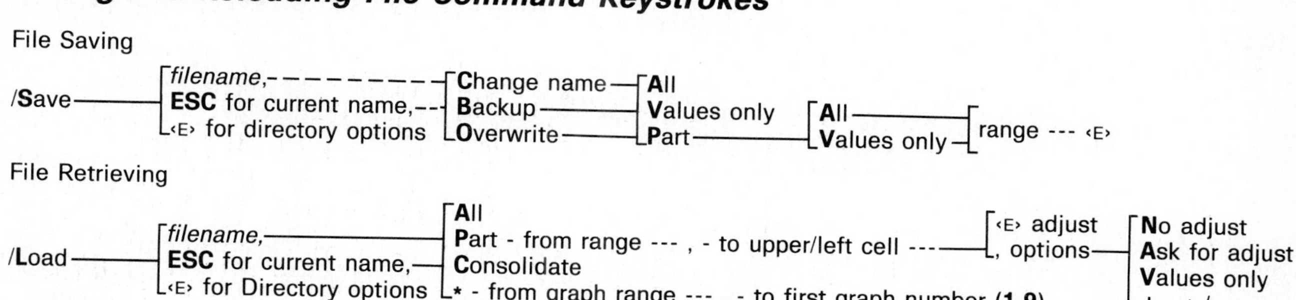

- *Bar charts or line plots:* These two plots are related. They both plot data points at a height above the horizontal axis proportional to the size of the data. Points from the same data set are typically spaced equally along the horizontal axis. The horizontal axis is sometimes referred to as the time line (it is assumed that all data sets share the same intervals along the time line). The bar chart draws bars vertically upward from the horizontal axis to each data point. The line plot connects adjacent data points with straight lines.
- *Stacked bar charts or area plots:* These are similar to the bar and line charts above, except that common time-line data points from each data set are stacked vertically on top of the previous data set, instead of being plotted adjacent to one another.
- *Hi-lo plots:* These graphs are similar to bar charts and line plots except that vertical lines are drawn at each time-line data point, but between the maximum data point and the minimum data point—thus the name "hi-low."

Topic 18-18 Lotus 1-2-3

Window: Open and Close Command Keystrokes

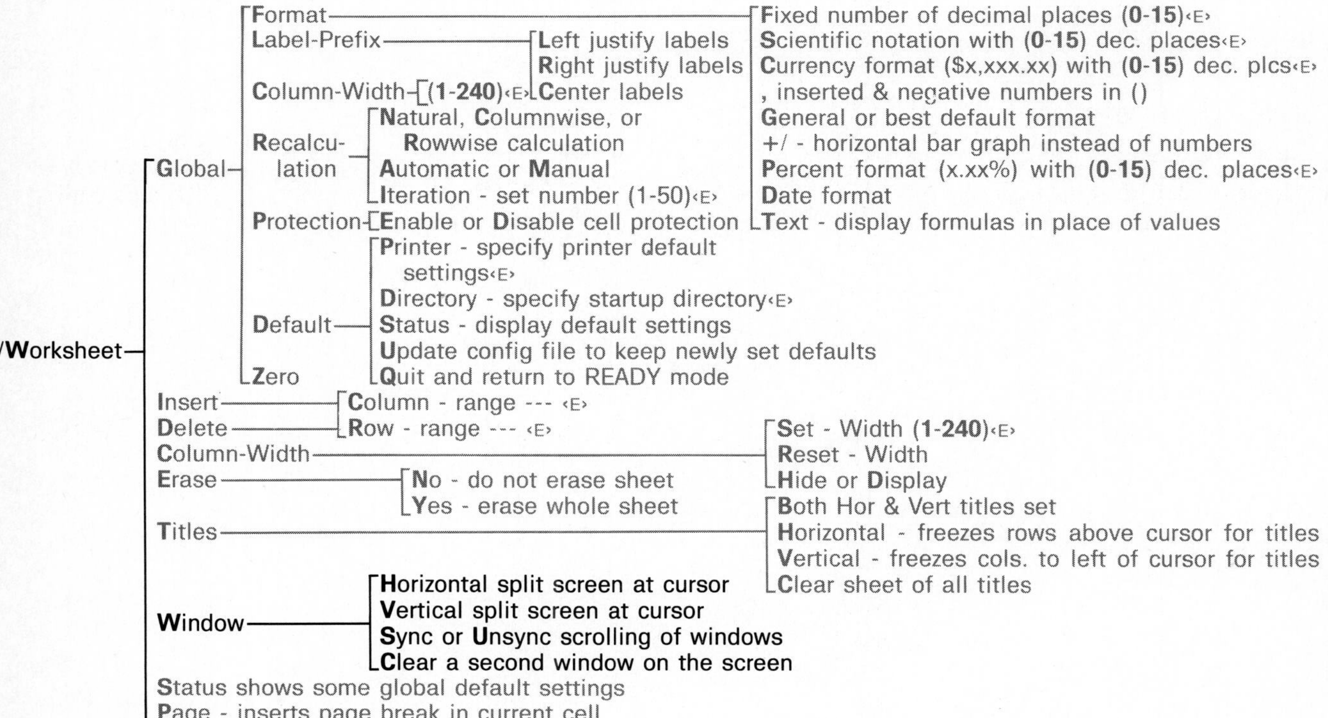

F6 FUNCTION KEY MOVES CURSOR TO NEXT WINDOW (IBM PC AND COMPATIBLES)

Of course, missing from this collection are semilog, log-log, and polar plots (we discussed graphing in Chapter 9). The log plotting can be accommodated by using the log function to compute the logarithms of the data points you wish to plot. The axis of a graph made with these new logarithms will not be labeled in the conventional manner,[16] but the plot will be a logarithmic one. Polar plots cannot be accommodated.

Choosing the Data to Be Graphed

Most integrated spreadsheet/graphics packages permit one or more sets of data to be displayed in graphical form, depending on the type of graph desired. Pie charts, for example, can display only one set of data, whereas line, bar, and *x-y* plots may accommodate several sets

[16]The conventional axes on logarithmic scales are labeled in terms of the numbers whose logarithms are plotted. In the approach to logarithmic plots suggested here for spreadsheets, the logarithmic axes will be labeled in terms of the logarithms of the original numbers.

Topic 18-18 — *SuperCalc3*

Window Open and Close Command Keystrokes

/**W**indow —
- **H**orizontal split
- **V**ertical split
- **C**lear to right or below split
- **S**ynchronize split-wise scroll
- **U**nsynchronize split-wise scroll

; KEY MOVES CURSOR TO NEXT WINDOW

(up to nine in SuperCalc3 and six in Lotus 1-2-3). Each set of data must reside in contiguous cells that form a rectangle, because they are identified to the program by their range (e.g., C3.C20, or C3.M3, or C3.F21).

Viewing the Graph

Once you have identified the data sets to be plotted and have chosen the type of graph, you simply actuate the *graphing key* to view the graph on the CRT (assuming that you are using a graphics monitor). In both SuperCalc3 and Lotus 1-2-3, the graphing key is a function key[17] on machines that have function keys [in SuperCalc3 it can also be a Ctrl T (i.e., hold down the key labeled Ctrl, push the T key, and then release both keys)]. With or without function keys, an <Enter> following selection of the / **V** option in SuperCalc3 will draw the graph on the CRT. The / **G**(raph **V**(iew) will execute a similar action in Lotus 1-2-3.

[17]For IBM PCs and compatibles, the F10 function key initiates display of the latest graph on the screen for either SuperCalc3 or Lotus 1-2-3.

Topic 18-19 — Lotus 1-2-3

Graphics Command Keystrokes

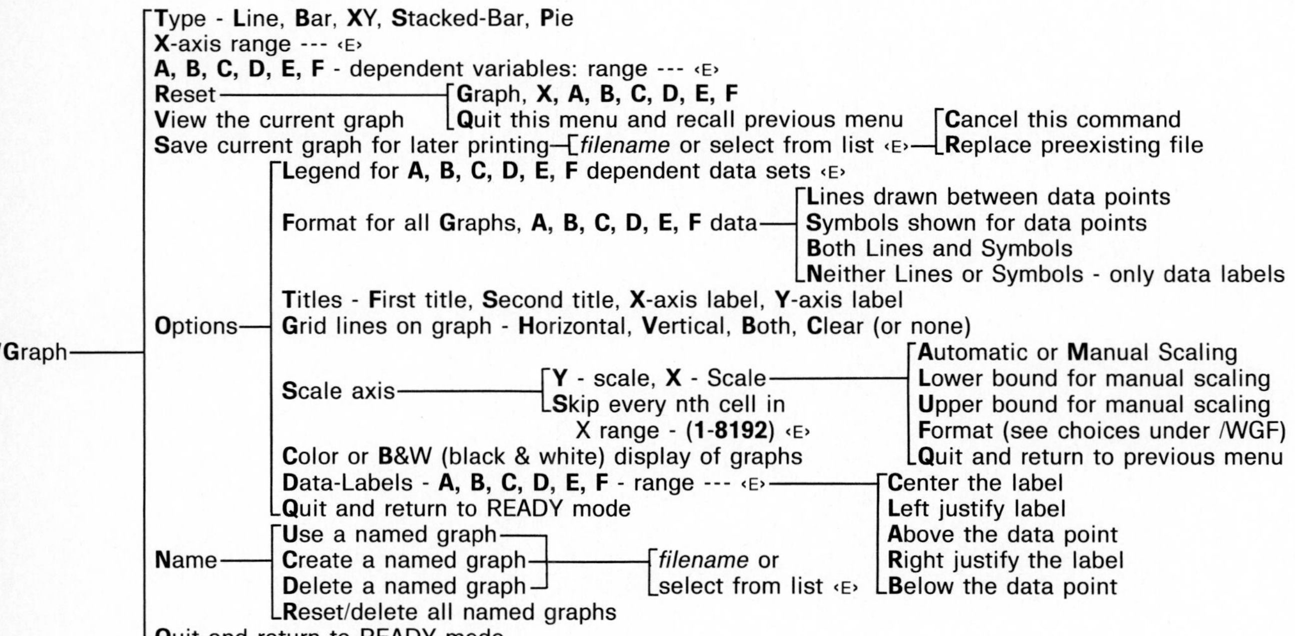

X Is the independent variable for Xy plots specified by range
A, B, C, D, E, F are 6 sets of dependent variables specified by range. Not all are necessary, but you must have one set. Pie charts are plotted from A; you must have only A defined for pie charts.

Topic 18-19 covers the graphing commands. Example plots are shown in Figure [18-12].

Formatting the Graph

Commonly buried in the graphing commands are the user choices for several formatting features for graphs. These features allow you to add titles, axis labels, legends, and data point labels to your graph. It also allows you to select, for appropriate types of graphs, whether you want data points without connecting straight lines, connecting straight lines without data points, or both data points and lines [these options in SuperCalc3 are chosen under / **G**(lobal) **G**(raphics) **O**(ptions)]. You also have the choice of permitting the program to scale your axes automatically or you may specify the scales.

Getting a Hard Copy of the Graph

Hard copy of graphs can be obtained on any printer or pen plotter which is supported by the package you are using. Once the device is

Topic 18-19 — *SuperCalc3*

Graphics Command Keystrokes

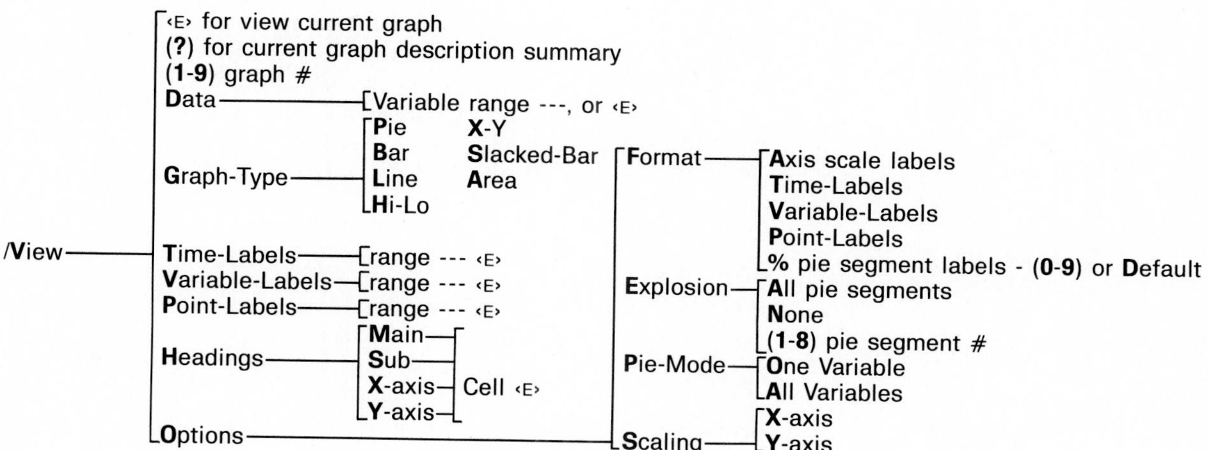

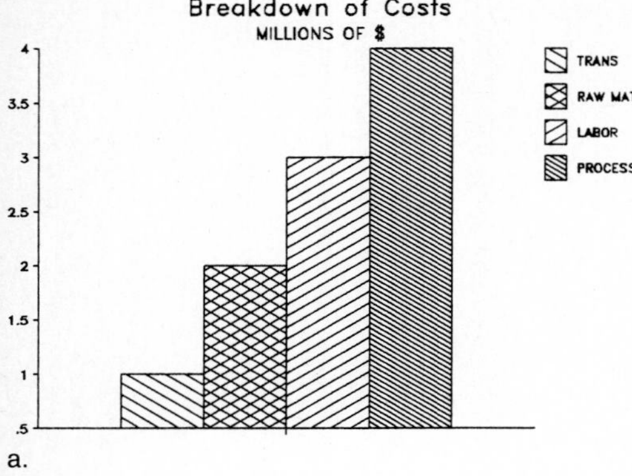

a.

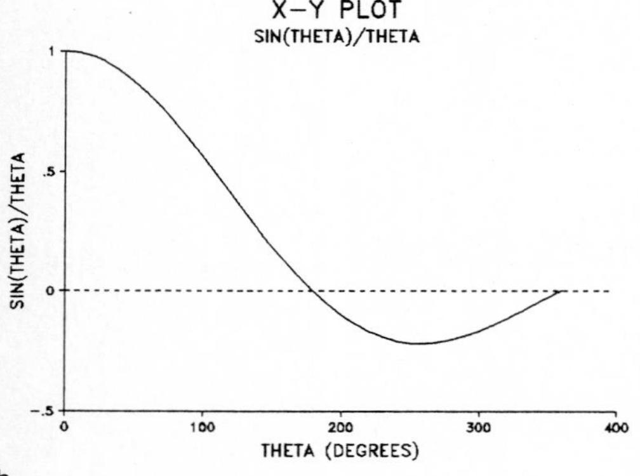

b.

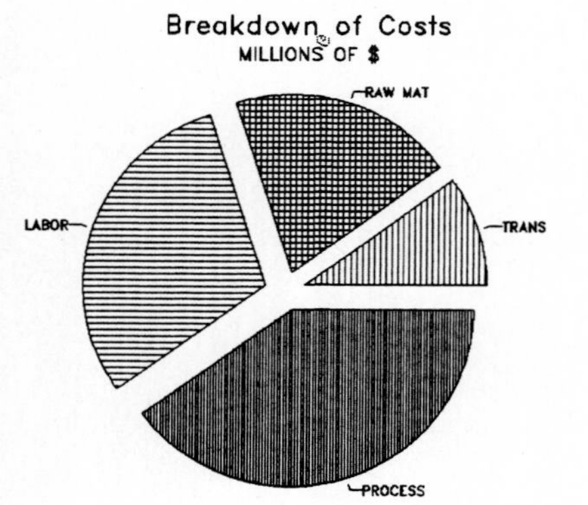

c.

[18-12] Example spreadsheet plots (SuperCalc3).

properly in place and you have defined a graph, it is a simple matter to make a plot from SuperCalc3. You simply actuate the plot key,[18] which is generally one of the function keys of the keyboard.

Lotus 1-2-3 requires a little more preparation on your part before hard copy can be obtained. First you must / **G(raph) S(ave)** the graph, giving it a name when requested. This writes a file with the chosen name and a .PIC extension to storage. You must then **Q(uit)** the graph menu and / **Q(uit) Y(es)** the 1-2-3 spreadsheet. This will then put you back into the Lotus Access Program, where you must **P(rintGraph) S(elect)** the .PIC file to be plotted (you will select the one you just saved), followed by **O(ptions) F(onts) (1)** and selection of the graph name font, followed by **O(ptions) F(onts)(2)** and selection of the fonts for the balance of the labels on the graph, followed by a **Q(uit)** of the option menu, followed by a selection of **G(o)** in the PrintGraph menu. When the graph has plotted, you must **Q(uit)** the PrintGraph menu to reenter the Lotus Access Program.

LOADING OTHER DATA INTO THE SPREADSHEET

When Is It Useful?

We have just shown that sophisticated spreadsheets can handle large amounts of data. They can do data reduction, statistical analysis, and plotting. Spreadsheets are also fairly fast and are extremely easy to use (we have not yet spent much time on the latter point, but it will be illustrated in the sample problems that follow).

All of the foregoing attributes make spreadsheets desirable for analyzing many kinds of data, whether they be generated and recorded in the field, in the laboratory, or from analysis (possibly on the computer).

How to Do It

A few spreadsheets do not provide for any means of "foreign" data entry except by hand. Fortunately, some of the better spreadsheet programs give you several options for loading "foreign" data, most of which are too lengthy to explain here. However, several spreadsheets, including SuperCalc3 and Lotus 1-2-3, provide an easy way to import ASCII files[19] into their sheets.

The ASCII files should consist of rows of data, with each row ending in a carriage return and line feed (ASCII 13_{10} and ASCII 10_{10}, respectively). Data within each row that are intended for indi-

[18]On IBM PCs and compatibles, the plot key is the F9 function key or Ctrl Y (i.e., hold down the key labeled Ctrl, push the "Y" key, and then release both keys).

[19]ASCII files consist of all ASCII characters between ASCII codes 32_{10}, and 127_{10}, and includes 8_{10}, 10_{10}, 12_{10}, and 13_{10}, which are the backspace, line feed, form feed, and carriage return, respectively. The absence of escape characters (ASCII 27_{10}) and other control characters makes these files rather inert when transferring them around from program to program or computer to computer.

vidual cells must be separated from each other by a comma (referred to as "comma delimited"). Data that are intended to be text or label entries must be enclosed in quotation marks "like this".

SuperCalc3, for example, supplies an auxiliary program called the Super Data Interchange Program for processing ASCII files of the foregoing type that terminate with an ASCII 26_{10}. This SDI Program refers to such files as Comma-Separated Value or CSV format files. The SDI program loads the entries of the CSV file row for row into a spreadsheet file (called a .CAL file by SuperCalc3), each entry in the row being given its own column. The SDI Program will, of course, handle several other "foreign" formats.

Lotus 1-2-3 makes loading these ASCII files even simpler if the ASCII file has a .PRN extension. While in the 1-2-3 spreadsheet, you take the cursor to where you would like the upper left-hand corner of the ASCII file data to be loaded and invoke the / **F(ile) I(mport)** commands. You will then be asked to supply the name of the ASCII file to be loaded. If all goes well, you will see your data appear on the worksheet. Like SuperCalc3, the file is loaded row for row, with each entry in each row being given its own column.

SOME FURTHER SPREADSHEET PROBLEMS

We will cover four additional spreadsheet problems to demonstrate the versatility of the spreadsheet as an engineering tool. The first is intended to demonstrate several commands that we have talked only briefly about in the preceding text. The example shows the construction of tables much like that in Figure [18-1]. We will partially duplicate the table in Appendix III.

The table was constructed with a BASIC program that solves the following equations (these equations were discussed in detail in Chapter 8; the BASIC program is also given in that chapter).

Example 18-6

Construct a multicolumn table with the following attributes:

Table title: "DECIMAL INTEREST RATE = ". *Interest rate:* i = ".02".

Column 1: titled "PERIOD", the independent variable, having values starting with 1, ending with 5, and incrementing by 1.

Column 2: titled "F/P" and listing dependent variable $(1 + i)^n$, where n is the period and i is the interest rate.

Column 3: titled "P/F" and listing the reciprocal of column 2.

Column 4: titled "F/A" and listing dependent variable $((1 + i)^n - 1)/i$.

Column 5: titled "A/F" and listing the reciprocal of column 4.

Column 6: titled "P/A" and listing dependent variable $((1 + i)^n - 1)/(i(1 + i)^n)$.

Column 7: titled "A/P" and listing the reciprocal of column 6.

Solution:

To give a little more space between the numbers in the table we are about to create, let's expand the column widths to 10 characters for at least columns A through G (seven columns). For SuperCalc3 and Lotus 1-2-3, the keystrokes are (Topic 18-7):

SuperCalc3: / **F C A.G,10** < Enter>
Lotus 1-2-3: / **W G C 10** < Enter>

Place the cursor in cell C2 and enter the table header. Place the cursor in cell F2 and enter the number .02. Choose row 4 for entering the column titles, entering PERIOD into cell A4, F/P into cell B4, and so on. If we decide that these labels would look better right justified, we can invoke the following keystrokes:[20]

SuperCalc3: / **F E A4.G4** < Enter> **TR**
Lotus 1-2-3: / **R L R A4.G4** < Enter>

Now start the table, beginning with the first column. Enter the number 1 in cell A6. Enter the formula A6 + 1 into cell A7. Replicate the formula in A7 into cells A8.A10 by placing the cursor in cell A7 and executing the following keystrokes[21] (Topic 18-16):

SuperCalc3: / **R A7 A8.A10** < Enter>
Lotus 1-2-3: / **C** <Enter > **A8.A10** < Enter>

In either case, note that the formulas are copied using relative (or "adjusted," in SuperCalc3 language) cell addresses. Each cell references the cell immediate above itself. Column A should now contain the numbers 1, 2, 3, 4, 5.

Enter the proper formulas in row 6 of the columns. For example, into cell B6, enter the formula (1 + F2) ^A1; into cell C6 enter the formula 1/B6; and so on. We have to be extremely careful here because the cells will be referencing a single cell (F2) for the value of the interest rate, i. If we replicate the formulas in cells B6.G6 down their respective columns (that is the easy way to complete the sheet), we will want absolute cell references to cell F2 where the interest rate, i, resides. Cell F2 is not separated by the same relative displacement from all cells. Therefore, relative cell addresses to cell F2 are not appropriate.

SuperCalc3: replicate the formula in B6 to cells B7.B10 by taking the cursor to cell B6 and executing the following commands (Topic 18-16):

/ **R B6,B7.B10,A**

Because we added the **A** (for Ask) at the end, the prompt line will ask us which cell references in the formula are to be "adjusted" (considered relative) and which are *not* to be "adjusted" (these will be considered absolute). Answer **N** for reference F2 and **Y** for Reference A6. The col-

---

[20]In Lotus, the " label prefix character could have been used when entering the label. This would have automatically right justified the label in the cell.

[21]In Lotus, a better alternative is provided by the Data Fill commands invoked by the keystrokes

/ D(ata) F(ill) A6.A10 < Enter > 1 < Enter > <Enter > <Enter >

There is no comparable command in SuperCalc3.

umn should fill with numbers nearly matching the table in Appendix III.

Lotus 1-2-3: In order to make F2 an absolute reference in cells B6.G6, it should be entered as F2 in the formulas found there. The formula in cell B6 should read (1 + F2) ^A6. The formula can be copied to cells B7.B10 by taking the cursor to cell B6 and invoking the following keystrokes (Topic 18-16)

/ **C** <Enter> **B7.B10** < Enter>

After copying all the formulas down their respective columns, the table should nearly match the table in Appendix III. However, the appendix table is given to four decimal places, so we should perhaps format cells B6.G10 (Topic 18-9).

SuperCalc3 keystrokes: First we should set up our format selection for general use:

/ **F D**

and change User-defined format #1 to **N**(o) dollar sign and four decimal places (use the cursor control keys and editing keys to make the changes) and exit that screen with the **F2** function key (or Ctrl Z). Then initiate

/ **F E B6.G10 U 1** < Enter>

Lotus 1-2-3 keystrokes:

/ **R F F 4** < Enter > **B6.G10** <Enter>

The resulting table appears as follows (from SuperCalc3):

	A	B	C	D	E	F	G
1							
2			DECIMAL INTEREST RATE =			.02	
3							
4	PERIOD	F/P	P/F	F/A	A/F	P/A	A/P
5							
6	1	1.0200	.9804	1.0000	1.0000	.9804	1.0200
7	2	1.0404	.9612	2.0200	.4950	1.9416	.5150
8	3	1.0612	.9423	3.0604	.3268	2.8839	.3468
9	4	1.0824	.9238	4.1216	.2426	3.8077	.2626
10	5	1.1041	.9057	5.2040	.1922	4.7135	.2122

In case you did not get a table that matched the above, we have shown the same spreadsheet with formulas being displayed instead of the formula results.[22] Check these formulas with what you have. This is for SuperCalc3. Lotus 1-2-3 will differ only slightly because all references to F2 should read F2 to be absolute cell references.

[22]In SuperCalc3, you can display the formulas stored in the cells by toggling the cell display with the keystrokes / G F (toggling means that if you issue these same keystrokes an second time, you will revert back to results being shown in the cells). A somewhat similar feature is available in Lotus. It is accessed with the commands: / **W**(orksheet **G**(lobal) **F**(ormat) **T**(ext). However, regaining the original worksheet is not as convenient as it is in SuperCalc3, if more than one numeric formula is used. A simple solution is to save the sheet, command / **WGFT**, and then reload the saved sheet after viewing the formulas. Formulas in individual cells or a range of cells in Lotus 1-2-3 can be displayed with the /RFT command and reset to values with the /RFR command.

	A	B	C	D	E	F	
1							
2				DECIMAL INTEREST RATE =		0.02	
3							
4	PERIOD	F/P	P/F	F/A	A/F	P/A	A/P
5							
6	1	(1+F2)^A6	1/B6	(B6-1.)/F2	1/D6	D6/B6	1/F6
7	A6+1	(1+F2)^A7	1/B7	(B7-1.)/F2	1/D7	D7/B7	1/F7
8	A7+1	(1+F2)^A8	1/B8	(B8-1.)/F2	1/D8	D8/B8	1/F8
9	A8+1	(1+F2)^A9	1/B9	(B9-1.)/F2	1/D9	D9/B9	1/F9
10	A9+1	(1+F2)^A10	1/B10	(B10-1.)/F2	1/D10	D10/B10	1/F10

Before we leave this problem, let us explore the graphing of some of the data on the spreadsheet we have just made. We will choose the data in the last column (G), labeled A/P, to plot as a function of the first column (A), labeled PERIOD.

In SuperCalc3, we set up the plot by defining the data on the sheet to plot and by choosing the type of plot. The command keystrokes are

/ **V**(iew) **D**(ata) **A6.A10** < Enter> **D**(ata) **space bar**

G6.G10 < Enter> **G**(graph-Type) **X**(-Y) **F10**

The keystrokes **/VD** allow you to enter sets of data, the default of which is set A (this shows right above the prompt line). Set A is the independent variable if an XY plot is requested. The keystrokes above have just defined the data in cells **A6.A10** to be the independent variable set A. In entering the second set of data (the second D in the keystroke sequence), the **[space bar]** toggles the line above the prompt line to define data set B, which we assign the data in cells **G6.G10.** We then define the **G**raph-Type to be an **X**-Y plot and request it to be displayed with the **F10** function key. The graph should magically appear. Remember what you formerly did to plot a data set?

We can define labels for this plot if we place the labels we wish to use in cells on the spreadsheet. The **/VT, /VV, /VP,** and **/VH** commands are then used to make the labels in these cells usable on the graph. The **/VO** command also allows the choice of many options. The /**GGO** command also allows more options to choose from, in order to customize the screen to your liking.

In Lotus 1-2-3, graphing is done just as easily, perhaps even more so. The keystrokes are

/ **G**(raph **X A6.A10** < Enter > **A G6.G10** < Enter > **V**(iew)

and, like magic, the plot should appear on your screen if you have a graphics monitor and capability. The **X** selects for definition the independent variable from your sheet (we have assigned X to be the data in cells A6.A10). The A B C D E F options under the main menu /G command allow you to define six sets of dependent variables (in our use we have defined data set **A** to be the data in cells **G6.G10**). The **V**(iew) command produces the plot on your screen. You can add labels under the /**G**(raph **O**(ptions) command. When you want to quit the **G**(raph commands, just select the **Q**(uit) option.

The next example problem is intended to highlight the use of the IF-THEN-ELSE function that is available on most good spreadsheets.

Example 18-7

Set up a spreadsheet to solve for the roots of a quadratic equation using the binomial theorem.

Solution:

Let the quadratic equation be designated by

$$ax^2 + bx + c = 0$$

The binomial theorem, given by

$$x = \frac{-b \pm (b^2 - 4ac)^{1/2}}{2a}$$

can be used to solve for the roots (the values of x that satisfy the quadratic equation). The number and type (real or imaginary) of roots, of course, depend on the value of the discriminant, $(b^2 - 4ac)$.

Given below is a spreadsheet, with formulas showing in the cells, that solves for the roots. (This is a SuperCalc3 sheet that has been toggled to display the formula.) You should enter the text, numbers, and formulas into the cells in the manner we have covered previously. Only after switching the

/ G F

toggle will your sheet look like the following one.

```
  |        A        ||    B   ||
 1| This sheet solves the quadratic equation:
 2|   aX^2+bX+c=0  Text
 3|
 4| Enter here    a            1
 5| Enter here    b           12
 6| Enter here    c            3
 7|
 8| The Discriminant    B5^2-4.*B4*B6
 9|
10|      See below for formula
11|     stored here in A10.B11
12|
13|
14|
```

Stored in cell A10 is the formula

IF(B8 < 0 , ("REAL_PART") , ("X1 = "))

Stored in cell A11 is the formula

IF(B4 < 1.E−10 , ("1_ROOT") ,)

IF(B8 < 0 , ("IMAGINARY"),)

IF(B8 > 0 , ("X2 = "), ("1_ROOT"))

Stored in cell B10 is the formula

IF(B4 < 1.E − 10 , B6/B5 ,)

IF(B8 < 0 ,-B5/(2*B4) ,)

IF(B8 > 0 , (−B5+SQRT(B8))/(2*B4) , −B5/(2*B4))

Stored in cell B11 is the formula

```
IF( B4 < 1.E−10 , (" ") ,  )
        IF( B8 < 0 , SQRT(−B8)/(2*B4) ,  )
      IF(B8 > 0 , B5 ^2−4.*B4*B6(−B5−SQRT(B8))/(2*B4) , (" "))
```

The IF-THEN-ELSE statements in cells A10.B11 get a little complicated, but basically they all have the form

> IF (*condition is satisfied, give the function IF this value, else give the function IF this value)*

where *condition is satisfied* is any mathematical expression that includes a comparison (=,<,>,<=,>=,<>) and *give the function IF this value* and *else give the function IF this value* are formulas or numbers. In SuperCalc3, these can also be text strings that are written to the cell if the strings are enclosed in double quotes and parentheses ("like this")[23] and if they are short (≤ nine characters). In Lotus 1-2-3 (version 2.0), text strings must be enclosed in double quotes; there is no reasonable limit on the length of the string.[24]

In both SuperCalc3 and Lotus, the *give the function IF this value* and *else give the function IF this value* can be another IF-THEN-ELSE. We have spread apart the IFs above for clarity to show the use of such *nested* structures. In the actual cells the statements are all on one line of code and do not need embedded spaces.

The IF-THEN-ELSE statements in column A cause messages to be displayed that tell the user the number of roots and whether the roots are real or imaginary. The statements in column B are used to display the numerical value of the root(s). The pseudocode for the statements in columns B11 and B12 are given below to aid you in understanding what they do. We will let you write the pseudocode for cells A11 and A12.

Cell B10:

```
If B4 <  a very small number ≈ 0
  display the single root B6/B5
Else
  If B8 < 0
    Display the real part of complex root
Else
    If B8 > 0
      Display one of two real roots
    Else
      Display the only real root
End of formula
```

Cell B11:

```
If B4 < is a very small number ≈ 0
  there is no second root - print nothing
Else
  If B8 < 0
```

---

[23]The ()s may be omitted here.

[24]Lotus (version 1A) users would have to replace the text strings that appear as ("text") with a numeric code of some sort and, perhaps, a legend for this code in another cell (e.g., cell A12).

```
        Display the imaginary part of complex
          root
      Else
        If B8 > 0
          Display second of two real roots
        Else
          There is no second root - print
            nothing
    End of formula
```

The actual sheet in operation looks like the following screen (from SuperCalc3):

```
   |       A         ||     B    ||
 1 This sheet solves the quadratic equation:
 2   aX^2+bX+c=0
 3
 4 Enter here    a                 1
 5 Enter here    b                12
 6 Enter here    c                 3
 7
 8 The Discriminant              132
 9
10                  X1= -.2554374
11                  X2= -11.74456
12
```

Whenever the value of a, b, or c is changed in cell B4, B5, or B6, respectively, the new roots appear almost immediately in cells B10 and B11.

The following example demonstrates another useful feature of spreadsheets, the table-lookup capability. This feature has many possibilities for use with tabular data that are not easily expressed in equation form.

Example 18-8

Assume that we want to easily determine the minimum-size copper wire that would be needed to carry a known direct current over a given distance with a maximum possible voltage drop. We want to pick a standard-size wire for economical reasons, but standard sizes increase in discrete steps. This means that the resistance of these standard wires will also vary in discrete steps. Expressing the relationship between standard size and resistance in a mathematical relationship is not satisfactory since permissible resistances are not continuous. Therefore, we need to incorporate a table lookup into our solution.

Solution:

We use Ohm's law to calculate the maximum resistance that will be required. This relationship is

$$R = \frac{V}{I}$$

where R is the resistance (in ohms), V is the voltage (in volts), and I is the current (in amperes). The resistance per foot of length for wire can be found in standard handbooks, so we will calculate the resistance per length that we will need to match our conditions. That is, we will calculate

$$\frac{R}{L} = \frac{V}{IL}$$

where L is the given length in appropriate units. We then look up this value in a table and let it determine the appropriate wire size.

A possible spreadsheet solution to this problem is shown below. The spreadsheet is Supercalc3 and the / **G F** toggle has been switched to show the formulas in the cells rather than the results of the formulas.

```
  |            A              ||    B     || C || D     ||    E    |
 1 This sheet uses Ohm's law to determine                Ohms/Mft WireGauge
 2   the size of wire to use in order to                 .09827    0
 3   not exceed a given voltage drop over                .1239     1
 4   a specified distance, at a given current.           .1563     2
 5 Inputs:                                               .197      3
 6   Current (amps)                      .5              .2485     4
 7   Max Voltage drop                    40              .3133     5
 8   Distance (specify units below)      20              .3951     6
 9   Units,enter 1-ft,2-mi,3-m,4-km      2               .4982     7
10 Outputs:                                              .6282     8
11         Resistance- Ohms/1000ft see below for formula .7921     9
12                                                       .9989     10
13              Wire gauge needed LOOKUP(B11,D2.D18)     1.588     12
14                                                       2.525     14
15                                                       4.016     16
16                                                       6.385     18
17                                                       10.15     20
18                                                       16.14     ERR
19
```

We are using column A to display labels that prompt the user as to what is needed and where the inputs go. Column B is the working column. Notice that columns D and E contain a table, with column D containing the resistance (in ohms) per unit length (per 1000 ft) for standard copper wire sizes, and column E containing the corresponding wire gauge sizes.

There are only two formulas stored in cells on this sheet. One can be found in cell B11 and one in B13. The formula in cell B11 calculates the resistance per foot of wire from the input information on current, length (including units), and maximum possible voltage drop, all of which must be supplied by the user. The formula is given by

IF(OR(B9 < 1,B9 > 4) = 1,ERROR,1000.*B7/B6/(B8*IF(B9 = 1,1,
  IF(B9 = 2,5280,IF(B9 = 3,1/.3048,1000/.3048)))))

Notice the use of the nested IF functions (@IF in Lotus 1-2-3) within the formula, just as if they were another constant or variable (see Example 18-7) if you have not done so already).

Pseudocode for this formula is given by the following:

If B9 < 1 or B9 > 4 then
  display ERROR for resistance and wire size
    (inappropriate units have been specified)
Else

Calculate R/L from V/I/(L*const) where the constant is determined from the input units:

```
If B9 = 1
  the const = 1
Else
  If B9=2
      the const = 5280
  Else
    If B9 = 3
      the const = 1/.3048
    Else
      the const = 1000/.3048
```

End of formula

If you are using Lotus 1-2-3, you will need to replace OR(B9 < 1,B9 > 4) = 1 with the Lotus equivalent B9 < 1#OR#B9 > 4, which is more conventional in its use of the logical OR function. You will also have to add the @ to the IF function and replace ERROR with @ERR (see Topic 18-3). The built-in help facility in Lotus 1-2-3 is very good (it is reached with the F1 function key on IBM PCs and compatibles). Do not hesitate to use it.

In SuperCalc3, the OR function returns a 1 if either expression in the argument list is met; otherwise, it returns a 0. The other logical functions in SuperCalc3 (see Topic 18-3) behave in a similar fashion. Use the program's built-in help screens to verify the details on each one.

The other formula, located in cell B13, is much simpler looking than the nested IFs. In SuperCalc3, its syntax is LOOKUP *(cell reference, table range),* where *cell reference* is the cell address of the independent variable to be used in entering the table (in our case, the value from cell B11 for the maximum resistance per unit length that can be tolerated). The *table range* second parameter of LOOKUP is the range of cells which defines where the table's independent variables are located. If this is a column range (e.g., D1.D12), the dependent variables are assumed to be located in the next column over (here, E1.E12). If *table range* is a row reference (e.g., A20.L20), the dependent variables are assumed to be in the next row down (correspondingly, A21.L21).

LOOKUP searches for the last value in the range of numbers that is less than or equal to the independent variable given (the *cell reference*) and returns the adjacent value to the right of the search column or below the search row. This is exactly what we want in this example problem, since we need the wire with the next smallest resistance.

The Lotus 1-2-3 table-lookup syntax allows a more sophisticated table structure than SuperCalc3 permits. in Lotus 1-2-3, there are two table-lookup functions: @VLOOKUP and @HLOOKUP. They both have arguments of *(cell reference, range, offset).* As implied in the names, @VLOOKUP is used for vertical tables where the dependent and independent variables are both in columns. For @HLOOKUP, both variables must be horizontal in rows.

The argument *cell reference* gives the location of the unique independent variable with which you want to enter the table in order to find the corresponding dependent variable (for our example it would be cell B11). The argument *range* encompasses at least the independent and dependent variables that constitute the table (our example would need a *range* of D2.E18). The *offset* is the number of columns (or rows) to move over (or down) from the independent variable column (or row) to find the dependent variable (in our problem it would be one).

Lotus 1-2-3 will permit text strings to be used for either the independent variables or the dependent variables. There is no need to enclose the text strings in quotes or parentheses. SuperCalc3 will also permit the use of textual values for either variable as long as the textual values are enclosed in both () and "", like ("this") and the textual strings contain no more than nine characters [excluding the () and ""].

The sheet below shows the working version with values showing, not formulas. Any time one of the four input variables is changed (cells B6 through B9), the resistance value (cell B11) and wire size output (cell B13) are immediately updated.

	A	B	C	D	E
1	This sheet uses Ohm's law to determine			Ohms/Mft	WireGauge
2	the size of wire to use in order to			.09827	0
3	not exceed a given voltage drop over			.1239	1
4	a specified distance, at a given current.			.1563	2
5	Inputs:			.197	3
6	Current (amps)	.5		.2485	4
7	Max Voltage drop	40		.3133	5
8	Distance (specify units below)	20		.3951	6
9	Units,enter 1-ft,2-mi,3-m,4-km	2		.4982	7
10	Outputs:			.6282	8
11	Resistance- Ohms/1000ft	.757575757576		.7921	9
12				.9989	10
13	Wire gauge needed	8		1.588	12
14				2.525	14
15				4.016	16
16				6.385	18
17				10.15	20
18				16.14	ERROR
19					

You should experiment with the table limits. That is, input values for current, voltage, and distance that yield values for the resistance per length to exceed the table limits. Also, note what happens when numbers less than 1 or greater than 4 are entered for the units key.

The table columns do not have to be located close to the cells that use the LOOKUP command. In our example we could have moved them to columns that were not in the window of the screen. There are two other ways of hiding a table in either SuperCalc3 or Lotus 1-2-3 (version 2.0). We could specify the width of the table columns (here, D and E) to be 0 (zero) in SuperCalc3 or the column-width to be hidden in Lotus 1-2-3. We could format them with the Hide option in either program.

The last spreadsheet problem involves the solution of an iterative problem. Such problems are often encountered in engineering and it is important to understand how they can be solved on a spreadsheet. The key is in initializing the problem to get the iteration going and then determining when it has converged to an answer sufficiently close to the correct one.

Example 18-9

Set up a spreadsheet to solve

$$f = \frac{\alpha - \sin \alpha}{2\pi}$$

for α, given a value of f. This equation represents the fraction, f, of a circle's total area that is contained in a segment whose central angle is α. It is extremely important to remember that the α's that appear in this equation must be in radians and not degrees. Note that when $\alpha = 0, f = 0$ and when $\alpha = 2\pi, f = 1$.

As demonstrated in the flowchart section of Chapter 16, an appropriate iteration scheme consists of making an initial guess of

$$\alpha_1 = (12\pi f)^{1/3}$$

and then getting new estimates of α, based on previous estimates of α. These new estimates can be based on the following equation derived using Newton's method:

$$\alpha_{new} = \alpha_{old} - \frac{\alpha_{old} - \sin \alpha_{old} - 2\pi f}{1 - \cos \alpha_{old}}$$

In terms of the difference between the desired f and the calculated f, the latter being based on the estimated α, this can be written as

$$\alpha_{new} = \alpha_{old} - \frac{2\pi(f_{calculated} - f_{desired})}{1 - \cos \alpha}$$

where $f_{calculated} = (\alpha_{old} - \sin \alpha_{old})/2\pi$.

The trick in doing iteration problems on the spreadsheet is to work in the initial guesses to start the calculations. Once an iteration has started, spreadsheets with iterative capability can keep it going, but there must be a way to initiate the process. This initiation can be done with the IF-THEN-ELSE statement. We show below a SuperCalc3 iteration sheet for this problem. Formulas are displayed in the cells where they are stored.

```
CIRC|             A            ||     B    ||
 1  Newton Iteration to Solve  ←Text
 2  f = (alf - sin (alf))/(2pi)     ←Text
 3    for alf, given f       ←Text
 4                        de/12/84←Text
 5
 6                Desired f .5
 7  Iter 0-reset,1-iterate 0
 8      Converg Tol on alf .00001
 9
10                Begin alf ((12*PI*B6)^(1/3))*180/PI
11
12             # Iterations IF(B7>0,B12+1,0)
13           alf, last iter IF(B7>0,B16,B10)
14      f(calc)-f(desired) ((B13*PI/180)-SIN(B13*PI/180))/(2*PI)-B6
15                          (1-COS(B13*PI/180))/(2*PI)
16          alf, this iter (B13*PI/180-B14/B15)*180/PI
17
```

Column A of this SuperCalc3 sheet is being used to prompt the user and contains various labels or text. Column B is where the work gets done.

There are three user-supplied numbers in column B. (On a Lotus 1-2-3 worksheet, the convergence tolerance that is requested in cell B8 cannot be used effectively. Thus, this cell would be left blank in Lotus 1-2-3.) Notice that the desired value of f is entered into cell B6. This value of f is then used in cell B10 to calculate the first value of α, and is also referenced in the iteration scheme.

Cell B7 contains what we will call the *iteration flag,* which the user gives the value of either 0 (actually, any number ≤ 0 will work) or 1 (any number ≥ 1 will work). If this flag is 0 (or less), the iterative problem is initialized and made ready for the iterative solution. For example, the IF function (@IF in Lotus 1-2-3) in cell B12 checks the iteration flag (B7) and if the flag is 0 (or less), sets the value of cell B12 to 0. The purpose of this is to initialize the value of cell B12 which is used for an iteration counter. Once the flag is set to 1 (or greater), cell B12 indexes its own count by 1. B12 continues to do this indexing each time the sheet is recalculated or until the flag in B7 is reset to 0 (or less).

Cell B13 also contains an IF function (@IF in Lotus 1-2-3) that examines the value of the iteration flag (B7). If the flag is 0 (or less), the value of cell B13 is set to the first estimate of α (from cell B10). After the flag in B7 gets set to 1 (or larger), cell B13 takes its value from cell B16, which, as we shall see, holds the result of the last updated α.

Of course, this reference of cell B16 by cell B13 is a circular one that must be resolvable by the spreadsheet you are using.[25] SuperCalc3 informs you that there is a circular reference on the sheet by displaying the CIRC in the upper left-hand corner (on Lotus 1-2-3 the warning appears in the lower right-hand corner). SuperCalc3 (Release 2 or later) will allow you to resolve circular references, as will Lotus 1-2-3. Many spreadsheets will not.

The formula in cell B14 calculates the numerator in the fourth equation at the beginning of this problem. It represents how close the f, based on the last value of α, is to the desired f (cell B6).[26] This difference is useful to you, because it gives you immediate knowledge about how close the true f is being approached.

In cell B15 is stored the denominator in the equation for α_{new}. This value is not useful to you and may be hidden from view, on SuperCalc3, with the command keystrokes (Topic 18-9)

/ F E B15,H

The comparable command in Lotus 1-2-3 is **/R F H B15** <Enter>.

Cell B16 displays the updated value of α (α_{new} in the equations at the beginning of this example). It is based on the last value that was calculated during the last iteration through the sheet (which is stored in cell B13).

[25]Actually, B12 contains the first circular reference on the sheet, since a cell referencing itself is considered a circular reference.

[26]Remember, arguments of trigonometric functions must be in radians, not degrees: thus the appearance of the term PI/180 (@PI/180 in Lotus 1-2-3) in the argument of the sine function. Also, note that if Lotus 1-2-3 were being used, the character @ would have to precede the functions SIN and COS that appear on this sheet (see Topic 18-3).

Now, in operation, you initialize the sheet by setting the iteration flag to 0 (or less) (you may have to push the Recalc key, which is ! in SuperCalc3 or the F9 function key in Lotus 1-2-3, depending on whether the recalculation mode is set to manual or automatic). This places the beginning value of α (from cell B10) in cell B13 and calculates a better estimate of α for display in cell B16.

When the iteration flag is set to 1 (or larger), the sheet has a good initial value and is free to iterate. How the iterations proceed, however, is determined by you, the user. Both SuperCalc3 and Lotus 1-2-3 programs can be set to either automatic or manual mode of recalculation with a fixed number of iterations—experiment with this by trying different numbers of iteration, starting with a small number and increasing it if needed. The command keystrokes for setting this are (Topic 18-15):

SuperCalc3: / **G** [**M** or **A**] / **G I** [supply number] < Enter >
Lotus 1-2-3: / **W G R** [**M** or **A**] / **W G R I** [no.] < Enter >

To operate the sheet, enter the desired f in cell B6 and enter a 0 in cell A7. If you are on SuperCalc3 and want the sheet to stop iterating automatically when your specified convergence tolerance is reached, enter the convergence tolerance in cell B8 (we will say more about this later). The sheet should reset itself (you will need to push the Recalc key if you are in the manual mode of recalculation). If the sheet does not reset itself, you probably have a bug in the program which you will have to fix. Carefully examine all formulas.

If the sheet did reset itself, enter a 1 in B7. The numbers in B12.B16 should change[27] (you will need to push the Recalc key if you are in the manual mode of recalculation). If the numbers do not change, you have an error to debug. If they did change, push the Recalc key to watch them converge even closer to the true answer (if possible). The sheet will undergo another set of iterations every time you push the Recalc key, even when you are in the automatic mode.

How do you know you are getting the correct answer? Simply do the calculations for a problem for which you know the answer. An α of 180° should produce an f of 0.5, for example. Does your α converge to 180 when $f = 0.5$?

SuperCalc3 will automatically iterate until the convergence criteria you specify has been met it you carry out the following command keystrokes:

/G I S D B8 R B16 < **Enter**> **F2**

With this "solve" mode on, you should be able to reset the sheet (B7 = 0), initiate the iteration (B7 = 1), and watch as the iteration occurs. The sheet will automatically stop iterating when the value in cell B8 changes by less than the tolerance set in cell B8. We hope you are not in an infinite loop—if you are, use the F2 function key to abort.

Correct sheets for $f = 0.5$ are given below for your reference. The first sheet shows after initialization, while the second sheet shows the results after iteration. Note that the answer converges in three itera-

---

[27]SuperCalc3 will display the numbers as they are being recalculated. Lotus 1-2-3 will only display the numbers at the end of the fixed number of iterations you have told it to complete. It displays a WAIT sign in the upper right-hand corner of the sheet while calculating.

tions to the correct one. Set the number of iterations to 1, and watch this converge iteration by iteration.

The sheet after initializing and before iterating:

```
CIRC|          A          ||    B    ||
 1  Newton Iteration to Solve
 2  f = (alf - sin (alf))/(2pi)
 3    for alf, given f
 4                       de/12/84
 5
 6              Desired f          .5
 7  Iter 0-reset,1-iterate          0
 8      Converg Tol on alf     .00001
 9
10              Begin alf 152.4835544
11
12           # Iterations           0
13         alf, last iter 152.4835544
14     f(calc)-f(desired) -.149964663
15                        .3003060022
16         alf, this iter 181.0955109
17
```

The sheet after converging:

```
CIRC|          A          ||    B    ||
 1  Newton Iteration to Solve
 2  f = (alf - sin (alf))/(2pi)
 3    for alf, given f
 4                       de/12/84
 5
 6              Desired f          .5
 7  Iter 0-reset,1-iterate          0
 8      Converg Tol on alf     .00001
 9
10              Begin alf 152.4835544
11
12           # Iterations           3
13         alf, last iter         180
14     f(calc)-f(desired) 1.11022e-16
15                        .3183098862
16         alf, this iter         180
17
```

Our purpose in presenting this material on spreadsheets has been to demonstrate their versatility and to give you some working knowledge of their capability. We have stressed here only small and relatively simple problems in order not to cloud the learning process. However, once you learn the fundamentals, you will be able to solve much more complicated problems.

There are many indications that spreadsheets will soon be discovered by the engineering community. For much of the engineer's work, they offer a welcome alternative to programming using one of the high-level languages of the type we discussed in Chapter 17.

PROBLEMS

18-1. Duplicate Table 18-1 by programming a spreadsheet.

18-2. Duplicate Figure [18-1], using the spreadsheet that you used in Problem 18-2 if it has graphics capability. If your spreadsheet does not have graphics capability, do you have access to any plotting programs that can load and plot data stored in a spreadsheet file? If so, plot a graph that is as similar to Figure [18-1] as possible.

18-3. Enter your social security number on a spreadsheet in cell B2 (R2C2 for Multiplan).

18-4. Enter the following numbers into a columnar range of cells.

3.0

3

1,238,876

0.0000586

10.568952

Copy this column of cells into the next five columns to the right. Then format each column for one of the following formats.

a. Integer.
b. Decimal with zero places after the decimal.
c. Decimal with three places after the decimal.
d. With embedded commas.
e. As dollar values to the penny and with dollar signs.
f. In scientific notation.

18-5. Enter the following formulas into the cells of a spreadsheet.

a. $y = mx + b$.
b. $y = 3(x + 2)^2$
c. $y = 3x^2 + 2z$
d. $x_1 = x_2 + vt + at^2/2$
e. $E = mc^2$
f. $F = ma$
g. $E = IR$
h. $y = \sin x$
i. $z = e^{y-2}$
i. $y\text{-}\tan x/\cos x$

18-6. Explore the behavior of your spreadsheet.
a. What happens to the value of $\tan x$ when x approaches 90°?
b. What happens to the value of $\arctan x$ when x becomes large and positive? when x becomes large but negative?
c. Does the square root function give the correct answers for all sizes of numbers, both large and small?

18-7. Enter a string of numbers in a column on your spreadsheet but leave one cell in the string blank. In the cell immediately below the last number in the string, enter the built-in function that averages this string of numbers (include the blank cell in the range to be averaged). Has the formula included the blank cell into the average? That is, was the blank cell averaged in as a zero, or was it ignored?

Enter into the cell below the one holding the average, the built-in function for counting the number of entries in a range of cells. With this function, count the number of entries in the string of numbers you have entered, including the blank cell. Was the blank cell counted or was it ignored?

18-8. Enter the formulas of Problem 17-11 into a spreadsheet and evaluate them for a few choices of independent variables.

18-9. Write expressions for the *condition* of an IF-THEN-ELSE function that will check for the items listed in Problem 17-18.

18-10. Enter some numbers in cells B2 and B3 of your spreadsheet. In cell B4 enter a formula that will display twice the value of the number in B2 if B2 is greater than B3. Otherwise, the value displayed in B4 should be 0.5 times the value of B3.

18-11. Solve Problem 17-22 using iteration on a spreadsheet.

18-12. Use the data in Table 9-1 to construct a table on a spreadsheet. Program the sheet so that when you enter a wire gauge in the range of the first column into some cell on the sheet, an appropriate value of cross-sectional area is returned in another cell.

18-13. (For spreadsheets that can load comma-delimited ASCII files with graphing capability.) Using an editor or word processor, create a comma-delimited file that has the following two lines of data:

14.5, 16., 17.5, 19., 20.5, 22., 23.5, 25.
1, 2, 3, 4, 5, 6, 7, 8

Store this file as an ASCII file (editors typically do this; most word processors can do this if specifically instructed), and load this into your spreadsheet. Assuming that the first row contains the dependent variable and the second row contains the independent variable, plot these data by pairing up the points using the first point in the first line with the first point in the second line, the second point in the first line with the second point in the second line, and so on.

Chapter 19

Analysis and Optimization in Design

This chapter continues to explore engineering problem solving by drawing on material developed in earlier chapters. In contrast to Chapter 11, which stressed problems that could be solved by manual calculations or by hand calculators, the current chapter stresses more involved or difficult problems that can be solved easily on the microcomputer.

We present the material in the form of example problems which are solved in a way that produces general applications programs. Some of these programs may be useful to you as you start into your engineering courses. For each problem we give the full solution, including the programs (either spreadsheet formulas or source code for programming languages), along with screen displays and data entry directions.

When we draw on material presented earlier in the text, we strongly recommend that you review the appropriate sections as needed. It is through this type of renewal and use that you become intimately familiar with the subject matter.

PROJECTILE MOTION

Projectile motion is an interesting physics problem that can lead to a deeper understanding of the equations of motion. Such phenomena as sports (throwing or hitting of objects such as balls, javelins, shot, and discus; "throwing of yourself, such as in the jumping events of long jump, high jump, and pole vault) and warfare (firing of artillery shells, dropping of bombs) depend on the physics of projectile motion.

If the effect of air resistance is neglected, the equations governing projectile motion become amazingly simple. In free fall (which adequately describes projectile motion), objects are under the influence of just gravitational attraction, or their weight. A free-body diagram of an object in such a situation is shown in Figure [19-1]. Newton's second law [10-1] applied to the object yields

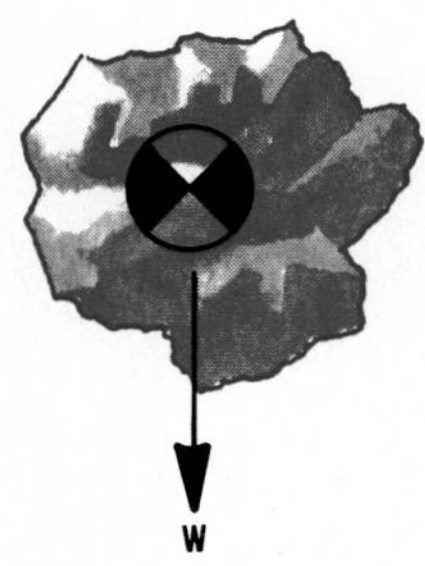

[19-1] A falling particle.

$$W = m_0 a \tag{19-1}$$

where a is the acceleration of the object in the direction of the force W and m_0 is its mass. From the law of universal gravitation [10-2], the force W is related to the masses of the two gravitating objects via

$$W = k_g \frac{m_0 m_m}{r^2} \tag{19-2}$$

Combining these two equations into one yields

$$a = k_g \frac{m_m}{r^2} \tag{19-3}$$

which tells us that the acceleration of any object gravitating with another is not dependent on the mass of the object, but depends only on the mass of the other gravitating body and on the distance of separation. That is, all masses are attracted to any common body with the same acceleration, assuming that all objects are removed from the common body by the same distance of separation.

For objects near the surface of the earth, this free-fall acceleration due to the attraction of the earth is approximately 32.2 ft/s^2 in English units or 9.8 m/s^2 in SI units.

Knowing the constant acceleration of a falling object, we can determine where the object is and how fast it is moving at any time if we know where (what place in space) the object began its fall and what initial velocity is had at that time. The equations relating these variables are:

1. Velocities:
   a. Horizontal (x direction assumed):

$$V_x = V_{x0} \tag{19-4}$$

   b. Vertical (y direction assumed, positive direction taken away from the gravitating body):

$$V_y = -g\,(t - t_0) + V_{y0} \tag{19-5}$$

2. Position:
   a. Horizontal:

$$x = x_0 + V_{x0}(t - t_0) \tag{19-6}$$

   b. Vertical:

$$y = y_0 + V_{y0}(t - t_0) - \frac{g(t - t_0)^2}{2} \tag{19-7}$$

In the equations above, t is the current time when the velocity has horizontal and vertical components of V_x and V_y, respectively, and the object is located at point x, y, while t_0 is the time when the velocity has horizontal and vertical components of V_{x0} and V_{y0}, respectively, and the object is located at point x_0, y_0. V_{x0}, V_{y0}, x_0 and y_0 are usually known, while V_x, V_y, x, and y are usually being sought. The constant g has the numerical value of 32.2 ft/s^2 or 9.8 m/s^2.

There are many interesting problems that can be solved using these equations, but we desire here to construct plots of a free-falling object's trajectory. Since most spreadsheets can do plotting of data, we will set up this problem on a spreadsheet. Figure [19-2] shows an example of such a sheet.

Note in Figure [19-2] that cells C4.C12 are reserved for data entry so that the user can input his or her own information. In fact, we prefer to protect all the cells on a spreadsheet except the data entry cells. You cannot overwrite the contents of a protected cell unless you purposely remove the protection before doing so. This prevents inadvertent and/or careless loss of important formulas.

```
 |          A          || B  ||  C  || D || E  || F|| G |
1  Projectile Motion - Plots        Data
2   Hor. vs Vert. Location          Entry        Hor. DistTime  Ver.Dist
3                                   !!!!!           20.00   .00   300.00
4  Units 1-Metric, 2-English            2           24.21   .17   306.84
5  Initial velocity                                 28.42   .34   312.76
6   Magnitude of Velocity   ft/sec     50           32.63   .51   317.77
7    Angle up from Horiz.  Degrees     60           36.84   .67   321.86
8  Initial Displacement                             41.05   .84   325.05
9     Horiz. Displacement      ft      20           45.26  1.01   327.32
10     Vert. Displacement      ft     300           49.47  1.18   328.67
11 Final Displacement                               53.68  1.35   329.11
12             Horizontal      ft     100           57.89  1.52   328.64
13                                                  62.11  1.68   327.26
14 Accel. of Gravity      ft/sec2    32.2           66.32  1.85   324.96
15   Possible?                Yes                   70.53  2.02   321.75
16 Time of Flight             sec    3.20           74.74  2.19   317.63
17                                                  78.95  2.36   312.59
18                                                  83.16  2.53   306.64
19                                                  87.37  2.69   299.77
20                                                  91.58  2.86   292.00
  C6          Form=50
Width:  9  Memory:217 Last Col/Row:G22     ? for HELP
  1>
```

[19-2] Projectile Motion (SuperCalc3).

The sheet shown in Figure [19-2] assumes that there will be some initial component of velocity in the horizontal direction. It is set up to give error readings if this is not the case. The actual formulas stored in the sheet are given in Figure [19-3].

Notice that most of the cells in any one column contain formulas that are quite similar to one another. This means that the contents can be copied from cell to cell, thus eliminating much meticulous formula entry work.[1]

Figure [19-4] is a plot of the trajectory of the projectile being described on the spreadsheet of Figure [19-2]. The object begins its motion at the point 20 horizontal feet from the origin and 300 ft above it. It ends its flight 3.2 seconds later, 100 horizontal feet from the origin and about 275 ft above it. The effect of the gravitational attraction is to reverse its initial vertical component of velocity of 50 cos (60) ft/s into a final downward velocity (not computed). The object reaches a maximum height of nearly 330 ft above the origin.

Figure [19-5] displays another set of conditions. The problem solved here is to construct the trajectory of a projectile having an initial velocity of 105 ft/sec directed 30° up from the horizontal as the object travels 300 ft in the horizontal direction (initial position at the origin assumed).

The trajectory is shown in Figure [19-6]. The object requires 3.3 s to travel the 100 ft. It reaches a maximum height of about 43 ft and ends up a little below its initial vertical location.

```
Column B
6   IF(C4=1,("m/sec"),("ft/sec"))
7                                      Degrees
8
9   IF(C4=1,("m"),("ft"))
10  IF(C4=1,("m"),("ft"))
11
12  IF(C4=1,("m"),("ft"))
13
14  IF(C4=1,("m/sec2"),("ft/sec2"))
15  IF(AND(C12-C9>0,C6*COS(C7*PI/180)>0)=1,
      ("Yes"),IF(AND(C12-C9<0;C6*COS(C7*PI/180)<0)=1,("Yes"),("No")))
16                                        sec

Column C
1                                                Data
2                                                Entry
3                                                !!!!!
4   2
5
6   50
7   60
8
9   20
10  300
11
12  100
13
15  IF(B15=("No"),("   Check"),(" "))
16  IF(B15=("Yes"),(C12-C9)/(C6*COS(C7*PI/180)),ERROR)

Column D
6   IF(B15=("No"),("---Check"),(" "))
7   IF(B15=("No"),("---Check"),(" "))
8   IF(B15=("No"),("   |   "),(" "))
9   IF(B15=("No"),("---Check"),(" "))
10  IF(B15=("No"),("   |   "),(" "))
11  IF(B15=("No"),("   |   "),(" "))
12  IF(B15=("No"),("---Check"),(" "))
13  IF(B15=("No"),("   |   "),(" "))
14  IF(B15=("No"),("   |   "),(" "))
15  IF(B15=("No"),("Entries "),(" "))

Column E
2   Hor. Dist
3   C9
4   E3+(C12-C9)/19
5   E4+(C12-C9)/19
6   E5+(C12-C9)/19
7   E6+(C12-C9)/19
8   E7+(C12-C9)/19
9   E8+(C12-C9)/19
10  E9+(C12-C9)/19
11  E10+(C12-C9)/19
12  E11+(C12-C9)/19
13  E12+(C12-C9)/19
14  E13+(C12-C9)/19
15  E14+(C12-C9)/19
16  E15+(C12-C9)/19
17  E16+(C12-C9)/19
18  E17+(C12-C9)/19
19  E18+(C12-C9)/19
20  E19+(C12-C9)/19
21  E20+(C12-C9)/19
22  E21+(C12-C9)/19
```

[19-3] Cell formulas (SuperCalc3).

[1]When copying formulas remember that one of the most common mistakes is to disregard the difference between absolute and relative cell addresses.

Figure [19-7] demonstrates the error-checking capability that we have built into this sheet. Here a positive horizontal velocity was programmed [105 cos (30) ft/s], but we sought to have the projectile undergo a negative total displacement (start at +200 ft from the origin and end +100 ft from the origin). The diagnostics tell us that we had better check our inputs.

THE IDEAL SOLAR CELL

In the work of Chapter 11 and in the example of projectile motion above, the object was to model the system with physics and mathematics. We will now look at the more complicated phenomenon exhibited by a photovoltaic cell and explore a model of its behavior. More commonly called "solar cells," photovoltaic cells convert light (usually sunlight) to direct-current electricity.

We will not attempt to explain the physics underlying their behavior, but rather, we will explore the electrical output behavior of solar cells. We might be interested, perhaps, in analyzing the performance of electrical circuits that contain solar cells.

A common simple model which is sometimes used for modeling electrical behavior of the cells is known as the "single-diode" model, which we will also call the "ideal" solar cell model. It is shown schematically in Figure [19-8]. Under illumination, the solar cell acts somewhat like a battery in that it will produce a voltage across its output leads and will support a current if the leads are connected

```
  |          A          ||    B    ||  C  ||  D  ||    E    || F  ||   G   |
 1  Projectile Motion - Plots           Data
 2   Hor. vs Vert. Location             Entry          Hor. DistTime  Ver.Dist
 3                                      !!!!!               .00   .00      .00
 4  Units 1-Metric, 2-English               2              15.79   .17     8.63
 5  Initial velocity                                       31.58   .35    16.29
 6   Magnitude of Velocity   ft/sec      105               47.37   .52    22.98
 7    Angle up from Horiz.  Degrees       30               63.16   .69    28.70
 8  Initial Displacement                                   78.95   .87    33.44
 9     Horiz. Displacement       ft        0               94.74  1.04    37.22
10      Vert. Displacement       ft        0              110.53  1.22    40.03
11  Final Displacement                                    126.32  1.39    41.86
12               Horizontal      ft      300              142.11  1.56    42.73
13                                                        157.89  1.74    42.62
14  Accel. of Gravity       ft/sec2     32.2              173.68  1.91    41.54
15    Possible?                 Yes                       189.47  2.08    39.49
16  Time of Flight              sec     3.30              205.26  2.26    36.47
17                                                        221.05  2.43    32.48
18                                                        236.84  2.60    27.52
19                                                        252.63  2.78    21.59
20                                                        268.42  2.95    14.69
 C6          Form=105
Width:  9  Memory:217 Last Col/Row:G22     ? for HELP
 1>
```

[19-5] A projectile motion problem (SuperCalc3).

```
Column F
2  Time
3   0
4   IF(B15=("Yes"),(E4-E3)/(C6*COS(C7*PI/180)),ERROR)
5   IF(B15=("Yes"),(E5-E3)/(C6*COS(C7*PI/180)),ERROR)
6   IF(B15=("Yes"),(E6-E3)/(C6*COS(C7*PI/180)),ERROR)
7   IF(B15=("Yes"),(E7-E3)/(C6*COS(C7*PI/180)),ERROR)
8   IF(B15=("Yes"),(E8-E3)/(C6*COS(C7*PI/180)),ERROR)
9   IF(B15=("Yes"),(E9-E3)/(C6*COS(C7*PI/180)),ERROR)
10  IF(B15=("Yes"),(E10-E3)/(C6*COS(C7*PI/180)),ERROR)
11  IF(B15=("Yes"),(E11-E3)/(C6*COS(C7*PI/180)),ERROR)
12  IF(B15=("Yes"),(E12-E3)/(C6*COS(C7*PI/180)),ERROR)
13  IF(B15=("Yes"),(E13-E3)/(C6*COS(C7*PI/180)),ERROR)
14  IF(B15=("Yes"),(E14-E3)/(C6*COS(C7*PI/180)),ERROR)
15  IF(B15=("Yes"),(E15-E3)/(C6*COS(C7*PI/180)),ERROR)
16  IF(B15=("Yes"),(E16-E3)/(C6*COS(C7*PI/180)),ERROR)
17  IF(B15=("Yes"),(E17-E3)/(C6*COS(C7*PI/180)),ERROR)
18  IF(B15=("Yes"),(E18-E3)/(C6*COS(C7*PI/180)),ERROR)
19  IF(B15=("Yes"),(E19-E3)/(C6*COS(C7*PI/180)),ERROR)
20  IF(B15=("Yes"),(E20-E3)/(C6*COS(C7*PI/180)),ERROR)
21  IF(B15=("Yes"),(E21-E3)/(C6*COS(C7*PI/180)),ERROR)
22  IF(B15=("Yes"),(E22-E3)/(C6*COS(C7*PI/180)),ERROR)
```

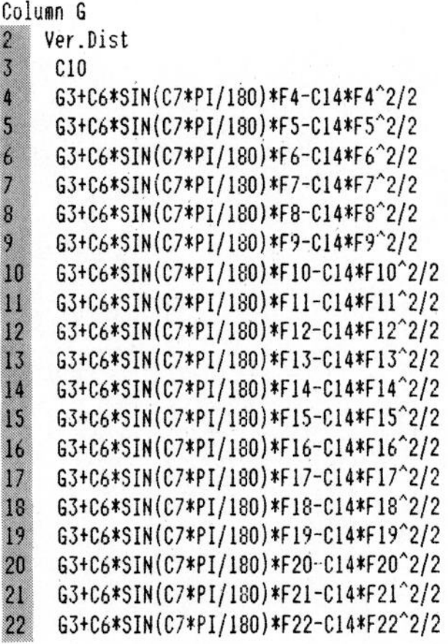

```
Column G
2  Ver.Dist
3   C10
4   G3+C6*SIN(C7*PI/180)*F4-C14*F4^2/2
5   G3+C6*SIN(C7*PI/180)*F5-C14*F5^2/2
6   G3+C6*SIN(C7*PI/180)*F6-C14*F6^2/2
7   G3+C6*SIN(C7*PI/180)*F7-C14*F7^2/2
8   G3+C6*SIN(C7*PI/180)*F8-C14*F8^2/2
9   G3+C6*SIN(C7*PI/180)*F9-C14*F9^2/2
10  G3+C6*SIN(C7*PI/180)*F10-C14*F10^2/2
11  G3+C6*SIN(C7*PI/180)*F11-C14*F11^2/2
12  G3+C6*SIN(C7*PI/180)*F12-C14*F12^2/2
13  G3+C6*SIN(C7*PI/180)*F13-C14*F13^2/2
14  G3+C6*SIN(C7*PI/180)*F14-C14*F14^2/2
15  G3+C6*SIN(C7*PI/180)*F15-C14*F15^2/2
16  G3+C6*SIN(C7*PI/180)*F16-C14*F16^2/2
17  G3+C6*SIN(C7*PI/180)*F17-C14*F17^2/2
18  G3+C6*SIN(C7*PI/180)*F18-C14*F18^2/2
19  G3+C6*SIN(C7*PI/180)*F19-C14*F19^2/2
20  G3+C6*SIN(C7*PI/180)*F20-C14*F20^2/2
21  G3+C6*SIN(C7*PI/180)*F21-C14*F21^2/2
22  G3+C6*SIN(C7*PI/180)*F22-C14*F22^2/2
```

[19-3] Cell formulas (SuperCalc3) (Continued)

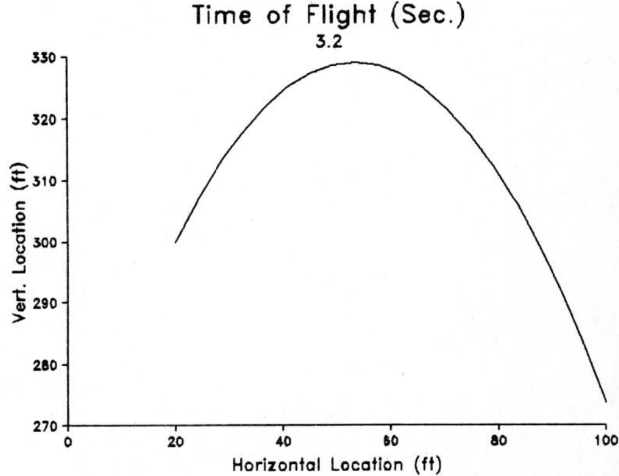

[19-4] A plot of the trajectory of the projectile described in Figure [19-2].

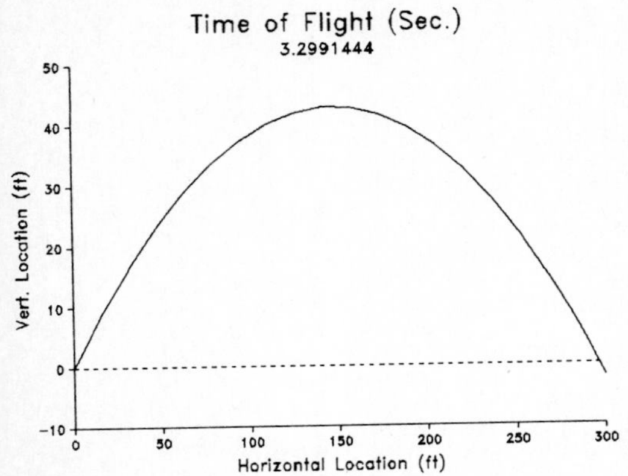

[19-6] A plot of the trajectory of the projectile described in Figure [19-5].

```
   |     A        || B || C  || D  || E  || F || G |
 1  Projectile Motion - Plots       Data
 2   Hor. vs Vert. Location         Entry        Hor. DistTime  Ver.Dist
 3                                  !!!!!          200.00   .00      .00
 4  Units 1-Metric, 2-English           2          194.74 ERROR    ERROR
 5  Initial velocity                               189.47 ERROR    ERROR
 6   Magnitude of Velocity  ft/sec    105 ---Check 184.21 ERROR    ERROR
 7    Angle up from Horiz.  Degrees    30 ---Check 178.95 ERROR    ERROR
 8  Initial Displacement                     !     173.68 ERROR    ERROR
 9     Horiz. Displacement      ft    200 ---Check 168.42 ERROR    ERROR
10      Vert. Displacement      ft      0      !   163.16 ERROR    ERROR
11  Final Displacement                         !   157.89 ERROR    ERROR
12               Horizontal     ft    100 ---Check 152.63 ERROR    ERROR
13                                             !   147.37 ERROR    ERROR
14  Accel. of Gravity       ft/sec2  32.2      !   142.11 ERROR    ERROR
15    Possible?                 No   Check Entries 136.84 ERROR    ERROR
16  Time of Flight              sec  ERROR         131.58 ERROR    ERROR
17                                                 126.32 ERROR    ERROR
18                                                 121.05 ERROR    ERROR
19                                                 115.79 ERROR    ERROR
20                                                 110.53 ERROR    ERROR
  C9            Form=200
Width:  9  Memory:217 Last Col/Row:G22      ? for HELP
  1>
```

[19-7] Error trapping on a spreadsheet (SuperCalc3).

through a load resistance (if the switch is closed in Figure [19-8]). The dc electrical power that the model produces is the product of its voltage (V) and current (I). In this "ideal" model, the relationship between the current and voltage is given by

$$I = \frac{\exp(V/V_T) - \exp(V_{0c}/V_T)}{1 - \exp(V_{0c}/V_T)} I_{sc} \qquad (19\text{-}8)$$

Here V_T is a temperature-dependent constant that is calculable from the physics involved.[2] Its value is $8.625 \times 10^{-5}T$ when T is in degrees Kelvin (K).

V_{0c} and I_{sc} in the equation above, are parameters that are determined by a combination of the physics involved, the way the solar cell is constructed, and the intensity of the solar radiation. These two parameters are easily determined by experimental tests. V_{0c} is known as the open-circuit voltage because it is the voltage that the cell produces across its output leads when the leads are not connected to one another and the current produced is zero (note in the equation above that $I = 0$ when $V = V_{0c}$). I_{sc} is known as the short-circuit current since this is the current that is produced when the output leads are shorted through a load that has no resistance (note in the equation above that $I = I_{sc}$ when $V = 0$).

The power produced by the cell is the product of V and I, or

$$P = V\frac{\exp(V/V_T) - \exp(V_{0c}/V_T)}{1 - \exp(V_{0c}/V_T)} I_{sc} \qquad (19\text{-}9)$$

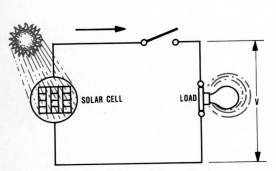

SOLAR CELL
LOAD
V

[19-8] A schematic of an idealized solar cell.

[2]The theory which yields a value for V_T is beyond the scope of this book.

A plot of current versus voltage is called the *IV* curve and its shape is useful in understanding the performance of the cell.

We can construct the *IV* curve on a spreadsheet by calculating I at several different values of V from $V = 0$ to $V = V_{0c}$ from Equation (19-9). We can also tabulate P for each V at the same time and plot P versus V in order to see the sensitivity of P with V. From Equation (19-9), we can see that P will be zero both when $V = 0$ and when $V = V_{0c}$. From this knowledge we can anticipate that P will have a maximum power (we will call this V_{mp}, for maximum some place between these two voltages.)

Finding the exact V for maximum power (we will call this V_{mp}, for maximum power voltage) cannot be done nicely in a table of P versus V since the V that maximizes the power will generally fall between two table entries. Short of interpolating between values, we could attempt to analytically solve for V_{mp}.

The use of calculus permits us to solve the following equation[3] for V_{mp}:

$$\left(1 + \frac{V_{mp}}{V_T}\right) \exp\left(\frac{V_{mp}}{V_T}\right) - \exp\left(\frac{V_{oc}}{V_T}\right) = 0 \qquad (19\text{-}10)$$

Unfortunately, this equation does not yield to a closed-form solution for V_{mp}. However, we can use Newton–Raphson iteration to solve it. That is, if V_{mp}, is an estimate of the maximum power voltage, a better estimate, $(V_{mp})_{new}$, is obtained from

$$(V_{mp})_{new} = V_{mp} - \frac{\left(1 + \frac{V_{mp}}{V_T}\right) \exp\left(\frac{V_{mp}}{V_T}\right) - \exp\left(\frac{V_{oc}}{V_T}\right)}{\exp\left(\frac{V_{mp}}{V_T}\right)\left(2 + \frac{V_{mp}}{V_T}\right)} \qquad (19\text{-}11)$$

where the denominator of the fraction is the derivative of the numerator (see Chapter 16 for a discussion of the Newton–Raphson iteration technique). Therefore, this equation can be solved by iteration. When you see your first *IV* curve, you will recognize that a first good guess for V_{mp} is simply V_{0c}.

Figure [19-9] is a spreadsheet for solving this problem. It accepts data input on the cell temperature (needed for calculating V_T), V_{0c}, and I_{sc}. It then produces a table of V, I, and P values over the range of V from $V = 0$ to $V = V_{0c}$ and produces a plot of the corresponding *IV* curve. An iteration is carried out on the same sheet in order to calculate V_{mp}, I_{mp}, and P_{mp}, the voltage, current, and power, respectively,where the electrical output power is a maximum.

Figure [19-10] is a listing of the formulas that make up the sheet for this ideal solar cell model. Figure [19-11] is a copy of the plot that was produced for the data input used in Figure [19-9].

Figure [19-12] is the spreadsheet for a temperature of 400 K, an open circuit voltage of 0.5 V, and a short-circuit current of 0.25 A. The *IV* curve for this set of parameters is shown in Figure [19-13]. All of the *IV* curves show that there is a condition that will yield maximum electrical power. The voltage for which the power out is a

[3]For those who know calculus, this equation is obtained by setting $dP/dV = 0$ and replacing V with V_{mp}. That is, we are after the V that maximizes P (and we call that voltage V_{mp}).

```
CIRC¦  A  ¦¦    B   ¦¦    C   ¦¦    D    ¦¦ E ¦¦   F   ¦¦   G   ¦¦   H  ¦
2   Ideal Solar Cell Model
3        300  Enter Temperature T here in degrees K
4        .75  Enter Open Circuit Voltage here (volts)
5        .25  Enter Short Circuit Current here (amps)
6
7    .025875  VT is computed here
8    V Volts  I (amps)     I x V Max P Amps
9          0       .25         0 .240634864Voltage
10      .075 .25000000 .01875000 .240634864Current
11       .15 .25000000 .03750000 .240634864Power
12      .225 .25000000 .05625000 .240634864Diode Model of Solar Cell
13        .3 .24999999 .07500000 .240634864MP Amps
14      .375 .24999987 .09374995 .240634864     Iterate for Max Power
15       .45 .24999770 .11249896 .240634864      8.536e10
16      .525 .24995817 .13122804 .240634864      1.588e14     .001
17        .6 .24924091 .14954455 .240634864      .6650144 Max Power Voltage
18     .6375 .24676627 .15731350 .240634864      .2406349 Max Power Current
19      .675 .23622424 .15945137 .240634864      .1600257 Max Power
20     .7125 .19131492 .13631188 .240634864
21       .75         0         0 .240634864
  A7          P Form=8.625E-5*A3
Width:  8  Memory:219 Last Col/Row:G21     ? for HELP
  1>
```

[19-9] A spreadsheet for exploring the current-voltage (IV) characteristics of a solar cell.

```
CIRC¦  A  ¦¦                         B                          ¦¦    C   ¦¦    D    ¦
2   Ideal Solar Cell Model
3    300      Enter Temperature T here in degrees K
4    .75      Enter Open Circuit Voltage here (volts)
5    .25      Enter Short Circuit Current here (amps)
6
7    8.625E-  VT is computed here
8    V Volts                                      I (amps)     I x V Max P Amps
9    0       A5*(EXP(A9/A7)-EXP(A4/A7))/(1-EXP(A4/A7))     A9*B9     F18
10   .1*A4   A5*(EXP(A10/A7)-EXP(A4/A7))/(1-EXP(A4/A7))    A10*B10   F18
11   .2*A4   A5*(EXP(A11/A7)-EXP(A4/A7))/(1-EXP(A4/A7))    A11*B11   F18
12   .3*A4   A5*(EXP(A12/A7)-EXP(A4/A7))/(1-EXP(A4/A7))    A12*B12   F18
13   .4*A4   A5*(EXP(A13/A7)-EXP(A4/A7))/(1-EXP(A4/A7))    A13*B13   F18
14   .5*A4   A5*(EXP(A14/A7)-EXP(A4/A7))/(1-EXP(A4/A7))    A14*B14   F18
15   .6*A4   A5*(EXP(A15/A7)-EXP(A4/A7))/(1-EXP(A4/A7))    A15*B15   F18
16   .7*A4   A5*(EXP(A16/A7)-EXP(A4/A7))/(1-EXP(A4/A7))    A16*B16   F18
17   .8*A4   A5*(EXP(A17/A7)-EXP(A4/A7))/(1-EXP(A4/A7))    A17*B17   F18
18   .85*A4  A5*(EXP(A18/A7)-EXP(A4/A7))/(1-EXP(A4/A7))    A18*B18   F18
19   .9*A4   A5*(EXP(A19/A7)-EXP(A4/A7))/(1-EXP(A4/A7))    A19*B19   F18
20   .95*A4  A5*(EXP(A20/A7)-EXP(A4/A7))/(1-EXP(A4/A7))    A20*B20   F18
21   A4      A5*(EXP(A21/A7)-EXP(A4/A7))/(1-EXP(A4/A7))    A21*B21   F18
  A7          P Form=8.625E-5*A3
Width:  8  Memory:219 Last Col/Row:G21     ? for HELP
  1>
```

[19-10] The formulas programmed into the spreadsheet of Figure [19-9]. (Continues on next page)

```
Column E
9   Voltage
10  Current
11  Power
12  Diode Model of Solar Cell
13  MP Amps

Column F
14  Iterate for Max Power
15   IF(ITER=1,EXP(A4/A7)*(1+A4/A7)-EXP(A4/A7),
           EXP(F17/A7)*(1+F17/A7)-EXP(A4/a7))
16   IF(ITER= .001
17   IF(ITER= Max Power Voltage
18   A5*(1.-( Max Power Current
19   F17*F18  Max Power

Column G  (The iteration tolerance for max power voltage)
16   .001
```

[19-10] The formulas programmed into the spreadsheet of Figure [19-9]. (Continued)

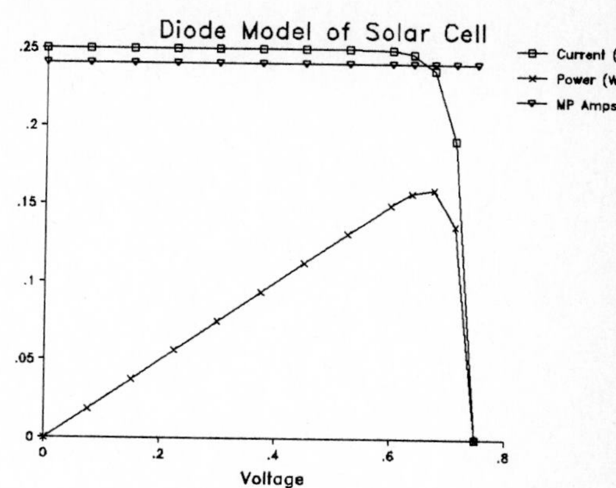

[19-11] The graph constructed from the spreadsheet of Figure [19-9].

```
CIRC|  A  ||   B  ||   C  ||   D   || E ||  F  ||  G  ||  H  |
2   Ideal Solar Cell Model
3        400  Enter Temperature T here in degrees K
4         .5  Enter Open Circuit Voltage here (volts)
5        .25  Enter Short Circuit Current here (amps)
6
7      .0345  VT is computed here
8    V Volts  I (amps)    I x V Max P Amps
9          0       .25        0 .230669665Voltage
10       .05 .24999959 .01249998 .230669665Current
11        .1 .24999782 .02499978 .230669665Power
12       .15 .24999031 .03749855 .230669665Diode Model of Solar Cell
13        .2 .24995830 .04999166 .230669665MP Amps
14       .25 .24982194 .06245548 .230669665     Iterate for Max Power
15        .3 .24924104 .07477231 .230669665      2736.156
16       .35 .24676640 .08636824 .230669665      61553025      .001
17        .4 .23622436 .09448975 .230669665      .4116876 Max Power Voltage
18      .425 .22156719 .09416605 .230669665      .2306697 Max Power Current
19       .45 .19131502 .08609176 .230669665      .0949638 Max Power
20      .475 .12887505 .06121565 .230669665
21        .5         0         0 .230669665
  A3            Form=400
Width:  8  Memory:219 Last Col/Row:G21     ? for HELP
  1>
```

[19-12] A spreadsheet using another set of parameters.

maximum is slightly smaller that the open-circuit voltage. If you were designing a solar power plant, you would want to control the electrical circuitry so that you operated the cell near its maximum power output. Otherwise, performance would suffer and you would have to install more cells in order to produce a fixed amount of power.

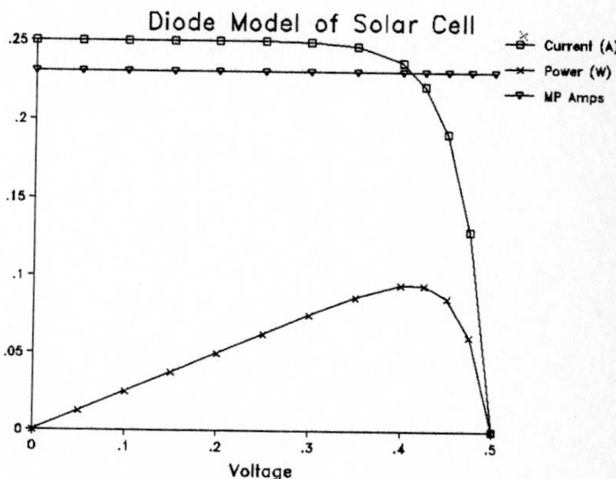

[19-13] The IV curve constructed from the spreadsheet of Figure [19-12].

Engineering Economic Analysis

As indicated in Chapter 8, engineering economic analysis is used to evaluate the worth of a single engineering project or to compare two or more engineering projects. A single project is deemed to be a worthy one if it earns a rate of return greater than or equal to the firm's minimum attractive rate of return (MARR). Equivalently the net present worth, annual worth, or future worth of the project's cash flows over the estimated life of the project will be greater than or equal to the MARR if the project is worth undertaking. The best among two or more engineering projects is the one with the largest rate of return or equivalently the one with the largest present, annual, or future worth.

The program listed in Figure [19-14] has been developed to compute for any cash flow, the net present worth, net annual worth, and net future worth using a specified rate of return, typically the MARR.

Example 19-1

To illustrate the use of the engineering economic analysis program, a problem which includes single, uniform, and gradient cash flows has been developed. Suppose that a $1,000 investment is expected to have a 12 year useful life and to return positive cash flows of $500 in years 1 and 9, and $200 in year 12. A $300 positive cash

```
10 DIM S(10), NS(10), A(5), B(5), E(5), U(5), G(5), R(5), F(5)
20 PW=0:AW=0:FW=0
30 CLS
40 PRINT " "
50 INPUT "ENTER THE NUMBER OF PERIODS IN THE PLANNING HORIZON";N
60 PRINT " "
70 INPUT "ENTER THE NUMBER OF SINGLE CASH FLOWS-IF NONE, ENTER 0";MS
80 PRINT " "
90 IF MS=0 THEN 170
100 FOR I=1 TO MS
110 PRINT "FOR SINGLE CASH FLOW NUMBER", I
120 PRINT "ENTER THE SINGLE CASH FLOW VALUE- USE A MINUS SIGN IF NEGATIVE"
130 PRINT "AND THE PERIOD IN WHICH THIS CASH FLOW OCCURS"
140 INPUT "SEPARATE WITH COMMAS";S(I), NS(I)
150 PRINT " "
160 NEXT I
170 PRINT "ENTER THE NUMBER OF UNIFORM SERIES CASH FLOWS"
180 INPUT "IF THERE ARE NONE, ENTER 0";NU
190 PRINT " "
200 IF NU=0 THEN 280
210 FOR I = 1 TO NU
220 PRINT "FOR UNIFORM SERIES NUMBER",I
230 PRINT "ENTER THE SERIES AMOUNT (USE A MINUS SIGN IF NEGATIVE)"
240 PRINT "AS WELL AS THE STARTING PERIOD AND ENDING PERIOD"
250 INPUT "SEPARATE WITH COMMAS";A(I), B(I), E(I)
260 PRINT " "
270 NEXT I
280 PRINT " "
290 INPUT "ENTER THE NUMBER OF GRADIENT SERIES- IF NONE, ENTER 0";NG
300 IF NG=0 THEN 410
310 FOR I = 1 TO NG
320 PRINT " "
330 PRINT "FOR GRADIENT SERIES NUMBER",I
340 PRINT "ENTER THE INITIAL AMOUNT (USE MINUS SIGN IF NEGATIVE)"
350 PRINT "THE GRADIENT AMOUNT (USE MINUS SIGN IF NEGATIVE),"
360 PRINT "AS WELL AS THE PERIOD AT WHICH THE GRADIENT SERIES BEGINS,"
370 PRINT "AND THE PERIOD AT WHICH THE GRADIENT SERIES ENDS"
380 INPUT "SEPARATE WITH COMMAS";U(I), G(I), R(I), F(I)
390 PRINT " "
400 NEXT I
410 PRINT " "
420 INPUT "ENTER THE RATE OF RETURN FOR THIS PROBLEM IN DECIMAL FORM";RR
430 CLS: PRINT " "
440 PRINT "THE NUMBER OF PERIODS IN THE PLANNING HORIZON IS"N
450 PRINT " "
460 INPUT "DO YOU WISH TO CHANGE THIS VALUE (Y OR N)";CN$
470 IF CN$="N" THEN 500
480 PRINT " "
490 INPUT "ENTER THE NEW VALUE OF THE PLANNING HORIZON PERIODS";N
```

[19-14] BASIC language program for engineering economic analysis. (Continues on next page)

[19-14] (Continued)

```
500 IF MS=0 THEN 690
510 CLS:PRINT "THE SINGLE CASH FLOWS FOR THIS PROBLEM ARE":PRINT " "
520 PRINT "   NUMBER        VALUE        PERIOD"
530 PRINT " "
540 FOR I = 1 TO MS
550 PRINT USING "#####.##      ";I,S(I),NS(I)
560 NEXT I
570 PRINT " "
580 INPUT "DO YOU WISH TO CHANGE ANY EXISTING VALUE (Y OR N)";C1$
590 IF C1$="N" THEN 630
600 PRINT "ENTER CASH FLOW NUMBER, VALUE AND PERIOD"
610 PRINT "TO DELETE A CASH FLOW ENTER 0 FOR ITS VALUE"
620 INPUT "SEPARATE WITH COMMAS";I,S(I),NS(I)
630 INPUT "DO YOU WISH TO ADD ANY SINGLE CASH FLOW (Y OR N)";A1$
640 IF A1$="N" THEN 680
650 PRINT "ENTER NEW CASH FLOW VALUE AND PERIOD"
660 INPUT "SEPARATE WITH COMMAS";S(MS+1),NS(MS+1)
670 MS=MS+1
680 IF C1$="N" AND A1$="N" THEN 690 ELSE 510
690 IF NU=0 THEN 880
700 CLS: PRINT "THE UNIFORM SERIES FOR THIS PROBLEM ARE":PRINT " "
710 PRINT "   NUMBER         VALUE        START          END   ":PRINT " "
720 FOR I = 1 TO NU
730 PRINT USING "#####.##      ";I,A(I),B(I),E(I)
740 NEXT I
750 PRINT " "
760 INPUT "DO YOU WISH TO CHANGE ANY EXISTING SERIES (Y OR N)";CU$
770 IF CU$="N" THEN 820
780 PRINT " "
790 PRINT "ENTER SERIES NUMBER, VALUE, STARTING AND ENDING PERIODS"
800 PRINT "TO DELETE A SERIES, ENTER 0 FOR ITS VALUE."
810 INPUT "SEPARATE WITH COMMAS  ",J,A(J),B(J),E(J)
820 PRINT " ":INPUT "DO YOU WITH TO ADD A UNIFORM SERIES (Y OR N)";AU$
830 IF AU$="N" THEN 870
840 PRINT "ENTER VALUE OF THE SERIES, STARTING PERIOD AND ENDING PERIOD"
850 INPUT "SEPARATE WITH COMMAS";A(NU+1),B(NU+1),E(NU+1)
860 NU=NU+1
870 IF CU$="N" AND AU$="N" THEN 880 ELSE 700
880 IF NG=0 THEN 1100
890 CLS: PRINT "THE GRADIENT SERIES FOR THIS PROBLEM ARE:":PRINT " "
900 PRINT "   NUMBER      INITIAL       GRADIENT        START         END "
910 PRINT " "
920 FOR I = 1 TO NG
930 PRINT USING "#####.##      ";I,U(I),G(I),R(I),F(I)
940 NEXT I
950 PRINT " "
960 INPUT "DO YOU WISH TO CHANGE ANY EXISTING GRADIENT SERIES (Y OR N)";CG$
970 IF CG$="N" THEN 1040
980 PRINT " "
990 PRINT "ENTER GRADIENT SERIES NUMBER, INITIAL VALUE, GRADIENT VALUE,"
1000 PRINT "BEGINNING PERIOD AND ENDING PERIOD OF THE GRADIENT SERIES."
1010 PRINT "TO DELETE A SERIES, ENTER 0 FOR BOTH THE INITIAL AND GRADIENT AMOUNT
S"
1020 INPUT "SEPARATE WITH COMMAS  ",J,U(J),G(J),R(J),F(J)
1030 PRINT " "
1040 INPUT "DO YOU WISH TO ADD A GRADIENT SERIES (Y OR N)";AG$
1050 IF AG$="N" THEN 1090
1060 PRINT "ENTER INITIAL VALUE, GRADIENT VALUE, STARTING AND ENDING PERIOD"
1070 INPUT "SEPARATE WITH COMMAS";U(NG+1),G(NG+1),R(NG+1),F(NG+1)
1080 NG=NG+1
1090 IF CG$="N" AND AG$="N" THEN 1100 ELSE 890
1100 CLS:PRINT " "
1110 PRINT "THE VALUE OF THE RATE OF RETURN FOR THIS PROBLEM IS"RR
1120 PRINT " "
1130 INPUT "DO YOU WISH TO CHANGE THIS VALUE (Y OR N)";CR$
1140 IF CR$="N" THEN 1170
1150 PRINT " "
1160 INPUT "ENTER THE NEW VALUE OF THE RATE OF RETURN IN DECIMAL FORM";RR
1170 IF MS=0 THEN 1210
1180 FOR I = 1 TO MS
1190 PW = PW + S(I) * (1/(1+RR)^NS(I))
1200 NEXT I
1210 IF NU=0 THEN 1260
1220 FOR I = 1 TO NU
1230 T=E(I)-B(I)+1
1240 PW = PW + (A(I)*((1+RR)^T-1)/(RR*(1+RR)^T))*(1/(1+RR))^(B(I)-1)
1250 NEXT I
1260 IF NG=0 THEN 1350
1270 FOR I = 1 TO NG
1280 Z=F(I)-R(I)+1
1290 PA = ((1+RR)^Z-1)/(RR*(1+RR)^Z)
1300 XW = U(I)*PA
1310 PF = 1/(1+RR)^Z
1320 YW = G(I)*(PA-Z*PF)/RR
1330 PW = PW + (XW+YW)*(1/(1+RR)^(R(I)-1))
1340 NEXT I
1350 AW = PW * (RR*(1+RR)^N)/((1+RR)^N-1)
1360 FW = PW * (1+RR)^N
1370 CLS:PRINT " "
1380 PRINT "NET PRESENT WORTH OF ALL CASH FLOWS = $",PW
1390 PRINT " "
1400 PRINT "NET ANNUAL WORTH OF ALL CASH FLOWS =  $",AW
1410 PRINT " "
1420 PRINT "NET FUTURE WORTH OF ALL CASH FLOWS =  $",FW
1430 PRINT " "
1440 END
```

[19-15] Cash flow diagram for Example 19-1.

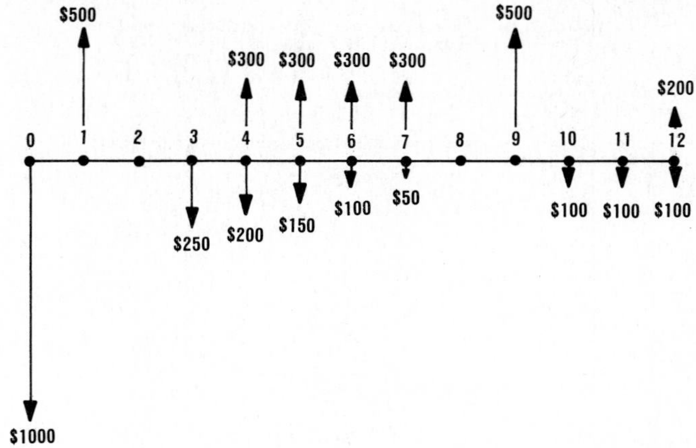

flow is expected in years 4 through 7. The projected negative cash flows include a gradient series in years 3 through 7 with the initial and final amounts $250 and $50 respectively. Furthermore, there is a uniform negative $100 cash flow in years 10 through 12. The MARR is 15 percent per year compounded yearly. The cash flow diagram associated with this situation is given in Figure [19-15].

The complete set of screen displays associated with entering data for this problem is given in Figure [19-16].

It is important that the user make a correct count of the number of single cash flows, uniform series, and gradient series (although an opportunity to correct errors is provided). It is also important to enter the correct sign associated with each cash flow (minus for negative or downward cash flows; a plus sign is not needed for positive cash flows). For uniform series, note that the series amount is followed by the starting period and then the ending period of the series separated by commas. With respect to gradient series, the initial amount, the gradient amount, the starting period, and then the ending period must be entered in that order, separated by commas. For this problem, note that although the cash flows are all negative, the gradient amount ($50) is actually a positive amount and is entered as such.

Next [19-17], the user is given an opportunity to change the number of periods in the planning horizon. The user must respond with a capital "N" or "Y" to indicate no change or change, respectively. The data for each of the three types of cash flows are then echoed to the user [19-18]. In each case, an opportunity to change a cash flow and add a cash flow of each type is provided. Once again, a capital "N" or "Y" is required. To delete a cash flow (not shown in this example), it is necessary only to respond "Y" to the question "DO YOU WISH TO CHANGE ANY EXISTING VALUE (SERIES or GRADIENT SERIES)" and then change only the cash values (single amount, uniform amount, gradient initial amount, and gradient value) to *zero* while entering the same period values as in the original cash flow (actually, with zero cash flows any values given for the periods are irrelevant). The output for this example problem is given in Figure [19-19].

```
ENTER THE NUMBER OF PERIODS IN THE PLANNING HORIZON? 12

ENTER THE NUMBER OF SINGLE CASH FLOWS-IF NONE, ENTER 0? 4

FOR SINGLE CASH FLOW NUMBER  1
ENTER THE SINGLE CASH FLOW VALUE- USE A MINUS SIGN IF NEGATIVE
AND THE PERIOD IN WHICH THIS CASH FLOW OCCURS
SEPARATE WITH COMMAS? -1000,0

FOR SINGLE CASH FLOW NUMBER  2
ENTER THE SINGLE CASH FLOW VALUE- USE A MINUS SIGN IF NEGATIVE
AND THE PERIOD IN WHICH THIS CASH FLOW OCCURS
SEPARATE WITH COMMAS? 500,1

FOR SINGLE CASH FLOW NUMBER  3
ENTER THE SINGLE CASH FLOW VALUE- USE A MINUS SIGN IF NEGATIVE
AND THE PERIOD IN WHICH THIS CASH FLOW OCCURS
SEPARATE WITH COMMAS? 500,9

FOR SINGLE CASH FLOW NUMBER  4
ENTER THE SINGLE CASH FLOW VALUE- USE A MINUS SIGN IF NEGATIVE
AND THE PERIOD IN WHICH THIS CASH FLOW OCCURS
SEPARATE WITH COMMAS? 200,12


SEPARATE WITH COMMAS? 200,12

ENTER THE NUMBER OF UNIFORM SERIES CASH FLOWS
IF THERE ARE NONE, ENTER 0? 2

FOR UNIFORM SERIES NUMBER     1
ENTER THE SERIES AMOUNT (USE A MINUS SIGN IF NEGATIVE)
AS WELL AS THE STARTING PERIOD AND ENDING PERIOD
SEPARATE WITH COMMAS? 300,4,7

FOR UNIFORM SERIES NUMBER     2
ENTER THE SERIES AMOUNT (USE A MINUS SIGN IF NEGATIVE)
AS WELL AS THE STARTING PERIOD AND ENDING PERIOD
SEPARATE WITH COMMAS? -100,10,12


ENTER THE NUMBER OF GRADIENT SERIES- IF NONE, ENTER 0? 1

FOR GRADIENT SERIES NUMBER    1
ENTER THE INITIAL AMOUNT (USE MINUS SIGN IF NEGATIVE)
THE GRADIENT AMOUNT (USE MINUS SIGN IF NEGATIVE),
AS WELL AS THE PERIOD AT WHICH THE GRADIENT SERIES BEGINS,
AND THE PERIOD AT WHICH THE GRADIENT SERIES ENDS
SEPARATE WITH COMMAS? -250,3,7


IF THERE ARE NONE, ENTER 0? 2

FOR UNIFORM SERIES NUMBER     1
ENTER THE SERIES AMOUNT (USE A MINUS SIGN IF NEGATIVE)
AS WELL AS THE STARTING PERIOD AND ENDING PERIOD
SEPARATE WITH COMMAS? 300,4,7

FOR UNIFORM SERIES NUMBER     2
ENTER THE SERIES AMOUNT (USE A MINUS SIGN IF NEGATIVE)
AS WELL AS THE STARTING PERIOD AND ENDING PERIOD
SEPARATE WITH COMMAS? -100,10,12


ENTER THE NUMBER OF GRADIENT SERIES- IF NONE, ENTER 0? 1

FOR GRADIENT SERIES NUMBER    1
ENTER THE INITIAL AMOUNT (USE MINUS SIGN IF NEGATIVE)
THE GRADIENT AMOUNT (USE MINUS SIGN IF NEGATIVE),
AS WELL AS THE PERIOD AT WHICH THE GRADIENT SERIES BEGINS,
AND THE PERIOD AT WHICH THE GRADIENT SERIES ENDS
SEPARATE WITH COMMAS? -250,50,3,7


ENTER THE RATE OF RETURN FOR THIS PROBLEM IN DECIMAL FORM? .15
```

[19-16] User inputs for Example 19-1.

```
THE NUMBER OF PERIODS IN THE PLANNING HORIZON IS 12

DO YOU WISH TO CHANGE THIS VALUE (Y OR N)? N
```

[19-17] Screen display—planning horizon, Example 19-1.

```
THE SINGLE CASH FLOWS FOR THIS PROBLEM ARE

   NUMBER          VALUE          PERIOD

    1.00         -1000.00          0.00
    2.00           500.00          1.00
    3.00           500.00          9.00
    4.00           200.00         12.00

DO YOU WISH TO CHANGE ANY EXISTING VALUE (Y OR N)? N
DO YOU WISH TO ADD ANY SINGLE CASH FLOW (Y OR N)? N


THE UNIFORM SERIES FOR THIS PROBLEM ARE

   NUMBER          VALUE          START          END

    1.00          300.00           4.00          7.00
    2.00         -100.00          10.00         12.00

DO YOU WISH TO CHANGE ANY EXISTING SERIES (Y OR N)? N

DO YOU WITH TO ADD A UNIFORM SERIES (Y OR N)? N


THE GRADIENT SERIES FOR THIS PROBLEM ARE:

   NUMBER       INITIAL        GRADIENT        START        END

    1.00        -250.00          50.00          3.00        7.00

DO YOU WISH TO CHANGE ANY EXISTING GRADIENT SERIES (Y OR N)? N
DO YOU WISH TO ADD A GRADIENT SERIES (Y OR N)? N
```

[19-18] Summary of inputs, Example 19-1.

```
NET PRESENT WORTH OF ALL CASH FLOWS = $    -302.7857

NET ANNUAL WORTH OF ALL CASH FLOWS =  $    -55.85815

NET FUTURE WORTH OF ALL CASH FLOWS =  $    -1619.98

Ok∅
```

[19-19] Output for Example 19-1.

Shortest-Route Algorithm

Consider the network shown in Figure [19-20]. The goal of this optimization problem is to find the path from the starting node (node 1 in this problem) to the ending node (node 7 in this problem) which results in the minimum total cost of travel (the cost of travel between any pair of nodes is indicated by numerical values with arrows in the direction of travel). It should be mentioned that

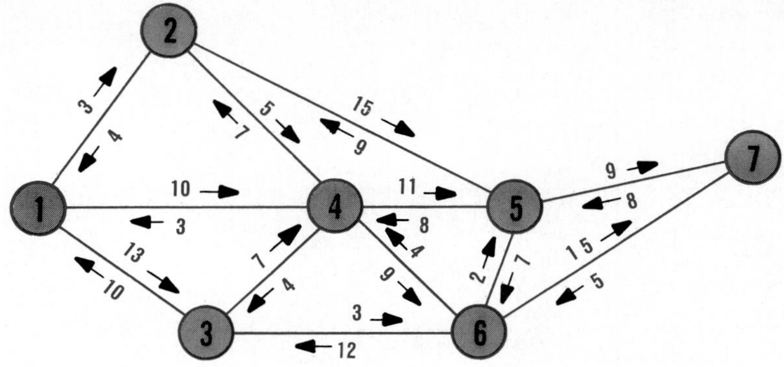

[19-20] Network used to illustrate shortest route algorithm.

the branch values could also represent a measure of the distance between any pair of nodes. This problem was introduced in Chapter 14 as an example of a problem that could be solved by "brute force" or the exhaustive enumeration of all possible routes or paths from the starting node to the ending node. The cost of travel (or total distance) would be totaled for each path and the path resulting in the minimal cost (or distance) would be selected. However, the program presented in [18-8] uses a algorithm (set of mathematical rules) which allows a solution without tracing all paths. The algorithm will not be described in detail here, but the solution procedure is based on remembering for each node only the path leading to that node which results in the smallest cost of travel (or distance) to that node. As each new node is examined, the program uses only the shortest-route information for the nodes which are connected to the new information for the nodes which are connected to the new node, adds the pairwise cost (or distance), and remembers the previous node resulting in the smallest route as part of the optimal path. Readers interested in learning more about this algorithm may consult any textbook on operations research. It should be noted that the procedure represented in the code of Figure [19-21] will find only a single route of minimal cost or distance. That is, if there is more than

```
10 CLS
20 DIM D(50,50), FT(50), A(50),B(50), C(50)
30 Z=1
40 INPUT "ENTER THE NUMBER OF NODES IN THE NETWORK";NN
50 FOR I = 1 TO NN
60      C(I) = 32767
70      A(I) = 0
80      B(I) = 32767
90         FOR J = 1 TO NN
100              D(I,J) = 32767
110        NEXT J
120 NEXT I
130 PRINT " "
140 PRINT " FOR EACH PAIR OF NODES WITH A DISTANCE ARC BETWEEN THEM"
150 PRINT " "
160 PRINT " ENTER THE SOURCE NODE NUMBER, THE DESTINATION NODE"
170 PRINT " NUMBER AND DISTANCE--SEPARATE WITH COMMAS"
180 INPUT " TO STOP, ENTER '1,1,1' ";I,J,D(I,J)
190 IF D(1,1) <> 1 THEN 150
200 D(1,1)=32767
210 CLS:PRINT:PRINT"  FROM          TO          DIST.          FROM          TO
 DIST.":PRINT
220 LL=0
230 FOR I=1 TO NN
240 FOR J=1 TO NN STEP 2
250 IF J=NN AND I <> J AND D(I,J) <> 32767 THEN PRINT USING "#####        ";I,J,D
(I,J):LL=LL+1
260 IF J <> NN AND I=J AND D(I,J+1) <> 32767 THEN PRINT USING "#####        ";I,J
+1,D(I,J+1):LL=LL+1
270 IF J <> NN AND I <> J AND D(I,J) <> 32767 AND D(I,J+1) <> 32767 THEN PRINT U
```

[19-21] Basic language program for the shortest-route algorithm. (Continued)

[19-21] (Continued from page 630)

```
SING "#####        ";I,J,D(I,J),I,J+1,D(I,J+1):LL=LL+1
280 IF J <> NN AND I <> J AND D(I,J) = 32767 AND D(I,J+1) <> 32767 THEN PRINT US
ING "#####        ";I,J+1,D(I,J+1):LL=LL+1
290 IF J <> NN AND I <> J AND D(I,J) <> 32767 AND D(I,J+1) = 32767 THEN PRINT US
ING "#####        ";I,J,D(I,J):LL=LL+1
300 IF LL MOD 15 <> 0 THEN 360
310 PRINT
320 PRINT "PLEASE NOTE ANY CHANGES, ADDITIONS OR DELETIONS THAT YOU"
330 PRINT "WISH TO MAKE AND THEN PRESS ANY KEY TO CONTINUE"
340 A$=INKEY$: IF A$="" THEN 340
350 CLS:PRINT:PRINT"  FROM          TO          DIST.          FROM          TO
 DIST.":PRINT
360 NEXT J
370 NEXT I
380 PRINT:INPUT"DO YOU WISH TO MAKE ANY CHANGES (Y OR N) ";C$
390 IF C$="N" THEN 420
400 PRINT:PRINT"ENTER SOURCE, DESTINATION AND DISTANCE"
410 INPUT "SEPARATE WITH COMMAS";I,J,D(I,J):GOTO 210
420 PRINT:INPUT"DO YOU WISH TO ADD A NEW DISTANCE ARC (Y OR N) ";A$
430 IF A$="N" THEN 460
440 PRINT:PRINT "ENTER NEW SOURCE, DESTINATION AND DISTANCE"
450 INPUT I,J,D(I,J):GOTO 210
460 PRINT:INPUT "DO YOU WISH TO DELETE A DISTANCE ARC ";D$
470 IF D$="N" THEN 510
480 PRINT:PRINT"ENTER SOURCE AND DESTINATION TO BE DELETED"
490 INPUT"SEPARATE WITH COMMAS";I,J
500 D(I,J)=32767 :GOTO 210
510 IF C$="N" AND A$="N" AND D$="N" THEN 520 ELSE 210
520 A(1)=1
530 B(1)=0
540 FOR I = 2 TO NN
550 FOR J = 1 TO NN
560 IF A(J) = 1 THEN 580
570 IF  C(J) > D(Z,J)+B(Z) THEN C(J) = D(Z,J) + B(Z)
580 NEXT J
590 Y = C(2)
600 R=2
610 FOR K = 3 TO NN
620 IF C(K) > Y THEN 650
630 R = K
640 Y = C(K)
650 NEXT K
660 C(R)= 32767
670 A(R)=1
680 B(R)=Y
690 Z=R
700 NEXT I
710 FOR L = 2 TO NN
720 V=L
730 FT(1)=V
740 S = 2
750 CLS
760 PRINT:PRINT:PRINT:PRINT:PRINT
770 PRINT "THE SHORTEST ROUTE FROM 1 TO "V
780 PRINT "HAS A TOTAL DISTANCE OF "B(V)
790 PRINT "  "
800 PRINT "AND THE ROUTE IS :"
810 FOR I = 1 TO NN
820 IF V=I THEN 840
830 IF D(I,V) = B(V)-B(I) THEN 850
840 NEXT I
850 FT(S)=I
860 IF I=1 THEN 900
870 V=I
880 S=S+1
890 GOTO 810
900 FOR I = 1 TO S-1
910 E=S-I
920 F=FT(E+1)
930 G=FT(E)
940 PRINT  F" TO "G"DISTANCE IS "D(F,G)
950 NEXT I
960 PRINT:PRINT
970 PRINT "PRESS ANY KEY TO CONTINUE VIEWING THE RESULTS"
980 A$=INKEY$: IF A$="" THEN 980
990 NEXT L
1000 END
```

one path from starting node to ending node with the same total travel cost (or total distance), the program will output only one of these paths or routes.

Example 19-2

This example obtains a solution for the network [19-20]. A portion of the user inputs is shown in Figure [19-22]. Numbering of the nodes is not critical—however, the starting node must be numbered "1" and the ending node number must equal the number of

```
ENTER THE NUMBER OF NODES IN THE NETWORK? 7

 FOR EACH PAIR OF NODES WITH A DISTANCE ARC BETWEEN THEM

 ENTER THE SOURCE NODE NUMBER, THE DESTINATION NODE
 NUMBER AND DISTANCE--SEPARATE WITH COMMAS
 TO STOP, ENTER '1,1,1' ? 1,2,3

 ENTER THE SOURCE NODE NUMBER, THE DESTINATION NODE
 NUMBER AND DISTANCE--SEPARATE WITH COMMAS
 TO STOP, ENTER '1,1,1' ? 2,1,4

 ENTER THE SOURCE NODE NUMBER, THE DESTINATION NODE
 NUMBER AND DISTANCE--SEPARATE WITH COMMAS
 TO STOP, ENTER '1,1,1' ? 1,4,10

 ENTER THE SOURCE NODE NUMBER, THE DESTINATION NODE
 NUMBER AND DISTANCE--SEPARATE WITH COMMAS
 TO STOP, ENTER '1,1,1' ? 4,1,3

 ENTER THE SOURCE NODE NUMBER, THE DESTINATION NODE
 NUMBER AND DISTANCE--SEPARATE WITH COMMAS
 TO STOP, ENTER '1,1,1' ? 1,3,13
```

[19-22] Portion of the user inputs, Example 19-2.

[19-23] Summary of data inputs, Example 19-2.

```
  FROM        TO        DIST.          FROM        TO        DIST.

   1          2           3
   1          3          13             1          4          10
   2          1           4
   2          4           5
   2          5          15
   3          1          10
   3          4           7
   3          6           3
   4          1           3             4          2           7
   4          3           4
   4          5          11             4          6           9
   5          2           9
   5          4           8
   5          6           7
   5          7           9

PLEASE NOTE ANY CHANGES, ADDITIONS OR DELETIONS THAT YOU
WISH TO MAKE AND THEN PRESS ANY KEY TO CONTINUE

  FROM        TO        DIST.          FROM        TO        DIST.

   6          3          12             6          4           4
   6          5           2
   6          7          15
   7          5           8             7          6           5

DO YOU WISH TO MAKE ANY CHANGES (Y OR N) ? N

DO YOU WISH TO ADD A NEW DISTANCE ARC (Y OR N) ? N

DO YOU WISH TO DELETE A DISTANCE ARC ? N
```

nodes in the network ("7" in this case). After the last distance arc and value is entered, the user simply enters "1, 1, 1" to indicate that no more data will follow.

The data as entered are shown in Figure [19-23]. For this problem, it is necessary to display the data on two consecutive screens. Thus the user should note on paper any changes, additions, or deletions to be made before viewing the second screen. The "N" or "Y" response requested should be capitalized. Output on the screen consists of a display of the shortest route to *each* node and the cost or distance value. The final result is shown in Figure [19-24].

```
THE SHORTEST ROUTE FROM 1 TO  7
HAS A TOTAL DISTANCE OF  26

AND THE ROUTE IS :
 1  TO  2 DISTANCE IS  3
 2  TO  4 DISTANCE IS  5
 4  TO  3 DISTANCE IS  4
 3  TO  6 DISTANCE IS  3
 6  TO  5 DISTANCE IS  2
 5  TO  7 DISTANCE IS  9


PRESS ANY KEY TO CONTINUE VIEWING THE RESULTS
```

[19-24] Final output, Example 19-2.

PERT Program for Planning Engineering Projects

The program evaluation and review technique (PERT) was introduced in Chapter 14. For any network of events and activities, the program will compute the earliest [$T(E)$] and latest [$T(L)$] event time and the slack [$T(L) - T(E)$] time for each event. The user may then identify the critical path through the network simply by listing the events that have a zero slack time. The PERT program is listed in Figure [19-25].

Example 19-3

To illustrate the use of the PERT program, it will be used with the network of Figure [19-26]. A portion of the screen displays for data input is given in Figure [19-27]. Assume that all activity times are given in months, and they exist as indicated on the proper activity branch. It is desired to find, for each event, the earliest time, T_E, the latest time, T_L, and the slack time. You will recall that for each event on the critical path, the slack time will be zero, and the critical path length or time is the T_E value for the last event, in this case, event 7. The user simply enters a count of the number of events in the network followed by the number of activities. Prompts then appear for the starting event, the ending event, and the time for the activity between those events for each activity in the network. All the

```
10 CLS
20 DIM E1(50),F1(50),E2(50),F2(50),H(50),T(50),A(50),SL(50),ER(20,5),TE(20),TL(2
0),C(20)
30 INPUT "ENTER NUMBER OF EVENTS IN THE NETWORK";NE
40 PRINT " "
50 INPUT "ENTER THE NUMBER OF ACTIVITIES ";NA
60 PRINT " "
70 PRINT " FOR EACH PAIR OF CONNECTED EVENTS"
80 FOR J = 1 TO NA
90 PRINT " "
100 PRINT " ENTER THE STARTING EVENT, THE ENDING EVENT,"
110 PRINT " AND THE ACTIVITY TIME--SEPARATE WITH COMMAS"
120 INPUT H(J), T(J), A(J)
130 NEXT J
140 LL=0
150 CLS:PRINT:PRINT" SET      FROM       TO       TIME       SET       FROM       TO
   TIME ":PRINT
160 FOR I = 1 TO NA STEP 2
170 IF I=NA THEN PRINT USING "##.#      ";I,H(I),T(I),A(I)
180 IF I <> NA THEN PRINT USING "##.#      ";I,H(I),T(I),A(I),I+1,H(I+1),T(I+1),A
(I+1):LL=LL+1
190 IF LL MOD 15 <> 0 THEN 240
200 PRINT:PRINT"PLEASE NOTE ANY CHANGES, ADDITIONS OR DELETIONS THAT YOU"
210 PRINT"WISH TO MAKE AND THEN PRESS ANY KEY TO CONTINUE"
220 A$=INKEY$: IF A$="" THEN 220
230 CLS:PRINT:PRINT" SET      FROM       TO       TIME       SET       FROM       TO
   TIME ":PRINT
240 NEXT I
250 PRINT:INPUT"DO YOU WISH TO MAKE ANY CHANGES (Y OR N) ";C$
260 IF C$="N" THEN 310
270 PRINT:PRINT"ENTER SET NUMBER, STARTING NODE, ENDING NODE AND ACTIVITY TIME"
280 PRINT "SEPARATE WITH COMMAS; DO NOT USE DECIMALS ON THE SET NO., STARTING"
290 PRINT "AND ENDING NODES- USE DECIMAL, ONLY IF NEEDED ON THE TIME"
300 INPUT K,H(K),T(K),A(K):GOTO 140
310 PRINT:INPUT "DO YOU WISH TO ADD A NEW ACTIVITY (Y OR N) ";A$
320 IF A$="N" THEN 350
330 PRINT:PRINT" ENTER NEW STARTING NODE, ENDING NODE, AND TIME"
340 INPUT "SEPARATE WITH COMMAS";H(NA+1),T(NA+1),A(NA+1):NA=NA+1:GOTO 140
350 PRINT:INPUT "DO YOU WISH TO DELETE AN ACTIVITY (Y OR N) ";D$
360 IF D$="N" THEN 450
370 PRINT:INPUT"ENTER SET NUMBER OF THE ACTIVITY TO BE DELETED";K
380 FOR M = K TO NA
390 H(M)=H(M+1)
400 T(M)=T(M+1)
410 A(M)=A(M+1)
420 NEXT M
430 NA=NA-1
440 GOTO 140
450 IF C$="N" AND A$="N" AND D$="N" THEN 460 ELSE 140
460 KB=1
470 FOR J=KB TO NA-1
480 L=0
490 FOR K=J+1 TO NA
500 IF H(K) > H(J) THEN 570
510 IF H(K) < H(J) THEN 530
520 IF T(J) < T(K) THEN 570
530 L=1
540 HD=H(K): SV=T(K): TP=A(K)
550 H(K)=H(J): T(K)=T(J): A(K)=A(J)
560 H(J)=HD: T(J)=SV: A(J)=TP
570 NEXT K
580 NEXT J
590 IF L=0 THEN 620
600 KB=KB+1
610 GOTO 150
620 SO=0
630 E1(1)=0
640 F1(1)=A(1)
650 FOR I=2 TO NA
660 TP=0
670 IF H(I)=H(1) THEN 750
680 FOR J=1 TO NA
690 IF H(I) < > T(J) THEN 720
700 IF F1(J) > TP THEN TP = F1(J)
710 E1(I)=TP
720 NEXT J
730 F1(I)= E1(I)+A(I)
740 GOTO 770
750 E1(I)=0
760 F1(I)=A(I)
770 NEXT I
780 FL=T(NA)
790 FOR I=NA TO 1 STEP -1
800 IF FL < > T(I) THEN 820
810 IF SO < F1(I) THEN SO = F1(I)
820 NEXT I
830 F2(NA)=SO
840 FOR I = NA TO 1 STEP -1
850 LO=32767
860 IF T(NA)=T(I) THEN 950
870 FOR J = NA TO 1 STEP -1
880 IF H(J) < T(I) THEN 930
890 IF H(J) < > T(I) THEN 920
900 IF LO > E2(J) THEN LO = E2(J)
910 F2(I)=LO
920 NEXT J
```

[19-25] Basic language program for PERT network. (Continued on page 634)

[19-25] (Continued)

```
930 E2(I)=F2(I)-A(I)
940 GOTO 970
950 F2(I)=SO
960 E2(I)=F2(I)-A(I)
970 NEXT I
980 TE(1)=0: TL(1)=0
990 FOR I=2 TO NE
1000 K=0
1010 FOR J=1 TO NA
1020 IF T(J) <> I THEN  1060
1030 K=K+1
1040 ER(I,K) = F1(J)
1050 TL(I) = F2(J)
1060 NEXT J
1070 C(I)=K
1080 NEXT I
1090 FOR I = 2 TO NE
1100 TE(I) = -1
1110 FOR J = 1 TO C(I)
1120 IF ER(I,J) > TE(I) THEN TE(I)=ER(I,J)
1130 NEXT J
1140 NEXT I
1150 CLS:PRINT
1160 PRINT "EVENT         T(E)           T(L)           SLACK"
1170 PRINT " "
1180 FOR I = 1 TO NE
1190 SL(I) = TL(I)-TE(I)
1200 PRINT I, TE(I), TL(I), SL(I)
1210 NEXT I
1220 END
```

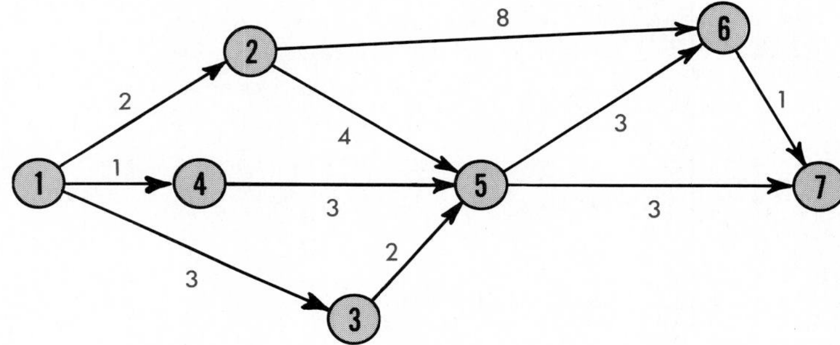

[19-26] Network used with PERT, Example 19-3.

data entries are then presented to the user in table form [19-28] and the user is given an opportunity to change any data element (starting event, ending event, and activity time) to add a new activity and to delete an activity. A capital "N" or "Y" is required when responding to the questions involving changes, additions, or deletions.

Output for this program [19-29] consists of a table much like that illustrated for the same problem in Chapter 14. For each event, the earliest and latest times to complete each event are presented together with the slack time. The critical path is identified as those events with a zero slack.

```
ENTER NUMBER OF EVENTS IN THE NETWORK? 7

ENTER THE NUMBER OF ACTIVITIES ? 10

 FOR EACH PAIR OF CONNECTED EVENTS

 ENTER THE STARTING EVENT, THE ENDING EVENT,
 AND THE ACTIVITY TIME--SEPARATE WITH COMMAS
? 1,2,2

 ENTER THE STARTING EVENT, THE ENDING EVENT,
 AND THE ACTIVITY TIME--SEPARATE WITH COMMAS
? 1,4,1

 ENTER THE STARTING EVENT, THE ENDING EVENT,
 AND THE ACTIVITY TIME--SEPARATE WITH COMMAS
? 1,3,3

 ENTER THE STARTING EVENT, THE ENDING EVENT,
 AND THE ACTIVITY TIME--SEPARATE WITH COMMAS
? 2,5,4
```

[19-27] Portion of the user input, Example 19-3.

```
SET      FROM      TO      TIME      SET      FROM      TO      TIME

1.0      1.0       2.0     2.0       2.0      1.0       4.0     1.0
3.0      1.0       3.0     3.0       4.0      2.0       5.0     4.0
5.0      4.0       5.0     3.0       6.0      3.0       5.0     2.0
7.0      2.0       6.0     8.0       8.0      5.0       6.0     3.0
9.0      5.0       7.0     3.0      10.0      6.0       7.0     1.0

DO YOU WISH TO MAKE ANY CHANGES (Y OR N) ? N

DO YOU WISH TO ADD A NEW ACTIVITY (Y OR N) ? N

DO YOU WISH TO DELETE AN ACTIVITY (Y OR N) ? N
```

[19-28] Summary of data input, Example 19-3.

```
EVENT      T(E)      T(L)      SLACK

1          0         0         0
2          2         2         0
3          3         5         2
4          1         4         3
5          6         7         1
6          10        10        0
7          11        11        0
Ok
```

[19-29] Output for Example 19-3.

Kirchhoff's Voltage Law

Kirchhoff's second law (voltage law) was explained and illustrated in Chapter 11. This law states that the algebraic sum of the voltages in any closed loop must be zero. This program applies the voltage law to solve for the value of the current in one, two, or three closed-loop networks. The program is listed in Figure [19-30].

```
10 CLS
20 M$="Y"
30 AD$="N"
40 H=0
50 DIM E(3), NP(10), R(3,10), C(5), SN(3), SP(3), Z(3)
60 FOR I=1 TO 3
70    FOR J= 1 TO 10
80       R(I,J)=0
90    NEXT J
100 NEXT I
110 INPUT "ENTER NUMBER OF LOOPS (1,2, OR 3)";N
120  FOR J=1 TO N
130     SN(J)=0
140     SP(J)=0
150     PRINT " "
160     PRINT " ENTER EMF VALUE IN VOLTS FOR LOOP";J
170     PRINT " USE A MINUS SIGN IF THE POLARITY ARROW"
180     PRINT " (MINUS TO PLUS) AT THE ENERGY SOURCE IS"
190     INPUT " OPPOSITE THE CLOCKWISE CURRENT FLOW ";E(J)
200     PRINT " "
210     PRINT " ENTER THE NUMBER OF RESISTANCES IN THIS"
220     PRINT " LOOP--PARALLEL RESISTANCES COUNT AS A "
230     INPUT " SINGLE RESISTANCE"; NP(J)
240        FOR K = 1 TO NP(J)
250        PRINT " "
260          PRINT " IS RESISTANCE "K
270          INPUT " A SINGLE , (S), OR PARALLEL, (P), RESISTANCE ?";L$
280          IF L$ = "S" THEN GOTO 380
290          INPUT "HOW MANY RESISTANCES ARE IN PARALLEL";P
300          SM=0
310             FOR M=1 TO P
320                INPUT " ENTER PARALLEL RESISTANCE VALUE IN OHMS";C(M)
330                SM = SM + 1/C(M)
340             NEXT M
350          R(J,K)= 1/SM
360          SN(J)=SN(J)+R(J,K)
370          IF J=N THEN 450 ELSE 420
380          INPUT " ENTER SINGLE RESISTANCE VALUE IN OHMS";R(J,K)
390          SN(J)=SN(J)+R(J,K)
400          IF J=N THEN 450
410          PRINT " "
420          PRINT " IS RESISTANCE "K
430          INPUT "SHARED WITH THE NEXT LOOP ? (Y OR N)";S$
440          IF S$="Y" THEN SP(J)=SP(J)+R(J,K)
450        NEXT K
460 NEXT J
470 CLS
480 PRINT "        SUMMARY OF INPUT DATA     "
490 CLS
500     PRINT "YOU WILL BE GIVEN AN OPPORTUNITY TO CHANGE THE INPUT DATA"
510     PRINT "THE EMF VALUES AND RESISTANCES WILL BE PRINTED ON THE SCREEN"
520     PRINT "AS YOU HAVE ENTERED THEM.  YOU WILL BE GIVEN AN OPPORTUNITY"
530     PRINT "TO CHANGE ANY VALUES THAT YOU WISH.  PARALLEL RESISTANCES"
540     PRINT "HAVE BEEN CONVERTED TO A SINGLE RESISTANCE."
550     PRINT "A MAXIMUM OF 9 RESISTANCES ARE ALLOWED PER LOOP"
560     PRINT "A GROUP OF PARALLEL RESISTORS COUNT AS ONE RESISTOR."
570     PRINT "IF AN ERROR WAS MADE ON A PARALLEL GROUP,"
580     PRINT "THEN THE PROGRAM MUST BE RERUN"
590     PRINT " "
600     PRINT " "
610     PRINT "PRESS ANY KEY TO CONTINUE"
620     A$=INKEY$:IF A$="" THEN 620
630     CLS
```

[19-30] BASIC language program, Kirchhoff's voltage law. (Continued on page 635)

[19-30] (Continued)

```
640 FOR I=1 TO N
650     PRINT " "
660    PRINT "FOR LOOP"I",THE EMF VALUE IS  "E(I)
670    PRINT "RESISTANCES 1, 2, AND 3 ARE (DISREGARD O'S)"
680    PRINT USING "#####.##"; R(I,1),R(I,2),R(I,3)
690    IF NP(I) <=3 THEN 730 ELSE PRINT "RESISTANCES 4, 5, AND 6 ARE (DISREGARD
O'S)"
700    PRINT USING "#####.##"; R(I,4),R(I,5),R(I,6)
710    IF NP(I) <=6 THEN 730 ELSE PRINT "RESISTANCES 7, 8, AND 9 ARE (DISREGARD
O'S)"
720    PRINT USING "#####.##"; R(I,7),R(I,8),R(I,9)
730 NEXT I
740 IF H=1 THEN INPUT "ANY MORE CHANGES (Y OR N) ";M$
750 IF M$="N" THEN 1130
760 PRINT " "
770 INPUT "DO YOU WISH TO CHANGE ANY EMF VALUE (Y OR N)";C$
780 IF C$ = "N" THEN 810
790 H=1
800 INPUT "ENTER THE LOOP NO. AND EMF VALUE--SEPARATE WITH COMMAS ",L,E(L)
810 INPUT "DO YOU WISH TO CHANGE ANY RESISTANCE VALUES (Y OR N)";D$
820 IF D$="N" THEN 880
830 PRINT "IF YOU WISH TO CHANGE A NUMERICAL VALUE ENTER N FOR"
840 PRINT "NUMERICAL; IF YOU WISH TO ADD OR DELETE A RESISTANCE"
850 INPUT "THEN ENTER AD FOR ADD/DELETE";TC$
860 IF TC$="AD" THEN 1000
870 IF D$="N" THEN 890 ELSE 920
880 IF C$="N" AND D$="N" THEN 890 ELSE 770
890 INPUT "WAS AN ERROR MADE ON A PARALLEL GROUP (Y OR N)";PE$
900 IF PE$="Y" THEN 1330
910 IF C$="N" AND D$="N" THEN 1080 ELSE 770
920 H=1
930 PRINT "ENTER LOOP NO., RESISTANCE NO. AND THE NEW VALUE"
940 INPUT "SEPARATE WITH COMMAS";I,J,NR
950 IF AD$="Y" THEN R(I,J)=0
960 SN(I)=SN(I)-R(I,J)+NR
970 INPUT "IS RESISTANCE SHARED WITH ANOTHER LOOP (Y OR N) ";S$
980 IF S$="Y" THEN SP(I)=SP(I)-R(I,J)+NR
990 R(I,J)=NR
1000 INPUT "DO YOU WISH TO ADD ANOTHER RESISTANCE (Y OR N)";AD$
1010 IF AD$="Y" THEN 920
1020 INPUT "DO YOU WISH TO DELETE A RESISTANCE (Y OR N)";DR$
1030 IF DR$="N" THEN 1070
1040 PRINT "ENTER THE LOOP NUMBER, RESITANCE NUMBER AND RESISTANCE"
1050 INPUT "TO BE DELETED--SEPARATE WITH COMMAS";I,J,R(I,J)
1060 NR=0: H=1: GOTO 960
1070 GOTO 770
1080 IF H=0 THEN 1130
1090 CLS
1100 PRINT "       CHANGED VALUES"
1110 PRINT " "
1120 GOTO 640
1130 IF N = 1 THEN II(I) = E(1)/SN(1):CLS: PRINT "I(1) = "II(I):GOTO 1330
1140 IF N = 3 GOTO 1230
1150 SN(2)=SN(2)+SP(1)
1160 DD = SN(1)*SN(2)-SP(1)^2
1170 II(1) = 1/DD*(SN(2)*E(1)+SP(1)*E(2))
1180 II(2) = 1/DD*(SP(1)*E(1)+SN(1)*E(2))
1190 CLS
1200 PRINT "I(1) = "II(1)
1210 PRINT "I(2) = "II(2)
1220 GOTO 1330
1230 SN(2)=SN(2)+SP(1)
1240 SN(3)=SN(3)+SP(2)
1250 DD = -SN(1)*SN(2)*SN(3)+SP(1)^2*SN(3)+SN(1)*SP(2)^2
1260 CLS
1270 II(1) = 1/DD*((SP(2)^2-SN(2)*SN(3))*E(1)-SP(1)*SN(3)*E(2)-SP(1)*SP(2)*E(3))

1280 II(2) = 1/DD*(-SP(1)*SN(3)*E(1)-SN(1)*SN(3)*E(2)-SN(1)*SP(2)*E(3))
1290 II(3) = 1/DD*(-SP(1)*SP(2)*E(1)-SN(1)*SP(2)*E(2)+(SP(1)^2-SN(1)*SN(2))*E(3)
)
1300 PRINT "I(1) = "II(1)
1310 PRINT "I(2) = "II(2)
1320 PRINT "I(3) = "II(3)
1330 END
```

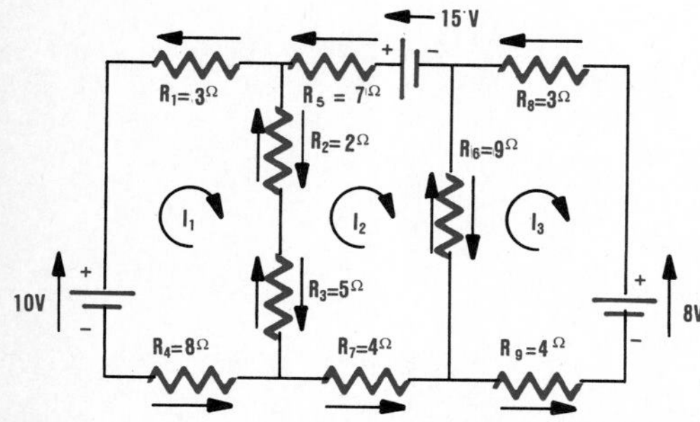

[19-31] Three-loop circuit used with Example 19-4.

Example 19-4

The use of the Kirchhoff's voltage law program is illustrated using the three-loop circuit of Figure [19-31]. It is desired to obtain numerical values for I_1, I_2, and I_3, the current flows in each of the three loops, Kirchhoff's voltage law states that the algebraic sum of the potential differences across the circuit elements in each loop must be zero. Thus, the program solves simultaneous equations for the I values. The program uses the conventions adopted in Chapter 11. Clockwise current flow is always positive, the sense or direction for the source voltage is from the negative to the positive pole, and the sense or direction of the voltage at each resistor is oriented so that the head of the arrow points in a direction opposite the current flow, which is always assumed to be positive. A negative result for any I value indicates that flow is in a counterclockwise direction. The complete set of user inputs is given in Figure [19-32]. The first prompt asks for the number of loops in the network—remember that the maximum is 3. For each loop, the emf value is entered, followed by the number of resistances in that loop. For each resistance, three responses are required. The first is the type of resistance—single (S) or parallel (P). A capital "S" or "P" is required. For the case of parallel resistances (not a part of this example), the user enters the value of each resistor and the program computes the resistance of a single equivalent resistor. The next prompt asks for the resistance value and then the user must indicate whether (capital "Y") or not (capital "N") this resistance is shared with the next loop. The message shown in Figure [19-33] then appears on the screen.

The next display [19-34] presents all values entered by the user as well as opportunities to change emf values and resistance values. The "Y" or "N" values should be capitalized. It is important to note that a "Y" response should be used for any change to a resistance. This includes a change in numerical value, an addition or a deletion. If an addition or deletion is desired, the user should respond "N" to the second prompt for a change. The next prompts will be for an addition and a deletion, respectively. The final prompt involves an error with any parallel group. If the user responds "Y", the program will terminate and must be rerun. Output for this example is given in Figure [19-35].

```
YOU WILL BE GIVEN AN OPPORTUNITY TO CHANGE THE INPUT DATA
THE EMF VALUES AND RESISTANCES WILL BE PRINTED ON THE SCREEN
AS YOU HAVE ENTERED THEM.  YOU WILL BE GIVEN AN OPPORTUNITY
TO CHANGE ANY VALUES THAT YOU WISH.  PARALLEL RESISTANCES
HAVE BEEN CONVERTED TO A SINGLE RESISTANCE.
A MAXIMUM OF 9 RESISTANCES ARE ALLOWED PER LOOP
A GROUP OF PARALLEL RESISTORS COUNT AS ONE RESISTOR.
IF AN ERROR WAS MADE ON A PARALLEL GROUP,
THEN THE PROGRAM MUST BE RERUN


PRESS ANY KEY TO CONTINUE
```

[19-33] Prompt used with Example 19-4.

```
FOR LOOP 1 ,THE EMF VALUE IS   10
RESISTANCES 1, 2, AND 3 ARE (DISREGARD O'S)
    3.00     2.00     5.00
RESISTANCES 4, 5, AND 6 ARE (DISREGARD O'S)
    8.00     0.00     0.00

FOR LOOP 2 ,THE EMF VALUE IS  -15
RESISTANCES 1, 2, AND 3 ARE (DISREGARD O'S)
    7.00     9.00     4.00

FOR LOOP 3 ,THE EMF VALUE IS  -8
RESISTANCES 1, 2, AND 3 ARE (DISREGARD O'S)
    3.00     4.00     0.00

DO YOU WISH TO CHANGE ANY EMF VALUE (Y OR N)? N
DO YOU WISH TO CHANGE ANY RESISTANCE VALUES (Y OR N)? N
WAS AN ERROR MADE ON A PARALLEL GROUP (Y OR N)? N
```

[19-34] Summary of input data, Example 19-4.

```
I(1) =  .2396097
I(2) = -.8124323
I(3) = -.9569931
Ok
```

[19-35] Output for Example 19-4.

[19-32] Complete user input, Example 19-4.

```
ENTER NUMBER OF LOOPS (1,2, OR 3)? 3

 ENTER EMF VALUE IN VOLTS FOR LOOP 1
 USE A MINUS SIGN IF THE POLARITY ARROW
 (MINUS TO PLUS) AT THE ENERGY SOURCE IS
 OPPOSITE THE CLOCKWISE CURRENT FLOW ? 10

 ENTER THE NUMBER OF RESISTANCES IN THIS
 LOOP--PARALLEL RESISTANCES COUNT AS A
 SINGLE RESISTANCE? 4

 IS RESISTANCE  1
 A SINGLE , (S), OR PARALLEL, (P), RESISTANCE ?? S
 ENTER SINGLE RESISTANCE VALUE IN OHMS? 3

 IS RESISTANCE  1
SHARED WITH THE NEXT LOOP ? (Y OR N)? N

 IS RESISTANCE  2
 A SINGLE , (S), OR PARALLEL, (P), RESISTANCE ?? S
 ENTER SINGLE RESISTANCE VALUE IN OHMS? 2
 A SINGLE , (S), OR PARALLEL, (P), RESISTANCE ?? S
 ENTER SINGLE RESISTANCE VALUE IN OHMS? 2

 IS RESISTANCE  2
SHARED WITH THE NEXT LOOP ? (Y OR N)? Y

 IS RESISTANCE  3
 A SINGLE , (S), OR PARALLEL, (P), RESISTANCE ?? S
 ENTER SINGLE RESISTANCE VALUE IN OHMS? 5

 IS RESISTANCE  3
SHARED WITH THE NEXT LOOP ? (Y OR N)? Y

 IS RESISTANCE  4
 A SINGLE , (S), OR PARALLEL, (P), RESISTANCE ?? S
 ENTER SINGLE RESISTANCE VALUE IN OHMS? 8

 IS RESISTANCE  4
SHARED WITH THE NEXT LOOP ? (Y OR N)? N

 ENTER EMF VALUE IN VOLTS FOR LOOP 2
 USE A MINUS SIGN IF THE POLARITY ARROW
 (MINUS TO PLUS) AT THE ENERGY SOURCE IS
 OPPOSITE THE CLOCKWISE CURRENT FLOW ? -15
 (MINUS TO PLUS) AT THE ENERGY SOURCE IS
 OPPOSITE THE CLOCKWISE CURRENT FLOW ? -15

 ENTER THE NUMBER OF RESISTANCES IN THIS
 LOOP--PARALLEL RESISTANCES COUNT AS A
 SINGLE RESISTANCE? 3

 IS RESISTANCE  1
 A SINGLE , (S), OR PARALLEL, (P), RESISTANCE ?? S
 ENTER SINGLE RESISTANCE VALUE IN OHMS? 7

 IS RESISTANCE  1
SHARED WITH THE NEXT LOOP ? (Y OR N)? N

 IS RESISTANCE  2
 A SINGLE , (S), OR PARALLEL, (P), RESISTANCE ?? S
 ENTER SINGLE RESISTANCE VALUE IN OHMS? 9

 IS RESISTANCE  2
SHARED WITH THE NEXT LOOP ? (Y OR N)? Y

 IS RESISTANCE  3
 A SINGLE , (S), OR PARALLEL, (P), RESISTANCE ?? S
 ENTER SINGLE RESISTANCE VALUE IN OHMS? 4
 IS RESISTANCE  3
 A SINGLE , (S), OR PARALLEL, (P), RESISTANCE ?? S
 ENTER SINGLE RESISTANCE VALUE IN OHMS? 4

 IS RESISTANCE  3
SHARED WITH THE NEXT LOOP ? (Y OR N)? N

 ENTER EMF VALUE IN VOLTS FOR LOOP 3
 USE A MINUS SIGN IF THE POLARITY ARROW
 (MINUS TO PLUS) AT THE ENERGY SOURCE IS
 OPPOSITE THE CLOCKWISE CURRENT FLOW ? -8

 ENTER THE NUMBER OF RESISTANCES IN THIS
 LOOP--PARALLEL RESISTANCES COUNT AS A
 SINGLE RESISTANCE? 2

 IS RESISTANCE  1
 A SINGLE , (S), OR PARALLEL, (P), RESISTANCE ?? S
 ENTER SINGLE RESISTANCE VALUE IN OHMS? 3

 IS RESISTANCE  2
 A SINGLE , (S), OR PARALLEL, (P), RESISTANCE ?? S
 ENTER SINGLE RESISTANCE VALUE IN OHMS? 4
```

PROBLEMS

19-1. Assume that interest is at 10 per cent per period, and use the program shown in Figure [19-14] to determine the equivalent present worth of the cash flow shown in Figure [19-P1].

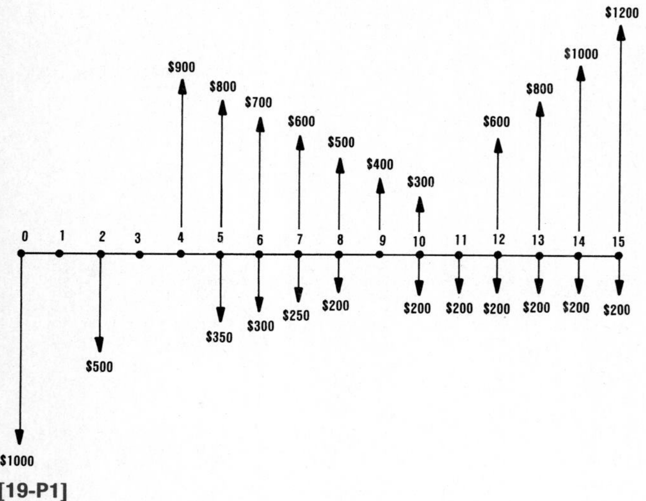

[19-P1]

19-2. Assume that interest is at 15 per cent per period. Where appropriate, use the program shown in Figure [19-14] to determine the equivalent annual worth of the cash flow shown in Figure [19-P2].

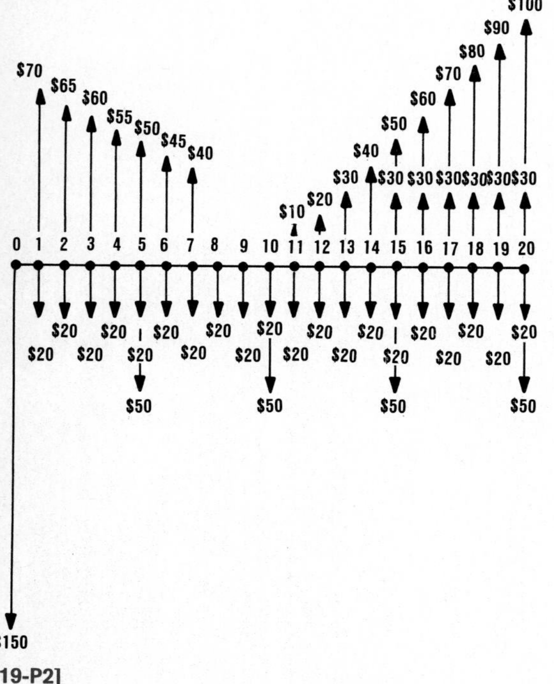

[19-P2]

19-3. The cash flow diagram shown in Figure [19-P3] represents projected receipts and disbursements for an engineering proposal. What is the rate of return for this proposal? *Hint*: The rate of return is the value of i that makes the net present worth equal to zero. Where appropriate, use the program shown in Figure [19-14] and solve by trial and error. Begin with any value of i, say 10 per cent. If the net present worth using this value of i is greater than zero, you have guessed too low; rerun the program with a larger value of i. If the new present worth is less than zero, you have used a value of i that is too large; rerun the program with a smaller value of i. Repeat this process until the net present worth is *approximately* zero. Express your result to the nearest tenth of a percent.

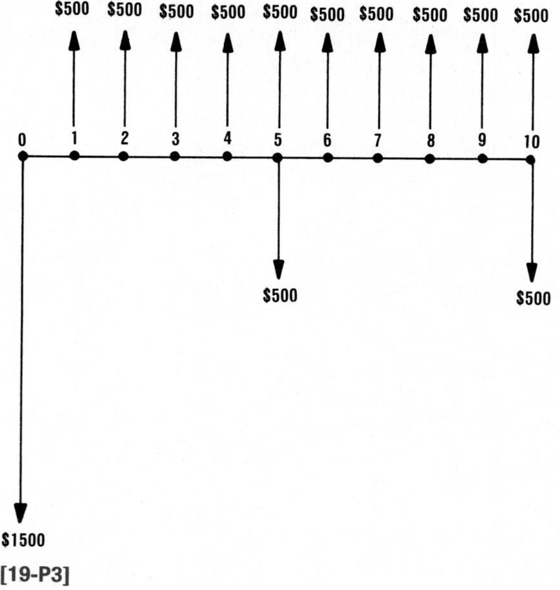

[19-P3]

19-4. Where appropriate, use the program shown in Figure [19-14] to solve Problem 8-9.

19-5. Where appropriate, use the program shown in Figure [19-14] to solve Problem 8-21.

19-6. Where appropriate, use the program shown in Figure [19-14] to solve Problem 8-25.

19-7. Where appropriate, use the program shown in Figure [19-14] to solve Problem 8-26.

19-8. Where appropriate, use the program shown in Figure [19-14] to solve Problem 8-27.

19-9. Where appropriate, use the program shown in Figure [19-14] to solve Problem 8-28.

19-10. There are hundreds of paths from node 1 to node 13 in the network shown in Figure [19-P10]. Imagine the difficulties and time required to determine the shortest route from node 1 to node 13 using the "brute force" approach (this approach involves evaluating the length of *all* of the paths from node 1 to node 13. Where appropriate use the program shown in Figure [19-21] to solve this problem.

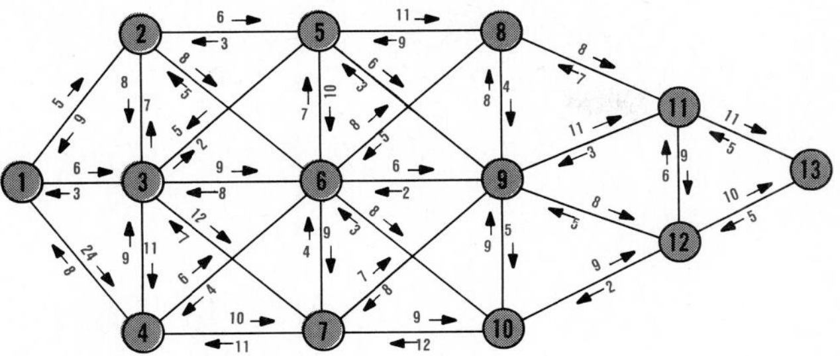

[19-P10]

19-11. Where appropriate, use the program shown in Figure [19-21] to find the shortest route from node 13 to node 1 for the network of Problem 19-10. *Hint:* Renumber the nodes so that node 13 becomes node 1, node 12 becomes node 2, and so on (i.e., node i becomes node 14-i).

19-12. Where appropriate, use the program shown in Figure [19-21] to find the shortest route from node 1 to node 9 in the network shown in Figure [19-P12].

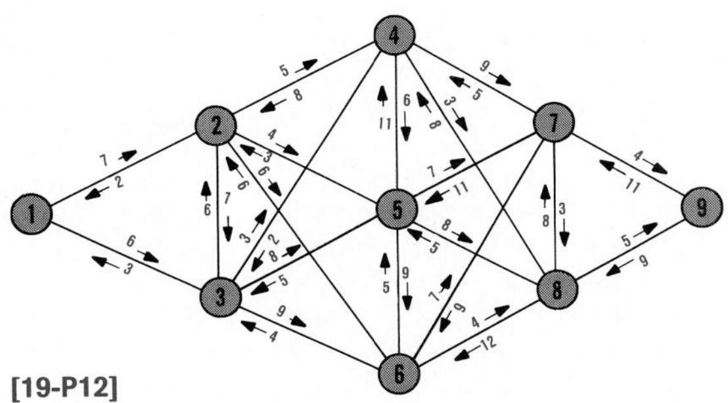

[19-P12]

19-13. Where appropriate, use the program shown in Figure [19-21] to find the shortest route from node 9 to node 1 in the network of Problem 19-12. *Hint:* Renumber the nodes so that node 9 becomes node 1, node 8 becomes node 2, and so on (i.e., node *i* becomes node 10-i).

19-14. Where appropriate, use the program shown in Figure [19-21] to find the shortest route from node 1 to node 9 in the network shown in Figure [19-P14].

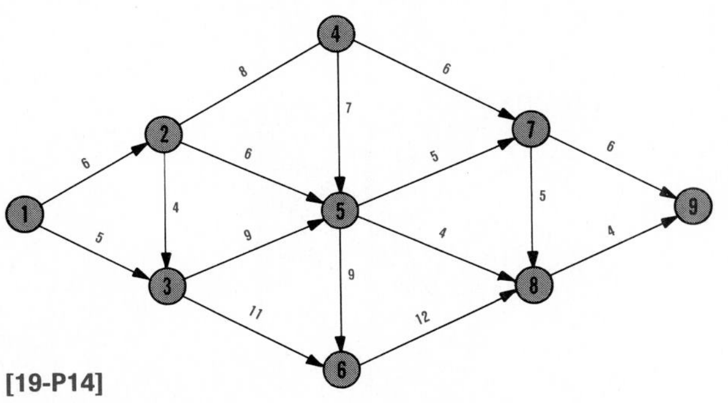

[19-P14]

19-15. Suppose that the network of Problem 19-14 is a PERT network rather than a shortest-route problem. Suppose that the numerical values for the branches of the network are expected times to complete activities rather than distances. Where appropriate use the program shown in Figure [19-25] to find the critical path through the network.

19-16. Where possible, use the program shown in Figure [19-25] to solve the PERT network shown in Figure [19-P16]. *Hint:* Make the activity from node 1 to node 2 a dummy activity by assigning to this activity a time of zero.

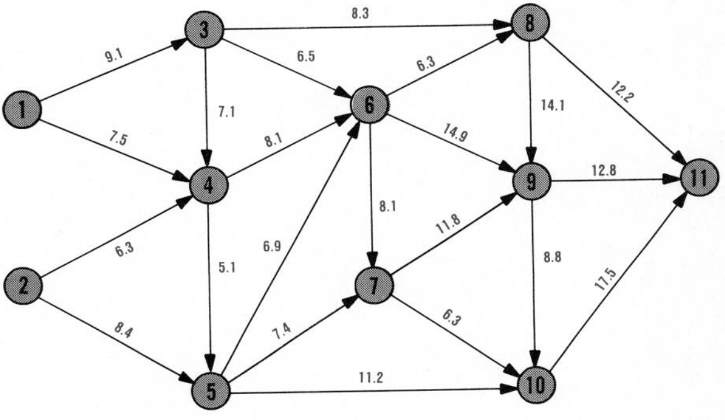

[19-P16]

19-17. Where possible, use the program shown in Figure [19-25] to solve Problem 14-1.

19-18. Where possible, use the program shown in Figure [19-25] to solve Problem 14-6.

19-19. Where possible, use the program shown in Figure [19-25] to solve Problem 14-8.

19-20. Where possible, use the program shown in Figure [19-30] to solve for I_1 and I_2 in the closed-loop system shown in Figure [19-P20].

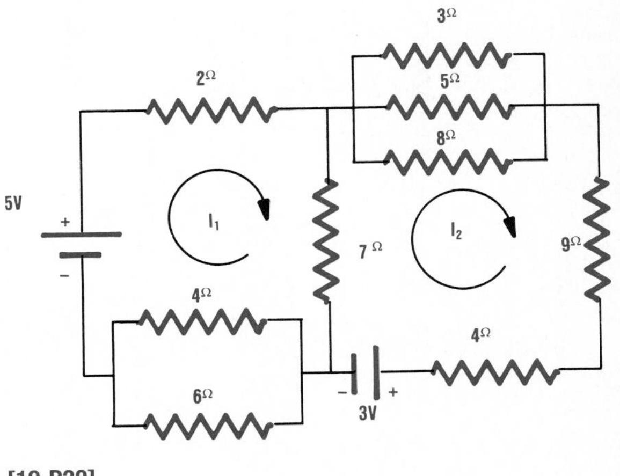

[19-P20]

19-21. Where possible, use the program shown in Figure [19-30] to solve for I_1, I_2, and I_3 in the closed-loop system shown in Figure [19-P21].

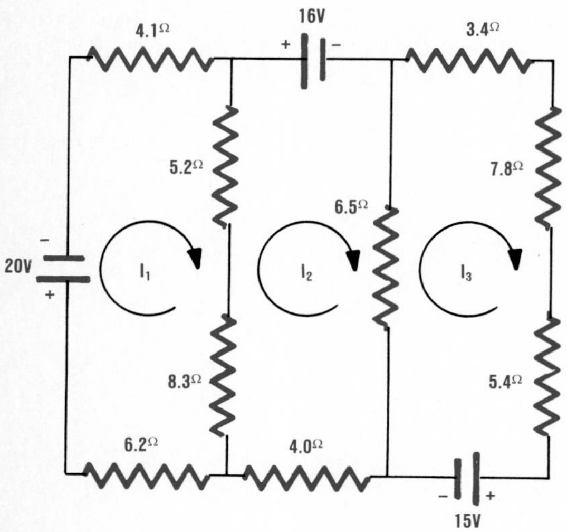

[19-P21]

19-22. Where possible, use the program shown in Figure [19-30] to solve for I_1, I_2, and I_3 in the closed-loop system shown in Figure [19-P22].

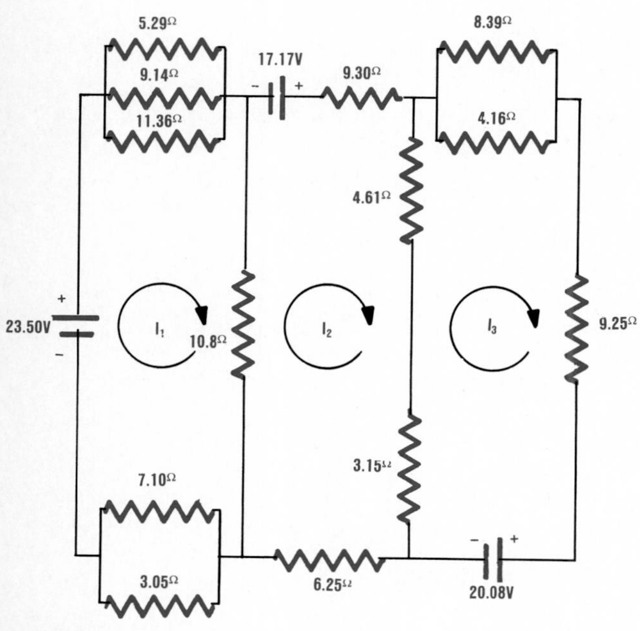

[19-P22]

19-23. Where possible, use the program shown in Figure [19-30] to solve Problem 11-93.

19-24. Where possible, use the program shown in Figure [19-30] to solve Problem 11-94.

19-25. Where possible, use the program shown in Figure [19-30] to solve Problem 11-95.

19-26. Where possible, use the program shown in Figure [19-30] to solve Problem 11-96.

19-27. Where possible, use the program shown in Figure [19-30] to solve Problem 11-97.

19-28. Where possible, use the program shown in Figure [19-30] to solve Problem 11-98.

19-29. Optimization plays an important role in engineering design. Most realistic problems are posed in terms of the "best," the "optimum," the "maximum," or the "minimum." Usually, cost is one of the most important variables that needs to be minimized, but there are always cost/benefit trade-offs that need to be evaluated.

There exists a large number of tools for solving optimization-type problems, depending on how the problems are stated. These tools are usually the subject of a senior- or graduate-level course in most engineering schools. They typically require some advanced mathematics and computer capability.

In optimization problems there is always some *objective function* that can be defined. This is nothing more than a statement of what is to be minimized or maximized. Any condition that is to be fixed or maintained during the process of finding the optimum solution is called a *constraint*.

Suppose that you were to build a container that has rectangular sides and bottom made of some impermeable material. The volume of the container must be 5 m^3. The application does not require that the container have a top. Due to the expense of the material that we have decided to use, we wish to minimize the amount of material.

In this problem the objective function is to *find the minimum surface area for a given volume*. The constraint is that the volume must be equal to 5 (m^3). With the labels from Figure [19-P29], this means that we must minimize the function

$$f = 2xz + 2yz + xy$$

subject to

$$xyz = 5$$

Set up this problem on a computer (FORTRAN, BASIC, or spreadsheet) to solve by what might be called "brute force" rather than by a sophisticated optimization scheme. That is, first reduce these two equations to one equation with three unknowns. Then program the computer to prompt the user for

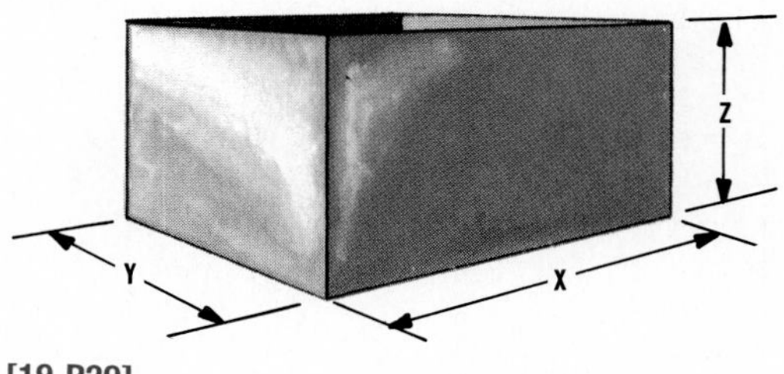

[19-P29]

two of the unknowns (the remaining two dimensions contained in the equation) and return to the user the value of the function f and the length of the third side. Exercise this program to explore the behavior of f.

19-30. Suppose that you are assigned the task of designing a solar power plant that used solar cells to convert sunlight into electricity. Solar cells act much as batteries except that they are activated by sunlight and can operate over a wide spectrum of voltage levels. This should be obvious from the problem explored earlier in this chapter.

Your requirement for the power plant is that is must produce a voltage of 1000 V and a current of 1000 A (dc or direct current). The nature of the load that is to be connected to this plant is such that it forces the voltage on the plant to be held constant at the 1000-V level.

There are many ways to connect the cells together such that the design requirements are met, but not all these ways use the same number of cells. However, your goal (imposed by your budget) is to require as few solar cells as possible.

You will find that neither the maximum voltage nor the maximum current that a single cell will produce can meet the design requirements for your problem. Therefore, you will have to wire or connect the solar cells into circuits in order to get the job done. This is identical to placing flashlight batteries in series (as in a two-cell flashlight) or in parallel arrangements to meet the voltage and/or current requirements of some electrical gadget.

Series connections: Wiring the cells in series is demonstrated in Figure [19-P30a]. In series strings, the voltages V of each cell add to give the total voltage V_{tot} across the string. The current I through each of the cells is the same as the total current I_{tot} in the string.

Parallel connections: Wiring the cells in parallel is demonstrated in Figure [19-P30b]. In parallel strings, the currents I through each cell add to give the total currents I_{tot} across the string. The voltage V across each of the cells is the same as the total voltage V_{tot} across the circuit.

Series–parallel connections: Wiring the cells in series–parallel combinations is demonstrated in Figure [19-P30c]. (For your application, you will want to keep the number of solar cells in each series string the same.) The voltage V_{tot} is the voltage across each series string in the circuit. The total current I_{tot} is equal to the sum of the currents in each series string. The current I through each cell is the same as the current through the series string of which the cell is a part.

The cells you have at your disposal have an open-circuit voltage, V_{oc}, of 1.0 V and a short-circuit current, I_{oc}, of 1.0 A under full sun (see the section on the ideal solar cell earlier in this chapter). The design temperature of the cells is to be 90° F. Their physical size is 7.5 cm in diameter and negligible thickness.

How many cells would be required for a good design?

19-31. Of course, it is impractical just to lay solar cells out on the ground, wire them together, and sit back and collect electrical energy. The cells need to be mounted on some surface and this surface needs to be supported by some structure. The term "array" is usually used to describe the cells mounted on some surface. These arrays then have to be supported by a structure.

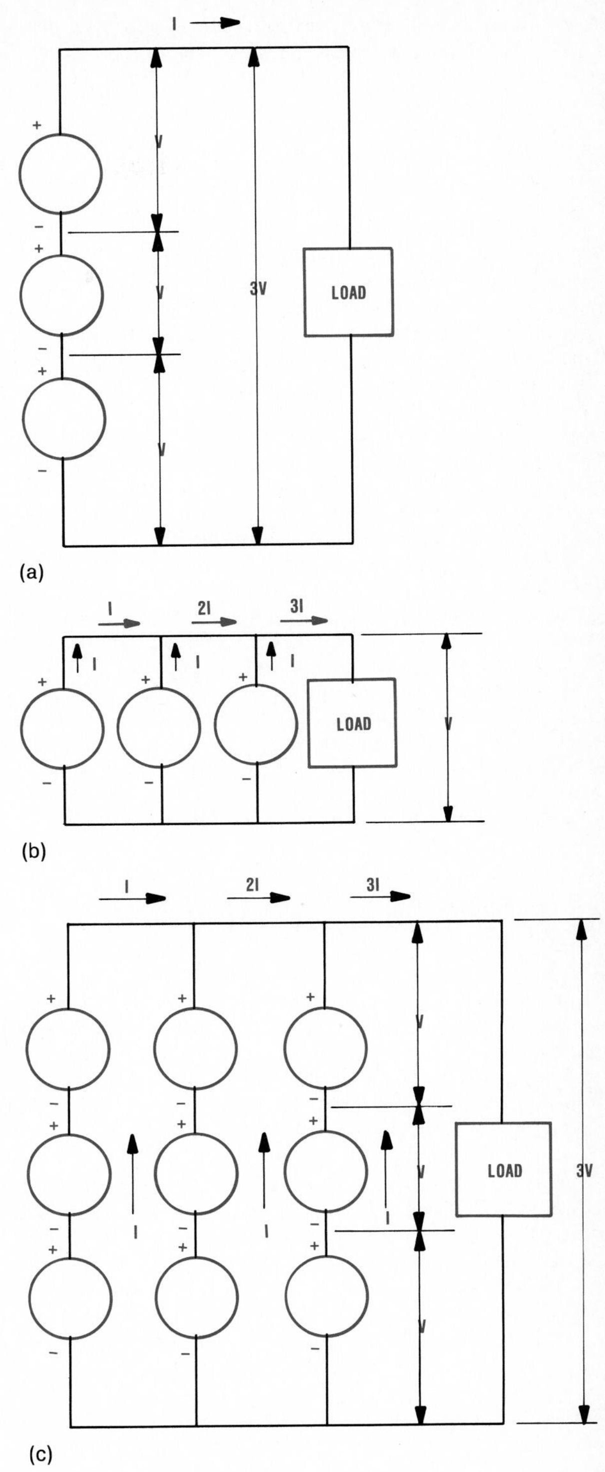

[19-P30]

You are to design the arrays and their supporting structure. Your design should include design sketches and lists of materials. It should specify the geometry for positioning the cells on an appropriate surface (you must also specify what that surface is made from). That is, the array configuration or geometry should be determined by you.

To prevent electrical arcing between cells, you should allow 5 mm separation between any two cells, regardless of how you configure the arrays.

The glue or actual mounting technique for holding the cells on this surface need not concern you. Similarly, weatherizing the cells should be of no concern to you.

Your design should specify the support structure that will hold the arrays. You may use wood or metal structural members, but these members should be made of standard-size materials (e.g., 2 × 4 wood member, etc.).

The support structure must orient the cells so that they are tilted up from the horizontal at an angle of 35° and generally point toward the south. The adjacent sketch should clarify this.

The land on which you are to build this power plant is flat and level. Land preparation is not a concern for you.

Assume that you already own equipment that will produce a foundation that consists of four concrete pads, each having one 30-mm-diameter anchor bolt. The four bolts with each foundation should be used to secure the structure. The bolts are located at the corners of an imaginary rectangle with sides of 1 × 2 m. Each bolt is located in the center of a 0.25 × 0.25 m pad. The top of all the pads are flat and in the same horizontal plane. The top of each pad is located 0.5 m above the ground. These foundations can be replicated over and over to hold as many arrays as you need.

Now conceive and document your "design."

19-32. You are the bombardier of an Air Force plane that has just been assigned a critical mission. Unfortunately, the equipment assigned to you is rather antiquated. As a result, you decide to computerize the calculations you need to make in order to hit your target. The equations of projectile motion apply here. The known quantities are the velocity (assumed to be horizontal) and altitude of the plane you are flying. Program the calculations (in BASIC, FORTRAN, or on a spreadsheet) that will compute the distance from the target at which you must release your bomb. (Hint: the bombs initial velocity is the same as the plane's velocity.)

19-33. You have just signed a lucrative contract to become a place kicker for a National Football League team. Using the equations of projectile motion, program (using BASIC, FORTRAN, or a spreadsheet) the computations that will yield the velocity (magnitude and direction) that you must impart to the ball as it leaves the ground in order to just clear the crossbar between the goal posts. Your input should be the distance from the goal line. Is there a unique solution here?

19-34. The following equations describe the temperature distribution in a rod that has one end maintained at temperature T_L and the other end maintained at temperature T_R while having its circumference exposed to air at temperature T_{inf}.

$$T_1 = T_L$$
$$T_1 - 2(\text{kS} + 1)T_2 + T_3 = -2\text{kS}T_{\text{inf}}$$
$$T_2 - 2(\text{kS} + 1)T_3 + T_4 = -2\text{kS}T_{\text{inf}}$$
$$T_3 - 2(\text{kS} + 1)T_4 + T_5 = -2\text{kS}T_{\text{inf}}$$
$$T_4 - 2(\text{kS} + 1)T_5 + T_6 = -2\text{kS}T_{\text{inf}}$$
$$T_6 = T_R$$

The rod is sketched below:

T_L 1 2 3 4 5 6 T_R

T_1 is the temperature at point 1, T_2 is the temperature at point 2, etc. When air is blown over the surface of the rod, it affects the constant k. The size of the rod affects the constant S. The true physical significance of these constants will become clear if you ever take a course in the subject of *heat transfer*.

The set of equations above, when solved yields the temperature distribution along the length of the rod. These are simultaneous equations and can be solved in the normal manual algebraic manner (with much perseverance) or by automated (computer) procedures. One way that is quite effective in this situation is the Gauss-Seidel iterative method. In this method a first guess is made for all the unknown temperatures ($T_2, \ldots, T_5$). Then each question is solved for one of the unknown temperatures. For example,

$$T_2 = F_2(T_1,T_3,T_4,T_5,T_6) \text{ from equation 2,}$$
$$T_3 = F_3(T_1,T_2,T_4,T_5,T_6) \text{ from equation 3,}$$
$$T_4 = F_4(T_1,T_2,T_3,T_5,T_6) \text{ from equation 4,}$$
$$T_5 = F_5(T_1,T_2,T_3,T_4,T_6) \text{ from equation 5.}$$

The solution procedure goes like this.

1. Use the guesses for T_3, T_4, and T_5, to calculate a new T_2, using the first equation.
2. Use this new T_2 and the guesses for T_4, and T_5, to calculate a new T_3 using the second equation.

3.,4.,5.,6.,etc. Continue this march through the equations using the latest values of the T's to get updated values until a new set of T_2, T_3, T_4, and T_5 is obtained. Calling this new set the old guesses, repeat the march through the process by starting over at step 1. Continue this until the set of numbers has "converged."

Solve this set of equations

a. Using a spreadsheet (and plot the temperatures).

b. By writing a program in either BASIC or FORTRAN. Make your solution methodology amenable to easy incorporation of user supplied T_{inf}, T_1, T_6, k, and S. In a written problem summary, explain what you have learned about the behavior of the temperature distribution by varying the input parameters. Is the distribution sensitive to k, S, or T_{inf}?

Appendixes

Appendix I

Conversion Tables

Example: Convert 87.5 cm to miles (see Table 1).

$$1 \text{ cm} = 6.214 \times 10^{-6} \text{ mile}$$
$$87.5 \text{ cm} = 87.5(6.214 \times 10^{-6}) \text{ mile}$$
$$= 5.44 \times 10^{-4} \text{ mile}$$

Example: Convert 14.7 lb/in.^{2} to pascals (see Table 7).

$$1 \text{ lb/in.}^2 = 6.895 \times 10^3 \text{ pascals}$$
$$14.7 \text{ lb/in.}^2 = 14.7(6.895 \times 10^3) \text{ pascals}$$
$$= 1.014 \times 10^5 \text{ pascals}$$

In the following tables, the E, minus or plus an exponent, following a value indicates the power of 10 by which this value should be multiplied. Thus 6.214 E − 06 means 6.214×10^{-6}.

1. LENGTH EQUIVALENTS

cm	*in.*	*ft*	*m*	*mi*[a]	
1	3.937 E − 01	3.281 E − 02	1.0 E − 02	6.214 E − 06	cm
2.540	1	8.333 E − 02	2.54 E − 02	1.578 E − 05	in.
3.048 E + 01	1.2 E + 01	1	3.048 E − 01	1.894 E − 04	ft
1.0 E + 02	3.937 E + 01	3.281	1	6.214 E − 04	m
1.609 E + 05	6.336 E + 04	5.280 E + 03	1.609 E + 05	1	mi

[a] Mile.

Additional measures

Metric: 1 km = 10^3 m
1 mm = 10^{-3} m
1 μm = 10^{-6} m (micrometer or micron)
1 Å = 10^{-10} m (angstrom)

English: 1 mil = 10^{-3} in.
1 yd = 3.0 ft
1 rod = 5.5 yd = 16.5 ft
1 furlong = 40 rod = 660 ft

2. AREA EQUIVALENTS

m^2	$in.^2$	ft^2	acres	mi^2	
1	1.55 E + 03	1.076 E + 01	2.471 E − 04	3.861 E − 07	m^2
6.452 E − 04	1	6.944 E − 03	1.594 E − 07	2.491 E − 10	$in.^2$
9.290 E − 02	1.44 E + 02	1	2.296 E − 05	3.587 E − 08	ft^2
4.047 E + 03	6.273 E + 06	4.356 E + 04	1	1.562 E − 03	acres
2.590 E + 06	4.018 E + 09	2.788 E + 07	6.40 E + 02	1	mi^2

Additional measures

1 hectare = 10^4 m^2

3. VOLUME EQUIVALENTS

cm^3	$in.^3$	ft^3	gal (U.S.)	
1	6.102 E − 02	3.532 E − 05	2.642 E − 04	cm^3
1.639 E + 01	1	5.787 E − 04	4.329 E − 03	$in.^3$
2.832 E + 04	1.728 E + 03	1	7.481	ft^3
3.785 E + 03	2.31 E + 02	1.337 E − 01	1	gal (U.S.)

Additional measures

Metric: 1 liter = 10^3 cm^3
1 m^3 = 10^6 cm^3

English: 1 quart = 0.250 gal (U.S.)
1 bushel = 9.309 gal (U.S.)
1 barrel = 42 gal (U.S.)
(petroleum measure only)
1 imperial gal = 1.20 gal (U.S.) approx.
1 board foot (wood) = 144 $in.^3$
1 cord (wood) = 128 ft^3

4. MASS EQUIVALENTS

kg	slug	lb_m[a]	g	
1	6.85 E − 02	2.205	1.0 E + 03	kg
1.46 E + 01	1	3.22 E + 01	1.46 E + 04	slug
4.54 E − 01	3.11 E − 02	1	4.54 E + 02	lb_m
1.0 E − 03	6.85 E − 05	2.205 E − 03	1	g

[a] Not recommended.

5. FORCE EQUIVALENTS

N[a]	lb_f[b]	dyn[c]	kg_f[d]	g_f[d]	Poundal[d]	
1	2.248 E − 01	1.0 E + 05	1.019 E − 01	1.019 E + 02	7.234	N
4.448	1	4.448 E + 05	4.54 E − 01	4.54 E + 02	3.217 E + 01	lb_f
1.0 E − 05	2.248 E − 06	1	1.02 E − 06	1.02 E − 03	7.233 E − 05	dyn
9.807	2.205	9.807 E + 05	1	1.0	7.093 E + 01	kg_f
9.087 E − 03	2.205 E − 03	9.087 E + 02	1.0 E − 03	1	7.093 E − 02	g_f
1.382 E − 01	3.108 E − 02	1.383 E + 04	1.410 E − 02	1.410 E + 01	1	poundal

Additional measures
1 metric ton = 10^3 kg_f = 2.205 × 10^3 lb_f
1 pound troy = 0.8229 lb_f
1 oz[b] = 6.25 × 10^{-2} lb_f
1 oz troy = 6.857 × 10^{-2} lb_f
1 grain = 0.0001429 lb_f

[a]Newton.
[b]Avoirdupois.
[c]Dyne.
[d]Not recommended.

6. VELOCITY AND ACCELERATION EQUIVALENTS

a. Velocity

cm/s	*ft/sec*	*mi/hr (mph)*	*km/h*	
1	3.281 E − 02	2.237 E − 02	3.60 E − 02	cm/s
3.048 E + 01	1	6.818 E − 01	1.097	ft/sec
4.470 E + 01	1.467	1	1.609	mi/hr
2.778 E + 01	9.113 E − 01	6.214 E − 01	1	km/h

b. Acceleration

cm/s²	*ft/sec²*	*ḡ[a]*	
1	3.281 E − 02	1.019 E − 03	cm/s²
3.048 E + 01	1	3.109 E − 03	ft/sec²
9.807 E + 02	3.217 E + 01	1	ḡ

Additional measures
1 knot = 1.152 mi/hr

[a]Standard acceleration of gravity.

7. PRESSURE EQUIVALENTS

dyn/cm²	*N/m² (pascal)*	*lb/in.² (psi)*	*lb_f/ft^2 (psf)*	*atm[a]*	*Head[b] in. (Hg)*	*Head[b] ft (H_2O)*	
1	1.0 E − 01	1.45 E − 05	2.089 E − 03	9.869 E − 07	2.953 E − 05	3.349 E − 05	dyn/cm²
1.0 E + 01	1	1.45 E − 04	2.089 E − 02	9.869 E − 06	2.953 E − 04	3.349 E − 04	N/m²
6.895 E − 04	6.895 E + 03	1	1.44 E + 02	6.805 E − 02	2.036	2.309	lb_f/in.²
4.788 E + 02	4.788 E + 01	6.944 E − 03	1	4.725 E − 04	1.414 E − 02	1.603 E − 02	lb/ft²
1.013 E + 06	1.013 E + 05	1.47 E + 01	2.116 E + 03	1	2.992 E + 01	3.393 E + 01	atm
3.336 E + 04	3.386 E + 03	4.912 E − 01	7.073 E + 01	3.342 E − 02	1	1.134	in. (Hg)
2.986 E + 04	2.986 E + 03	4.331 E − 01	6.237 E + 01	2.947 E − 02	8.819 E − 01	1	ft (H_2O)

Additional measure
1 bar = 1 dyne/cm²

[a]Standard atmospheric pressure.
[b]At standard gravity and 0°C for Hg, 15°C for H_2O.

8. WORK AND ENERGY EQUIVALENTS

J[a]	*ft-lb$_f$*	*W-h*	*Btu*[b]	*kcal*[c]	*kg-m*	
1	7.376 E − 01	2.778 E − 04	9.478 E − 04	2.388 E − 04	1.020 E − 01	J
1.356	1	3.766 E − 04	1.285 E − 03	3.238 E − 04	1.383 E − 01	ft-lb$_f$
3.60 E + 03	2.655 E + 03	1	3.412	8.599 E − 01	3.671 E + 02	W-h
1.055 E + 03	7.782 E + 02	2.931 E − 01	1	2.520 E − 01	1.076 E + 02	Btu
4.187 E + 03	3.088 E + 03	1.163	3.968	1	4.269 E + 02	kcal
9.807	7.233	2.724 E − 03	9.295 E − 03	2.342 E − 03	1	kg-m

Additional measures

1 newton-meter = 1 J

1 erg = 1 dyne-cm = 10^{-7} J

1 cal = 10^{-3} kcal

1 therm = 10^{-5} Btu

1 million electron volts (MeV) = 1.602×10^{-13} J

[a] Joule.
[b] British thermal unit.
[c] Kilocalorie.

9. POWER EQUIVALENTS

J/s	*ft-lb$_f$/s*	*hp*[a]	*kW*	*Btu/h*	
1	7.376 E − 01	1.341 E − 03	1.0 E − 03	3.412	J/s
1.356	1	1.818 E − 03	1.356 E − 03	4.626	ft-lb$_f$/s
7.457 E + 02	5.50 E + 02	1	7.457 − 01	2.545 E + 03	hp
1.0 E + 03	7.376 E + 02	1.341	1	3.412 E + 03	kW
2.931 E − 01	2.162 E − 01	3.930 E − 04	2.931 E − 04	1	Btu/h

Additional Measures

1 W = 10^{-3} kW

1 cal/s = 14.29 Btu/h

1 poncelet = 100 kg-m/s = 0.9807 kW

1 ton of refrigeration = 1.2×10^4 Btu/h

[a] Horsepower.

Time

1 week	7 days	168 hours	10,080 minutes	604,800 seconds
1 mean solar day		24 hours	1440 minutes	86,400 seconds
1 calendar year	365 days	8760 hours	5.256 E + 05 minutes	3.1536 E + 07 seconds

1 tropical mean solar year 365.2422 days (basis of modern calendar)

Temperature

Δ1° Celsius (C) = Δ1° Kelvin (K) = Δ1.8° Fahrenheit (F) = 1.8° Rankine (R)
0°C = 273.15°K = 32°F = 491.67°R = 0°R 0°C = $\frac{5}{9}$(0°F − 32)
0°K = −273.15°C = −459.67°F 0°F = $\frac{9}{5}$(°C + 32)

Electrical

1 coulomb	1.036×10^5 faradays	0.1 abcoulomb	2.998×10^9 statcoulombs
1 ampere		0.1 abampere	2.998×10^9 statcoulombs
1 volt	10^3 millivolts	10^8 abvolts	3.335×10^{-3} statvolt
1 ohm	10^6 megohms	10^9 abohms	1.112×10^{-12} statohm
1 farad	10^6 microfarads	10^{-9} abfarad	8.987×10^{11} statfarads

Appendix II

Special Tables and Formulas

SPECIFIC GRAVITIES AND SPECIFIC WEIGHTS

Material	Average specific gravity	Average specific weight, lb_f/ft^3
Acid, sulfuric	1.80	112
Air, S.T.P.	0.001293	0.0806
Alcohol, ethyl	0.790	49
Aluminum, cast	2.65	165
Asbestos	2.5	153
Ash, white	0.67	42
Ashes, cinders	0.68	44
Asphaltum	1.3	81
Babbitt metal, soft	10.25	625
Basalt, granite	1.50	96
Brass, cast-rolled	8.50	534
Brick, common	1.90	119
Bronze	8.1	509
Cedar, white, red	0.35	22
Cement, portland, bags	1.44	90
Chalk	2.25	140
Clay, dry	1.00	63
Clay, loose, wet	1.75	110
Coal, anthracite, solid	1.60	95
Coal, bituminous, solid	1.35	85
Concrete, gravel, sand	2.3	142
Copper, cast, rolled	8.90	556
Cork	0.24	15
Cotton, flax, hemp	1.48	93
Copper ore	4.2	262
Earth	1.75	105
Fir, Douglas	0.50	32
Flour, loose	0.45	28
Gasoline	0.70	44
Glass, crown	2.60	161
Glass, flint	3.30	205
Glycerine	1.25	78
Gold, cast-hammered	19.3	1205
Granite, solid	2.70	172
Graphite	1.67	135
Gravel, loose, wet	1.68	105
Hickory	0.77	48
Ice	0.91	57
Iron, gray cast	7.10	450
Iron, wrought	7.75	480
Kerosene	0.80	50
Lead	11.34	710
Leather	0.94	59
Limestone, solid	2.70	168
Limestone, crushed	1.50	95
Mahogany	0.70	44
Manganese	7.42	475
Marble	2.70	166
Mercury	13.56	845
Monel metal, rolled	8.97	555
Nickel	8.90	558
Oak, white	0.77	48
Oil, lubricating	0.91	57
Paper	0.92	58
Paraffin	0.90	56
Petroleum, crude	0.88	55
Pine, white	0.43	27
Platinum	21.5	1330
Redwood, California	0.42	26
Rubber	1.25	78
Sand, loose, wet	1.90	120

(Continued on next page)

Material	Average specific gravity	Average specific weight, lb_f/ft^3	Material	Average specific gravity	Average specific weight lb_f/ft^3
Sandstone, solid	2.30	144	Tungsten	19.22	1200
Seawater	1.03	64	Turpentine	0.865	54
Silver	10.5	655			
Steel, structural	7.90	490	Water, 4°C	1.00	62.4[a]
Sulfur	2.00	125	Water, snow, fresh fallen	0.125	8.0
Teak, African	0.99	62			
Tin	7.30	456	Zinc	7.14	445

[a] The value for the specific weight of water which is usually used in problem solutions is 62.4 lb_f/ft^3 or 8.34 lb_f/gal.

FORMULAS USEFUL FOR SOLVING UNIFORM MOTION PROBLEMS

Legend
V velocity V_2 final velocity t time a acceleration
V_1 initial velocity V_{av} average velocity S distance

Given:	To find:	Suggested formulas	Given:	To find:	Suggested formulas
V_1, V_2, t	S	$S = \left(\frac{V_1 + V_2}{2}\right)t$	V_1, a, S	V_2	$V_2 = \sqrt{V_1^2 + 2aS}$
V_1, V_2, a	S	$S = \frac{V_2^2 - V_1^2}{2a}$	V_1, S, t	V_2	$V_2 = \frac{2S}{t} - V_1$
V_1, a, t	S	$S = V_1 t \frac{at^2}{2}$	V_1, V_2, S	t	$t = \frac{2S}{V_1 + V_2}$
V_1,V_2	V_{av}	$V_{av} = \frac{V_1 + V_2}{2}$	V_1, a, S	t	$t = \frac{-V_1 \pm \sqrt{V_1^2 + 2aS}}{a}$
S, t	V_{av}	$V_{av} = \frac{S}{t}$	V_1, V_2, a	t	$t = \frac{V_2 - V_1}{a}$
V_2, a, t	V_1	$V_1 = V_2 - at$	V_1, V_2, t	a	$a = \frac{V_2 - V_1}{t}$
V_2, a, S	V_1	$V_1 = \sqrt{V_2^2 - 2aS}$			
S, a, t	V_1	$V_1 = \frac{S}{t} - \frac{at}{2}$	V_1, V_2, S	a	$a = \frac{V_2^2 - V_1^2}{2S}$
V_1, a, t	V_2	$V_2 = V_1 + at$	V_1, S, t	a	$a = \frac{2S}{t^2} - \frac{2V_1}{t}$

Appendix III

Interest Factors

DECIMAL INTEREST RATE = .005

PERIOD	F/P	P/F	F/A	A/F	P/A	A/P	P/G	A/G
1.0000	1.0050	0.9950	1.0000	1.0000	0.9950	1.0050	-0.0002	-0.0002
2.0000	1.0100	0.9901	2.0050	0.4988	1.9851	0.5038	0.9911	0.4992
3.0000	1.0151	0.9851	3.0150	0.3317	2.9702	0.3367	2.9589	0.9962
4.0000	1.0202	0.9802	4.0301	0.2481	3.9505	0.2531	5.9027	1.4942
5.0000	1.0253	0.9754	5.0503	0.1980	4.9259	0.2030	9.8033	1.9902
6.0000	1.0304	0.9705	6.0755	0.1646	5.8964	0.1696	14.6597	2.4862
7.0000	1.0355	0.9657	7.1059	0.1407	6.8621	0.1457	20.4485	2.9799
8.0000	1.0407	0.9609	8.1414	0.1228	7.8230	0.1278	27.1781	3.4741
9.0000	1.0459	0.9561	9.1821	0.1089	8.7791	0.1139	34.8251	3.9668
10.0000	1.0511	0.9513	10.2280	0.0978	9.7304	0.1028	43.3886	4.4591
11.0000	1.0564	0.9466	11.2792	0.0887	10.6770	0.0937	52.8547	4.9503
12.0000	1.0617	0.9419	12.3356	0.0811	11.6190	0.0861	63.2181	5.4409
13.0000	1.0670	0.9372	13.3973	0.0746	12.5562	0.0796	74.4646	5.9305
14.0000	1.0723	0.9326	14.4643	0.0691	13.4887	0.0741	86.5894	6.4194
15.0000	1.0777	0.9279	15.5365	0.0644	14.4166	0.0694	99.5741	6.9069
16.0000	1.0831	0.9233	16.6142	0.0602	15.3399	0.0652	113.4268	7.3942
17.0000	1.0885	0.9187	17.6973	0.0565	16.2586	0.0615	128.1250	7.8804
18.0000	1.0939	0.9141	18.7858	0.0532	17.1728	0.0582	143.6676	8.3660
19.0000	1.0994	0.9096	19.8797	0.0503	18.0824	0.0553	160.0369	8.8504
20.0000	1.1049	0.9051	20.9791	0.0477	18.9874	0.0527	177.2373	9.3344
21.0000	1.1104	0.9006	22.0840	0.0453	19.8880	0.0503	195.2454	9.8173
22.0000	1.1160	0.8961	23.1945	0.0431	20.7841	0.0481	214.0701	10.2997
23.0000	1.1216	0.8916	24.3104	0.0411	21.6757	0.0461	233.6800	10.7808
24.0000	1.1272	0.8872	25.4320	0.0393	22.5629	0.0443	254.0886	11.2613
25.0000	1.1328	0.8828	26.5591	0.0377	23.4457	0.0427	275.2728	11.7409
26.0000	1.1385	0.8784	27.6919	0.0361	24.3240	0.0411	297.2325	12.2197
27.0000	1.1442	0.8740	28.8304	0.0347	25.1980	0.0397	319.9551	12.6976
28.0000	1.1499	0.8697	29.9746	0.0334	26.0677	0.0384	343.4387	13.1749
29.0000	1.1556	0.8653	31.1245	0.0321	26.9331	0.0371	367.6720	13.6513
30.0000	1.1614	0.8610	32.2801	0.0310	27.7941	0.0360	392.6403	14.1268
31.0000	1.1672	0.8567	33.4414	0.0299	28.6508	0.0349	418.3365	14.6012
32.0000	1.1730	0.8525	34.6087	0.0289	29.5033	0.0339	444.7663	15.0751
33.0000	1.1789	0.8482	35.7817	0.0279	30.3515	0.0329	471.9098	15.5481
34.0000	1.1848	0.8440	36.9606	0.0271	31.1956	0.0321	499.7631	16.0203
35.0000	1.1907	0.8398	38.1454	0.0262	32.0354	0.0312	528.3165	16.4916
36.0000	1.1967	0.8356	39.3361	0.0254	32.8710	0.0304	557.5638	16.9622
42.0000	1.2330	0.8110	46.6066	0.0215	37.7983	0.0265	747.1963	19.7680
48.0000	1.2705	0.7871	54.0979	0.0185	42.5804	0.0235	959.9282	22.5439
54.0000	1.3091	0.7639	61.8168	0.0162	47.2214	0.0212	1194.2380	25.2902
60.0000	1.3489	0.7414	69.7701	0.0143	51.7256	0.0193	1448.6560	28.0065
66.0000	1.3898	0.7195	77.9650	0.0128	56.0970	0.0178	1721.8060	30.6934
72.0000	1.4320	0.6983	86.4089	0.0116	60.3396	0.0166	2012.3570	33.3506
78.0000	1.4755	0.6777	95.1093	0.0105	64.4570	0.0155	2319.0510	35.9783
84.0000	1.5204	0.6577	104.0740	0.0096	68.4531	0.0146	2640.6770	38.5764
90.0000	1.5666	0.6383	113.3110	0.0088	72.3313	0.0138	2976.0870	41.1452
96.0000	1.6141	0.6195	122.8287	0.0081	76.0953	0.0131	3324.1980	43.6847
102.0000	1.6632	0.6013	132.6354	0.0075	79.7482	0.0125	3683.9640	46.1949
108.0000	1.7137	0.5835	142.7400	0.0070	83.2935	0.0120	4054.3900	48.6760
114.0000	1.7658	0.5663	153.1517	0.0065	86.7342	0.0115	4434.5430	51.1280
120.0000	1.8194	0.5496	163.8795	0.0061	90.0735	0.0111	4823.5200	53.5510

DECIMAL INTEREST RATE = .01

PERIOD	F/P	P/F	F/A	A/F	P/A	A/P	P/G	A/G
1.0000	1.0100	0.9901	1.0000	1.0000	0.9901	1.0100	-0.0001	-0.0001
2.0000	1.0201	0.9803	2.0100	0.4975	1.9704	0.5075	0.9803	0.4975
3.0000	1.0303	0.9706	3.0301	0.3300	2.9410	0.3400	2.9212	0.9933
4.0000	1.0406	0.9610	4.0604	0.2463	3.9020	0.2563	5.8043	1.4875
5.0000	1.0510	0.9515	5.1010	0.1960	4.8534	0.2060	9.6099	1.9800
6.0000	1.0615	0.9420	6.1520	0.1625	5.7955	0.1725	14.3200	2.4709
7.0000	1.0721	0.9327	7.2135	0.1386	6.7282	0.1486	19.9166	2.9602
8.0000	1.0829	0.9235	8.2857	0.1207	7.6517	0.1307	26.3807	3.4477
9.0000	1.0937	0.9143	9.3685	0.1067	8.5660	0.1167	33.6948	3.9335
10.0000	1.1046	0.9053	10.4622	0.0956	9.4713	0.1056	41.8435	4.4179
11.0000	1.1157	0.8963	11.5668	0.0865	10.3676	0.0965	50.8063	4.9005
12.0000	1.1268	0.8874	12.6825	0.0788	11.2551	0.0888	60.5682	5.3814
13.0000	1.1381	0.8787	13.8093	0.0724	12.1337	0.0824	71.1123	5.8607
14.0000	1.1495	0.8700	14.9474	0.0669	13.0037	0.0769	82.4215	6.3383
15.0000	1.1610	0.8613	16.0969	0.0621	13.8650	0.0721	94.4805	6.8143
16.0000	1.1726	0.8528	17.2579	0.0579	14.7179	0.0679	107.2728	7.2886
17.0000	1.1843	0.8444	18.4304	0.0543	15.5622	0.0643	120.7828	7.7613
18.0000	1.1961	0.8360	19.6147	0.0510	16.3983	0.0610	134.9955	8.2323
19.0000	1.2081	0.8277	20.8109	0.0481	17.2260	0.0581	149.8946	8.7017
20.0000	1.2202	0.8195	22.0190	0.0454	18.0455	0.0554	165.4654	9.1693
21.0000	1.2324	0.8114	23.2392	0.0430	18.8570	0.0530	181.6940	9.6354
22.0000	1.2447	0.8034	24.4716	0.0409	19.6604	0.0509	198.5649	10.0998
23.0000	1.2572	0.7954	25.7163	0.0389	20.4558	0.0489	216.0654	10.5625
24.0000	1.2697	0.7876	26.9735	0.0371	21.2434	0.0471	234.1789	11.0236
25.0000	1.2824	0.7798	28.2432	0.0354	22.0231	0.0454	252.8923	11.4830
26.0000	1.2953	0.7720	29.5256	0.0339	22.7952	0.0439	272.1953	11.9409
27.0000	1.3082	0.7644	30.8209	0.0324	23.5596	0.0424	292.0689	12.3970
28.0000	1.3213	0.7568	32.1291	0.0311	24.3164	0.0411	312.5038	12.8515
29.0000	1.3345	0.7493	33.4504	0.0299	25.0658	0.0399	333.4856	13.3044
30.0000	1.3478	0.7419	34.7849	0.0287	25.8077	0.0387	355.0013	13.7556
31.0000	1.3613	0.7346	36.1327	0.0277	26.5423	0.0377	377.0386	14.2052
32.0000	1.3749	0.7273	37.4941	0.0267	27.2696	0.0367	399.5846	14.6531
33.0000	1.3887	0.7201	38.8690	0.0257	27.9897	0.0357	422.6277	15.0994
34.0000	1.4026	0.7130	40.2577	0.0248	28.7027	0.0348	446.1562	15.5441
35.0000	1.4166	0.7059	41.6603	0.0240	29.4086	0.0340	470.1575	15.9871
36.0000	1.4308	0.6989	43.0769	0.0232	30.1075	0.0332	494.6198	16.4285
42.0000	1.5188	0.6584	51.8790	0.0193	34.1581	0.0293	650.4510	19.0424
48.0000	1.6122	0.6203	61.2226	0.0163	37.9740	0.0263	820.1443	21.5976
54.0000	1.7114	0.5843	71.1410	0.0141	41.5687	0.0241	1001.5720	24.0944
60.0000	1.8167	0.5504	81.6696	0.0122	44.9550	0.0222	1192.8040	26.5333
66.0000	1.9285	0.5185	92.8460	0.0108	48.1451	0.0208	1392.0940	28.9145
72.0000	2.0471	0.4885	104.7099	0.0096	51.1504	0.0196	1597.8650	31.2386
78.0000	2.1730	0.4602	117.3037	0.0085	53.9814	0.0185	1808.6980	33.5059
84.0000	2.3067	0.4335	130.6722	0.0077	56.6484	0.0177	2023.3130	35.7170
90.0000	2.4486	0.4084	144.8632	0.0069	59.1609	0.0169	2240.5660	37.8724
96.0000	2.5993	0.3847	159.9272	0.0063	61.5277	0.0163	2459.4270	39.9727
102.0000	2.7592	0.3624	175.9180	0.0057	63.7573	0.0157	2678.9840	42.0184
108.0000	2.9289	0.3414	192.8925	0.0052	65.8578	0.0152	2898.4190	44.0103
114.0000	3.1091	0.3216	210.9113	0.0047	67.8365	0.0147	3117.0080	45.9488
120.0000	3.3004	0.3030	230.0386	0.0043	69.7005	0.0143	3334.1120	47.8348

DECIMAL INTEREST RATE = .015

PERIOD	F/P	P/F	F/A	A/F	P/A	A/P	P/G	A/G
1.0000	1.0150	0.9852	1.0000	1.0000	0.9852	1.0150	-0.0001	-0.0001
2.0000	1.0302	0.9707	2.0150	0.4963	1.9559	0.5113	0.9703	0.4961
3.0000	1.0457	0.9563	3.0452	0.3284	2.9122	0.3434	2.8828	0.9899
4.0000	1.0614	0.9422	4.0909	0.2444	3.8544	0.2594	5.7090	1.4812
5.0000	1.0773	0.9283	5.1522	0.1941	4.7826	0.2091	9.4218	1.9700
6.0000	1.0934	0.9145	6.2295	0.1605	5.6972	0.1755	13.9942	2.4564
7.0000	1.1098	0.9010	7.3230	0.1366	6.5982	0.1516	19.4004	2.9403
8.0000	1.1265	0.8877	8.4328	0.1186	7.4859	0.1336	25.6140	3.4216
9.0000	1.1434	0.8746	9.5593	0.1046	8.3605	0.1196	32.6105	3.9006
10.0000	1.1605	0.8617	10.7027	0.0934	9.2222	0.1084	40.3652	4.3770
11.0000	1.1779	0.8489	11.8632	0.0843	10.0711	0.0993	48.8546	4.8510
12.0000	1.1956	0.8364	13.0412	0.0767	10.9075	0.0917	58.0544	5.3224
13.0000	1.2136	0.8240	14.2368	0.0702	11.7315	0.0852	67.9425	5.7915
14.0000	1.2318	0.8118	15.4503	0.0647	12.5433	0.0797	78.4962	6.2580
15.0000	1.2502	0.7999	16.6821	0.0599	13.3432	0.0749	89.6943	6.7221
16.0000	1.2690	0.7880	17.9323	0.0558	14.1312	0.0708	101.5145	7.1837
17.0000	1.2880	0.7764	19.2013	0.0521	14.9076	0.0671	113.9365	7.6428
18.0000	1.3073	0.7649	20.4893	0.0488	15.6725	0.0638	126.9399	8.0995
19.0000	1.3269	0.7536	21.7966	0.0459	16.4261	0.0609	140.5047	8.5537
20.0000	1.3469	0.7425	23.1236	0.0432	17.1686	0.0582	154.6113	9.0055
21.0000	1.3671	0.7315	24.4704	0.0409	17.9001	0.0559	169.2408	9.4548
22.0000	1.3876	0.7207	25.8375	0.0387	18.6208	0.0537	184.3750	9.9016
23.0000	1.4084	0.7100	27.2250	0.0367	19.3308	0.0517	199.9961	10.3460
24.0000	1.4295	0.6995	28.6334	0.0349	20.0304	0.0499	216.0852	10.7879
25.0000	1.4509	0.6892	30.0629	0.0333	20.7196	0.0483	232.6259	11.2274
26.0000	1.4727	0.6790	31.5138	0.0317	21.3986	0.0467	249.6013	11.6644
27.0000	1.4948	0.6690	32.9866	0.0303	22.0676	0.0453	266.9950	12.0990
28.0000	1.5172	0.6591	34.4813	0.0290	22.7267	0.0440	284.7899	12.5311
29.0000	1.5400	0.6494	35.9986	0.0278	23.3760	0.0428	302.9718	12.9608
30.0000	1.5631	0.6398	37.5385	0.0266	24.0158	0.0416	321.5247	13.3881
31.0000	1.5865	0.6303	39.1016	0.0256	24.6461	0.0406	340.4340	13.8129
32.0000	1.6103	0.6210	40.6881	0.0246	25.2671	0.0396	359.6845	14.2353
33.0000	1.6345	0.6118	42.2984	0.0236	25.8789	0.0386	379.2626	14.6553
34.0000	1.6590	0.6028	43.9329	0.0228	26.4817	0.0378	399.1538	15.0728
35.0000	1.6839	0.5939	45.5919	0.0219	27.0755	0.0369	419.3452	15.4880
36.0000	1.7091	0.5851	47.2758	0.0212	27.6606	0.0362	439.8232	15.9007
42.0000	1.8688	0.5351	57.9229	0.0173	30.9940	0.0323	568.0120	18.3265
48.0000	2.0435	0.4894	69.5649	0.0144	34.0425	0.0294	703.5374	20.6665
54.0000	2.2344	0.4475	82.2948	0.0122	36.8305	0.0272	844.2087	22.9215
60.0000	2.4432	0.4093	96.2142	0.0104	39.3802	0.0254	988.1572	25.0928
66.0000	2.6715	0.3743	111.4342	0.0090	41.7120	0.0240	1133.7950	27.1815
72.0000	2.9211	0.3423	128.0765	0.0078	43.8446	0.0228	1279.7830	29.1891
78.0000	3.1941	0.3131	146.2739	0.0068	45.7949	0.0218	1424.9960	31.1169
84.0000	3.4926	0.2863	166.1717	0.0060	47.5786	0.0210	1568.5020	32.9666
90.0000	3.8189	0.2619	187.9288	0.0053	49.2098	0.0203	1709.5320	34.7397
96.0000	4.1758	0.2395	211.7189	0.0047	50.7016	0.0197	1847.4600	36.4379
102.0000	4.5660	0.2190	237.7321	0.0042	52.0659	0.0192	1981.7880	38.0630
108.0000	4.9926	0.2003	266.1760	0.0038	53.3137	0.0188	2112.1220	39.6169
114.0000	5.4592	0.1832	297.2778	0.0034	54.4548	0.0184	2238.1660	41.1014
120.0000	5.9693	0.1675	331.2858	0.0030	55.4984	0.0180	2359.6990	42.5183

DECIMAL INTEREST RATE = .0175

PERIOD	F/P	P/F	F/A	A/F	P/A	A/P	P/G	A/G
1.0000	1.0175	0.9828	1.0000	1.0000	0.9828	1.0175	0.0001	0.0001
2.0000	1.0353	0.9659	2.0175	0.4957	1.9487	0.5132	0.9662	0.4958
3.0000	1.0534	0.9493	3.0528	0.3276	2.8980	0.3451	2.8649	0.9886
4.0000	1.0719	0.9330	4.1062	0.2435	3.8310	0.2610	5.6640	1.4785
5.0000	1.0906	0.9169	5.1781	0.1931	4.7479	0.2106	9.3318	1.9655
6.0000	1.1097	0.9011	6.2687	0.1595	5.6490	0.1770	13.8374	2.4495
7.0000	1.1291	0.8856	7.3784	0.1355	6.5347	0.1530	19.1516	2.9308
8.0000	1.1489	0.8704	8.5076	0.1175	7.4051	0.1350	25.2448	3.4091
9.0000	1.1690	0.8554	9.6564	0.1036	8.2605	0.1211	32.0884	3.8846
10.0000	1.1894	0.8407	10.8254	0.0924	9.1012	0.1099	39.6550	4.3571
11.0000	1.2103	0.8263	12.0149	0.0832	9.9275	0.1007	47.9179	4.8268
12.0000	1.2314	0.8121	13.2251	0.0756	10.7396	0.0931	56.8508	5.2936
13.0000	1.2530	0.7981	14.4566	0.0692	11.5377	0.0867	66.4282	5.7575
14.0000	1.2749	0.7844	15.7096	0.0637	12.3220	0.0812	76.6247	6.2185
15.0000	1.2972	0.7709	16.9845	0.0589	13.0929	0.0764	87.4172	6.6767
16.0000	1.3199	0.7576	18.2817	0.0547	13.8505	0.0722	98.7817	7.1320
17.0000	1.3430	0.7446	19.6017	0.0510	14.5951	0.0685	110.6953	7.5844
18.0000	1.3665	0.7318	20.9447	0.0477	15.3269	0.0652	123.1357	8.0340
19.0000	1.3904	0.7192	22.3112	0.0448	16.0461	0.0623	136.0811	8.4806
20.0000	1.4148	0.7068	23.7017	0.0422	16.7529	0.0597	149.5112	8.9245
21.0000	1.4395	0.6947	25.1165	0.0398	17.4476	0.0573	163.4046	9.3655
22.0000	1.4647	0.6827	26.5560	0.0377	18.1303	0.0552	177.7417	9.8036
23.0000	1.4904	0.6710	28.0208	0.0357	18.8013	0.0532	192.5034	10.2388
24.0000	1.5164	0.6594	29.5111	0.0339	19.4607	0.0514	207.6708	10.6713
25.0000	1.5430	0.6481	31.0276	0.0322	20.1088	0.0497	223.2252	11.1009
26.0000	1.5700	0.6369	32.5706	0.0307	20.7458	0.0482	239.1489	11.5276
27.0000	1.5975	0.6260	34.1405	0.0293	21.3718	0.0468	255.4251	11.9515
28.0000	1.6254	0.6152	35.7380	0.0280	21.9870	0.0455	272.0363	12.3726
29.0000	1.6539	0.6046	37.3634	0.0268	22.5917	0.0443	288.9667	12.7909
30.0000	1.6828	0.5942	39.0173	0.0256	23.1859	0.0431	306.1998	13.2063
31.0000	1.7123	0.5840	40.7001	0.0246	23.7699	0.0421	323.7208	13.6189
32.0000	1.7422	0.5740	42.4124	0.0236	24.3439	0.0411	341.5145	14.0287
33.0000	1.7727	0.5641	44.1546	0.0226	24.9080	0.0401	359.5662	14.4358
34.0000	1.8037	0.5544	45.9273	0.0218	25.4624	0.0393	377.8617	14.8400
35.0000	1.8353	0.5449	47.7310	0.0210	26.0073	0.0385	396.3875	15.2414
36.0000	1.8674	0.5355	49.5663	0.0202	26.5428	0.0377	415.1302	15.6400
42.0000	2.0723	0.4826	61.2726	0.0163	29.5679	0.0338	531.4420	17.9736
48.0000	2.2996	0.4349	74.2631	0.0135	32.2939	0.0310	652.6118	20.2085
54.0000	2.5519	0.3919	88.6787	0.0113	34.7504	0.0288	776.5419	22.3463
60.0000	2.8318	0.3531	104.6757	0.0096	36.9640	0.0271	901.5028	24.3886
66.0000	3.1425	0.3182	122.4276	0.0082	38.9589	0.0257	1026.0790	26.3375
72.0000	3.4872	0.2868	142.1270	0.0070	40.7565	0.0245	1149.1260	28.1949
78.0000	3.8698	0.2584	163.9874	0.0061	42.3764	0.0236	1269.7280	29.9631
84.0000	4.2943	0.2329	188.2461	0.0053	43.8362	0.0228	1387.1670	31.6443
90.0000	4.7654	0.2098	215.1659	0.0046	45.1517	0.0221	1500.8880	33.2410
96.0000	5.2882	0.1891	245.0389	0.0041	46.3371	0.0216	1610.4800	34.7557
102.0000	5.8683	0.1704	278.1890	0.0036	47.4053	0.0211	1715.6470	36.1910
108.0000	6.5121	0.1536	314.9758	0.0032	48.3680	0.0207	1816.1930	37.5495
114.0000	7.2265	0.1384	355.7983	0.0028	49.2354	0.0203	1912.0050	38.8339
120.0000	8.0192	0.1247	401.0990	0.0025	50.0171	0.0200	2003.0340	40.0470

DECIMAL INTEREST RATE = .02

PERIOD	F/P	P/F	F/A	A/F	P/A	A/P	P/G	A/G
1.0000	1.0200	0.9804	1.0000	1.0000	0.9804	1.0200	-0.0000	-0.0000
2.0000	1.0404	0.9612	2.0200	0.4951	1.9416	0.5151	0.9609	0.4949
3.0000	1.0612	0.9423	3.0604	0.3268	2.8839	0.3468	2.8456	0.9867
4.0000	1.0824	0.9238	4.1216	0.2426	3.8077	0.2626	5.6168	1.4751
5.0000	1.1041	0.9057	5.2040	0.1922	4.7134	0.2122	9.2397	1.9603
6.0000	1.1262	0.8880	6.3081	0.1585	5.6014	0.1785	13.6793	2.4421
7.0000	1.1487	0.8706	7.4343	0.1345	6.4720	0.1545	18.9025	2.9207
8.0000	1.1717	0.8535	8.5829	0.1165	7.3255	0.1365	24.8769	3.3959
9.0000	1.1951	0.8368	9.7546	0.1025	8.1622	0.1225	31.5707	3.8679
10.0000	1.2190	0.8203	10.9497	0.0913	8.9826	0.1113	38.9538	4.3366
11.0000	1.2434	0.8043	12.1687	0.0822	9.7868	0.1022	46.9964	4.8020
12.0000	1.2682	0.7885	13.4120	0.0746	10.5753	0.0946	55.6695	5.2641
13.0000	1.2936	0.7730	14.6803	0.0681	11.3483	0.0881	64.9460	5.7229
14.0000	1.3195	0.7579	15.9739	0.0626	12.1062	0.0826	74.7982	6.1785
15.0000	1.3459	0.7430	17.2934	0.0578	12.8492	0.0778	85.2000	6.6307
16.0000	1.3728	0.7284	18.6392	0.0537	13.5777	0.0737	96.1267	7.0798
17.0000	1.4002	0.7142	20.0120	0.0500	14.2918	0.0700	107.5533	7.5255
18.0000	1.4282	0.7002	21.4122	0.0467	14.9920	0.0667	119.4558	7.9680
19.0000	1.4568	0.6864	22.8405	0.0438	15.6784	0.0638	131.8116	8.4072
20.0000	1.4859	0.6730	24.2973	0.0412	16.3514	0.0612	144.5976	8.8431
21.0000	1.5157	0.6598	25.7832	0.0388	17.0112	0.0588	157.7933	9.2759
22.0000	1.5460	0.6468	27.2989	0.0366	17.6580	0.0566	171.3768	9.7053
23.0000	1.5769	0.6342	28.8449	0.0347	18.2922	0.0547	185.3280	10.1316
24.0000	1.6084	0.6217	30.4218	0.0329	18.9139	0.0529	199.6275	10.5546
25.0000	1.6406	0.6095	32.0302	0.0312	19.5234	0.0512	214.2560	10.9743
26.0000	1.6734	0.5976	33.6708	0.0297	20.1210	0.0497	229.1956	11.3909
27.0000	1.7069	0.5859	35.3442	0.0283	20.7069	0.0483	244.4279	11.8042
28.0000	1.7410	0.5744	37.0511	0.0270	21.2812	0.0470	259.9358	12.2143
29.0000	1.7758	0.5631	38.7921	0.0258	21.8443	0.0458	275.7031	12.6213
30.0000	1.8114	0.5521	40.5679	0.0247	22.3964	0.0447	291.7130	13.0250
31.0000	1.8476	0.5412	42.3793	0.0236	22.9377	0.0436	307.9500	13.4255
32.0000	1.8845	0.5306	44.2269	0.0226	23.4683	0.0426	324.3996	13.8229
33.0000	1.9222	0.5202	46.1114	0.0217	23.9885	0.0417	341.0468	14.2171
34.0000	1.9607	0.5100	48.0336	0.0208	24.4985	0.0408	357.8778	14.6081
35.0000	1.9999	0.5000	49.9943	0.0200	24.9986	0.0400	374.8786	14.9960
36.0000	2.0399	0.4902	51.9942	0.0192	25.4888	0.0392	392.0361	15.3807
42.0000	2.2972	0.4353	64.8620	0.0154	28.2347	0.0354	497.5964	17.6235
48.0000	2.5871	0.3865	79.3532	0.0126	30.6731	0.0326	605.9608	19.7555
54.0000	2.9135	0.3432	95.6726	0.0105	32.8382	0.0305	715.1760	21.7788
60.0000	3.2810	0.3048	114.0510	0.0088	34.7608	0.0288	823.6918	23.6960
66.0000	3.6950	0.2706	134.7480	0.0074	36.4681	0.0274	930.2944	25.5098
72.0000	4.1611	0.2403	158.0562	0.0063	37.9840	0.0263	1034.0500	27.2233
78.0000	4.6861	0.2134	184.3050	0.0054	39.3301	0.0254	1134.2590	28.8394
84.0000	5.2773	0.1895	213.8654	0.0047	40.5255	0.0247	1230.4130	30.3615
90.0000	5.9431	0.1683	247.1552	0.0040	41.5869	0.0240	1322.1640	31.7928
96.0000	6.6929	0.1494	284.6449	0.0035	42.5294	0.0235	1409.2920	33.1369
102.0000	7.5373	0.1327	326.8643	0.0031	43.3663	0.0231	1491.6800	34.3972
108.0000	8.4882	0.1178	374.4104	0.0027	44.1095	0.0227	1569.2970	35.5773
114.0000	9.5591	0.1046	427.9548	0.0023	44.7694	0.0223	1642.1780	36.6808
120.0000	10.7651	0.0929	488.2546	0.0020	45.3554	0.0220	1710.4110	37.7113

DECIMAL INTEREST RATE = .025

PERIOD	F/P	P/F	F/A	A/F	P/A	A/P	P/G	A/G
1.0000	1.0250	0.9756	1.0000	1.0000	0.9756	1.0250	-0.0000	-0.0000
2.0000	1.0506	0.9518	2.0250	0.4938	1.9274	0.5188	0.9518	0.4938
3.0000	1.0769	0.9286	3.0756	0.3251	2.8560	0.3501	2.8090	0.9835
4.0000	1.1038	0.9060	4.1525	0.2408	3.7620	0.2658	5.5268	1.4691
5.0000	1.1314	0.8839	5.2563	0.1902	4.6458	0.2152	9.0621	1.9506
6.0000	1.1597	0.8623	6.3877	0.1566	5.5081	0.1816	13.3735	2.4280
7.0000	1.1887	0.8413	7.5474	0.1325	6.3494	0.1575	18.4212	2.9013
8.0000	1.2184	0.8207	8.7361	0.1145	7.1701	0.1395	24.1664	3.3704
9.0000	1.2489	0.8007	9.9545	0.1005	7.9709	0.1255	30.5721	3.8355
10.0000	1.2801	0.7812	11.2034	0.0893	8.7521	0.1143	37.6030	4.2965
11.0000	1.3121	0.7621	12.4835	0.0801	9.5142	0.1051	45.2244	4.7534
12.0000	1.3449	0.7436	13.7955	0.0725	10.2578	0.0975	53.4035	5.2062
13.0000	1.3785	0.7254	15.1404	0.0660	10.9832	0.0910	62.1085	5.6549
14.0000	1.4130	0.7077	16.5189	0.0605	11.6909	0.0855	71.3088	6.0995
15.0000	1.4483	0.6905	17.9319	0.0558	12.3814	0.0808	80.9755	6.5401
16.0000	1.4845	0.6736	19.3802	0.0516	13.0550	0.0766	91.0797	6.9766
17.0000	1.5216	0.6572	20.8647	0.0479	13.7122	0.0729	101.5948	7.4091
18.0000	1.5597	0.6412	22.3863	0.0447	14.3534	0.0697	112.4945	7.8375
19.0000	1.5986	0.6255	23.9460	0.0418	14.9789	0.0668	123.7540	8.2619
20.0000	1.6386	0.6103	25.5446	0.0391	15.5892	0.0641	135.3492	8.6823
21.0000	1.6796	0.5954	27.1832	0.0368	16.1845	0.0618	147.2569	9.0986
22.0000	1.7216	0.5809	28.8628	0.0346	16.7654	0.0596	159.4550	9.5110
23.0000	1.7646	0.5667	30.5844	0.0327	17.3321	0.0577	171.9225	9.9193
24.0000	1.8087	0.5529	32.3490	0.0309	17.8850	0.0559	184.6384	10.3237
25.0000	1.8539	0.5394	34.1577	0.0293	18.4244	0.0543	197.5838	10.7241
26.0000	1.9003	0.5262	36.0117	0.0278	18.9506	0.0528	210.7398	11.1205
27.0000	1.9478	0.5134	37.9120	0.0264	19.4640	0.0514	224.0881	11.5130
28.0000	1.9965	0.5009	39.8598	0.0251	19.9649	0.0501	237.6117	11.9015
29.0000	2.0464	0.4887	41.8563	0.0239	20.4535	0.0489	251.2942	12.2861
30.0000	2.0976	0.4767	43.9027	0.0228	20.9303	0.0478	265.1197	12.6668
31.0000	2.1500	0.4651	46.0002	0.0217	21.3954	0.0467	279.0731	13.0436
32.0000	2.2038	0.4538	48.1502	0.0208	21.8492	0.0458	293.1400	13.4165
33.0000	2.2588	0.4427	50.3540	0.0199	22.2919	0.0449	307.3064	13.7856
34.0000	2.3153	0.4319	52.6128	0.0190	22.7238	0.0440	321.5593	14.1508
35.0000	2.3732	0.4214	54.9282	0.0182	23.1451	0.0432	335.8861	14.5122
36.0000	2.4325	0.4111	57.3014	0.0175	23.5562	0.0425	350.2743	14.8697
42.0000	2.8210	0.3545	72.8397	0.0137	25.8206	0.0387	437.2888	16.9357
48.0000	3.2715	0.3057	90.8595	0.0110	27.7731	0.0360	524.0365	18.8685
54.0000	3.7939	0.2636	111.7568	0.0089	29.4568	0.0339	608.9409	20.6723
60.0000	4.3998	0.2273	135.9914	0.0074	30.9086	0.0324	690.8646	22.3518
66.0000	5.1024	0.1960	164.0960	0.0061	32.1606	0.0311	769.0186	23.9119
72.0000	5.9172	0.1690	196.6888	0.0051	33.2401	0.0301	842.8876	25.3576
78.0000	6.8622	0.1457	234.4864	0.0043	34.1709	0.0293	912.1698	26.6943
84.0000	7.9580	0.1257	278.3201	0.0036	34.9736	0.0286	976.7278	27.9276
90.0000	9.2288	0.1084	329.1537	0.0030	35.6658	0.0280	1036.5490	29.0629
96.0000	10.7026	0.0934	388.1050	0.0026	36.2626	0.0276	1091.7140	30.1058
102.0000	12.4118	0.0806	456.4705	0.0022	36.7773	0.0272	1142.3700	31.0619
108.0000	14.3938	0.0695	535.7535	0.0019	37.2210	0.0269	1188.7130	31.9366
114.0000	16.6924	0.0599	627.6975	0.0016	37.6037	0.0266	1230.9710	32.7354
120.0000	19.3581	0.0517	734.3243	0.0014	37.9337	0.0264	1269.3890	33.4634

DECIMAL INTEREST RATE = .03

PERIOD	F/P	P/F	F/A	A/F	P/A	A/P	P/G	A/G
1.0000	1.0300	0.9709	1.0000	1.0000	0.9709	1.0300	-0.0000	-0.0000
2.0000	1.0609	0.9426	2.0300	0.4926	1.9135	0.5226	0.9426	0.4926
3.0000	1.0927	0.9151	3.0909	0.3235	2.8286	0.3535	2.7728	0.9803
4.0000	1.1255	0.8885	4.1836	0.2390	3.7171	0.2690	5.4383	1.4631
5.0000	1.1593	0.8626	5.3091	0.1884	4.5797	0.2184	8.8887	1.9409
6.0000	1.1941	0.8375	6.4684	0.1546	5.4172	0.1846	13.0761	2.4138
7.0000	1.2299	0.8131	7.6625	0.1305	6.2303	0.1605	17.9547	2.8818
8.0000	1.2668	0.7894	8.8923	0.1125	7.0197	0.1425	23.4806	3.3450
9.0000	1.3048	0.7664	10.1591	0.0984	7.7861	0.1284	29.6119	3.8032
10.0000	1.3439	0.7441	11.4639	0.0872	8.5302	0.1172	36.3087	4.2565
11.0000	1.3842	0.7224	12.8078	0.0781	9.2526	0.1081	43.5329	4.7049
12.0000	1.4258	0.7014	14.1920	0.0705	9.9540	0.1005	51.2481	5.1485
13.0000	1.4685	0.6810	15.6178	0.0640	10.6350	0.0940	59.4195	5.5872
14.0000	1.5126	0.6611	17.0863	0.0585	11.2961	0.0885	68.0141	6.0210
15.0000	1.5580	0.6419	18.5989	0.0538	11.9379	0.0838	77.0001	6.4500
16.0000	1.6047	0.6232	20.1569	0.0496	12.5611	0.0796	86.3476	6.8742
17.0000	1.6528	0.6050	21.7616	0.0460	13.1661	0.0760	96.0279	7.2936
18.0000	1.7024	0.5874	23.4144	0.0427	13.7535	0.0727	106.0136	7.7081
19.0000	1.7535	0.5703	25.1169	0.0398	14.3238	0.0698	116.2787	8.1179
20.0000	1.8061	0.5537	26.8704	0.0372	14.8775	0.0672	126.7985	8.5229
21.0000	1.8603	0.5375	28.6765	0.0349	15.4150	0.0649	137.5495	8.9231
22.0000	1.9161	0.5219	30.5368	0.0327	15.9369	0.0627	148.5093	9.3186
23.0000	1.9736	0.5067	32.4529	0.0308	16.4436	0.0608	159.6565	9.7093
24.0000	2.0328	0.4919	34.4265	0.0290	16.9355	0.0590	170.9709	10.0954
25.0000	2.0938	0.4776	36.4593	0.0274	17.4131	0.0574	182.4335	10.4768
26.0000	2.1566	0.4637	38.5530	0.0259	17.8768	0.0559	194.0259	10.8535
27.0000	2.2213	0.4502	40.7096	0.0246	18.3270	0.0546	205.7306	11.2255
28.0000	2.2879	0.4371	42.9309	0.0233	18.7641	0.0533	217.5318	11.5930
29.0000	2.3566	0.4243	45.2188	0.0221	19.1885	0.0521	229.4135	11.9558
30.0000	2.4273	0.4120	47.5754	0.0210	19.6004	0.0510	241.3612	12.3141
31.0000	2.5001	0.4000	50.0027	0.0200	20.0004	0.0500	253.3607	12.6678
32.0000	2.5751	0.3883	52.5027	0.0190	20.3888	0.0490	265.3992	13.0169
33.0000	2.6523	0.3770	55.0778	0.0182	20.7658	0.0482	277.4640	13.3616
34.0000	2.7319	0.3660	57.7302	0.0173	21.1318	0.0473	289.5435	13.7018
35.0000	2.8139	0.3554	60.4621	0.0165	21.4872	0.0465	301.6265	14.0375
36.0000	2.8983	0.3450	63.2759	0.0158	21.8323	0.0458	313.7027	14.3688
42.0000	3.4607	0.2890	82.0232	0.0122	23.7014	0.0422	385.5022	16.2650
48.0000	4.1323	0.2420	104.4084	0.0096	25.2667	0.0396	455.0253	18.0089
54.0000	4.9341	0.2027	131.1374	0.0076	26.5777	0.0376	521.1155	19.6073
60.0000	5.8916	0.1697	163.0534	0.0061	27.6756	0.0361	583.0524	21.0674
66.0000	7.0349	0.1421	201.1626	0.0050	28.5950	0.0350	640.4405	22.3969
72.0000	8.4000	0.1190	246.6671	0.0041	29.3651	0.0341	693.1224	23.6036
78.0000	10.0301	0.0997	301.0018	0.0033	30.0100	0.0333	741.1120	24.6955
84.0000	11.9764	0.0835	365.8804	0.0027	30.5501	0.0327	784.5432	25.6806
90.0000	14.3005	0.0699	443.3487	0.0023	31.0024	0.0323	823.6300	26.5667
96.0000	17.0755	0.0586	535.8500	0.0019	31.3812	0.0319	858.6376	27.3615
102.0000	20.3890	0.0490	646.3011	0.0015	31.6985	0.0315	889.8592	28.0726
108.0000	24.3456	0.0411	778.1858	0.0013	31.9642	0.0313	917.6011	28.7072
114.0000	29.0699	0.0344	935.6629	0.0011	32.1867	0.0311	942.1697	29.2720
120.0000	34.7110	0.0288	1123.6990	0.0009	32.3730	0.0309	963.8634	29.7737

DECIMAL INTEREST RATE = .04

PERIOD	F/P	P/F	F/A	A/F	P/A	A/P	P/G	A/G
1.0000	1.0400	0.9615	1.0000	1.0000	0.9615	1.0400	-0.0000	-0.0000
2.0000	1.0816	0.9246	2.0400	0.4902	1.8861	0.5302	0.9245	0.4902
3.0000	1.1249	0.8890	3.1216	0.3203	2.7751	0.3603	2.7025	0.9738
4.0000	1.1699	0.8548	4.2465	0.2355	3.6299	0.2755	5.2669	1.4510
5.0000	1.2167	0.8219	5.4163	0.1846	4.4518	0.2246	8.5546	1.9216
6.0000	1.2653	0.7903	6.6330	0.1508	5.2421	0.1908	12.5062	2.3857
7.0000	1.3159	0.7599	7.8983	0.1266	6.0021	0.1666	17.0657	2.8433
8.0000	1.3686	0.7307	9.2142	0.1085	6.7327	0.1485	22.1805	3.2944
9.0000	1.4233	0.7026	10.5828	0.0945	7.4353	0.1345	27.8011	3.7391
10.0000	1.4802	0.6756	12.0061	0.0833	8.1109	0.1233	33.8812	4.1772
11.0000	1.5395	0.6496	13.4863	0.0741	8.7605	0.1141	40.3770	4.6090
12.0000	1.6010	0.6246	15.0258	0.0666	9.3851	0.1066	47.2476	5.0343
13.0000	1.6651	0.6006	16.6268	0.0601	9.9856	0.1001	54.4544	5.4533
14.0000	1.7317	0.5775	18.2919	0.0547	10.5631	0.0947	61.9616	5.8658
15.0000	1.8009	0.5553	20.0236	0.0499	11.1184	0.0899	69.7353	6.2721
16.0000	1.8730	0.5339	21.8245	0.0458	11.6523	0.0858	77.7439	6.6720
17.0000	1.9479	0.5134	23.6975	0.0422	12.1657	0.0822	85.9579	7.0656
18.0000	2.0258	0.4936	25.6454	0.0390	12.6593	0.0790	94.3495	7.4530
19.0000	2.1068	0.4746	27.6712	0.0361	13.1339	0.0761	102.8930	7.8341
20.0000	2.1911	0.4564	29.7781	0.0336	13.5903	0.0736	111.5645	8.2091
21.0000	2.2788	0.4388	31.9692	0.0313	14.0292	0.0713	120.3411	8.5779
22.0000	2.3699	0.4220	34.2479	0.0292	14.4511	0.0692	129.2022	8.9406
23.0000	2.4647	0.4057	36.6179	0.0273	14.8568	0.0673	138.1281	9.2973
24.0000	2.5633	0.3901	39.0826	0.0256	15.2470	0.0656	147.1009	9.6479
25.0000	2.6658	0.3751	41.6459	0.0240	15.6221	0.0640	156.1037	9.9925
26.0000	2.7725	0.3607	44.3117	0.0226	15.9828	0.0626	165.1209	10.3312
27.0000	2.8834	0.3468	47.0842	0.0212	16.3296	0.0612	174.1381	10.6640
28.0000	2.9987	0.3335	49.9675	0.0200	16.6631	0.0600	183.1421	10.9909
29.0000	3.1186	0.3207	52.9662	0.0189	16.9837	0.0589	192.1202	11.3120
30.0000	3.2434	0.3083	56.0849	0.0178	17.2920	0.0578	201.0615	11.6274
31.0000	3.3731	0.2965	59.3283	0.0169	17.5885	0.0569	209.9553	11.9371
32.0000	3.5081	0.2851	62.7014	0.0159	17.8735	0.0559	218.7921	12.2411
33.0000	3.6484	0.2741	66.2095	0.0151	18.1476	0.0551	227.5631	12.5395
34.0000	3.7943	0.2636	69.8578	0.0143	18.4112	0.0543	236.2604	12.8324
35.0000	3.9461	0.2534	73.6521	0.0136	18.6646	0.0536	244.8765	13.1198
36.0000	4.1039	0.2437	77.5982	0.0129	18.9083	0.0529	253.4049	13.4018
42.0000	5.1928	0.1926	104.8195	0.0095	20.1856	0.0495	302.4367	14.9828
48.0000	6.5705	0.1522	139.2630	0.0072	21.1951	0.0472	347.2443	16.3832
54.0000	8.3138	0.1203	182.8451	0.0055	21.9930	0.0455	387.4433	17.6167
60.0000	10.5196	0.0951	237.9903	0.0042	22.6235	0.0442	422.9964	18.6972
66.0000	13.3107	0.0751	307.7666	0.0032	23.1218	0.0432	454.0844	19.6388
72.0000	16.8422	0.0594	396.0558	0.0025	23.5156	0.0425	481.0167	20.4552
78.0000	21.3108	0.0469	507.7699	0.0020	23.8269	0.0420	504.1692	21.1597
84.0000	26.9649	0.0371	649.1237	0.0015	24.0729	0.0415	523.9430	21.7649
90.0000	34.1193	0.0293	827.9814	0.0012	24.2673	0.0412	540.7368	22.2825
96.0000	43.1717	0.0232	1054.2940	0.0009	24.4209	0.0409	554.9310	22.7236
102.0000	54.6260	0.0183	1340.6500	0.0007	24.5423	0.0407	566.8775	23.0979
108.0000	69.1193	0.0145	1702.9830	0.0006	24.6383	0.0406	576.8948	23.4145
114.0000	87.4580	0.0114	2161.4500	0.0005	24.7142	0.0405	585.2667	23.6814
120.0000	110.6622	0.0090	2741.5560	0.0004	24.7741	0.0404	592.2427	23.9057

DECIMAL INTEREST RATE = .05

PERIOD	F/P	P/F	F/A	A/F	P/A	A/P	P/G	A/G
1.0000	1.0500	0.9524	1.0000	1.0000	0.9524	1.0500	-0.0000	-0.0000
2.0000	1.1025	0.9070	2.0500	0.4878	1.8594	0.5378	0.9070	0.4878
3.0000	1.1576	0.8638	3.1525	0.3172	2.7232	0.3672	2.6346	0.9675
4.0000	1.2155	0.8227	4.3101	0.2320	3.5459	0.2820	5.1027	1.4390
5.0000	1.2763	0.7835	5.5256	0.1810	4.3295	0.2310	8.2368	1.9025
6.0000	1.3401	0.7462	6.8019	0.1470	5.0757	0.1970	11.9678	2.3579
7.0000	1.4071	0.7107	8.1420	0.1228	5.7864	0.1728	16.2319	2.8052
8.0000	1.4775	0.6768	9.5491	0.1047	6.4632	0.1547	20.9697	3.2445
9.0000	1.5513	0.6446	11.0265	0.0907	7.1078	0.1407	26.1265	3.6758
10.0000	1.6289	0.6139	12.5779	0.0795	7.7217	0.1295	31.6517	4.0990
11.0000	1.7103	0.5847	14.2068	0.0704	8.3064	0.1204	37.4985	4.5144
12.0000	1.7959	0.5568	15.9171	0.0628	8.8632	0.1128	43.6237	4.9219
13.0000	1.8856	0.5303	17.7129	0.0565	9.3936	0.1065	49.9876	5.3215
14.0000	1.9799	0.5051	19.5986	0.0510	9.8986	0.1010	56.5534	5.7133
15.0000	2.0789	0.4810	21.5785	0.0463	10.3796	0.0963	63.2876	6.0973
16.0000	2.1829	0.4581	23.6574	0.0423	10.8378	0.0923	70.1592	6.4736
17.0000	2.2920	0.4363	25.8403	0.0387	11.2741	0.0887	77.1400	6.8423
18.0000	2.4066	0.4155	28.1323	0.0355	11.6896	0.0855	84.2038	7.2033
19.0000	2.5269	0.3957	30.5389	0.0327	12.0853	0.0827	91.3270	7.5569
20.0000	2.6533	0.3769	33.0659	0.0302	12.4622	0.0802	98.4879	7.9029
21.0000	2.7860	0.3589	35.7192	0.0280	12.8211	0.0780	105.6667	8.2416
22.0000	2.9253	0.3419	38.5051	0.0260	13.1630	0.0760	112.8455	8.5729
23.0000	3.0715	0.3256	41.4304	0.0241	13.4886	0.0741	120.0081	8.8970
24.0000	3.2251	0.3101	44.5019	0.0225	13.7986	0.0725	127.1397	9.2139
25.0000	3.3863	0.2953	47.7270	0.0210	14.0939	0.0710	134.2270	9.5237
26.0000	3.5557	0.2812	51.1133	0.0196	14.3752	0.0696	141.2579	9.8265
27.0000	3.7334	0.2678	54.6690	0.0183	14.6430	0.0683	148.2220	10.1224
28.0000	3.9201	0.2551	58.4024	0.0171	14.8981	0.0671	155.1095	10.4114
29.0000	4.1161	0.2429	62.3225	0.0160	15.1411	0.0660	161.9120	10.6936
30.0000	4.3219	0.2314	66.4386	0.0151	15.3724	0.0651	168.6220	10.9691
31.0000	4.5380	0.2204	70.7606	0.0141	15.5928	0.0641	175.2327	11.2381
32.0000	4.7649	0.2099	75.2986	0.0133	15.8027	0.0633	181.7386	11.5005
33.0000	5.0032	0.1999	80.0635	0.0125	16.0025	0.0625	188.1345	11.7565
34.0000	5.2533	0.1904	85.0667	0.0118	16.1929	0.0618	194.4162	12.0063
35.0000	5.5160	0.1813	90.3200	0.0111	16.3742	0.0611	200.5801	12.2498
36.0000	5.7918	0.1727	95.8360	0.0104	16.5468	0.0604	206.6231	12.4872
37.0000	6.0814	0.1644	101.6278	0.0098	16.7113	0.0598	212.5428	12.7185
38.0000	6.3855	0.1566	107.7091	0.0093	16.8679	0.0593	218.3372	12.9440
39.0000	6.7047	0.1491	114.0946	0.0088	17.0170	0.0588	224.0048	13.1636
40.0000	7.0400	0.1420	120.7993	0.0083	17.1591	0.0583	229.5446	13.3774

DECIMAL INTEREST RATE = .06

PERIOD	F/P	P/F	F/A	A/F	P/A	A/P	P/G	A/G
1.0000	1.0600	0.9434	1.0000	1.0000	0.9434	1.0600	-0.0000	-0.0000
2.0000	1.1236	0.8900	2.0600	0.4854	1.8334	0.5454	0.8900	0.4854
3.0000	1.1910	0.8396	3.1836	0.3141	2.6730	0.3741	2.5692	0.9612
4.0000	1.2625	0.7921	4.3746	0.2286	3.4651	0.2886	4.9455	1.4272
5.0000	1.3382	0.7473	5.6371	0.1774	4.2124	0.2374	7.9345	1.8836
6.0000	1.4185	0.7050	6.9753	0.1434	4.9173	0.2034	11.4593	2.3304
7.0000	1.5036	0.6651	8.3938	0.1191	5.5824	0.1791	15.4496	2.7676
8.0000	1.5938	0.6274	9.8975	0.1010	6.2098	0.1610	19.8415	3.1952
9.0000	1.6895	0.5919	11.4913	0.0870	6.8017	0.1470	24.5766	3.6133
10.0000	1.7908	0.5584	13.1808	0.0759	7.3601	0.1359	29.6022	4.0220
11.0000	1.8983	0.5268	14.9716	0.0668	7.8869	0.1268	34.8701	4.4213
12.0000	2.0122	0.4970	16.8699	0.0593	8.3838	0.1193	40.3367	4.8112
13.0000	2.1329	0.4688	18.8821	0.0530	8.8527	0.1130	45.9628	5.1920
14.0000	2.2609	0.4423	21.0150	0.0476	9.2950	0.1076	51.7127	5.5635
15.0000	2.3966	0.4173	23.2759	0.0430	9.7122	0.1030	57.5544	5.9260
16.0000	2.5403	0.3936	25.6725	0.0390	10.1059	0.0990	63.4591	6.2794
17.0000	2.6928	0.3714	28.2128	0.0354	10.4773	0.0954	69.4009	6.6240
18.0000	2.8543	0.3503	30.9056	0.0324	10.8276	0.0924	75.3567	6.9597
19.0000	3.0256	0.3305	33.7599	0.0296	11.1581	0.0896	81.3060	7.2867
20.0000	3.2071	0.3118	36.7855	0.0272	11.4699	0.0872	87.2302	7.6051
21.0000	3.3996	0.2942	39.9927	0.0250	11.7641	0.0850	93.1133	7.9151
22.0000	3.6035	0.2775	43.3922	0.0230	12.0416	0.0830	98.9410	8.2166
23.0000	3.8197	0.2618	46.9957	0.0213	12.3034	0.0813	104.7005	8.5099
24.0000	4.0489	0.2470	50.8155	0.0197	12.5504	0.0797	110.3810	8.7951
25.0000	4.2919	0.2330	54.8644	0.0182	12.7834	0.0782	115.9730	9.0722
26.0000	4.5494	0.2198	59.1563	0.0169	13.0032	0.0769	121.4682	9.3414
27.0000	4.8223	0.2074	63.7057	0.0157	13.2105	0.0757	126.8598	9.6029
28.0000	5.1117	0.1956	68.5280	0.0146	13.4062	0.0746	132.1418	9.8568
29.0000	5.4184	0.1846	73.6397	0.0136	13.5907	0.0736	137.3094	10.1032
30.0000	5.7435	0.1741	79.0580	0.0126	13.7648	0.0726	142.3586	10.3422
31.0000	6.0881	0.1643	84.8015	0.0118	13.9291	0.0718	147.2862	10.5740
32.0000	6.4534	0.1550	90.8896	0.0110	14.0840	0.0710	152.0899	10.7987
33.0000	6.8406	0.1462	97.3430	0.0103	14.2302	0.0703	156.7679	11.0165
34.0000	7.2510	0.1379	104.1835	0.0096	14.3681	0.0696	161.3189	11.2275
35.0000	7.6861	0.1301	111.4345	0.0090	14.4982	0.0690	165.7425	11.4319
36.0000	8.1472	0.1227	119.1206	0.0084	14.6210	0.0684	170.0385	11.6298
37.0000	8.6361	0.1158	127.2678	0.0079	14.7368	0.0679	174.2070	11.8212
38.0000	9.1542	0.1092	135.9039	0.0074	14.8460	0.0674	178.2489	12.0065
39.0000	9.7035	0.1031	145.0581	0.0069	14.9491	0.0669	182.1650	12.1857
40.0000	10.2857	0.0972	154.7616	0.0065	15.0463	0.0665	185.9566	12.3590

DECIMAL INTEREST RATE = .07

PERIOD	F/P	P/F	F/A	A/F	P/A	A/P	P/G	A/G
1.0000	1.0700	0.9346	1.0000	1.0000	0.9346	1.0700	0.0000	0.0000
2.0000	1.1449	0.8734	2.0700	0.4831	1.8080	0.5531	0.8735	0.4831
3.0000	1.2250	0.8163	3.2149	0.3111	2.6243	0.3811	2.5061	0.9549
4.0000	1.3108	0.7629	4.4399	0.2252	3.3872	0.2952	4.7948	1.4155
5.0000	1.4026	0.7130	5.7507	0.1739	4.1002	0.2439	7.6467	1.8650
6.0000	1.5007	0.6663	7.1533	0.1398	4.7665	0.2098	10.9784	2.3032
7.0000	1.6058	0.6227	8.6540	0.1156	5.3893	0.1856	14.7149	2.7304
8.0000	1.7182	0.5820	10.2598	0.0975	5.9713	0.1675	18.7890	3.1466
9.0000	1.8385	0.5439	11.9780	0.0835	6.5152	0.1535	23.1405	3.5517
10.0000	1.9672	0.5083	13.8165	0.0724	7.0236	0.1424	27.7156	3.9461
11.0000	2.1049	0.4751	15.7836	0.0634	7.4987	0.1334	32.4666	4.3296
12.0000	2.2522	0.4440	17.8885	0.0559	7.9427	0.1259	37.3507	4.7025
13.0000	2.4098	0.4150	20.1407	0.0497	8.3577	0.1197	42.3303	5.0649
14.0000	2.5785	0.3878	22.5505	0.0443	8.7455	0.1143	47.3719	5.4167
15.0000	2.7590	0.3624	25.1291	0.0398	9.1079	0.1098	52.4462	5.7583
16.0000	2.9522	0.3387	27.8881	0.0359	9.4467	0.1059	57.5272	6.0897
17.0000	3.1588	0.3166	30.8403	0.0324	9.7632	0.1024	62.5924	6.4110
18.0000	3.3799	0.2959	33.9991	0.0294	10.0591	0.0994	67.6221	6.7225
19.0000	3.6165	0.2765	37.3790	0.0268	10.3356	0.0968	72.5992	7.0242
20.0000	3.8697	0.2584	40.9955	0.0244	10.5940	0.0944	77.5092	7.3163
21.0000	4.1406	0.2415	44.8652	0.0223	10.8355	0.0923	82.3395	7.5990
22.0000	4.4304	0.2257	49.0058	0.0204	11.0612	0.0904	87.0794	7.8725
23.0000	4.7405	0.2109	53.4362	0.0187	11.2722	0.0887	91.7203	8.1369
24.0000	5.0724	0.1971	58.1768	0.0172	11.4693	0.0872	96.2546	8.3923
25.0000	5.4274	0.1842	63.2491	0.0158	11.6536	0.0858	100.6766	8.6391
26.0000	5.8074	0.1722	68.6766	0.0146	11.8258	0.0846	104.9815	8.8773
27.0000	6.2139	0.1609	74.4840	0.0134	11.9867	0.0834	109.1657	9.1072
28.0000	6.6488	0.1504	80.6978	0.0124	12.1371	0.0824	113.2266	9.3290
29.0000	7.1143	0.1406	87.3467	0.0114	12.2777	0.0814	117.1623	9.5427
30.0000	7.6123	0.1314	94.4609	0.0106	12.4090	0.0806	120.9720	9.7487
31.0000	8.1451	0.1228	102.0732	0.0098	12.5318	0.0798	124.6551	9.9471
32.0000	8.7153	0.1147	110.2184	0.0091	12.6466	0.0791	128.2121	10.1381
33.0000	9.3254	0.1072	118.9336	0.0084	12.7538	0.0784	131.6436	10.3219
34.0000	9.9781	0.1002	128.2590	0.0078	12.8540	0.0778	134.9509	10.4987
35.0000	10.6766	0.0937	138.2371	0.0072	12.9477	0.0772	138.1354	10.6687
36.0000	11.4240	0.0875	148.9137	0.0067	13.0352	0.0767	141.1991	10.8321
37.0000	12.2236	0.0818	160.3377	0.0062	13.1170	0.0762	144.1442	10.9891
38.0000	13.0793	0.0765	172.5614	0.0058	13.1935	0.0758	146.9731	11.1398
39.0000	13.9948	0.0715	185.6407	0.0054	13.2649	0.0754	149.6884	11.2845
40.0000	14.9745	0.0668	199.6355	0.0050	13.3317	0.0750	152.2929	11.4234

DECIMAL INTEREST RATE = .08

PERIOD	F/P	P/F	F/A	A/F	P/A	A/P	P/G	A/G
1.0000	1.0800	0.9259	1.0000	1.0000	0.9259	1.0800	0.0000	0.0000
2.0000	1.1664	0.8573	2.0800	0.4808	1.7833	0.5608	0.8573	0.4808
3.0000	1.2597	0.7938	3.2464	0.3080	2.5771	0.3880	2.4450	0.9487
4.0000	1.3605	0.7350	4.5061	0.2219	3.3121	0.3019	4.6501	1.4040
5.0000	1.4693	0.6806	5.8666	0.1705	3.9927	0.2505	7.3725	1.8465
6.0000	1.5869	0.6302	7.3359	0.1363	4.6229	0.2163	10.5233	2.2764
7.0000	1.7138	0.5835	8.9228	0.1121	5.2064	0.1921	14.0242	2.6937
8.0000	1.8509	0.5403	10.6366	0.0940	5.7466	0.1740	17.8061	3.0985
9.0000	1.9990	0.5002	12.4876	0.0801	6.2469	0.1601	21.8081	3.4910
10.0000	2.1589	0.4632	14.4866	0.0690	6.7101	0.1490	25.9769	3.8713
11.0000	2.3316	0.4289	16.6455	0.0601	7.1390	0.1401	30.2657	4.2395
12.0000	2.5182	0.3971	18.9771	0.0527	7.5361	0.1327	34.6340	4.5958
13.0000	2.7196	0.3677	21.4953	0.0465	7.9038	0.1265	39.0463	4.9402
14.0000	2.9372	0.3405	24.2149	0.0413	8.2442	0.1213	43.4723	5.2731
15.0000	3.1722	0.3152	27.1521	0.0368	8.5595	0.1168	47.8857	5.5945
16.0000	3.4259	0.2919	30.3243	0.0330	8.8514	0.1130	52.2641	5.9046
17.0000	3.7000	0.2703	33.7503	0.0296	9.1216	0.1096	56.5884	6.2038
18.0000	3.9960	0.2502	37.4503	0.0267	9.3719	0.1067	60.8426	6.4920
19.0000	4.3157	0.2317	41.4463	0.0241	9.6036	0.1041	65.0134	6.7697
20.0000	4.6610	0.2145	45.7620	0.0219	9.8181	0.1019	69.0899	7.0370
21.0000	5.0338	0.1987	50.4230	0.0198	10.0168	0.0998	73.0630	7.2940
22.0000	5.4365	0.1839	55.4568	0.0180	10.2007	0.0980	76.9257	7.5412
23.0000	5.8715	0.1703	60.8933	0.0164	10.3711	0.0964	80.6726	7.7786
24.0000	6.3412	0.1577	66.7648	0.0150	10.5288	0.0950	84.2998	8.0066
25.0000	6.8485	0.1460	73.1060	0.0137	10.6748	0.0937	87.8042	8.2254
26.0000	7.3964	0.1352	79.9545	0.0125	10.8100	0.0925	91.1842	8.4352
27.0000	7.9881	0.1252	87.3509	0.0114	10.9352	0.0914	94.4391	8.6363
28.0000	8.6271	0.1159	95.3389	0.0105	11.0511	0.0905	97.5687	8.8289
29.0000	9.3173	0.1073	103.9660	0.0096	11.1584	0.0896	100.5739	9.0133
30.0000	10.0627	0.0994	113.2833	0.0088	11.2578	0.0888	103.4558	9.1897
31.0000	10.8677	0.0920	123.3460	0.0081	11.3498	0.0881	106.2163	9.3584
32.0000	11.7371	0.0852	134.2137	0.0075	11.4350	0.0875	108.8575	9.5197
33.0000	12.6761	0.0789	145.9508	0.0069	11.5139	0.0869	111.3820	9.6737
34.0000	13.6901	0.0730	158.6269	0.0063	11.5869	0.0863	113.7925	9.8208
35.0000	14.7854	0.0676	172.3170	0.0058	11.6546	0.0858	116.0920	9.9611
36.0000	15.9682	0.0626	187.1024	0.0053	11.7172	0.0853	118.2839	10.0949
37.0000	17.2456	0.0580	203.0706	0.0049	11.7752	0.0849	120.3714	10.2225
38.0000	18.6253	0.0537	220.3162	0.0045	11.8289	0.0845	122.3579	10.3440
39.0000	20.1153	0.0497	238.9415	0.0042	11.8786	0.0842	124.2470	10.4598
40.0000	21.7245	0.0460	259.0569	0.0039	11.9246	0.0839	126.0422	10.5699

DECIMAL INTEREST RATE = .09

PERIOD	F/P	P/F	F/A	A/F	P/A	A/P	P/G	A/G
1.0000	1.0900	0.9174	1.0000	1.0000	0.9174	1.0900	0.0000	0.0000
2.0000	1.1881	0.8417	2.0900	0.4785	1.7591	0.5685	0.8417	0.4785
3.0000	1.2950	0.7722	3.2781	0.3051	2.5313	0.3951	2.3861	0.9426
4.0000	1.4116	0.7084	4.5731	0.2187	3.2397	0.3087	4.5113	1.3925
5.0000	1.5386	0.6499	5.9847	0.1671	3.8897	0.2571	7.1111	1.8282
6.0000	1.6771	0.5963	7.5233	0.1329	4.4859	0.2229	10.0924	2.2498
7.0000	1.8280	0.5470	9.2004	0.1087	5.0330	0.1987	13.3746	2.6574
8.0000	1.9926	0.5019	11.0285	0.0907	5.5348	0.1807	16.8877	3.0512
9.0000	2.1719	0.4604	13.0210	0.0768	5.9952	0.1668	20.5711	3.4312
10.0000	2.3674	0.4224	15.1929	0.0658	6.4177	0.1558	24.3728	3.7978
11.0000	2.5804	0.3875	17.5603	0.0569	6.8052	0.1469	28.2481	4.1510
12.0000	2.8127	0.3555	20.1407	0.0497	7.1607	0.1397	32.1590	4.4910
13.0000	3.0658	0.3262	22.9534	0.0436	7.4869	0.1336	36.0732	4.8182
14.0000	3.3417	0.2992	26.0192	0.0384	7.7862	0.1284	39.9634	5.1326
15.0000	3.6425	0.2745	29.3609	0.0341	8.0607	0.1241	43.8069	5.4346
16.0000	3.9703	0.2519	33.0034	0.0303	8.3126	0.1203	47.5850	5.7245
17.0000	4.3276	0.2311	36.9737	0.0270	8.5436	0.1170	51.2821	6.0024
18.0000	4.7171	0.2120	41.3014	0.0242	8.7556	0.1142	54.8860	6.2687
19.0000	5.1417	0.1945	46.0185	0.0217	8.9501	0.1117	58.3868	6.5236
20.0000	5.6044	0.1784	51.1602	0.0195	9.1285	0.1095	61.7770	6.7675
21.0000	6.1088	0.1637	56.7646	0.0176	9.2922	0.1076	65.0510	7.0006
22.0000	6.6586	0.1502	62.8734	0.0159	9.4424	0.1059	68.2048	7.2232
23.0000	7.2579	0.1378	69.5320	0.0144	9.5802	0.1044	71.2360	7.4357
24.0000	7.9111	0.1264	76.7899	0.0130	9.7066	0.1030	74.1433	7.6384
25.0000	8.6231	0.1160	84.7010	0.0118	9.8226	0.1018	76.9265	7.8316
26.0000	9.3992	0.1064	93.3241	0.0107	9.9290	0.1007	79.5863	8.0156
27.0000	10.2451	0.0976	102.7233	0.0097	10.0266	0.0997	82.1241	8.1906
28.0000	11.1672	0.0895	112.9684	0.0089	10.1161	0.0989	84.5420	8.3571
29.0000	12.1722	0.0822	124.1355	0.0081	10.1983	0.0981	86.8423	8.5154
30.0000	13.2677	0.0754	136.3077	0.0073	10.2737	0.0973	89.0280	8.6657
31.0000	14.4618	0.0691	149.5754	0.0067	10.3428	0.0967	91.1025	8.8083
32.0000	15.7634	0.0634	164.0372	0.0061	10.4062	0.0961	93.0691	8.9436
33.0000	17.1821	0.0582	179.8006	0.0056	10.4644	0.0956	94.9315	9.0718
34.0000	18.7284	0.0534	196.9827	0.0051	10.5178	0.0951	96.6935	9.1933
35.0000	20.4140	0.0490	215.7111	0.0046	10.5668	0.0946	98.3590	9.3083
36.0000	22.2513	0.0449	236.1251	0.0042	10.6118	0.0942	99.9320	9.4171
37.0000	24.2539	0.0412	258.3764	0.0039	10.6530	0.0939	101.4163	9.5200
38.0000	26.4367	0.0378	282.6303	0.0035	10.6908	0.0935	102.8158	9.6172
39.0000	28.8160	0.0347	309.0670	0.0032	10.7255	0.0932	104.1345	9.7090
40.0000	31.4095	0.0318	337.8831	0.0030	10.7574	0.0930	105.3762	9.7957

DECIMAL INTEREST RATE = .1

PERIOD	F/P	P/F	F/A	A/F	P/A	A/P	P/G	A/G
1.0000	1.1000	0.9091	1.0000	1.0000	0.9091	1.1000	0.0000	0.0000
2.0000	1.2100	0.8264	2.1000	0.4762	1.7355	0.5762	0.8264	0.4762
3.0000	1.3310	0.7513	3.3100	0.3021	2.4869	0.4021	2.3291	0.9366
4.0000	1.4641	0.6830	4.6410	0.2155	3.1699	0.3155	4.3781	1.3812
5.0000	1.6105	0.6209	6.1051	0.1638	3.7908	0.2638	6.8618	1.8101
6.0000	1.7716	0.5645	7.7156	0.1296	4.3553	0.2296	9.6842	2.2236
7.0000	1.9487	0.5132	9.4872	0.1054	4.8684	0.2054	12.7631	2.6216
8.0000	2.1436	0.4665	11.4359	0.0874	5.3349	0.1874	16.0287	3.0045
9.0000	2.3579	0.4241	13.5795	0.0736	5.7590	0.1736	19.4215	3.3724
10.0000	2.5937	0.3855	15.9374	0.0627	6.1446	0.1627	22.8914	3.7255
11.0000	2.8531	0.3505	18.5312	0.0540	6.4951	0.1540	26.3963	4.0641
12.0000	3.1384	0.3186	21.3843	0.0468	6.8137	0.1468	29.9012	4.3884
13.0000	3.4523	0.2897	24.5227	0.0408	7.1034	0.1408	33.3772	4.6988
14.0000	3.7975	0.2633	27.9750	0.0357	7.3667	0.1357	36.8005	4.9955
15.0000	4.1772	0.2394	31.7725	0.0315	7.6061	0.1315	40.1520	5.2789
16.0000	4.5950	0.2176	35.9497	0.0278	7.8237	0.1278	43.4164	5.5493
17.0000	5.0545	0.1978	40.5447	0.0247	8.0216	0.1247	46.5820	5.8071
18.0000	5.5599	0.1799	45.5992	0.0219	8.2014	0.1219	49.6396	6.0526
19.0000	6.1159	0.1635	51.1591	0.0195	8.3649	0.1195	52.5827	6.2861
20.0000	6.7275	0.1486	57.2750	0.0175	8.5136	0.1175	55.4069	6.5081
21.0000	7.4003	0.1351	64.0025	0.0156	8.6487	0.1156	58.1095	6.7189
22.0000	8.1403	0.1228	71.4028	0.0140	8.7715	0.1140	60.6893	6.9189
23.0000	8.9543	0.1117	79.5431	0.0126	8.8832	0.1126	63.1462	7.1085
24.0000	9.8497	0.1015	88.4974	0.0113	8.9847	0.1113	65.4813	7.2881
25.0000	10.8347	0.0923	98.3471	0.0102	9.0770	0.1102	67.6964	7.4580
26.0000	11.9182	0.0839	109.1818	0.0092	9.1609	0.1092	69.7940	7.6187
27.0000	13.1100	0.0763	121.1000	0.0083	9.2372	0.1083	71.7773	7.7704
28.0000	14.4210	0.0693	134.2100	0.0075	9.3066	0.1075	73.6495	7.9137
29.0000	15.8631	0.0630	148.6310	0.0067	9.3696	0.1067	75.4146	8.0489
30.0000	17.4494	0.0573	164.4941	0.0061	9.4269	0.1061	77.0766	8.1762
31.0000	19.1944	0.0521	181.9435	0.0055	9.4790	0.1055	78.6396	8.2962
32.0000	21.1138	0.0474	201.1379	0.0050	9.5264	0.1050	80.1078	8.4091
33.0000	23.2252	0.0431	222.2517	0.0045	9.5694	0.1045	81.4856	8.5152
34.0000	25.5477	0.0391	245.4768	0.0041	9.6086	0.1041	82.7773	8.6149
35.0000	28.1025	0.0356	271.0245	0.0037	9.6442	0.1037	83.9872	8.7086
36.0000	30.9127	0.0323	299.1270	0.0033	9.6765	0.1033	85.1194	8.7965
37.0000	34.0040	0.0294	330.0397	0.0030	9.7059	0.1030	86.1781	8.8789
38.0000	37.4044	0.0267	364.0437	0.0027	9.7327	0.1027	87.1673	8.9562
39.0000	41.1448	0.0243	401.4480	0.0025	9.7570	0.1025	88.0908	9.0285
40.0000	45.2593	0.0221	442.5928	0.0023	9.7791	0.1023	88.9525	9.0962

DECIMAL INTEREST RATE = .12

PERIOD	F/P	P/F	F/A	A/F	P/A	A/P	P/G	A/G
1.0000	1.1200	0.8929	1.0000	1.0000	0.8929	1.1200	0.0000	0.0000
2.0000	1.2544	0.7972	2.1200	0.4717	1.6901	0.5917	0.7972	0.4717
3.0000	1.4049	0.7118	3.3744	0.2963	2.4018	0.4163	2.2208	0.9246
4.0000	1.5735	0.6355	4.7793	0.2092	3.0373	0.3292	4.1273	1.3589
5.0000	1.7623	0.5674	6.3528	0.1574	3.6048	0.2774	6.3970	1.7746
6.0000	1.9738	0.5066	8.1152	0.1232	4.1114	0.2432	8.9302	2.1720
7.0000	2.2107	0.4523	10.0890	0.0991	4.5638	0.2191	11.6443	2.5515
8.0000	2.4760	0.4039	12.2997	0.0813	4.9676	0.2013	14.4715	2.9131
9.0000	2.7731	0.3606	14.7757	0.0677	5.3283	0.1877	17.3563	3.2574
10.0000	3.1058	0.3220	17.5487	0.0570	5.6502	0.1770	20.2541	3.5847
11.0000	3.4785	0.2875	20.6546	0.0484	5.9377	0.1684	23.1288	3.8953
12.0000	3.8960	0.2567	24.1331	0.0414	6.1944	0.1614	25.9523	4.1897
13.0000	4.3635	0.2292	28.0291	0.0357	6.4235	0.1557	28.7024	4.4683
14.0000	4.8871	0.2046	32.3926	0.0309	6.6282	0.1509	31.3624	4.7317
15.0000	5.4736	0.1827	37.2797	0.0268	6.8109	0.1468	33.9202	4.9803
16.0000	6.1304	0.1631	42.7533	0.0234	6.9740	0.1434	36.3670	5.2147
17.0000	6.8660	0.1456	48.8837	0.0205	7.1196	0.1405	38.6973	5.4353
18.0000	7.6900	0.1300	55.7497	0.0179	7.2497	0.1379	40.9080	5.6427
19.0000	8.6128	0.1161	63.4397	0.0158	7.3658	0.1358	42.9979	5.8375
20.0000	9.6463	0.1037	72.0524	0.0139	7.4694	0.1339	44.9676	6.0202
21.0000	10.8038	0.0926	81.6987	0.0122	7.5620	0.1322	46.8188	6.1913
22.0000	12.1003	0.0826	92.5026	0.0108	7.6446	0.1308	48.5543	6.3514
23.0000	13.5523	0.0738	104.6029	0.0096	7.7184	0.1296	50.1776	6.5010
24.0000	15.1786	0.0659	118.1552	0.0085	7.7843	0.1285	51.6929	6.6406
25.0000	17.0001	0.0588	133.3339	0.0075	7.8431	0.1275	53.1046	6.7708
26.0000	19.0401	0.0525	150.3339	0.0067	7.8957	0.1267	54.4177	6.8921
27.0000	21.3249	0.0469	169.3740	0.0059	7.9426	0.1259	55.6369	7.0049
28.0000	23.8839	0.0419	190.6989	0.0052	7.9844	0.1252	56.7674	7.1098
29.0000	26.7499	0.0374	214.5828	0.0047	8.0218	0.1247	57.8141	7.2071
30.0000	29.9599	0.0334	241.3327	0.0041	8.0552	0.1241	58.7821	7.2974
31.0000	33.5551	0.0298	271.2926	0.0037	8.0850	0.1237	59.6761	7.3811
32.0000	37.5817	0.0266	304.8477	0.0033	8.1116	0.1233	60.5010	7.4586
33.0000	42.0915	0.0238	342.4295	0.0029	8.1354	0.1229	61.2612	7.5302
34.0000	47.1425	0.0212	384.5210	0.0026	8.1566	0.1226	61.9612	7.5965
35.0000	52.7996	0.0189	431.6635	0.0023	8.1755	0.1223	62.6052	7.6577
36.0000	59.1356	0.0169	484.4632	0.0021	8.1924	0.1221	63.1970	7.7141
37.0000	66.2318	0.0151	543.5987	0.0018	8.2075	0.1218	63.7406	7.7661
38.0000	74.1797	0.0135	609.8306	0.0016	8.2210	0.1216	64.2394	7.8141
39.0000	83.0812	0.0120	684.0102	0.0015	8.2330	0.1215	64.6968	7.8582
40.0000	93.0510	0.0107	767.0914	0.0013	8.2438	0.1213	65.1159	7.8988

DECIMAL INTEREST RATE = .15

PERIOD	F/P	P/F	F/A	A/F	P/A	A/P	P/G	A/G
1.0000	1.1500	0.8696	1.0000	1.0000	0.8696	1.1500	-0.0000	-0.0000
2.0000	1.3225	0.7561	2.1500	0.4651	1.6257	0.6151	0.7561	0.4651
3.0000	1.5209	0.6575	3.4725	0.2880	2.2832	0.4380	2.0712	0.9071
4.0000	1.7490	0.5718	4.9934	0.2003	2.8550	0.3503	3.7864	1.3263
5.0000	2.0114	0.4972	6.7424	0.1483	3.3522	0.2983	5.7751	1.7228
6.0000	2.3131	0.4323	8.7537	0.1142	3.7845	0.2642	7.9368	2.0972
7.0000	2.6600	0.3759	11.0668	0.0904	4.1604	0.2404	10.1924	2.4498
8.0000	3.0590	0.3269	13.7268	0.0729	4.4873	0.2229	12.4807	2.7813
9.0000	3.5179	0.2843	16.7858	0.0596	4.7716	0.2096	14.7548	3.0922
10.0000	4.0456	0.2472	20.3037	0.0493	5.0188	0.1993	16.9795	3.3832
11.0000	4.6524	0.2149	24.3493	0.0411	5.2337	0.1911	19.1289	3.6549
12.0000	5.3503	0.1869	29.0017	0.0345	5.4206	0.1845	21.1849	3.9082
13.0000	6.1528	0.1625	34.3519	0.0291	5.5831	0.1791	23.1352	4.1438
14.0000	7.0757	0.1413	40.5047	0.0247	5.7245	0.1747	24.9725	4.3624
15.0000	8.1371	0.1229	47.5804	0.0210	5.8474	0.1710	26.6930	4.5650
16.0000	9.3576	0.1069	55.7175	0.0179	5.9542	0.1679	28.2960	4.7522
17.0000	10.7613	0.0929	65.0751	0.0154	6.0472	0.1654	29.7828	4.9251
18.0000	12.3755	0.0808	75.8364	0.0132	6.1280	0.1632	31.1565	5.0843
19.0000	14.2318	0.0703	88.2118	0.0113	6.1982	0.1613	32.4213	5.2307
20.0000	16.3665	0.0611	102.4436	0.0098	6.2593	0.1598	33.5822	5.3651
21.0000	18.8215	0.0531	118.8101	0.0084	6.3125	0.1584	34.6448	5.4883
22.0000	21.6447	0.0462	137.6316	0.0073	6.3587	0.1573	35.6150	5.6010
23.0000	24.8915	0.0402	159.2764	0.0063	6.3988	0.1563	36.4988	5.7040
24.0000	28.6252	0.0349	184.1679	0.0054	6.4338	0.1554	37.3023	5.7979
25.0000	32.9190	0.0304	212.7930	0.0047	6.4641	0.1547	38.0314	5.8834
26.0000	37.8568	0.0264	245.7120	0.0041	6.4906	0.1541	38.6918	5.9612
27.0000	43.5353	0.0230	283.5688	0.0035	6.5135	0.1535	39.2890	6.0319
28.0000	50.0656	0.0200	327.1041	0.0031	6.5335	0.1531	39.8283	6.0960
29.0000	57.5755	0.0174	377.1697	0.0027	6.5509	0.1527	40.3146	6.1541
30.0000	66.2118	0.0151	434.7452	0.0023	6.5660	0.1523	40.7526	6.2066
31.0000	76.1436	0.0131	500.9570	0.0020	6.5791	0.1520	41.1466	6.2541
32.0000	87.5651	0.0114	577.1005	0.0017	6.5905	0.1517	41.5006	6.2970
33.0000	100.6998	0.0099	664.6655	0.0015	6.6005	0.1515	41.8184	6.3357
34.0000	115.8048	0.0086	765.3653	0.0013	6.6091	0.1513	42.1033	6.3705
35.0000	133.1755	0.0075	881.1701	0.0011	6.6166	0.1511	42.3586	6.4019
36.0000	153.1519	0.0065	1014.3460	0.0010	6.6231	0.1510	42.5872	6.4301
37.0000	176.1246	0.0057	1167.4970	0.0009	6.6288	0.1509	42.7916	6.4554
38.0000	202.5433	0.0049	1343.6220	0.0007	6.6338	0.1507	42.9742	6.4781
39.0000	232.9248	0.0043	1546.1660	0.0006	6.6380	0.1506	43.1374	6.4985
40.0000	267.8636	0.0037	1779.0900	0.0006	6.6418	0.1506	43.2830	6.5168

DECIMAL INTEREST RATE = .18

PERIOD	F/P	P/F	F/A	A/F	P/A	A/P	P/G	A/G
1.0000	1.1800	0.8475	1.0000	1.0000	0.8475	1.1800	0.0000	0.0000
2.0000	1.3924	0.7182	2.1800	0.4587	1.5656	0.6387	0.7182	0.4587
3.0000	1.6430	0.6086	3.5724	0.2799	2.1743	0.4599	1.9355	0.8902
4.0000	1.9388	0.5158	5.2154	0.1917	2.6901	0.3717	3.4828	1.2947
5.0000	2.2878	0.4371	7.1542	0.1398	3.1272	0.3198	5.2313	1.6728
6.0000	2.6996	0.3704	9.4420	0.1059	3.4976	0.2859	7.0834	2.0252
7.0000	3.1855	0.3139	12.1415	0.0824	3.8115	0.2624	8.9670	2.3526
8.0000	3.7589	0.2660	15.3270	0.0652	4.0776	0.2452	10.8292	2.6558
9.0000	4.4355	0.2255	19.0859	0.0524	4.3030	0.2324	12.6329	2.9358
10.0000	5.2338	0.1911	23.5213	0.0425	4.4941	0.2225	14.3525	3.1936
11.0000	6.1759	0.1619	28.7552	0.0348	4.6560	0.2148	15.9717	3.4303
12.0000	7.2876	0.1372	34.9311	0.0286	4.7932	0.2086	17.4811	3.6470
13.0000	8.5994	0.1163	42.2187	0.0237	4.9095	0.2037	18.8765	3.8449
14.0000	10.1473	0.0985	50.8181	0.0197	5.0081	0.1997	20.1577	4.0250
15.0000	11.9738	0.0835	60.9653	0.0164	5.0916	0.1964	21.3269	4.1887
16.0000	14.1290	0.0708	72.9391	0.0137	5.1624	0.1937	22.3885	4.3369
17.0000	16.6723	0.0600	87.0681	0.0115	5.2223	0.1915	23.3482	4.4708
18.0000	19.6733	0.0508	103.7404	0.0096	5.2732	0.1896	24.2123	4.5916
19.0000	23.2145	0.0431	123.4137	0.0081	5.3162	0.1881	24.9877	4.7003
20.0000	27.3931	0.0365	146.6281	0.0068	5.3527	0.1868	25.6813	4.7978
21.0000	32.3238	0.0309	174.0212	0.0057	5.3837	0.1857	26.3000	4.8851
22.0000	38.1421	0.0262	206.3450	0.0048	5.4099	0.1848	26.8506	4.9632
23.0000	45.0077	0.0222	244.4872	0.0041	5.4321	0.1841	27.3394	5.0329
24.0000	53.1091	0.0188	289.4949	0.0035	5.4509	0.1835	27.7725	5.0950
25.0000	62.6687	0.0160	342.6039	0.0029	5.4669	0.1829	28.1555	5.1502
26.0000	73.9491	0.0135	405.2727	0.0025	5.4804	0.1825	28.4935	5.1991
27.0000	87.2599	0.0115	479.2219	0.0021	5.4919	0.1821	28.7915	5.2425
28.0000	102.9667	0.0097	566.4817	0.0018	5.5016	0.1818	29.0537	5.2810
29.0000	121.5007	0.0082	669.4485	0.0015	5.5098	0.1815	29.2842	5.3149
30.0000	143.3709	0.0070	790.9493	0.0013	5.5168	0.1813	29.4864	5.3448
31.0000	169.1776	0.0059	934.3202	0.0011	5.5227	0.1811	29.6638	5.3712
32.0000	199.6296	0.0050	1103.4980	0.0009	5.5277	0.1809	29.8190	5.3945
33.0000	235.5630	0.0042	1303.1280	0.0008	5.5320	0.1808	29.9549	5.4149
34.0000	277.9643	0.0036	1538.6910	0.0006	5.5356	0.1806	30.0736	5.4328
35.0000	327.9979	0.0030	1816.6550	0.0006	5.5386	0.1806	30.1773	5.4485
36.0000	387.0375	0.0026	2144.6530	0.0005	5.5412	0.1805	30.2677	5.4623
37.0000	456.7044	0.0022	2531.6910	0.0004	5.5434	0.1804	30.3465	5.4744
38.0000	538.9111	0.0019	2988.3950	0.0003	5.5452	0.1803	30.4152	5.4849
39.0000	635.9152	0.0016	3527.3070	0.0003	5.5468	0.1803	30.4749	5.4941
40.0000	750.3800	0.0013	4163.2220	0.0002	5.5482	0.1802	30.5269	5.5022

DECIMAL INTEREST RATE = .2

PERIOD	F/P	P/F	F/A	A/F	P/A	A/P	P/G	A/G
1.0000	1.2000	0.8333	1.0000	1.0000	0.8333	1.2000	0.0000	0.0000
2.0000	1.4400	0.6944	2.2000	0.4545	1.5278	0.6545	0.6944	0.4545
3.0000	1.7280	0.5787	3.6400	0.2747	2.1065	0.4747	1.8519	0.8791
4.0000	2.0736	0.4823	5.3680	0.1863	2.5887	0.3863	3.2986	1.2742
5.0000	2.4883	0.4019	7.4416	0.1344	2.9906	0.3344	4.9061	1.6405
6.0000	2.9860	0.3349	9.9299	0.1007	3.3255	0.3007	6.5806	1.9788
7.0000	3.5832	0.2791	12.9159	0.0774	3.6046	0.2774	8.2551	2.2902
8.0000	4.2998	0.2326	16.4991	0.0606	3.8372	0.2606	9.8831	2.5756
9.0000	5.1598	0.1938	20.7989	0.0481	4.0310	0.2481	11.4335	2.8364
10.0000	6.1917	0.1615	25.9587	0.0385	4.1925	0.2385	12.8871	3.0739
11.0000	7.4301	0.1346	32.1504	0.0311	4.3271	0.2311	14.2330	3.2893
12.0000	8.9161	0.1122	39.5805	0.0253	4.4392	0.2253	15.4667	3.4841
13.0000	10.6993	0.0935	48.4966	0.0206	4.5327	0.2206	16.5883	3.6597
14.0000	12.8392	0.0779	59.1959	0.0169	4.6106	0.2169	17.6008	3.8175
15.0000	15.4070	0.0649	72.0351	0.0139	4.6755	0.2139	18.5095	3.9588
16.0000	18.4884	0.0541	87.4421	0.0114	4.7296	0.2114	19.3208	4.0851
17.0000	22.1861	0.0451	105.9306	0.0094	4.7746	0.2094	20.0419	4.1976
18.0000	26.6233	0.0376	128.1167	0.0078	4.8122	0.2078	20.6805	4.2975
19.0000	31.9480	0.0313	154.7400	0.0065	4.8435	0.2065	21.2439	4.3861
20.0000	38.3376	0.0261	186.6880	0.0054	4.8696	0.2054	21.7395	4.4643
21.0000	46.0051	0.0217	225.0256	0.0044	4.8913	0.2044	22.1742	4.5334
22.0000	55.2061	0.0181	271.0307	0.0037	4.9094	0.2037	22.5546	4.5941
23.0000	66.2474	0.0151	326.2369	0.0031	4.9245	0.2031	22.8867	4.6475
24.0000	79.4969	0.0126	392.4843	0.0025	4.9371	0.2025	23.1760	4.6943
25.0000	95.3962	0.0105	471.9811	0.0021	4.9476	0.2021	23.4276	4.7352
26.0000	114.4755	0.0087	567.3773	0.0018	4.9563	0.2018	23.6460	4.7709
27.0000	137.3706	0.0073	681.8529	0.0015	4.9636	0.2015	23.8353	4.8020
28.0000	164.8447	0.0061	819.2233	0.0012	4.9697	0.2012	23.9991	4.8291
29.0000	197.8136	0.0051	984.0681	0.0010	4.9747	0.2010	24.1406	4.8527
30.0000	237.3764	0.0042	1181.8820	0.0008	4.9789	0.2008	24.2628	4.8731
31.0000	284.8516	0.0035	1419.2580	0.0007	4.9824	0.2007	24.3681	4.8908
32.0000	341.8219	0.0029	1704.1100	0.0006	4.9854	0.2006	24.4588	4.9061
33.0000	410.1863	0.0024	2045.9310	0.0005	4.9878	0.2005	24.5368	4.9194
34.0000	492.2236	0.0020	2456.1180	0.0004	4.9898	0.2004	24.6038	4.9308
35.0000	590.6683	0.0017	2948.3410	0.0003	4.9915	0.2003	24.6614	4.9406
36.0000	708.8019	0.0014	3539.0090	0.0003	4.9929	0.2003	24.7108	4.9491
37.0000	850.5623	0.0012	4247.8120	0.0002	4.9941	0.2002	24.7531	4.9564
38.0000	1020.6750	0.0010	5098.3740	0.0002	4.9951	0.2002	24.7894	4.9627
39.0000	1224.8100	0.0008	6119.0490	0.0002	4.9959	0.2002	24.8204	4.9681
40.0000	1469.7720	0.0007	7343.8580	0.0001	4.9966	0.2001	24.8469	4.9728

Appendix IV

Anthropometric Tables*

Anthropometric Table IV-1 (all dimensions in inches)

Measurement	Range	Mean	Standard Deviation	Percentiles 1st	5th	50th	95th	99th
Weight								
1. Weight (pounds)	104. - 265.	163.66	20.86	123.1	132.5	161.9	200.8	215.9
Body Lengths								
2. Stature	59.45 - 77.56	69.11	2.44	63.5	65.2	69.1	73.1	74.9
3. Nasal root height	56.30 - 73.23	64.95	2.39	59.4	61.0	65.0	68.9	70.7
4. Eye height	56.30 - 73.23	64.69	2.38	59.2	60.8	64.7	68.6	70.3
5. Tragion height	54.72 - 74.41	63.92	2.39	58.4	60.0	64.0	67.8	69.6
6. Cervicale height	50.39 - 66.93	59.08	2.31	53.7	55.3	59.2	62.9	64.6
7. Shoulder height	47.24 - 64.17	56.50	2.28	51.2	52.8	56.6	60.2	61.9
8. Suprasternale height	48.03 - 63.78	56.28	2.19	51.3	52.7	56.3	59.9	61.5
9. Nipple height	42.13 - 57.09	50.41	2.08	45.6	47.0	50.4	53.9	55.3
10. Substernale height	41.34 - 55.51	48.71	2.02	44.0	45.6	48.7	52.1	53.5
11. Elbow height	36.61 - 49.21	43.50	1.77	39.5	40.6	43.5	46.4	47.7
12. Waist height	34.65 - 48.82	42.02	1.81	37.7	39.1	42.1	45.0	46.4
13. Penale height	27.95 - 41.34	34.52	1.75	30.6	31.6	34.5	37.4	38.7
14. Wrist height	27.56 - 39.76	33.52	1.54	30.1	31.0	33.6	36.1	37.1
15. Crotch height (inseam)	26.77 - 38.19	32.83	1.73	29.3	30.4	32.8	35.7	37.0
16. Gluteal furrow height	25.20 - 37.01	31.57	1.62	27.9	29.0	31.6	34.3	35.5
17. Knuckle height	24.80 - 35.04	30.04	1.45	26.7	27.7	30.0	32.4	33.5
18. Kneecap height	15.75 - 23.23	20.22	1.03	17.9	18.4	20.2	21.9	22.7

*Adapted from H. T. E. Hertzberg, G. S. Daniels, and E. Churchill, *Anthropometry of Flying Personnel*—1950, WADC Technical Report 52-321, USAF, Wright Air Development Center, Wright-Patterson AFB, Ohio, September, 1954. It should be noted that these data represent measurements made on approximately 4,000 male USAF personnel and thus do not specifically represent the U.S. population at large.

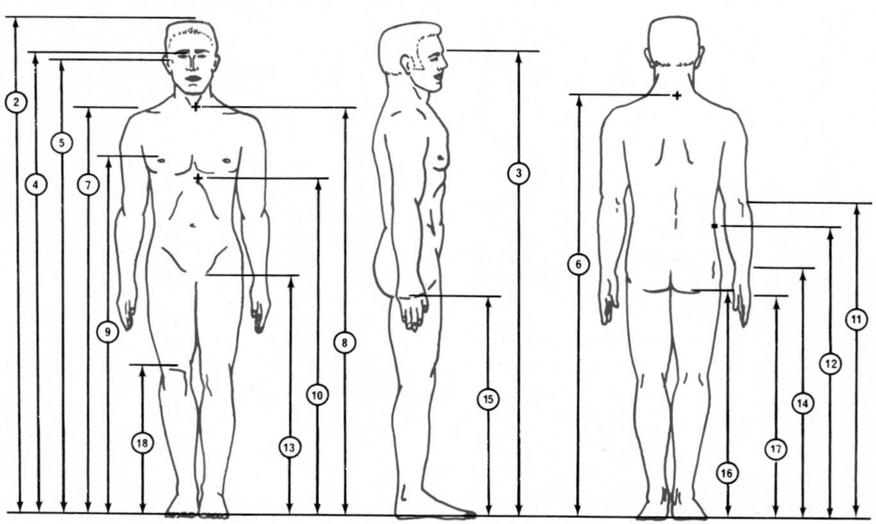

Anthropometric Table IV-2 (all dimensions in inches)

Measurement	Range	Mean	Standard Deviation	Percentiles 1st	5th	50th	95th	99th
19. Sitting height	29.92 - 40.16	35.94	1.29	32.9	33.8	36.0	38.0	38.9
20. Eye	26.38 - 36.61	31.47	1.27	28.5	29.4	31.5	33.5	34.4
21. Shoulder	18.90 - 27.17	23.26	1.14	20.6	21.3	23.3	25.1	25.8
22. Waist height, sitting	6.30 - 12.99	9.24	0.76	7.4	7.9	9.3	10.4	10.9
23. Elbow rest height, sitting	4.33 - 12.99	9.12	1.04	6.6	7.4	9.1	10.8	11.5
24. Thigh clearance height	3.94 - 7.09	5.61	0.52	4.5	4.8	5.6	6.5	6.8
25. Knee height, sitting	17.32 - 24.80	21.67	0.99	19.5	20.1	21.7	23.3	24.0
26. Popliteal height, sitting	14.17 - 19.29	16.97	0.77	15.3	15.7	17.0	18.2	18.8
27. Buttock-knee length	18.50 - 27.56	23.62	1.06	21.2	21.9	23.6	25.4	26.2
28. Buttock-leg length	35.43 - 50.00	42.70	2.04	38.2	39.4	42.7	46.1	47.7
29. Shoulder-elbow length	11.42 - 18.11	14.32	0.69	12.8	13.2	14.3	15.4	15.9
30. Forearm-hand length	15.35 - 22.05	18.86	0.81	17.0	17.6	18.9	20.2	20.7
31. Span	58.27 - 82.28	70.80	2.94	63.9	65.9	70.8	75.6	77.6
32. Arm reach from wall	27.56 - 39.76	34.59	1.65	30.9	31.9	34.6	37.3	38.6
33. Maximum reach from wall	31.10 - 46.06	38.59	1.90	34.1	35.4	38.6	41.7	43.2
34. Functional reach	26.77 - 40.55	32.33	1.63	28.8	29.7	32.3	35.0	36.4

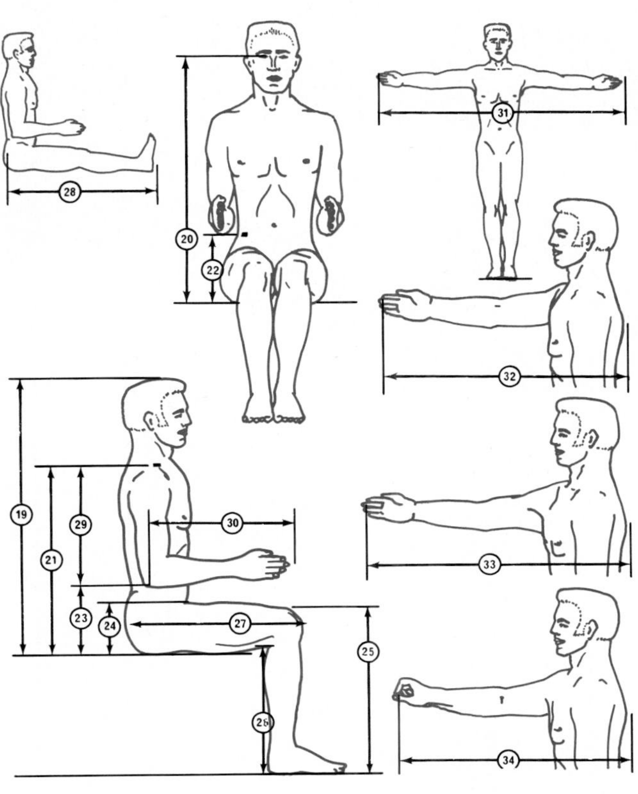

Anthropometric Table IV-3 (all dimensions in inches)

Measurement	Range	Mean	Standard Deviation	Percentiles 1st	5th	50th	95th	99th
Body Breadths and Thicknesses								
35. Elbow-to-elbow breadth	11.42 - 23.62	17.28	1.42	14.5	15.2	17.2	19.8	20.9
36. Hip breadth, sitting	11.42 - 18.11	13.97	0.87	12.2	12.7	13.9	15.4	16.2
37. Knee-to-knee breadth	6.30 - 10.24	7.93	0.52	7.0	7.2	7.9	8.8	9.4
38. Biacromial diameter	12.60 - 18.50	15.75	0.74	14.0	14.6	15.8	16.9	17.4
39. Shoulder breadth	14.57 - 22.83	17.88	0.91	15.9	16.5	17.9	19.4	20.1
40. Chest breadth	9.45 - 15.35	12.03	0.80	10.4	10.8	12.0	13.4	14.1
41. Waist breadth	7.87 - 15.35	10.66	0.94	8.9	9.4	10.6	12.3	13.3
42. Hip breadth	8.27 - 15.75	13.17	0.73	11.3	12.1	13.2	14.4	15.2
43. Chest depth	6.69 - 12.99	9.06	0.75	7.6	8.0	9.0	10.4	11.1
44. Waist depth	5.51 - 11.81	7.94	0.88	6.3	6.7	7.9	9.5	10.3
45. Buttock depth	6.30 - 11.81	8.81	0.82	7.2	7.6	8.8	10.2	10.9
Circumferences and Body Surface Measurements								
46. Neck circumference	10.24 - 19.29	14.96	0.74	13.3	13.8	14.9	16.2	16.8
47. Shoulder circumference	35.43 - 56.69	45.25	2.43	40.2	41.6	45.1	49.4	51.5
48. Chest circumference	31.10 - 49.61	38.80	2.45	33.7	35.1	38.7	43.2	44.8
49. Waist circumference	24.41 - 47.24	32.04	3.02	26.5	27.8	31.7	37.5	40.1
50. Buttock circumference	29.92 - 46.85	37.78	2.29	33.0	34.3	37.7	41.8	43.5
51. Thigh circumference	14.57 - 28.74	22.39	1.74	18.3	19.6	22.4	25.3	26.4
52. Lower thigh circumference	11.81 - 23.23	17.33	1.41	14.2	15.1	17.3	19.6	20.9
53. Calf circumference	9.84 - 18.50	14.40	0.96	12.2	12.9	14.4	16.0	16.7
54. Ankle circumference	7.09 - 12.99	8.93	0.57	7.8	8.1	8.9	9.8	10.5

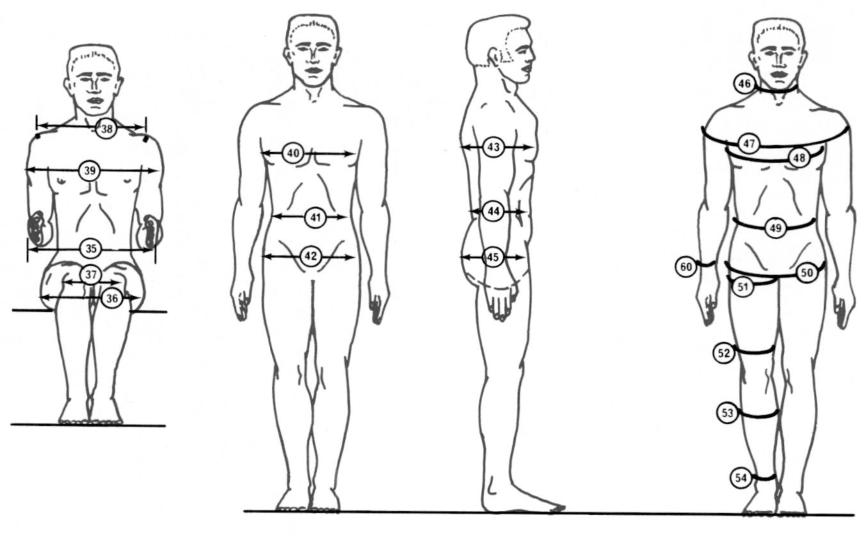

Anthropometric Table IV-4 (all dimensions in inches)

Measurement	Range	Mean	Standard Deviation	Percentiles				
				1st	5th	50th	95th	99th
55. Scye circumference	11.02 - 22.83	18.09	1.38	15.1	16.1	18.0	20.5	21.8
56. Axillary arm circumference	7.87 - 16.54	12.54	1.10	10.2	10.9	12.4	14.4	15.2
57. Biceps circumference	8.27 - 16.93	12.79	1.07	10.5	11.2	12.8	14.6	15.4
58. Elbow circumference	8.27 - 15.35	12.26	0.80	10.7	11.1	12.2	13.6	14.3
59. Lower arm circumference	8.66 - 15.35	11.50	0.73	9.9	10.4	11.5	12.7	13.3
60. Wrist circumference	3.94 - 8.27	6.85	0.40	6.0	6.3	6.8	7.5	7.8
61. Sleeve inseam	15.35 - 24.80	19.83	1.14	17.1	18.0	19.8	21.7	22.6
62. Sleeve length	27.56 - 38.98	33.64	1.50	30.2	31.3	33.7	36.0	37.3
63. Anterior neck length	1.38 - 5.31	3.40	0.64	1.8	2.3	3.4	4.4	4.9
64. Posterior neck length	1.57 - 6.10	3.64	0.61	2.3	2.7	3.6	4.7	5.2
65. Shoulder length	4.33 - 8.66	6.77	0.56	5.5	5.9	6.8	7.7	8.1
66. Waist back	11.81 - 22.83	17.72	1.07	14.8	16.1	17.7	19.4	20.2
67. Waist front	10.63 - 21.26	15.24	1.12	12.3	13.5	15.2	17.0	18.1
68. Gluteal arc	7.87 - 17.32	11.71	0.92	9.7	10.4	11.7	13.1	14.8
69. Crotch length	20.08 - 38.19	28.20	2.00	23.7	25.1	28.2	31.6	33.5
70. Vertical trunk circumference	54.72 - 74.41	64.81	2.88	58.3	60.2	64.8	69.7	71.7
71. Interscye	12.20 - 24.41	19.62	1.40	16.3	17.3	19.6	22.0	22.9
72. Interscye maximum	17.72 - 27.17	22.85	1.33	19.8	20.7	22.9	25.1	26.0
73. Buttock circumference	33.46 - 52.36	41.74	2.82	36.1	37.4	41.5	46.7	49.3
74. Knee circumference	11.42 - 20.47	15.39	0.92	13.5	14.0	15.4	16.9	17.7

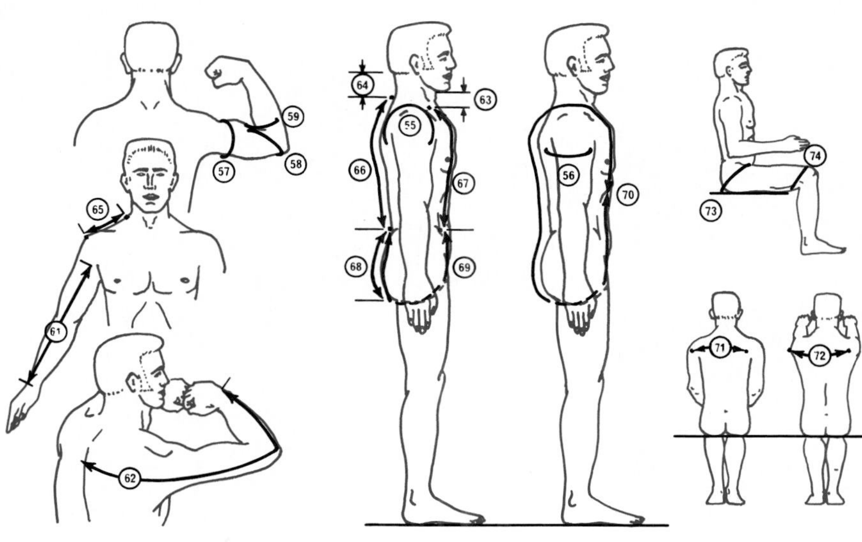

Anthropometric Table IV-5 (all dimensions in inches)

Measurement	Range	Mean	Standard Deviation	Percentiles 1st	5th	50th	95th	99th
The Foot								
75. Foot length	8.86 - 12.24	10.50	0.45	9.5	9.8	10.5	11.3	11.6
76. Instep length	6.42 - 8.86	7.64	0.34	6.9	7.1	7.6	8.2	8.4
77. Foot breadth	3.19 - 4.65	3.80	0.19	3.40	3.50	3.78	4.10	4.36
78. Heel breadth	2.13 - 3.27	2.64	0.15	2.30	2.40	2.63	2.87	3.01
79. Bimalleolar breadth	2.44 - 3.58	2.95	0.15	2.61	2.70	2.95	3.19	3.32
80. Medial malleolus height	2.60 - 4.29	3.45	0.21	3.0	3.1	3.5	3.8	4.0
81. Lateral malleolus height	2.01 - 3.70	2.73	0.22	2.2	2.4	2.7	3.1	3.3
82. Ball of foot circumference	7.87 - 12.60	9.65	0.48	8.6	8.9	9.6	10.4	10.8
The Hand								
83. Hand length	5.87 - 8.74	7.49	0.34	6.7	6.9	7.5	8.0	8.3
84. Palm length	3.39 - 5.04	4.24	0.21	3.77	3.89	4.24	4.60	4.74
85. Hand breadth at thumb	3.23 - 4.76	4.07	0.21	3.59	3.73	4.08	4.42	4.57
86. Hand breadth at metacarpale	2.99 - 4.09	3.48	0.16	3.12	3.22	3.49	3.74	3.86
87. Thickness at metacarpale III	0.75 - 1.54	1.17	0.07	1.00	1.05	1.17	1.28	1.35
88. First phalanx III length	2.21 - 3.07	2.67	0.12	2.40	2.49	2.67	2.85	2.95
89. Finger diameter III	0.75 - 1.00	0.86	0.05	0.77	0.79	0.85	0.93	0.96
90. Grip diameter (inside)	1.37 - 2.63	1.90	0.14	1.52	1.62	1.83	2.05	2.16
91. Grip diameter (outside)	3.15 - 4.72	4.09	0.21	3.58	3.72	4.09	4.44	4.57
92. Fist circumference	7.09 - 13.39	11.56	0.57	10.2	10.7	11.6	12.4	12.8

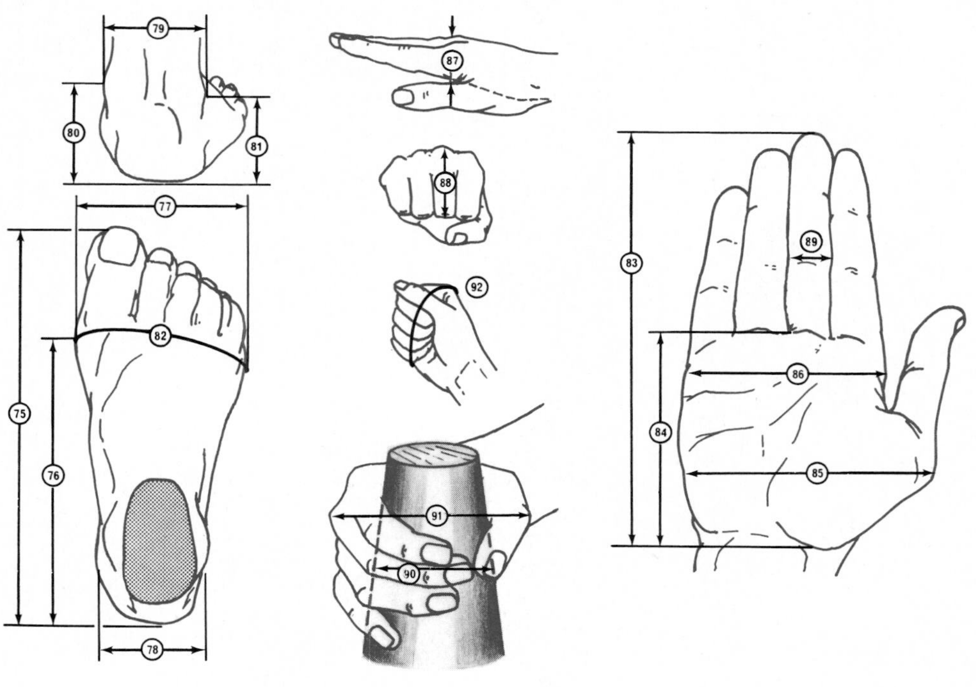

Anthropometric Table IV-6 (all dimensions in inches)

Measurement	Range	Mean	Standard Deviation	Percentiles				
				1st	5th	50th	95th	99th
The Head and Face								
93. Head length	6.89 - 8.78	7.76	0.25	7.2	7.3	7.7	8.2	8.3
94. Head breadth	5.35 - 6.89	6.07	0.20	5.61	5.74	6.05	6.40	6.56
95. Minimum frontal diameter	3.54 - 5.00	4.35	0.19	3.88	4.04	4.35	4.68	4.80
96. Maximum frontal diameter	4.02 - 5.47	4.71	0.20	4.26	4.39	4.72	5.05	5.20
97. Bizygomatic diameter	4.72 - 6.22	5.55	0.20	5.07	5.21	5.54	5.88	6.02
98. Bigonial diameter	3.50 - 5.08	4.27	0.22	3.8	3.9	4.3	4.6	4.8
99. Bitragion diameter	4.76 - 6.30	5.60	0.21	5.1	5.3	5.6	5.9	6.1
100. Interocular diameter	0.87 - 1.65	1.25	0.10	1.03	1.09	1.25	1.42	1.50
101. Biocular diameter	3.19 - 4.45	3.78	0.17	3.38	3.48	3.78	4.06	4.19
102. Interpupillary distance	2.01 - 2.99	2.49	0.14	2.19	2.27	2.49	2.74	2.84
103. Nose length	1.46 - 2.56	2.01	0.14	1.69	1.79	2.00	2.23	2.33
104. Nose breadth	0.91 - 1.85	1.31	0.11	1.09	1.16	1.31	1.49	1.58
105. Nasal root breath	0.28 - 0.91	0.61	0.08	0.42	0.48	0.61	0.74	0.81
106. Nose protrusion	0.43 - 1.42	0.89	0.11	0.63	0.72	0.90	1.08	1.17
107. Philtrum length	0.35 - 1.46	0.77	0.14	0.48	0.54	0.76	0.98	1.09
108. Menton-Subnasale length	1.81 - 3.54	2.63	0.27	2.05	2.19	2.62	3.07	3.28
109. Menton-Crinion length	6.18 - 8.58	7.36	0.36	6.6	6.8	7.4	8.0	8.2
110. Lip-to-Lip distance	0.16 - 1.26	0.64	0.12	0.35	0.44	0.63	0.83	0.94
111. Lip length (Bichelion Dia.)	1.34 - 2.64	2.03	0.14	1.72	1.81	2.02	2.27	2.38
112. Ear length	1.69 - 3.15	2.47	0.16	2.08	2.21	2.47	2.73	2.85
113. Ear breadth	1.10 - 1.93	1.44	0.11	1.20	1.27	1.44	1.61	1.70
114. Ear length above tragion	0.79 - 1.61	1.17	0.11	0.92	0.99	1.17	1.35	1.42
115. Ear protrusion	0.31 - 1.54	0.84	0.14	0.55	0.63	0.83	1.10	1.23

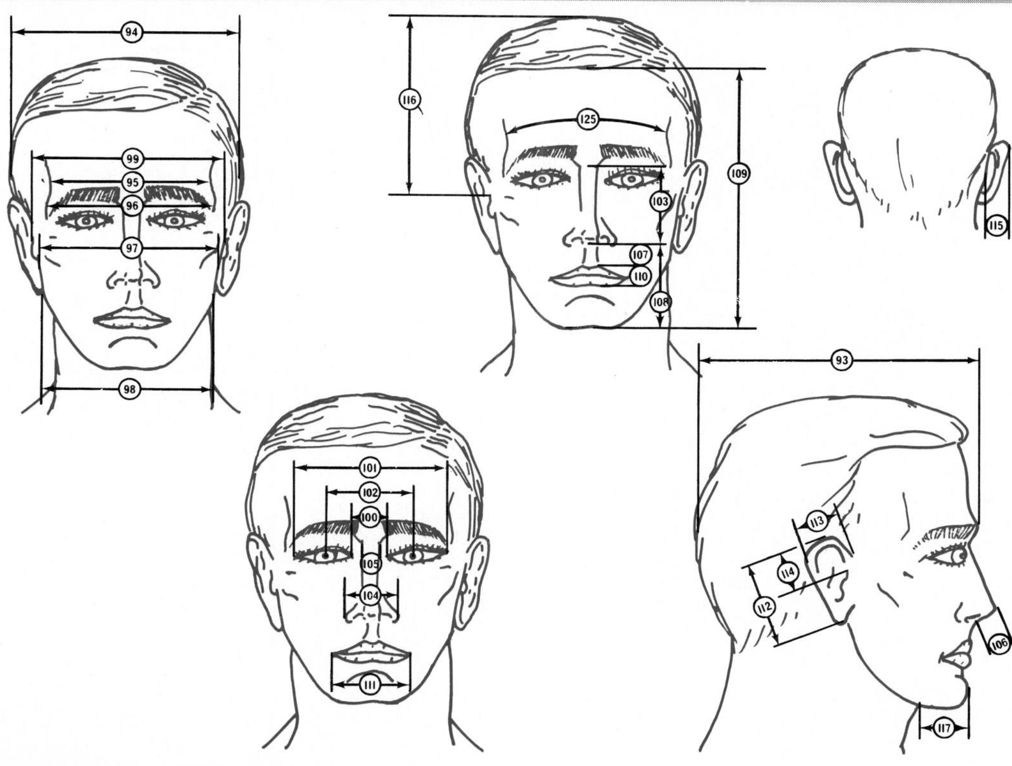

Index

C

D